Essays on Sigma

Simulation and Implementation of General Mathematical Activity

John S N Elvey

ISBN–13: 978–1502869531

This book is dedicated to my beloved wife Annie

I would like to thank my friend and colleague Emeritus Professor Ray d'Inverno without whose tireless efforts this book would not have appeared in print

Contents

Preface		(iii)
Chapter 0	Introduction	1
Chapter 1	The Sigma Mathematical Processor	9
Chapter 2	Fundamental SAM Algorithm	22
Chapter 3	Mathematical Information	58
Chapter 4	Operational Theorems	67
Chapter 5	Approximation Framework	85
Chapter 6	Abstraction and Analogy as Approximation	113
Chapter 7	Approaches to Mathematical Information	128
Chapter 8	Proofs as Combinations of Operational Schemes	174
Chapter 9	An Approach View of Mathematics	199
Appendix 1	Some Quasi-Operational Schemes	226
Appendix 2	Diverse Examples of Approach	250
Appendix 3	Analogue Representation	303
Appendix 4	Analogue Synthesis and Analysis	351
Appendix 5	Analogue Exploration	369
Appendix 6	Analogue Examples	409

Further Sigma Development 426

Conventions and Notations 430

Conspectus 436

Exercises 451

References Web Site 461

Preface

These Essays embody a comprehensive scheme for the development of a database for research in all areas of Mathematics. The underlying mathematical computer processor proposed, called SIGMA or Σ for short [1] is seen as providing an interactive research environment.

The application of arbitrary theorems to well-formed expressions constitutes the basic activity within this environment and the goal is to produce theorems which are formulated in maximally operational and constructive modes. This presents a huge challenge to mathematical ingenuity and technique and it is proposed that it is pursued internationally through a dedicated Institute. The Essays on Σ furnish a substantial foundation for this crucially important enterprise.

A motivation for Σ is the imminently irreversible disintegration of Mathematics into substantially independent subtheories. The scheme adumbrated in these extended Essays addresses 'the disintegration phenomenon' in multifariously interconnected ways. The main goal is to fabricate a collection of frameworks and procedures collectively encompassing and unifying M_t, the content of Σ_t, on the plausible premise that $M_t \sim' M_t^*$, as $t \to \infty$. Here, M_t^* denotes 'Mathematical Knowledge at time t', and the asymptotic quasi-convergence ($\sim'$) is realized by 'approach' which, as we shall see, subsumes all standard forms of deterministic and non-deterministic approximation, abstraction, direct analogy and 'inverse analogy' (via quasi-reflection or refraction over scientific or mathematical theories).

The systematic description, let alone preliminary implementation, of this scheme is of such vast scope that even these Essays only furnish a base-camp for the longterm project. This crucial initial phase, however, arguably presents the most formidable challenge. As with any pioneering scheme spanning many years, there will surely be room for improvement in formulations and implementations, but such modifications may be pursued it is suggested within a dedicated international Institute. To some extent this reflects the Bourbaki paradigm for parts of Pure Mathematics, albeit in a rarefied style, and without the 'approach' content.

The main aim is to promote and facilitate a constructive attitude to research in all areas of Mathematics and its applications. The overriding concern is, therefore, to enhance nontrivial calculability in arbitrary contexts, no matter how 'ostensibly uncomputational' they may appear at first in their original or conventional interpretations. In this way, what we will call the 'Σ -Operational Information Base' (Σ -OIB) may be extended indefinitely as a central research resource. Expository accounts of all nontrivial results will also be covered in the Σ -OIB. The Σ -References contain diverse examples of 'approach phenomena' which may be investigated by using the concepts, structure and techniques developed in the Essays. Thus the Essays should be viewed as key foundational documents in the effective representation of Mathematical Knowledge, rather than as expositions of various combinations of mathematical topics. This distinction is of fundamental importance.

The starting point of this work is Symbolic / Algebraic Manipulation (SAM) packages. These have existed for over 50 years, with intensive activity since 1980, yet their overall effect on general research in Mathematics so far has been very small. This is mainly on account of the profound mismatch between Input/Output (I/O) facilities and modes of formulation and application of theorems in virtually all areas of Mathematical Knowledge.

In these Essays, the consequences of harmonizing the SAM I/O with the conventions and notations of standard research schemes are examined in considerable depth. It is demonstrated that the systematic use of operational paradigms in the formulation of arbitrary theorems, together with the imposition of 'approach' criteria for closeness of Mathematical Entities (subsuming the most general forms of deterministic and non-deterministic approximation, abstraction, and direct and quasi-inverse analogy from Mathematics and the Sciences) cumulatively yields the Simulation and Implementation of General Mathematical Activity, or SIGMA for

[1] An early precursor of Σ is outlined in the paper: 'Varieties of Approximation' by JSN Elvey (pp 77 – 94) in : Jeffrey Johnson, et al., eds., 'AI in Mathematics' (OUP, 1994).

short – also written symbolically as Σ .

Not surprisingly, this program involves many highly innovative concepts and procedures which are to be deployed within maximally general frameworks. The resulting structures plainly have the potential for interactive and collaborative improvement. The harmonization problem requires precise correspondences between handwritten input, using Roman and Greek alphabets, and symbols for arbitrary mathematical operations. Moreover, every mathematical procedure or process occurring in a formal proof must be reproducible in these packages (the so-called 'Reduction Problem', discussed in Chapter 2), provided that such procedures have been formulated operationally with standard I/O. All of these problems are pragmatically solvable through combinations of Optical Recognition Software and associated SAM subroutines, for the 'quasi-unit-operators' involved. The specification and representation of 'Approach Environments' is discussed in detail in the Essays (especially, in Chapter 5 and 6). The 'Operational-Formulation problems' constitute the principal research activity of Σ . All of the key issues arising from this vast project are substantially resolved in the Essays.

In parallel with the formal developments there is a vast collection of deliberately speculative structures, notions and propositions, designed to stimulate research in many often novel directions. This is in part to blur or remove most of the apparent boundaries between and within Pure, Applied, Applicable and Scientific Mathematics. In this way, we shall show that Mathematical Knowledge (as a 'quasi-space') is effectively rendered 'calculationally connected. Even this material, however, is outlined with due regard for potential rigour through diverse references and many innovative structures are explored. The resulting calculational milieux which essentially characterize Σ as a vast collection of 'Environments for Investigation'. These will provide rich settings for quasi-formalized heuristics, where all nontrivial mathematical entities are exhibited in 'approach' settings. Multifarious examples of 'approach phenomena' are identified through several tagged and partially-annotated reference lists, totalling some 5,000 items.

In short: we plan to show that the crucial importance of the Σ -agenda is incontestable, where the component definitions and processes are all rigorously specified, and the overall structure, uniting 'approach' and 'operationally', is pellucid.

We finish this Preface with some introductory comments on the Σ -Essays :

(i)They constitute a radically innovative scheme for genuinely mathematical computation through reformation of arbitrary theorems, in operational modes, for interactive application.

(ii) The underlying calculational procedures are based on maximally constructive forms of theorems via enhanced computer packages.

(iii) The basic assumption in the Σ -Essays is that Input/Output is in conventional mathematical notation. Handwritten input on paper-like tablets' is allowed and initially, at least, a WYSIWYG (what you see is what you get) input facility with matching printed output is envisaged.

(iv) The overall aim is to produce an Operational Information Base asymptotically covering the whole of Mathematics which emphasises the 'the' (subsuming deterministic and non-deterministic approximation, abstraction and analogy).

(v) Σ is inevitably incomplete, but the proposal is that it will be maintained and enlarged as new results are obtained.

(vi) Σ is inherently nontrivial. This leads to difficulties of comprehension because, on the one hand, the unavoidably intricate formalism (with its pan-mathematical scope) makes it very abstract and, on the other, it is computationally and operationally complex.

(vii) It is hoped that the Essays will act as a blueprint for the longterm collective creation of an interactive Operational Information Bank for the Simulation and Implementation of General Mathematical Activity.

Editor's note

The book comprises 10 essays in chapters numbered from 0 (the Introduction) to 9. The 10 essays are also correspondingly referred to in the text as E_0, E_1, ..., E_9. The essays are supported by 6 extensive Appendices. Appendix 1 and 2 were oiginally considered to be two parts of a tenth essay (called Diverse Examples of Approach) and are referred to in the book as $E_{10/I}$ and $E_{10/II}$. Similarly Appendices 3, 4, 5 and 6 are considered to be four parts of an eleventh essay (called Studies in Mathematical Analogy) and are referred to in the book as $E_{11/I}$, $E_{11/II}$, $E_{11/III}$ and $E_{11/IV}$ respectively. After the Appendices there are 3 sections on Further Σ -Development, Conventions and Notations, and Conspectus, the last of which essentially provides a highly detailed contents of the book. This is followed by a section called Exercises which offer opportunities to elucidate ideas found in the book. Finally, References Web Site gives the web address where the full set of References may be found.

Ray d'Inverno
October, 2014

CHAPTER 0
Introduction

These Essays are concerned with fundamental issues in the taxonomic organisation, and effective implementation of Mathematics – viewed as a potentially unlimited system of logical deduction / derivation, formal calculation, and computation (in various forms). In this enterprise, **the maximisation of approximative content and explicit representation, in all theorems**, is combined with **the development of algorithms for their use in the generation of proofs**. Here, the term '**approach**' is introduced, to subsume **standard approximation** (of all types), **abstraction**, and **analogy** – all broadly interpreted. The algorithms are to be implemented over a greatly enhanced amalgam of existing computer-based SAM[2] packages. The general system so constructed is called SIGMA (**Simulation / Implementation of General Mathematical Activity**)[3].

The taxonomic representation of Mathematics involves detailed classification of 'categories', formulation of 'species', identification of 'types', and the development of systematic descriptive notation – together with a study of the 'structures' (often overlapping) into which Mathematics may by subdivided.

The ultimate aim is **to produce a comprehensive Operational Information Bank for Mathematics and its applications**. Plainly, no 'static' collection of results (however extensive) can achieve this aim. Rather, regular maintenance / modification, and 'interpolation of new material' within the initial framework / package, will allow asymptotic progress towards the desired goal. In this connection, it is useful to think of '**Mathematics at time t**' (say, M_t)[4] as comprising all notation, definitions, axioms, and proven theorems 'validated at time t'. The operational forms of information are assumed to be 'algorithmic over Σ_t'.

Is is supposed that 'M_t is closed relative to all admissible approach environments' – but that M_t is still 'essentially finite' (as a set of axioms, definitions, results, ...), and so may be 'embedded within a sufficiently large / versatile computer network', with sophisticated search software, etc.. 'Mathematical activity over a time-interval from t_1 to t_2' means interaction with Σ_t from t_1 until t_2.[5] In these Essays, the implications for research of adopting Σ_t as a model of mathematical activity are explored from many points of view, with copious illustrations. **The primary intention** is to provide paradigms for the global organisation, and local applications, of the operational theorems constituting the '**calculational base**' (CB) of Σ_t. Without such organisation, the accumulated information will mushroom uncontrollably.

The nature of 'The Σ-project' obviously precludes completion; but the urgency of the tasks addressed requires that a start be made to address the fundamental problems. The construction of Σ_{t_0} – the initial framework[6], with a preliminary selection of results – will be a massive undertaking, even after most of the 'foundational problems' have been tackled. It is assumed here that packages for symbolic computation will have been developed to the levels of sophistication necessary for the implementation of the operational algorithms in Σ_{t_0}; and, moreover, that the I/O may be presented in conventional mathematical notations. Recent progress in these areas augurs well for such assumptions. Consequently, only the barest details (but **not** the algorithms) may be found in the various manuals. The key mathematical procedures on which these systems are based are discussed briefly in E_2.

[2]Symbolic / Algebraic Manipulation

[3]The symbol Σ is also used, and the underlying processor is denoted by P_Σ

[4]The associated 'system' is denoted by Σ_t. The 'full version', containing 'all mathematical information 'identified' at time t – but not necessarily all rigorously proven', is denoted by M_t^*. The aim is to establish that $\mathrm{M}_t \sim \mathrm{M}_t^*$, as $t \to \infty$!

[5]This must be carefully interpreted, since (formally) Σ_t 'varies with t over continuum intervals'. One solution is to partition such intervals into subintervals, over each of which Σ_t remains essentially unchanged, etc..

[6]The abbreviation (O)IB is used for Operational Information Bank

Comments on 'Time-Dependence' in M_t and M_t^*

It is important to understand, from the outset, that 'the extent of M_t' is just 'the content of Σ_t'. Mathematical entities are accepted 'at face value' – no matter what their degree of constructivity or realisation (at t) may be. In this context, M_t^* encompasses not only M_t, but all mathematical phenomena discovered (up to t), or **inherent** (**independent** of t!). Of course, some phenomena may be **gradually** revealed (as t increases) without ever being fully comprehended; while, others may never be discovered at all (on the Platonist view, at least). These subtle issues are peripheral to the present work[7], since Σ_t comprises only entities discovered (by t) and used – either purely existentially, or else, at any attainable level of constructivity.

The evolution of $\Sigma_{t''}$ from $\Sigma_{t'}$ may be represented formally as:

$$\Sigma_{t''} = \Phi_{t''t'}(\Sigma_{t'}),$$

where $\Phi_{t''t'}$ incorporates all modifications of $\Sigma_{t'}$ between times t' and t''. For a computer-based system, in which all operations, whatever their sources, may be 'recorded', the 'history operator', Φ_I over intervals $\mathrm{I} := [\mathrm{I}^-, \mathrm{I}^+]$, may be regarded as (retrospectively) well-defined, in principle. The aim is to construct from Σ_{t_0} (with its initial mathematical content, M_{t_0}) the system Σ_t (with mathematical content M_t) in such a way that M_t is – in some sense(s), to be made precise – **asymptotic to** M_t^* (the '**full** representation of Mathematics at time t'). Systematic adherence to the **operational** paradigm, combined with the most general species of approximation (subsuming abstraction, analogy and 'inverse analogy' from all classes of 'scientific theories') will characterize Σ as an environment of unparaleled richness for the pursuit of mathematical research.

One of the key aspects of this work involves **a detailed discussion of the most general forms of approximation environments; and, of their interrelations with variants of abstraction and analogy**. The role of '**inverse**-analogy', from arbitrary scientific theories, in suggesting novel mathematical structures, is also explored – along with various species of '**stability**' of mathematical entities, under classes of perturbations of their defining frameworks. Above all, the requirement that all objects be viewed as 'products' of (sufficiently versatile) **approximation** processes – so that they may be calculated / estimates / approximated / perturbed / ... – dominates the treatment for Σ. In a sense, **a mathematical object is fully comprehended only when it is located in conceptual space**', as well as being accessible to analyses within conventional frameworks (the **operational aspect**). This combination of general approximation and inverse analogy enlarges greatly the scope of investigations and raises many fascinating issues that are scarcely even hinted at in the current literature. In the context of SIGMA, such topics are of central importance, and involve qualitatively new viewpoints in standard domains of both 'pure' and 'applied' mathematics.

Indeed, this approximative / analogical scheme essentially obliterates the apparent boundaries – not only among 'pure mathematics', 'applied mathematics' and all 'applications of mathematics', but also, **within** each of these three basic categories. For instance, algebra and analysed are merged in the theory of algebraic numbers; 'abstract differential geometry' is essential in modern theories of gravitation; and (non)-linear functional analysis underlies much of control theory, mathematical economics and many other fields. Any overall framework capable of accommodating all of these facets of mathematics must necessarily be abstract and general, with many linked levels of information / operation. This structure will be achieved partly through the routing use of a whole range of generic 'spaces' – each diversely exemplified throughout the book, in various 'realizations'.

The spaces selected are: **Linear (vector), Topological, Uniform, Proximity, Contiguity, and Nearness** spaces – to which are added: **Statistical(-Metric), Fuzzy**, and **Measure** (μ) spaces. Notice that all of the metric / normed / inner-product / function / ... spaces may be fitted into this classification; and, that the possibility of introducing other generic spaces (without invalidating the existing framework of Σ_t) remains open at all times. Moreover, each of the generic

[7] An extensive discussion is given in: Penrose,R., *The Emperor's New Mind* (OUP, 1989), where many references may be found.

spaces may be endowed with extra (e.g., algebraic, differential, geometrical) structure – also without affecting results over 'simpler spaces'. Even more: to each of the generic spaces there correspond classes of the other generic spaces – so there is no 'absolutely preferred space' for any topic considered, though some environments are 'richer than others' for the treatment of particular mathematical phenomena, especially for the variety of approximations. The treatment of direct and inverse analogues is developed within this fundamental framework which may undergo sporadic extension, or modification, as a result of attempts to realise reflections of 'strange' scientific phenomena.

The **goal** is **the longterm, cumulative construction of an interactive IB** unifying, and promoting, all of these – largely complementary – facets of mathematical information, some of which have barely been recognised up to now. A computer-based system, suitable enhanced, furnishes the ideal vehicle for achieving this goal in a practicable way, as an organic repository of operational theorems embedded in a 'medium' possessing many qualitative properties of physical / chemical / biological / . . . media – in that it admits multifarious deformations or interactions of (collections of) its constituents. Ultimately, some of these processes may occur autonomously within Σ, in ways analogous to fermentation of other forms of chemical reactions! Even without such transformations, the user-driven implementations of Σ facilities will require rigorous classification and static / dynamic organization, to eliminate redundancies, maximize scope, speed and efficiency of combination, for the potentially huge collection of algorithms realising the operational versions of theorems – 'new', or well-known in less constructive forms. This organisation devolves on the creation of several information structures, among which **taxonomic skeletons, pre-taxonomies, partial taxonomies**, and **full taxonomies**, play basic roles.

Although the term 'taxonomy' originally applied only to the classification / formation of living organisms, it is now interpreted more broadly in the construction and operation of very large DBMS[8], for arbitrary fields of application. The implementation of any taxonomic scheme usually entails an underlying **numerical labelling** of all items classified – with concomitant criteria for 'precedence', 'derivation', 'stages of evolution' (of some species from others), etc.. All of this may be formulated mathematically, in terms of trees, lattices, order-relations, order-preserving mappings and so on[9]; but this does not seem to be fruitful, since it contributes almost nothing to the **content** of the taxonomy. At best, it could be used to refine of extend already-substantial taxonomies – which must, apparently, be produced on the basis of general knowledge of the field(s) considered, combined with 'interpolatory procedures'. The design and implementation of DBMS, however, is highly relevant for the 'system aspects' of Σ, and has been cultivated intensively for many years.[10] As this book is concerned almost entirely with the mathematical content of Σ – rather than with problems of implementation – only the mathematical issues of DBMS will be discussed and even these, briefly. It should be noted that all SAM packages have built-in procedures for data control, so that further DBMS facilities must satisfy various compatibility requirements. No problems of principle seem to arise here.

It should be stressed, also, that all issues of taxonomic / DBMS design are peripheral to the main objective: **the effective implementation of 'Mathematical Knowledge'** – in the sense that no useful purpose is served be seeking genera, species, evolutionary mechanisms, etc., for mathematical concepts. Rather, these categories are valuable primarily as guiding principles in the cumulative construction of a very-largescale operational information bank (OIB), where only forms of '**asymptotic** completeness' are even **conceivable**. The extensive use of multivariate statistical analysis in 'numerical taxonomy' is hardly relevant to the hierarchical structure of Σ (except, possibly, where very detailed/specialized data classification is concerned). Similarly, the more intricate 'recognition algorithms', dependency analyses, partitioning procedures, etc., of **DBMS-design** are not apparent at the 'Σ-user level'. Consequently, all of these matters, often presenting recondite and challenging problems, are mentioned only when they bear directly

---

[8]Data-Base Management System(s)

[9]See, e.g. Benzecri, JP, **L'analyse des données, 1. Le taxonomy; 2.**[e] **L'analyse des correspondences** (Dunod, 1976)

[10]See, e.g., Adv. Comp. Research Vol 3 (1986): **The Theory of Databases** (ed. P.C. Kanellakis) JAI Press

on the mathematical content of Σ – though the associated **algorithms** will, of course, appear as results in the OIB!

This apparent circularity effectively precludes demonstrations of logical consistency; it is, however, endemic to any large, intricate OIB, though it does not detract critically from the soundness or power of the OIB, 'in practice'. In particular, the 'overview of the system' requires some **meta**-mathematical (even some k-meta-mathematical!) constructions; but this diversity of descriptive levels may be monitored 'intuitively' by users, and should not produce contradictory or paradoxical results. For a fully-automated ('theorem-proving') **deductive** package, serious problems of 'interpretation' would certainly arise.[11] Similar obstacles limit the use of all 'inference engines' – and general AI facilities – in Σ, for the foreseeable future.[12] Decision procedures, in all areas, **are** suitable for SAM-implementation, and such procedures will be incorporated systematically into the OIB. This Introduction (E_0) concludes with outlines of the topics covered in the main Essays (designated as $E_1, \ldots, E_11$).

E_1: The SIGMA Mathematical Processor, P_Σ

The SAM packages MATHEMATICA (C-based, 1988–) and AXIOM (LISP-based, 1992–) offer the most extensive and sophisticated resources currently available for symbolic mathematical computation; both combine symbolic, numerical and graphics environments. MATHEMATICA is '**rule-based**' (expressions of particular forms always transform in the same way) but not '**typed**'. By contrast, AXIOM is **strongly typed** (every object is identified by its type, e.g. 'integer', 'rational', 'polynomial'), but computations may be done over essentially arbitrary mathematical structures. The range of facilities covered by these systems is considerable (and regularly augmented) for numerical / algebraic / analytical / graphical computations; but SIGMA – as a **research environment for 'general mathematical activity'** – must possess several other features that have (so far) never been developed in SAM packages.[13] These include **the capacity to accommodate 'operational theorems'** (of arbitrary complexity / sophistication) for 'application' by users to expressions generated in interactive computation – and **the construction of a comprehensive taxonomy**, potentially adequate (with extensions/modifications) for the detailed classification of 'all mathematical knowledge' (in particular, of the collections of results already designated as M_t and M_t^*).

Various categories of information ('**descriptive**', '**partially operational**', '**t-max-operational**' and '**fully operational**') are introduced, and their characteristics and interrelations are discussed. The connections with 'approximation' – broadly interpreted to cover abstraction and analogy – are linked with the view of SIGMA as a (regularly expanded) '**operational information bank**'. The issues raised in this introductory essay pervade the rest of the book, where they are treated in some depth, and diversely exemplified. On this basis, the essential tasks for the development of P_Σ from existing SAM packages are identified, and possible approaches to these tasks are outlined.

E_2: Fundamental SAM Algorithms

At the basic / intermediate levels these algorithms have been carefully treated in the book: '**Algorithms for Computer Algebra**' (Kluwer, 1992), by Geddes / Czapor / Labahn . . . where many references are given. There, too, a brief **historical sketch** (spanning 1961 - 1992) may be found.

Most of the algorithms covered are 'purely algebraic', but some aspects of Newton iteration and symbolic integration are included. Even for Σ_{t_0}, however, a much broader collection of pro-

[11] See, e.g., Siekmann, J., Wrightson, G. (eds.), **The Automation of Reasoning**, Vols 1,2, Springer, 1983

[12] The structure of Σ allows such routines to be added, without invalidating the existing form of the system

[13] The work of PAGER (CACM, 1972, 1973) is concerned mainly with the efficient **communication** of proofs, so the overlap with Σ is slight, but some of the basic ideas are valuable for the DBMS aspects of the OIB

cedures is required – especially, in Analysis; so a discussion of these extra topics is essential – though many of them have now been (partially) implemented, as 'library routines', mainly in REDUCE and MACSYMA (for the original versions). Other calculational schemes / facilities necessary for any adequate (basic) specification of Σ_{t_0} are also considered, along with **structures tailored for the development of operational theorems** (OT), in the performance of calculations and the construction of proofs. Only the **mathematical** forms of these algorithms are covered here; issues of **implementation** / **optimisation** / ... are examined in G/c/L, in the book by Zippel, and in many papers cited in these books.

E$_3$: Mathematical Information

The characterisation of Σ as an IB emphasises the central role of '**information**'. Accordingly, a broad discussion of **species of mathematical information** opens this Essay. The other dominant features of Σ involve **calculation**, and **approximation** – in their most general forms (subsuming **abstraction** and **analogy**). This leads to the concepts of **t-calculational bases** and **t-max.-operational information** – with associated **algebraic** / **analytical** / **topological** / ... **structures**, which jointly determine the **super**structure of Σ as a research environment. In this context, the **meta**mathematical aspects of IB design are immediately apparent, and a discussion of the principles involved, and of their longterm potential, is given. More detailed treatments of these fundamental issues are developed in subsequent Essays.

E$_4$: Operational Theorems

For the overall organization of Σ, the basic IB entities are **theorems** – interpreted here to cover **'definitions', 'constructions', 'procedures', 'calculational schemes',** ..., as well as the standard deductive format: '$p \Rightarrow q$', with **premise(s)** p and **conclusion(s)** q. Indeed, all types of 'mathematical objects' may be represented, **formally**, as 'theorems' – the artificiality in some cases being more than compensated by the uniformity of notation. The distinctions among **'operational', 'effective', 'constructive', 'implementable', 'realisable',** ..., are considered, and these terms are '**specified** for use in Σ'. Clases of **canonical representation of operational theorems** are introduced – along with **general transformations of operational theorems, formal structures over sets of operational theorems**, and **levels of operationality**. All of this contributes to the establishment of a framework within which Σ_{t_0} may be extended cumulatively to cover arbitrary domains of mathematics and its applications: an **Operational-Theorem Calculus** (OCT).

E$_5$: Approximation Frameworks

All processes of approximation must be located / realized within suitable environments. Although diverse applications may entail intricate constructions, it appears that al cases of significance for research may be covered be combinations of: **Linear(-vector)** / **Topological** / **Uniform** / **Proximity** / **Contiguity** / **Nearness** / **Statistical(-metric)** / **Measure** / **Fuzzy** spaces – all with extra **algebraic** / **analytical** / **topological** / **geometrical** / **combinatorial** /... imposed structure(s), as required. These '**Basic Spaces**' are multifariously interrelated, so a broad treatment of their general properties constitutes an adequate foundation for 'the descriptive aspects of Σ'. Detailed accounts of these spaces are available in books; the aim here is to give a unified presentation, with copious references for the more recondite results – so that each ME may be associated with 'optimally related approximation-environments' / approach-frameworks (and thus realised in all admissible contexts).

E_6: Abstraction and Analogy as Approximation

The subsumption of 'abstraction' and 'analogy' under 'general approximation' (or, 'approach') is much more than a linguistic convenience; rather, it highlights essential characteristics of the processes commonly interpreted as abstraction and analogy, but more pertinently viewed as varieties of approximation. For Σ, which is built around a comprehensive taxonomy (where conceptual nuances are significant), combined with maximal calculational facilities in **all** domains of mathematics and is applications, all possible extensions of estimation or quantification are important. Accordingly, a 'scientific', rather than 'philosophical', approach is adopted – to minimize vagueness and to maximize connections with the standard criteria of 'Approximation Theory'. Inevitably, the precision attainable for such abstraction / analogy aspects of general approximation depends strongly on the domains and problems treated; but the enlargement of analytical scope is essential for the Σ-scheme. **Representations of abstractions / analogues of ME in terms of operations on their 'specification relations'** allows connections with conventional species of approximation to be discussed in a unified way.

E_7: Approaches to Mathematical Information

Several aspects of 'approximation' arise for arbitrarily given 'information'. First: is there a sufficiently broad scheme of **representation** to cover all types of mathematical information? If so, **how** can this scheme be **specified** (with minimal ambiguity)? Second: each element of **imprecision** in the information $\mathcal{I}$ corresponds to some **mode(s) of approximation** – other modes being associated with **modifications** of $\mathcal{I}$. The basic idea is that **all mathematical entities (ME) are determined by collections of logical predicates** (of various orders and classes), and that modified ME correspond to suitably modified sets of **implicatory / non-implicatory deformations of ME, general deformations of ME** (including **generalizations, abstractions** and **analogues**); and so, to species of '**conceptual approximation**' – all with associated notions of '**stability** under predicate-modification'. The possible choices of **approximation frameworks**, and the analysis of **abstraction and analogy as types of approximation**, are discussed in other Essays.

The association of quasi-**diagrams** with (arbitrary) ME (c.f., E_4) is extended here, systematically, to consider (sub)**taxonomies** of special / general **ME-diagrams, via** typical **components** (-settings). This is an enormous task (even in preliminary form); but it constitutes an indispensable part of the effective construction of the OIB.

E_8: Proofs as Combinations of Operational Theorems

The fundamental conception here is that **proofs may be regarded as (finite) compositions of suitably transformed operational theorems acting on 'initial data' (the premises).** The compositions may contain repetitions of some generic theorem(s) – with any admissible mixture of transformations. Infinite compositions yield proofs through various limit processes (within the allowable approximation framework(s)). For this model of proof in Σ, a range of interesting criteria and problems may be considered, involving '**quasi-geodesics over sets of theorems**', '**topological conditions**', and forms of '**perturbation of proofs**'. Questions of this nature would be either trivial, or else, intractable, 'by hand'; but they become both tractable, and heuristically suggestive, over a computer-based IB with sophisticated search software. Various notions of '**semantic (quasi-)distance(s)**' arise naturally, here – and may be heuristically valuable, despite their inherent limitations.

E_9: An 'Approach View' of Mathematics

For Σ, Mathematics has been characterized (in E_0) as: a **potentially unlimited system of rigorous deduction and (formal) computation / calculation.** It is convenient to extend 'calculation' to cover both **derivation** (in proofs) and **systematic transformation / manipulation of w.d. numerical / symbolic expressions.** All calculations may be performed **partially**, but only 'small sub-classes' of calculations may be **completed**. Thus: '**typical calculations**' are **in**complete – and so must be represented as species of 'approximations'. This mirrors one of the fundamental goals of Σ: **to maximize calculability in all domains of Mathematics and its applications** (not just in the most obvious cases). Here, it is argued that (if 'approximation' is augmented to cover abstraction / analogy / inverse-analogy) **every ME is representable either, as (part of) a general ' approach scheme', or else, in terms of results of such schemes.**

As in the designation of all items of mathematical information as 'theorems', this approximative view of ME will be highly artificial in certain situations – but 'generically nontrivial', and productive for effective implementation or representation or determination of the ME generated in 'typical research activity'. Further, this incompleteness of 'almost all calculations' suggests several sources of approximation – for instance, **definitions**; conditions imposed by various '**limitations**'; realizations of '**quasi-varieties**' (based on collections of generalized equations and inequalities). In this context, identities and equations may be viewed as 'singular' or 'extremal' elements (among the sets of all **in**equalities of '**in**complete statements'). Next, there are the classes of **constructive** and **non-constructive** approximations; and finally, the analysis of **speed, efficiency, optimality** and **scope** of approximation – which raise qualitatively new issues / problems in many established branches of mathematics where 'approximation' has so far played at most a minor role.

E_{10}: Diverse Examples of Approach

In this Essay, the problem of representing essentially arbitrary mathematical objects in terms of general approximation procedures / schemes is explored through a wide range of examples. The closely related task of developing operational versions of theorems – in any domain – is also illustrated, with discussion of **the interrelations among (in) effective / (non-)constructive / (partially) operational versions of 'the same theorem(s)'**. The idea is to produce **paradigms** for the Σ-formulation(s) of theorems – in the hope that 'new results' will be framed for almost-routine incorporation in the IB. The challenge of systematically modifying existing **non**operational results to t-max.operational forms (and the relation of this to current 'schools' of Constructive Mathematics) must be fully appreciated before significant progress can be made.Numerous approximative phenomena are exhibited through references, with brief explanations.

E_{11}: Studies in Mathematical Analogy

For any ME, e, there may exist a whole range of analogues – from transcriptions to the 'vaguest imitations'. In the present **Studies**, a collection of (about sixty) 'basic concepts' has been made – some, 'standard'; some, less familiar; and some, highly individual (but important for Σ).

One aim is **to show how progressively vaguer analogues correspond to weaker modes of approximation** – mostly, starting from the 'standard forms'. A broad discussion of the aims of analogical analysis in general is given, along with various procedures for generating analogues of e, or for testing 'suspected analogues of e'.

Processes of 'inverse analogy' (or, '**reflection**') from arbitrary 'scientific theories' are also considered. These differ fundamentally from 'mathematical models' (as used extensively in 'Applied Mathematics' and 'Applications of Mathematics'), and they form a very natural element of Σ. A large collection of diverse **illustrations** (with references and capsule descriptions) concludes this Essay.

CHAPTER 1
The SIGMA Mathematical Processor P_Σ

The synopses given in the **Introduction** indicate that the entire SIGMA scheme devolves on the development of a '**calculational milieu**' (CM)[14] adequate for the representation, manipulation, transformation and extension of **operational theorems** (OT) of arbitrary complexity and sophistication. This entails the provision of both '**logical**' and'**conceptual**' **substructures**, facilitating the analysis of '**theorems**' and '**proofs**' – both **as entities and as compound objects** whose **components** are variously interrelated. Such facilities are almost entirely lacking in the current SAM packages. These packages do have the potential to form a solid algorithm c basis, from which the more complicated capabilities may be fashioned. Part of this evolutionary process depends on the wholesale employment of very general 'modes of **approach**' (extensions of motions of **approximation**) for the construction or decomposition of mathematical entities (ME). This is illustrated, again and again, in the later Essays.

In spite of the long period over which SIGMA has evolved – with several different treatments of some fundamental issues – the essential structure has remained unchanged. In any innovative, large-scale scheme it is inevitable that certain problems are only partially solved, some formulations are far from optimal, and many techniques require clarification or extension. This is not surprising here, since the primary aim is to establish a durable framework for the organization and implementation of 'mathematical knowledge'! From this point of view the Essays are intended to furnish a sound foundation for systematic, long-term development, as well as a moderately detailed specification – which will be modified and enlarged when the OIB is implemented as a research resource. As already mentioned, **graphics / AI / theorem-proving / logic packages** may be introduced at any time, without undue disruption of the existing system. **Here, attention is focussed entirely on mathematical aspects of classification, approximative formulation, and operational realization over** P_Σ – all within sufficiently general calculational frameworks, exemplified by the '**basic spaces**', designated here as: **L**inear / **C**losure / **T**otological / **U**niform / **P**roximity / **C**contiguity / **N**earness / **S**tatistical(-metric) / **F**uzzy / **M**easure.[15]

The fact that these spaces may be distinguished (almost!) by their initial letters allows one to write '**X-space**', with X $\equiv$ L/Γ/T/U/P/C/N/S/F/μ. (Here, Γ denotes **closure** spaces, and μ, **measure** spaces, since M is used for **metric** spaces). The symbol ${}_SX$ stands for '**space of type X**'; and **realizations of any** ${}_SX$ **in terms of some class,** ξ **of ME** (e.g., matrices, differential / integral operators) may be denoted by ${}_SX(\xi)$, while ${}_{\{A\}S}X$ denotes **the space** ${}_SX$ **subject to extra axioms** $\{A\}$. A **cumulative axiom-list** for Σ_t – permanently ordered as, say, $(A)_{k_t^A}$ – may be maintained / augmented, as necessary. In this scheme, the axiom-sets for particular types of spaces are just examples of axiom-sets for arbitrary Σ_t-objects. They need not be either logically or mathematically self-evident (as they are intended to be for, say, Euclidean geometry); rather, **they serve to define the meaning(s) / possible-use(s) of the object concerned** – in relation to the 'environments' into which it is introduced.

A typical realisation of e within ${}_{AS}X(\xi)$ may be denoted by $e\langle {}_{AS}X(\xi)\rangle$, where all of e, A, ξ are chosen from appropriate Σ_t-lists, say, ${}^Q\Lambda_t$, for $Q \equiv e,\ A,\ \xi$. More generally, these (finite) lists may be replaced by **ordered, (un)countable families**; the formal notation covers all cases, and is convenient for taxonomic purposes. The 'index' t is essential if comparisons are made between $\Sigma_{t'}$ and $\Sigma_{t''}$; indeed, even mathematical operations involving t may be exploited! Nevertheless, dependence on t is always suppressed unless there is a danger of misunderstanding.

This representation of objects within approximative frameworks depends on the existence of a sufficient variety of **constituent-lists** (with no operational content) which may be produced **quasi-recursively** (i.e., by defining 'later items' in terms of 'earlier items'). This obvious proce-

[14]or, CMD, if Deduc(k)tion is included

[15]Measure-based formulations cover *stochastic variants of ME*, as discussed in later Essays

dure would entail regular 'interpolation of extra definitions', in practice; but this does not matter, in principle! The purpose of using 'lists', ${}^{Q}L_{\Lambda_t}$, for each class, Q, of objects, is to permit the systematic representation of **approximative forms of Q over suitable enhanced species of the basic calculational frameworks**.Such representations may be **parsed** efficiently, **modified** (via their 'components'), and **converted** into a range of variously operational versions. For any generic object, e, the challenge is to develop forms $e\langle_S X\rangle$, for every generic space ${}_S X$. The incorporation of abstraction of basic spaces may be accomplished by adding approximately chosen conditions to the characterizing axiom-sets, as may be necessary. Plainly, all analogues / abstractions of e must be 'located' in some framework(s), obtainable by 'modification' from the standard spaces. This process, though not algorithmic, may, nevertheless, be systematized, once the formal descriptions of 'abstraction' and 'analogy' within Σ have been developed.

The task of compiling '**initial lists**' for the introduction of various classes, Q, of mathematical objects (of increasing complexity / sophistication) os comparable with that of producing a **dictionary** (or fundamental terms and results). The crucial requirement is that all definitions / theorems should be 'covered' by others – with the 'most basic therms' regarded as 'self-explanatory'. Processes of 'interpolation' may be used to refine and extend the entrees, without affecting the underlying structure, so that adequate levels of consistency / completeness are attained. It must be emphasized again that shortcomings in the design and content of Σ_t are inescapable; but, that the effectiveness of the OIB as a research resource need not be significantly reduced on this account. A parallel my be drawn with Gödel's 'incompleteness theorem' (in Mathematical logic), where the existence of 'undecidable propositions' (within 'deductive systems containing arithmetic') has virtually no practical effect on the scope or methodology of research.

A fairly recent example, '**Dictionary of Mathematics**', by E.J. Borowsky / J.M. Borwein (pp659, Collins, 1989) illustrates how a considerable amount of 'core information' may be compressed, without undue loss of clarity, into a book of paperback size – covering material up to Master's Degree level. As a working basis for essentials this is useful, but more substantial dictionaries / handbooks are required for the initial list in Σ_0^D. These larger works, however, are not lexicographically ordered, making it harder to assess content. Nevertheless, gradual accumulation of material from diverse sources seems to be the only viable scheme for producing Σ_0^D.

The natural choices of source-books include: NAS / SCHMIDT '**Mathematische Wortenbuch**' (Pergamon); FREIBERGER, W.F., et al. (eds.), '**International Dictionary of Applied Mathematics**' (Illiffe); SNEDDON, I.M., et al. (eds.), '**Encylopedia of Applied Mathematics**' (Pergamon); PEARSON, E.S. (ed.), '**Handbook of Applied Mathematics**' (Van Nostrand); and, above all, the two major works: IYANAGA, S. / KAWADA, Y. (eds.), '**Encyclopedic Dictionary of Mathematics**' (1st Ed., MIT Press, 1977, pp1750; 2nd Ed. (K.ITO, ed.) MIT Press, 1993, pp2260), and PROKHOROV, A.M. / HAZEWINKEL, M. (eds.), '**Encyclopedia of Mathematics**' (translated from Russian; Reidel, 1987, pp c.4600).

These works comprise alphabetically arranged articles (varying from less than a single column to several pages) on the main parts of 'Pure Mathematics', and a range of topics from 'Applied Mathematics', including most of the principal theorems and substantial lists of references. All of this information, however, is 'static'; there are no operational elements and the material cannot be updated / maintained effectively. Moreover, in spite of publishers' claims of 'exhaustive coverage', it is obvious that many important results on standard topics (let along more obscure topics) are missing. Consequently, such compendia, though very useful, only furnish a detailed **guide**, say ${}_0\Sigma_0^D$, for the descriptive part of Σ_{t_0} – to be obtained primarily by combining results from all of these (and other suitable) sources. This constitutes the **preliminary phase** in the development of Σ_0^D: schematically,

$$\psi_1({}_0\Sigma_0^D) =: {}_1\Sigma_0^D.$$

The **second phase** – depicted as

$$\psi_2({}_1\Sigma_0^D) =: {}_2\Sigma_0^D$$

– involves the basic construction of partial and total **(pre-)taxonomies**, corresponding to some quasi-partition(s) of the aggregated information into broad fields / subfields / . . . , for both 'pure', and 'applicable', mathematics, as conventionally characterised. The 'Σ-view' is that there is, fundamentally, no significant distinction here!

In the **third phase**, interpretations / versions of all items in ${}_2\Sigma_0^D$ over the basic frameworks / spaces, $\{_SX\}$, are recorded (if they already exist), or else, developed – to produce

$$\psi_3({}_2\Sigma_0^D) =: {}_3\Sigma_0^D \equiv \{_SX\}_2\Sigma_0^D \equiv {}_2^1\Sigma_0^D$$

realized over $\{_SX\}$.

One of the pioneering aspects of Σ lies in ψ_3, where **variants of standard results over unfamiliar frameworks, $_SX$, are to be formulated – and proven**. This is a very longterm aim, but the great adaptability of any computer-based IB allows the system to function at a mixture of levels of completeness (beyond certain 'threshold states') without precluding further progress.

Notice that all results considered so far are '**exact**' (even where they involve standard approximation procedures). Perhaps the most radical feature of Σ is **the search for maximal calculability** in **all** results – no matter how ostensibly 'uncomputational' they may appear. This entails **the derivation of 'approach-formulations' of arbitrary 'standard theorems', over all of the frameworks $_SX$**. A Pre-requisite for this is the (general-)approximative representation of all entities occurring as 'constituents' of such theorems. The **forms** of approach may be characterised as **analytical**, **abstract**, and **(inverse-)analogical** – conveniently associated with the phases $\psi_{4.1}$, $\psi_{4.2}$, $\psi_{4.3}$ and $\psi_{4.4}$ (respectively). Thus, one has (again, schematically!):

$$\Sigma_0^{Apr} \equiv \psi_4({}_3\Sigma_0^D) \equiv \{\psi_{4.k}({}_3\Sigma_0^D) : 1 \leqslant k \leqslant 4\}.$$

Here, Σ_0^{Apr} denotes the 'approach-IB', and one may put: ${}_4\Sigma_0^D \equiv \Sigma_0^D \cup \Sigma_0^{Apr}$, where, for any M.E., e, **Apr(e) comprises all M.E. that approach ϵ – in any sense covered by the $\psi_{4.k}$. In these terms, it is natural to write**:

$$\delta(e) \equiv \{\text{descriptive forms of } e \text{ in } \Sigma_t\}$$

$$\text{Apr} \equiv \{\text{Apr}(e) :\in \Sigma\}$$

$$\text{Apr} =: \text{Ap} \cup \text{Ab} \cup \text{An} \cup -\text{An}$$

$$D(e) \equiv \{\text{determinations of } e\}$$

Here, the **determinations** of e comprise the (exact) **evaluation(s)** of e in all admissible context / frameworks. Formally, $D(\epsilon) \subset A(\epsilon) := \{\text{approximations to } e\} \equiv \text{Ap}(e)$. Here, the sets Ap($e$), Ab($e$), An($e$), -An($e$) correspond (respectively) to all (standard-)**approximative, abstract, analogical, inverse-analogical approaches** to e within Σ_t. These abbreviations will be used from now on.

The next stage in the formation of Σ_t involves **the derivation / construction of t-max(imally) -operational realizations of all elements of $\Sigma^D \cup \Sigma^{Apr}$** – a gargantuan task which (for any range of t) will proceed quasi-hysterically, as new results facilitate the effective realisation of others (often long-known in less effective forms). These realizations constitute '**pre-algorithms**', potentially implementable as '**operational forms**' over P_Σ (the processor underlying Σ), to produce Σ^{Imp} in phase ψ_5 – and thence, Σ^{Op} in phase ψ_6. These steps may be depicted as:

$$\Sigma^{Imp} \equiv {}_5\Sigma^D =: \psi_5({}_4\Sigma^D);$$

$$\Sigma^{Op} \equiv {}_6\Sigma^D =: \psi_6({}_5\Sigma^D).$$

In this evolutionary / cumulative scheme, successive phases, $\psi_2, \ldots, \psi_6$ will overlap, and evolve 'in parallel' – after a staggered start. This (inescapably cumbersome) notation allows 'the operational state' (from 'purely descriptive' to 'fully operational') to be monitored. **Notice that** every item in Σ^{Op} has some precursor(s) in Σ^D, whereas (successively larger) classes of items in Σ^D may have no images in $\Sigma^{Apr}, \Sigma^{Imp}, \Sigma^{Op}$. One may say, figuratively, that these classes are '**non-extendable**' beyond certain levels of operationally – which is reminiscent of of natural boundaries (in Function Theory) / inaccessible (or, unattainable) states in Thermodynamics / Probability-Theory) and various '**limitation-conditions**' in other fields, to be specified in later Essays.

From this point of view, Σ should be regarded as an extremely complex, quasi-organic system, functioning simultaneously on several levels – each item at the 'lowest' (purely descriptive) level interacting with others at all levels, and with its own images at higher levels. Various results from **General Systems Theory** may be useful in studying the local / global behaviour of Σ as a hierarchical structure. Moreover, the **meta**mathematical aspects are ubiquitous, and cannot be separated from the operational aspects. Logical circles / self-references / formal-inconsistencies will abound, as the edifice emerges from hazy, shaky foundations – requiring extensive underpinning! Yet the overall scheme just adumbrated **is inherently sound** – and susceptible of unlimited refinement / consolidation.

Since the longterm goal is to construct a maximally complete research resource, with an international body of users, the problem of basic **nomenclature** / **notation** is acute. In particular, the I.B. should be thought of as containing **generic** items, along with all of their **variants** (for other frameworks / forms of approach). The selection / specification of generic items must therefore, be made, as far as possible, in line with 'universal practice'. The inevitable ambiguities / clashes / obscurities / ... entailed may be minimized by providing a running **inventory of alternative notations / terminologies / ...** backed by **references to other parts of the I.B.** where details may be found. All of this is already incorporated in DBMS design, and is indispensable for the viability of such packages. The term '**entity**' will stand for '**generic mathematical entity**' (typically, e), and the set of all Σ_t-variants of e (say, $V_t(e)$) will comprise all descriptive forms of e and all elements of Apr(e) – where, as usual, the subscript t will be understood, unless ambiguities arise. Thus:

$$V(e) \doteq \delta(e) \cup \text{Apr}(e) \cup \text{Imp}(e) \cup \text{Op}(e),$$

where Apr(e) $\doteq$ Ap(e) $\cup$ Ab(e) $\cup$ An(e) $\cup$ $-$An(e), and -An(e) contains forms of e derived by **inverse analogy** from arbitrary theories / 'phenomena' pertaining to domains other than 'Pure Mathematics' or 'Applied Mathematics', as commonly interpreted.

This terminology allows a near-exhaustive classification of the items in the (O)IB; extra items may be interpolated, whenever necessary, without invalidating the material already selected. Of course, 'absolute exhaustiveness' is unattainable; but also irrelevant, since Σ is interactive and dynamic in structure. The existence of an OIB on this scale raises qualitatively new mathematical issues of basic importance for research and applications. Only a 'global view' of Mathematics founded on the concept of approach, can offer adequate scope for 'general mathematical activity'. Within the collection $\{_S X\}$, of approximation environments, there are many subdivisions. For instance, the compendium of results in the book 'Function Spaces' be **A. Kufner, et al.** (Noordhoff, 1977) covers properties of diverse function spaces, mostly Banach spaces – and hence, 'topological spaces', $_S T$, with 'the norm topology'. Some of these may also be regarded as basic spaces of other types (e.g., 'uniform', $_S U$). The choice of $\{_S X\}$ is intended to include, as 'special cases', all spaces found in Mathematical practice. The linear spaces are ubiquitous, but mostly possess additional structure. An interesting question arises here: To what extend can approximation be defined for **arbitrary** linear spaces? This excludes spaces with inner products, norms, etc., as well as **topological** linear (vector) spaces. Finally, one must consider whether there exist approximation environments **other than** those already specified, since Σ aims to encompass **all** substantive results, within approximative frameworks. It would appear that only **measure spaces** ($_S \mu$, here) offer possibilities for approximation on otherwise arbitrary sets; but the minimal conditions for an initially arbitrary set to admit a w.d. **measure** must be investigated before this question can be resolved. Exceptional /'pathological' cases are of only marginal interest as Σ deals primarily in general procedures. It should be realized, however, that, if the combined scope of the basic approximation environments, $\{_S X\}$ proves to be inadequate, then some **extra space(s)** may be added without disrupting the previous classifications. The selection made so far seems very broad, and it should suffice for all of the developments to be discussed here. The built-in flexibility of DBMS / IB packages is ideally suited to the accommodation of both evolution and revolution in Mathematics!

Up to now, the treatment of (detailed) **proofs** in Σ has not been discussed. This is mainly due to the typical strategy for deriving proofs of some theorem, (say, T) through sequences of references to (the proofs of) other theorems (say, $T_1, \ldots, T_n$), so that the proof of T is a sort of composition of the proofs of $T_1, \ldots, T_n$. This scheme is used in the proof-checking package, **AUTOMATH** (implemented, in **typed λ-calculus,** by **N.G. de Bruijn** and coworkers – mainly in Eindhoven (see, e.g., **A. Rezus, 'Abstract AUTOMATH'**, Mathematisch Centrum, Amsterdam, 1983 – and references there)). For pedagogical / aesthetic quality, proofs should be discursive and elegantly phrased; such proofs will be called **descriptive**, here. They form an important part of Σ^D. If the totality of proofs in Σ is denoted by Σ^π, then it is useful to decompose Σ^π as:

$$\Sigma^\pi = \Sigma^{\pi^D} \cup \Sigma^{\pi^F}$$

– covering, respectively, all descriptive, and all formal, proofs (in the sense of compositions of transformed operational theorems). A further decomposition, corresponding to **constructive**, and **partially constructive**, proofs, may be represented as:

$$\Sigma^{\pi^D} = \Sigma^{\pi^{D+}} \cup \Sigma^{\pi^{D-}}; \Sigma^{\pi^F} = \Sigma^{\pi^{F+}} \cup \Sigma^{\pi^{F-}},$$

the '+' signifying full constructively, and the '-', partial (including '**nil**') constructively. In this context, every theorem may be identified with its formal proof(s) – i.e. with combinations of proofs of certain other theorems, say, $T = \Gamma_T((T_\sigma)_n)$, where $(T_\sigma)_n := (T_\sigma, \ldots, T_{\sigma_n})\ (\subset \Sigma_t)$ is an appropriate selection of n theorems from the 'current stock' in Σ_t. This representation raises interesting questions about the stability of proofs of T under perturbations of the T_{σ_k}, and about the class of deformations of T resulting from such perturbations. Moreover, analogues of the 'response functions' of physical systems may be formulated (for 'changes in $(T_\sigma)_n$', or under augmentations of $(T_\sigma)_n$ by specified sets of OT – say, $(T'_{\sigma'})_m$). In the present SAM packages, investigations of this kind are not even feasible, but for Σ they are both viable and heuristically valuable. Indeed, these wide-ranging / speculative studies will complement – but not replace – the more familiar research methods, raising qualitatively novel issues. This has already occurred for aspects of numerical computation (e.g., in Number Theory and in Dynamical Systems Theory). Numerical Analysis, Graphics, and Simulation are often combined to generate conjectures / hypotheses. Soon, these techniques will be extended to include types of response analysis for general symbolic calculation, and it will be possible to estimate the 'centrality' of theorems – through their frequency of use in proofs, and other criteria to be formulated in Σ. Of course, the most basic results in Algebra / Analysis are cited (explicitly or implicitly) in almost all nontrivial proofs, so more subtle conditions must be found to obviate this effect. One possibility involves the length of proofs: strong results tend to produce relatively short (and 'sharp') proofs, etc .... All of these factors arise in the design of the 'Σ-research-environment' which will offer unparalleled resources for the exploration of hypotheses and the development of intricate / large-scale proofs.

Notational Recap

The sign 'Σ' is used in many ways (with various indices). The entire system is denoted by Σ; classes ζ of ME, by Σ^ζ states of evolution of Σ, by ${}_j\Sigma \equiv \psi_j({}_{j-i}\Sigma)$. All of the 'mathematical activity' is located within combinations / variants / extensions of the **basic frameworks / spaces**, namely: $\{_S X : x \approx L/\Gamma/T/U/C/P/N/S/\mu\}$, possibly with additional ('imposed') **combinatorial / algebraic / analytical / geometrical / topological / ... structure**. All of these (augmented) frameworks are characterised by collections of **conditions** – which may be viewed as **quasi-axioms** (regardless of their complexity).

To every ME ϵ in Σ there corresponds a set of '**approaches to** ϵ', comprising all forms of **approximations** to ϵ over the basic (augmented) spaces, together with all admissible **abstractions / (inverse-)analogues** of ϵ. A distinction is maintained among **descriptive, formal**, and **operational** entities, especial, for proofs (typically, only partially operational). Every theorem T has a set, $\pi(T)$ of valid proofs, over Σ; and, conversely, every derivation, δ, comprising a chain of deductions, corresponds to a see $\{T(\delta)\}$ of theorems, all of which are 'provable from δ. In other words: several formal chains of deductions may contain T, and every valid chain of deductions (potentially) contains several theorems. These obvious remarks, though somewhat academic in standard mathematical practice, are nevertheless significant for Σ, where the detailed properties of proofs will be analysed and compared. The interactive facilities in Σ have been largely ignored, so far, but they are of fundamental importance in providing maximally transparent interfaces with P_Σ – ultimately, through handwritten input on suitable 'tablets' (producing conventional output after processing by 'recognition software'). Criticism of this aim as 'Luddite' is wholly misplaced. On the contrary, the combination of optimal use of range / power with a psychologically familiar research environment – where all forms of visual patterns play a profound role – offers a vast improvement on the unstimulating technological milieux typical of current SAM systems. It is undeniable that, unless humans mutate significantly towards robots, the discovery of proofs through a mixture of heuristic and formal procedures will flourish in 'natural surroundings' (here coupled to the OIB for information and interactive calculation). There is no doubt that such user / system links are now within reach; only commercial pressures impede further progress. As a prerequisite for Σ, alone, this R&D can be strongly justified – it is enough to consider the potential impact on diverse applications of mathematics. The probably restrictions on handwriting / style / 'diagrams' / ... will not significantly limit user-OIB communication – even the monitors may be made to merge into the furniture!

For the first time it had become practicable **to treat Mathematics as a unified body of knowledge**, from 'pure there' to 'the most practical applications'. Further, this **should** be done, to revers the fragmentation so apparent from a perusal of journals. The ideas underlying Σ constitute a response to this problem. The sheer volume and heterogeneity of the material raise profound issues of organisation – let alone implementation. Mathematics is especially suitable for this kind of treatment – which has deep implications for all scientifically-based fields. With the advent of Σ, mathematicians will be liberated from routine calculation, bibliographical search and the frequent re-derivation of (sometimes insignificant!) results. In this way, the superficial research output will be reduced – and the vast majority of published results will be both valid and nontrivial.

As stated in E_0, one aim of this Essay is to characterize $P_\Sigma$qualitatively as an enhanced extension of an amalgam of current SAM systems – notably, MACSYMA / SYMBOLICS (MIT, 1971–); REDUCE 2 (Utah, 1970 –); SCRATCHPAD / AXIOM (IBM, 1970– / 1992–); SMP / MATHEMATICA (Caltech, 1998–). Since AXIOM[16] and MATHEMATICA[17] have near-comprehensive User-Manuals, no discussion of format/I/O-languages/scope need be given here – beyond remarking that the I/O is far from adequate for Σ ; and, that almost no details are given of the underlying algorithms (which are outlined in E_2, with many references)[18]. Further, **theorems / proofs** (as such) are not formally represented at all, whereas, for Σ , these objects are of central importance. Again, notions of **approximation** are treated only incidentally in SAM packages, in contrast to Σ , where the concept of '**approaches to the entity** ϵ' (combining all standard forms of approximation with abstraction, analogy and inverse analogy) is absolutely **fundamental**.

It follows that the current systems – no matter how much they may be enlarged along existing lines – will (partially) meet only the **basic** computational requirements of Σ (as a Universal Operational Information Bank) for the cumulative implementation of (suitable reformulated) arbitrary theorems. Another extension of SAM procedures, enlarging their analytical scope,

[16]JENKS, R.D. / SUTOR, R.s., '*Axiom ...*' (NAG/SPRINGER, 1st ed, 1992)

[17]WOLFRAM, S., '*Mathematica ...*' (Addison-Wesley, 1st ed, 1998)

[18]For the most basic algorithms, see Geddes, KO, et al., '*Algorithms for Computer Algebra*' (Kluwer, 1992). Selected 'advanced algorithms' are also outlined in E_2, with additional comments / refs.

involving the use of **set-based algorithm**. This has been systematized in the package **SETL**[19] (initially, by J.T. Schwartz, et al., at NYU), where intrinsically analytical arguments can be handled. It is convenient to denote the programming language associated with Σ_t by L_{Σ_t}. This will combine desirable features of the (I/O) languages of existing SAM packages – with additional syntactic / semantic facilities for contracting / processing procedures / proofs (as 'structured objects') and for analysing 'deductive chains'.

These programming / calculational operations are to be embedded in a DBMS at (probably) unprecedented levels of size / information-resolution – to be attained through the systematic use of **(pre-)taxonomies.** In this scheme, definitions / constructions / procedures / proofs / ... are all treated as species of 'theorems'. Thus, '**mathematical information**' consists (entirely) of **theorems**, of various kinds – with a body of '**laws of combination**', and a range of 'operationally' from **nil** (descriptive) to **full**; but, mostly, partial/t-maximal. Although it is not possible at this stage to specify L_Σ precisely, no problems of principle seem to arise in **assuming that L_Σ is w.d.**, and governs the behaviour (internal / interactive) of P_Σ, with **conventional mathematical** I/O.

At any time t, the OIB consists of descriptive, partially-operational, t-max-operational, and (fully) operational parts – say, $\Sigma_t \doteq \Sigma^D \cup \Sigma^{Op}$, where all theorems that are not purely descriptive are considered as being operational (of some degree). Plainly, Σ^D contains **descriptive** forms of **all** species of theorems.[20] The partially operational theorems either contain inherently (logically) nonoperational elements, or else, have not been (re-)formulated operationally in Σ_t. The t-max-operational theorems represent the best reformulations up to time t (where no inherent obstacles exist). These subdivisions are useful in assessing the level of effectiveness of the IB at any time, by helping to identify domains particularly 'resistant' to operational treatment.

The main tasks here involve:

(i) the **representation(s)** of theorems of all species;

(ii) the **transformation(s)** of theorems under classes of mappings; and

(iii) the '**action**' of theorems on (arbitrary) well-formed expressions (wfe).

These tasks are aimed at introducing one of the characteristics of Σ : **calculational bases** (collections of operational theorems of various species). The overall effectiveness of Σ depends strongly on the range of (partially) implemented theorems. The **structure** of such bases raises several questions about '**calculational span**', '**operational realisation**' (of theorems) and **derivation** (of proofs). The other crucial constituent is the availability of all admissible **approaches** to (arbitrary) mathematical entities[21] (ME), e – in relation to the basic spaces, $_SX$, and to the associated abstractions, analogues, and inverse analogues of e. Preliminary discussions of these 'extended forms of approximation' are essential – to establish notation, and to develop representations / procedures. The detailed treatment of all these matters will be pursued in later Essays.

Representation of Theorems in Σ

Although there is room in the design of Σ for **automated theorem-proving** capabilities, they will become significant only when their scope is greatly enlarged; so far, such packages have produced either 'trivial', or else, 'uninteresting' results.[22] Nevertheless, the **terminology** of 'automatic deduction' has been developed to study the syntax (and aspects of semantics) of the-

---

[19] SCHWARTZ, J.T., et al., '*Programming With Sets*', Springer, 1986. An **interactive** version, ISETL, may be linked with AXIOM or MATHEMATICA (on some basis).

[20] The abbreviation **OT** is used for operational theorem. The symbol T **denotes** OT but is not short for theorem!

[21] A generic ME is denoted by e, and a generic wfe, by ϵ. The **ingredients** of any ϵ (which **denote** (non)generic ME) are usually written as x_k, u_k, θ_k, etc..

[22] The volumes edited be Siekmann/Wrightson: '*The Automation of Reasoning*' (Springer, 1983) cover the key foundational papers.

orems and proofs **as 'linguistic objects'**, so the notation of Σ should at least take these techniques into account. In particular, grammar(s) and proof-procedures are established through various algorithms in Mathematical Logic. The aims in Σ are far more pragmatic: theorems are viewed as **tools** to be applied in the transformation of 'expressions' (from '**premises**' to '**conclusions**') The continuous **monitoring** of this process by the user obviated the need for detailed rules to assess relative values of procedures and to avoid the 'explosions' in computer storage or memory requirements that make the effective design and implementation of algorithms so problematical.

In the book, '**Logic, Form and Function**' (Edinburgh U.P., 1979), J.A. Robinson gives a detailed account of the logical schemes / algorithms underlying the representation and deduction of mathematical statements. **The aim**, of implementing such deductions in computer routines, necessitates the use of techniques that are precise, systematic and essentially exhaustive – to produce **decision procedures** for (variants of) the (first-order) **Predicate Calculus** (PC1). Since this system encompasses virtually all proofs encountered in 'general mathematical activity', it should be adequate as a basis for Σ , including eventual automated-deduction (AD)/AI facilities.

An essential difference between any PC1-development of Mathematics, and Σ , is that Σ **deals only with relatively high-level theorems**; all 'elementary results' are 'assumed' (but **stated**, for completeness, in Σ^D, the descriptive part of Σ). The **structure** of logical formulae / expressions, however, remains relevant at all levels of sophistication – so the AD-classifications are highly pertinent for Σ .

The characterization of formulae in PC1 is straightforward. The set F of all all formulae may be decomposed as [23] $F = I + S$, where $I \equiv \{$ingredients$\}$ and $S \equiv \{$sentences$\}$. The descriptive content of F is encompassed by the universe, U, of individuals, and the ingredients denote individuals. The sentences express propositions – about / properties – of (collections of) individuals – and, as such, have only two possible values: t (truth) or f (falsehood). The set I has its own decomposition as $I \equiv U+V+C+X$, where $U(\neq \phi) \equiv \{$individuals$\}$, V(countable) $\equiv$ $\{$variables$\}$, C(countable) $\equiv$ $\{$(operator, operand) ordered pairs$\}$ $\equiv$ $\{$(n-any-constructor, n-element set of ingredients)$\}$ $\equiv \sum_{n\geqslant 0} C_n$. Hence $C = \sum_{n\geqslant 0}(C_n \times I^n)$, with $I^n \equiv I \times \cdots \times I$ (n copies).

Each **construction** is also representable as a parenthesised (finitely-nested) **list**, whose **initial component / entry** is the **operator** – the remaining components being its **immediate constituents**. Thus, constructions form a type of **applicative expressions**.

The set X of **exemplifications** is obtained from $G \equiv \{$**generalisations**$\}$. Here, G is part of the decomposition of S as: $S = \{\underline{t}, \underline{f}\} + P + B + G$, where $P \equiv$$\{$**predicati- ons**$\}$ and $B \equiv \{$**Boolean combinations**$\}$. The **predications** (or, **atomic** sentences) are distinguished from **Boolean combinations** (or, **molecular** sentences) in that the (immediate constituents of the former are simple ingredients, whereas those of the latter are **sentences**). In Σ , the notions of '**elementary**' and '**compound' expressions** are used (with 'atomic' and 'molecular' expressions as special cases). **Chemical imagery** is used widely in developing various aspects of Σ -structure. In particular, species of **arity** (e.g., for lists / operators / connectives) mirror basic ideas for **valency**. The list of Boolean connectives may be taken as: $\{$**not, and, or, if, iff**$\}$; in symbols: $\{\neg, \wedge, \vee, \rightarrow, \leftrightarrow\}$. Sometimes, '**not**' is denoted by one of '-', '$\sim$'; '**if**', by one of '$\supset$', '$\Rightarrow$'; '**iff**', by '$\Leftrightarrow$', '$\equiv$'. The **infix notation**' is often used for **conjunctions**: $(\wedge S_1 S_2 \ldots S_n) \equiv S_1 \wedge S_2 \cdots \wedge S_n$; for **disjunctions**, one has: $(\vee S_1 S_2 \ldots S_n) \equiv S_1 \vee S_2 \cdots \vee S_n$. Consistency requires that, for empty lists, $(\wedge) = (\underline{t})$; $(\vee) = (\underline{f})$. Formally, G may be represented as: $G \equiv ((\{\forall\} \times \vee) \times S) + ((\{\exists\} \times \vee) \times S)$ – corresponding to a partition into the sets of **universal**, and, **existential** generalizations, with **quantifiers** $\forall$, $\exists$. The crucial property relating X and G is that: the truth values of any $g \in G$ and of its X-instance(s) (say, x_g), **coincide**. The set P of **predications** (or atomic sentences) has the form: $P = \sum_{k\geqslant 0}(P_k \times I^k)$, where $P_k \equiv \{$predicate symbols of arty $k\}$ is effectively countable, while the set of molecular sentences ($\equiv$ Boolean combinations) is given by $B = \sum_{k\geqslant 0}(B_k \times S^k)$, where 'arities must match' for each element of

[23] $A + B$ is used for the union of A, B; it is usual to put $T \equiv \{$terms$\}$, but $I \equiv \{$theorems$\}$

the (Boolean) sums (recall that **and** / **or** take all possible arities). The P_k are also referred to as sets of **k-peace predicated** (interrelating k individuals (or predicate variables)). These are the constituents out of which arbitrarily intricate proofs are produces, through combinations of **construction, direct deduction**, and '**indirect deduction**' ($\equiv$ '**proof by contradiction**'). These basic structures, for the representation / proof of mathematical statements – familiar from Analysis / Algebra / Euclidean Geometry / ...– must be exploited in all proofs (however complex or arcane) within Σ , where the fundamental methods of proof involve **the action of operational theorems** (appropriately '**focussed**) **on** (suitable **approaches** to) **well-formed expressions (wee)** generated in interactive computation – these 'approaches' covering all of the designated '**basic approximation frameworks**', together, with species of **abstraction, analogy** and **inverse analogy**. Accordingly, schemes for the classification, representation and application of (partially) operational theorems must now be developed.

The simplest classification of (generic) theorems is into purely **descriptive** and partially **operational**; say, as: $\mathcal{J} = \mathcal{J}^D \cup \mathcal{J}^{Op}$, where $\mathcal{J}^D$ contains (as a subset) **descriptions** of theorems in $\mathcal{J}^{Op}$, while $\mathcal{J}^{Op}$ covers all degrees of operationally from '**nil**' ($\equiv$ 'initially operational') to '**full**'. It is supposed that all ME in Σ_t (including all theorems) are labelled by successive (or, at least, strictly increasing) elements of some totally ordered set(s) – as the ME are introduced / recorded. This labelling is arbitrary, but, once adopted, is **fixed** for all time. The set of positive integers, $\mathbb{P}$, may be used, together with (subsets of) the real numbers, $\mathbb{R}$ to cover uncountable **facilities** of generic ME. The obvious difficulties of producing such a labelling scheme do **not** present a major problem of **principle**. This is because a computer-based IB allows more of less unlimited 'interpolation' and reorganization, before it assumes a viable 'initial form'l and even after this, material may be added or modified, subject only to **consistency criteria** – new classifications / codes being added to the taxonomic scheme as necessary. This book is concerned with possible **models** of Σ_t, and their implications for research; consequently, the details of classificatory labelling and other aspects of software implementation need not be considered here – beyond the claim of 'cumulative feasibility', which is justified by the increasing power of DBMS packages.

It follows that the **objects** of Σ_t (of all kinds) may be represented as **an effectively countable set**, some of whose members are themselves represented by (ordered) **families** of elements. If Σ_T^E stands for 'the set of ME in Σ^t, then one may put $\Sigma_t^E \equiv X^{k_t} = \{x_k : 1 \leqslant k \leqslant K_t\}$, where K_t is the label of 'the latest ME added to Σ_t^E. It is convenient to use the notation

$$ {}_nP = (P, \ldots, P+n-1); {}_{\{n\}}P \equiv \{P, \ldots, P+n-1\}; $$

$$ (x)_n := (x_1, \ldots, x_n); \{x\}_n := \{x_1, \ldots, x_n\}; $$

$$ (x_\sigma)_n := (x_{\sigma_1}, \ldots, x_{\sigma_n}); \{x_\sigma\}_n := \{x_{\sigma_1}, \ldots, x_{\sigma_n}\}; $$

$$ ({}_kx_\sigma)_n := (\underbrace{x_1, \ldots, x_1}_{k_1}, \ldots, \underbrace{x_n, \ldots, x_n}_{k_n}); $$

$$ \{{}_kx_\sigma\}_n := \{\underbrace{x_1, \ldots, x_1}_{k_1}, \ldots, \underbrace{x_n, \ldots, x_n}_{k_n}\}; $$

where $(\ldots)$ denotes **ordered lists**, and $\{\ldots\}$ denotes **un**ordered **sets**, and the function $\sigma : {}_{(n)}1 \to \mathbb{P}$ is strictly increasing, while $k : {}_{\{n\}}1 \to \mathbb{P}$ is unrestricted. In this way it will be possible to represent theorems in terms of the entities x_k appearing in their formulations.

Theorems in Σ are denoted typically by T, with (sets of) subscripts, superscripts or other 'labels', as required. For instance, $\{T_k\}\{T_{l_k}\}, \{\tilde{T}_{pq}\}, \{T_\lambda\}\{T_{\psi(\mu)}\}$, where k, l_k, p, q vary over countable sets, and λ, $\psi(\mu)$, over uncountable sets.

The notations:

$$ f(x)_m := f((x)_m) = f(x_1, \ldots, x_m); $$

$$ f\{x\}_m := f(\{x_1, \ldots, x_m\}); $$

$$ f[x]_{(m)} := (f(x_1), \ldots, f(x_m)); $$

$$f[x]_{\{m\}} := \{f(x_1), \ldots, f(x_m)\}$$

are also used from now on – with the further conventions

$$f((x_1, \ldots, x_m)) \equiv f(x_1, \ldots, x_m);$$

$$f(\{x_1, \ldots, x_m\}) \equiv f\{x_1, \ldots, x_m\},$$

provided there is no danger of ambiguity.

For **ordered** families Λ, define: $(x)_\Lambda \equiv (x_\lambda : \lambda \nearrow \Lambda)$; and for **all** families Λ, $\{x\}_\Lambda \equiv \{x_\lambda : \lambda \in \Lambda\}$. Here, '$\lambda \nearrow \Lambda$' means '$\lambda$ increases though Λ' (with $\lambda \searrow \Lambda$ for **de**crease through Λ). **NOTE that**, aspects of **randomness** are introduced in $f\{x\}_m$ (and $mf\{x\}_\Lambda$, etc.); and that **invariance** criteria may also arise.

For distinguishing **species** of theorems[24] (upper / lower) prefixes may be used; for instance, $\{_r T_\lambda\}, \{^s T'_{pq}\}$. Lastly, **powers** of OT may be denoted as $T_k^{<m>}, {}_r T_\lambda^{<n>}$, etc., where $T^{<1>} \equiv T$ and $T^{<n>} \equiv T \circ T^{<n-1>} (n \geqslant 2)$. The possible definitions of $T^{<0>}$ and $T< -n >$ involve identity mappings and operational inverses (based on **converse** OT). These matters are considered in later Essays.

Since **theorems** are composed of **sentences** (either formal, or else written / spoken – in suitably augmented languages) it follows that theorems are equivalent to (collections of) **propositions** about (sets of) **individuals** (denoted by the **ingredients** of these theorems). In the context of Σ , I corresponds to $\Sigma\ ^E$, and so: **theorems express propositions about collections of objects / entities.** By combining the notations for ME / sentences /sequences of theorems / ..., one obtains representations of the form:

$$T(x)_n; T^{<q>}\{x\}_n; T(_k x_a)_n; \{_{r_\lambda} T_\lambda(_{k_\lambda} x_{\sigma_\lambda})_{n_\lambda} : \lambda \in \Lambda\}; \ldots$$

and so on – where the abbreviated notation $T(x)_n := T((x)_n)$, etc., is understood in all cases. Thus, the family $\{_{r_\lambda} T_\lambda(_{k_\lambda} x_{\sigma_\lambda})_{n_\lambda} : \lambda \in \Lambda\}$ has as its typical member

$$_{r_\lambda} T_\lambda(\underbrace{x_{\sigma_\lambda 1}, \ldots, x_{\sigma_\lambda 1}}_{k_{\lambda 1}}, \ldots, \underbrace{x_{\sigma_\lambda n_\lambda}, \ldots, x_{\sigma_\lambda n_\lambda}}_{k\lambda n_\lambda}).$$

Here, r_λ covers possible **species** of theorems, σ_λ, the **choice-functions**, k_λ, the '**repetition functions**', and n_λ, the number of '**blocks**' (of copies of generic objects). The symbol λ takes every value in Λ exactly once (where Λ may be finite, countable, or uncountable of any order). Some extensions of this notation are specified as they are needed.

It is useful to introduce the symbol $^{[r]}q$ for 'r copies of q'; the expounded representation just given then takes the form:

$$r_\lambda T_\lambda \left[{}^{k_{\lambda 1}} x_{\sigma_{\lambda_1}}, \ldots, {}^{[k_{\lambda_n}]} x_{\sigma_{\lambda_n}} \right].$$

This notation will be used (as appropriate) from now on.

The application of OT to wfe entails at least two types of transformations – to make the OT **compatible** with the wfe; and to **modify** the wfe in accordance with the results embodied in the OT. In these formal processes, all copies of any generic entity must be similarly transformed within the (original) wfe/OT – each regarded as a collection of '**formulae**' in PC1 (which are, in turn, denoted by equivalent **expressions**, say, e, involving sets of ingredients $^{[k_1]}x_{\sigma_1}, \ldots, {}^{[k_n]}x_{\sigma_n}$ – no matter how the copies of the x_{σ_k} may be distributed in e). Thus: **theorems**, and the **data** to which they are applied, are species of **expressions**, with **syntactic structure** and **semantic content**. As such, they may be put into '**standard forms**' (to be specified) for parsing and other procedures.

Notice that, as the generic ME in $\Sigma\ ^E$ are permanently **labelled**, every ME may be **identified in every context** – in particular, in the modification of **expressions**, e. One scheme is to convert e initially into the unordered form $\{_k x_\sigma\}_n$, and thence into the ordered list $(_k x_\sigma)_n$ – all

[24]Other symbols (e.g. S,U,V,W) may be used in place of T if theorems of the same species / type must be considered together.

of σ, k, n being chosen to 'reproduce e'. It is necessary here to explain the distinction between the (purely) **descriptive**, and **nil-operational** forms of expressions – especially, of **theorems**. For a theorem to qualify as '**operational**', all of its **ingredients** must be (labelled) **symbols**, and all designated interrelationships among (collections of) ingredients must be **symbolically represented**. This covers classifications / constructions / proofs / ... (even where there is no possibility of implementing them over Σ_t). Expressions, e, may be called operational whenever the could constitute parts of OT. An **initially-operational theorem** is, therefore, one in operational **syntax**, but with **nil-implementation** For Σ, all expressions are assumed to be, either descriptive, or else (at least), initially-operational; there are no 'intermediate forms'.

Representations such as $T({}_kx_\sigma)_n$ have all constituents of T identified with labelled symbols (i.e., with ${}_{k_1}x_{\sigma_1}, \ldots, {}_{k_n}x_{\sigma_n}$ (appearing blocks of $k_i(\geqslant 1)$ copies)). All **functions** of $T({}_kx_\sigma)_n$ – say, f, must be of the same '**arty**' as $T({}_kx_\sigma)_n$. In other words, each occurrence of any x_{σ_i} is to be 'matched' by the occurrence in f of some component(s), f_{σ_i} acting on x_{σ_i}. Variants of these representations (e.g., with $({}^{[k]}x_\sigma)_n$) may also be used. The definition of $f(T)$ is discussed fully in later Essays. Here, it is enough to observe that such definitions may be formulated; and that they are fundamental in the overall structure of Σ.

From the point of view of 'basic content', any theorem $T(\theta)$ may be associated with its ingredients (collectively denoted[25] by θ, ignoring here all details of structure). The totality of 'admissible forms of T in Σ' in generated through transformations f, where $f(T(\theta)) := T(f(\theta))$, provided only that $f(\theta)$ and $T(f(\theta))$ are w.d.. Recall that θ represents some (finite, or (un)countable) family of ME – based on a finite collection of **generic** ME – identified by their permanent labels / numbers; and, that $T(f(\theta))$ is obtained from $T(\theta)$ by replacing each ME in θ by its image under (a specified component of) f. Consequently, $f(T(\theta))$ is always **formally** w.d. if $f(\theta)$ is – through $f(T(\theta))$ need not 'make sense' as a theorem!

There are several possibilities:

(i) $f(T(\theta))$ is a recognizable variant of $T(\theta)$;

(ii) $f(T(\theta))$ corresponds to another theorem, say, $T'(\theta)$;

(iii) $f(T(\theta))$ is provably '**false**' (even though $T(\theta)$ is 'true');

(iv) The validity of $f(T(\theta))$ is '**unclear**' – but the use of $f(T(\theta))$ may be of (at least) **heuristic** value. (Occasionally, this strategy has produced highly significant discoveries – as illustrated later).

Note: Other possible definitions of $f(T)$ and $f(T)_m$ are discussed in E_4/E_8.

The **action** of (transformed) operational theorems on wfe, e, is of three basic types: (a) '**informative**'; (b) '**substitutive**'; and, (c), '**constructive**'. Typically, one has a combination of these types of action, since 'most theorems in Σ' are only partially operational, and are 'substantially incompatible with almost all wfe'. This is not as disastrous as it may appear, since Σ is used **interactively**, and so broad matching of theorems and expressions is monitored by the **user**. At all events, it may be convenient to define 'the action of T on e' **formally** for arbitrary pairs (T, e), even though significant results will be obtained only when T and e are (at least, partially) 'matched'. On this basis one may represent **any** theorem, T, as: $t =: \bigcup_{1\leqslant j\leqslant 3} u_j\tau_j[T(\theta_T)]$, where the u_j have possible values 0 or 1, and the τ_j correspond to the 'decomposition of T into sub theorems of the basic types'. For any wfe e, the τ_j may be defined as follows:

$$\tau_1\{T\}e := e \cup T^D \cup \pi_T(e);$$

$$\tau_2\{T\}e := e^*(T) \cup R_T(e) \cup \rho_T(e^*, e);$$

$$\tau_3\{T\}e := C_T(v_e, w_e).$$

[25]Strictly, one must write $T(\theta_T)$, etc., since (unlike the x_{σ_k} in the ordered set U of underlying ME) θ has no unambiguous interpretation. This basic T-dependence must always be remembered.

Here, is is assumed that $e \cap$ **dom** $(T) \neq \phi$; $0\tau_j(T)e = \phi$; $1\tau_j(T)e = \tau_j(T)e$'. The 'operator' τ_1 corresponds to the descriptive form, T^D, of T, with $\pi_T(E)$ being the set of properties of (parts of) e that are directly deducible from T. For τ_2, the wfe, $e^*(T)$ is obtained by 'tuning T to e' and then replacing appropriate parts of e by their T-associates, while $\rho_T(e, e^*)$ specifies all relations of e to e^* induced by this process, and $R_T(e)$ contains all parts of e **dom**(T). Lastly, if the 'logical form of T' i given by $T(\theta) \Leftrightarrow p(v) \Rightarrow q(w)$, then the set $C_T(v_e, w_e)$ comprises all objects **constructed** through T from the e-realizations of v and w. In all of these notations, $T \equiv T(\theta) \equiv T(\theta_T) \equiv [p(v) \Rightarrow q(w)]$. Furthermore, one has $T({}_k x_\sigma)_n \equiv T({}_{k_T} x_{\sigma_t})_{n_T}$, etc., for the 'component forms', where, again, T-dependence is 'understood' (unless clarity requires that it be shown explicitly). This does not affect the Σ -formalism significantly.

The aim in **this** Essay is just to outline the **apparatus** for representation / transformation / manipulation / application of operational theorems, within the 'processor', P_Σ, of Σ . More detailed explanations and illustrations are given later – along with the introduction of **calculational bases** 'spanned' be a given set of operational theorems, and other structures involving algebraic / analytical / . . . combinations of theorems. The possible notions of '**neighbourhood** of ME' are also explored, to develop environments for the ubiquitous exploitation of generalised approximation procedures.

The problem of representing (quasi-)**neighbourhoods** of ME, say $\mathcal{N}(y)$, is basic for Σ , since approximation / abstraction /analogy / inverse analogy are involved. If the ME in Σ_t are listed as $(x)_{k_t}$, then any ME, say, $y := x_j$, may be characterised by (for instance) a set of **conditions** interrelating y and subsets of some (finite) collection of other ME. This association is denoted in two ways. **First**, the ME associated with y in this way are identified, and one puts (say) $y; ; \{x_{\sigma^y}\}_l$. **Second**, the relation of y to the simultaneous **conditions characterizing** y **in terms of the ME in** $\{x_{\sigma^y}\}_l$ may be denoted as $y :: \underset{\alpha \in A}{\Lambda} C_\alpha((x_{\sigma^y})_l, y)$, where (in principle), A may be finite, countable or uncountable of any cardinality.

Every representation of y in such a form (with some choice of selection function σ^y, and of l) constitutes a permissible quasi-**definition** of y; but the 'proper' definition(s) of y are commonly expected to satisfy other criteria as well – so that y appears 'as naturally as possible' as an object within some **theory**, related to other objects already defined. It may happen, however, that 'y in terms of $\{x_{\sigma^y}\}_l$' is especially appropriate in certain contexts – even though the 'basic definitions of y' involve different collections of ME. In any case, the 'fundamentality' of definitions depends strongly on the 'environment' within which they are viewed; so 'basic' / 'proper' definitions will be so designated only for specified mathematical domains. This topic is elaborated in other Essays, but the essential point here is that **every** ME, y, may be **characterized** through some **set(s) of conditions** involving y and associated ME, $\{x_{\sigma^y}\}_l$. There is no attempt to construct 'maximally primitive characterisations' (as is often done in Mathematical Logic). Consequently, for quasi-neighbourhoods, $\mathcal{N}(y)$, y may be 'generated', **either**, be **varying** the **forms** of some of the C_α, **or**, by **modifying** some of **the** $s_{\sigma^y_k}$ from their (initial) 'y-determinations'. Of course, these basic processes may be combined, in any consistent way. Notice that not all of the C_α need depend on all of the $x_{\sigma^y_k}$; but that every $x_{\sigma^y_j}$ must appear in at least one condition C_β. It is further supposed that each $s_{\sigma^y_k}$ ranges over the 'space' – say, $X_{\sigma^y_k}$ – of its direct exemplifications; for instance, square matrices of order n with elements in $\mathbb{R}$ may be associated with the space $\mathbb{R}^{n^2}$. Variations of the $x_{\sigma^y_k}$ are, therefore, to be specified (initially) within the corresponding spaces $X_{\sigma^y_k}$. For the conditions C_α, however, no such simple reductions seem to be possible. Both modes of modification are present, in general. A broader view of variation of the $x_{\sigma^y_k}$ includes 'contiguous species', as well as the above variation of types within a single species. The conceivable varieties of **conditions** encompass both **syntactic** and **semantic** deformations of the C_α – the latter admitting at least two forms: imprecatory and non-implicatory. These matters raise many fundamental issues, which are dealt with in other Essays. Here, is suffices simply to introduce these 'conceptual neighbourhoods', and to note that: **all w.d. approaches** to the constituents of the $C_\alpha((x_{\sigma^y})_l, y)$ are involved in determining the possible forms of $\mathcal{N}(y)$ over the approximative environments in $\{{}_S X\}$, together with abstraction, analogy and inverse-analogy.)

Sets of OT may be viewed as objects – **Calculational Bases** (CB). Their general (structural) properties are explored later, in considering the application of (almost-)arbitrary **combinations** of OT in the development of **constructions / proofs**. These 'generalized compositions of OT' are blended with the diverse modes of **approach** to the ingredients of expressions – in particular, of OT, themselves. In this context it is useful to introduce **modifications of the universal and existential quantifiers** – reflecting **the approach environments**. In particular, the symbol ${}_S\forall$ means: 'for ${}_SX$-almost al'; and, ${}_S\exists$ means: 'there exists at least one ${}_SX$-approximation (to)'. **Notice that** both of these modified quantifiers may have several interpretations / realizations, so further clarification is necessary before they may be used in the formulation of results.

CHAPTER 2
Fundamental SAM Algorithms

The purpose of this Essay is **to assess the scope of current SAM packages; to identify and outline the algebraic theorems underlying the key SAM algorithms;** and, **to indicate how the sophisticated or procedures typical of 'general mathematical activity' may be synthesized from these elements.** The essential point is that **proofs** – no matter how intricate or recondite – no matter how intricate or recondite – are composed of comparatively simple 'deductive blocks' ('in series', or 'in parallel') whose validation depends substantially on basic techniques (algebraic / analytic / combinatorial / ...) already covered in existing systems. In particular the flexibility of **Axiom** allows calculations to be performed in a wide range of **algebraic** structures – within which **analytical** facilities may subsequently be developed.

Examination of the 'central theorems of Analysis' over 'progressively general paces', indicates that the proofs are mostly reducible to results for 'simpler objects' over $\mathbb{R}$ or $\mathbb{C}$. Further, the situation is comparable sophisticated **algebraic** procedures in relation to existing SAM facilities over 'basic domains'. The deductive nature of Mathematics make these claims 'intuitively obvious' – or, at least, very plausible. The aim here however, is to justify (if not, 'prove') them, on the basis of accepted synopses of Algebra, and Analysis but before this is considered, the principal built in capabilities of existing SAM systems must be outlined along – along with the algorithms through which they are realized. Since essentially all of 'Pure Mathematics' – and thence, of 'Applicable Mathematics' – is ultimately derivable from Analysis and Algebra, the above claims apply, with minor modifications, to the entire Σ scheme.

The diversity of possible algebraic environments – and the range of implemented functions – may be conveniently summarised through **Axiom**[26] (by far the richest package for structural graduations). Indeed, the present discussion of scope is based primarily on Axiom, since it is the most recent system, and probably the most comprehensive. In any case, the processor, P_Σ (with I/O language, L_Σ) is to be developed from all current systems, so there is no question of detailed comparisons (which may be found in SIGSAM Bulletin and the Journal of Symbolic Computation, for instance).

In Axiom, the Basic Algebraic Hierarchy (over the specified category of sets) is generated (roughly) by combining **'Fundamental Frameworks'**, $\mathcal{F}$; (**G**roups/ **R**ings/**Mod**ules/**Mon**oids/ **V**ector-Spaces/**A**lgebras/ **D**omains) with **Structural Properties**, $\mathcal{S}$: [**O**rder/ **Comm**utativity/ **Can**cellation/**Lin**earity/**D**ifferential-operations/**Char**acteristic forms/**E**xtension/**F**actorization].

Typical 'structured frameworks' (SF) formed in this way are: '**ordered cancellation abelian monoids**', '**commutative rings**', '**partial differential rings**', '**unique factorization domains**', '**ordered abelian semigroups**', and '**fields of prime characteristic**'. The full list is given (as an '**Interdependence Chart**') in the Axiom Manual. Most of these SF are defined in (for instance) the book: '**Algebra**', by **S.Maclane** and **G.Birkhoff** (Macmillan, 1967)[27], where calculational procedures for each SF are specified, together with the principal modes of combination of the SF[28]. It should be noted that the frameworks may be classified into three primary collections – corresponding to **additive**, **multiplicative** and **additive / multiplicative** properties with **the (non)existence of zero / unit / inverse elements, (non)validity of cancellation, (non)dependence on sets of 'parameters'** (as for example, in vector space); and so on. In each collection of frameworks, calculational scope varies from 'minimal' to 'substantial', according to the range of 'extra conditions' imposed. A further **sub**classification involves the **arities** of the operations within each (structured) framework. In the simplest forms there are just unary and binary operations; more generally, one has m**-ary** operations, for $1 \leqslant m \leqslant n$ (say). This

[26]See: 'axiom – the Scientific Computation System', by RD JENKS, R.S SUTOR, et al., pp 742 (Springer, 1992) referred to have as (the) axiom manual

[27]This is used as a basic reference here(cited as 'M/B') – along with Geddes / Czapor / Labahn (G/G/L), op cit. Other references are introduced later, to cover basic Analysis, and synopses of Algebra and of Analysis.

[28]Here, 'SF' means *algebraic* structured framework 'GSF' $\equiv$ *general* SF

increases the number of possible extra conditions considerably, but the classifications remain essentially unchanged. Another variation on this theme allows certain operations to be w.d. only over proper **subsets** of the underlying framework sets – giving rise to **partial SF** of various sorts. Other classes of SF appear naturally in treatments of the foundations of Computer Science, and in Universal Algebra. These variants are identified as they arise in the main discussion of Σ-facilities, or in the treatment of **approaches** to general ME, where **G**SF play a basic role.

In Axiom, every object has a definite **type**, which determines how it is handled in the system, as well as the possible choice of operations open to the user. Every SF is associated with elements of certain types; conversely every choice of type corresponds to only 'the right sort of SF', and in this way consistency is maintained. For Mathematica, by contrast, all operations are **rule**-based: objects are not typed. Consequently, every structure has to be defined 'from scratch' and so cannot be invoked 'dynamically'. In spite of this fundamental difference the notion of 'synthesizing P_Σ / L_Σ from Axiom, Mathematica, and other sources' is not absurd, since one is concerned with **methods**, **procedures**, **algorithms**, **layout** / **format**, **user-facilities**, ..., not merely formal structure. Moreover, the objects (O.T.) encountered in Σ cannot be handled in any extension of existing systems along present lines. Rather, a **blending** of attributes from several sources with some innovative ideas – and the addition of extra 'types' – will be required.

The essence of 'algebraic activity' appears to encompass the following **tasks**.

(i) Identification / definition / classification of SF.

(ii) Specification of '**elementary computation**'

(iii) Representation of all types of SF

(iv) Study of **interrelations** among classes of SF (and, between **members** of any single class of SF)

(v) Determination of (**near**-)**morphisms** between **classes** of SF (and, between **members** of any single class of SF)

(vi) Study of other types of **basic mappings** involving SF

(vii) Study of **quotient** mappings, for SF

(viii) Study of **inverse mappings**, for SF

(ix) Definition / determinations of **universal elements**

(x) Study of **canonical forms**, over SF

(xi) Unification of items (i) - (x) [via **categories**]

In this list there is no mention of (quasi-)**metrical**, or **topological** conditions - or, of the **cardinality** of the underlying sets, and of the associated SF - sets. Such considerations pertain mainly (though not exclusively) to **analytical** / **topological** / **geometrical** / ... GSF **environments**, founded on the **algebraic** SF. The Theories of **topological groups**, and of **differentiable manifolds**, provide obvious illustrations of this process. Indeed, it is apparent that: **every analytical/topological structure maybe realized over suitable SF**, on this basis[29]. Further, the fundamental analytical operations within these 'richer structures' may also be realized from those already defined for Axiom and some other systems. From this point of view, the possibility of implementing the explicit calculations in the OT of Σ becomes transparent – at least in principle!

Each of the SF in Axiom has been multifariously studied in its own right. For the simpler SF, some of these studies tend towards the arid and pedantic; whereas, the more complex

[29] Many *examples* of such realizations may be recognized from the subsequent discussions of the basic spaces, ${}_SX$

SF (such as **abelian groups**, and **commutative rings**) have intricate theories, replete with elaborate techniques and constructions – even when attention is confined to strictly algebraic (as opposed to analytical) processes. The degrees of coverage of these SF theories vary in Axiom, but are all at least adequate for 'routine computations'; some are quite sophisticated. There is a sound basis for upgrading these facilities, and for **embedding them within various** (more abstract) **structures typical of** Σ-calculations. The diverse **interrelations** among SF are exploited in the efficient implementation of basic SAM algorithms (which are often far from direct realizations of the original algebraic procedures).

The seminal capabilities for algebraic manipulation may be cited (following, e.g.., G/C/L) as: **routines for basic symbolic operations on polynomials**; on **rational functions**; and, on **formal power series**. This, apparently very restricted, scope is actually surprisingly broad – provided that all entities are defined over general SF, and potential constructions[30] are taken into account. Among the key tasks are the following.

(i) Optimization of numerical/formal field operations.

(ii) Use of **modular representations**

(iii) Power-series division via Newton iteration

(iv) General use of Newton iteration

(v) Computation with **homomorphic images**

(vi) Use of **Hemsel-type constructions** (including univariate and multivariate **lifting**

(vii) Determination of **polynomial GCD**

(viii) Polynomial factorization

(ix) Use of **partial-fraction expansions**

(x) Treatment of **simultaneous linear equations**

(xi) Treatment of sets of **linear equations**

(xii) Techniques for **polynomial ideals/Gröbner bases**

(xiii) Algorithms for **rational-function integration**

(xiv) The Risch algorithm for **purely transcendal** integrands

(xv) The Risch algorithm for **purely algebraic** integrands

(xvi) The Risch algorithm for **mixed transcendental/algebraic**($\equiv$ **elementary**) integrands. (The term 'transcendental' covers both 'logarithmic' and 'exponential'; all of the Trigonometrical' and 'Hyperbolic' functions are included here)

(xvii) Most of the tasks (i)-(xvi) are formulated initially for the **uni**variate case'; extensions to classes of **multi**variate object are required.

Notice that, although '**integration**' is usually classed as 'Calculus' depending on 'guesswork', 'tricks', and a few simple methods ('integration by parts','substitution', 'partial fractions', 'recurrence formulae', etc) the **general solution** (ultimately based on Hermite's decision procedure for rational integrands, and Liouville's representation theorems) involves results from **Differential Algebra** and **Algebraic Geometry**[31]. The crucial point however is that 'most' Elementary integrands are **not 'integrable in finite terms'**, and so analytical (approximative)

[30] For example: **Multilinear Algebra** is founded on these procedures, if Standard Linear Algebra is assumed. More generally, all ME obtainable through species of **approximated** over polynomial/rational-function domains are covered

[31] By contrast, **differentiation** is almost trivially algorithmic – and was the first symbolic operation implemented for computers

techniques are 'the rule', not 'the exception', in studying the properties of functions defined (or otherwise encountered) as indefinite integrands. Indeed, 'exactly calculable objects' are **exceptional** in mathematics[32]. Consequently, a systematic development of generalized **approximation** processes ('approaches', in Σ) is indispensable, if the power of existing SAM packages is to be effectively usable in mathematical research. This holds as much for thee axiomatic formulations, typified by Bourbaki, as for the problems routinely met in 'applicable analysis', for instance. The central goal of Σ is to embed exact calculations, i all mathematical domains, 'as naturally as possible', within diverse **approach-frameworks**. In this way, arbitrary ME generated in any 'theory' become **approachable** (to varying 'levels of precision'), qualitatively new issues are raised, and unchartered terrain is discovered! These matters are discussed, from several angles, in subsequent Essays. Here, they are raised mainly to indicate the algebraic aspects of 'integration'; and, to emphasize that, from a calculational viewpoint, Analysis constitutes an extension of Algebra[33].

Many of the above tasks amount to the implementation of **computational routines in Commutative Algebra**. The considerable differences between formal algorithms, and their implementations over SAM systems are undefined, if one compares the routines in G/C/L with the corresponding theorems in the book :'Ideals, Varieties and Algorithms', by **C**ox/**L**ittle/**O'**shea (Springer, 1992)[34]. Tasks (i)-(vi) are tacitly taken for granted in C/L/O'S (where SAM packages are used) to produce succinct formulations; yet it is precisely these quasi-arithmetic procedures that render the algorithms viable – especially, minimizing the devastating 'intermediate expression swell' and maximizing 'garbage collection', during computations[35]. The pure theory, in other words, requires radical modification before it can be used effectively. Thus it is appropriate to consider **two classes of fundamental theorems**: those **underlying** the formal algorithms; and, those necessary for developing efficient **implementations** – call these classes C^F, C^I, respectively. It happens frequently that results in C^I are developed – or modified from other contexts – to improve the implementation of some formal procedures in C^F. The principal theorems in C^F constitute formal procedures for:

(i) Canonical representation of 'basic algebraic objects'

(ii) Calculation of the GCD for n-variable polynomials (n $\geqslant$ 1) over various SF

(iii) Partial-fraction expansions for n-variable rational functions (n $\geqslant$ 1)

(iv) factorization of n-variable polynomials (n $\geqslant$ 1) over specified SF

(v) Resultant calculation for n-variable polynomials

(vi) Calculations for **PS-ideals**, and for **polynomial ideals** (over n-variable domains, n $\geqslant$ 1)

(vii) Elimination, and related problems in commutative algebra

(viii) Basic calculations in **group theory**

(ix) Basic calculations over**other SF**

(x) Basic results in **matrix theory**

(xi) Solution of sets **nonlinear** (polynomial) **equations** in n variables (n $\geqslant$ 1)

---

[32]This point will be elaborated often in later Essays

[33]Purely algebraic techniques ,ay be used within combined algebraic/analytical GSFI but analytical operations are simply undefined over most of the basic SF

[34]An even greater contrast is formed with the classic treatise by O.ZARISKI and P. SAMUEL $\equiv$ Z/S

[35]That is, the explosive growth of coefficients during computations (often in cases where the final results are 'small') and the removal of expressions as they become redundant

Notice that many of the later facilities make use of earlier ones – and, that the distinction between a built-in procedure and a 'library routine' is not always clear-cut. In particular the **GCD**, and **factorization**, procedures involve the use of p-**adic expansions**, **Netwon iteration**, and **Hensel lifting** (all within C^F), while efficient implementation of such formal procedures requires, for instance, **arithmetically optimal schemes** (to reduce 'intermediate expression swell') and the determination of **bounds** for the maximal number of steps informal procedures (tasks within C^I).

The overall scope of Axiom maybe assessed, on a very basic level, from the **'partial list of operations'** given on pp 627-290 of the Manual. This reinforces the claim that large tracts of quasi-computational procedures (in all areas of mathematics) are potentially realizable from a relatively small 'core' of fundamental algorithms over the (algebraic) SF – through extension to GSF for 'standard approximation environments'. The dominant aim for Σ is a natural **continuation** of this scheme, to cover GSF over all possible **approach** environments, corresponding to the 'basic spaces' $_SX$, together with abstraction, analogy and inverse analogy. Another aspect of the design of Axiom contributing to its efficiency and scope, is the complex **hierarchy of data structures** over which it operates. The range of admissible data **types**, and their interdependencies, determine the variety of ME that may be handled efficiently, as well as (substantially) the '**expressive power**' of the associated programming language[36]. This combination of flexible data structures, versatile SF and 'strong typing' (covering both algebraic and data types) with associated domain/category structures[37],may be incorporated in the specification of P_Σ/L_Σ. Indeed, many of the quasi-effective results in **Zoriski/Samuel (op. cit.)** appear to be implementable within Axiom – as has already happened in various algebraic - geometric investigations.

The classes C^F, C^I, of basic results, correspond principally to C/L/O'S (or Z/S), and G/C/L, respectively – as far as style of proof is concerned – but the identification of key theorems in **both** classes is best made from G/C/L alone. This is because the other books scarcely even mention the crucial constructive algebraic algorithms – even though their proofs are formal and non-numerical. This reflects the different objectives of books on 'pure' and 'applied' algebra; which, in turn exemplifies the fundamental confusion over 'pure' and 'applied' **mathematics** as a whole. The 'Σ -view' is that this dichotomy (a comparatively recent phenomenon[38]) proves, on informed examination, to be largely spurious; and, that SAM packages have already inaugurated a re-unification of Mathematics, along with a rapprochement between Mathematics and The Sciences. It should be remarked that, even the decision procedures for (indefinite) integration and for the (closed form) solution of (ordinary) differential equations have now been given essentially **algebraic** formulation – the results on bounds for the torsion on Jacobian varieties (used by Davenport, for purely algebraic integrands) being avoided in the 'rationalization procedures' of Rothstein/Trager/Bronstein, involving 'resultant - based representations'. Nevertheless, the Risch 'Jacobi-variety scheme' will be outlined later, on account of its elegance, and, of the inobvious links that it forges between ostensibly disparate domains. For this **outline of the key results** underlying SAM packages (as well as for later reference) it is useful to list some important books/papers – with brief comments.[39]

Maclane/Birkhoff (M/B) : Algebra

Most of the SF are treated here – mainly, from a non-constructive (Category-theoretic) perspective many of the associated structural developments are covered via 'forms' and 'morphisms'. Lattices are included but field extensions (crucial for 'symbolic' integration) are barely mentioned (but cf. 2nd Ed. for extra topics).

[36]There is also interactive 'user' language, but routines are implemented in the programming language.

[37]There 'categories' are not to be confused with algebraic ones

[38]For Euler, Cauchy, Gauss and Riemann, Mathematics was indivisible. The gradual rise of 'Axiomatics' (now somewhat in decline) has promoted this schism

[39]See the main Bibliography for detailed references – and, for selected references taken from the bibliographies of the books/papers cited here.

Cox/Little/O'Shea (C/L/O'S): Ideals, Varieties and Algorithm

This covers the rudiments of computational commutative-algebra / algebraic-geometry – but it does this in a comparatively ineffective way, though references is made to SAM texts/articles for details of some procedures. A correspondence ('dictionary') is established between 'algebraic' and 1geometrical' objects. Many pre-algorithms are treated - usually with the aim of SAM-implementation.

Zariski/Samuel (Z/S): Commutative Algebra

This two-volume treatise covers all of the commutative algebra considered necessary (in 1960) for a rigorous development of Algebraic Geometry subsequent formulations (in terms of '**Sche mes**' (see, e.g., **Harte Horne, R.**, '**Algebraic Geometry**', Springer, 1977), or in terms of **complex manifolds** (see, e.g., **Griffiths, P, and J. Harris**, '**Principles of Algebraic Geometry**', Wiley, 1978)) already offer possibilities for effective SAM routines; but these are not directly related to the basic algorithms for Axiom or other packages. All of the background results for **Differential Algebra** (including FPS) are given in Z/S; but the body of theorems/procedures of particular relevance for SAM packages is to be found mainly in various papers (and some PhD theses)[40], though much of this work is ultimately based on the treatment of **Ritt** in his book '**Differential Algebra**' (AMS, 1950; repr. Dover 1966). See also: **E.R.Kolchin**: '**Differential Algebra and Algebraic Groups**', AP, 1977.

Geddes/Czapor/Labahn (G/C/L): 'Algorithms for Computer Algebra'

This is the most thorough account of the **basic** algebraic procedures underlying SAM systems. Although it is concerned more with implementation than with 'pure algebra', most of the fundamental theorems are carefully stated (with references) and some of them are proved! The primary emphasis however m is on detailed presentation of the key algorithms – in a 'descriptive/Pascal-like language' covering the essential steps economically.

Connell(C): Modern Algebra: A Constructive Introduction

The basic procedures are treated in great detail for a variety of SF; an excellent complement to G/C/L, with much extra material – some analytical.

Knuth(K): 'The Art of Computer Programming Volume 2: Semi numerical Algorithms

Detailed analyses of the basic arithmetic and quasi-arithmetic (e.g., GCD) processes, including estimates of space/time requirements, are given. SAM procedures, as such, barely existed when the first edition was written, but this material is highly pertinent for effective methods/efficient implementations.

Davenport(D): 'On the Integration of Algebraic Functions'

This is a detailed account of Risch's approach to the integration problem with the (partial) implementation of algorithms for purely-algebraic integrands. The algebraic-geometric background is covered, and procedures based on fundamental results of Bliss/Coates, and of Manin, are used to derive (non)termination-criteria, in cases where 'elementary representations' of the integrands may not exist.

[40] See, e.g., papers by Risch; Rosenlicht; Suiger; Epstein/Caviness; Rothstein/Caviness; Bronstein and the theses of Rothstein; Trager

Zippel(Z): 'Computer Algebra'

Here, the algebraic algorithms covered by G/C/L, and some procedures for power-series, differential equations, integration, etc, are treated within a broad framework – combining historical, mathematical and computational content. The algorithms are implemented in MACSTMA. It is less systematic than G/C/L, but more stimulating (and, in places, at a higher level).

Henrici (H_1, H_2, H_3): 'Applied and Computational Complex Analysis' (Vols 1,2,3)

These books contain a vast collection of quasi-algorithmic procedures ranging over many aspects of Complex Analysis. The broad headings are:

H_1 Power Series / Integration / Conformal Mapping / Location of Zeroes

H_2 Special Functions / Integral Transforms / Asymptotics / Continued Fractions

H_3 Discrete Fourier Analysis / Cauchy Integrals / Construction of Conformal Maps / Univalent Functions

Most of this material is directly relevant for the development (of extension) of SAM routines. It complements the algebraic schemes that have, so far, dominated SAM packages – though it also depends on them, for implementation Henrici's approach furnishes a bridge between existing systems and the foothills of SIGMA.

Further progress towards the processor P_Σ and its associated programming language, L_Σ, may be made through another seminal reference:

Bishop/Bridges (B/B): 'Constructive Analysis'

This is a revision and extension **'Foundations of Constructive Analysis'**, (B), by **Bishop** alone – where an object is constructively defined (or 'B-Constructive') if, and only if, there is a procedure for its determination to any desired degree of accuracy – a deceptively simple criterion, for results in domains far from the directly numerical! In spite of many problems in formulation and interpretation, a surprisingly large body of fundamental mathematical results has been rendered constructive, in the above sense, and additional results appear regularly.[41]

It is useful to introduce here the symbol M_t^C for the **collection of all ME, in M_t 'ratified as B-constructive over Σ_t**, with the associated collections, M_t^{*C}, of **all** B-constructive ME over Σ_t (ratified, or not), and M^{*C} (corresponding to M^*). The 'status' of ME in M_t / M_t^C is logically clear: such ME may be used in calculations, without loss of consistency; but (B-) constructiveness will in general, be reduced as a result.

The extension of these ideas to Σ leads to the notion that: an ME is Σ **-constructive** IFF **there is a CF within which it is arbitrarily closely approachable.**[42]

This type of constructiveness is characteristic of Σ, and pervades these Essays. In some contexts it is equivalent to B-constructiveness – but it is more general, allowing estimates or comparison to be made (so far) ultimately-numerical criteria have not been obtained (though they need not be precluded). Thus, **there is a 'longterm trend' from purely existential definitions of ME to B-constructivity, via Σ -Constructivity** – and this is the tendency of mathematical information in general. One of the purposes of Σ is to furnish an environment conducive to the promotion and exploitation of this trend. In fact, B-constructivity is somewhat too restrictive for some facets of mathematical investigation; so the ultimate goal also involves the accommodation of certain **heuristic / speculative ingredients**. This is, of course, in line with the way in which research normally proceeds; it is just a matter of filling it into an 'IB/SAM setting'.

---

[41] The Bibliography from B/B is 'incorporated' here, for references in later Essays – in the sense that items from it are cited, as appropriate, without further comment.

[42] CD $\equiv$ Calculational Framework – as introduced in E_0.

Key papers on the basic SAM algorithms are listed at the ends of chapters in G/C/L. Apart from several PhD theses, most of these papers appear in J. Symbolic Computation and in various 'Proceeding volumes' (in the *LNCS series*). More advanced (theoretical) developments – especially, for elementary representation of solutions of (O)DE, may be found in **BAMS**, **Pacific J Math**., **American J. Math**.,etc., while more practically orientated work appears in SIAM K. Computing, or in some journals for 'applied algebra'.[43]. The G/C/L Bibliography is also 'incorporated' here[44], so **it suffices to add items of a more advanced character** – many of which are also pertinent to the Σ -developments discussed in other Essays.

Rothstein/Caviness (R/C): 'A Structure Theorem for Exponential and Primitive Functions'

Epstein/Caviness (E/C): 'Elementary Proofs of Algebraic Relationships for the Exponential and Logarithmic Functions'

Singer (S$_1$): 'Liouvillian Solutions of n-th Order Linear Differential Equations'

Risch (R): Algebraic Properties of the Elementary Functions of Analysis'

Mack (M): 'Integration of Affine Forms Over Elementary Functions

Singer (S$_2$): 'Elementary Solutions of Differential Equations'

Ax (A): 'On Schanuels Conjectures'

Rothstein (R)L 'Aspects of Symbolic Integration and Simplification of Exponential and Primitive Functions'

Caviness (C): 'On Canonical Forms and Simplification'

Risch(R): 'Implicitly Elementary Integrals'

Bronstein(B$_1$ - B): 'A Unification of Liouvillian Extensions','The Transcendental Risch Differential Equation', 'Integration and Differential Equations in Computer Algebra'

Bronstein (B): 'On Solutions of Linear Ordinary Differential Equations in their Coefficient Field', 'Integration of Elementary Functions'

Rosenlicht (R): 'Liouville's Theorem on Functions with Elementary Integrals'

Risch (R): 'The Problem of Integration in Finite Terms'

Coates (C): 'Construction of Rational Functions on a Curve'

Manin (M): 'Algebraic Curves over Fields with Differentiation'

Manin (M): 'Rational Points of Algebraic Curves over Function Fields'

It should be evident by now that the crucial capabilities for most SAM routines involve the **manipulation of polynomials and rational functions (in n variables, over classes of algebraic structures**. Whilst it may be surprising that a considerable body of computational procedures can be developed from such a narrow base, the inherent **limitations** of this approach should also be appreciated. Indeed, the entire Σ -scheme has evolved from the realization that **current SAM packages maybe 'embedded' in a broader environment, at a higher level of abstraction, in which generalized approximation schemes are well defined and arbitrary theorems may be deployed 'operationally'.** The manipulative (and numerical) facilities (appropriately extended) may then be used to implement theorems for application to expressions

[43]Some of these items are also in G/C/L

[44]As for B/B, this is an acknowledgement. The Bibliography of the present book is very large. Most entries result from work on the basic projects, on 'General Approximations', and 'Mathematical Analogy'; but (as is usual), many have been selected from a variety of existing bibliographies – each of which is identified at its first appearance.

of generated in interactive computation. Consequently, it suffices, here, to give relatively formal statements of the theorems central to SAM procedures for fields of rational functions in n variables (mainly, following G/C/L).

These theorems are concerned principally with the inversion of (classes of) homomorphisms between 'computational environments'. In particular, the classes of modular homomorphisms, and of, evaluation homomorphisms play basic roles. These mappings are defined as follows.

Modular:

$$\phi_m : \mathbb{Z}[x]_v \rightarrow \mathbb{Z}_m[x]_v$$
$$\phi_m(x_i) := x_i(1 \leqslant i \leqslant v)$$
$$\phi_m(a) := \text{rem}(a, m)$$

for all coefficients $a \in \mathbb{Z}$

Evaluation:

$$\phi_{x_i-\alpha} : \mathbb{D}[x]_v \rightarrow \mathbb{D}[(x)_v \backslash x_i]$$
$$\phi_{x_i-\alpha}(a(x)_v) := a(x)_v|_{x_i=\alpha}$$

Here, $\mathbb{Z}[x]_v$ comprises all v-variate polynomials over $\mathbb{Z}$, and $\mathbb{D}[x]_v$, all v-variate polynomials. The first inversion procedures involve so-called **C**hinese-**R**emainder-**T**heorem (CRT), and Interpolation(Int) Algorithms – which entail the use of several prime moduli, p_i, with associated sets of sub-problems in $\mathbb{Z}_p[x_1]$, and subsequent inversion (via Int/CRT algorithms) to obtain solutions of the original problem(s) in $\mathbb{Z}[x]_v$.

If π^k_m denotes a generic problem over $\mathbb{Z}_m[x]_k$, and σ^k_m, its solution(s), then the above process may be depicted succinctly as:

$$\pi^v \xrightarrow[\phi_{p_i}]{\text{mod}} \{\pi^v_{p_i} : 0 \leqslant i \leqslant n\} \xrightarrow[\phi_{I,l}]{\text{eval}} \{\pi^1_{p_i,l} : 1 \leqslant l \leqslant N_i\}$$

$$\rightarrow \{\sigma^1_{p_i,l} : 0 \leqslant i \leqslant n; 1 \leqslant l \leqslant N_i\} \xrightarrow{\text{int}} \{\sigma^v_{p_i,l} : 0 \leqslant i \leqslant n; 1 \leqslant l \leqslant N_i\} \xrightarrow{\text{CRT}} \sigma^v_1$$

NOTE that various forms of interpolation may be used to invert mappings of the mod/eval types.

Another 'inversion scheme' depends on p-adic/id-adic iteration, and Hensel lifting – which has significant computational advantages over the CRT-based techniques. Further variants are, mainly, refinements/extensions of these two schemes. The restriction of Newton-Iteration to the solution of single, univariate polynomial equations is overcome by the introduction of **bi**variate Newton iteration (via the Hensel Construction). The **linear p-adic** (Newton) iteration guarantees convergence (for any initial solution)l indeed, only a finite number of steps is required. The exact solution (to $F(u) = 0$) is assumed to have the form $u(\underline{x}) = u_0(\underline{x}) + u_1(\underline{x})p + \cdots + u_n(\underline{x})p^n \in \mathbb{Z}[\underline{x}]$, if n is taken 'sufficiently large'. The **iteration** generates an order -k approximation from an order - $(k-1) approximation - -intheassumedp - adicrepresentationof$u((x)).

The **id-adic iteration** may be used to invert multivariate **evaluation** – homomorphisms – for instance, $\phi_I : \mathbb{Z}_p[x]_v \rightarrow \mathbb{Z}_p[x_1]$, with 'kernel'
$I \equiv \langle x_2 - \alpha_2, \ldots, x_v - \alpha_v \rangle$, corresponding to the 'evaluations' (i.e., substitutions of α_i for $x_i, 2 \leqslant i \leqslant v$). If I is viewed as an **ideal** in $\mathbb{Z}_p[x_1] = \mathbb{Z}_p[x]_v/I$. Thus: **given** $u^{(1)} := \phi_I(\tilde{u}) \in \mathbb{Z}_p[x_1]$ as an order -1 id-adic approximation to $\tilde{u}$, where $\tilde{u}$ satisfies $F(\tilde{u}) = 0$, and $F(u) \in \mathbb{Z}_p[x]_v[u]$, the desired iteration produces $u^{(k+1)}$ from $u^{(k)}$ – through the finite quasi-Taylor Series representation: $a(x+y) = a(x) + a'(x)y + b(x,y)y^2$, which holds for any univariate polynomial $a \in D[x]$, if it is viewed (for $a(x+y)$) over $D[x][y] = D[x,y]$. Here, this representation is applied to $F(u^{(k)} + \Delta u^{(k)})$, where $\Delta u^{(k)} := u^{(k+1)} - u^{(k)}$.

Two basic problems which motivate **Hensel - type lifting procedures** are (1) **factorization** of multivariate polynomials, and (2) **GCD calculation** for pairs of multivariate polynomials. In **case**(1), the problem is equivalent to that of finding polynomials $u, w,$ in$\mathbb{Z}[x]_v$ such that

$$F(u, w) := a - uw = 0,$$

$$u(x)_v \equiv u_0(x_1)(\text{mod}\langle I, p\rangle)$$
$$w(c)_v \equiv w_0(x_1)(\text{mod}\langle I, p\rangle)$$

where $a(x)_v$ is the polynomial to be factorized. For the GCD, a composite homomorphism of the form $\phi_{I,p} : \mathbb{Z}[x]_v \to \mathbb{Z}_p[x_1]$ yields the basic GCD problem for a_0, b_0in$\mathbb{Z}_p[x_1]$ – which is readily solved by the standard Euclidean algorithm. Accordingly, the main task now becomes that of 'lifting' the image polynomials $u_0 = G \in D(a_0, b_0)$ and $w_0 := a_0/u_0$ from $\mathbb{Z}_p[x_1]$ to u, w in $\mathbb{Z}[x]_v$. This process is known as multivariate Hensel lifting – which is a generalization of the basic **uni**variate 'Hensel Construction'. Both of these schemes are **realizable** as (SAM) algorithms.

Many variants of the above interpolation CRT, iteration, and Hensel procedures corresponding to particular computational problems (especially, **coefficient-growth**); but the underlying mathematical structure is not affected. Estimates of running times or computer storage requirements are central for the development of efficient algorithms – and entail careful analysis of the basic algorithmic schemes. For Σ , however such considerations are peripheral, since it is **assumed** that P_Σ is obtained by embedding some combination of (augmented) current SAM packages within a 'calculational environment' capable of accommodating mathematical processes at arbitrary levels of abstraction/sophistication. This is exemplified by the formal treatment of '**approaches to ME**', and, of **operational theorems**. Problems of **elimination** of variables from (finite) sets of multivariate polynomial conditions – and, in particular, of the solution of simultaneous multivariate polynomial equations – have been partially solved through the use of so called **Gröbner bases** of polynomial ideals. more general 'canonical bases', whose **existence** was established by **Hironaka**, for ideals of formal power series[45], make it possible to decide (in principle) whether a given FPS (say, 1) belongs to an arbitrarily specified FPS **idea** (say, $I \equiv \langle p_1, \ldots, p_n\rangle$), by transforming I into canonical form. In the special case where the p_k are polynomials (rather than FPS) it was conjectured by Gröbner that a **decision procedure** could be established – and such an algorithm was derived by Buchberger[46]. Collections of 'Gröbner-basis routines' are now implemented in several SAM packages. Among the **applications** of these routines are several procedures in **constructive commutative algebra**[47], and alternative schemes for **factorization** and **GCD-computation**. The 'iteration', and 'k-variate lifting' procedures may be depicted (respectively) as follows.

$$\pi^v \xrightarrow[\phi_p]{\text{mod}} \pi_p^v \xrightarrow[\phi_I]{\text{eval}} \pi_p^1 \to \sigma_p^1 \xrightarrow[\text{iter}]{id\text{-adic}} \sigma_p^v \xrightarrow[\text{iter}]{\text{p-adic}} \sigma^v$$

$$\pi^v \xrightarrow[\phi_I]{\text{eval}} \pi^1 \xrightarrow[\phi_p]{\text{mod}} \pi_p^1 \to \sigma_p^1 \xrightarrow[\text{iter}]{ip\text{-adic/Hensel}} \sigma_{p^l}^1 \xrightarrow[\text{iter}]{id\text{-adic}} \sigma^v$$

The most significant **analytical** SAM procedures are concerned with **indefinite integration**, (closed-form) **solution classification for ODEs**, and the treatment of **Special/Elementary functions**. In fact, these procedures are also based on **algebraic** structural relationships, formal properties of FPS, and on the routines for GCD, factorization, partial fractions, resultants; and so on. The integration decision procedure derived by Risch for purely transcendental integrands (and further indicated for purely algebraic integrands) has now been modified and extended to cover **arbitrary Elementary integrands**. The basic principle is to consider the integrand as a member of a tower of (successive) transcendental or algebraic extensions of the 'ground field', $K[x]$, of the integrand – in such a way that the integrand is rationally expressible

[45]This was part of a solution of the 'resolution of singularities' problem for algebraic varieties: Ann.Math. 79(1964) 109-326

[46]PhD Thesis (Innsbruck, 1965)

[47]See, e.g., 'Computational Aspects of Commutative Algebra (L.Robbiano, ed.) (AP, 1989), and C/L/O'S

over the tower. There after, a combination of Hermite's algorithm (for rational integrands), 'square-free factorization', and a resultant - based finite representation (originally due to Trager, extended to transcendental integrands by Rothstein, then to algebraic integrands, by Trager – and finally, generalized to arbitrary Elementary integrands, by Bronstein!) has produced a complete decision procedure – provided that **the (non-)solvability of Risch's (associated) ODE system is decidable.**

This highly condensed description[48] omits the initial development by Davenport, based on an outline by Risch) off the scheme for purely algebraic integrands. Here, several results from Algebraic Geometry are required, to guarantee that the alleged decision procedure is indeed finitary; and the subsequent implementation is incomplete. Nevertheless, this approach offers insights into general problem that are less evident – or hidden – in the later formulations[49].

In the calculation of resultants, one scheme involves (finite) Polynomial Remainder Sequences (obtained in GCD computation by Euclid's algorithm). The Trager/Rothstein/Bronstein procedure may be specified (for a rational integrand, $f \equiv A/D$) initially through the formulae:

$$\int A/D = \sum_{\{\alpha = R(\alpha)=0\}} \alpha \log \gamma(\alpha) = \sum_{i=1}^{m} \sum_{R(\alpha)=0} \alpha \log S_i(x,\alpha)$$

$$R(z) := \text{res}_x(A - zD', D) \equiv \text{res}_x(B(z), D)$$

$$\gamma(z) := \text{gcd}_x(B(z), D)$$

$$S_i(x,z) := \text{remainder of degree} i \text{in PRS for} R(z).$$

For the more general integrands purely transcendental, purely algebraic, and Elementary – it is found that, if the integrands are expressed **rationally** (over suitable towers of field extensions) then various elaborations of Trager's representation in terms of the PRS for $R(z)$, suffice for a complete decision procedure for Elementary integrands.

The Risch ODE system associated with the above integration algorithms is ostensibly move complicated than the original integration problem. It is found, however, that (in the transcendental case, at least) a recursive solution procedure may be used. In the algebraic/Elementary cases, some apparatus from Algebraic Geometry is required (Puiseux series, divisors, residues, ...), but forms of Trager's scheme remain adequate. For the '**ODE Problem**', the primary aim is to obtain closed - form solutions for particular classes of equations – for instance, those with 'Liouvillian coefficients'. The main results in this area due to Singer, who proved that Liouvillian solutions occur in pairs, $y, \exp\{\int u\} =: z$

Theorem (Singer)

For an OD-operator, L, of order n, over F (a differential field whose field of constants is algebraically closed), existence of an Elementary y (in a Liouvillian extension of F) satisfying $L(y) = 0$, implies that of z, with z'/z algebraic over F and $[F(z'/z) : F] \leqslant I(n) \leqslant \max(b_n, n!I(n-1))$. Here, $b_n := (\sqrt{8n}+1)^{2n^2} - (\sqrt{8n}-1)^{2n^2}$, and $[F(w) : F]$ denotes the algebraic degree of $F(w)$ over F.

NOTES

(i) Not **all** solutions of $L(v) = 0$ have to be of this form

(ii) If the coefficient of L are **rational** functions, then there is a classical algorithm for finding all exponential solutions, z, with $\bar{z}/z \in \bar{K}F$. Singer has generalized this to cover certain classes of extensions of fields of rational functions.

[48]Parts of this section are based on a review by Bronstein (transl., Plenum, 1993, of: programmirovanie No5, Sept/Oct,1992, pp 26-44), where full references are given

[49]Complete/efficient solutions of the associated Risch ODE system are still an obstacle to full implementation

(iii) In practice, **rational** solutions are sought before exponential solutions; a SAM algorithm has been implemented for some lower-order cases of this process[50].

If $L(y) := \sum_{0 \leqslant k \leqslant n} A_{n-k}(x) y^{n-k}$, then it is found that $u := z'/z$ satisfies a non linear ODE[51] of the form $\sum_{0 \leqslant k \leqslant n} A_{n-k}(x) P_{n-k} = 0$, where the P_i are recursively defined polynomials in $u \equiv u^{(0)}, u^{(1)}, \ldots, u^{(n)}$. Singer's algorithm is based on this representation.

Although an effective procedure for rational coefficient functions in $L(v)$ was obtained in the Nineteenth Century, Singer's technique is valid for more general coefficients, and forms the basis for SAM procedures. Even here, however, the (practical) requirement of **optimal** bounds on $I(n)$ has restricted full implementation to $n = 2, 3$ (with possibilities up to $n = 11$). In Bronstein's review (**op. cit.**), Singer's algorithm and the corresponding algorithm for the associated Riccati equation (for the rational, and exponential, solutions, respectively) are discussed in the case where $L(v)$ has rational coefficient-functions. The overall scheme is analogous for more general operators, $L(v)$, but presents serious problems over full implementation. Several other types of ODE solutions (e.g., in **series** forms, or for **systems** of ODE) have now been implemented – or are being tested – but, as might be expected, corresponding facilities for **PDEs** are considerably less developed, apart from various specialized schemes (e.g., in General Relativity) that are listed as '**library routines**'.

From this sketch of the central features of advanced SAM packages it emerges that the core of fundamental (mostly, algebraic) theorems, from which such packages are developed, is surprisingly small (though some of these theorems have several computational variants). They correspond to basic mathematical procedures, namely:

(a) partial-fraction decomposition;

(b) **GCD** calculation;

(c) **factorization** (over various domains);

(d) **resultant-calculation**, with applications;

(e) **elimination** (from sets of polynomial conditions);

(f) solution of (simultaneous) **polynomial equations**;

(g) **integration** of (arbitrary) **rational** functions;

(h) **integration of Elementary functions** (with special cases);

(i) solution of classes of ODE's;

(j) procedures for **finite** groups;

(k) procedures for **infinite/continuous** groups

NOTE that all of the algebraic procedures apply to k-variate objects (K $\geqslant$ 1) over specified domains; and, that general operations on **rational functions** are covered here. On this basis, the corresponding key theorem(s) for procedures (a)-(k), above may be exemplified, schematically, as follows (in **effective** version)[52].

T$_a$: For polynomials a, b, c in a Euclidean domain $F[x]$, with GCD $(a, d) = g$, g/c, deg$(c) <$ deg(ab/g), there is a unique constructive representation of c as $\sigma a + \tau b, \sigma, \tau \in F[x]$, deg$(\sigma)$ $<$ deg(b/g), deg$(\tau) <$ deg(a/g)

[50] This bound on $I(n)$ is not optimal, but a group-theoretic criterion-resolvable, so far, for $n \leqslant 11$ - **can** produce optimal bounds. See: Applic. Alg. Engrg. Commun. Comp 2 (1992) 171-1193(F. Ulmer)

[51] A so-called associated Riccati equation (of order n-1, here)

[52] There are several algorithms for many of these procedures: the essential point is that they can now be performed **efficiently** in most situations met in 'applicable mathematics'. Full references for these theorems are given in the Bibliography.

T$_{b1}$: Application of modular/evaluation homomorphisms to reduce problems in $\mathbb{Z}[x]_v$ to problems in $\mathbb{Z}_p[x_1]$.

T$_{b2}$: Repeated use of variants of the standard Euclidean algorithm for pairs of univariate polynomials over finite coefficient fields.

T$_{b3}$: Inversion of the homomorphisms of T$_b1$ through a combination of CRA, interpolation, Newton/p-adic iteration and forms of Hensel lifting.

T$_{c1}$: Factorization of polynomials in $\mathbb{Z}_p[x]$ – the 'case $m = 1$' of Berlekamp's algorithm for polynomials in GF(p^m)

T$_{c2}$: Reduction of factorization of polynomials in $\mathbb{Z}[x]_v$ to that for $\mathbb{Z}_p[x_1]$

T$_{c3}$: Factorization in $\mathbb{Z}_p[x_1]$

T$_{c4}$: Hensel-type lifting/p-adic iteration in $\mathbb{Z}[x]_v$

T$_{c5}$: Factorization of polynomials in **algebraic** extensions, $F[\alpha]_k$ through **Trager's** algorithm, or the LLL algorithm

T$_{d1}$: **Univariate resultant calculation** through PRS procedures

T$_{d2}$: Collins' **modular resultant algorithm** (for pairs of s-variate polynomials, $f(x)_s$, $g(x)_s$, relative to and x_i, with the other x_j 'fixed'). **NOTE:** this produces several univariate resultant calculations (see T$_{d1}$, above)

T$_{e1}$: Resultant-based procedures

T$_{e2}$: Methods involving the **construction of Gröbner bases**

T$_{f1}$: Fraction-free Gaussian elimination

T$_{f2}$: Minor-expansion of determinants

T$_{f3}$: Homomorphism-reductions/lifting

T$_{f4}$: Elimination routines

T$_{f5}$: Basic procedures from commutative algebra/algebraic geometry

T$_{g1}$: **Hermite's algorithm**

T$_{g2}$: **Horowitz' algorithm** (for the rational part)

T$_{g3}$: **Rothstein/Trager** (resultant-based) **representation** (for the logarithmic part)

T$_{h1}$: **Risch(transcendental) procedure**

T$_{h2}$: **Risch/Davenport/Coates/Marin (algebraic) procedure**

T$_{h3}$: **Rothstein/Trager representations** extended to **transcendental**, or else **algebraic**, cases

T$_{h4}$: **Bronstein's theorems**, generalizing T$_{h3}$ to **combined** transcendental/algebraic ($\equiv$ **Elementary cases**

T$_{i1}$: **Singer's theorem** (as above) on Liouvillian solutions of Liouvillian ODEs

T$_{i2}$: Implementations for **Frobenius-types series solutions** of classes of ODEs

$\{\mathbf{T}_j\}$, $\{\mathbf{T}_k\}$: Extensive facilities for diverse calculations in both finite and infinite continuous groups have been developed sine the late Seventies (when a computer-based Bibliography[53] of over 600 items was first circulated). Citations of reviews in Mathematical Reviews, Zentralblatt, are given wherever possible most of these (and other facilities have now been implemented.[54]

The Axiom system is probably close to optimal in overall design. Other packages – especially, Mathematica and MACSYMA/Symbolics– have extensive facilities and a broadly comparable scope, but they lack the versatility and coherence of Axiom's elaborate **domain structure.** MAPLE and REDUCE have relatively small 'kernels' of basic procedures, extended by large collections of library routines. Other specialized, packages cover **finite/infinite/continuous groups, algebraic number theory** and various aspects of **differential geometry**, including schemes for **generalized 'separation of variables' (through Lie-group transformations)** for classes of PDEs. Additional 'applications packages' will certainly continue to appear regularly. all of these developments may be followed in the 'dedicated journals'; **J.Symbolic Computation (J.S.C)** and, **Applicable Algebra in Engineering, Communication and Computing (AAECC)**; in **conference proceedings** published by **ACM**, and **Springer (LNCS)**, and in scattered articles in journals for **Approximation Theory, Constructive Mathematics**, and diverse areas of applications.

In view of this vigorous research activity, it is inappropriate to describe in more detail any of the current SAM packages. Subject to the limitations inherent in exact calculation, without the operational approximative use of general theorems, these systems will attain the progressively higher levels of completeness and efficiency that are essential for the successful realization of any Σ –type OIB. The aim of this Essay, as emphasized at the outset is to support the claim that (in principle, at least) **arbitrary theorems may be formulated in various operational versions, potentially implementable over** P_Σ. These various **versions** correspond to the **underlying spaces,** ${}_SX$, specified (in E_1) as 'calculational frameworks'. Although this claim is not susceptible of formal proof, one may (as was argued earlier) justify it substantially by considering **general synopses of Algebra and Analysis** from which all mathematical theories, and their rigorous applications, are ultimately derived.

An equivalent statement of this claim – more directly related to SAM routines – is that: **arbitrary calculations encountered in general theorems may be represented as combinations of 'procedures' over a suitably enhanced SAM system** – in this case, the system P_Σ and its associated OIB. In this context, **'calculation'** involves manipulation and computation (formal or numerical); for Σ , this term also cover **derivation/deduction/estimation**, through the application of compositions of operational theorems. Two synopses of Algebra (Al_1: '**Lectures on General Algebra**', by **A.G.Kurosh** (Pergamon Press, 1965), and Al_2: '**Basic Notions of Algebra**', by **I.R. Shafarevich** (Springer, 1990)[55]) provide an outstanding, near-comprehensive overview of (mainly) '**theoretical** algebra'. A diverse selection of '**applications**' (developed from the underlying theories) may be found in: Al_3: '**Applied Modern Algebra**' by **L.L. Dornhoff and F.E. Hohn** (Macmillan, 1977); Al_4: '**Applied Abstract Algebra**', by **R.Lidl and G.Pilz**(Springer, 1984); Al_5: '**Discrete Mathematical Structures for Computer Science**', by **R.E. Prather** (Houghton Mifflin, 1976). As **additional general references**, the books by **S.Lang: Algebra** (Addison-Wesley, 1965) $\equiv Al_6$, and **N.Bourbaki: Algebra,** I, II (Springer, 1974, 1989, 1990) $\equiv Al_7$, and **Commutative Algebra** (Springer (1972), 1989) $\equiv Al_8$, may be used.

For **Analysis**, the most obvious choice is '**Foundations of Modern Analysis**' ($\equiv$ Vol I of **Dieudonne's 'Treatise on Analysis'**, Vols $I - VIII$($\equiv A_{n_1}$) Academic Press, 1969 - 1978), supplemented with citations of topics from Volumes $II - VIII$ – in order to indicated how the more abstruse theories are essentially reducible to fundamental schemes in Real Complex

[53]'A Bibliography on the Use of Computers in Group Theory and Related Topics: Algorithms, Implementations and Applications' compiled by **V. Felsch** (Aacheu)

[54]The crucial point is that they are implementable - and so may be incorporated in P_Σ

[55]Encyclopaedia of Mathematical Sciences, Vol.11

Analysis, as covered in 'Foundations'.
NOTE that the treatment in Volumes $II - VIII$ is largely orientated towards **maximal applicability** (both within Mathematics proper and for problems in the theoretical sciences). This entails, for instance, the use of **separable spaces**, wherever possible – with cognate simplifying conditions in all of the domains covered. In this way, the hierarchical nature of the theories falling broadly under the heading 'Analysis' is exhibited – along with the dependence on progressively 'elementary' calculational frame works, most of which come within the purview of 'advanced SAM packages'. Other excellent sources for the fundamental development of analytical techniques are furnished by the books:

A$_{n_2}$: '**Analyse: Fondements, Techniques, Evolution**' by **J. Mawhim** (De Boeck-Wesinoel, 1992)

A$_{n_3}$: '**Trigonometric Series**' Vols. $I - II$, by **A. Zygmund** (2nd ed., Cambridge U.P, 1959; repr. 1979)

A$_{n_4}$: **Analysis**, Vols $I - IV$, (Encyclo. of Math. Sciences, vols 13, 14, 26, 27 (Springer, 1989, 1990, 1991, – various authors))

A$_{n_5}$: (**Bourbaki**) Integration.

A$_{n_6}$: (**Bourbaki**) Functions of a Real Variable

A$_{n_7}$: (**Bourbaki**) **General Topology**

A$_{n_8}$: (**Bourbaki**) **Topological Vector Spaces**

A$_{n_9}$: (**Dumford/Schwartz**) **Linear Operators**

A$_{n_10}$: (Deimling) Non linear **Functional Analysis**

A$_{n_11}$: (Zeidlier) Non linear Functional Analysis

For an (algebraic/analytical) subtheory considered, it should be possible to construct a '**decomposition scheme**', indicating how calculations 'at the top level of abstraction' are realizable through a (finite) succession of calculations at a lower levels – culminating in procedures for CF potentially implementable over specifiable extensions of (say) Axiom. The initial program, therefore consists in **listing** (in skeletal outline) **the principal subtheories constituting Algebra/Analysis**, in their 'research manifestations' – and then **producing 'calculational decompositions'**, for several (sample) non trivial subtheories. This process comes as close as is possible to a **proof** of the (potential) adequacy of extended SAM systems to produce viable foundations for Σ , since arbitrary subtheories may be 'tested in this way. In any case, it is indisputable that the books just cited do give a representative overview for the intended assessments.

Since these synopses have been written from long experience by noted authorities, no paraphrase of the contents lists seems appropriate. Accordingly, the (main) entries are now reproduced (with some abbreviations).

Al_1 (Kinosh)

Chapter One: Relations

1. Sets 2. Binary relations 3. Equivalence relations 4. Partial ordering
5. The minimum condition 6. Theorems equivalent to the axiom of choice

Chapter Two: Groups and rings

1. Groupoids, semigroups, groups 2. Rings skew fields, fields 3. Subgroups, subrings 4. Isomorphism 5. Embedding of semigroups in groups and rings in skew fields 6. Nonassociative skew fields, quasigroups Isotopy 7. Normal subgroups, ideals 8. Gaussian semigroups 9. Gaussian rings 10. Dedekind rings

Chapter Three: Universal algebras, Groups with multi operators

1. Universal algebras, Homomorphisms 2. Groups with multi-operators 3. Automorphisms, endomorphisms. The field of p-adic numbers 4. Normal and composition series 5. Abelian, nilpotent and soluble Ω-groups 6. Primitive classes of universal algebras 7. Free universal algebras 8. Free products of groups

Chapter Four: Lattices

1. Lattices, complete lattices 2. Modular lattices 3. Direct unions. The Schmidt-Ore theorem 4. Direct decompositions of Ω-groups 5. Complete direct sums of universal algebras 6. Distributive lattices

Chapter Five: Operator groups and rings modules. Linear algebras

1. Operator groups and rings 2. Free Modules. Abelian groups 3. Vector spaces over skew fields 4. Rings of linear transformations 5. Simple rings. Jacobson's theorem 6. Linear algebras. The algebra of quaternions and the Cayley algebra 7. Alternative rings. Artim's theorem 8 A generalization of Frobenius' theorem 9. The Birkhoff-Witt theorem on Lie algebras 10. Derivations. Differential rings.

Chapter Six: Ordered and topological groups and rings. Normal rings

1. Ordered groups 2. Ordered rings 3. Archimedean groups and rings 4. Normed rings 5. Valuated fields 6. Albert's theorem on normed algebras 7. Closure. Topological spaces 8. Special types of topological spaces 9. Topological groups 10. The connection between topologies and norms in rings and skew fields 1. Galois correspondences. The fundamental theorem of Galois theory.

Al_2 (Shaforevichi)

1. What is Algebra? 2. Fields 3. Commutative Rings 4. Homomorphisms and Ideals 5. Modules 6. Algebraic Aspects of Dimension 7. The Algebraic View of Infinitesimal Notions 8. Noncommutative Rings 9. Modules over Noncommutative Rings 10. Semisimple Modules and Rings 11. Division Algebras of Finite Rank 12. The Notion of a Group 13. Examples of Groups: Finite Groups 14. Examples of Groups: Infinite Discrete Groups 15. Examples of Groups: Lie Groups and Algebraic groups. A. Compact Lie Groups. B. Complex Analytic Lie Groups. C. Algebraic Groups 16. General Results of Group Theory 17. Group Representations A. Representations of Finite Groups. B. Representations of Compact Lie Groups. C. Representations of the Classical Complex Lie Groups. 18. Some Applications of Groups. A. Galois Theory. B. The Galois Theory of Linear Differential Equations (Picard Vessiot Theory). C. Classification of Unramified Covers. D. Invariant Theory. E. Group Representations and the Classification of Elementary Particles 19. Lie Algebras and Nonassociative Algebra. A. Lie Algebras. B. Lie Theory. C. Applications of Lie Algebras. Other Non associative Algebras 20. Categories 21. Homological Algebra. A. Topological Origins. B. Cohomology of Modules and Groups. C. Sheaf Cohomology 22. K-Theory. A. Topological K-Theory. B. Algebraic K-Theory.

Al_3 (Dornhoff/Hohm)

Chapter 1: Sets and Functions

Chapter 2: Relations and Graphs

Chapter 3: Rings and Boolean Algebras

Chapter 4: Semigroups and Groups

Chapter 5: Applications of Group Theory

Chapter 6: Lattices

Chapter 7: Linear Algebra and Field Theory

Chapter 8: Linear Machines

Chapter 9: Algebraic Coding Theory

Al_4 (Lidl/Pilz)

Chapter 1: Lattices

Chapter 2: Applications of Lattices

Chapter 3: Finite Fields and Polynomials

Chapter 4: Coding Theory

Chapter 5: Further Applications of Fields and Groups

Chapter 6: Automata

Chapter 7: Further Applications of Semigroups

Chapter 8: Solutions to the Exercises

Al_5 (Prather)

Chapter 0: Preliminaries

Chapter 1: Algebras and Algorithms

Chapter 2: Graphs and Digraphs

Chapter 3: Monoids and Machines

Chapter 4: Lattices and Boolean Algebras

Chapter 5: Groups and Combinatorics

Chapter :6 Logic and Languages

Al_6 (Lang)

Chapter I Groups. **Chapter II** Rings. **Chapter III** Modules. **Chapter IV** Homology. **Chapter V** Polynomials. **Chapter VI** Noetherian Rings and Modules. **Chapter VII** Algebraic Extensions. **Chapter VIII** Galois Theory. **Chapter IX** Extensions of Rings. **Chapter X** Transcendental Extensions. **Chapter XI** Real Fields. **Chapter XII** Absolute Values. **Chapter XIII** Matrices and Linear Maps. **Chapter XIV** Structure of Bilinear Forms. **Chapter XV** Representation of One Endomorphism. **Chapter XVI** Multilinear Products. **Chapter XVII** Semisimplicity. **Chapter XVIII** Representation of Finite Groups.

$\mathbf{A}l_7$ / I, II (Bourbaki) 'Algebra'

Chapter I Algebraic Structures. **Chapter II** Linear Algebra. **Chapter III** Tensar Algebra, Exterior Algebras, Symmetric Algebras. **Chapter IV** Polynomials and Rational Fractions. **Chapter V** Fields. **Chapter VI** Ordered Groups and Fields. **Chapter VII.** Modules over Principal Ideal Rings. **Chapter VIII** Semisimple Modules and Rings. **Chapter IX** Sesquilinear and Quadratic Forms.

$\mathbf{A}l_8$ (Bourbaki) Commutative Algebra

Chapter I Flat Modules. **Chapter II** Localization. **Chapter III** Graduations, Filtrations and Topologies. **Chapter IV** Associated Prime Ideals and Primary Decomposition. **Chapter V** Integers. **Chapter VI** Valuations. **Chapter VII** Divisors.

$\mathbf{A}l_9$ (Bourbaki) 'Lie Groups and Lie Algebras'

Chapter I Lie Algebras. **Chapter II** Free Lie Algebras. **Chapter III** Lie Groups. **Chapter IV** Coxeter Groups and Tits Systems. **Chapter V** Groups Generated by Reflections. **Chapter VI** Root Systems.

$\mathbf{A}l_{10}$ (Bourbaki) 'Spectral Theory'

Chapter I Normed Algebras. **Chapter II** Locally Compact Groups.

$\mathbf{A}n_1/I$ (Dieudonné)

Chapter I: Elements of the Theory of Sets

Chapter II: Real Numbers

Chapter III: Metric Spaces

Chapter IV: Additional Properties of the Real Line

Chapter V: Normed Spaces

Chapter VI: Hilbert Spaces

Chapter VII: Spaces of Continuous Functions

Chapter VIII: Differential Calculus (in Banach Spaces)

Chapter IX: Analytic Functions (with Applications to Plane Topology)

Chapter X: Existence Theorems

Chapter XI: Elementary Spectral Theory

Appendix: Elements of Linear Algebra

A$n_1/II - VI$

Chapter XII: Topology and Topological Algebra

Chapter XIII: Integration

Chapter XIV: Haar Measure and Convolution

Chapter XV: Normed Algebras and Spectral Theory

Chapter XVI: Differential Manifolds

Chapter XVII: Distributions and Differential Operators

Chapter XVIII: Differential Systems

Chapter XIX: Lie Groups

Chapter XX: Riemannian Geometry

Chapter XXI: Compact Lie Groups

Chapter XXII: Harmonic Analysis

Chapter XXIII: Linear Functional Equations

Chapter XXIV: Elementary Differential Topology

Chapter XXV: Non-linear Problems

An_2 **(Mawhin)**

1. Ensembles, graphes, fonctions

2. Limites et continuité

3. Dérivabilité

4. Fonctions continues ou dérivables

5. Fonctions implicites

6. Fonctions monotones

7. D/'eveloppments de Taylor et séries

8. Equations différentielles linéaires

9. Fonctions primitivables

10. Fonctions inteégrables

11. Intégrale sur un intervalle et séries

12. Suites et séries de fonctions

13. Fonctions et ensembles mesurables

14. Représentations et transformations

15. Analyse vectorielle et extérieure

16. Analyse complexe

17. Analyse fonctionnelle

18. Index historique

An_3 (Zygumund)

Chapter I: Trigonometric Series and Fourier Series. Auxiliary Results

Chapter II: Fourier Coefficients. Elementary Theorems on the Convergence of $S[f]$ and $\tilde{S}[f]$

Chapter III: Summability of Fourier Series

Chapter IV: Classes of Functions and Fourier Series

Chapter V: Special Trigonometric Series

Chapter VI: The Absolute Convergence of Trigonometric Series

Chapter VII: Complex methods in Fourier Series

Chapter VIII: Divergence of Fourier Series

Chapter IX: Riemann's Theory of Trigonometric Series

An_4 (I: M.A. Evgravovv; II, III: M.V. Fedorynk)

I. Series and Integral Representations

Chapter 1: The Evolution of the Concept of Convergence

Chapter 2: The Techniques of Operating with Series and Integrals

IIL Asymptotic Methods in Analysis

Chapter 1: Integrals and Series

Chapter 2: Linear Ordinary Differential Equations

III: Integral Transforms

An_4/II (V.M. Tikhomirov)

I: Convex Analysis

Chapter 1: The Basic Ideas of Convex Analysis

Chapter 2: Convex Calculus

Chapter 3: Some Applications of Convex Analysis

Chapter 4: Extensions of the sphere of Convex Analysis and Generalizations of Convexity

II: Approximation Theory

Chapter 1: Classical Approximation Theory

Chapter 2: Classical Methods of Approximation

Chapter 3: Best Methods of Approximation and Recovery of Functions

An_4/III

I: Spaces of Differentiable Functions of Several Variables and Imbedding Theorems

Chapter 1: Function Spaces

Chapter 2: Soboleu Spaces

Chapter 3: The Imbedding Theorems of Nikol'skü

Chapter 4: Nikol'skü - Besov Spaces

Chapter 5: Sobolev-Liouville Spaces

Chapter 6: Spaces of Functions Defined in Domains

Chapter 7: Weighted Function Spaces

Chapter 8: Interpolation Theory of Nikol'skü - Besov and Lizorkin - Triebel Spaces

Chapter 9: Orlicz and Orliz-Sobolev Spaces

Chapter 10: Symmetric and Nonsymmetric Banach Function Spaces

An_4/IV

I: Linear Integral Equations

Chapter 1: Some Facts from Abstract Operator Theory

Chapter 2: Foedholm Integral Equations

Chapter 3: One-Dimensional Singular Equations

Chapter 4: Multi-dimensional Singular Equations

II: Boundary Integral Equations

Chapter 1: Theory of Harmonic Potentials

Chapter 2: Integral Equations for the Equations of Launé and Stokes

Chapter 3: Some More Applications of the Boundary Integral Equations Method

Chapter 4: The Integral Equations of Potential Theory in the Spaces C and L_p

Chapter 5: Boundary Integral Equations on Piecewise Smooth Surfaces

An_5 (Bourbaki) 'Integration'

Chapter I: Convexity inequalities

Chapter II: Riész Spaces

Chapter III: Measures on Locally Compact Spaces

Chapter IV: Extension of a Measure L^p spaces

Chapter V: Integration of Measures

Chapter VI: Vectorial Integration

Chapter VII: Haar Measure

Chapter VIII: Convolution and Representation

$\mathbf{A}n_6$ (Bourbaki) 'Functions of a Real Variable'

Chapter I: Derivatives

Chapter II: Primitives and Integrals

Chapter III: Elementary Functions

Chapter IV: Differential Equations

Chapter V: Local Study of Functions

Chapter VI: Generalized Taylor Expansions. The Euler-Maclaurin Summation Formula

Chapter VII: The Gamma Function Dictionary

$\mathbf{A}n_7$ (Bourbaki) 'General Topology'

Chapter I: Topological Structures

Chapter II: Uniform Structures

Chapter III: Topological Groups

Chapter IV: Real Numbers

Chapter V: One Parameter Groups

Chapter VI: Real Number Spaces, Affine and Projective Spaces

Chapter VII: The Additive Groups $\mathbb{R}^n$

Chapter VIII: Complex Numbers

Chapter IX: Use of Real Numbers in General Topology

Chapter X: Function Spaces

$\mathbf{A}n_8$ (Bourbaki) 'Topological Vector Spaces'

Chapter I: Topological Vector Spaces over a Valued Field

Chapter II: Convex Sets and Locally Convex Spaces

Chapter III: Spaces of Continuous Linear Mappings

Chapter IV: Duality in Topological Vector Spaces

Chapter V: Hilbert Spaces: Elementary Theory. Dictionary

$\mathbf{A}n_9$ (Dunford/Schwartz) 'Linear Operators' (Parts I, II, III)

Chapter I: Preliminary Concepts (Set-theoretic, topological, algebraic)

Chapter II: Three Basic Principles of Linear Analysis (Uniform boundedness/Interior Mapping/ Hahn-Banach)

Chapter III: Integration and Set Functions

Chapter IV: Special Spaces

Chapter V: Convex Sets and Weak Topologies

Chapter VI: Operators and Their Adjoints

Chapter VII: General Spectral Theory

Chapter VIII: Applications

Chapter IX: B-Algebras

Chapter X: Bounded Normal Operators in Hilbert Space

Chapter XI: Miscellaneous Applications

Chapter XII: Unbounded Operators in Hilbert Space

Chapter XIII: Ordinary Differential Operators

Chapter XIV: Linear Partial Differential Equations and Operators

Chapter XV: Spectral Operators

Chapter XVI: Spectral Operators: Sufficient Conditions

Chapter XVII: Algebras of Spectral Operators

Chapter XVIII: Unbounded Spectral Operators

Chapter XIX: Perturbations of Spectral Operators with Discrete Spectrum

Chapter XX: Perturbations of Spectral Operators with Continuous Spectrum

An_{10} (Deimling) Nonlinear Functional Analysis

Chapter 1: Topological Degree in Finite Dimensions

§1. Uniqueness of the Degree

§2. Constructions of the Degree

§3. Further Properties of the Degree

§4. Borsuk's Theorem

§5. The Product Formula

§6. Concluding Remarks

Chapter 2: Topological Degree in Infinite Dimensions

§7. Basic Facts about Banach Spaces

§8. Compact Maps

§9. Set Contractions

§10. Concluding Remarks

Chapter 3: Monotone and Accretive Operators

§11. Monotone Operators on Hilbert Spaces

§12 Monotone Operators on Banach Spaces

§13. Accretive Operators

§14. Concluding Remarks

Chapter 4: Implicit Functions and Problems at Resonance

§15. Implicit Functions

§16. Problems at Resonance

Chapter 5: Fixed Point Theory

§17. Metric Fixed Point Theory

§18. Fixed Point Theorems Involving Compactness

Chapter 6: Solutions in Cones

§19. Cones and Increasing Maps

§20. Solutions in Cones

Chapter 7: Approximate Solutions

§21

§22

Chapter 8: Multis

§23 Monotone and Accretive Multis

§24 Multis and Compactness

Chapter 9: External Problems

§25 Convex Analysis

§26 Extrema Under Constrains

§27 Critical Points of Functionals

Chapter 10: Bifurcations

§28 Local Bifurcation

§29 Global Bifurcations

§30 Further Tops in Bifurcation Theory

A$n_{11/I}$ (Zeidler) Fundamental Fixed Point Theorems

Chapter 1: The Banach FP Theorem and Iterative Methods

Chapter 2: The Schauder FP Theorem and Compactness

Chapter 3: Ordinary Differential Equations in B-Spaces

Chapter 4: Differential Calculus and the Implicit Function Theorem

Chapter 5: Newton's Method

Chapter 6: Continuation with Respect to a Parameter

Chapter 7: Positive Operators

Chapter 8: Analytic Bifurcation Theory

Chapter 9: Fixed Points of Multivalued Maps

Chapter 10: Nonexpansive Operators and Iterative Methods

Chapter 11: Condensing Maps and the Bourbaki-Kneser FP Theorem

Chapter 12: The Leray-Schauder FP Index

Chapter 13: Applications of the FP Index

Chapter 14: The FP Index of Differentiable and Analytic Maps

Chapter 15: Topological Bifurcation Theory

Chapter 16: Essential Mappings and the Borsuk Antipodal Theorem

Chapter 17: Asymptotic FP Theorems

A$n_{11/II}$ (Zeidler) Monotone Operators

Chapter 18: Variation Problems, the Ritz Method and the Idea of Orthogonality

Chapter 19: The Galerkin Method for Differential and Integral Equations and the Idea of Self-Adjointness

Chapter 20: Difference Methods

Chapter 21: Auxiliary Means for the Hilbert Space Methods and the Unique Approximative Solvability of Equations

Chapter 22: Hilbert Space Methods and Elliptic Differential Equations

Chapter 23: Hilbert Methods and Parabolic Differential Equations

Chapter 24: Hilbert Space Methods and Hyperbolic Differential Equations

Chapter 25: Projection-Iteration Methods and Monotone Operators

Chapter 26: Monotone Operators and Quasi-Linear Elliptic Differential Equations

Chapter 27: Pseudomonotone Operators and Quasi-Linear Elliptic Differential Equations

Chapter 28: Monotone Operators and Hammerstein Integral Equations

Chapter 29: Generalized Fredholm Alternatives

Chapter 30: Evolution Equations of the First Order and The Galerkin Method

Chapter 31: Evolution Equations of the First Order and Semigroups

Chapter 32: Maximal Monotone Operators and Their Applications

Chapter 33: Evolutions Equations of Second Order and the Galerkin Method

Chapter 34: Internal Approximation Schemes and the Galerkin Method

Chapter 35: External Approximation Schemes and the Difference Method

Chapter 36: Mapping Degree for A-Proper Maps

A$n_{11/III}$ (Zeidler) Variational Methods and Optimization

Chapter 37: Introductory Typical Example

Chapter 38: Compactness and Extremal Principles

Chapter 39: Convexity and Extremal Principles

Chapter 40: Free Local Extrema of Differentiable Functionals and the Calculus of Variations

Chapter 41: Potential Operators

Chapter 42: Free Minima for Convex Functionals, Ritz Method and the Gradient Method

Chapter 43: Lagrange Multipliers and Eigen Value Problems

Chapter 44: Ljusternik-Schnirelman Theory and the Existence of Several Eigenvectors

Chapter 45: Bifurcation for Potential Operators

Chapter 46: Differential Functionals on Convex Sets

Chapter 47: Convex Functionals on Convex Sets and Convex Analysis

Chapter 48: General Lagrange Multipliers (Dubovickü-Miljutin Theory)

Chapter 49: General Duality Principle by Means of Lagrange Functions

Chapter 50: Duality and the Generalized Kuhn-Tucker Theory

Chapter 51: Duality, Conjugate Funtionals, Monotone Operators and Elliptic Differential Equations

Chapter 52: General Duality Principle by Means of Perturbed Problems and Conjugate Functions

Chapter 53: Conjugate Functionals and Orlicz Spaces

Chapter 54: Elliptic Variational Inequalities

Chapter 55: Evolution Variational Inequalities

Chapter 56: Evolution Variational Inequalities of the Second Order in H-Spaces

Chapter 57: Accretive Operators and Multivalued First Order Evolution Equations in B-Spaces

A$n_{11/IV}$ (Zeidler) Applications to mathematical Physics

Chapter 58: The Fundamental Equations of Mechanics

Chapter 59: The Dualism Between particles and Waves: a Glance at Quantum Theory and Elementary Particles

Chapter 60: The Elasto-Plastic Wire

Chapter 61: The Fundamental Equations of Nonlinear Elasticity

Chapter 62: Monotone Potential Operators and a Class of Models with Nonlinear Hooke's Law

Chapter 63: Variational Inequalities and the Signormi Problem for Nonlinear Materials

Chapter 64: Bifurcation for Variational Inequalities and the Beaim

Chapter 65: Pseudomonotone Operators, Bifurcation and Von Kármán's Nonlinear Plate Equation

Chapter 66: Convex Analysis, Maximal Monotone Operators, and Elasto plastic Material with Hysteresis

Chapter 67: Phenomenological Thermodynamics of Quasi-Equilibrium States

Chapter 68: Statistical Physics

Chapter 69: Continuation with Respect to a Parameter and Carleman's Radiation Problem

Chapter 70: The Fundamental Equations of Hydrodynamics

Chapter 71: Permanent Gravity Waves and Bifurcation

Chapter 72: Viscous Fluids and the Navier Stokes Equations

Chapter 73: Banach Manifolds

Chapter 74: Classical Surface Theory and the Theorems Egregium of Gauss

Chapter 75: Special Relativity

Chapter 76: General Relativity and Cosmology

Chapter 77: Simplicial Methods, FP Theory and Mathematical Economics

Chapter 78: Homotopy Methods

Chapter 79: Dynamical Stability and Bifurcation in B-Spaces

A$n_{11/V}$ (Zeidler) Applications to Mathematical Physics

Chapter 80: Stable Manifolds of Dynamical Systems

Chapter 81: The Methods of Ljapumov Function in Stability Theory

Chapter 82: The Calculus of Differential Forms on B-Manifolds

Chapter 83: Maxwell's Theory of Electromagnetism

Chapter 84: Symplectic Manifolds, Mechanics and Statistical Physics

Chapter 85: Riemannian Manifolds

Chapter 86: Irreversible Thermodynamical processes

Chapter 87: Applications to Mathematical Biology

Chapter 88: Symmetry, Group Theory and Bifurcation

Chapter 89: The Fundamental Equations of Nonrelativistic Quantum Mechanics and Quantum Chemistry

Chapter 90: The Helium Atom with Nonlinear Interaction

Chapter 91: Relativistic Field Theories and the Relativistic Electron

Chapter 92: Quantum Electrodynamics

Chapter 93: The Free Quantum Field

Chapter 94: Quantum Statistics, Phase Transitions and Supra Fluidity

Chapter 95: Group Theory and the Quark Model for Elementary Particles

Chapter 96: Connections in Principle Fibre Bundles and Gauge Field Theories

Chapter 97: Homology

Chapter 98: Homotopy

Chapter 99: Cohomology

Chapter 100: Algebraic Topology and the Mapping Degree

$\langle$Brief Outline of the History of Nonlinear Functional Analysis $\rangle$

An_{12}: Maurin, K., Analysis (Vol II) (Reidel, 1980)

Chapter XII: Topology, Uniform Structures, Function Spaces

Chapter XIII: Theory of the Integral

Chapter XIV: Tensor Analysis, Harmonic Forms, Cohomology, (Applications to Electrodynamics)

Chapter XV: Elementary Properties of Holomorphic Functions of Several Variables. Harmonic Functions

Chapter XVI: Complex Analysis in one Dimension. Riemann Surfaces

Chapter XVII: Normal and Paracompact Spaces. Partition of Unity

Chapter XVIII: Measurable Mappings. The Transport of a Measure. Convolutions of Measures and Functions

Chapter XIX: The Theory of Distributions. Harmonic Analysis.

Comments on the Reference Sources

It is evident, from the profusion of texts, treatises and monographs written over the past 30 years or so, that the possible choice of reference sources for Σ is extremely wide. Nevertheless, a selection must be made especially, to support the claim that suitably enhanced SAM packages are potentially adequate for the development of Σ . This claim is itself based on the (implicit) argument that arbitrary results in all other areas of mathematics are, ultimately, convertible into (collections of) theorems in Algebra or Analysis broadly interpreted. For '**Pure Mathematics**', this entails consideration of **Topology** and **Geometry** (each of several species), **Function Theory** (of k complex variables; $k \geqslant 11$), **Number Theory** (algebraic, geometric, analytic), **Functional Analysis**; and so on. A sufficient classification may be obtained from the full enumeration of mathematical topics used for **M**athematical **R**eviews.

The crucial point is that a relatively small collection of theories may be identified as **fundamental** for the systematic construction of all the rest. This is not a subjective assessment; rather, it reflects the essentially hierarchical nature of Modern Mathematics, in its axiomatic development.

Thus: **Topology** has several species, **General**, **Combinatorial**, **Algebraic**, **Differential** or **Geometric** (with comparatively slight 'overlaps'); Geometry may be viewed as **Synthetic**, **Analytical**, **Algebraic** or **Differential**

dots to cite only the most obvious examples. It would appear that such considerations motivated the choice of Subject matter in the, **Bourbaki 'Elements of Mathematics'**, series, where the emphasis is strongly on the more basic aspects of **Algebra and Analysis**(rather than on the various 'applications', even within 'pure mathematics' – of which Topology and

Algebraic Geometry are examples). As far as '**Applicable Mathematics**' (as this is usually understood) is concerned, no fundamentally 'new' (generic) **mathematical** concepts are introduced – and hence the 'reduction' of Pure Mathematics suffices. Somewhat more problematical issues arise from the **inverse analogues** derived from scientific theories; for, here, essentially novel structures may be generated, whose incorporation within the overall 'reduction' may not be straightforward. In spite of such possible complications, the general strategy of 'reduction to Algebra and Analysis' seems highly plausible and potentially realizable within all mathematical subdomains. Moreover, since the content of 'Advanced Algebra Analysis' is demonstrably obtainable through elaboration and extension of the 'nuclei of essential results' it is enough to show that: **every calculation is equivalent to some combination(s) of basic algebraic and analytical calculations**. The feasibility of the cumulative construction of Σ from the SAM package, P_Σ, is predicated on this basis. For the **computational** aspects of calculations, fairly direct enhancements of existing SAM routines will suffice. The more abstract constructions/manipulations/derivations, however, present challenges of an altogether different – and substantially new –character. The key tools for the resolution of these difficulties are species of **operational information** and, in particular, **operational theorems**.

The interactive nature of Σ ensures that sequences of noncomputational transformations of **wfe** (through the application of OT) may be interpreted by users 'at the top level of abstraction', without the 'active participation' of the underlying OIB. In other words, the system must display/record the results of all abstract transformations (as images of wfe under compositions of OT), but it need not 'understand' these transformations or their justifications. The primary task for Σ is, therefore, **the formulation/proof/implementation of operational versions of theorems**,in all mathematical domains (**over the CF**,s^X, **designated in** E_1). This (gargantuan) task, and its manifold implications for Mathematics, form most of the subject matter of the remaining Essays.

Calculational Decomposition Schemes

The following general prescription seems to cover the essential steps in any calculational reduction process. Let ${}_1\zeta$ denote an arbitrary calculation over the set $\{{}_1\mathcal{F}\}_{j_1}$ of GSF, and suppose that the ${}_1\mathcal{F}_k$ are specified in terms of operations over (other) sets $\{{}_2\mathcal{F}\}_{j_2}$ of GSF – at levels of generality/abstraction not exceeding those in $\{{}_1\mathcal{F}\}_{j_1}$. By iterating this representation, one obtains a (finite) sequence $\{\ \{{}_m\mathcal{F}\}_{j_m}\ :\ 1 \leqslant m \leqslant N\}$, where each ${}_m\mathcal{F}_k$ is specified through operations over $\{{}_{m+1}\mathcal{F}\}_{m+1}$, and calculations over $\{\ {}_N\mathcal{K}\}_{j_N}$ are assumed to be effectively implementable within P_{Σ_0}, the processor for $\S_0$, corresponding to some 'enhanced amalgam of current SAM packages'. The underlying assumption here is that all GSF are definable through axioms/rules over existing (G)SF, at (ultimately) decreasing levels of complexity – culminating in the (G)SF covered by SAM systems. This assumption reflects the typical path of development for 'new' mathematical structures, with their associated calculational procedures. For Σ , the standard approximation techniques are augmented by very general approximation criteria – including abstraction/analogy/inverse analogy – along with cognate formulations of operational theorems. Consequently, it is enough to demonstrate the validity of this scheme in representative examples taken from all of the main divisions of 'pure mathematics' (as classified, for instance, by Mathematical Reviews). The aim is (in each example) to identify the sets, $\{{}_m\mathcal{F}\}_{j_m}$, of (G)SF – especially, $\{{}_N\mathcal{F}\}_{j_N}$; and then, to obtain the associated decompositions (say, $\delta({}_1\zeta)$) for the calculation(s) ${}_1\zeta$ over $\{\ {}_1\mathcal{F}\}_{j_1}$.

If one regards each ${}_m\mathcal{F}_k$ as a 'realization of prescribed axioms and basic rules of calculation', then these rules (say, $\{{}_m\mathcal{P}_{kl}\ :\ 1 \leqslant l \leqslant L_{mk}\}$ constitute a sort of **basis** for arbitrary calculations within ${}_m\mathcal{F}_k$. Moreover, since the $m\mathcal{P}_{kl}$ are themselves (elementary) calculations over ${}_m\mathcal{F}_k$, the decompositions $\delta({}_m\mathcal{P}_{kl})$ are w.d. and may (collectively) furnish a basis for the general decompositions $\delta({}_1\zeta)$ of calculations at the top level.

In other words, the set $\{\delta({}_m\mathcal{P}_{kl}) 1 \leqslant m \leqslant N; 1 \leqslant k \leqslant j_m; 1 \leqslant l \leqslant L_{mk}\}$ may be adequate for the (iterative) generation of representations of the form

$$\delta({}_1\zeta) = \Phi[\{{}_N\mathcal{F}\}_{j_N}] = \Psi\langle\{{}_N\mathcal{P}_{kl} : 1 \leqslant k \leqslant j_N; 1 \leqslant l \leqslant L_{NK}\}\rangle$$

, where the notation $\Phi[\ldots], \Psi\langle\ldots\rangle$ connotes extended functional relationship. Here, the ${}_N\mathcal{P}_{kl}$ include all of the basic calculational rules governing the SF covered by $\mathcal{P}_{\Sigma_0}$ there are also rules corresponding to fundamental analytical operations, combinatorial results; and so on. Eventually, all $ME, e,$ inΣ_t will have '**calculational profiles**'for every known mode of calculation–exhibiting the essential constituents of each algorithm for the determination of (all possible approaches to) e. Such profiles themselves furnish approaches to the complete algorithms concerned!

This view of mathematical development and calculational decomposition does not exclude **identities** (or '**formulae**') for the closed-form representation of one (set of) ME in terms of another (set of) ME; or else, in some different form(s) over the same (set of) ME. On the contrary, such alternative representations may be viewed as 'quasi–extremal elements', where inequality is typical. They are particular value in the assessment of approximation schemes–at all levels of sophistication.

Some Sophisticated Algorithms: On the Role of SAM Σ

The achievements of SAM packages over the past 30 years are remarkable, but is only recently that some basic quasi-algorithms at the highest levels of sophistication have been encompassed.[56].The task remains gargantuan[57]. The interplay of 'general' and 'special' theorems; and, of 'abstract' and 'concrete' calculations – especially in the formulation/realization of OT, in all parts of Mathematics – forms a dominant theme in these Essays. Here is has been argued that: **arbitrary mathematical calculations are effectively reducible to collections of relatively basic quasi-computational routines, combined with a range of (often subtle/intricate)** Σ -facilities – to be developed in other Essays. Notable advances have been made in **computational commutative algebra** (useful, e.g., in **Algebraic Geometry**). Other obviously useful areas include: **group-representation procedures (for (in)finite/continuous groups)**, with wide application in high-energy/solid-state-physics/quantum-chemistry; and, computational aspects of complex analysis [58]. It is rarely emphasized that many ostensibly 'abstract' theorems, in all domains, have crucial **enumerative/calculational facets** (at least, in their initial proofs; often inescapably); for instance, in determining the '**best constants**' in classical approximation theory; or, in **covering all possible cases** (as in the computer - assisted proof of the 4-colour conjecture). Equally, in many applications of 'abstract' theorems, **formal criteria** (e.g., for **continuity, summability, coerciveness** must be satisfied, to justify the applications. **It is these multifarious blends of abstract criteria and concrete procedures that make the development of Σ so challenging – and essential!** It is clear, therefore, that SAM algorithms themselves are rarely involved in **general** mathematical activity, where one deals with 'abstract ME and their (non)implicatory inter-relationships' – i.e., with the derivation of general theorems. Even the **OT** are not specific in their actions, except in particular applications. Hence, the implementation of OT becomes 'targeted' only over the collections of 'particular elements (e.g., polynomials, matrices, Markov chains, PDO, ...) among the totalities of such elements as abstractly definable. **In Short:** Most of the SAM procedures are used for computation in concrete situations. When OT are applicable to 'abstract data', the results are primarily transformational – but (e.g.) the use of 'general (in)eqs may still be covered, since no special computations need be done.[59] Although detailed accounts of the **basic** SAM routines are given in (e.g.) the book by Geddes, et al. (1992), some **more advanced procedures** (especially the factorization, and for indefinite integration) seem to be treated only[60] in papers – for instance

[56]e.g., in parts of Group Theory, and Algebraic Geometry

[57]A wide range of examples may be found in Elvey (1980); a few of these have new been (partially) implemented by various people

[58]References are given in E_5 (for 'diagrams of ME')

[59]See, e.g., D.S. Mistrimovic, et al., 'Analytic Inequalities (Springer, 1970)

[60]A partial exception in Risch's 'formal integration procedure'. The outline given here is based largely on Davenport's account (LNM\ 102, Springer, 1981), where trial algorithms are given but mathematical details are (mostly omitted here)

in the Journal of Symbolic Computation. It is therefore worthwhile to discuss a few of these algorithms briefly, here, to give the flavour of the deductive/decisions schemes involved.

1. Polynomial Factorization

One of the key facilities is the **factorization of polynomials in** $\mathbb{Q}[x]_v$ – and, **in various algebraic - number fields.** The '**CRT - Hensel**' routines suffice for $\mathbb{Z}_p[x]_v$ and $\mathbb{Z}[x]_v$; further, 'the case of $\mathbb{Q}[x]_v$' is equivalent to that for 'primitive polynomials in $\mathbb{Z}[x]_v$'. Nevertheless, issues of **complexity/cost** are also important so alternative algorithms have been developed. The best versions – of various types – run in (essentially) 'polynomial time in the degrees of the variables'. The extensions to algebraic - number fields have broadly comparable efficiencies. Procedures originating in results from the **Geometry of Numbers** are of special interest [61].

The 'L.L.L. algorithm' depends on the fact that, given a **primitive polynomial** $f \in \mathbb{Z}[x]$, of degree n, **it is possible to construct**[62] **(in polynomial time in** l**) a polynomial** $h \in \mathbb{Z}[x]$, **of degree** l such that (**for a prime** p **and an integer** k – **to be specified**) h is **monic**; $h_{pk}|f_{pk}$ in $(\mathbb{Z}/p^k\mathbb{Z})[x]$; h_p is **irreducible** in $F_p[x] := (\mathbb{Z}/p\mathbb{Z})[x]; (h_p)^2 \nmid f_p$ in $F_p[x]$. Here, f_{p^j} stands for '$f \text{mod} p^j$', etc., and the quotients $(\mathbb{Z}/p^j\mathbb{Z})[x]$ have their usual meanings.

From these properties of h it follows that f **has an irreducible factor** h_0, and that $h_p \backslash h_{0_p}$. Within this framework it may be proven that **the shortest vector(s) of a certain lattice** L **must equal some integer multiple(s) of** h_0, where, for $m \geqslant l, L$ **comprises all polynomials** q **in** $\mathbb{Z}[x]$ **of degree** $\leqslant m$ **for which** $h_{p^k}|q_{p^k}$ **in** $(\mathbb{Z}/p^k\mathbb{Z})[x]$.

This result, and the subsequent arguments, depends on the (proven) equivalence of the following conditions (for any **factor** g of f in $\mathbb{Z}[x]$): $h_p|g_p$ in $F_p[x]; h_{p^k}|g_{p^k}$ in $(\mathbb{Z}/p^k\mathbb{Z})[x]; h_0|g$ in $\mathbb{Z}[x]$.

An algorithm for finding the shortest vector(s), say, $\hat{b}_1$, in L – or, in a reduced form of L – was given by Dieter (1975); but this is not (in general) 'in polynomial time' (in various senses). The fact that $h_0|\hat{b}_1$ depends on the (purely algebraic) bound

$$\| h_0 \| \equiv \| \sum_{i=0}^{m} a_i x_i \| \leqslant (\sum_{i=0}^{m} \binom{m}{i}^2 \| f \|^2)^{3/2}$$

(since $\| a_i \| \leqslant \binom{m}{i} \| f \|$, because[63] $h_0|f), andhence\$parallelh_0 \| \leqslant \binom{2m}{m}^{1/2} \| f \| =: B --$ from which it follows that $[p^{kl} > B^{2m} \wedge h_0|b\,] \Rightarrow \| b \| < B$.

This establishes that $h_0|\hat{b}_1$. Another estimate, valid for $p^{kl} > B^{2M}2^{m(n-1)/2}$, implies that all multiples of h_0 in L have length $\leqslant 2^{(n-1)/2}B$ – so that, for a reduced basis (say, $(b)_n$ in an n-dimensional lattice L) the general bound, $\| b_1 \|^2 \leqslant 2^{n-1} \| x \|^2$, valid for all nonzero $x \in L$, shows that $h_0|b_1$. Hence: the LLL-algorithm (which is 'polynomial') may be used (instead of Dieter's algorithm) to determine h_0.

To find a suitable $h \in \mathbb{Z}[x]$ one may WLOG assume that f has only simple factors (otherwise, consider $\tilde{f} \equiv f/gcd(f, f')$. If p is chosen so that $p \mid R(f, f')$ – this **resultant**[64] of f and f' – and f_p is decomposed into **irreducible factors** (using Berlkam['s algorithm[65]) then it follows that f_p has degree n, with only simple factors in $F_p[x]$[66]. Next, via 'Hensel's Lemma h may be modified

[61] See: Lenstra, A.K/Lenstra, HW/Lovaász, Math.Ann 261 (1982) 513-534; Kannan, R/Lenstra,AK/Lovás L., Report CMU-CS-84-111(1984) Bachem, A/Kannan,R, Report CMU CS-84-112 (1984); Mulimiley, K Report CMU - CS-84-126 (1984); Lenstra, A.K, EUROJAM 84, LNCS 174 (Springer, 1984) 272-276

[62] This construction depends on several known procedures mentioned later. The LLL algorithm requires the prior determination of h, for specified f. The associated 'shortest-lattice-vector' criterion has applications in several other forms of factorization, and in some 'integer programming' problems (see: Mulmuley, op cit. also, Bachem/Kannan, op.cit, for references).

[63] See Mignolte, M., Math.Comp. 28 (1974) 1153=1157, and DE Knuth, 'The Art of Computer Programming' Vol 2 (A-W, 2nd ed., 1981) p.627. For Dieter's algorithm, see Math.Comp. 29 (1975) 827-833

[64] The **subresultant algorithm** may be used. See, e.g., **Knuth D., 'The Art of Computer Programming' Vol.2 pp 410-418**

[65] See, e.g. **Knuth, op.cit, Sec.4.6.2**

[66] Hence, '$(h_p)^2 \nmid f_p$ in $F_p[x]$' is always valid

(leaving h_p fixed) so that $h_{p^k}|f_{p^k}$ in $(\mathbb{Z}/p^k\mathbb{Z})[x]$[67]. Finally, h_p may be made monic, since F_p is a field.

An **alternative scheme** (to find L) involves the 'sufficiently close approximation' of a root of $f(x) = 0$ – via a constructive form of FTA[68]. Let $f \in \mathbb{Z}[x]$ be primitive, $f(\alpha) = 0 (\alpha \in \mathbb{C})$; $h \in \mathbb{Z}[x]$ is the minimum polynomial of α, then (plainly) $h|f$ and h is irreducible.

$\models$ If $\tilde{\alpha}$ 'approximates α suitably', in $\mathbb{Q}(i)$, then h is determinable from $\tilde{\alpha}$, in polynomial time, in terms of the shortest element of a lattice, L, constructible from $\tilde{\alpha}$

This **procedure** depends on the **estimate:**

$$[s \in \mathbb{Z}^+ \wedge \tilde{\alpha} \in \mathbb{C} \wedge |\alpha - \tilde{\alpha}| < 2^{-s}] \Rightarrow |h(\tilde{\alpha}| < K_h(s, f) \text{ where}$$
$$K_h(s, f) := 2^{-s}\delta_h|f|(2 + |f|)^{\delta_h - 1}, \delta_h \equiv \text{deg}(h)$$

and, on the introduction of the **linear form** $\tilde{g} : \tilde{g}(\tilde{\alpha}) := \sum_{1 \leqslant j \leqslant \delta_g} g_j\tilde{\alpha}_j$, where the $\tilde{\alpha}_j$ suitably approximate the powers, $\tilde{\alpha^j}$, for $0 \leqslant j \leqslant m \geqslant \delta_g$[69].Then the **mapping** $\hat{v} : \mathbb{Z}^{m+1} \to \mathbb{Q}^{m+3}$ may be defined[70] by: $v := (v)_{0,m} \to ((v)_{0,m}, cRe(\tilde{v}(\tilde{\alpha})), cIm(\tilde{v}(\tilde{\alpha})))$, and $L := \hat{v}(\mathbb{Z}^{m+1})|_{\mathbb{Z}^{m+1}}$. The **motivation** for these definitions is that (for certain bounds, $\Lambda_1(m, f), \Lambda_2(m, f)) \models [\Lambda_1 \leqslant c \leqslant 2^s\Lambda_2] \Rightarrow |\hat{h}|^2 < B \wedge |\hat{g}|^2 \geqslant 2^m B \; (B \equiv 1 + \binom{2m}{m}|f|^2)$; from which it may be deduced that , **if** $(b^*)_{m+1}$ **is a reduced basis of** L **then** (provided $\delta_h = m)h = b_1^*|_{\mathbb{Z}^{>+\times}}$ (up to sign, $\pm$). Here the LLL-algorithm may be used to obtain $(b^*)_{m+1}$.

2.The Rothstein/Trager/Bronstein Integration Scheme

Risch's (1969) solution of PIFT[71] fore **purely transcendental** extension fields was implemented in several SAM packages – through his outline solution (1970) for **purely algebraic** extension fields has, apparently been implemented, so far, only for **simple radical**[72] extension fields (**Davenport** (1981)). **Modifications** of RIsch's procedure (to minimize**factorization** of integrand denominators), by **Rothstein**(1976), produced a **resultant-based representation** of elementary integrals. Next, **Trager** (1979) generalized Risch's 'transcendental algorithm' to cover integrands containing **un-nested radicals** – essentially the same class of integrands for which implementations of Davenport's algorithm have been developed. In **Trager** (1984), the generalization of Risch's scheme was **combined** with Rothstein's approach, to admit **arbitrary algebraic** integrands. (**NOTE that** the case of arbitrary transcendental extensions of a 'given' algebraic extension field is essentially within the scope of Risch's original method).[73]

Finally, **Bronstein** (1990,1987)[74] proved that **all of Trager's**[75] **representations may be extended** to allow for **general elementary integrands** – in a (potentially fully implementable) decision procedure. The general 'resultant formula' is given in this Essay, but its derivation is somewhat involved – in all of its versions. Details may be found in Rothstein (1976)[76], Trager (1984) and Bronstein(1987, 1990).

3. Risch's Formal Solution. Davenport's Algorithm

In spite of the apparent practical limitations of Risch's formal scheme[77] (1970), its mathematical depth, and elegance, justify the inclusion, here, of a sketch – along with an indication of its

[67] See, e.g., Geddes, et al (op. cit) p230

[68] See A.K. Lenstra, LNCS\ 174, 272-276 (Springer, 1984)

[69] See KNCS \ 174, 272-276, for the details

[70] This construction is used generally, to define $\hat{h}, \hat{g}$, etc ...

[71] Problem of Integration in Finite Terms

[72] i.e, for fields $F(y)$, where $y^n \in F, n \in \mathbb{Z}^+$

[73] The only extra problem is to solve the associated Risch DE over algebraic extension fieldsL see Bronstein's thesis (Berkeley, 1987) for Risch's procedure

[74] J. Symbolic Comp. 9(1990) 117-173, Thesis (Berkeley.1987)

[75] B.Trager, Thesis, MIT, 1984

[76] M.Rothstein, Thesis, U.Wisconsin,1976

[77] R.H. Risch, BAMS 76 (1970) 605-608

partial implementation[78], based on some highly nontrivial results from Algebraic Geometry.[79] Only the barest outline is given by Davenport, but details may be found in the references cited.

Risch's , modernized version of Liouville's characterization theorem for elementary integrals $\int w$ – where w is a differential of

$$\mathcal{F} := [K(z,w) : F(z,w) = 0],$$

may be stated as follows.[80]

Theorem (Risch/Liouville)

Let the $\mathbb{Z}$ - module generated by Res(w) have basis $(T)_k$, so that

$$\text{Res}_p(w) = \sum_{1 \leqslant i \leqslant k} a_{ip} r_i, a_{ip} \in \mathbb{Z}$$

and define $S_i := \prod_p p^{a_{ip}}$; then

$\exists (v)_{0,k} \subset \mathcal{F}\ \exists (j)_k \subset \mathbb{Z} : S_i^{j_i} = (v_i)$, and $w = v_0 + \sum_{1 \leqslant i \leqslant k} (r_i/j_i)(dv_i/v_i)$.

NOTE: $\models$ elements in $K[x,y] | F(x,y) = 0]$ may be represented as $g(x,y)/h(x)$, where g, h are polynomials. [81]

It follows[82] that **the R/L Theorem yields a decision procedure for elementary integrability – provided that effective bounds on the j_i may be obtained.** The residues are just the terms of order -1 in the Puiseux expansions in local parameters at the place p.

In the typical case, $w \equiv f(x,y)dx$, where $y(x)$ is determined by $F(x,y) = 0$, for a specified rational F, so **fields of algebraic functions** are involved.[83] The basic problem is to **to identify divisors that are rationally (but not linearly)** ~ 0 (i.e., those $\delta \ni \delta \neq (h)$ but $\delta^j = (h), etc)$. **Coates' algorithm determines rational functions on algebraic curves** [84] **over algebraic number fields.** More precisely, suppose given D, a divisor of poles over finite places – all unramified (i.e., admitting $x - a$, or x^{-1}, as local parameter for the Puiseuk expansion(s) in question)[85]. The sum of all the multiplicities in δ is called the say **degree** of δ. The order of a function at a place is defined to be (the index of the first (non zero) term in its Puiseux expansion that place) $\div$ (the ramification index of the relevant local parameter). Then the algorithm produces an integral basis [86], V, for the VS, Γ, over K of all functions f such that $(f) \geqslant D$ and f is regular at ∞. This is achieved in three stages - starting from any collection, C, of elements having no finite poles:

(i) 'insert' all of the poles specified in D;

(ii) ensure that the resulting basis is (quasi-) integral;[87]

(iii) modify the basis further to eliminate all elements with poles at ∞ ('normality'). **NOTE that,** if $\deg_y F \equiv \beta_y$, then an initial (non-integral) form of V is given (over $K[x]$) by $\{y\}^{0,\beta_y - 1}$ – to be transformed into an **integral/ normal** basis of Γ.

The rôle of **torsion** arises as follows.

(a) Torsion-divisors (i.e. those $\sim_l 0$ form a subgroup, D_l, of the (abelian) group, say D; of all divisors.

[78] J.H. Davenport, LNM\102 (Springer,1981)

[79] The key algorithm (for rational functions over an algebraic curve) is due to J.Coates (Proc Camb. PhilSoc 68 (1970) 105-123)

[80] See R.H. Risch BAMS 76 (1970) 605-608

[81] See R.H. Risch BAMS 76 (1970) 605-608

[82] Here, Res$_p(w)$ is essentially the coefficient of t^{-1} in the Puiseux series for w at P. Since the form of $\int w$ is known – apart from the exact values of the j_i (which may, in principle be found by trial, if bounds are known.)

[83] See, e.g., Walker, R.J., 'Algebraic Curves' (Princeton UP, 1950; Dover 1962)

[84] at places P on the curve –say, $\delta \sim_{l_m} 0/\delta \sim rat0$

[85] If $t^p = x - a$, or $t^p = x^{-1}$ then p is called the ramification index of t (at P). Zeros could be included in D.

[86] i.e., $(v)_k \subset K \ni v \in V \Rightarrow v = \Sigma r_i v_i, r_i \in K$

[87] i.e. 'within' a non zero common factor of integral'

(b) $\models (D_l \cup D_r)/D_l =: D_{t,l}$ is a **finitely generated abelian group**[88] (the**Jacobian divisor group** of the underlying curve).[89]

(c) Hence

$$D_{r,l} =: \bigotimes_{q=1}^{n} C_q \otimes \mathcal{J}$$

, where the C_q are infinite cyclic groups, and the abelian group $J \equiv$ the **torsion part** of $D_{r,l}$ is finite.

(d) By 'Lagrange's theorem', $\delta \in D_r \Rightarrow$ ordδ/tor$D_{r,l}$ where tor$D_{r,l} := |\mathcal{J}| \equiv$ the group-theoretic order of $\mathcal{J}$.

Consequently, a 'universal' bound, B, on tor$D_{r,l}$ would establish the decision procedure, since – for arbitrary algebraic curves[90] – all divisors, δ, could be tested to discover whether, for some $j \in \mathbb{Z}, j \leqslant B$, and $w.d.$ function, h, $\delta^j = (h)$. Here the curves may be of arbitrary $w.d.$ **genus**, but most of the detailed results have been derived only for elliptic curves (of genus 1).[91] A scheme applicable to **curves of any genus,** g, may be based on 'modular procedures' – where the problem is reduced to some associated problem(s) over finite fields (for which detailed results may be obtained) of several orders. The combined estimates from these finite fields eventually yield the possible value(s) for B. For curves of genus 1, adequate algorithms have been produced from theorems proven only for $g = 1$. The modular approach – involves **specialization**[92] of **Jac(ζ)**, and introduction[93] of **p-part**(Jac(ζ)):= $\{\alpha \in$ Jac(ζ) : ord$\alpha = p^m$, for some $m \in \mathbb{Z}^+\} =: p - pt(J(\zeta))$– a subgroup of $J(\zeta)$ of order a power of p.

The **non -p part** of $J(\zeta)$, say $n - p - pt(J(\zeta))$, comprises those elements of $J(\zeta)$ that are of **finite order prime to** p. If all elements of $J(\zeta)$ have finite orders, then $J(\zeta) = p - pt(J(\zeta)) \otimes n - p - pt(J(\zeta))$ – for **any** prime p. Within this framework, with both of ζ, D (a divisor) w.d. over K, let p' be a **prime ideal** of K lying over the prime $p \in \mathbb{Z}$, and $K' := K$modp' (generated by {integers of K modp'});[94] and let, A' be the be the associated **specialization** of the abelian variety A (taken to be $J(\zeta)$, here)[95]. Then – **by definition** – A has **good reduction** at

$$p' \Leftrightarrow A \in \langle \otimes_{p'}; A', K \rangle$$

$\equiv$ {objects constructible, from A', K, through the use of tensor products over the valuation-ring at p'}. On this basis:

(i) $\models |n - p - pt(J_A)|$ divides $|A'|$ (with A' regarded as a group);[96]

(ii) $\models$ **for fixed K, A only finitely many primes p have bad reduction.**

(iii) $\models$**WLOG p may be taken as unramified (over K)**, since, by a theorem of Dedekind, K contains only finitely many ramified primes.[97] For finite fields, several **implementable** results may be proven[98]. For instance

(iv) $\models$ If ζ **has genus** g **then** $J(\zeta)$ **has dimension** $g(v)$ Let $|A|_k := \#\{w \in A : w$ is $w.d.$ over K }. Then $\models$ [dim$A = r \wedge \#K = q] \Rightarrow |A|_k = \prod_{1 \leqslant j \leqslant 2t} (\lambda_j - 1)$, **where the λ_j are the eigenvalues of the matrix (say ϕ) realizing the Frobenius automorphism** on K – and inducing an automorphism on A. $(v) \models |\lambda_j| = q^{1/2}(1 \leqslant j \leqslant 2r)$.

---

[88] by the Mordell-Weil Theorem: see, e.g., S.Lang, 'Elliptic Curves and Diophantine Analysis' (Springer, 1978) §4.2

[89] also known as the **Jacobian Variety)(Jac) of the curve**

[90] i.e. for all finitely generated fields L, and all rational F.

[91] See, e.g., Simple, J.G./Kneebone, G.T., 'Algebraic Curves' (OUP, 1959) Ch X

[92] See, e.g. Hodge & Pedoe, 'Methods of Algebraic Geometry' (CUP, 1952;1994)Vol 2 Essentially, this amounts to substitution of specific values for variables from the field.

[93] These definitions hold for any abelian variety A, but only Jac(ζ) $\equiv J()$ is relevant here.

[94] See,e.g., E. Grosswold, 'Topics from the Theory of Numbers'(Macmillen, 1966),Ch10

[95] See,e.g. Hodge & Pedoe, 'Methods of Algebraic Geometry' (CUP 1952) Vol 2 and Weil,A., 'Foundations of Algebraic Geometry'. (AMS, rev.ed. 1962)Ch2

[96] In fact: $\models n - p - pt(J_A)$ is mapped ineffectively into $n - p - pt(J_{A'})$

[97] See,e.g., E.Weiss 'Algebraic Number Theory' (McG-H, 1963)p157

[98] Most of these results may be found in 'Abelian Varieties', by S.Lang (Wiley/Interscience, 1959)

(v) $\models |J(\zeta)|_k \leqslant (q^{1/2}+1)^{2g}$, independent of r.

(vi) $\models$ If ζ **has good reduction at** p**, then so does** $J(\zeta)$**.**

(vii) Let ζ' **be a specialization of** ζ **(over** K'**), and** F'**, of** F. $\models$ **If** ζ **and** ζ' **have the same genus, and** F' **is irreducible in all algebraic extensions of** K'**, then there is good reduction**.

These results furnish possible **tests for good reduction.**[99]

Within this framework, it may be shown that **bounds on** $|J_{J(\zeta)}|$ **may be obtained from bounds on the torsions**

$$\mathcal{T}^{(\alpha)} := |\mathcal{J}_{J(\zeta)\text{mod}P_\alpha}|, \mathcal{T}^{(\beta)} := |\mathcal{J}_{J(\zeta)\text{mod}Q_\beta}|$$

, **provided that** rational primes, p, q, have been determined – with **prime - ideal factorizations** [100]

$$p = \prod_{i \in S_p} P_i^{r_i}, q = \prod_{j \in S_q} Q_j^{s_j}$$

such that

$$J^{(\alpha)}(\zeta), J^{(\beta)}(\zeta)$$

have good reduction at P_α, Q_β, respectively. Here $J^{(\alpha)}(\zeta), J^{(\beta)}(\zeta)$ denote the specializations of $J(\zeta)$, through $F \equiv F(x,y)$, for the residue - class fields K modP_α, K modQ_β, with $\alpha \in S_p, \beta \in S_q$. The test embodied in (vii) may be used (with prime-ideal factorization routines) to search the set of all rational primes for the smallest admissible prime p, q. **The torsion bounds then have the form** $B(p,q)$. If $\mathcal{T}^{(\alpha)}, \mathcal{T}^{(\beta)}$ are known exactly (rather than merely having known bounds) then, typically, $B(p,q)$ may be lowered significantly.

For **parametrized curves**, ζ, governed by the equation $F(u;x,y) = 0$, defined over K with u transcendental over some field k, and K an algebraic extension of $k(u)$, an alternative procedure may be used to attack the problem – though some 'good reduction' criteria are still involved (at least, for tractable implementation). With this method it is sometimes possible to prove that some divisor is not of finite order (and hence that the function is not elementarily integrable). Otherwise the dependence on u may be removed, by [101] finding $u_0 \in k$ such that $D(u)$ has finite order on '$F(u,x,y) = 0 \Leftrightarrow D(u_0)$' has finite order on '$F(u_0,x,y) = 0$' – after which, the previous scheme (producing bounds $B(p,q)$) may be used. **NOTE that**, if D depends essentially on u (i.e., if u-dependence of D cannot be eliminated) then D is of order ∞, and elementary integration is impossible.

Manin's technique, however, does yield a different algorithm for obtaining the order of

$$D \equiv \sum_1^M n_j P_j$$

(i.e. additively represented).

The algorithm produces either: ∞, or else, an integer, N, dividing the order of D – i.e., the least multiple of D that is linearly equivalent to 0, and so is the divisor of a function. The actual **estimate** of ord$D$ may be made by determining $B(p,q)$, as above; but direct testing of mND, for $m = 1,2,3\ldots$, will eventually generate a divisor $m * ND = (h)$, for some w.d. function, h, over ζ.

NOTE that Bronstein's scheme (op.cit), bases on Trager's rationalization procedures, avoids the most recondite aspects of Risch's approach – except for decision/solution problems for the associated systems of ODE introduced by Risch.[102] The elegance and breadth of the 'Jacobi

---

[99] A full discussion of various tests for good reduction is given in **J-P Serre/J.Tate, 'Good Reduction of Abelian Varieties', Ann.Math 88(1968) 492-5**

[100] In this context, the r_i, s_j are ramification indices, and p (or q) is unramified if all r_i (or all s_i)=1; otherwise, ramified cf. the definition of ramification for places (e..g., on algebraic curves)

[101] Good reduction is used here $\models$ only finitely many u have bad reduction

[102] R.H. Risch, TAMS 139(1969) 1670189; see also, M.Bronstein, J.Symbolic Comp. 9 (1990)49-60

variety framework', however, are highly illustrative for Σ – even though the resulting procedures may not be fully/efficiently implementable. More recently, research on enumerative/effective techniques in Algebraic Geometry has become widespread.

These algorithms, for **factorization integration** and (O)**DE**[103], illustrate the extent to which existing SAM procedures depend on sophisticated results of **algebraic-geometric** character – involving the most general structures in nontrivial ways. This tendency will become more pronounced – to include, for instance, fundamental results for **topological**, or **differentiable, structures** such classes of results must be reflected in the specification of Σ_0. As indicated in E_0/E_1, the underlying results may be incorporated as **subtheories** (with their own nomenclature/definitions/axioms/theorems). The consequent 'opacity' of the system is offset by an overall gain i coherence – furnishing enhanced tools for the realization of maximally effective ME – viewed as OT.

In particular, all of the formal 'apparatus' for investigating **algebraic, analytic varieties, schemes manifolds**, in the utmost generality, may be represented in Σ_0 – along with the subtheories of the **Basic Spaces**, ${}_XS$, introduced in E_1, and taken as **approach environments**, in E_6. Possible sources for some of this material are given earlier in this Essay for **Algebra** and **Real Analysis**. Sources for other subdomains of 'Pure Mathematics' are given in E_5 – in relation to the formation of 'diagrams' **for 'arbitrary sub theories'**– as well as, for some aspects of 'Applied Mathematics'.

[103]e.g.,Singer's results, mentioned earlier in this Essay

CHAPTER 3
Mathematical Information

An IB may be regarded as a (very large) collections of **items** of information arranged hierarchically, and diversely interconnected – with extra structure for the implementation of versatile search routines. The general classification of information given to rise to **taxonomic schemes.** Here, **genera** (or, **categories**) cover **species** (or, **types** and each species comprises **representatives** (or, **examples**). The principles for the formation of these groupings may be systematized (even, partially formalized) mainly, through **lattices** and other p.o. systems. This scheme, however, is ill-suited to Σ , since mathematical information is, typically, both more varied and more intricately interrelated, than that characteristic of the Life Sciences (where taxonomic methods originated). Somewhat paradoxically, then, a combination of the intuitive and heuristic seems appropriate for the construction of mathematical **pretaxonomies**, and for **skeletal/partial/full taxonomies**. Of course all of these forms of classification are (inevitably) incomplete; refinements is always possible.

The complexity of mathematical information stems partly from the high level of abstraction, but also from the great variety of categories /species/examples. Moreover, within any species, the distinctions between examples may be very subtle – so, attempts at rigorous formalization will inevitably produce turgid schemes that are hard to use (let alone, implement) and unstimulating. All of this has serious implications for the development of 'genuinely interesting theories (promoting invention/innovation), as opposed to arid exercises in axiomatics. Since Σ is an **interactive** OIB, it is convenient to view every item of information as a (potentially operational) **theorem**. This includes [104] **Nomenclature, Definitions, Axioms,** and the '**Rudiments**' in all domains. These sets are denoted (respectively) by $\mathcal{N}, \mathcal{D}, \mathcal{A}, \mathcal{R}$. Also included (as a cumulative bodies of deductions) are all recognized **subdomains of** $\mathcal{M}_t$. The set of $\mathcal{J}_\Sigma$, given by $\mathcal{R} \cap (\mathcal{N} \cup \mathcal{D} \cup \mathcal{A})$ is regarded as the **foundation** for the developments of Σ . This foundation, $\mathcal{J}_\Sigma$, must be specified (once and for all) in great detail for subsequent development (though the results in $\mathcal{J}_\Sigma$ are 'assumed', here). The content of $\mathcal{J}_\Sigma$ is a matter of judgement; all that is relevant here is that some adequate form(s) of $\mathcal{J}_\Sigma$ have been specified for P_Σ. An excellent example of this process is furnished by the book: '**Retracing Elementary Mathematics**', by L.Henkin, et al.,[105] where the main goal is to establish the Real Number System axiomatically – with careful discussing of cognate topics in Set Theory, Logic, and Algebra (as a basis for Real Analysis). All proofs are given as sequences (lists) of numbered steps, with some extra labels (—— >> for essential structure, $\rightarrow$ for key deductions). Although this scheme would be unbearable tedious for P_Σ, it does seem appropriate for the embedding of $\mathcal{J}_\Sigma$ within P_Σ[106]. Once $\mathbb{R}$ has been constructed (as a complete, ordered field) the constructions of $\mathbb{C}$ complete but **un**orderable[107], field – in which every algebraic equation over $\mathbb{C}$ has at least one root – is straightforward. The fact that $\mathbb{C}$ is **algebraically closed** permits the development of Complex Analysis (initially, for functions

$$f : \mathbb{C} \supseteq A \rightarrow B \subseteq \mathbb{C}, \text{and thence, for}$$

$$g : \mathbb{C}^k \supseteq A \rightarrow B \subseteq \mathbb{C}, k = 2, 3, \dots).$$ [108]

[104] An arbitrary (sub) theory is denoted by $\overline{\mathcal{T}} \equiv \mathcal{R}(\mathcal{T}) \cup \mathcal{T}$ – so that $\mathcal{T}$ is the 'non rudimentary part of $\overline{\mathcal{T}}$'. Then one has $\mathcal{N} =: \mathcal{B} \cup \{\mathcal{N}(\overline{\mathcal{T}}) : \overline{\mathcal{T}} \in \overline{\mathcal{J}}\}$, where $\mathcal{B} \equiv$ 'basic nomenclature'.

[105] '**Retracing Elementary Mathematics**', by L.Henkin/W N Smith/ V J Varineau/ M J Walsh (pp418, Macmillan, 1962)

[106] Much briefer developments of $\mathbb{R}$ (and $\mathbb{C}$) are given by: H.A. Thurston, 'The Number System' (pp134, Blackie, 1956 (Dover, 1967)), and by: L W Cohen/ G. Ehrlich, '**The Structure of the Real Number System** (pp116, Van Nostrand, 1963), but they lack the discussions of Set Theory, Logic, and Foundations.

[107] $\models$ No ordered field extension of $\mathbb{Q}$ contains a root of the equation $z^2 - 1 = 0$. Further, $\models$ Every algebraic extension of $\mathbb{R}$ is isomorphic to $\mathbb{C}$ (and so is also unorderable)

[108] Various topological properties of planar domain are also crucial (f, above); and, of higher – dimensional domains (for g).

It should be understood that the sets $\mathcal{N}, \mathcal{D}, \mathcal{A}, \mathcal{R}$, are far from disjoint/independent. Definitions are framed in terms of accepted nomenclature; axioms are formulated on the basis of existing definitions; and, basic results are derived from axioms, through rules of inference/ deduction. On the other hand, as the sets $\mathcal{N}, \mathcal{D}, \mathcal{A}, \mathcal{R}$, evolve, each generates new elements in the others (in intuitively obvious, but formally tricky, ways!), in symbiotic processes. it is impossible, therefore to characterize these sets formally, but their essential properties will be used from now on!

The **Nomenclature** of any scientific subdomain encompasses it terminology, including the accepted (range of) symbols for the objects/processes/... involved. In mathematics and its applications, the choice of notation is of great importance, since it often suggests further developments of topics considered. The symbiotic growth of the sets $\mathcal{N}, \mathcal{D}, \mathcal{A}, \mathcal{R}$, precludes their full specification. Indeed, the terms listed in $\mathcal{N}$ are introduced only as required in some (sub)theory. All that can be said is that such criteria as logical consistency, calculational effectiveness, 'suggestiveness', and 'elegance', play key roles in determining 'the steady state of any theory'. Perusal of mathematical journals reveals a wide range of accepted terminologies/notations; collectively, these constitute the nomenclature. Certain terms/objects/processes/... are customarily denoted in established forms (to be listed in $\mathcal{N}$) while other may have several representations. Since a concept is semantically independent of the symbol(s) used to depict it, arbitrary (but definite) assignment of symbols to 'nonestablished entities' in $\mathcal{N}$ will be adequate. Ultimately **all** ME will be listed as 'dictionary entries' – to fix their notations and basic interpretations.

Definitions are of central importance for Σ – even though 'the deductive span' of any subtheory is invariant under their introduction/ They are fundamental in the formulation of 'new' theories, and in the extension of existing ones, both for '**compression** without loss of information', and for **explication** – corresponding to **nominal**, and **explanatory**, defunctiodefinitionsns, respectively. Although several **modes** of definition may be identified, no exhaustive classification seems to be attainable. Nevertheless, the classes of e**explicit, inductive/recursive,** and **implicit** definitions, appear to suffice Σ . Indeed, it may be shown that, within certain general/frameworks, alls of these modes are essentially equivalent! In 'mathematical practice', however, this is definitely, not the case, since each mode lends itself 'naturally' to particular classes of investigations, but is relatively ineffectual otherwise.

Explicit ($\equiv$ nominal/explanatory) definitions are ubiquitous in mathematics. A **nominal definition** (**of** A **by** B) is simply a declaration of the (possibly, conditional) interchangeability of A and B in any wfe. The effect is to **compress** the wfe – often facilitating calculations that were intractable, or exhibiting hidden structural features. one of the most remarkable illustrations is provided by the definition, and literal denotation, of 'function', and the subsequent development of analysis. A subtly different example is found in the **abstraction** of the entity 'matrix' (as an array of coefficients), followed by the literal representation of this array, initiating the study of **matrix algebra** (and, later, of **matrix analysis**).

An **explanatory definition (of** A **by** B **for context** C**)** normally prescribes the admissible interpretations/uses of A, as well as providing an identity for A in terms of B, where B is 'already understood and w.d', for C. An obvious **example** is the definition of any function $f : A \to B, x \mapsto f(x)$, by a specified **power-series, within the circle of convergence** – where the interpretation and use of power-series are assumed 'known'.

The **inductive/recursive definitions** yield **sequences** of objects, by specifying the initial object and providing an **algorithm** for the calculation of any other object in the sequence from its predecessor(s). Typical examples include the solutions of finite -difference equations with coefficients in some given domain. The prescription just given corresponds to a very general class of difference relations. Even more generally, one may consider procedures for the generation of 'new' objects from other already w.d. (over some theory); if these w.d. objects are primitive notions of the theory then the inductive definitions is called **fundamental** otherwise it is **non-fundamental**. A more restricted form requires the algorithm to be strictly constructive (usually, **finitary**)

The class of **implicit definitions** covers all cases where there is **an identity between two**

wfe, only one of which contains the definiendum(i.e., the entity to be defined). **Implicit and generalized functions,** along with **eigenvalues and eigenfunctions** (in Algebra/Analysis); the prescription is quite general, however. Such definitions do not produce interchangeable entities unless additional rules are specified (through suitable algorithms). Another example of implicit definition involves '**primitive objects/relations,** which are given 'coherence' through sets of **axioms** (e.g., as in axiomatic formulations of Geometry (see, e.g., '**The Foundations of Geometry**' (U. of Toronto Press, 3rd Ed, 1952), by **G.de B. Robinson** and, **'Axiomatic Projective Geometry'** (leicester U.Press, 1962), by **R.L. Goodstein/EJF Primrose**. Other variants include **prescriptive definition** (containing criteria to 'test' the presence of prescribed properties), of relevance for operational information, but probably not consistently applicable in strictly axiomatic theories; and, **definition by abstraction** (where 'essential properties' are abstracted). This process of abstraction yields classes of objects interrelated through a form of 'restricted equality', say aRb, where R is symmetric, transitive and satisfies the condition (for the class C): $\forall x \in C \, \exists \, y : xRy$ (i.e. there is at least one y related to each x in C by R). An **example** is given by **similarity** (for any class of geometrical figures) and thus also illustrates, in an elementary context, the notion of **'abstraction as approximation'**, which – together with analogy/inverse analogy – is discussed in E_7. This is a fundamental process in Σ .

So far, the classes of notations/terminologies (Nomenclature), and of Definitions have been discussed as instances of 'Mathematical Information'. The class of **Axioms** comprises fundamental propositions involving the primitive elements of theories – each of which is fully determined (in principle) by its underlying set of (assumed) axioms. This process is especially transparent for **algebraic** structures; but it is widely held that **all** of Pure Mathematics may (and should!) be developed axiomatically. For Applicable Mathematics (of all types), a more pragmatic approach seems unavoidable, though various subtheories may still be treated axiomatically.[109] At all events, the mathematical results upon which applications are based, are ultimately derived from Analysis and Algebra; so, to this extent, **all** applications also have an axiomatic basis. The prime exemplar of this scheme is **Bourbako**, whose 'Elements of Mathematics' is intended as a demonstration of the rigorous formulation of theories.

The diversity (in complexity and sophistication) of axioms is considerable, but, in fruitful theories, the axioms tend to have a degree of (retrospective!) inevitability – often coupled with the elusive quality of elegance. The axioms of a theory, $\overline{\mathcal{T}}$, are simply sets of mutually consistent propositions which (together) 'generate $\overline{\mathcal{T}}$. As mathematical information they have the extra characteristic that they are **postulated**, and cannot be proved or disproved within $\overline{\mathcal{T}}$ (or, in most cases, within any other theory). This justifies treating them as a separate type of information.

The class of Rudiments (for a theory $\overline{\mathcal{T}}$) comprises 'the most basic theorems derivable from the $\overline{\mathcal{T}}$ - axioms. The choice of material for $\mathcal{R}(\overline{\mathcal{T}})$ is partially subjective, but the exact form is immaterial here provided that **some** choice is made, and that the items of information involved are suitably labelled in P_Σ. In Arithmetic/(basic) Analysis, the treatment given by Henkin, et al. (op cit.) will suffice; in Algebra, it may be supplemented from any thorough going introduction[110]. Comparable presentations for Euclidean[111] and non Euclidean[112] Geometries may be envisaged on similar lines. The main purpose in distinguishing between $\mathcal{R}(\overline{\mathcal{T}})$ and $\mathcal{T}$ is to make the operational results in P_Σ relatively nontrivial – but properly underpinned by elementary properties from $\mathcal{R}(\overline{\mathcal{T}})$. So far, only basic Arithmetic, Analysis, Algebra and Geometry have been considered; but the intention is to produce $\mathcal{R}\overline{\mathcal{T}}$ for every (sub)theory, $\overline{\mathcal{T}}$, in P_Σ. The (inescapable) nonuniqueness/nondisjointness of the elements in $\{\mathcal{R}(\overline{\mathcal{T}})\}$ raises no problems of

[109]Many examples are discussed in: '**The Axiomatic Method**', with special reference to Geometry and Physics', **L. Henkin, P.Suppes and A.Tarski**, Eds. (N.Holland, 1959) For Continuum Mechanics, see various books/articles by C.Truesdell (and Co-workers)

[110]Possible texts include: 'Elementary Abstract Algebra', by E M Patterson/D E Rutherford (Oliver & Boyd, 1965); 'An Introduction to Abstract Algebra', by T A Whitelaw (Blackie, 1978); 'A First Course in Abstract Algebra', by P J Higgins (van Nostrand, 1975); 'Rings, Fields and Groups', by R B J R Allenby (Arnold, 1983)

[111]Such material may be compiled from (e.g.) 'Introduction to Plan Geometry', by H.F.Baker (CUP,1943); 'Analytical Geometry', by A.Robson (CUP,1940); 'Coordinate Geometry', by RJT Bell (Macmillan, 3rd ed, 1956).

[112]See, e.g., 'Euclidean and Non Euclidean Geometries ...', by M J. Greenburg (2nd ed. Freeman, 1980)

principle, and will be helpful in effective presentation of the results. The construction of **pre-/skeletal/partial/full taxonomies** depends critically on this initial specification of (sub)theories, $\overline{\mathcal{T}}$, with associated collections, $\mathcal{R}(\overline{\mathcal{T}})$ - a tedious but crucial task. If the totality of (sub) theories is demonstrated by $\mathcal{J} \equiv \{\mathcal{R}(\overline{\mathcal{T}}) \cup \mathcal{T}\}$, then the cumulative mathematical information in P_Σ will comprise $\mathcal{N}, \mathcal{D}, \mathcal{A}, \mathcal{R}, \mathcal{J}$, where $\overline{\mathcal{J}} =: \mathcal{R} \cup \mathcal{J}, \mathcal{N} = \{\mathcal{N}(\overline{\mathcal{T}})\} \cup \mathcal{B}, \mathcal{D} = \mathcal{D}(\mathcal{B}) \cup \mathcal{D}(\overline{\mathcal{J}}), \mathcal{A} = \mathcal{A}(\overline{\mathcal{J}})$ and $\mathcal{R} = \mathcal{R}(\overline{\mathcal{J}})$. In other words, Σ -information is either **basic** or else, $\overline{\mathcal{T}}$-related for some (collection of) $\overline{\mathcal{T}}$. Of course, every $\overline{\mathcal{T}}$ involves many items of Σ -information; indeed, each $\overline{\mathcal{T}}$ is itself such an item!

Lastly, in this outline of the types of Σ -information comes the set $\mathcal{J}$, of nonrudimentary (sub)**Theories**. As already explained, $\mathcal{J}$ comprises (sub)theories, $\mathcal{T}$ may be represented as a collection of (interrelated) nontrivial (quasi-) **operational theorems**. For purposes of taxonomy **approaches to** $\mathcal{T}$ are treated as separate (sub)theories. In general, these approaches correspond to the approximation-environments, $_SX$, introduced in E_1, together with the processes of abstraction, analogy and inverse analogy. From now on, the word 'theory' will be understood to include all alternative representations a a **sub**theory designations will be reserved for situations where structural embeddings have significant implications. In the above example, Group Theory would be developed in its own right, as a self contained body of theorems; and similarly for the theory of Abelian Groups (in relation to the general theory of groups).

On this basis, every theory $\mathcal{T}$m is characterized by its nomenclature, definitions, axioms, rudiments, and operational theorems – along with the basic terminology, B. On this basis, coherent taxonomic structures and search software may be produced, guided by the combined aims of scope/key-topics/effectiveness, for every established mathematical subdomain (taken here from the lists in Mathematical Reviews (MR)). The MR scheme is designed to locate the content of papers/books accurately enough for specialized use. This is indispensable 'first step' in specifying a collection of fields which -properly interpreted = encompasses 'all of mathematics'. This collection is modified /augmented regularly to ensure that new fields, and 'topographical changes', in Mathematics are identified, and covered.

although every entry in the 'MR Subject Classification' could be a root of further subclassifications, experience indicates that such extra refinement is likely to be counter-productive at this level of use. For P_Σ, by contrast, this initial identification of subjects is just the start of the design process! Each MR subject entry corresponds to a Σ -subdomain; the totality of entries 'covers M_t' (including applications). These subdomains yield **ore-taxonomies**: lists of terminology, definitions, axioms, rudiments, and theories. The essential structural items, systematically arranged, constitute **skeletal taxonomies** – each item of which may be expanded (with the skeleton classification consistently embedded in a more refined 'tagging scheme'), to produce **full taxonomies** for the MR -subdomains. On account of the 'universality of Mathematics', any procedure devised to organize 'Mathematical Knowledge' may itself be treated mathematically! Taxonomy, and Data-Base-Management have established methodologies[113] – which are not, however, pertinent to the fundamental issues which these Essays are concerned (though they may be relevant for the design of search software). The **structural** aspects of taxonomy involve partially ordered algebraic/analytical systems – in particular, for the generation of classification - refinements with preservation of complex ordering inter relations. Other facets correspond to problems of multivariate (statistical) cluster analysis. As indicated in E_1, the **production** of taxonomies requires, primarily, wide-ranging knowledge of the subjects, and of their interconnections; and, of links with other fields. Subsequent elaboration and checking may benefit from formal taxonomic techniques – but not for the selection of material.

The abstract study of data-base theory involve **relational logic, transaction-processing,** and **access methods,** with special relevance to **updating**, serial and parallel **algorithms,** and **data-structures** (respectively). The (declarative) **relational calculus** has a dual (procedural) counterpart: **relational algebra**, for data manipulation. This set-up has been extended to allow consistent 'knowledge input', and incomplete information/updating. The 'transactions' correspond to serial/concurrent programs as they affect the integrity/security of (existing) data; a

[113]See, e.g., the references given in E_1

theory of reliable transactions is sought. Lastly, 'access methods' refer mainly to information retrieval and so are directly related to IB design. These techniques of DB analysis are intricate and sophisticated often blending Mathematical Logic with Algebra. Essentially, they furnish foundations for the efficient/consistent I/O of general classes of data, where 'input' covers updating, and 'output' covers data retrieval. On the scale of the Σ -IB, the use of effective data procedures is crucial – and taxonomic structures must be designed to exploit these procedures. As already indicated, the principal ingredients of any taxonomic scheme are genera ($\equiv$ categories), speciies ($\equiv$ types), and examples ($\equiv$ representatives), and the overall schemes may be realized **cumulatively**, without invalidation of existing form when new items are added. Thus, each 'heading' in the MR subject list will initiate 'its own' pre/skeletal/full taxonomy, which at any time, will be only partially complete. **This MR list is taken as the basis for all illustrations/discussions of taxonomic issues in** Σ .

The most novel aspects of Σ as an IB are embodied in the emphasis on **maximal calculation**, in all 'situations', and the consequent introduction of **approaches** – subsuming the most general 'approximation procedures', together with abstraction/analogy/inverse analogy. This leads naturally to the concepts of: t**-max.-operational information** and t**- calculational bases**(the t-dependence being suppressed in most contexts), with associated, structural characteristics – which determine the **super** structure of Σ as a research environment for mathematical **investigations**.

The representation of **all** items of information as 'theorems' requires each of them to be expressible as '$p \Rightarrow q$', for suitable wfe p, q. It suffices, therefore, to consider notation, terminology, definitions , and axioms, from this point of view. Since notation and terminology are essentially equivalent to nominal definition, only definitions and axioms need to be dealt with.

The basic types (explicit, recursive, implicit) of definitions over first-order logical theories, have been characterized succinctly by Rasiowa/Sikorski[114]as follows (where it is assumed that recursive definitions are reducible to explicit definitions).[115]

Let $\mathcal{J} \equiv \{\mathcal{L}, \zeta, \mathcal{A}\}$ be any first-order formalized theory, with $\mathcal{L}$ a **formalized language**, and ζ, the consequence operation in $\mathcal{L}$ (so that $\zeta(\mathcal{A})$ comprises the set of **theorems** of $\mathcal{J}$). Notice that the First-Order Predicate Calculus (PC-1) is covered here – and this suffices for almost all proofs in 'general mathematical activity'.[116] A w.f.f(ormula) in $\mathcal{J}$ is characterized by the conditions:

(i) every sentence is a wff;

(ii) if α, β are wff then so are $\sim \alpha, \sim \beta, \alpha \wedge \beta, \alpha \vee \beta, \alpha \Rightarrow \beta, \alpha \Leftrightarrow \beta$

Any set S containing all sentences of $\mathcal{J}$, and satisfying (i) and (ii), is called **inductive Propositional functions** have the structure of sentences, but contain 'variables' ranging over a fixed set of objects; they yield sentences whenever these variables are instantiated'.

Predicates represent **relations** (among constituents), and **elementary formulae** are formed from terms, predicates and auxiliary signs (e.g., parentheses). Finally, **formulae** are obtained by linking elementary formulae through propositional connectives quantifiers and auxiliary signs. A formula is **closed** if it contains no free variable (for which 'values' may be substituted)

With this background, one may specify the processes of explicit and implicit definition.[117]

Let $\mathcal{T}(x)_m$ be a term in $\mathcal{J} = \{\mathcal{L}, \mathcal{G}, \mathcal{A}\}$, for $m \geqslant 0$, having precisely the free variables $x_1, \ldots z_m (m \geqslant 1)$, or **no** free variable $(m = 0)$. Let: $\psi_\mathcal{T}(x)_m \notin \mathcal{A} \cup \mathcal{T}(x)_m, L' := L \cup \psi_\mathcal{T}(x)_m; \mathcal{A}' := \mathcal{A} \cup$ eq$(\psi_\mathcal{T}(x)_m, \mathcal{T}(x)_m)$. Then $\mathcal{J}' := \{\mathcal{L}', \mathcal{G}, \mathcal{A}'\}$ extends $\mathcal{J}$ through the explicit definition eq$(\psi_\mathcal{T}(x)_m, \mathcal{T}(x)_m)$ of the 'functor' $\psi_\mathcal{T}$. Intuitively: $\psi_\mathcal{T}$ is 'accepted' as a shorter notation for the function $\mathcal{T}(x)_m$.

---

[114]'The Mathematics of Metamathematics' (P W N, Warsaw, 1963, §15

[115]See, e.g., 'The Foundations of Mathematics', by E.W. Beth (North-Holland, 1968, Sec.53)

[116]See, e.g., 'A Mathematical Introduction to Logic' (HB Enderton)(AP, 1972), Ch.2.

[117]See Suppes, P., 'Introduction to logic' (Van Nostrand, 1957) Ch. for a brief study of definitions'.

Next, let $(\#) \equiv (\forall_\xi)_m \exists_\eta \beta((\xi)_m, \eta), m \geqslant 0$, be a closed theorem in $\mathcal{J}, \mathcal{L}' := \mathcal{L} \cup \varphi_\beta, \varphi_\beta \notin A \cup (\#)$; $A' := A \cup (\forall_\xi)_m \beta((\xi)_m, \varphi_\beta(\xi)_m)$. Then $\mathcal{J}' := (\mathcal{L}', \mathcal{G}, \mathcal{A}')$ is obtained from $\mathcal{J}$ through use of the **implicit** definition of φ_β (and φ_{beta} may be taken as unique if eq $\in \mathcal{J}$).

Intuitively, φ_β instantiates η in $(\#)$ **Note that: nominal** definitions may be specified formally through the conditions:
Let $\alpha(x)_m \in \mathcal{L}(m > 0)$, with precisely $(x)_m$ as free variables. Take $\mathcal{L}' : \mathcal{L} \cup \mathcal{P}_\alpha, \mathcal{P}_\alpha(x)_m \notin A \cup \alpha(x)_m$, and $A' := A \cup [\mathcal{P}_\alpha(x)_m \Rightarrow \alpha(x)_m, \alpha(x)_m \Rightarrow \mathcal{P}_\alpha(x)_m]$

Then $\mathcal{J}' := (\mathcal{L}', \mathcal{C}, \mathcal{A}')$ allows the **replacement** of $\alpha(x)_m$ by $\mathcal{P}_\alpha(x)_m$, and **vice versa**; i.e., $\alpha(x)_m$ and $\mathcal{P}_\alpha(x)_m$, are **interchangeable.**

If follows that the above formal definitions may be represented as
theorems:
Nominal: $\alpha(x)_m \Rightarrow \beta(x)_m \wedge \beta(x)_m \Rightarrow \alpha(x)_m$
Explicit: $\psi_\mathcal{T}(x)_m \Rightarrow \mathcal{T}(x)_m \wedge \mathcal{T}(x)_m \Rightarrow \psi_\mathcal{T}(x)_m$
Implicit: $(\forall\xi)_m\ \beta((\xi)_m, \varphi_\beta(\xi)_m)$.
Recursive/inductive: reducible to **explicit.**

For **axioms,** $a(x)_m$, it is enough to observe that, as postulates (delivered to be true in $\mathcal{J}$) they may be represented as:
$\{1 = 1\} \Rightarrow a(x)_m$
In this way, all items of information may be represented by **theorems** as claimed, so it is now possible to consider the concepts of t**-max operational information,** t**-max implemented** information, and t**-calculational bases** – which together embody the principal mathematical challenges for the development of P_Σ.

The fundamental class of Σ -objects is that of **operational theorems** (OT) – which, in view of the above representation of all Σ - information by theorems, appears to cover **all** entities in Σ . Although this is so, in principle, there is a sharp distinction between the **rudiments** and the mainstream **theorems** in any theory, $\mathcal{J}$. Indeed, such theorems are expressed in terms of 'elementary formulae' through logical connective, recursion and other devices. Here, attention is confined to preliminary observations, a more systematic discussions of OT is given in E_4.

The degree to which any wfe is **operational** measures the extent to which its constituents are realizable (over some **calculational milieu** (CM))[118]. Thus, operational procedures are clos- ely related to effective/constructive/algorithmic procedures – the distinction is mainly a matter of context, in general, but questions of approximation/approach also arise in Σ .

A further distinction must be drawn between representations of $\mathcal{P}_k(e)$ – of whatever operationality – and possible implementations, say, $i_l(e)$, of e over ζ with in Σ_t. Such implementations typically differ radically from the mathematical procedures they embody, since they must be realized in terms of the fixed functions and library routines of the computer system involved. Even for SAM packages this difference in form is often considerable. The notion of t**-max implemented wfe** is, therefore, system - dependent, but this raises no problems of principle. For any wfe e, its representations, $\mathcal{P}_k(e)$ may have implementations ranging from 'nil' to Σ_t – maximal (which may still be far short of 'full', unless the corresponding procedures are available within Σ_t). As in the case of operationality, one has to compare the actual and potentially maximal 'states' of implementation of the $\mathcal{P}_k(e)$, over Σ_t. In general, wfe may 'exist' at many levels of operationality and implementation – these two facets of the wfe being mutually (almost) independent.

At this point, an informal description may be given of the concept of **Calculational Base** (CB) for Σ_t – viewed as a collection of t-max implemented procedures available for application in Σ_t. Here, 'calculation' covers both inference and computation. To avoid circularity, it is assumed that an initial CB, say, B_0, is 'given at $t = 0$'. The system builder's task becomes that of extending B_0 to B_t $(t > 0)$; and, the user's task, that of constructing elements of the $\mathcal{A}$ -closure of B_t – where (roughly) $\mathcal{A}$ stands for the set of all admissible constructions. Although the t-dependence is fundamental, it is not to be taken literally, in most contexts– rather, it serves as a reminder that new results may be derived 'at any time', and that procedures may

[118]The term Calculational **framework**(CF) is also used.

be synthesized only from simpler procedures already implemented. Occasionally, however, the symbols $\Sigma_t/\Sigma(t), B_t/B_it)$, etc., may be interpreted as 'functions of t, upon which various operations may be performed; in such cases, great care must be exercised to ensure that meaningful results are obtained!

A simple example of '$\mathcal{A}$ - closure' is given by the construction of $\mathbb{R}$ (the field of real numbers) from the rationals through the use of (limits of) Cauchy sequences – and, more generally, of the isometric embedding of any metric space within a **complete** metric space. Informally, any CB may be represented as a set, say, $\{P\}_L$, of procedures , P_k, supplemented by 'descriptive material', $(D)_n$ and 'rules of application/combination, $(r)_m$, covering the valid use of the P_k on (classes of) wfe. Since Σ is an **interactive** system modelled on 'standard mathematical activity', there is no need for exhaustive sets of rules such as those for elementary logical systems, or for programming languages. Rather, the user is encouraged to deploy the P_k 'imaginatively' – especially, where the 'usual conditions' are 'almost realized'; or else, where the rules r_j are applicable in unfamiliar frameworks.

Various representations/terminologies may be introduced to highlight particular aspects of the CB. For instance, one may define the (compositional) product P_iP_k – and , thence, the product of any finite collection of P_k, with or without repetitions. Again, the formal additive and multiplicative combinations used to define (respectively) the group of n-chains over a triangulated surface, and the group of divisors of a meromorphic function over a Riemann surface, suggest possible interpretations in the general study of CB. It should be noted that, 'with suitable conventions, the system $\langle (P)_L, E\rangle$, where $E \equiv \{wfe\}$, constitutes a (non commutative) semigroup, under multiplication; and, that theorems/procedures with converses have (quasi-) inverses in this system – allowing a full (sub)group structure. Further, the possibility of associating metrical/topological/geometric structures with CB should also be explored.

Some extension of the definition of CB may be made by allowing for (partially) operational information $(O)_q$ as well as the descriptive items $(D)_n$ and the (t-max implemented) procedures, $(P)_L$. This yields: $B \equiv \{(D)_n, (O)_q, (P)_L, (r)_m\}$, a finite (discrete) set, provided that entities associated with continuously varying 'parameters' are counted as single objets – in which event the notation $B \equiv \{\}_{\underline{J}}$ may be used, where $J := n + q + L + m$. Usually, however, it is necessary to differentiate between qualitatively distinct types of CB elements. Here all of n, q, L, m depend on t, but – subject to this – the numbering of the b_k (but **not** the **form**!) is permanently **fixed**. In other words, 'new' elements of types D, O, P, r, may be added - as n, q, L, m increase with t – and the forms of the 'previous elements' may be changed because of new algorithms, improved implementations, generalizations; and so on. All of this corresponds to the 'maintenance/updating' of IB.

In the study of CB, the sets $(D)_n, (O)_q, (P)_L, (r)_m$ – **always in that order** – are treated as separate entities; the 'collective form', $B \equiv \{b\{_J$, above, is viable only if n, q, L and m are regarded as **fixed** – so that the logical nature of any b_k is ascertainable from k alone. In considering a 'specified CB', such fixed numbering may be useful. In general, the abbreviated representation $B \equiv \{D, O, P, r\}$ seems opposite, unless detailed manipulation are involved (when the 'components', P_k, etc., must be introduced). The key idea is to 'fabricate' operational theorems from 'simpler' ingredients with in the CB and then to convert these theorems, as far as possible, into implemented forms over P_{Σ_t}.

In all that follows, it is essential to recognize that the distinction between operational and implemented versions of any item of information is more 'technological' than mathematical. The status of operational information is as 'potentially implementable' (to the extent/degree of its operationality). The current limitations of P_Σ may restrict the implementation; but the assumption is that this is only a temporary deficiency. Thus: 'operational' means 'implementable in principle'. The terms 'effective', 'realizable', 'constructive', . . . normally refer to particular species of operationality – and when they are used in these Essays, contexts should clarify any particular emphasis intended. Information, then, may be (purely) descriptive, (partially) operational, non-implementable, or (partially) implementable. The mathematical context of any wfe determines as its operationality; and the algorithms through which it may be (partially) implemented are themselves instances of operational procedures. Consequently, the discussion of the CB, its components, and the theorems synthesized from them, centres on operational information – and this emphasis is maintained.

The notions of 'elementary' and 'compound' wfe (with special cases, respectively, 'atomic' and 'molecular' wfe) were mentioned in E_1. Recall that e is **atomic** IFF e contains none of the (logical) connective $\sim, \vee, \wedge, \Rightarrow, \Leftrightarrow$. A useful extension defines e to be k-atomic IFF e is irreducibly comprised of k atomic wfe. This means that e cannot be transformed into any combination of l atomic wfe, where $l < k$. Other variants of 'near-atomicity' will be encountered in E_4, for operational **theorems** and their P_{Σ_t} – implementations. Analogously, one may call a **theorem simple** if it has the form $p \Rightarrow q$, and neither of p, q contains the connectives $\Rightarrow$, or, $\Leftrightarrow$; and so on. For **parsing** purposes, wfe are viewed as certain combinations of descriptive/ expository, and (partially) operational components – typically, k-**atomic predicate-functions of** n **variables.** This is especially pertinent for the interactive application of operational theorems to wfe. On this basis every wfe may be represented schematically as:

$$e \approx \Gamma^e(u_\sigma)_h$$

where the interpretation is as follows. It is supposed that Σ_t is (partially) characterized by a (finite) set, U_t, of underlying objects. The set U_t (regularly augmented) is permanently **ordered**: $U_t \equiv (u)_{L_t}$. All wfe must be 'specifiable over U_t. Typically, one has: $e \equiv e(\theta)_h \equiv \Gamma^e(u_\sigma)_h$, with $(u_\sigma)_h \subset U_t$, and all details varying as e varies. More succinctly, one has $e(\theta)_h = \Gamma^e(U^e)$, for suitable $U^e \subset U_t$. Intuitively, one regards $e(\theta)_h$ as 'being about' $(\theta)_h \equiv (u_\sigma)_h \subset U_t$; thee syntactic and semantic forms of e are embodied in Γ^e. Formally, the set U_t limits the scope of discussion over Σ_t; in practice, however, all of the accepted content of Mathematics may be covered, cumulatively, as t is increased. This formalism makes it possible to work **finite** systems, at all times. Indeed, one has (again, schematically)

$$U_{t+t'} \equiv \{(u)_{L(t)}, u_{L(t)+1}, \ldots, u_{L(t+t')}\}$$

for $t_0 < t < t + t'$, where it is assumed that Σ_t evolves from Σ_{t_0} with overall information I_{t_0} containing the initial CB, B_{t_0} – all stemming from an 'initial set', U_{t_0}, of 'underlying objects'. The system $P_{\Sigma_{t_0}}$ corresponds to some enhanced amalgam of existing SAM packages – with which $U_{t_0}, I_{t_0}, B_{t_0}, \ldots$ are associated. The full specification of B_t includes both 'rules of application' for the procedures P_k and cognate **modes of approach** governing the representations of arbitrary entities in terms of (other) classes of entities.

Although this model for evolution of Σ_t may appear simplistic, it does embody the essence of the process, in ways that facilitate more rigorous studies of particular aspects – especially, of the formulation, proof and implementation of operational theorems, as elaborated in E_4, where the overall aim is to explore the possibilities of creating an **operational-theorem calculus** (OTC) – covering the formal manipulation and application of OT to wfe within Σ_t.

Notice that: **the** CB for Σ_t denotes 'the largest set of t-max-implemented information over Σ_t; but, that any subset of this B_t constitutes a sub-CB for Σ_t. Hence, the intersection and union of sub-CB form (other) sub-CB, as does the $\mathcal{A}_t$ - closure of any sub CB. The algebraic and other, structures that can be associated with CB are considered in E_4. Ultimately, such structural properties will play an important part in the organization of Σ as a largescale research environment.

Even at this stage it is worthwhile to discuss briefly how this research environment could function, in relation to arbitrary 'users' of the system. The development of Σ proceeds through contributions from mathematicians and other scientists – these contributions being 'processed' in a Universal Editorial Facility (UEF), where some results are rejected, others, modified and the rest accepted into I_t (and subsequently into $B_{t+t'}$, if appropriate). It is apparent that there is a complex, symbiotic relationship involving any particular user, other users, the UEF, $B_{t+t'}$, and additional Σ -structure. The UEF is fundamental for the soundness of Σ ; it comprises of a panel of referees (and, eventually, extensive AI/IT software) whose tasks include the following activities

(a) The assessment of information for basic corrections

(b) Checking to minimize duplication

(c) 'Storage' of formally correct but apparently trivial results – for further consideration

(d) Estimation of the operationality (over Σ_t) of results received 'after t'

(e) Rectification of inconsistencies in terminology/notation (in relation to Σ_t)

(f) Suggestion of possible ways of increasing operationality over Σ_t

(g) Iteration of steps (a)-(f), after contributions have been re-submitted to the UEF.

Of course, all of this corresponds closely to the procedures for soliciting, editing and publishing material for printed journals. The essential difference lies in the scale, consistency of criteria, and the widespread canvassing of opinions – since **all** users may offer comments on accepted material at any time (as part of the maintenance/updating processes). Moreover, these tasks can be monitored globally over the system – so that all of its subdomains are updated, not just the basic lists, etc,. The introduction of 'environments for investigation (EfI), covering specified classes of problems is practicable only through extensive use of taxonomies/tagging/ IT-Software. The view of Σ as an interactive (dynamic) IB also admits concepts of **network analysis, information flows,** and **system-optimization** – with associated notions of 'progress towards proofs', 'obstacles to proofs' (e.g., counter-examples!); and so on. Of course, an obstacle may (potentially) be surmounted .circumvented by modifying the 'proof-path' (as discussed in later Essay). Some of these aspects of 'Σ as a system' may be (partially) formalized – and the extensive results of **General Systems Theory**[119] may then be applied to Σ itself! Most of these ideas are 'natural' for Σ ; they, too, are discussed subsequently.

[119] See, especially, '*GST*' by MD Mesarovic / Y Takahara (AP, 1975)

CHAPTER 4
Operational Theorems

Although all items of Σ -information are formally representable as 'theorems', the designation 'operational theorem' (OT) is reserved for nontrivial wfe of the forms $T : p \Rightarrow q$; $\sum nT : p_1 \Rightarrow q_1 \wedge \cdots \wedge p_n \Rightarrow q_n$ $(n = 1, 2, \ldots)$, where none of the P_k, q_k contains the symbols $\Rightarrow$, $\Leftrightarrow$.[120] The OT of type ${}_1T$ are called simple, while those of type $nT (n \geqslant 2)$ are compound[121], of order n. A useful notation for compound OT is ${}_nT =: (T)_n$, with all T_k simple. Alternatively, one had: ${}_nT = \bigwedge_{1 \leqslant k \leqslant n} T_k$. Notice that a simple OT is 'irreducible', in the sense that it cannot be split into a conjunction of $m (\geqslant 2)$ simple OT. The 'operational aspect' is embodied in the action of OT on wfe – as outlined later in this Essay, and considered further in E_8.

Next, following the notation for general Σ -information, one puts:

$$T \equiv T(\theta)_h =: \Gamma^T (u_\sigma)_h$$

for the **the formulation of T in terms of the underlying objects** $u_{\sigma_1}(T), \ldots, u_{\sigma_h}(T)$, selected from the basic set U_t. The $\theta_i(T) \equiv u_{\sigma_i}(T)$ may be called the **ingredients of T** (from U_t, over Σ_t), and the function $\Gamma \equiv \Gamma^T$ determines the **formulation** of T over the $\theta_i(T)$ – while, Γ also embodies the **specification** of T (in Σ_t). These two aspects of Γ^T correspond (respectively) to the '**stylistic form** of T', and to a list of the **mathematical tasks** undertaken in T. From the latter angle, T may be represented as $T \equiv T(\tau)_k$, where $\tau_i \equiv \tau_i(\theta)_{h_i}$ are the tasks involved. These tasks range from the purveying of pure (i.e., nil-operational) information to the performance of procedures (at various levels of operationally). Another partial description of T, say, c(T), may be called the **cast of** T (i.e., a list of the non-trivial ME that appear in any expository formulation of T). Thus: $C(T)$ **comprises the primary ingredients of** T.

In this representation of T, as a set of tasks, the convention is adopted that $i \leqslant j \Rightarrow \mathrm{op}(\tau_i) \leqslant \mathrm{op}(\tau_j)$. The **descriptive index** of T is defined as: $i_d(T) := \mathrm{Max}\{i : 1 \leqslant i \leqslant k(T), \mathrm{op}(\tau_i) = 0\}$, with the corresponding **operational index** given by: $i_{\mathrm{op}}(T) := k(T) - i_d(T)$, and the (total) operationally of T, by: $\mathrm{op}(T) := \sum_{1 \leqslant i \leqslant k(T)} \mathrm{op}(\tau_i)$. These 'parameters of T' are all formulation / specification-dependent – but they do indicate, very broadly, the 'overall operational states' of **theories**, comprising various collections of theorems. Moreover, this formalism (with extensions to be considered shortly) is amenable to refinement and also well suited to IB applications where comparisons of theories entail the identification of qualitative, as well as quantitative, features – especially through the use of (in)equivalence of representations of theorems, or, of theories ($\mathcal{J}_1$ and $\mathcal{J}_2$ being equivalent, in this context, IFF their sets of indices coincide). All of the above definitions may be extended (additively) to multiple theorems, ${}_nT$ (with obvious modifications for 'equivalence', etc.).

It has already been noted (for arbitrary wfe) that 'information e, applied to e_2' produces a combination of **exposition, substitution, calculation** and **construction**. For any operational theorem T, the alternative forms: $T \equiv T(\theta)_{n(T)} = \Gamma^T (u_\sigma)_k$, and $T \equiv T(\tau)_{k(T)}$, $\tau_i = \tau(\theta)_{h(T)}$, have been introduced – to emphasise 'ingredients' ('what T is about'), and 'tasks' ('what T accomplishes'). Frequently, it is convenient to use $T(\theta_T)$, $T(\tau_T)$ in place of $T(\theta)_{h(T)}$, $T(\tau)_{k(T)}$, and this is done from now on, as appropriate. From the above remarks on 'application of e_1 to e_2' it follows, in particular, that an arbitrary OT, say, T, may be 'resolved into elementary actions', α_j, so that:

$$T(\theta_T) = \bigvee_{1 \leqslant j \leqslant 3} b_j \alpha_j (T(\theta_T))$$

$$T(\theta_T) e = \bigcup_{1 \leqslant j \leqslant 3} b_j \alpha_j (T(\theta_T)) e,$$

[120] In E_8, the basic form is: $T : p \Rightarrow q \wedge r$, where r is the 'residual information'.

[121] There are also other sorts of OT viewed in Σ as compound – as specified in later Essays.

where $\alpha_1, \alpha_2, \alpha_3$ correspond (respectively) to the **informative, substitutive**, and **constructive** 'components' of T, the b_j take possible values 0 or 1, independently, and (by convention):

$$0\alpha_j(T)e = \phi(\text{the 'empty wfe'})$$

$$1\alpha_j(T)e = \alpha_j(t)e.$$

The α_j may be specified by the conditions (for $T(\theta_T) \Leftrightarrow p(v) \Rightarrow q(w)$):

$$\alpha_1(T)e := e \cup \tilde{T}$$

$$\alpha_2(T)e := e^*(T) \cup R_T(e) \cup \rho_T(e^*, e)$$

$$\alpha_3(T)e := C_T(v_e, w_e).$$

Here, $\tilde{T}$ denotes the **descriptive** part of T (possibly, a full statement of T); $e^*(T)$ is the **transformed** part of e (under the substitution associated with α_2); and, C_T comprises objects / wfe **constructed** from e-realizations of v, w. The action of $\alpha_2(T)$ – producing $e^*(T)$, with '**residue**' $R_T(e)$ – could be modified by ignoring R_T and defining $\alpha_2 e$ to be e^*; but the point is that $R_T(e)$ is the part of e **un**affected by the substitutions inherent in $\alpha_2(T)$, and so 'invariant under $\alpha_2(T)$'; and this is a significant mathematical property. Lastly, $\rho_T(e^*, e)$ specifies all relations between e and e^*, in this process.

As before, these definitions may be extended to multiple theorems, ${}_nT$, by combining the definitions for the (simple) 'component theorems', $T_k \equiv {}_1T_k$, $1 \leqslant k \leqslant n$. Recall that, for $T \equiv T(\tau)_k$, the τ_i are ordered so that $i \leqslant j \Rightarrow \text{op}(\tau_i) \leqslant \text{op}(\tau_j)$. There is, however, no obvious way of ordering the 'ingredients', θ_i in $T \equiv T(\theta)_h$, by reference to T alone. On the other hand, the representation of $T(\theta)_h$ as $T(u_\sigma)_h$ (ignoring the details of formulation / specification embodies in Γ, as above), and the recognition that $(u_\sigma)_h \subset U_t$, with U_t **permanently ordered**, 'induces' an ordering of $(\theta)_h$. This is vital for the simplest definition of $\hat{f}(T)$, for (almost) arbitrary functions $(\hat{f})$, of operational theorems, T; and, for the systematic treatment of such transformations of (families of) OT – a model for 'general mathematical activity'. It is, in this context, better to call the $\theta_i \equiv u_{\sigma_i}$ **arguments**[122] of $T \equiv T(\theta)_h$.

The notational conventions introduced in E_1 may now be used to write

$${}_nT({}_n^\theta T) = (T(\theta)_h)_n = (T_1(\theta_1)_{h_1}, \ldots, T_n(\theta_n)_{h_n}),$$

where the $(\theta_k)_{h_k}$ are subsets of U_t given by $(u_{\sigma_k})_{h_k} = (u_{\sigma_{k1}}, \ldots, u_{\sigma_{kh_k}})$, and any $u_{\sigma_{kj}}$ may occur v_{kj} times in the representation of ${}_nT$, with $1 \leqslant v_{kj}$, $1 \leqslant j \leqslant h_k$. The main purpose of this formalism is to allow definitions of $f({}_nT(\theta_{{}_nT}))$ to be formulated consistently and with adequate generality. For any simple OT, one defines

$$\hat{f}(T(u_\sigma)_h) := T(\hat{f}(u_\sigma))_h;$$

and, for the compound theorem, ${}_nT$:

$$\hat{f}({}_nT(\theta_{{}_nT})) \equiv \hat{f}((T(\theta)_h)_n) := (T(\hat{f}(u_\sigma)_h))_n.$$

Here, it is worthwhile to give the full versions, namely:

$$\hat{f}(T(u_\sigma)_h) = T(\hat{f}_{11}(u_{\sigma_{11}}), \ldots, \hat{f}_{1h_1}(u_{\sigma_1 h_1})); \text{ and,}$$

$$\hat{f}({}_nT(\theta_{{}_nT})) = (T_1(\hat{f}_{11}(u_{\sigma_{11}}), \ldots, \hat{f}_{1h_1}(u_{\sigma_{1h_1}})), \ldots,$$
$$T_n(\hat{f}_{n1}(u_{\sigma_{n1}}), \ldots, \hat{f}_{nh_n}(u_{\sigma_n h_n}))).$$

[122]The more evocative term **ingredients** is also used

Here, for $1 \leqslant k \leqslant n$, the lists $(u_{\sigma_k})_{h_k}$ correspond to the sets of arguments $(\theta)_{h_k}$ of the OT constituting ${}_nT \equiv (T)_n$, and $\hat{f} \equiv ((\hat{f})_h)_n = ((\hat{f}_1)_{h_1}, \ldots, (\hat{f}_n)_{h_n})$ exhibits the collections of h_k function(al)s to be applied (in order) to the arguments of T_k.

For this to 'make sense' in general, one may define **the action of $\hat{f}$ on ${}_nT$**, through **the actions of the $\hat{f}_{ik}$** on θ_k, as follows:

$$\hat{f}_{ik}(\theta_k) := \begin{cases} f_{ik}(\theta_k), & \text{if } (\theta_k)n \text{ is even} \\ \theta_k, & \text{otherwise.} \end{cases}$$

Thus, an alternative form is:

$$\hat{f}_{ik} := f_{ik}\chi_{ik} + (1 - \chi_{ik}),$$

where, χ_{ik} is the **characteristic function** of **dom** f_{ik}, $\hat{f}_{ik} : PU_t \to U_t$, PX being the **power-set** of any set X, and it is assumed that U_t has been 'specified-in-principle at t'.

<u>NOTE</u>: Extensions of the concept of compound theorem, ${}_nT$, to voce families of theorems – say, as ${}_\Lambda T \equiv \{T_\lambda : \lambda \in \Lambda\}$ are discussed in E_8.

A convenient form for $f(T)$, where f is 'arbitrary' and $T \equiv T(\theta)_h \equiv T(\underline{\theta})$, may be obtained in terms of the '**function-matrix**' ${}^k\tilde{f}$ given by:

$${}^k\tilde{f}_{ij} := \delta_{ij}\delta_{ik}f + \delta_{ij}(1 - \delta_{ik}),$$

$$\tilde{f} := {}^h(\tilde{f}) \equiv ({}^1\tilde{f}, \ldots, {}^h\tilde{f}),$$

for $1 \leqslant i, j, k \leqslant h$ (no summation over i). Thus: ${}^k\tilde{f}$ 'focuses f' on $\theta_k \in \theta_T \equiv (u_\sigma)_h \subset \mathcal{U}_t$. Here, f may denote any w.d. operation potentially applicable to the θ_l, and one may define ${}^kf(T(\underline{\theta}))$ as

$${}^kf(T(\underline{\theta})) := T({}^k\tilde{f}\underline{\theta}), \qquad \theta \equiv (\theta)_{h_T},$$

with the convention that – in this context – $f\theta_k$ **is interpreted as** $f(\theta_k)$.

<u>NOTE</u> that this convention is already well established for formal treatments of PS expansions, and in many other procedures.[123] An alternative scheme – based on vectors f_i rather than matrices $\tilde{f}$ – is also introduced here. There is scope for both forms in later developments.

If the notation

$$({}^kf^p)_r \equiv {}^{k_1}f_1^{p_1} \circ \cdots \circ {}^{k_r}f_r^{p_r},$$

where f_p is the p-fold composition of f, is now used, then the most general form of modification of any generic OT in a proof / procedure / construction / ... takes the form:

$$({}^kf^p)_r(T(\underline{(\theta)})), \text{ with } 1 \leqslant k_l \leqslant h,\ p_l \geqslant 1,\ r \geqslant 1.$$

Observe that, in this representation , all occurrences of any f_l are combined. In operational proofs / ..., several generically distinct OT may be involved – and the application(s) of various functions $f_l, f_m, \ldots$ are distributed non-uniformly, according to the structure(s) of the proof(s). These matters are discussed extensively in E_8.

It may appear that the class of transformations of OT constructible in this way is very specialised. Then the following question arises: **What is the most general class of transformations ψ : {OT} → {OT}, where {OT} denotes the class of all (non-generic) OT that are w.d. over Σ_t?**

The simplest candidate is the class of **compositions of type-preserving modifications of single arguments of OT**. Next, comes: **the 'closure' of this class under combinations of 'limit processes' on 'sequences of modifications of individual arguments'** (which may undergo '**type transitions**'). A partial extension is obtained by introducing **families** of modifications; and so on.

The (primarily **semantic**) constraints on argument types seem to rule out **genuinely** more general classes of transformations – since apparent extensions examined so far are reducible to the simpler cases, for which the definitions given here are formulated. Nevertheless, a

[123]See, e.g., Turnhall, H.W., '*The theory of Determinants, Matrices, and Invariants*' (Blackie, 1945; Dover, 1960)

generalisation of this scheme is introduced in E_8, to allow for nontrivial extensions that may eventually be obtained.

NOTE Another possible definitions of $f(T)$ involves the identification of 't' with 'v', and 'x', with 'o' (composition). This allows a direct interpretation of Taylor-series expansions of '$f(x)$', and $f(T)$. **Notice that**, in this form, extensions to functions $H = H(T)_m$ may be given at once; whereas, the 'transformation version' $f(T(\theta)) := T(f(\theta))$, appears to permit only somewhat formal extensions to $H(T)_m$: **see** E_8 for further discussion of this issue.

Plainly, this definition of $\hat{f}$ is somewhat contrived! Nevertheless, it permits the application of essentially arbitrary functions to collections of multiple OT - within a consistent framework. (**Notice that** functionals, in the standard sense are covered – as 'functions into $\mathbb{R}$'). In practice, modifications / transformations of wfe are effected **sequentially** – as part of (human) mathematical activity. When IT/AI facilities are involved, however, the possibility of (some) **parallel** activities must be considered – and the above definitions seem to allow for all admissible situations. In spire of this requirement, it appears that – as far as the overall **results** are concerned – the 'compound operations' $\hat{f}({}_nT(\theta_{{}_nT}))$ is equivalent to a sequence of individual operations of the form

$$f_{\sigma_i}T(u_\sigma)_h := T(u_\sigma)_h|_{u_{\sigma_i}=f_{\sigma_i}(u_{\sigma_i})}.$$

More succinctly, if $\underline{(U)}^h := (u_\sigma)_h$ and $\underline{f}^{hi}$ is the 'operator vector' given by $f_j^{hi} := 1 + \delta_{ij}(f_{\sigma_j} - 1)$, then one may define $\underline{f}^{hi} : \underline{u}^h := (f^{hi}u^h)_h$, so that $f_{\sigma_i}T(u_\sigma)_h = T(\underline{f}^{hi} : \underline{u}^h)$. This notation will be used from now on, interchangeably with the form $\hat{f}(T(u_\sigma)_h) = T(\hat{f}(u_\sigma))_h = T(\hat{f}_1(u_{\sigma_1}), \ldots, \hat{f}_h(u_{\sigma_h}))$, as given above, where the reduction of $\hat{f}({}_nT(\theta_{{}_nT}))$ to operations involving simple OT is also exhibited.[124] The (apparently inescapable) intricacy of this formalism stems from the diversity of admissible instantiations of any generic OT – corresponding to its possible applications to wfe 'generated in interactive investigations'. This is, of course, how proofs are forged – often from a morass of conjectures, false starts, and incomplete calculations. The 'final version' may be (retrospectively) elegant, but the disjointed groupings from which proofs (sometimes) crystallise are usually far more instructive – maybe even suggestive additional problems or other lines of investigations. The sets [125] I_t, B_t and U_t are represented within Σ_t through taxonomic structures, with near-exhaustive tagging, to allow the development of versatile search routines. The unification of this vast pool of mathematical information depends upon the efficient and high-resolution manipulation of 'formulae', for the general modification of symbolic expressions. In particular, **the exploration of hypotheses**, through the application of OT and the transformations of associated wfe, is of fundamental importance.

In treating the transformation of mathematical information, all of the terms 'effective', 'constructive', 'realisable', 'operational', 'implementable', ..., are pertinent, but the distinctions among them are by no means clear-cut. It is useful, therefore, to characterise such terms in relation to Σ.

For **processes**, π, **over** (or **within**) a **system**, S, the following characterisations are made:

→ π is **realisable** over S IFF π can be **modelled** within S.

→ π is **effective** over S IFF π can be 'realised arbitrarily accurately' (within S) in a finite number of 'steps'.

→ π is **constructive** over S IFF π is realisable **exactly** (over S) in a finite number of **explicit** steps.

→ π is **implementable** over S IFF there exists an **algorithm** for π over S.

→ π is **operational** over S IFF π is **formally applicable** within S.

[124] The alternative form, based on ${}^kf$ and $({}^kf^p)_r$, involves matrices ${}^kf$, rather than vectors f^i. **Note that** superscripts h are dropped unless there are ambiguities of interpretation.

[125] $I_t \equiv$ {all information in Σ_t}; $B_t \equiv$ {all calculational procedures in Σ_t}; $U_t \equiv$ {all 'nontrivial ingredients of OT' over Σ_t}. See E_1

→ π is **applicable** within S IFF π has some implementation(s) over S.

These compact definitions may be elucidated by explaining the terms 'modelled', 'explicit', and, 'formally applicable' – all of which 'should', however, be 'intuitively clear'! It is supposed that S is obtained from some set of axioms, through w.d. procedures (possibly, numerous / intricate) where all entities are 'endowed with meaning' through appropriate **interpretations**. If all specific notions embodied in the process π have interpretations in S, then S **is a model for** π. This relationship is developed multifariously in Mathematical Logic, and in all of the 'Mathematical Sciences'. Next, an explicit step (over S) is any operations that is precisely specified and w.d. (over S). Lastly, a process is **formally applicable** (within S), whenever it is implementable, **and** the associated (pre-)algorithm(s) over S may be exhibited, but **not used** – since no implementation is involved.

In the context of Σ as an OIB all of these definitions refer, ultimately, to P_Σ, the mathematical **processor** – whose basic capabilities stem from those of existing SAM packages. In this sense, the **degree of implementation** of π, over S, increases (weakly) with t. For book-keeping purposes within Σ , it is worthwhile to introduce symbols for implementations of π (say, $\text{imp}_S(\pi)$), and, for the 'order of implementation' (say, $\text{ord}_S(\pi)$). Although these symbols will appear in calculations only very exceptionally, they facilitate the production of 'snapshots of the state of Σ_t' – in relation to the variety of available procedures and their orders of applicability. Analogous estimates for the effectiveness, constructiveness, realizability, and operationally of informations, e, also contribute to assessments of the scope of Σ_t. In particular, such estimates may be made for the elements of the calculational base, B_t – which contains all of the OT that are fully operational, at t.

In E_5, the **approximation** of mathematical information is treated, from several points of view – making it possible to consider (e.g.) **perturbations** of CB, and quasi-invariant sets of wfe under the action of CB. Notice that, **any** collection of applicable procedures constitutes a CB; the notation B_t is reserved for **the** CB of Σ_t – essentially, the t-maximal collection of applicable procedures. Although B_t is not uniquely specified by this description, it is still viable for the development of Σ , provided that consistency is maintained. In principle, all ambiguities could be removed: but the crucial idea is that the evolution of Σ_t is viewed at a sequence of times, say, $\{t_k\}$, at each of which the (finite) sets I_{t_k}, B_{t_k}, U_{t_k}, are specified by **enumeration, listing** in taxonomic structures, and cognate tagging. As discussed in E_0, one regards Σ_t as a **model** of $M_t^* \equiv$ 'Mathematics at time t'. The **aim** is for $M_t \equiv$ 'Mathematics as embodied in Σ_t' to 'approach' M_t^* as t increases: symbolically, $M_t \sim M_t^*$ as $t \to \infty$. This entails that M_t becomes 'progressively more complete' (compared with the abstract ideal M_t^*), but it does not prescribe uniquely the content of any M_{t_k}. Of course, t varies continuously, and M_t has an interpretation for **any** value of t. For most purposes, however, it suffices to consider a fixed sequence, $\{t_k\}$ – for instance, with t_k representing the kth day after t_0 (the day on which Σ was 'inaugurated'). This corresponds to taking snapshots of Mathematics on a daily basis. The 'approach' of M_t to M_t^* is governed by the systematic application of all 'standard modes of approximation', augmented by other modes associated with abstraction, analogy and inverse analogy. This generates a highly complex body of information, comprising many symbiotically interrelated subsystems, some of which are recursively determined. The extreme complexity of this information scheme must be matched by the taxonomic and calculational facilities – which may themselves be regarded as mathematical subsystems! In spite of these severe challenges, it is possible to determine local and global structures within which the evolution of Σ_t may be monitored coherently – and even accelerated! Moreover, this comprehensive representation (only viable through IT/AI/SAM developments) produces a total **unification** of Mathematics and its applications. The remaining constituents of this framework are discussed – with diverse illustrations – in subsequent Essays. The rest of this Essay is devoted to some general issues involving CB.

Every CB is an ordered collection of applicable procedurals as such, it is an ordered subset of B_t – and hence, an element of B_t, the **power-set**[126] of $\mathcal{P}B_t$. Thus, $\mathcal{P}B_t$ comprises all

[126] Recall that B_t, though 'large', is a **finite** set, provided that (parametrized) families of entities are treated as single

possible CB in Σ_t, so it is instructive to consider structural features of B_t. If B_t is finite, then the boolean algebra $\mathcal{P}B_t$ has significant algebraic properties, but only trivial topological ones[127]. If B_t is regarded as (un)countable infinite, however,[128] then the general theory of boolean algebras becomes applicable to Σ_t. It is quite tenable to treat B_t as finite, for 'basic book-keeping', but also to view it as **in**finite in other contexts (where the symbol B_t^{∞} is used). If the cardinality of B_t is not relevant, then the symbol B_t is used to cover both possibilities. Similar conventions hold for **all** infinite (sub)sets, from now on – especially, here, for (un)countable subsets of B_t.

The interest centres on (quasi-)metrical properties of 'topological boolean algebras' (for the set of CB over Σ_t); and, on **cylindrical / polyadic algebras** – including the **Lindenbaum - Tarski algebras** of formalised theories.[129] These algebras have been introduced as generalisations of boolean algebras (with an extra set of operations); they are considered in E_0, in the discussion of **proofs** as combinations of OT. Since every OT **acts on wfe** (as indicated above) the CB may be assumed (without loss of generality) to consist of collections of (**labelled, generic**) OT – 'procedures' being special types[130] of OT. The applications of elements from a CM to transform wfe – typically, in the course of **proofs** – may, therefore, be represented in terms of the transformed OT, $\hat{f}(T(\sigma)_h)$ and $\hat{f}(T(\theta)_h)_n$, acting on **collections**, $(e)_k$, of wfe. The choice of **generic** OT in any CB ensures that all CB are represented as **finite** sets – for which 'canonical forms' may be sought. Alternative (infinite-set) forms treat **variants** of generic OT as distinct elements of the CB.

An obvious partial classification of CB corresponds to **fields of application**. For instance, B^1 may contain OT on finite groups; B^2, on matrix algebra; B^3, on Lie groups – and so on. Mostly, such specialised CB are not mutually disjoint. As subsets of B_t, they may in some cases be viewed, either as finite (for generic OT), or else, as infinite (when variants of generic OT are included). Progressive specialisation yields collections of interrelated CB, covering particular mathematical areas – for example, those domains specified by the principal / secondary MR headings. The overall taxonomic structure(s) must reflect these correspondences. The application of any selection of OT from[131] the 'compound CB', $(B)^k$, may be handled by forming multiple theorems with the selected OT as 'components' – and then transforming / applying these multiple theorems as already indicated for $\hat{f}(T(\theta)_h)_n$.

Several classes of **fundamental operations** may be applied to (collections of) elements of any B^k in $\mathcal{P}B_t$. These operations are associated (respectively) with the families **V, G, Ab, ∓An, Ap**, of **Variants, Generalizations, Abstractions, (Inverse-)Analogues**, and **(Standard) Approximations** – which, collectively, comprise the family, **Ar**, of **Approaches**. All of these operations are discussed in detail in other Essays. They are mentioned here because of their relevance to the systematic use of CB (generic or otherwise), and to the corresponding structures deniable over $\mathcal{P}B_t$. Each of the above species of approach is governed by complex 'internal relations'; while, all of these species are intricately **inter**related. Consequently, for any ME, e (which may be a set of other ME, etc..), the set, $A(e)$, of all admissible **approaches** to e, has itself a complicated structure. The possibility of defining (G)SF (in the sense of E_1)[132] over CB, and of forming various types of 'compound CB' from selected 'components', are of considerable interest for the efficient design of P_Σ.

For any boolean algebra, say A_B, the assignment: $d(a,b) := ab' + a'b$ determines a **metric** – since, $d(a,b) = 0 \Leftrightarrow a = b; d(a,b) = d(b,a)$, and $d(a,b) + d(b,c) \supset d(a,c)$, where + denotes boolean multiplication, and $r \supset s$ means 'r contains s' (with the possibility of $r = s$). If A_B is a **finite** set, then the associated 'distance geometry' of the boolean MS is comparatively trivial. Fortunately, it may be shown[133] that: every boolean MS is iso**morphemically** (hence,

elements, etc..

127 See, e.g., Rasiowa-Sikorski, '*The Mathematics of Metamathematics*' (PWN, Warnisaw, 1963; pp518)§8

128 by treating members of families of entities as distinct elements of B_t, etc..

129 See Rasiowa - Sikorski, op.cit., §10

130 Though every OT may be formulated as a procedure – with some artificiality, if necessary

131 Here, $(B)^k := (B^1, \ldots, B^k)$, as indicated in the notational section preceding E_0.

132 **G**eneralized **S**structural **F**rameworks. Recall that the term **Calculational Milieu (CM)** is also used (in E_1)

133 See '**Boolean Geometry II**', by LM Blumenthal - CJ Penning, Rend. Cire - Mat.Palermo. Ser. II Vol 1(1952)

isometrically) **embeddable** in a **convex** boolean MS – where A_B is **metrically convex** IFF $\{a, c \in A_B \wedge a \neq c\} \Rightarrow \exists b \in A_B : \{a \neq b \neq c\} \wedge d(a,b) + d(b,c) = d(a,c)$ – so that, b is **metrically between** a and c. **Thus:** the boolean algebra $\mathcal{P}B_t$ may be assumed metrically convex – in the sense of isometric embedding.

On this basis, a **sequential (order) topology** may be defined for A_B – for which the order relation, the distance function, and all boolean operations are **continuous**. Moreover, $\models \lim_{n\to\infty} x_n = x \Leftrightarrow \lim_{n\to\infty} d(x_n, x) = 0$; so that this **sequential topology** is **equivalent** to the **metric topology**, generated by d. Since $\mathcal{P}B_t$ is a boolean algebra, these topological constructions – and divers approximative properties derivable from them – have important implications for the assessment of 'relative power' among collections of CB selected from $\mathcal{P}B_t$. In particular, if a **proof** is obtained through the application of a multiple OT to 'initial data' (the **premises**), then the component simple OT may be viewed as elements of a CB – and various CB containing these simple OT may, in turn, be viewed as elements of the TS with the metric topology, allowing quasi-metrical criteria to be developed.

Analogously, if $L \equiv$ the Lindenbaum-Tarski algebra of the (formalised) theory[134] $\mathcal{J}$, then sequential and metric topologies may be defined on L. Here, the algebra L is specified as follows. Let $\mathcal{J} \equiv \{\mathcal{L}, C, A\}$ have **language $\mathcal{L}$, consequence relation C, and axiom set,** A, where $C(A)$ is closed under **modus ponens** (i.e., $\{\alpha \in \mathcal{J} \wedge \alpha \Rightarrow \beta \in \mathcal{J}\} \Rightarrow \beta \in \mathcal{J}$). Then $\models \alpha \leqslant \beta$ IFF $\{\alpha \Rightarrow \beta\} \in \mathcal{J}$ determines a **quasi-ordering** of $F \equiv$ {formulae in $\mathcal{L}$}, **provided that** the conditions $\alpha \leqslant \alpha$, and $\{\{\alpha \Rightarrow \beta\} \Rightarrow \{\{\beta \Rightarrow \gamma\} \Rightarrow \{\alpha \Rightarrow \gamma\}\}\}$ hold, for arbitrary α, β, γ in F (i.e., these conditions represent theorems in $\mathcal{J}$). For, it follows that $\leqslant$ is reflexive and transitive. If one now defines $\alpha \approx \beta \Leftrightarrow \alpha \leqslant \beta \wedge \beta \leqslant \alpha$ in $\mathcal{J}$, then α and β are called **equivalent** in $\mathcal{J}$. This induces an ordering of $F/\approx$, where (for elements η, λ, of F with equivalence classes $\| \eta \|$, $\| \lambda \|$, respectively) $\| \eta \| \leqslant' \| \lambda \|$ IFF $\eta \leqslant \lambda$. Here, $\leqslant'$ refers to $F/\approx$, and $\leqslant$, to F. It may be shown[135] that $F/\approx$ (denoted here, also, be L) is a boolean algebra (of formulae), and so admits the metric structure – and metic **topology** – specified by d, where $d(a,b) := ab' + a'b$, for $a, b \in L$. This has implications for gauging 'the closeness of formulae to one another', and so, for monitoring the progress of proofs corresponding to compositions of OT applied to wfe (premises). This general manipulation of transformed / multiple OT is referred to in these Essays as 'the **O**perational-**T**heorem **C**alculus. Attempts to provide rigorous foundations for this OTC are outlined in E_8. The principal challenge is: **to furnish consistent interpretations of the basic processes of algebra / analysis, for families of OT**. It is found that, although such interpretations cannot be established for arbitrary families of OT, valuable results may be obtained for various classes of OT – which, fortunately, seem to be among the most important for 'research over Σ '. In particular, it is fruitful to consider all forms of **converses** of OT as OT-**inverses**, in various senses.

Diagrams as Operational Information

An important part of 'the Σ -environment', complementing the operational development of theorems, corresponds to various classes of (modifiable) **diagrams**. These diagrams range over **exact** representations (e.g., 'technical / scale drawings' in CAD), geometrical / topological 'figures', schematic depictions of 'general processes' (often figurative) and diverse types of calculational representations for series / integral / function(al) **expansions** in many domains of Mathematics and The Sciences. Such pictorial schemes play a crucial role in the discovery, formulation and proof of theorems – yet these aids to visualisation are usually suppressed in formal (axiomatic) presentations of new results / theories. One aim for Σ is to reverse this trend, by exhibiting the pictorial motivation underlying conjectures / formulations / proofs / structures, wherever possible.

343-361, for most of the results cited here; also, Rasiowa - Sikorski, op.cit., and Siroski, '**Boolean Algebras**' (2nd Ed., Springer, 1964, pp237)

[134] Of the first order, here (predicate calculus . . .)

[135] See Rasiowa - Sikorski, op.cit., §10

The ideas of Pager[136] (for 'a computer-based Mathematical Community') have already been mentioned. Although they are relevant for Σ , the overlap is comparatively small, since Pager is concerned mainly with the efficient **communication** of results / proofs, rather than with the construction of operational / approach versions – the primary task in Σ .

Among the commonly occurring classes of diagrams are: (1) CAD representations (especially in most branches of engineering); (2) control-theoretic procedures; (3) geometrical / topological configurations (of dimensions 2 or 3); (4) indications of higher-dimensional configurations (in all fields); (5) diagrams for General Algebra; (6) diagrams for Homological Algebra, (7) diagrams for other aspects of Algebraic / Geometric Topology; (8) diagrams for Algebraic Geometry; (9) diagrams for Differential Topology / Geometry; (10) diagrams for Real Analysis; (11) diagrams for Complex Analysis (12) diagrams for Functional Analysis; (13) diagrams for Lie-Groups / Lie-Algebras; (14) diagrams for many-body calculations in (Chemical) Physics; (15) diagrams in High-Energy Physics; (16) (un)labelled / (un)directed graphs (in all fields); (17) generally illustrative diagrams (in all fields); (18) diagrams in all areas of 'pure Geometry'.

NOTE that there are many (more specialised) classes of diagrams not covered in this list. **These may be identified as they arise** (especially in illustrations for E_{10} / E_{11}).

The plan here is **to exemplify each class of diagrams**; and, **to indicate typical forms of modifications producible through the actions of suitable (classes of) OT. The general notation for diagrams** is: $D_\lambda, 1 \leqslant \lambda \leqslant 19$, where $D_{\lambda,j}(e)$ denotes the j-th type of diagram of species λ (for the ME e), and D_{19} corresponds to the '**Miscellaneous**' class of diagrams not included so far (in $\{D_1, \ldots,$
$D_{13}\}$). Here, e is called the **source** of the diagram $D_{\lambda,j}(e)$.

(1) CAD Diagrams

The development of **C**omputer-**A**ided **D**esign packages since the 1970s – especially, in the automotive / aerospace industries – has been intensive. Each package has its own software / user-interface, with the associated programming language(s). There is no problem of principle in incorporating such systems into the OIB – alongside general **graphics** facilities. Comparable packages exist for the specification / control of **robotic mechanisms**.

The CAD diagrams represent (sections of) 2-dimensional / 3-dimensional 'objects', whose specifications are stored as lists of points / lines / curves / .... **Operations** on the diagrams may be performed interactively. In current CAD systems, arbitrarily complicated objects (e.g., new car designs) may be represented / modified in this way. For Σ , such diagrams occur only in various applications for engineering / scientific routines. Even within these setting, CAD diagrams have many facets – exemplified by (a) interactive graphics[137]; (b) digital image processing[138]; (c) computational visions[139]; (d) image analysis / morphology[140]. Many concepts / problems of seminal mathematical importance arise in these domains, though their motivation has been primarily technological. Several examples are given in E_{10}, E_{11}. Combinations of CAD and 'numerical control' (especially for machine tools) are dominant in forming / shaping procedures[141]. The software underlying interactive diagram-modification is itself based on coordinate geometry, but a variety of more subtle theories are also involved.

Remark: A great variety of diagrams for **Civil / Structural / Mechanical / Chemical / ... Engineering** may be covered here.

---

[136] op. cit.

[137] e.g., PARK, C.S., '*Interactive Microcomputer Graphics*' (A-W, 1985)

[138] e.g., PRATT, W.K., '*DigitalImage Processing*' (Wiley, 1978)

[139] e.g., WECHSLER, H., '*Computational Vision*' (AP, 1990)

[140] e.g., SERRA, J., '*Image Analysis and Mathematical Morphology*' (AP, 1982)

[141] e.g., CHANG, C.H. / MELKANOFF, M.A., '*NC Machine Programming and Software Design*' (P-H, 1989)

(2) Control-Theoretical Diagrams

A somewhat broader designation is: '**system** diagrams', covering everything from 'single devices' to industrial plants.[142] This is an area potentially rich in inverse-analogues, since it involves 'processes' from all of 'The Sciences'. The diagrams are (mostly) schematic, but have precise interpretations. The range of admissible modifications includes changes in some input(s) / component(s) / environment(s). Special symbols are used for classes of components, and (un)labelled / (un)directed graphs are introduced to monitor types / rates / directions of flows. Every diagram has some interpretation(s) in terms of systems of (coupled) operator equations. Questions of **sampling / feedback / stability / realizability / separability / optimality / ...** arise naturally, here. Possible operations on such diagrams may be of many types, reflecting the performance of component procedures, etc.. Moreover, collections of diagrams may be **combined** in various ways, and **reduced** to 'canonical forms'. All of these processes are embodied in the **block-diagram algebra**[143] associated with the class(es) of systems concerned.

<u>**NOTES**</u> (i) Diagrams for **Electrical Engineering** of all types[144] – including Electronics – may be covered here, since aspects of 'control' (e.g., 'feedback') are basic to these areas (and also, to parts of **Chemical Engineering**)[145].

(ii) Diagrams for (theoretical) **Computer Science**[146] (including program (flow-)dioramas) may be viewed as controlling computation – from the most formal / abstract to the purely numerical; and from concurrent / pipeline / . . . / supercomputer **hardware** to the associated **software**. Various aspects of **Automata Theory** fall broadly under this heading.

(3) Geometrical / Topological Configurations

These diagrams furnish qualitatively accurate representations of the geometrical / topological properties 'observable' in planar or spatial objects. This includes all wd 'figures' for theorems in Euclidean Projective Geometry[147] and all decompositions of planar / spatial objects into undoing of 'elementary objects' (e.g., triangles, squares, rectangles, circles, ellipses) with appropriate 'remainders' whenever exact decomposition is unattainable. These forms are especially pertinent to Finite-Element[148] / Boundary-Element[149] / Multigrid[150] procedures for the numerical solution of differential equations. It also covers the topological shape / connectivity of objects[151], and the relative location of all types of 'singular points'[152], and various non-Euclidean geometries.[153]

(4) Diagrams for 'Higher-Dimensional Objects' / Proofs

These diagrams refer quasi geometrical / topological configurations (direct or figurative) that cannot be realised in low-($\leqslant 3$)dimensional spaces. The primary aim is to exhibit locations /

[142]A wide range of examples may be found in: NETUSHIL, A., et al., eds. '*Theory of Automatic Control*' (MIR Publ., Moscow (transl. of 2nd ed, 1976), 1987)

[143]e.g., DISTEFANO, J.J., et al., '*Feedback and Control Systems*', (schaum/McGr-H, 1967/1976) Ch.7. Other general properties of system equations / diagrams are considered in Ch. 8 of 'emphIntroduction to System Dynamics', by J.L. SHEARER, et al., (A-W, 1971)

[144]e.g., SHEARER, et al., op. cit.

[145]eg. SHEARER, et al., op. cit.

[146]See, e.g., LNCS, #1313 (J. Fitzgerald, et al., ads) Springer, 1997

[147]e.g., COXETER, HSM, '*Intro. To Geometry*' (Wiley, 1969, 2nd ed., repr. 1981); BERGER, M., '*Geometry*', Vols. 1,2 (Springer, 1987)

[148]e.g., ZIENKIEWICZ, OC, '*The Finite Element Method . . .*' (McGr-H, 2nd ed., 1971)

[149]e.g. BREBBIA, CA, (ed.) '*Topics in Boundary Element Research*', Vol 3 (Springer, 1987)

[150]e.g., HACKBUSCH, W., et al., (ads) '*Multigrid Methods*' (LNM #960, Springer, 1982)

[151]esp., GRIFFITHS, HB, '*Surfaces* (CUP, 2nd ed., 1981), ref 6 (below) and KOSNIOWSKI, C, '*A First Course in Algebraic Topology*' (CUP, 1980)

[152]e.g., FOMENKO, A, '*Visual Geometry and Topology*' (Springer, 1994)

[153]BERGER, M, (ref 1., above)

interrelations for 'the components of compound objects' – in all domains of Mathematics. Such diagrams may be introduced fruitfully in virtually all contexts, as precursors to more detailed investigations[154], since they may indicate, broadly, how a rigorous theory should be formulated – simply by highlighting underlying structures. In spite of this, the use of schematic representations in 'Pure Mathematics' seems to be comparatively rare – at least in 'final versions'.[155] In Σ , an effort is made to emphasise quasi-geometric content, especially in lengthy / intricate proofs where complicated transformations / estimates are involved. Indeed, the 'Σ -view' is that every (nontrivial) proof may be given a (valuable!) **pictorial outline** – to be stored alongside the other forms of the proofs.

(5) Diagrams for General Algebra

This covers essentially all parts of Algebra except for Lie Groups / Algebras and Homological Algebra. For instance: Group[156] / Ring[157] / Field[158] Theory; Lattice[159] Theory; Boolean / Cylindrical Algebras[160]; Universal Algebras[161]; Linear Algebras[162]; Topological Groups / Rings / Fields / Algebras[163]; Operator Algebras[164] – all, broadly interpreted; other classes of algebras[165].

Although the scope for diagrammatic representation varies greatly among these structures, ranging from (Multi-)Linear Algebras (which admit comprehensive geometric interpretations) to Universal Algebras (which are intrinsically formal / axiomatic), progress in all of these areas may be enhanced (at least) by depicting the (internal / external) interactions among key results.

(6) Diagrams for Homological Algebra

This has been satirised as 'the science of diagram-chasing' (mostly, unnecessary)[166] for (non-)commutative diagrams / in(exact) sequences. Some extensions / complements of such techniques are embodied in Category Theory[167] and Topos Theory[168]. In spite of the wealth of diagrammatic representation here, motivational pictorial schemes are uncommon[169], and would be valuable.

(7) Diagrams for Other Aspects of Algebraic / Geometric Topology

A typical example here is provided by Shape Theory – a form of 'Applied Category Theory' – which allows the development of subtle forms of approximation[170]. Another example (in-

---

[154] See, e.g., the citations in E_5, where 'diagram-modification' is treated.

[155] See FOMENKO, A, op. cit. for abundant 'pictorial motivation'.

[0] [2-10] All of these structures are covered briefly (and inter-related) in KUROSH, AG, '*Lectures in General Algebra*' (Pergamon Press, 1965)

[156] Bourbaki, N, '*Algebra*', Vols 1,2, op. cit.; Lang, S., '*Algebra*', op. cit.

[157] Rowen, LH, '*Ring Theory*' (Student Edition; AP, 1991)

[158] Winter, DJ, '*The Structure of Fields*' (Springer, 1974); ROMAN, S, '*Field Theory*' (Springer, 191)

[159] Birkhoff, G, '*Lattice Theory*' (AMS, 3rd ed., 1967)

[160] SIKORSKI, R, '*Boolean Algebras*' (Springer, 2nd ed., 1964)

[161] BOURBAKI (op. cit, CH III)

[162] BOURBAKI (op. cit, CH II)

[163] BOURBAKI, N, '*General Topology*' (op. cit.) Vol 1 Ch III

[164] SCHWARTZ, JT, '*W*-Algebras*' (Nelson, 1967); RICKART, CE, '*Banach Algebras*' (Van Nostrand, 1960); PEDERSEN, GK, '*C*-Algebra and Their Automorphism Groups*' (AP, 1979); NAIMARK, MA, '*Morned Rings*' (Noordhoff, 1959)

[165] See, e.g., PIERCE, RS, '*Associative Algebras*' (Springer, 1982); FARB, B / DENNIS, RK, '*Noncommutative Algebra*' (Springer, 1993); WÖRZ-BUSEKROS, A, '*Algebras in Genetics*' (Springer, 1980)

[166] e.g., LANG, S, '*Algebra*' (A-W, 1965)

[167] e.g., STROOKER, JR, '*Introduction to Categories, Homological Algebra and Sheaf Cohomology*' (CUP, 1978)

[168] e.g., JOHNSTONE, PT, '*Topos Theory*' (AP, 1977)

[169] See, however, many papers in the *Reports of the Midwest Category Seminar* (Springer LNM # 47, 61, 106, 137, 195, . . .), where a wide range of structural systems is exhibited diagmatically.

[170] Cordier, JM, Porter, T, '*Shape Theory: categorical methods of approximation*' (Ellis Harwood, 1989)

volving Feyman diagrams in QED, etc.) emphasises homology calculations[171]. In Statistical Mechanics, various (power-)series expansions may be derived in terms of 'sums over classes of graphs' (as characterised by certain graph-theoretic / topological criteria)[172]. Again, topological features of chemical forms may be exhibited diagrammatically, in many contexts[173]. A wide range of illustrations –from the above fields and others – are covered in E_{10}/E_{11}. For **Geometric** Topology[1], the **problems** are geometrical, but the techniques may be combinatorial, algebraic, differential – or geometrical / 'pictorial'. Each of these frameworks has its graphical aspects.

(8) Diagrams for Algebraic Geometry

Even the most abstruse / general formulations of Algebraic Geometry are motivated by the theory of algebraic (plane) **curves** – where 'natural geometrical questions' have (gradually) produced profound theories[174]. Although most treatments of (plane) curves are heavily algebraic / analytical[175], genuinely geometrical accounts **have** been given[176], with high pictorial content. As in several other 'abstract domains', schematic diagrams would help to elucidate the intricate structures of 'modern' Algebraic Geometry – whether based on complex manifolds[177], or on schemes[178].

NOTE that the book by **A. Formenko** (op. cit.) is pertinent to visual representations in **all** areas of Geometry and Topology.

(9) Diagrams for Differential Topology / Geometry

Differential Topology and Differential Geometry are considered here together, since the overlaps in both methods and content seem to be especially large. Again, in many expositions, formal analytical / algebraic techniques predominate[179] even though most of the key results have suggestive quasi-graphical interpretations[180].

(10) Diagrams for Real Analysis

Most of the diagrammatic information here elucidates the conditions involving class- es of mappings for local / global configurations; for instance, inverse / composite mappings[181]. A detailed grasp of these basic results is essential for the development of more intricate / abstract theories (or, Calculus in Banach Spaces). Other areas benefiting greatly from 'illustrations' include vector / tensor analysis / differential forms[182].

---

[171] Hwa, RC, Teplitz, VL, '*Homology and Feynman Integrals*' (Benjamin, 1966)

[172] e.g. Stell, G, '*Cluster Expansions...*' ppll 171 – II 266, in Frisch, HL, Lebowitz, JL (ads) '*The Equilibrium Theory of Classical Fluids*' (Benjamin, 1964) and references there

[173] e.g. Haken, W, in '*Studies in Modern Topology*' (P. Hilton, ed.) (MAA/P-H, 1968) pp 39-98; Mandelbaum, R, '*4-D Topology: An Intro*' (BAMS 2 (1980) pp 1-159)

[174] e.g., Semple, JG, Roth, L, '*Intro to Algebraic Geometry*' (OUP, 1949)

[175] e.g., Coolidge, JL, '*A Treatise on Algebraic Plane Curves*' (Dover, 1959)

[176] e.g., Brieskorn, E, Knörrer, H, '*Plane Algebraic Geometry*' (Wiley, 1978, repr. 1994)

[177] e.g., Griffiths, P, Harris, J, '*Plane Algebraic Curves*' (1981, transl. Birkhauser, 1986)

[178] e.g., Hartshorne, R, '*Algebraic Geometry*' (Springer, 1977)

[179] e.g., Golubitsky, M, Guillemin, V, '*Stable Mappings and Their Singularities*' (Springer, 1973, repr. 1986)

[180] e.g., Berger, M, Gostiaux, B, '*Differential Geometry...*' (Springer, 1988); Spivak, M, '*Differential Geometry*', Vols 1-5, (Publish or Perish, Inc., 2nd ed., 1979)

[181] e.g., Takenchi, Yu, '*Analysis Matematico*' (Spanish) (Universidad Nacional de Colombia, 1974)

[182] e.g., Bamberg, P, Sternberg, S, '*A Course in Mathematics for Students of Physics*', Vol 2 (CUP, 1990); Dodson, CTJ, Poston, T, '*Tensor Geometry*' (Pitman, 1977); Buck, RC, '*Advanced Calculus*' (Mc-Gr-H, 2nd ed., 1965); Price, GB, '*Multivariable Analysis*' (Springer, 1984)

(11) Diagrams for Complex Analysis

This is an area of Mathematics with obvious geometrical / topological content[183], though purely axiomatic accounts, with scarcely a single diagram of any kind (aimed at exhibiting the formal structures without distractions!) are not uncommon[184]. The most stimulating treatments **combine** the analytical and geometrical features inherent in the subject – especially, for functions of one variable[185].

For functions of n ($\geqslant 2$) complex variables, there are certainly some pictorial facets, but they are very rarely found in developments of the subject (mainly, one assumes, because of problems of visualisation / representation)[186].

(12) Diagrams for Functional Analysis

Since this is essentially the study of (non)linear mappings among various classes of spaces, the potential for illustration is evident[187]. In particular, 'the geometry of Banach spaces' is a well established field of study[188]. Moreover, detailed investigations of specific mappings often benefit from graphical representation, especially when **non**linearities are involved[189]. Again, in **applications** of Functional Analysis to The Sciences, the underlying abstractions may be greatly clarified through pictorial indications of the actions / interrelations of the operators characterising mathematical models[190].

(13) Diagrams for Lie-Groups / Lie-Algebras

The most basic forms here are the so-called **Dynkin diagrams** for root systems – originally, for the classification (up to isomorphism) of complex simple Lie Algebras[191]. This formalism has been modified / extended / reflected into several other mathematical domains[192], most notably, in the theory / application of **Building**[193] – for instance, in studying geometrical aspects of **reflection groups**[194], and in the investigation of **Linear Algebraic Groups**[195]. Although such topics are treated primarily in formal algebraic / analytical ways[196], the underlying diagrammatic content is considerable. Since Lie **groups** have associated Lie **algebras**, both structures are involved here.

---

[183]e.g., Carleson, L, Gamelin, TW, '*Complex Dynamics*' (Springer, 1993); Newman, MHA, '*... Topology of Plane Sets of Points*' (CUP, 2nd ed., 1951)

[184]e.g., Macker, GW, '*... Functions of a Complex Variable*' (V. Nestrand, 1967); Esterman, T, '*Complex Numbers and Functions*' (Athlone P, London, 1962); Burckel, RB, '*... Classical Complex Analysis*' (Birkhäuser, 1979)

[185]e.g., Palka, BP, *... Complex Function Theory*' (Springer, 1991); Jones, GA, Singerman, D, '*Complex Functions*' (CUP, 1987)

[186]e.g., for 'domains of holomorphy', and for results on complex manifolds

[187]This ranges from representations of the Unit Ball to the study of boundaries / contractions / iterations / compositions associated with classes of mappings

[188]e.g., '*The Volume of Convex Bodies and Banach Space Geometry*', by G. Pisier (CUP, 1989) and references there.

[189]e.g., Naylor, AW, Sell, GR, '*Linear Operator Theory ...*' (HRW, 1971, Springer, 1982); Ziedler, E, '*Nonlinear Functional Analysis*' Vols 1-5 (Springer, 1986-88)

[190]e.g., Porter, WA, '*Modern Foundations of Systems Engineering*' (Macmillan, 1966); Takayama, A, '*Mathematical Economics*' (CUP, 2nd ed., 1985); Zeidler, op.cit.; Ellickson, B., '*Competitive Equilibrium*' (CUP, 1993)

[191]e.g., Sattinger, DH, Weaver, OL, '*Lie Groups and Algebras, with Applications ...*' (Springer, 1986); Jacobson, N, '*Lie Algebras*' (Wiley, 1962; repr. Dover, 1979)

[192]e.g. Brown, KS, '*Buildings*' (Springer, 1989); Carter, RW, (a) '*Simple Groups of Lie Type*' (Wiley, 1972; repr. 1989); Carter, RW, (b) '*Finite Groups of Lie Type*' (Wiley, 1985; repr. 1993)

[193]e.g., Brown, KS, op.cit.

[194]Carter, RW (a), op.cit.

[195]e.g., Humphreys, JE, '*Linear Algebraic Groups*' (Springer, 1987)

[196]e.g., Kac, VG, '*Infinite Dimensional Lie Algebras*' (CUP, 2nd ed., 1985)

(14) Diagrams for Many-Body Calculations in (Chemical) Physics

Under this heading, several areas of theoretical physics / chemistry are considered; the 'statistical' element corresponds to calculations for many-particle systems. Thus, the following topics are among those covered here:

(a) classical / quantum statistical mechanics of:
(i) fluids;[197] **(ii)** solids;[198] **(iii)** hybrid phenomena[199]

(b) dynamics of chemical systems;[200]

(c) solid-state physics;[201]

(d) nuclear physics;[202]

(e) plasma physics;[203]

(f) Renormalization Theory (of critical phenomena).[204]

(15) Diagrams in High-Energy Physics

The formalism of Feynman diagrams, and various extensions, are central to calculations in Quantum Electro-dynamics[205] – and to all theories of elementary particles[206]. Here, the topological and graph-theoretical properties are often combined with additional 'perturbation criteria'[207]. The corresponding diagrammatic expansions are characterised by the forms of diagrams involved[208]. Another class of (elaborate) diagrams may be found in 'twistor diagram theory', especially for scattering problems[209].

(16) (Un)labelled / (Un)directed Graphs, In All Fields

These diagrams differ from most of the others considered in that they **do not** represent (steps / terms in) detailed calculations; rather, overall **interconnections**[210] / **flows**[211] / **structured-networks**[212], etc.. All of the resources of Graph Theory[213] / Combinatorial Optimization[214] may be deployed for the investigation of such 'objects' – which may (in the context of generalized / compound networks) encompass all of the more specialized diagrams already considered. Indeed, it is even possible to develop theories of **nonlinear** networks[215].

---

[197] e.g., Stell, G, '*Cluster Expansions*', pp II 171 - II 266 in Frisch, HL, Lebowitz, JL (eds.), '*... Classical Fluids*' (Benjamin, 1964);

[198] e.g., Montroll, EW, '*... The Ising Model...*', in Chrétien, M et. al., eds., '*Statistical Physics ...*' (Gordon & Breach, 1968); Syozi, I, '*Transformation of Ising Models*', in Dourb, C, Green, MS, eds. '*Phase Transitions ...*' Vol 1 (AP, 1972); Moraal, H, '*Classical Discrete Spin Models*' (LNP #214, Springer, 1984)

[199] e.g., Fetter, AL, Walecka, JD, '*Quantum Theory of Many-Particle Systems*' (McGr-H, 1971)

[200] e.g., King, RB, ed., '*Chemical Applications of Topology and Graph Theory*' (Elsevier, 1983), and references there.

[201] e.g., Cracknell, AP, Wong, KC, '*The Fermi Surface*' (CUP, 1973); Fetter, op.cit.

[202] e.g., Rung, P, Schuck, P, '*The Nuclear Many-Body Problem*' (Springer, 1980); Fetter, op. cit.

[203] Fetter, op. cit.

[204] e.g., Itzykson, C, JM, Drouffe, '*Statistical Field Theory*', Vols 1, 2 (CUP, 1989)

[205] e.g., Ryder, LH, '*Quantum Field Theory*' (CUP, 1985); Fradkin, ES, et. al., '*Quantum Electrodynamics ...*' (Springer, 1991)

[206] e.g., Cheng, TP, Li, LF, '*Gauge Theories of Elementary Particle Physics*' (OUP, 1984); Yndurain, FJ, '*The Theory of Quark and Gluon Interactions*' (Springer, 2nd ed., 1992)

[207] Hwa, RC, et. al., '*Homology and Feynman Integrals*' (Benjamin, 1966); Ryder, op. cot.

[208] e.g., Yndurain, op. cit.

[209] e.g., Hodges, AP, Proc. Roy. Soc. Lond. A397, 341-374/375-396

[210] e.g., Busacker, RG, Saaty, TL, '*Finite Graphs and Networks*' (McGr-H, 1965)

[211] e.g., Carré, B, '*Graphs and Networks (OUP, 1979); Busacker, op. cit.*

[212] e.g., Šiljak , D, '*Largescale Dynamic Systems*' (Elsevier/N-H, 1978)

[213] e.g., Bollabás, B, '*Graph Theory ...*' (Springer, 1979); Temperley, HNV, '*Graph Theory ...*' (Horwood, 1981); Busacker, op. cit.; Carré, op. cit.

[214] e.g., Papadimitriou, CH, Steiglitz, K, '*Combinatorial Optimization ...*' (P-H, 1982)

[215] e.g., Dolozal, V, '*Nonlinear Networks*' (Elsevier, 1977)

(17) Generally Illustrative Diagrams, In All Fields

The diagrams in this class are motivational / explanatory / suggestive / . . . of the 'quintessential ME' in the field(s) involved. Although there are no obvious sources to cite as examples, the value of such pictorial representations is considerable – and reflects the heuristic aspects of research investigations. One aim in Σ is to accumulate a collection of illustrative diagrams from diverse parts of Mathematics – to act as paradigm in the longterm evolution of the OIB.

(18) Diagrams in all Areas of Pure Geometry

These diagrams are (almost always) depictions of geometrical figures in planar / spatial euclidean geometry[216]; or else, partially schematic configurations from types of **non**euclidean geometries[217]. Their interpretations are usually direct and unambiguous – through care must be taken to avoid plausible-but-impossible constructions[218]! At the other extreme are the intricate 'engineering drawings' ('blue-prints') in which (sections of) complicated objects are reproduced to scale[219]. Further refinements include quasi-three-dimensional views of objects (through use of suitable optical filters)[220] and – more recently – holographic images[221]. This cast range of 'diagrams' may be integrated into the IB as part of a comprehensive graphics package.

(19) Miscellaneous Types of Diagrams

This 'class' of diagrams is primarily intended to cover inverse analogues from all of The Sciences (Engineering / Physics / Chemistry / Biology / Economics / . . .), together with inadvertent omissions from the list of Mathematically-based diagrams. In this way, the coverage will become progressively complete. It suffices to give headings, with minimal key references – for eventual elaboration as the IB evolves.

(19.1) Chemical Reaction Diagrams[222]

(19.2) Electrical Circuit Diagrams[223]

(19.3) Communication System Diagrams[224]

On Facets of Mathematical Information in Σ

Investigations are enhanced – especially in the heuristic phase(s) – if all **facets** of the 'central ME' are recognised. The simplest classification of 'basic facets' – into **arithmetical / combinatorial / algebraic / analytical / geometrical / topological** – seems adequate for Σ . Substantially arbitrary **combinations** of these basic facets appear to encompass 'all foreseeable mathematical activity' within current logical / calculational frameworks. If these basic facets are denoted as $\psi_1, \ldots, \psi_6$ (respectively) and $\psi_k(e)$ denotes the kth basic facet of the ME, e, then subtle nuances of the interrelations among items in the IB may be monitored systematically.

[216] e.g., Berger, M '*Geometry*' (Vols 1,2; Springer 1987, transl.); Berger, M, et. a., '*Problems in Geometry*' (Springer, 1984), Coxeter, HSM, '*Introduction to Geometry*' (2nd ed., Wiley, 1969; repr. 1989)

[217] e.g., Greenberg, MJ, '*Euclidean and Non-Euclidean Geometries . . .*' (freeman, 2nd ed., 1980); Berger (et al.), op. cit.

[218] This applies to all forms of geometries; but see: Maxwell, EA, '*Fallacies in Mathematics*' (CUP, 1963) for simple examples

[219] e.g., in designs of industrial machines, where CAD packages are used

[220] An interesting example is: '*Descriptive Geometry with '3-D' Figures*', by I Pál (Hungarian Technical Publishers, Budapest, 1962)

[221] Several architectural graphics packages use holographic images

[222] e.g., Norman, ROC, '*Principles of Organic Synthesis*' (Methuen, 1968); Purcell, KF, Kotz, JC, '*Inorganic Chemistry*' (Saunders, 1977)

[223] e.g., Desoer, CA, Kuh, ES, '*Basic Circuit Theory*' (McGr-H, 1969)

[224] e.g., Kim, WH, Chen, RT-W, '*Topological Analysis and Synthesis of Communication Networks*' (Columbia UP, 1962)

Before such matters can be addressed, however, a crucial issue must be considered – namely: **Is $\psi_k(e)$ nontrivially / consistently definable, for arbitrary e, with $1 \leqslant k \leqslant 6$?**

More precisely, **if $^k\mathcal{S}$ denotes the 'space' of ME definable (primarily) in relation to ψ_k, and e is arbitrary, is it possible to 'locate' e within $^k\mathcal{S}$** – and so 'to represent e as $e \approx \underset{1\leqslant k\leqslant 6}{\langle + \rangle} {}^k e$, where $\langle + \rangle$ stands for the combination of facets (in some w.d. sense)? Here, $^k e$ is 'the projection of e onto $^k\mathcal{S}$'.

NOTE that although this question may still appear dangerously vague, it is, nevertheless, of fundamental importance for Σ , as a 'structured OIB with high-resolution search facilities'. This is especially so because of the central role of **approach processes**, including (inverse) analogy. Essentially, what is involved is **the t-maximal recognitions of ψ-structure in each (collection of) ME; or, in each theory**. Of course, the terms 'representation' and 'projection' are used here figuratively, since such 'versions of e' are neither exact nor unique: the projections are qualitative. Thus: every ME is stored in the OIB as a collection of its qualitative basic facets – any of which may (in principle) be realised t-max.-operationally. The $^k e$ may also be viewed as **reflections** of e in the 'spaces' $^k\mathcal{S}$. As such, their **existence** (for arbitrary e) seems intuitively evident – though, attempts to provide formal existence proofs (without various restrictions of e) could produce, at best, only results of extreme generality, but little substance. A more fruitful path, followed here, is characterised by the gradual accumulation of **examples** (of diverse types / depths) from which viable frameworks – each covering substantial classes of ME – will emerge, and may be incorporated within the OIB. Pertinent examples are identified in the commentaries in E_{10}/E_{11}, as part of Σ_0, the foundation for the longterm evolution of Σ_t.

In spite of the schematic element in the form $e \approx \underset{1\leqslant k\leqslant 6}{\langle + \rangle} {}^k e$, it is possible – and useful – to introduce associated operations of e, in terms of operations of the $^k e$ within $^k\mathcal{S}$. Similarly, one may specify 'elementary operations of typical diagrams' from each of the classes considered above. These collections of qualitative transformations greatly enhance the scope of heuristic investigations, and the discovery of unsuspected interrelations among seemingly disparate ME.

Remark: The **existence** of geometrical / topological facets of e **implies that** e has some **diagrammatic content**. Moreover, other facets of e may 'generate' schematic / explicatory diagrams. It is therefore highly plausible (for **all** e), and rigorously provable (for 'large subclasses' of ME) that modifications may be effected by 'perturbing (combinations of) facets / diagrams associated with the ME concerned.

Modifications of Diagrams / Facets of ME

In the scheme of Σ , there are (so far) eighteen species of diagrams, D_i (plus an extra 'Miscellaneous' species covering possible 'new species' / extensions), and six species of facets ψ_k (plus ψ_7, to cover unforeseen developments!). **Modifications** of arbitrary ME, e, may therefore be investigated in terms of changes in the $D_{i,j}(e)$ and the $\psi_k(e)$ – provided that appropriate account is taken of the corresponding modifications of all relevant **combinations** of diagrams $D_{i_1,j_1}(e_1)$, $D_{i_2,j_2}(e_2)$, and of facets $\psi_{k_1}(e_1)$, $\psi_{k_2}(e_2)$, etc.. **The next task**, therefore, is to introduce an initial list of '**elementary modifications**' of (single or combined) diagrams / facets of (admissible classes of) ME. To accomplish this aim one must specify abstract interpretations of 'diagram', 'facet', 'combination', ..., applicable to all of the species pertinent to Σ .

The following definitions, for **diagrams**, involve finite / (un)countable **collections of** substantially arbitrary (active / passive) **elements** within prescribed **environments**. These elements are subject to **rules** of several types:

(i) Rules of (relative) **location / orientation**;

(ii) Rules of (directed) **interconnection**;

(iii) Rules of r-element **interaction** ($r \geqslant 1$).

The environment(s), element(s), and rule(s) – of all types – comprise the **components** of the diagram. Variations / modifications of (any subset of) the components of $D_{i,j}(e)$ yield corresponding overall changes in $D_{i,j}(e)$ itself.

In a similar vein, **facets** may be defined within prescribed environments, in terms of '**partial models**' – that is, of **partial realizations** over the underlying **theories**. Thus, each facet, $\psi_k(e)$, of the ME e, is defined in relation to environments, sets of 'primitive objects', axioms satisfied by these objects – and appropriate results / theorems from the associated theory.

Another view of this process may be expressed by regarding $\psi_k(e)$ as a mathematical reflection of e in $\mathcal{J}_k$ – the theory characterizing the kth 'mathematical mode' in Σ , where the designated modes have been listed (Above) as: arithmetical / combinatorial / algebraic / analytical / geometrical / topological :=: $\psi_1/\psi_2/\psi_3/\psi_4/\psi_5/\psi_6$.

The range of admissible facets, $\psi_k(e)$, may be explored – heuristically, at least – through variants / extensions / generalisations of the basic theories, $\mathcal{J}_k$. Such 'elaborations of the $\mathcal{J}_k$' are generated by modifications of the axions, A, of $\mathcal{J}_k$ – to produce (say) $\mathcal{J}'_k$, with axiom-set A' (which may also be called '**deformations** of $\mathcal{J}_k$'). This amounts to the investigation of variants of Arithmetic / Combinatorics / Algebra / Analysis / Geometry / Topology – each of these 'modes' being taken in 'Standard Form(s)', upon which the deformations may be based. As always, at these levels of generality, there are potential ambiguities; but they may be effectively removed by imposing definite (almost universally acceptable) choices of the $\mathcal{J}_k$ as a basis for the exploratory processes.

The next step is to develop generalized 'Addition Theorems' for various combinations of diagrams / facets. Here, $\langle + \rangle$ denotes 'amalgamation' of diagrams, and $\{+\}$, of facets. The combination of ME is denoted by $[+]$. Of course, these 'summation operations' are w.d. only under special conditions on all of the 'arguments' involved, but it is important to have uniform notations for them. Moreover, several possible 'versions' of $a\langle + \rangle b$, $c\{+\}d$, $p[+]q$ are pertinent to Σ , and will be specified as they arise; but the aim here is to comment on the structural forms involved. In these terms, one seeks results of the forms:

$$D_{i_1,j_1}(e_1)\langle + \rangle D_{i_2,j_2}(e_2) =' D_{i_3,j_3}(e'_3);$$

$$\psi_{k_1}(e_1)\{e_1\}\psi_{k_2}(e_2) ='' \psi_{k_3}(e''_3);$$

$$e_1[+]e_2 =''' e'''_3.$$

Here, all information on the left side of the 'quasi-equality signs' $(=', ='', =''')$ is supposed 'given' – in various admissible forms. The aim is to determine $i_3, j_3, e'_3/k_3,$
e''_3/e'''_3 in terms of these data. In general, such determinations cannot be formulated algorithmically; rather, it is a matter of the gradual accumulation of (sufficiently divers) results, in forms readily accessible within the OIB. There are, however, special cases allowing somewhat more systematic treatment.

NOTES (i) Although this notation is superficially clear, and will be used in certain contexts, it is also misleading, since i_3 need not be an integer (e.g., when $i_1 \neq i_2$). Similarly, if $k_1 \neq k_2$, then k_3 need not be an integer.

(ii) In these 'hybrid cases', the symbols D_{i_s}, ψ_{k_3} represent 'mixes' diagrams / facets – for which suitable interpretations must be found.

(iii) Where two diagrams are of different species $(i_1 \neq i_2)$ **and** involve 'arbitrary ME' e_1, e_2, the 'amalgamation' of these diagrams can be a formal juxtaposition, with no representation as a single 'composite diagram'.

(iv) If $i_1 = i_2$ $(= i$, say), then one has

$$D_{i,j_1}(e_1)\langle + \rangle D_{i,j_2}(e_2) =' D_{i,j_3}(\alpha),$$

where α is (somehow) describable in terms of e_1 and e_2 – each of which admits a diagrammatic representation of species i.

(v) /if $e_1 = e_2$ $(= e$, say), then one has

$$D_{i_1,j_1}(e)\langle + \rangle D_{i_2,j_2}(e) =' D_{v,j}(e),$$

where υ is defined in terms of i_1, i_2 (possibly, as an 'indicator vector', (i_1, i_2)). In exceptional cases – when D_{i_1}, D_{i_2} refer to aspects of e representable as compatible parts of a single diagram, the form of $D_{\upsilon,j}(e)$ may be realized explicitly.

(vi) Diagrams with matching species and subjects may be combined in a natural way to form more detailed diagrams of the same kind. (Classes of electrical circuits, each modelling some phenomenon, give rise to such behaviour.)

(vii) Plainly, possible combinations $e_1[+]e_2$ may have many, diverse forms, according to the nature of e_1, e_2; but 'almost always', only the formal pairing: $e_1[+]e_2 =$
$'''(e_1, e_2)$ will be w.d.. Nevertheless, it is useful to have a notation covering all cases, since there are important classes where $e_1[+]e_2$ is nontrivially w.d. – as many examples in $\mathrm{E}_{10}/\mathrm{E}_{11}$ demonstrate.

(viii) Notice that **mappings**, in (commutative) diagrams may be viewed as types of **interactions** (between 'vertices'), etc..

For combinations of **facets**, matters are more straightforward – amounting to **(a)** the **aggregation of facets** of a fixed ME:

$$\psi_{k_1}(e)\{+\}\psi_{k_2}(e) ='' \psi_{(k_1,k_2)}(e);$$

or else, **(b) extension of some fixed facet** from two ME to their 'amalgamations':

$$\psi_k(e_1)\{+\}\psi_k(e_2) ='' \psi_k(e_1[+]e_2),$$

where $e_1[+]e_2$ stands for any admissible combination of the ME e_1, e_2 (whose precise form varies with e_1 and e_2). **(c)** Lastly, there is the case: $k_1 \neq k_2$ and $e_1 \neq e_2$; and here, it seems that only formal pairing is w.d., in general – though special relationships between e_1 and e_2 (to be discussed in examples) are considered in $\mathrm{E}_{10}/\mathrm{E}_{11}$, in which $\psi_{k_1}(e_1)\{+\}(e_2)$ has the form $\psi_\alpha(\eta)$, 'suitably interpreted'! Here, ψ_α may be some 'blend' of the facets ψ_{k_1},, ψ_{k_2}, and η, an ME 'suggested by' e_1, e_2 – without necessarily being directly 'obtainable' from them. In other words, such processes are essentially heuristic / (inverse-)analogical, and so highly pertinent to the 'approach framework' underlying Σ . Indeed, the emphasis on facets / diagrams – at all levels of sophistication – promotes unification among ostensibly disparate domains of Mathematics – as exemplified by the (asymptotic) blurring of the boundaries between 'Pure', and 'Applied' Mathematics.

With all of these (unavoidable) preliminaries, one may introduce classes of modifications of diagrams / facets, covering all w.d. **combinations** of **basic** diagrams / facets – involving finite collections of 'arguments' (and even, with due caution, certain (un)countable collections, provided that appropriate quasi-convergence criteria may be formulated). **Notice that**, for **facets**, 'modification' entails **either**, the replacement of one facet (of e) by another; **or else**, the amalgamations of two (basic) facets, by partial realisation of e within a framework encompassing both facets – these operations being iterated in arbitrary combinations.

Let E* be any set of ME, and define $\mathcal{D}^{(\lambda)}_{E*}$, $\mathcal{J}^{(\mu)}_{E*}$ by:

$$\mathcal{D}^{(\lambda)}_{E*} := \{D_\lambda(e) : e \in E^*\};$$

$$\mathcal{J}^{(\mu)}_{E*} := \{\psi_\mu(e) : e \in E^*\}.$$

Further, introduce the classes of **modification operators** by:

$$\mathbb{M}^{(\lambda',\lambda'')}_{D;E*} := \left\{M_D : \mathcal{D}^{(\lambda')}_{E*} \to \mathcal{D}^{(\lambda'')}_{E*}\right\};$$

$$\mathbb{M}^{(\mu',\mu'')}_{F;E*} := \left\{M_F : \mathcal{J}^{(\mu')}_{E*} \to \mathcal{J}^{(\mu'')}_{E*}\right\}.$$

Here, all diagrams / facets (simple or compound, basic or hybrid) are covered by this notation – with $\lambda = 1, 2, \ldots, 18$, for basic diagrams, and $\mu = 1, 2, \ldots, 6$, for basic facets. **Note that** the species may be conserved ($\lambda' = \lambda''$, or $\mu' = \mu''$) in any of these operations, though, in general, this will not happen.

Although the heterogeneity of possible diagrams / facets precludes the development of general representations for the modification operators, M_D, M_F, some progress may be made within each class of diagrams / facets of any (fixed) basic species. For instance, Homological Algebra may be viewed as an elaborate framework for the systematic transformation of various types of commutative diagrams / exact sequences; and, 'at the other end of the spectrum', Electric Circuit Analysis / Synthesis is highly diagrammatic in content. More generally, is is 'the Σ -view' that 'the collection of diagrams / facets (of all species) encompasses $\mathcal{M}_t$' – in the sense that every ME exhibits some (basic) facet(s), and admits some (basic) diagrammatic representation(s). It follows that 'the quasi-graphical content of Σ ' (say, Σ^G) furnishes a 'global approach to Σ '. These ideas are developed extensively in E_5.

The detailed forms of these aspects of (arbitrary) ME, and the use of such forms in general investigations, depend on the establishment of systematic calculational procedures, to be implemented (progressively) over the OIB. This, in turn, depends on the gradual accumulation of diverse **examples** – many of which may be found in E_{10}/E_{11}. Moreover, powerful operational calculi already exist for the manipulation of 'diagrams' in many parts of Mathematics and The Sciences, and these resources will be codifies / extended within Σ .

CHAPTER 5
Approximation Frameworks

In E_5, the emphasis is on the definition and estimation of **conceptual vicinities** of (sets of) specified ME, where **qualitative** differences are dominant. Here, by contrast, the main problem is to approximate e **quantitatively** by some family $\{e_\lambda : \lambda \in \Lambda\}$, with all of e and the e_λ lying in the **same** 'environment'. Of course, there are, typically, structural differences between e and the e_λ – for instance, e may be merely continuous over a real interval, I, whereas the e_λ are polynomials, etc.. All of the ME concerned, however, are of the same broad type ('functions of one real variable' in the instance just cited). For Σ , the admissible environments – or approximation frameworks – are those introduced in E_1 as spaces, $_SX$, of **L**inear / **T**otological / **U**niform / **P**roximity / **C**ontiguity / **N**earness / **S**statistical / **F**uzzy / **M**easure types[225]. All of these spaces have been studied intensively, and their principal properties may be found in a wide range of books – at all levels[226]. Hence no attempt is made here to re-derive these results; rather, the aim is to **characterize** each of the spaces $_SX$, and to analyse their **interrelations**. It is asserted (with relevant references) that, from any of the (classes of) spaces Γ, T, U, P, C, N, corresponding 'versions' of all the rest may be constructed[227]. In this (formal) sense, the possibilities for approximation are (with some qualifications) independent of the choice of 'initial framework' – though, in practice, some choices are more efficacious than others. Moreover, it is demonstrated in E_9 / E_{10} / E_{11} that the spaces $_SX$cover all standard (and many nonstandard) applications, through multifarious types of spaces obtainable from the basic ones as 'special cases'. A preliminary list is given here; the more exotic constructions are identified as they occur, in examples.

As an initial step, it is necessary to specify primary reference sources for each of the basic spaces – so that proofs of the fundamental properties / characterizations / interrelations may be formulated succinctly. The general theories of these spaces are also covered in these sources, which are chosen for clarity and consistency, rather than exhaustiveness. Before this is done, it should be emphasized that (a) **L**inear (vector) spaces constitute a general **background** for many of the other space constructions; (b) **M**easure spaces are also different in kind from the other spaces, $_SX$– insofar as they do not necessarily carry other structures, but (rather) admit 'set functions', etc.. Consequently, these spaces ($_SL$ and $_SM$) are discussed separately from the rest – which are closely interrelated, as already indicated. Observe, also, that the statistical-metric spaces are obtained by combining **metric** spaces (derived from, e.g., topological spaces) with the (probabilistic) distribution function F, where $F(p, q; x) := \underline{\text{prob}}\{d(p, q) < x\}$, for $x > 0$, and the underlying metric function, d. Similar remarks apply to the **fuzzy** spaces – whose relations to S-MS, and to the basic spaces Γ/T/U/P/C/N, are far from transparent – and are elucidated, as far as possible, later.

On this account, references are listed separately for: (i) L-spaces, (ii) S-spaces, (iii), μ-spaces, (iv) Γ/T/U/P/C/N-spaces, and (v) **Fuzzy** spaces (F-spaces). This arrangement displays the (dis)similarities among these 'basic spaces', while leaving scope for the 'interpolation' of examples / modifications. Several attempts have been made to produce 'unified theories of topological structures'; some of these theories are mentioned here, too. It appears that C/N-spaces have not been treated in books (so far), but the (survey) articles cited are adequate for Σ . **Syntopogenous** spaces **have** been treated in the book, '*Foundations of General Topology*', by A. Csaszar (Pergamon, 1963). The resulting theory, however, is somewhat obscure, and does not seem to fit naturally with the basic spaces. Alternative possibilities are considered later.

---

[225] The validity / distinctiveness and potential relevance of **fuzzy** versions of these spaces are considered later. The prefix 'F-' may be used, where appropriate.

[226] References are given later.

[227] There are some limitations: e.g., only 'uniformizable' TS yield US; and, distinct US may yield the same TS; while, only certain classes of TS yield compatible PS – etc..

Background References for the Basic Spaces

1. Linear Spaces (LS)

1.1 Halmos, PR, '*Finite-Dimensional Vector Spaces*' (repr. 2nd ed. (1958); Springer, 1974)

1.2 Axler, S, '*Linear Algebra Done Right*' (Springer, 1996)

1.3 Shilov, GE, '*An Introduction to the Theory of Linear Spaces*' (Prentice-Hall, 1961)

1.4 Cater, FS, '*Lectures on Real and Complex Vector Spaces*' (Saunders, 1966)

1.5 Curtis, ML, '*Abstract Linear Algebra*', (Springer, 1990)

1.6 Greub, WH, '*Linear Algebra*' (2nd ed., Springer, 1963)

1.7 Blyth, TS, '*Module Theory – An Approach to Linear Algebra*' (Clarendon Press, Oxford, 1977)

1.8 Bourbaki, N, '*Algebra I*' (Springer, 1989; esp. Chapter II)

1.9 Noll, W, '*Finite-Dimensional Spaces*' (Nijhoff, 1987)

1.10 Schaefer, HH, '*Topological Vector Spaces*' (Macmillan, 1966 / Springer, 1986)

1.11 Smirnov, VI (ed. Silverman, RA), '*Linear Algebra and Group Theory*' (McGraw Hill, 1961)

1.12 Jameson, GJO, '*Topology and Normed Spaces*' (Champan and Hall, 1974)

1.13 Day, MM, '*Normed Linear Spaces*' (Springer, 3rd ed., 1973)

1.14 Naylor, AW / Sell, GR, '*Linear Operator Theory in Engineering and Science*' (Springer, 1982)

1.15 Gohberg, I / Goldberg, S, '*Basic Operator Theory*' (Birkhäuser, 1980)

1.16 Dunford, N / Schwartz, JT, '*Linear Operators*' (Vol. 1, 1956; Vol. 2, 1962, Interscience)

1.17 Kahn, PJ, '*Introduction to Linear Algebra*' (Harper & Row, 1967)

2. Statistical-Metric Spaces (S-MS)

2.1 Schweizer, B / Sklar, A, '*Probabilistic Metric Spaces*' (Elsevier, 1982)

2.2 Frank, MJ, '*Probabilistic Topological Spaces*' (J. Math. Anal. Applic. 34 (1971) 67–81)

3. Measure Spaces (μS)

3.1 Halmos, PR, '*Measure Theory*' (Van Nostrand, 1950)

3.2 Hewitt, E / Stromberg, K, '*Real and Abstract Analysis*' (Springer, 1965)

3.3 Taylor, AE, '*General Theory of Functions and Integration*' (Blaisdell, 1966; Dover, 1985)

3.4 Royden, HL, '*Real Analysis*' (macmillan, 3rd ed., 1988)

3.5 Brown, A / Pearcy, C, '*Introduction to Operator Theory I*'

3.6 Federer, H, '*Geometric Measure Theory*' (Springer, 1969)

3.7 Doob, JL, '*Measure Theory*' (Springer, 1994)

3.8 Kelley, JL / Srinivasan, TP, '*Measure and Integration, Vol 1*' (Springer, 1988)

3.9 Zaanen, AC, '*Integration*' (North-Holland, 1967)

3.10 Cotlar, M / Cignoli, R, '*An Introduction to Functional Analysis*' (North-Holland, 1974)

3.11 Rogers, CA, '*Hausdorff Measures*' (Cambridge UP, 1970)

3.12 Bichteler, K, '*Integration Theory (with special attention to Vector Measures)*' LNM #315 (Springer, 1970)

3.13 Hewitt, E / Ross, KA, '*Abstract Harmonic Analysis*' Vol 1 (Springer, 1963)

3.14 Shilov, GE / Gurevich, BL, '*Integral Measure and Derivative: A Unified Approach*' (Prentice-Hall, 1996 / Dover, 1977)

3.15 Cornfeld, IP / Fomin, SV / Sinai, Ya. G., '*Ergodic Theory*' (Springer, 1982)

3.16 Parthasarathy, KR, '*Introduction to Probability and Measure*' (Macmillan, 1977)

4. The Main (Γ/T/U/P/C/N) Basic Spaces

4.1 Kelley, JL, '*General Topology*' (Van Nostrand, 1955)

4.2 Čech, E / Frolik, Z / Katétor, M, '*Topological Spaces*' (Wiley, 1966)

4.3 Csaszar, A, '*Foundations of General Topology*' (Pergamon Press, 1963)

4.4 Csaszar, A, '*General Topology*' (Hilger, 1978)

4.5 Dugundji, J, '*Topology*' (Allyn & Bacon, 1966)

4.6 Gaal, SA, '*Point Set Topology*' (AP, 1964)

4.7 Kowalsky, H-J, '*Topological Spaces*' (AP, 1964)

4.8 Kuratowski, K, '*Topology, Vol 1*' (AP, 1966)

4.9 Mamuzić, ZP, '*Introduction to General Topology*' (Noordhoff, 1963)

4.10 Naimpally, SA / Warrack, BD, '*Proximity Spaces*' (CUP, 1970)

4.11 Bourbaki, N, '*General Topology*' (Springer, 1989)

4.12 James, IM, '*Topological and Uniform Spaces*' (Springer, 1987)

4.13 Csaszar, A (ed.), '*Topics in Topology*' Colloq. Mat. Soc. J. Bolyai **8** (North-Holland, 1974)

4.14 Herrlich, H, '*A Concept of Nearness*' Gen. Topol. and its Appl. **5** (1974) 191–212

4.15 Herrlich, H, '*Topological Structures*' (pp59–122 in Math. Centre Tracts No **52**, Amsterdam, 1974)

4.16 Naimpally, S, '*Nearness in Topology and Elsewhere*' Lec. Notes P&A Math. **#24** pp77–86

4.17 Thron, WJ, '*Proximity Structures and Grills*' Math. Ann. **206** (1973) 35–62

4.18 Bentley, HL, '*The Role of Nearness Spaces in Topology*' (pp1–22 in LNM **#540**, Springer)

4.19 Carlson, JW, '*Topological Properties in Nearness Spaces*' Gen. Topol. Appl. **8** (1978) 111–118

4.20 Reed, EE, '*Nearnesses, Proximities and T_1-Compactifications*' Trans. AMS **236** (1978) 193–207

4.21 Morrel, B / Nagata, J, '*Statistical Metric Spaces as Related to Topological Spaces*' Gen. Topol. Appl. **9** (1978) 233–237

4.22 Aleksandrov, PS / Fedorchuk, VV / Zaitsev, VI, '*The Main Aspects in the Development of Set-Theoretical Topology*' Russian Math. Surveys **33** (1978) 1–53

4.23 Schmid-Zartner, R / Reichel, H-C / Pötscher, B, '*Non-Numerical Distance Functions in Topology – A Survey and Bibliography, Furnished With Short Notes and Comments on Most of the Papers Listed Herein*' (University of Vienna, Dept. of Mathematics, 1984)

4.24 Cristescu, R, '*Ordered Vector Spaces and Linear Operators*' (Abacus Press, 1976)

4.25 Wong, Y-C / Ng, K-F, '*Partially Ordered Topological Vector Spaces*' (Oxford UP, 1973)

4.26 Wilansky, A, '*Modern Methods in TVS*' (McGraw-Hill, 1978)

4.27 Treves, F, '*TVS, Distributions and Kernels*' (AP, 1967)

4.28 Schaefer, HH, '*Banach Lattices and Positive Operators*' (Springer, 1974)

NOTE Some references (5.1–) for **fuzzy structures** are given later in this Essay. A subclass of this class (say, $\underline{F}$) is formally equivalent to a subclass of $\underline{S}$ – but the two theories have been pursued largely independently.

Brief Comments on the Background References

1.1 A systematic axiomatic treatment (to intermediate level).

1.2 This concentrates on the structure of linear maps between vector spaces, mainly via determinant-free methods.

1.3 Here, the aim is to establish linear spaces as environments for **analysis** / **geometry** – including infinite-dimensional euclidean spaces.

1.4 The stress here is on structural aspects of linear operators – including **algebras** of operators.

1.5 Algebraic aspects of linear operators dominate this book, in an abstract setting.

1.6 A comprehensive development for linear spaces of (in)finite dimension, combining axiomatic treatment of both algebraic and analytical topics (many, for arbitrary ground fields).

1.7 A thoroughly algebraic approach, emphasising structural aspects (canonical representations / decompositions / similarity / . . .).

1.8 The definitive axiomatic account, covering all algebraic, and some analytical, results (mainly for finite-dimensional spaces).

1.9 An extremely detailed presentation of linear algebra and calculus for finite-dimensional spaces (motivated by applications to 'rigorous continuum mechanics') with much unusual – often near-pedantic – terminology. Most of the formulations are coordinate-free.

2.1 This is the first book written on S-M spaces. The coverage is thorough, and the scope broad enough to treat probabilistic forms of all the other basic spaces (as well as various correspondences with '**fuzzy** versions' of standard results).

2.2 The underlying space is now a TS, so a theory of S-TS is outlined.

3.1 This is the standard reference for applications to other parts of Mathematics (though few applications are given). The (formal) **construction** of measures on various species of structures covers **M-spaces**.

3.2 Detailed applications in (real) analysis (and some other domains) are developed within a general framework.

3.3 The 'theory of real functions' is treated here through 'integrals' based either on (abstract) measures, or else, on classes of linear functionals.

3.4 The construction of M-spaces for operator theory forms the focus of this book; the scope is wide, including real / functional analysis.

3.5 The M-spaces are obtained in several contexts relevant to operator theory / Banach spaces.

3.6 Here, measures are constructed for diverse classes of manifolds, with a vast range of applications: **estimation** is the primary aim.

3.7 The emphasis here is on **pseudo**metric spaces, and on measures occurring in **probability theory**.

3.8 The measures in this treatment are obtained from 'stone integrals'; considerable ground is covered succinctly.

3.9 This is a detailed presentation of several types of integrals – with associated measures. The integrals are defined over various types of spaces (with some applications).

3.10 The measure-theoretic results are applied, here, to a broad selection of areas in functional analysis – providing some comparatively uncommon examples of M-spaces.

3.11 The emphasis is on the construction of measures as a means of **estimation** (of 'size' / 'context' / ...); as such, it is highly pertinent to the aims of Σ .

4.1 This is a standard reference for topological / uniform spaces. Many other topics are covered in Exercises.

4.2 For Σ , Čecho's book is a **key** reference (covering TS/US/PS, in great generality).

4.3 The theory of **syntopogenous spaces** is intended to subsume **all** topological structures. Although is has been accepted as formally correct, the resulting unification is somewhat unnatural; so the parallels in Σ are only pointed out – rather than used as a foundation.

4.4 This is a very thorough treatment of TS/US/PS (including many variants). As such, it is fundamental for Σ .

4.5 The scope here includes both General Topology (TS/US in many types) and Homotopy Theory. There are no literature citations (though some theorems are 'named').

4.6 The stress here is on topology for Analysis / Geometry, through TS/US. Many variants are discussed.

4.7 This covers TS/US. Most proofs are formulated in terms of **filters**.

4.8 The presentation is scholarly, with copious references and notes – as well as interesting illustrations, and many uncommon topics. All of TS/US/PS are covered (but some facets, e.g., connectedness, are deferred to Vol. 2).

4.9 This (brief) formulation is excellently attuned to Σ – since the most general kinds of approximation environments are considered – including TS/US/PS.

4.10 This is a tract on PS; their interrelations with TS/US are well covered, with some material on CS.

4.11 This detailed account includes topological groups and function spaces – but still confines itself to TS/US and many variants.

4.12 The treatment is introductory, but nontrivial. Many interesting connections between TS and US are exhibited (even though there exist topological structures **not** obtainable from any uniformity . . .).

4.13 This is a collection of papers on several aspects of Topology – but with some emphasis on General Topology.

4.14 This is a key reference for NS – a relatively successful scheme for the unification of all standard topological structures. Links with other basic spaces are discussed.

4.15 The stress, in this extended survey, is on NS and their interrelations with other standard spaces. There are 82 references.

4.16 This is another (brief) review of NS and some applications.

4.17 The 'grill formulation' of PS is discussed here.

4.18 Discussions of unification, extensions, homology, and connectedness (in terms of **NS** and **CS**) are presented here – with 54 references.

4.19 Links between NS and TS are treated here.

4.20 Further embeddings / equivalences are treated here.

4.21 This provides a useful link for understanding the interrelations among the basic spaces (especially, for **S** and **F**).

4.22 In this extensive review (with 369 references, mainly to **Russian** papers), the scope is very broad. The influence of algebraic / geometrical techniques on GT forms the principal theme.

4.23 This is very useful for Σ , though the notes / comments are brief, and of variable value. There are about 600 references. The role of S-MS (and **fuzzy spaces**) is covered. Many examples of non-numerical metrics are listed. **MR citations** are given for most items.

Interrelations Among Analytical Structures

Whereas the multifarious embedding / extensions / constructions / . . . involving **algebraic** structures have been studied extensively – and mirror the development of ('abstract') Algebra[228], the comparable results for **analytical** structures have been unified only relatively recently[229]. This has been accomplished, most perspicuously, through the introduction of **newness spaces**[230], and the use of category-theoretic representations, to establish that all of the standard analytical structures may be associated with subcategories of N, the category of NS and N-preserving maps. The table of 'secondary structures' induced / derived / constructed from any specified 'primary structure', must be **carefully interpreted** – in the light of a maze of qualifications, reflecting the tortuous progress towards a viable general theory. Even more general are the theories based on syntopogenous spaces[231], and on generalized uniform open covers[232].

[228] See, e.g., The Bourbaki development (in several volumes) op. cit.
[229] A good survey is given by H. Herrlich, in Math. Centre Tracts **52** (1974) 59–122 – where many references (and a hierarchy of pre-N-structures) are given
[230] The original paper (Herrlich) is Gen. Topol. Appl. **5** (1974) 191–212
[231] See Császár, A, '*Foundations of General Topology*', (Pergamon, 1963)
[232] See Harris, D, '*Structures in Topology*', Mem. AMS **115** (1971)

Interrelations Among The Basic Spaces

Let **L**, Γ, **T, U, P, C, N, S, M** denote the **classes** of (all) **Linear, Closure, Topological, Uniform, Proximity, Contiguity, Nearness, Statistical, Fuzzy**, and **Measure** spaces – defined over collections of subsets of some underlying set (say, W). All of the basic frameworks are characterised by sets of axioms, and the initial task is to represent these frameworks as far as possible in terms of any selected generic one – as depicted in the following table (where primed entries are qualified in various ways...).

$Y\backslash^X$	Γ	T	U	P	C	N
Γ	Γ	T_Γ	U'_Γ	P_Γ	C_Γ	N_Γ
T	Γ_T	T	U'_T	P_T	C_T	N_T
U	Γ_U	T_U	U	P_U	C_U	N_U
P	Γ_P	T_P	U_P	P	C_P	N_P
C	Γ_C	T_C	U_C	P_C	C	N_C
N	Γ_N	T_N	U_N	P_N	C_N	N

In this table, X_Y denotes the space of type X represented in terms of Y. Note that the class, $\underline{\Gamma}$, of **closure spaces**, with closure operators γ (say), properly contains $\underline{T}$; in fact, $\underline{T}$ corresponds to those members of $\underline{\Gamma}$ for which γ **is idempotent** (i.e., $j^2=\gamma$). Mostly, this distinction is not emphasized, but the extra generality is important for Σ. The various procedures for constructing X_Y from Y are expounded in treatises / papers on General Topology – especially, those listed above as references 4.1 – 4.16, which are cited, as necessary, to cover each entry in the table. Although many of these procedures are not constructive[233] they are explicit. The aim is to make it possible to reprint arbitrary mathematical entities within **any** chosen framework[234] – the optimal choice depending on context. In particular, diverse **approach**-procedures may be formulated in various environments, to cover all species of applications. It is not always at all obvious how to identify the 'optimal environment(s)' for a given application[235]. An infelicitous choice may greatly complicate subsequent calculations – or obscure crucial structural features.

Denote by Ω_1 the set $\{\underline{\Gamma}, \underline{T}, \underline{U}, \underline{P}, \underline{C}, \underline{N}\}$; by Ω_2, the set[236] $\{L, \mu, S, F\}$ and by Ω, the set $\Omega_1 \cup \Omega_2$. All constructions in Σ are based on elaborations / combinations of members of Ω – these elaborations including imposition of additional (order / combinatorial / geometric / differentiable / topological / ...) structures over the basic spaces. In order to systematise the process of representing all members of Ω_1 in terms of any chosen member, one may consider formal pairs $< W, \sigma >$, where W is a generic **set** (the **base set**) with an associated (collection of) structure(s), σ (**framework structures**) over appropriate families of sets (say, Φ_σ) constructed from W. The classes $\underline{L}$, $\underline{\mu}$, $\underline{S}$, $\underline{F}$ (comprising Ω_2) correspond to **extra** structures[237] over the members of Ω_1 – as well as constituting classes of spaces in their own right. This ambiguity motivates the distinction drawn between Ω_1 and Ω_2 – which is maintained from now on. The families Φ_σ are exemplified by $\mathcal{P}W$ (the **power-set** of W) and $W^{(k)}$ (the cartesian product of k copies of W).

This 'representation scheme' is realized (for Γ, T, U, P) in refs. 4.2 / 4.4 / 4.7 / 4.10 and thence (for all of Ω_1) in refs. 4.14 / 4.15 / 4.16 / 4.18 / 4.20. The various types of 'randomized spaces' may be (more tenuously) related to those in Ω_1; for $\underline{S}$, this is done in ref. 2.1, where it is also shown that the theories of **fuzzy sets**, and of **probabilistic semi-topological spaces**[238] are **equivalent**.

[233]There are possibilities for constructive versions, but this is not crucial to the Σ scheme. Relevant references may be found in the book by Bishop / Bridges, op. cit.

[234]The links between $\underline{\Gamma}$ and $\underline{T}$ are covered in 4.2/4.3

[235]For instance, it has been claimed that stability / recurrence phenomena in local dynamical systems are best studied over US/PS (see, Hajek, O, '*Dynamical Systems in the Plane*' (Academic Press, 1968, preface))

[236]The class $\underline{F}$ comprises **fuzzy** versions of other spaces in Ω

[237]Equally, **topologies**, etc., may be defined over elements in $\underline{S}$, but it is not possible to obtain Ω, from $\underline{S}$ under general conditions (though there are some links)

[238]See ref. 2.1, §15.3

Nevertheless, the two theories have developed independently (and have not been proven fully equivalent), so it is necessary to include the class, $\underline{F}$, of **fuzzy (analytical) structures**, in Ω_2, to produce $\Omega_2 = \{\underline{L}, \underline{\mu}, \underline{S}, \underline{F}\}$. Although the equivalence of certain subclasses of $\underline{F}$ to subclasses of $\underline{S}$ may appear to make the inclusion of F in Ω_2 partly formal, the **scope** of research on $\underline{F}$ has involved several detailed studies of fuzzy versions of elements in Ω_1 – which are pertinent for Σ . Consequently, some basic references for $\underline{F}$ are given now; more recent results may be found, especially, in J. Math. Anal. & Applications, from 1968, onwards – and, to a lesser extent, in (General) Topol. & Applications[239].

5.1 Chang, CL, '*Fuzzy Topological Spaces*' JMAA **24** (1968) 182–190

5.2 Lowen, R, '*Fuzzy Topological Spaces and Fuzzy Compactness*' JMAA **56** (1976) 621–633

5.3 Hutton, B, '*Uniformities on Fuzzy Topological Spaces*' JMAA **58** (1977) 559–571

5.4 Lowen, R, '*Initial and Final Fuzzy Topologies and the Fuzzy Tychonoff Theorem*' JMAA **58** (1977) 11–21

5.5 Lowen, R, '*A Comparison of Different Compactness Notions in Fuzzy Topological Spaces*' JMAA **64** (1978) 446–454

5.6 Hutton, B, '*Compactification Theory for Fuzzy Topological Spaces, Part I*' (Report No. 130, University of Auckland, April, 1978)

5.7 Lowen, R, '*Convergence in Fuzzy Topological Spaces*', General Topol. & Appl. **10** (1979) 147–160

5.8 Lowen, R, '*Compact Hausdorff Fuzzy Topological Spaces are Topological*' (Pre-print, Free Univ. Brussels, 1979)

5.9 Hutton, B, '*Products of Fuzzy Topological Spaces*', Topol. & Appl. **11** (1980) 59–67

5.10 Hutton, B / Reilly, I, '*Separation Axioms in Fuzzy Topological Spaces*', Fuzzy Sets and Systems **3** (1980) 93–104

5.11 Lowen, R, '*Fuzzy Uniform Spaces*', Preprint, Free Univ. Brussels, 1980

5.12 Lowen, R, '*Fuzzy Neighbourhood Spaces*, Preprint, Free Univ. Brussels, (?1980)

5.13 Lowell, R / Whyts, P, '*The Relation Between Completeness, Compactness and Pre-Compactness in Fuzzy Uniform Spaces*', Preprint, Free Univ. Brussels (?1980)

5.14 Lowen, R, '*I^X, The Hyperspace of Fuzzy Sets, A Natural Non-Topological Space*', Preprint, Free Univ. Brussels (?1980)

Remarks

(a) The relevance of $\underline{S}$ and $\underline{F}$ to Σ may be assessed only in terms of possible applications – where the extra scope allows stochastic phenomena to be treated within a unified scheme[240]. In view of the principal aim in Σ – to encompass the most general forms of **approach** to arbitrary ME – the inclusion of $\underline{S}$ and $\underline{F}$ is essential.

(b) None of the classes $\underline{L}$, $\underline{\mu}$, $\underline{S}$, $\underline{F}$ need to be considered in extensions of the 'representation table', since the links between spaces from Ω_2 and spaces from Ω_1 are incomplete. There are, however, many 'partial representations' – to be noted separately.

[239] It has been claimed that every result for F **could** be derived within standard topological structures; there is some parallel with nonstandard Analysis, here, where analogous results have been established (for 'non-exotic SF'). See J. Symbolic Logic, **51** (1986) 377–386 (CW Henson / HJ Keisler)

[240] See, e.g., the Bibliography in ref. 2.1, here.

(c) A major issue for the formulations in $\underline{F}$ is **the retrieval of non-fuzzy structures from the fuzzy versions**. This gives the subject a somewhat experimental flavour! A similar situation obtains for theories of '**calculus in TVS**', for which there are several models[241].

(d) All types of **stochastic convergence** are covered by $\underline{\mu}$ (the class of **measure spaces**); see, e.g., '*Stochastic Convergence*', by E. Lukacs (D.C. Heath and Co., 1968), and '*Measure Theory*', by JL Doob (Springer, 1994).

(e) If some radically new class of spaces is discovered, then appropriate changes / enlargements may be made in Ω_1, Ω_2, and the basic representation table – without invalidating any of the (then current) results in Σ . Thus, the collection Ω, which appears to encompass all structures currently recognised, may be augmented as necessary, within a **fixed** basic (calculational) framework.

(f) Many of the formulations in $\underline{F}$ may be viewed as fuzzy 'reflections' (in the sense introduced in Σ for analogues) of standard structures in Ω. The relevant comparisons are most aptly made within the framework of **category theory** – which is also a natural scheme for the interpretation of the representation table. There is a bewildering range of variants of almost every concept reflected into $\underline{F}$ from Ω, and only applications (e.g., to the modelling of stochastic analogues of deterministic entities) will identify and 'rank' these variants for their mathematical significance. Meanwhile, a formalism must be sought in which all variants are treated 'neutrally'. This situation is, of course, typical in Mathematics: fundamental notions tend to proliferate by spawning tributaries which may eventually meander all over the mathematical landscape! For fuzzy structures, the variants correspond (roughly) to modified versions of operations / procedures / structures / . . . of fundamental importance in the **non** fuzzy theories; this imparts some coherence to the classifications. To a lesser degree, these remarks also apply to the class $\underline{S}$.

(g) The most general uses of measure spaces are associated with 'estimation problems' over sets of 'nonstandard ME', for which the structures in Ω_1 are (at best) ill-defined. A selection of such problems (mainly pertaining to **differential geometry**) is treated extensively in ref. 3.6.

(h) For **probabilistic applications** where measure spaces are constructed / introduced in specified contexts (and called **probability spaces**), there is an abundance of sources, of use in Σ ; for instance:

Feller, W, '*Introduction to Probability Theory and its Applications*' (Vols. 1 (2nd ed., 1957) and 2 (1966), Wiley);

Rényi, A, '*Probability Theory*' (North Holland, 1970);

New, P (et al.), eds., '*Advances in Probability and Related Topics*', Vols 1–6 (Dekker, 1970–1979);

Bharucha-Reid, AT, '*Probabilistic Methods in Applied Mathematics*', Vols 1–3, (Academic Press, 1968–1973).

In all of these references, **measures** are introduced (e.g., through associated **probability densities**, or **distribution functions**) over **underlying sets** (often, spaces from Ω_1, but also sets with **no** topological classification). It follows that there are particular spaces within Ω_2 having no structural characteristic from Ω_1, and **vice versa**. Nevertheless, the view of Ω_1 as 'primary structures', and of Ω_2 as 'secondary structures', seems appropriate for Σ .

The further classification (in the taxonomic hierarchy of **genera** / **species** / **types** / **instance** (or **examples**) of **paces** (as genus) into the species (L/Γ/T/U/P/C/N/S/F/μ), delineates the range of possible **approach environments**[242] – provided that **abstraction**

[241] See, e.g., (i) Averbukh, VI / Smolyanov, OG, Russian Math. Surveys **22** (1967) 201–258; **23** (1968) 67–113; (ii) Frölicher, A / Bucher, W, LNM #30 (Springer, 1966); Yamamuro, S, LNM #374 (Springer, 1974)

[242] abbreviated as AE from now on

and (direct or inverse) **reflection** (from mathematical scientific theories) are taken to be included. Most of the **instances / examples** are exhibited in $E_9/E_{10}/E_{11}$; but a preliminary listing of the most prominent **types** (of each **species** of AE), with associated key references, is given here, to underpin the subsequent general developments.

A. Linear Spaces

(1) Vector spaces over (arbitrary) fields, K

(2) Spaces of linear maps between finite dimensional spaces

(3) Spaces of linear maps between infinite-dimensional spaces

(4) Inner-product vector spaces of finite dimension

(5) Infinite-dimensional inner-product spaces

(6) Quotient spaces (of vector spaces)

(7) Spaces of multilinear maps

(8) Tensor-product spaces

(9) Tensor spaces

(10) p-fold exterior-power spaces

(11) Spaces of semi-linear maps

(12) Ordered linear spaces

(13) Dual linear spaces

B. Closure Spaces (ClS)

The class, $\underline{\Gamma}$, of ClS (with associated closure operators $\underline{\gamma}$) has typical form $< W, \gamma >$, where $\gamma : \beta W \to \beta W$ is single-valued, and satisfies the **axioms**:

$$\gamma\varphi = \varphi; X \subset W \Rightarrow X \subset \gamma X;$$

$$\{X \subset W, Y \subset W\} \Rightarrow \gamma(X \cup Y) = \gamma X \cup \gamma Y$$

If, **in addition**, γ is **idempotent** (i.e., $X \subset W \Rightarrow \gamma\gamma X = \gamma X$) then it may be shown that the CS $< W, \gamma >$ is (in fact) a **TS**. Examples of **non**topological CS include spaces of maps with pointwise convergence of sequences, quotient spaces, and certain representations of spaces of maps between linear TS. Henve, it is worthwhile (for Σ) to preserve the distinction between $\underline{\Gamma}$ and $\underline{T}$, even though, formally, $\underline{T} \subset \underline{\Gamma}$. The known **types** of **non**topological CS include (as already mentioned):

(1) maps with pointwise convergence of sequences

(2) quotient spaces

(3) sequentially continuous maps between **TLS** (viewed as **standard** continuity of maps between **non**topological CS)

C. Topological Spaces

(1) Kolmogrov spaces

(2) Kuratowski spaces

(3) Hausdorff spaces

(4) Regular spaces

(5) Completely regular spaces

(6) Normal spaces

(7) Completely normal spaces

(8) Perfectly normal spaces

(9) Metrizable spaces

(10) First-countable spaces

(11) Second-countable spaces

(12) Compact spaces

- **(12.1)** sequentially
- **(12.2)** countably
- **(12.3)** pseudo
- **(12.4)** locally

(13) separable spaces

(14) Lindelöf spaces

(15) Paracompact spaces

(16) Strongly paracompact spaces

(17) Metric spaces

(18) Normed spaces

D. Uniform Spaces

Since uniform spaces 'lie between metric spaces and general topological spaces', there are few 'types' of US. Indeed, one possible **definition** of a US is as: **a TS whose topology is induced by a uniformity** (a class of **filter**). Thus the possible types of US correspond precisely to **the types of uniformizable TS** (for instance, T_1-CR spaces). **Criteria** for a TS to be uniformizable are cited later. The topological character of a US is identified with that of its induced TS. Nevertheless, several types may be listed:

(1) (hyper)complete;

(2) locally (sub)fine;

(3) Čech;

(4) coarse;

(5) fine;

(6) proximally coarse;

(7) proximally fine;

(8) Hewitt;

(9) pseudometrizable;

(10) topological;

(11) discrete;

(12) Hausdorff;

(13) p-adic;

(14) Mozzochi;

(15) quasi-;

(16) symmetric generalized;

(17) total;

(18) universal;

(19) finite-partition-;

(20) finite-open-cover-.

E. Proximity Spaces

The following **types** may be distinguished – the word ‘proximity’ being understood in each case.

(1) Uniform;

(2) Topological;

(3) Discrete / indiscrete;

(4) Efremović;

(5) Generalized;

(6) Hausdorff;

(7) Leader;

(8) Maximum / minimum;

(9) Lodato;

(10) Pervin;

(11) Induced;

(12) Local;

(13) Pseudo;

(14) Quasi;

(15) Separation;

(16) Sequential.

F. Contiguity Spaces[243]

The CS are essentially NS subject to certain finiteness conditions.

G. Nearness Spaces

This class of spaces (with category N) encompasses 'most' TS, and all US/PS/CS. The exceptional TS admit a nontrivial **order relation**, namely:

$$x \leqslant y \Leftrightarrow x \in \gamma\{y\} \text{ (but } y \notin \gamma\{x\}).$$

These TS **are** included in the category of **syntopogenous** spaces / maps; or, alternatively in theories involving **generalized uniform open covers**. It may be shown that the following concepts (for (arbitrary) **collections** of sets) are **equivalent**: (a) nearness; (b) existence of uniform covers; (c) existence of arbitrarily small members. (**Note** The existence of **generalized** open uniform covers is discusses, briefly, later).

The principal **types** of NS are designated as elements of the categories λ-**Near**, where 'λ =T/U/C/Pr/R/Q/P' – corresponding (respectively) to symmetric-**topological** / **uniform** / **continual** / **proximal** / **regular** / **quasi-near** / **pre-near spaces** (and associated **maps**). The **syntopogenous** spaces (Sn) and the **merotopic** spaces (Mr) have already been mentioned. The categories for NS and MS are equivalent; but they may be embedded in the categories for **Sn** of **gen-U** spaces. It follows that the NS may be listed as: (1) T-N; (2) U-N; (3) C-N; (4) Pr-N; (5) R-N; (6) Q-N; (7) P-N.

H. Statistical(-Metric) Spaces

This class ($\underline{S}$), also known as the class of **probabilistic**-metric spaces, yields stochastic versions of TS/US/PS (at least!) though their relations to the standard forms require careful analysis. The main **types** of S-MS are listed later.

I. Fuzzy Spaces

It may be shown that the theories of 'fuzzy sets', and, 'probabilistic semi-topological spaces', are **equivalent** – but the 'overall equivalence problem' seems to be open (and, again, the links with standard forms require careful interpretation). Here, too, fuzzy forms of TS/US/PS have been investigated.

For both $\underline{S}$ and $\underline{F}$, perhaps **the** crucial problem is to formulate criteria for the effective **use** of these classes of spaces in the realisation of constructions / calculations in general settings.

J. Measure Spaces

From one point of view, the class ($\underline{M}$) of measure saves encompasses all of the other classes of 'basic spaces' in Σ . This is so because (with due care over details!) some measure(s) may be defined over essentially arbitrary sets of ME – of any cardinality (finite / countable / uncountable). One way to classify measure spaces is through the various types of measure in use. A partial list is as follows (with the word 'space' understood in each case):

(1) Caratheodory;

(3) Lebesgue;

(4) Baire;

(5) Haar;

---

[243] Recall that **closure** spaces are denoted by CIS

(6) Radon;

(7) Hausdorff;

(8) Jordan;

(9) Markov;

(10) Spectral;

(11) Probability;

(12) Daniell-Stone;

(13) Vector;

(14) Geometric;

(15) Invariant.

Plainly, this is not exhaustive, but it allows an initial list of key references to be made.

Key References for the Main Types of Spaces

The references are given in sets (with at least one member!) labelled as $x_1, \ldots, x_{L_x}$, for $x = A, \ldots, J$, $L_x \geqslant 1$. **NOTE that**: the references already identified for **species** of spaces are also used (**in greater detail**) here – and **supplemented**, as necessary, with papers from the more specialized literature for each type of space. Previous lists are cited in their earlier forms.

A. Linear Spaces

A_1	1.1	1.6				
A_2	1.1	1.6	1.7	1.9		
A_3	1.3	1.4	1.11	1.14	1.15	1.16
A_4	1.6	1.9	1.17			
A_5	1.3	1.11	1.13	1.14	1.15	1.16
A_6	1.1	1.6	1.8	1.17		
A_7	1.6	1.7	1.8			
A_8	1.6	1.7	1.8			
A_9	1.6	1.7	1.8			
A_{10}	1.6	1.8				
A_{11}	1.8					
A_{12}	4.24	4.25				
A_{13}	1.1	1.6	1.7	1.8		

B. Closure Spaces

B_1	4.2
B_2	4.2
B_3	4.2

C. Topological Spaces

C_1	4.5	4.6	4.7	4.11				
C_2	4.1	4.4	4.5	4.7				
C_3	4.1	4.2	4.4	4.5	4.8			
C_4	4.1	4.4	4.5	4.7	4.8			
C_5	4.2	4.4	4.5	4.6	4.7	4.8		
C_6	4.1	4.2	4.4	4.5	4.6	4.7	4.8	4.11
C_7	4.1	4.4	4.5	4.6	4.7			
C_8	4.1	4.4	4.5	4.6	4.8			
C_9	4.1	4.4	4.5	4.6	4.7	4.8		
C_{10}	4.1	4.5	4.6	4.7	4.8			
C_{11}	4.1	4.5	4.6	4.7	4.8			
C_{12}	4.1	4.4	4.5	4.6	4.7	4.8	4.11	
C_{13}	4.1	4.4	4.5	4.6	4.7	4.8		
C_{14}	4.1	4.4	4.5	4.6	4.7	4.8	4.11	
C_{15}	4.1	4.2	4.4	4.5	4.6	4.7	4.8	4.11
C_{16}	4.4							
C_{17}	4.1	4.2	4.4	4.5	4.6	4.7	4.8	
C_{18}	4.5	4.6	4.7	4.8	1.12	1.13	1.14	1.16

D. Uniform Spaces

D_1	4.2		
D_2	4.2		
D_3	4.2		
D_4	4.2		
D_5	4.2		
D_6	4.2		
D_7	4.2		
D_8	4.2		
D_9	4.2		
D_{10}	4.2		
D_{11}	4.2		
D_{12}			4.16
D_{13}		4.11	
D_{14}			4.16
D_{15}			4.16
D_{16}			4.16
D_{17}			4.16
D_{18}			4.16
D_{19}		4.11	
D_{20}		4.11	

E. Proximity Spaces

All of the types here are covered in the Bambridge Tract (4.16) by Naim-Pally / Warrack – but room is left for other references, as appropriate.

E_1	4.10	4.2
E_2	4.10	
E_3	4.10	
E_4	4.10	
E_5	4.10	
E_6	4.10	
E_7	4.10	
E_8	4.10	
E_9	4.10	
E_{10}	4.10	
E_{11}	4.10	
E_{12}	4.10	
E_{13}	4.10	
E_{14}	4.10	
E_{15}	4.10	
E_{16}	4.10	

F. Contiguity Spaces

The possible **types** of contiguity spaces correspond to specializations of nearness spaces. **Thus:** it suffices to identify the known types of NS – together with the procedure(s) for obtaining associated CtS from NS. Hence: the **types** of CtS ($F_1, \ldots, F_7$) are derived from $G_1, \ldots, G_7$ (for NS), and they are covered by references 4.14/4.15.

G. Nearness Spaces

NOTE All of the types of NS are covered in references 4.14/4.15 (both by H. Herrlich); they have already been listed as 'λ-N', where 'λ=T/U/C/Pr/R/Q/P'. Here they are listed as $G_1, \ldots, G_7$.

G_1	4.14	4.15
G_2	4.14	4.15
G_3	4.14	4.15
G_4	4.14	4.15
G_5	4.14	4.15
G_6	4.14	4.15
G_7	4.14	4.15

H.Probabilistic / Statistical(-Metric) Spaces

The interrelations among the types of S-MS and the standard forms require careful analysis; but the following types have been recognised as 'distinct': PI / PI-P / PM / PN / PPM / PPN / Pre-PM / Pre-RM / PSM / PST / PT / URPM.

Here, P ≡ probabilistic, PP ≡ probabilistic pseudo, PS ≡ probabilistic semi, M ≡ metric, N ≡ normed, I-P ≡ inner-product, UR ≡ uniform-random – and the word 'space' is understood to follow all acronyms! Thus there are **12 types**, say, $\mathbf{H}_1, \ldots, \mathbf{G}_{12}$, and all of these types of S-MS are discussed in reference 2.1 (and, further, in works listed in its Bibliography).

NOTE The various connections between S-MS and **fuzzy** spaces ($\underline{F}$, S) have yet to be fully elucidated.

I. Fuzzy Spaces (FS)

Most of the recognized types of FS are obtained by considering the standard spaces within some version of 'fuzzy set throw'. The importance of these FS as approach environments (AE) has not been fully assessed yet; but the main types parallel those in Ω_1 (i.e., Γ/T/U/P/C/N). These FS are treated in references 5.1–5.14 (and in J. Math. Anal. Applic., from about 1965 onwards). These types (each of which has some variant(s)) are designated here as (respectively) $I_1, \ldots, I_6$.

I_1									
I_2	5.1	5.2	5.3	5.4	5.5	5.6	5.7	5.8	5.9
I_3	5.3	5.11	5.13	5.14					
I_4	5.3	5.11	5.13	5.14					
I_5	5.3	5.11	5.13	5.14					
I_6									

NOTE This will be amplified later.

J. Measure Spaces

Here, fifteen types have been identified, but some of these actually cover several 'subtypes'.

J_1	3.1	3.2	3.3	3.4	3.5	3.6	3.7	3.8	3.10	3.12
									3.13	3.14
J_2	3.3	3.4	3.7	3.8	3.10	3.11	3.13			
J_3	3.1	3.2	3.3	3.4	3.5	3.6	3.7	3.8	3.9	3.10
							3.11	3.12	3.13	3.14
J_4	3.1	3.4	3.5	3.12						
J_5	3.1	3.4	3.6	3.8	3.10	3.13				
J_6	3.2	3.3	3.6	3.7	3.10	3.11	3.12			
J_7	3.1	3.2	3.6	3.11						
J_8	3.2	3.3	3.5	3.9	3.13					
J_9	3.15									
J_{10}	3.10									
J_{11}	3.1	3.7	3.12							
J_{12}	3.6	3.8	3.9	3.10	3.14					
J_{13}	3.9									
J_{14}	3.2	3.6								
J_{15}	3.2	3.4	3.6	3.7	3.10	3.11	3.13			

Some Extra Structures Over the Basic Spaces

The basic spaces constitute fundamental **approach environments** (AE) for Σ – with minimal 'extra structure'. The need for additional structural properties is, however, ubiquitous within the theories of typical use in research investigations. Consequently, it is important to specify the nature and scope, of these 'imposed properties', which may be grouped (broadly) under the headings: **arithmetical; algebraic; combinatorial; analytical; geometrical; topological** (and, **combinations** of these types)[244]. In accordance with the procedure followed so far, the principal aim is to identify the extra structures, and then, to indicate (with adequate references) the extent to which they may be imposed on (each of) the AE. One way (perhaps, the simplest) of accomplishing this task, is to consider, in turn, sets (say, W) carrying any of these extra structures, as possible base-sets for the fundamental AE. In favourable cases, collections of 'φ-admissible AE' are obtained (where φ labels the imposed structure).

In many instances, the base set for an AE appears to be (essentially) **arbitrary** – and the various possible AE structures are defined through collections of **axioms** (usually, involving 'points' and 'subsets' of the base set, but otherwise not presuming this set to have any particular

[244]These are just the 'facets', φ_k (already introduced in E_4)

form). This is most obviously so for the definition of **topological** spaces, by specifying their collections of 'open sets'; but broadly comparable specifications will be found for all of the AS in Σ. The typical problem may therefore be stated as follows:

> For any structural framework (SF), W (regarded as a **set**[245]) and any AE, S (with axiom set Ax(S)): Is Ax(S) realisable over W?

Notice that, every structure (say, W*) is realized over **some** base-set (say, W) with its **own** structure (trivial or otherwise). If W is nonempty, then the collection of all topologies, say, $(\tau)_k$, $k \geqslant 1$, ordered by class inclusion, is a **complete lattice**. Moreover, corresponding to each τ_i, there is a **T-N structure** (say, ξ_{τ_i}) defined by:

$$\xi_{\tau_i} = \{A \subset P(W) : \cap\{\overline{A}^{\tau_i} : A \in \mathcal{A}\} \neq \varphi\},$$

where $\overline{A}^{\tau_i}$ denotes the closure of A for τ_i. This is the **N-structure induced by** τ_i. Further, a **U-N structure** may be induced through the given US, (W,μ), by:

$$\xi_\mu = \{\mathcal{A} \subset P(W) : \forall B \in \mu : B \cap \text{sec}\mathcal{A} \neq \varphi\},$$

where sec$A := \{B \subset W : \forall A \in \mathcal{A} : B \cap A \neq \varphi\}$. Conversely, starting from certain classes of N-spaces, associated T-spaces and U-spaces may be obtained. Moreover, similar results hold for C-spaces and P-spaces. Details may be found in references 4.14, 4.15, 4.19, 4.20.

Representations of the Basic Spaces

Let P be any set, with power-set $\mathcal{P}$(P). A **semi-pseudo-metric** on P^2 $\equiv$ PxP is defined as: d : P^2 $\rightarrow \mathbb{R}; d(x,x) = 0,\ d(x,y) = d(y,x) \geqslant 0$. If, further, $d(x,y) = 0 \Rightarrow x = y$, then d is a **semi-metric**; and if, also, $(\forall x, y, z \in$ P$)\ d(x,z) \leqslant d(x,y)$, then d is a **metric**. All of these forms are called '**distances**'. If $X \cup Y \subset$ P then: dist$_d(X,Y) := +\infty$ (if $X \cup Y = \varphi$), inf $d(X \times Y)$, otherwise; and, diam$_d(X) := 0$ (if $x = \varphi$), sup $d(X^2)$ (if X is bounded $\neq \varphi$), $+\infty$ (if X is unbounded). Then $\langle$P$, d\rangle$ is a space (of type given by that of d), and $f : \langle$P$, d\rangle \rightarrow \langle$P$_1, d_1\rangle$ is **distance-preserving** IFF $(x,y) \in$ P$^{(2)} \Rightarrow d_1(f(x), f(y)) = d(x,y)$.

New, denote by u, τ, $\mathcal{U}$, p, c, ξ (respectively) generic **closure, topological, uniform, proximity, contiguity, nearness** structures – over P; and, denote by α_β **the structure** α **induced by** β (over P), for $\alpha, \beta \equiv$ u, τ, $\mathcal{U}$, p, c, ξ. It may be proven that (with mild restrictions for some pairings) formal **representations of the structures** α_β may be obtained.

NOTE For (exhaustive!) details, see Čech, '*TS*' (op. cit.), and (for NS, etc.) Herrlich, '*T. Structures*' (op. cit.).

NOTE In some cases, these structures may be **induced** by a **distance**, d (of specified type: **s**(emi) / **p**(seudo) **m**(etric)) – **via** the d-induced closure, (semi-)uniformity, proximity. Such constructions / representations may be added to the main 'table' of forms, α_β (as u_d, $\mathcal{U}_d$, p_d). Moreover, since **contiguity is a restricted form of nearness**, the representations for α_c, or for c_β, are derivable from those for α_ξ, ξ_β – and so need not be given in detail.

(i) For $X \subset$ P, $u_d X := \{x \in$ P : dist$_d(x, X) = 0\}$.

(ii) d_1, d_2 are **topologically equivalent** IFF $u_{d_1} = u_{d_2}$.

(iii) X is a **neighbourhood** of x, in $\langle$P$, u_d\rangle$, IFF dist$_d(x,$ P$\backslash X) \neq 0$.

(iv) $\langle$P$, u\rangle$ is **s-p-metrizable** (by d) IFF $\exists u_d$ on P.

(v) $\langle$P$, u\rangle$ is **s-p-metrizable** IFF $\exists\{V_n\} \subset \mathcal{P}$(P^2) : $(\forall n \in \mathbb{N}, \forall x \in$ P$)[V_n = V_n^{-1}]\ [\{V_n[x] : n \in \mathbb{N}\}$ is a local base at x in $\langle$P$, u\rangle$].

[245] i.e. W is here identified with its base set; structural relations over W (as an SF) are ignored, in this context. Alternatively, the SF may be denoted by W*, and it base set, by W.

(vi) The metrization of arbitrary 'sums', and countable 'products', for families of s-p-m spaces, may be achieved in several (possible, equivalent) forms. Here, (uniform) **Lipschitz equivalence** – of $\langle \mathsf{P}, d_1 \rangle$, $\langle \mathsf{P}, d_2 \rangle$ – requires that the identity map, $i : \rangle \mathsf{P}, d_1 \rangle \to \langle \mathsf{P}, d_2 \rangle$, and its inverse be **Lipschitz-continuous.**

(vii) A **semi-uniformity**, for P, is a filter for P^2, say, $\mathcal{U}$, such that $\cap \mathcal{U} \supset \Delta_p$, and, $U \in \mathcal{U} \Rightarrow U^{-1} \in \mathcal{U}$. If, in addition, $\forall U \in \mathcal{U}, \exists V \in \mathcal{U} : V \circ V \subset U$, then $\mathcal{U}$ is a uniformity. For the following results $\mathcal{U}$ is a (semi-)uniformity, unless a distinction must be made.

(viii) $\models \mathcal{U}$ is s-p-metrizable $\Leftrightarrow \mathcal{U}$ has a countable base.

(ix) Given d, the family $\{U_+ \equiv \{y : d(x, y) < r, r > 0\}\}$ is a **base** for $\mathcal{U}$ – **induced by** d (but this does not yield all possible $\mathcal{U}$).

(x) If $U_x \equiv$ nhd of x in $\langle \mathsf{P}, u \rangle$, for $x \in \mathsf{P}$, and $U := \{\Sigma\{U_x : x \in \mathsf{P}\}\} =: \{\Sigma V_P\}$ ($\Sigma \equiv$ disjoint union) **then** $\models \exists! u : [\mathcal{U}][x] := \{U - x \subset \mathsf{P}, U \in \mathcal{U}\}$ is a **local base** at x in $\langle \mathsf{P}, u \rangle$, with u this induced by $\mathcal{U}$. Here, $\mathcal{U}$ is the set of all U w.d. via choices of families V_p.

(xi) $\models U$ is **semi-uniformizable** IFF $x \in U\{y\} \Rightarrow y \in U\{x\}$.

(xii) If $\mathcal{U}$ is p-m-(rather than s-p-m) induced then $\mathcal{U}$ is a ('full') uniformity.

(xiii) $\models \langle \mathsf{P}, u \rangle$ **uniformizable** $\Rightarrow \exists \tau_u : \langle \mathsf{P}, u \rangle \equiv \langle \mathsf{P}, \tau_u \rangle$.

(xiv) $\models d$ induces a full $\mathcal{U}$ IFF $\forall r > 0 \exists s > 0 : \max(d(x, y), d(y, z)) < s \Rightarrow d(x, z) < r$.

(xv) If $p_{\mathcal{U}}$ (on P^2) is given by XpY IFF $U_x \cap Y \neq \varphi$ (for some $x \in X, U \in \mathcal{U}$) then p **is induced by** $\mathcal{U}$.

(xvi) If U is the (unique) **closure induced by** $\mathcal{U}$ then $\models u_p X = \{x : \{x\} p X\}$, with p induced by $\mathcal{U}$.

(xvii) $\models p_d = \{(X, Y) : \mathrm{dist}_d(X, Y) = 0\}$.

(xviii) $\models u_p X = \{x : \{x\} p X, X \subset \mathsf{P}\}$.

(xix) $\langle \mathsf{P}, \xi \rangle$ is **topological** $\Leftrightarrow \{\mathcal{A} \in \xi = \cap \overline{\mathcal{A}} \neq \varphi\}$.

(xx) $\langle \mathsf{P}, \xi \rangle$ is **contigual** $\Leftrightarrow \{\mathcal{A} \in \xi \Rightarrow \exists B \subset A : \#B < \infty \wedge B$
$notin \xi\}$.

(xxi) $\langle \mathsf{P}, \xi \rangle$ is **totally bounded** $\Leftrightarrow \{\mathcal{A} \subset \mathcal{P}(\mathsf{P}) \wedge \mathcal{A}$ has a finite intersection prop$\} \Rightarrow \mathcal{A} \in \xi$. Here, $\overline{\mathcal{A}} := \{\overline{A} : A \in \mathcal{A}\} = \{\{x \in \mathsf{P} : x \xi A, A \in \mathcal{A}\}$.

(xxii) $\models \xi \mathcal{A} \Rightarrow \xi(\pi \mathcal{A})$, for any permutation, $\tilde{u} A$, of $\mathcal{A}$.

(xxiii) $\models \xi_{\mathcal{U}} = \{\mathcal{A} \subset \mathcal{P}(\mathsf{P}) : \cap \{U[A] : A \in \mathcal{A}\} \neq \varphi (\forall U \in \mathcal{U})\}$.

(xxiv) $\models \exists \mathcal{U}_U \Rightarrow \exists \tau_U \Leftrightarrow u \circ u = u$.

(xxv) $\models \xi_U = \{\mathcal{A} \subset \mathcal{P}(\mathsf{P}) : \cap \{UA : A \in \mathcal{A}\} \neq \varphi\}$.

(xxvi) $\models [p_u := p_{\mathcal{U}_u}]$ yields 'p **induced by** u'.

(xxvii) $\models u_\xi A = \{x \in \mathsf{P} : \{x\} \xi A\}$.

(xxviii) u_ξ is **topological** IFF $\cap \{u_\xi A : A \in \mathcal{A}\} \neq \varphi$.

(xxix) $\models [\mathcal{U}_\xi := \mathcal{U}_{u_\xi}]$ yields '$\mathcal{U}$ induced by ξ'.

(xxx) $\models [p_\xi := p_{u_\xi}]$ yields 'p induced by ξ'.

On this basis, **all of the entries in the 'α_β-Table' have definite representations**, which may be used to formulate essentially arbitrary results 'within the most convenient environment(s)'. Moreover, **statistical / fuzzy / measure** structures may be imposed on the spaces α_β – whenever such procedures are w.d.. The degrees of contractility of these representations depend critically on the particular realizations involved, as attested by various examples in E_{10} / E_{11}.

NOTE The interrelationships among the α_β are developed in detail (in various notations!) in the books: '*TS*' by E. Čech (Wiley, 1966); '*Structures in Topology*', by D. Harris ($\equiv$ Mem. AMS #115, 1971); and in: '*Topological Structures*', by H. Herrlich, pp59–122 of Math. Centre Tracts #52 (Amsterdam, 1974).

The essential issue, for Σ , is the extent to which the **intrinsic** structures of SF may be matched compatibly with the AE defined (or induced) over their base sets. In this connection, it is useful to differentiate the SF (say, W*) from its base set (say, W), and to consider the 'compound structure', $\langle W^*, \sigma(W)\rangle =: W^*_\sigma$, where the operations associated with W* and σ are to be combined, as far as possible. For some examples of W*, the relevant results are standard – notably, for **topological-algebraic** structures (see, e.g., reference 4.2, Section 19). Here, there seem to be few restricts on the allowable types of topologies; but their **realizability** over W depends on the nature of W as a set. For instance, W* may be a ring of functions of some class, or a group of matrices, etc. – over which not only TS/US structure, but even M(etric)S structure may be defined. On the other hand, the axiomatic specifications of many SF presume nothing about the properties of the elements of W, provided only that they satisfy the required axioms. Hence: **in general**, there are no guarantees of either compatibility or realizability[246], for '**arbitrary** pairings of SF and AE'. Rather, it is a matter of matching each type (say, φ) of SF, with each type (say, α) of AE: to some extent this is just a process of trial and error!

Before examining the range of possible SF, it is worthwhile to consider the inverse procedure – in which each AE is examined as a potential 'carrier' of various SF, to be realized over the chosen AE. Plainly, the only way of conducting such examinations is to test whether appropriate combinations of the elements (of the AE base sets) are well-defined, and satisfy the axioms characterizing the SF in question.

Under this scheme, there appear to be no general results at all! This is so because a **general** form of AE is not necessarily endowed with any extra structure whatever (whereas, for instance, a topology, of some kind, is definable over any nonempty set). Consequently, no matter whether the SF is chosen first and then matched with the AE, of **vice versa**, the question of compatibility of the resulting 'combined structure' is decidable only on a case-by-case basis. On reflection, this is not surprising, since the axiom-sets for the two species ($\{SF\}$, and $\{AE\}$) are largely independent of each other[247]. What is most relevant for Σ , however, is whether there exists **at least one** combined structure of every (admissible) mixed type[248], say, φ/α. Evidently, the special forms of φ and α would have to be tailored to satisfy several conditions combining the 'φ' and 'α' properties appropriately. This will be designated as **the existence problem for combined structures** (EPCS). This problem may be settled affirmatively (for specified generic lists $\{SF\}$, $\{AE\}$) simply by producing an example for every φ/α combination[249]. The broader questions – especially, those involving 'maximal domains of validity' for particular $\varphi//\alpha$ pairs – will be considered later. For the moment, it is necessary to specify the collections $\{SF\}$, $\{AE\}$ as completely as possible, so that the EPCS may be investigated. Denote $\{SF\}$ by Φ.

The **fundamental AE** have already been specified as a collection, Ω, of classes of spaces, where $\Omega =: \Omega_1 \cup \Omega_2$, $\Omega_1 = \{\underline{\Gamma}, \underline{T}, \underline{U}, \underline{P}, \underline{C}, \underline{N}\}$ and $\Omega_2 = \{\underline{L}, \underline{\mu}, \underline{S}, \underline{F}\}$, comprising (respectively) Closure, Topological, Uniform, Proximity, Contiguity, Nearness Spaces (for Ω_1); and, Linear, Measure, Statistical(-Metric), Fuzzy (for **preliminary** lists).

---

[246]Notation: '$q//p$' means 'q realized over p', for any SF, q, and AE, p. Hence: in general, $p//q \neq q//p$ (as – possible, empty – sets, even when both forms are w.d.

[247]The SF are mainly algebraic; the AE (quasi-) analytical

[248]Notation: CMS

[249]Not **so** simple, in practice!

The variety of **admissible SF** is considerable greater than it is for AE.

(A) For **algebraic** SF, a basic list is given (as a 'hierarchy') in the Axiom Manual. This may be amplified to include **non-associative** forms of all the objects listed (rather than merely not ruling them out). In particular, certain forms of algebraic SF **must** be associative, while the rest need not be so: this should be stated explicitly. Moreover, the classification should leave scope for enlargement of the list – and this applies to **all** categories of SF to be considered. Since Σ aspires to (at least) a level of completeness for 'everyday mathematical activity', these requirements are far from pedantic – indeed, probably minimal. The possibility of 'compatible extension' (maintaining the status of all items of information already in the IB) is absolutely fundamental for the construction of Σ .

The generic ME for algebraic SF are: **groups / rings / fields / mono ids / algebras / domains / modules / vector-spaces** (listed in no particular order of precedence). Variants correspond to the use of qualifying adjectives (QA), including:

(A′) **semi / abelian / commutative / associative / ordered / left / right / bi- / cancellation / partial / differential / extension / unique-factorization / characteristic-zero / division / polynomial-factorization / prime-characteristic / Euclidean / principal-ideal / non / explicit / linear(ly) / full(y)**.

Of course, neither of these lists is 'complete', nor are its members disjoint (as structured ME). Further, only 'relatively few' (ordered!) sets of QA are validly applicable to any selected generic algebraic SF – to produce recognised 'compound ME'. The ostensible choice is presented without extra conditions, to allow for the introduction of 'new combinations', should this become appropriate. The same scheme is used for **all** categories of qualified, generic SF.

(B) **Combinatorial** structures (especially, in Algebra and Topology) are characterized by the representation of ME in terms of **finite collections** of 'basic objects' – **generators** (in Algebra); **simplexes** (in Topology). The 'combinatorial' aspects encompass all valid interrelations among objects corresponding to admissible selections / combinations of the elements of sets of generators or simplexes. Thus, for Σ , the study of all possible **order relations** over sets is also relevant to the determination of their combinatorial characterizations. Thus, a minimal list of ME possessing combinatorial structure includes:

SF with **finite bases** (in some sense(s))

SF that are **triangulable**

SF that are **orientable**

SF that admit **partial order(s)**

SF that admit **total order(s)**

(C) For Σ , the **analytical structures** must support (at least, some of) the following capabilities:

determination of **bounds** on ME

calculation of **limits** of ME

differentiation of ME

integration of ME

definition of (quasi-)**manifolds** over ME

derivation of **inequalities** between ME

solution of **operator-(in)equations** over ME

(D) Among the **geometrical structures** that could (in principle) be imposed on the AE in Σ , a minimal list includes structures corresponding to:

convexity criteria

curvature criteria

affine transformations

projective transformations

non-euclidean geometries

finite geometries

isometric transformations, etc..

(E) Since the AE are themselves **general**-topological frameworks, a minimal list of extra structures realizable over AE must (here) pertain to the other branches of topology – **combinatorial / algebraic / differential / geometric**.

Thus, a minimal list should include frameworks for:

Homology structures

Cohomology structures

Homotopy structures

Cohomotopy structures

Contraction criteria

Orientation criteria

Transversality criteria

Topological-degree criteria

It has been remarked that (in some sense, to be made precise) 'Ω_2 lies between Ω_1 and Φ'. It is therefore useful to introduce the notation:

$$Q//P := \{q//p : p \in P, q \in Q\},$$

with **reverse**, $P//Q$, where (as before) $q//p$ denotes q **realized over** p. In these terms, one may consider separately the sets:

$$\Phi//\Omega \text{ and } (\Phi \cup \Omega_2)//\Omega_1,$$

where the elements of Ω_2 are viewed (respectively) as 'possible AE', and as 'additional SF'. The distinction is important, since, even for a single pairing, the **sets**[250] $\varphi//\alpha$ and $\alpha//\varphi$, or $\alpha_1//\psi$ and $\psi//\alpha_1$, need not coincide.

These scheme for generating[251] CSF as 'structures AE' is complemented by the definitions (given in E_5) for **conceptual vicinities**, $V(e)$, of e – the **points** of $V(e)$ being **variants** of e.

Moreover, further development of these 'CV criteria' yields a basis for the discussion (in E_6) of **abstraction / analogy / inverse analogy as species of approximation**; and (to a lesser degree) for the investigation of '**proofs as compositions of operational theorems**' (in E_8). The term **G**eneralized **CSF** $\equiv$ **GCSF** encompasses all admissible generalized approach environments involving the above CSF with all standard modes of approximation, together with abstraction / analogy / inverse analogy. This prescription extends the notion of 'approximation' very considerable, in a controlled way.

Although ME may be approached within GCSF environments on many different levels – from the absolutely quantitative to the highly qualitative – all possible modes of approach are equally valid! This statement is in no way paradoxical. Rather, it illustrates the fact that – even in Mathematics – objects have many characterizations; some, precisely representable others, only 'vaguely realizable', yet of critical importance for full descriptions of the objects.

[250]Here, $\alpha \in \Omega, \varphi \in \Phi, \alpha_1 \in \Omega_1$, and $\psi \in \Psi := \Phi \cup \Omega_2$

[251]CSF $\equiv$ combined structural framework(s)

Indeed, it may be reasonably claimed that the frontiers of Mathematics are advanced substantially through the identification, and partial realisation, of 'elusive facets of established ME'. In this sense, every ME has a **spectrum** of facets – each optimally observable over a certain bandwidth of approaches – and the researcher's task has astronomical parallels[252]!

In full generality, the EPCS covers all possible structures of the form $p//q$, where p and q are (essentially) **arbitrary** structures (i.e. (G)CSF); but this seems to be **reducible** to repeated forms of the **basic** cases, '$\alpha//\varphi$' and '$\varphi//\alpha$', already mentioned – together with the imposition of conditions to handle abstraction / analogy / inverse analogy. **In short**: general structures, of any complexity, may be constructed (quasi-)recursively from simpler structures.

If one disregards the qualifying adjectives (listed above for **algebraic** structures, but comparably numerous for many other classes of structures), then there are 10 AE (in Ω), and 35 SF (in Φ). Hence, **there are 350 basic combined SF pairings** (and so, **700 pairings if the order of α and φ is taken into account**). In view of the multifarious interdependencies among the AE, and among the SF, the irreducible minimum number of fully distinct pairings is comparatively small. It appears, therefore, that the question of existence of representative CSF, $\alpha//\varphi$, and $\varphi//\alpha$, may be resolved as follows.

Since each AE class, say, α', corresponds categorically to (restricted) classes of all α'' in $\Omega_1\backslash\{\alpha'\}$, it suffices to consider (initially) $\underline{T}//\Phi$ and $\Phi//\underline{T}$. Further, one may use the simplest adequate representatives of $\underline{\mathrm{T}}$ and the various classes comprising Φ, as the aim is merely to exhibit a w.d. example for each of the pairings, no matter how basic it may be.

For the structures in $\Omega_2//\Omega_1$, however, no such 'global procedure' seems to be available. Rather, it is necessary to treat the elements of Ω_2 one by one – to obtain:

$$\Omega_2//\Omega_1 = \Omega_2//\underline{L} \cup \Omega_2//\underline{\mu} \cup \Omega_2//\underline{S} \cup \Omega_2//\underline{F}.$$

The conditions sufficient for membership of $\Omega_2//L$ and of $\Omega_2//\mu$ (corresponding, respectively, to the imposition of linear-space, and measure-space, structures, over the AE in Ω_1) are straightforward, in principle, and so may be **assumed** (for now). By contrast, the combined SF associated with $\Omega_2//\underline{S}$ and $\Omega_2//\underline{F}$ must be considered largely on a case-by-case basis – though it is still possible to confine attention (formally) to the sets $\underline{T}//\underline{S}$, $\underline{S}//\underline{T}$, and $\underline{T}//\underline{F}$, $\underline{F}//\underline{T}$, because of the interdependencies among the AE in Ω_1. Proposed **definitions** for $\underline{\mathrm{S}}$ (see reference 2.1, chapter 12) correspond to $\underline{T}//\underline{S}$, and furnish several examples of TS structures defined over statistical spaces – and, thence, of the **other** AE structures over S-MS. In the 'opposite direction', if (W,σ) is any S-MS and t_σ is any topology over (W,σ), then the TS, (W,t_σ) is ('by construction') an element of $S//T$. It follows, from the AE-interrelations, that the other sets comprising $\Omega_1//\underline{S}$ and $\underline{S}//\Omega_1$ are also nonempty. **Notice that**, here, the definitions of the basic AE in Ω_1 are the standard ones.

For $\underline{F}$, the class of **fuzzy** SF, the situation is somewhat different from that governing $\underline{S}$, since the standard governing $\underline{S}$, since the standard definitions of basic AE structures are not directly applicable. Instead, there are definitions designed to blend the established AE 'as naturally as possible' with characteristic properties of $\underline{F}$. Nevertheless, it is known (see reference 5.8) that: **All fuzzy TS that are also F-compact and F-Hausdorff are (in fact) standard TS**. **More generally**: it is one aim of the F-definitions that their restrictions to the relevant categories of standard objects coincide with the standard definitions. These constructions ensure that (again, at least formally) the sets comprising $\Omega_2//\Omega_1$ are nonempty. The above result for compact, Hausdorff FTS shows that there are non-trivial examples – though care must be taken, since there are several F-versions of some 'topological phenomena'; and some versions of 'the same phenomenon' may be inequivalent! The fundamental results in this area are covered in references 5.1 to 5.14 – excepting general connections between $\underline{S}$ and $\underline{F}$, which are discussed in ref. 2.1, §15.3[253]. **Note that**: reference 2.2 treats **statistical-TS**, which have interesting links with F-spaces.

[252]Even the distinction between 'optical' and 'radio' astronomy is relevant, in this context.

[253]The main sources are: Frank, MJ, JMAA 34 (1971) 67–81; Nguyen, HT, Thèse, Univ. Lille (1975); Höhle, U. (i) Z. Wahrsch. V Geb. 36 (1976) 179–188, (ii) Manu-Math 26 (1979) 345–356, (iii) ibid. 26 (1978) 223–245 – as cited in ref 2.1

Definitions of the Basic AE and SF

In all of the following definitions, the base set is W (with power-set, $\mathcal{P}$W, relative complements $U' := W \backslash U$, typical elements x, y, z, of W, and A, B, C, of $\mathcal{P}$W). The 'field of scalars', k appears (for VS, etc.); and all structures have the generic forms $\langle \mathrm{W}, \theta \rangle$, or $\langle \mathrm{W}, \mathcal{H}, \varphi \rangle$, where θ, φ characterize the structural operations, and $\mathcal{H}$, special subsets of $\mathcal{P}$W. Here, θ, φ, range over **closures, topologies, uniformities, proximities, contiguities, nearnesses** (and analogues for statistical (-M)S / fuzzy spaces). All of these generic structures are of a single type – analytical or algebraic. The **combined SF**, and **generalized** combined SF are obtained by the compatible superposition of single SF – together with imposition of additional criteria / conditions for **abstraction / analogy / inverse analogy**.

Closure Spaces: $\langle W, \gamma \rangle$

$$\gamma\phi = \phi$$

$$A \subset W \Rightarrow A \subset \gamma A$$

$$A \cup B \subset W \Rightarrow \gamma(A \cup B) = \gamma A \cup \gamma B$$

NOTE In general, $\gamma^2 \neq \gamma$, here.

Topological Spaces: $\langle W, \tau \rangle$

$$\tau \subset \mathcal{P}W$$

$$\cup\{S_\lambda \in \tau : \lambda \in \Lambda\} \in \tau$$

$$\underset{1 \leqslant j \leqslant n}{\cap} \{S_j \in \tau\} \in \tau$$

$$\Rightarrow \phi \in \tau; W \in \tau$$

Uniform Spaces: $\langle W, \mathcal{U} \rangle$

$$\mathcal{U} \subset \mathcal{P}(W \times W)$$

$$W \supset A \subset B \in \mathcal{U} \Rightarrow A \in \mathcal{U}$$

$$\underset{i \leqslant j \leqslant n}{\cap} (B_j \in \mathcal{U}$$

$$B \in \mathcal{U} \Rightarrow \Delta \subset B$$

$$\forall V \in \mathcal{U}, \exists Y \in \mathcal{U} : Y \circ Y \subset V$$

Proximity Spaces: $\langle W, p \rangle$

$$ApB \Rightarrow BpA$$

$$A \cap B = \phi \Rightarrow ApB$$

$$(A \subset A_1 \wedge B \subset B_1 \wedge ApB) \Rightarrow A_1 p B_1$$

$$\phi \overline{p} W$$

$$(A\overline{p}C \wedge B\overline{p}C) \Rightarrow (A \cup B)\overline{p}C$$

$$A\overline{p}B \Rightarrow \exists R, S \in \mathcal{P}W, R \cap S = \phi : A\overline{p}R' \wedge B\overline{p}S'$$

Nearness Spaces: $\langle W, \xi \rangle$

$\mathcal{A} \subset \mathcal{P}W \wedge \cap\mathcal{A} \neq \phi \Rightarrow \xi\mathcal{A}$

$\mathcal{A}, B \subset \mathcal{P}W \wedge [\xi B \wedge \forall A \in \mathcal{A}, \exists B_A \in B : B \in cl_\xi A] \Rightarrow \xi\mathcal{A}$

$\bar{\xi}\mathcal{A} \wedge \bar{\xi}B \Rightarrow \bar{\xi}(\mathcal{A} \cup B)$

$\phi \in \mathcal{A} \Rightarrow \xi\mathcal{A}$

$(N_1) : \xi\{\{x\}, \{y\}\} \Rightarrow x = y$

Contiguity Spaces: $\langle W, \xi^* \rangle$

The NS $\langle W, \xi \rangle$ is **contigual** IFF $(\mathcal{A} \subset \mathcal{P}W \wedge \bar{\xi}\mathcal{A}) \Rightarrow \exists B$ (finite) $\subset \mathcal{A} : \bar{\xi}B$.

Linear Spaces: $\langle W, \lambda \rangle$

$x\lambda y = y\lambda x$

$(x\lambda y)\lambda z = x\lambda(y\lambda z)$

$\exists\theta \in W : x\lambda\theta = x(\forall x \in W)$

$\forall x \in W, \exists y_x : x\lambda y_x = \theta$

$\forall\alpha, \beta \in k, \alpha(\beta x) = (\alpha\beta)x$

$(\alpha + \beta)x = \alpha x\lambda\beta x$

$\alpha(x\lambda y) = \alpha x\lambda\alpha y$

NOTE: Here, λ denotes any operation satisfying the given conditions; usually, λ is taken to be $+$ (addition), and k, to be $\mathbb{R}$ or $\mathbb{C}$.

Measure Spaces: $\langle W, \mathcal{B}, \mu \rangle$

$\mathcal{B} \subset \mathcal{P}W; \phi \in \mathcal{B}$; $\mathcal{B}$ **closed under complementation and countable union**

$\mu : \mathcal{B} \to \mathbb{R}^+$

$\mu(\phi) = 0$

$\forall$ disjoint sequences $(E_i) \subset \mathcal{B}, \mu(\underset{i \in \mathbb{N}}{\cup} E_i) = \underset{i \in \mathbb{N}}{\Sigma} \mu(E_i)$

$\mathcal{B} \equiv$ **sigma-algebra** of sets; $\mu \equiv$ **measure**

Groups: $\langle W, g \rangle$

Set, W, endowed with an associative, binary, uniquely invertible operation, g, under which W is closed.

Semigroups: $\langle W, s \rangle$

Set, W, closed under an associative binary operation, s.

Monoids: $\langle W, s; e \rangle$

Semi-group with identity element, e (but no general inverses)

Groupoids: $\langle W, g^* \rangle$

Set, W, closed under a binary operation, g^*.

Modules: $\langle W, g, R \rangle$

Commutative group, $\langle W, g \rangle$, allowing two-sided, associative, distributive 'multiplication' of group elements by elements from a ring, R.

Rings: $\langle W, g, h \rangle$

Abelian groups $\langle W, g \rangle$ such that $\langle \langle W, g \rangle, h \rangle$ is also a **semi**group, with h two-sided / distributive over g.

Fields: $\langle W, g, h \rangle$

$\langle W, g \rangle$ is an abelian group with 'zero element', θ; $\langle W \backslash \{\theta\}, h \rangle$ is also an abelian group – with h distributive over g.

Algebras: $\langle W, g, h \rangle$

Rings that are (also) modules over fields.

Categories: $\langle \mathcal{O}, \mathcal{M} \rangle$ ($\mathcal{O} \equiv$ objects; $\mathcal{M} \equiv$ morphisms)

$\mathcal{M} := \bigcup_{i<j}^* \{f_{ij} : \mathcal{O}_i \to \mathcal{O}_j\} =: \bigcup_{i<j}^* \mathcal{M}_{ij}$,
with associative compositions (with identity) $c_{ink} : \mathcal{M}_{ij} \times \mathcal{M}_{jk} \to \mathcal{M}_{ik}$.

The 'objects' are generalizations of 'sets' (and include all of the AE/SF), and the 'morphisms' give corresponding generalizations of 'functions between sets'.

Functors: $\mathcal{F}$

Suitable generalisations of 'functions' for the mapping of one category into another.

The formulations of **abstraction, analogy** and **inverse analogy** are developed in E_7 – mainly, through rules for the reduction / deformation of sets of defining relations for (collections of) ME. These procedures may be incorporated into the schemes for classification of GCSF by starting from specifications of the CSF and then applying the rules to appropriate sets of relations – typically represented as predicate functions over various domains. The extremes to which this program may be accomplished with adequate rigour depends strongly on the nature of the ME; and, of the theories into (or from) which reflections are constructed. Nevertheless, the overall **techniques** for generating abstractions / analogues appear to be generally valid – provided only that the ME concerned may be characterized by some set(s) of w.d. predicate functions. This claim is investigated in E_7, where the underlying formalism is explained.

To conclude this Essay, an indication is given of the types of (generalized) AE/SF obtainable from specalization of the basic (generic) AE/SF. Most of the **examples** of these processes may be found in E_9/E_{10}/E_{11}. The selection of items here is designed mainly to illustrate the diversity of AE/SF (derived from the basic ones) that are prominent both within Mathematics and in applications to 'the sciences'. This great variety of spaces in especially important for Σ , where synthesis and unification are primary objectives. For the (algebraic) SF, the dominant variants are realizations of the fundamental frameworks, with sets of extra conditions relevant in calculation – at all levels of abstraction.

The majority of significant AE are (quasi- / semi-)**normed linear** (topological) **spaces**[254] (NL(T)S); or else, **non-normable L(T)S**[255]. Of the NLS, the class of **function spaces** (FS) is ubiquitous in applications – for instance, to **partial differential equations** (PDE), and to **integral equations** (with extensions, in both cases, to more general types of **operator equations**). Recall that **metric spaces** and NLS are closely related: every NLS yields an equivalent MS, with $d(x, y) :=\| x-y \|$; and, every MLS yields an NLS, with $d(x-y, \theta) =:\| x-y \|\equiv\| z \|$, say. On the other hand, metrics may be defined over sets that are **not** LS. Again, every **inner-product space** (I-PS) determines a norm, through: $\| x \|^2= (x, x)$; and hence, also, an equivalent MS. Within this specification lie **Banach Spaces** (BS): **complete** NLS; and; **Hilbert Spaces** (HS): **complete pre-HS** – where a **pre-HS** is a NLS whose norm is determined by an inner ('scalar') product. This scheme covers a vast array of special types of spaces, of which examples are cited now, for (n-)NL(T)S, metric FS, and other classes of spaces:

Sequence spaces – and their generalizations, **FK spaces**

Metrizable TS

Normable TLS

General NLS

General MS

All of these spaces appear in 'abstract' or 'concrete' forms – with cognate versions corresponding to the other basic AE (through the various alternative representations), where this is possible. The classes of Banach Spaces (BS), Hilbert Spaces (HS) and their multifarious realizations as FS – along with the remaining NLS and MS – are diversely exemplified in E_9/E_{10}/E_{11}. Here, it suffices to cite examples of special relevance to the study of operator equations.

Lebesgue Spaces (typically, $L_p(\Omega)$)

Orlicz Spaces (typically, $L_\Phi(\Omega)$)

Riesz Spaces ($\equiv$ **vector lattices**)

Campanato / Morrey Spaces (typically, $L_{p,\lambda}(\Omega)$)

Sobelov / Sobelov-Orlicz Spaces (e.g., $W^k_O L_\Phi(\Omega)$)

Nikol'skii Spaces (typically, $\hat{\gamma}^{k,p}(\Omega)$)

Besov Spaces (typically, $B^{k,p,s}(\Omega)$)

Lizorkin Spaces (typically, $L^{k,p}(\Omega)$)

Triebel Spaces (typically, $F^{k,p,s}(\Omega)$)

[254]Notation: (q-/s-)NL(T)S
[255]Notation: n-NL(T)S

Anisotropic Spaces (several of the above types)

NOTE For MS, one may have $d(x,y) = 0$ with some $x \neq y$ (pseudo-MS); $d(x,y) \neq d(y,x)$ for some x, y (quasi-MS); $d(x,y) \neq d(y,z) < d(x,y)$, for some x, y, z (semi-MS). For NLS, one may have $\| w \| = 0$, with $w \neq \theta$ (semi-NLS). Many variants of the FS just listed correspond to the definition of a semi-metric, or semi-norm, rather than a metric, or norm (as for the 'generic spaces').

With regard to the basic (algebraic) SF, the variants tend to be defined through the imposition of extra structural conditions / criteria. It is enough to cite a few examples for each of the generic SF characterised above[256].

Abelian groups

Nilpotent groups

Finitely presented groups

Symplectic groups

Free semigroups

Holomorphic semigroups

Flat modules

Quasi-coherent modules

Graded rings

Noetherian rings

Fields of formal power-series

Local class fields

Boolean algebras

Hopf algebras

Category of rings

Category of analytic manifolds

Contravariant functors

Universal functors

These types of algebraic SF are cites simply to illustrate the range of specializations of the basic SF. All of these modified SF (probably several hundred[257] in total!) may be located in Σ as single structures – together with the specializations of basic AE – from which CSF, and GCSF, may be constructed. In this way, arbitrarily complicated compound / combined structural frameworks may be introduced to realize / model mathematical / scientific phenomena – a fundamental aim for the Σ IB.

[256] The formal definitions of these specialized SF may be found (for instance) in the references cited in E_2.

[257] A good idea of the total may be obtained from a perusal of the index of the '*Encyclopedic Dictionary of Mathematics*' (MIT Press, 1977; 2nd ed., 1989)

CHAPTER 6
Abstraction and Analogy as Approximation

One of the most distinctive features of the Σ -scheme is the incorporation of abstraction and analogy as possible modes of 'generalized approximation', or, '**approach**', to (essentially arbitrary) ME – '**located** in conceptual space' through some set(s) of **defining relations / conditions**. These notations have already been considered – mainly from other points of view – as 'locational conditions' (LC), in E_5. Here, the principal objective is to develop qualitative - t-max.-quantitative formulations of procedures for 'generating' elements of Ab(e), An(e), -An(e) – the classes of (respectively) **abstractions, analogues**, and **inverse-analogues**, for the ME e. Types of quasi-**reflection** / **refraction** are suggestive in this context – as possible transformations of arbitrary ME/SE[258] (ranging from 'simple' / 'atomic' objects to mathematical / scientific theories). These aspects of analogy are also discussed.

In the following development, M^* represents $\{$mathematical entities$\}$, and S^*, $\{$scientific entities$\}$; s, σ are typical elements of S^*, and, m and μ, of M^*. **Processes** of the forms

(i) $s \to m \to \sigma$;

(ii) $m \to s \to \mu$;

(iii) $m_1 \to \mu \to m_2$;

(iv) $s_1 \to \sigma \to s_2$,

are called **reflections**; while, the processes

(v) $s \to m \to \mu$;

(vi) $m \to s \to \sigma$,

are called **refractions**. More precisely:

(i) σ is a mathematical reflection of s in m;

(ii) μ is a scientific reflection of m in s;

(iii) m_2 is an internal reflection of m_1 in μ;

(iv) s_2 is a scientific reflection of s_1 in σ;

(v) μ is a refraction of s through m;

(vi) σ is a refraction of m through s.

The detailed nature of the ME m_1, m_2, m, μ, and of the SE, s_1, s_2, s, σ, need not be specified at this stage, since all of these processes are schematic. The aim is to propose a general framework within which all of these transformations may be accommodated, through **interactions**, whose precise structures may be defined in each case considered. Another suggestive formalism for these transformations corresponds to **scattering**, of some (collection(s) of) ME/SE by others. Again, this is left vague to cater for a wide range of possibilities; specifications may be made as required.

The classes **(i)–(vi)** may be described alternatively as:

(i)′ Mathematical reflections of SE

(ii)′ Scientific reflections of ME

[258] SE/P stands for scientific entity / phenomena (or, for the plural), typical, α

(iii)′ Mathematical reflections of ME

(iv)′ Scientific reflections of SE

(v)′ Mathematical refractions of SE

(vi)′ Scientific refractions of ME.

In these terms, {mathematical analogues} is a subclass of **(ii)**∪**(iii)**; {scientific analogues}, of **(i)**∪**(iv)**; {inverse-mathematical-analogues}, of **(v)**, and {inverse-scientific analogues}, of **(vi)**. The extra conditions / criteria distinguishing analogues from reflections / refractions are investigated later in this Essay.

This broad distinction between ME and SE (extended to **analogues**, as MA and SA) does not produce mutually exclusive classes. Rather, it serves to identify objects with (respectively) predominantly 'mathematical', or 'scientific', interpretations / characteristic. Mostly, this is an unambiguous prescription; in exceptional cases, objects are designated as lying in {ME}, or {SE}, or both, according to the context involved. Indeed, a generic object may be viewed, **formally**, as an ME, or, an SE – simply by 'embedding it ins a suitable theory'. This claim is justified in diverse examples given in $E_9/E_{10}/E_{11}$.

Since the emphasis for Σ is strongly on **mathematical** properties, the classes **(ii), (iii)**, and **(v)**, are or principal concern – corresponding to **reflections** of ME, and (for **(v)**), **refractions** of SE, with associated **analogues** (from **(ii)/(iii)**) and **inverse analogues** (from **(v)**). These are the **mathematical** analogues / inverse analogues; those associated with classes **(i)/(iv)/(vi)** are labelled as 'scientific', and may be used subsequently to produce 'new' MA or inverse MA; and so on!

The basic notions of reflection, refraction, and scattering, for ME and SE, are purposely schematic; but the subclasses of MA and the inverse MA must be adequately specified to allow (at least) qualitative analyses to be conducted. Before this task is undertaken, however, the class, Ab(e), of **abstractions**, will be discussed. It turns out to be substantially distinct from An(e) – in that Ab(e) involves 'essential conceptual similarity', of e and $e' \in$ Ab(e); whereas, An(e) identifies / compares the 'internal organisation / behaviour' and 'external interaction with environment', of e and $e'' \in$ An(e). The relevant criteria may be based on the LC-representations[259] of e, e', e'' – as introduced in E_5.

Thus: suppose given[260] $e :: l(e) \equiv \underset{\alpha \in S_e}{\wedge} {}^{e}_{E}c_\alpha(x_{\sigma^e_\alpha})_{n_\alpha}$, which may be abbreviated to:

$$e :: l(e) \equiv \underset{\alpha \in S_e}{\wedge} {}^{e}_{\xi}c_\alpha({}^{e}\xi^{(\alpha)}), \text{ where}$$

$${}^{e}\xi^{(\alpha)} := (x_{\sigma^e_\alpha})_{n_\alpha}; \text{ or, even to:}$$

$$e :: l(e) \equiv \underset{\alpha \in S}{\wedge} c_\alpha(\xi^{(\alpha)}),$$

if dependence on e and ξ is suppressed, for S, c_α and ξ. On this basis, one prescription for the formation of $e' \in$ Ab(e), from e, is as follows.

For any set X, put $c^X := \{c_\alpha : \alpha \in X\}$. Then:

(a) Identify the set, say, c^F of **fundamental conditions for** e (from c^S);

(b) discard the conditions in $c^{S/F}$;

(c) postulate (usually, **weaker**) **variants** of the conditions in c^F – to obtain $(v(c))^F$;

(d) define $v(e)$ by: $v(e) :: (v(c))^F$.

Then $v(e)$ is an **abstraction** of e (over $\xi := \underset{\alpha \in F}{\cup} \xi_\alpha$).

---

[259]The (non unique) Locational Conditions 'fix' the ME within their structural environments (say, ξ); notation: $l_\xi(e)$, often abbreviated to $l(e)$.

[260]Recall that this denotes the conjunction of conditions ${}^{e}c_\alpha(x_{\sigma^e_{\alpha 1}}, \ldots, x_{\sigma^e_{\alpha n_\alpha}})$, within environment ξ.

Remarks

(1) Denote by Gen(e), Spec(e), respectively, the sets of generalizations, and, specializations, of e (over ξ). For abstractions $v(e)$, that are **not** generalizations, one has: $e \notin$ Spec($v(e)$), so that: $v(e) \in$ Gen(e) IFF $v(e) \neq e \wedge e \in$ Spec($v(e)$). In this sense, **generalizations** may be regarded as **'reversible abstractions'**. A related set is Sc(e), where Sc(e) $:= \{w : e \models w\}$, the **scope** of e (over ξ).

(2) The selection of $c^F(e)$ from $c^S(e)$, for given e, cannot be prescribed – still less, rendered 'algorithmic'. Rather, each choice of F(CS), and of $v(e)^F$, corresponds to some **level of abstraction** of $e' \in$ Ab(e), say, $\lambda(e'; e)$. Although no 'obvious' definitions of $\lambda(e'; e)$ are apparent, it is mainly the **comparison** of levels, $\lambda(e'_1; e)$ and $\lambda(e'_2; e)$ that seems relevant for Σ . This issue is pursued later.

(3) The 'admissible' types of **variants,** $v(c)$, have not been specified. For the moment, it suffices to leave this problem too open.

(4) In fact, the best plan for tackling these sorts of problems is to provide (context-dependent) 'working definitions'.

The description of **analogues** devolves on the specification of **imitative behaviour** – 'internal' and 'external' – of ME, within certain **'contexts'** ζ, or **environments,** ξ. Full imitation corresponds to faithful **simulation**, or **transcription**. This is of relatively low mathematical interest, though it may be useful in detailed calculation. Of far greater importance for Σ is the study of **partial** simulations of e through 'internal structure' and 'external interactions'. This representation of e, as an 'active element' of the IB, must (for consistency) be equivalent to the 'LC form' just considered for abstractions. Nevertheless, the emphasis is different.

This time, one has the formulation:

$$e :: \bigwedge_{\alpha \in I} {}^e c_\alpha^-(\xi^{(\alpha)}) \bigwedge_{\beta \in O} {}^e c_\beta^+(\eta^{(\beta)}),$$

where the c_α^+ denote 'inner conditions', and the c_β^+, 'outer conditions'. That is, the ${}^e c_\alpha^-$ account for 'the inner structure of e', while the ${}^e c_\beta^+$ determine 'how e interacts with its environment[261] (say, ξ)'. A 'good analogue', $e'' \in$ An(e), must 'closely reproduce the dominant (inner and outer) e-conditions', in terms of entities, in ξ'', that are ('in most contexts') 'far from those of ξ'. Evidently, this is somewhat imprecise, but its essential content seems intuitively clear. More precise formulations will emerge from the diverse examples considered in $E_9/E_{10}/E_{11}$, but the immediate task is to elucidate all of the terms in quotation marks.

(a) Notice, first, that some e are fully determined by internal structural conditions **alone**; this includes all objects axiomatically defined over some basic set, W (for instance, all of the algebraic SF underlying P_Σ). Thus, **inner conditions** are exemplified by laws of composition / combination and their precedence relations.

(b) A broad class of **outer conditions** on e is furnished by **optimality criteria** (relative to frameworks, $\mathcal{J}$), regardless of the nature of the **generic** defining conditions for e (which could be only of 'inner type'). Another natural source involves **representations of non-abelian groups with prescribed commutator relations** – to which the outer conditions correspond.

(c) The identification of **dominant conditions** for e is somewhat problematical since subjective judgements appear to be involved. In practice, however, the subjective element is of comparatively slight importance, because each choice of {dominant conditions} is associated with its own analogues, and, in principle, the choices may cover all possibilities.

[261] **Note that** e forms part of its own environment, etc..

(d) A fuller understanding of inner conditions may be obtained by considering the **interrelations among constituents of** e, where it is assumed that e is somehow 'composed' of various elements whose mutual interactions / interdependencies / dots **characterize** e. Thus one has[262]: $e \approx (\rho)_{m_e}$, for constituents, ρ_j, of essentially arbitrary types – for instance, a collection, $(\rho)_k$, ordered by a relation ρ_{k+1}, and subject to certain 'constraints', $\rho_{k+2}, \ldots, \rho_{k+q}$ (for $m \geqslant k+q$); and so on. Not only the collection $(\rho)_k$, but also, the ordering / constraint relations, must be suitably 'mirrored' in any adequate analogue of e. The analogues of the ρ_k must be ordered (over ξ'') by relations 'similar to' those for the ρ_l (over ξ), etc.. The apparent circularity of this prescription may be dealt with by recognizing the difference between **simple** and **compound** analogues.

Here, simple analogues amount to prescribed correspondences between single elements $\gamma'_j \in \xi'$ and $\gamma''_j \in \xi''$; while, the compound analogues are **constructed** on the basis of the simple ones. Again, for any set, S, of constituents, certain **prescribed** analogues over subsets, V, of S, may be **designated** as V-simple; after which, other analogues **constructed** from V_k-simple analogues (for $k \in I$, say) are regarded as S**-compound**. These, and cognate, notions are diversely illustrated in E_{10}/E_{11}. Meanwhile, the crucial idea is that: analogues correspond to **mappings** (say, i) between sets of conditions / relations[263]. Consequently, the **comparison** of e, defined over ξ, with $e'' \in \mathrm{An}(e)$, defined over e'' – or, of $e''_1, e''_2 \in \mathrm{An}(e)$ – amounts to the comparison of **relations**, ρ_k, ρ''_k (or, of $\rho''_{k_1}, \rho''_{k_2}$) where $\rho''_{k_1}, \rho''_{k_2}$ are possible images of ρ_k under i.

Initially, it should be mentioned that, for any '**envisaged** condition / relation', say, η, there are (basically) only 4 possibilities (over ξ):

(i) η is fully formulable / exactly realizable;

(ii) η is fully formulable / approximately realizable;

(iii) η is partially formulable / exactly realizable;

(iv) η is partially formulable / approximately realizable.

The variants correspond to interpretations of 'approximately', and 'partially', in cases **(ii)/(iii)/(iv)**. The questions of full / partial **formulation** are reflections of the investigator's capabilities, and of the logical nature of the conditions / relations concerned. Problems of exact / approximate **realization** of the proposed conditions, however, should be solvable as 'exercises in technique along' – subject to current limitations, or to relevant 'non-existence results'. All of the forms **(ii)/(iii)/(iv)** contain elements of imprecision; even **(i)** may cease to be exactly realizable when it occurs with other conditions.

It follows that any specified approach to e may, in principle, require some of the other approaches, for its implementation. In other words, exact calculation in M_t is (relatively) extremely rare! This observation – and the fundamental objective for Σ , of maximising calculability in all contexts – may be regarded as distinguishing **formal representation** from **effective determination**. Thus, many representations of ME are logically correct / consistent, yet they offer no possibility for estimation or application. The replacement of such ME by combinations of their admissible approaches may entail **loss of information, permutation of processes** (to be performed in some particular order, for the 'approach-formulation' to make sense), and **careful interpretation** of the (resulting) formulation(s) – especially, for abstractions or analogues.

It is useful to introduce the following sets. Let D/S/G/A/Q/I $\equiv$ **D**etermination / **S**pecialization / **G**eneralization / (Standard-)**A**pproximation / **Q**uintessence / **I**mitation, and put $M \equiv \{$D,S,G, A,Q,I$\}$. Further, let $\mathcal{D}/\mathcal{S}/\mathcal{G}/\mathcal{A}/\mathcal{Q}/\mathcal{I}$ stand for the classes of all w.d. **realizations** of D/S/G/A/Q/I (in appropriate contexts), and put $\mathcal{M} \equiv \{\mathcal{D}, \mathcal{S}, \mathcal{G}, \mathcal{A}, \mathcal{Q}^{\pm}\mathcal{I}\}, \mathbb{A} \equiv \mathcal{D} \cup \mathcal{S} \cup \mathcal{G} \cup \mathcal{A} \cup \mathcal{Q} \cup {}^{\pm}\mathcal{I}$. For $\mathcal{W} \subset \mathcal{M}$, define $\mathcal{W}e$ by: $\mathcal{M}e := \{we \in \mathcal{W}\}$. Then $\mathcal{W}e$ comprises all w.d. $\mathcal{W}$-modifications of e, and $\mathbb{A}e$, all w.d. approaches to e. Next, denote by ${}^{=}c({}^{\neq}c)/{}^{=}\rho({}^{\neq}\rho)$ any w.d. (in)equational

[262]Here, $e \approx (\rho)_{m_e}$ means: e is characterized by $\rho_1, \ldots, \rho_{m_e}$.

[263]i stands for imitation; other forms are introduced later.

condition / relation involving specified collections of ME. Here, the class ${}^{+}\mathcal{I}$ corresponds to $\mathcal{I} \equiv$ 'direct mathematical analogy', while ${}^{-}\mathcal{I}$ corresponds to 'inverse-analogy'. Unless there is a danger of confusion, direct analogy is denoted by ${}^{+}\mathcal{I} \equiv \mathcal{I}$ (and ${}^{+}I \equiv I$). The notation $\overset{(+)}{-}$ An is also used.

The classes: $\mathcal{D}, \mathbb{A}, \mathcal{S}, \mathcal{G}, \mathcal{A}, \mathcal{Q}, \mathcal{I}$, correspond to all admissible **determinations, approaches, specializations, generalizations, approximations, abstractions**, and **analogues** (respectively), over Σ_t. If D_e is the disjoint union of sets L, M, N, T, then the most general characterization of e is given by:

$$e \approx \bigwedge\{\rho_h : h \in D_e\} \equiv \bigwedge_{l \in L} {}^{=}\rho_l^- \bigwedge_{m \in M} {}^{\neq}\rho_m^- \bigwedge_{n \in N} {}^{=}\rho_n^+ \bigwedge_{t \in T} {}^{\neq}\rho_t^+ .$$

When D_e is finite, this representation comprises a finite conjunction of inner / outer (in)eqs.; otherwise, the conjunction is (un)countable. In any case, the formation of analogues involves the **modification** / **deformation** of the (in)eqs. in specified subsets of D_e, (non-)implicatorily; or, the **dropping** of some (in)eq(s); or, the **imposition** of some '**new**' (in)eq(s). All of these possibilities are envisaged in E_7, where a general view of the approximation of mathematical information is developed. Here, this scheme is applied to the ranking of analogues.

Remarks

(1) The underlying assumption (for Σ) is that: there is a 'fundamental set', say, $\mathcal{R}_\Sigma$, comprising all distinct, (in)equationally specified relations, ρ, over countable and uncountable cartesian products of 'spaces' whose elements are regarded – for ρ – as independently well-defined ME. Each of the ${}^{=}\rho_i^{\pm}$, ${}^{\neq}\rho_j^{\pm}$, is therefore itself an element of $\mathcal{R}_\Sigma$.

(2) This is equivalent to the claim that: every ME may be (in)equationally characterized in terms of others (taken as already w.d.), subject only to adequate **consistency criteria**. In other words: 'All of the world's a stage, and one ME, in its time, plays many parts'!

(3) In spite of the apparently ephemeral status of ME, in this 'picture of Mathematics', the existence of a 'background framework', for each definition, imposes order on the potential chaos – by demonstrating that such 'locally defined ME' may be combined to form increasingly inclusive structures, by amalgamating their background frameworks in essentially 'seamless' ways.

(4) The representations of e – as

$$e :: \bigwedge_{\alpha \in S} c_\alpha(\xi^{(\alpha)}),$$ for **abstractions**; or, as

$$e \approx \bigwedge\{\rho_h : h \in D\},$$ for **analogues** –

are all conjunctions of elements of $\mathcal{R}_\Sigma$.

(5) Several (largely equivalent) notations – with varying emphases – are useful in the attempt to analyse abstraction and analogy as modes of 'incomplete determination' of ME. These are elusive concepts, with many possible interpretations – each optimally reflected in some associated representation.

(6) As a further representation of e, one may introduce the notation: $e; [K_e, P_e]$, where K_e comprises the **key** (inner / outer) relations for e, and P_e, **the peripheral** relations – all within some **context(s)**, C, or **framework(s)**, F. Thus, a fuller representation is $e; [K_e, P_e/F]$, or $e; [K_e, P_e/C]$. Here, it must be understood, once and for all, that the nature of the abstraction / analogue is initially characterized by the selection of K_e from the full collection of defining relations for e. Subsequent deformations further influence the precision / scope

/ ... of the resulting variant of e – as do the choices of C or F. Once these basic decisions have been taken, however, to produce 'preliminary abstractions / analogues', say $q_m(e)/i_n(e)$, it is possible to formulate criteria for the comparison of $q_{m_1}(e)$ and $q_{m_2}(e)$; or, of $i_{n_1}(e)$ and $i_{n_2}(e)$; and also of $q_m(e)$ and e; or, of $i_n(e)$ and e.

(7) For any ME, u, with $u \approx \bigwedge\{\rho_h : h \in D_u\}$, defined R_u to be $\{\rho_h : h \in D_u\}$. **Then**: all of the sets R_v, for $v = Se/Ge/Ae/Qe/Ie$, are w.d., along with R_e itself – so their cardinalities may be compared. The analogue $u \in Ie$ is called full if $\underline{\text{card}R_u} \geqslant \underline{\text{card}R_e}$ – and, partial, if $\underline{\text{card}R_u} < \underline{\text{card}R_e}$. Similarly, $u \in Ie$ is called K**-full** if $\underline{\text{card}K_u} \geqslant \underline{\text{card}K_e}$; and, P**-full**, if $\underline{\text{card}P_u} \geqslant \underline{\text{card}P_e}$ – and, K**-partial** / P**-partial**, otherwise.

The analogue, u, may be called m**-partial** when $\underline{\text{card}H_e} - \underline{\text{card}H_u} = m$ (for $H \equiv R/K/P$). The '**quality**' of a partial analogue may be (and often is) 'superior to' that of another fully analogue (of some e). This depends on the relative 'degrees of replication' of key conditions in the full and partial analogues.

(8) The formulation (from e) of $Se/Ge/Ae/Qe/Ie$ exhibits S/G/A/Q/I both, as symbolic **operators** on ME, and, as symbolic mappings: $\mathcal{M}_t \to \mathcal{M}_t$. Each specialisation / generalisation / approximation / abstraction / analogue (i.e., each **approach**) corresponding to e embodies the action of some associated operator(s) / mapping(s). Consequently, it makes sense to consider, also, symbolic **compositions** of the form $(H_1 \circ \cdots \circ H_L)e =: (\circ H)_L e$, where all of the H_j are chosen (possibly, with repetitions) from the set $\mathbb{A}(e)$, of all Σ -admissible approaches to e.

(9) The 'operators', S/G/A/Q/I, cannot be given general representations; rather, they symbolize the operations of **forming** specializations / ... / analogues. Nevertheless, the **actions** of compositions, $(\circ H)_L$, on e still make sense, representing the **formation** of (classes of related) ME through prescribed sequences of operations. Each of the symbols $Ge/\ldots/Ie$ signified a generic element of the class concerned. Moreover, combinations such as $G \circ A)e$, $A \circ I)e$, $Q \circ A)e$, arise regularly in the course of investigations / calculations – both as aids in evaluation, and as means for enlarging scope. More complicated compositions are readily interpretable in comparable ways. In this formalism, I covers all forms of analogues (direct or inverse), and A, all types of (standard) approximation. Limitations of technique and range may be (partially) removed through the skilful use of such compositions, to maximize effective calculability in **all** mathematical domains – perhaps the central aim for Σ .

The criteria for **comparison** of two analogues – say, $i_1(e)$, $i_2(e)$, with one another, and with e itself – are considered now. It is clear that no universally valid, or exhaustive, criteria can be found – since the representations $e; [K_2, P_e]$, based as they are on the selection of 'key' and 'peripheral' characteristics (of e), must be partially subjective. Once e, K_e, P_e have been prescribed, **ordering-relations**, over K_e and P_e, may be imposed, so that each of the K_e, P_e is internally ordered (say, by $\gtrsim$), and $K_e \gtrsim P_e$, where '$k' \gtrsim k'''$' means 'k' is at least as important (for e) as k''' (etc.). Plainly, this process, too, is partially subjective – the components of k', k'',... and, of p', p'',..., must (somehow) be **weighted**, so that differences in importance (for e) can be 'measured' – and used 'to **rank** the elements of $I(e)$'.

Suppose that K_e is **partitioned**, say, as: $K_e =: \bigcup_{1 \leqslant l \leqslant L_e} K_{el}$, where K_{el} comprises those conditions of weight w_i; and, similarly: $P_e =: \bigcup_{1 \leqslant j \leqslant J_e} P_{ej}$. The primary 'analogical representation of e' – corresponding to $e; [K_e, P_e]$ – is (as before):

$$e \approx \bigwedge_{1 \leqslant l \leqslant L} \{\rho \in K_{el}\} \bigwedge_{1 \leqslant j \leqslant J} \{\rho \in P_{ej}\}.$$

To cater for the possible imposition of 'new conditions' in $i(e)$ which are **not** modifications of the basic LC (e), it is useful to insert the formal (tautological) conditions k_{e0}, p_{e0}, which are always satisfied (independently of e), and to put $K_{e0} := \{k_{e0}\}, P_{e0} := \{p_{e0}\}$ – singleton sets. The final representation now becomes:

$$e \approx \bigwedge_{0 \leqslant l \leqslant L} \{\rho \in K_{el}\} \bigwedge_{0 \leqslant j \leqslant J} \{\rho \in P_{ej}\},$$

and one may define $i(K_{e0})$ and $i(P_{e0})$ as follows (with $A \backslash B := A \cap B'$), for sets A, B with complements A', B':

$$i(K_{e0}) := (R_{i(e)} \backslash R_e) \cap K_{i(e)}; \text{ and,}$$

$$i(P_{e0}) := (R_{i(e)} \backslash R_e) \cap P_{i(e)}.$$

In other words, all 'new conditions' are regarded as 'created from the formal e-conditions K_e, P_e'. In the 'other direction', conditions may be 'destroyed' simply by mapping them to tautologies in $R_{i(e)}$. In a full analogue, no conditions are created or destroyed; but assessment of the **quality** of the analogue involved many other considerations.

NOTE Since analogical formation may be **repeated** – taking e first, to $i_1(e)$, and thence, to $i_2(i_1(e))$ – the formal representation of $i(e)$ also includes tautologies in $i(K_e)$ and $i(P_e)$; and so on.

For purposes of comparisons (among admissible analogues of e) a more uniform notation is necessary. Let $K_{el}^{(1)} \equiv K_{el}, P_{ej}^{(1)} \equiv P_{ej}; i(k_{el}^{(m)}) =: k_{el}^{(m+1)}, i(p_{ej}^{(m)}) := p_{ej}^{(m+1)}$, for $m = 1, 2, \ldots$. Then the basic representations become:

$$e \approx \bigwedge_{0 \leqslant l \leqslant L^{(1)}} \bigwedge_r k_{elr}^{(1)} \bigwedge_{0 \leqslant j \leqslant J^{(1)}} \bigwedge_s p_{ejs}^{(1)}; \text{ and}$$

$$i(e) \approx \bigwedge_{0 \leqslant l \leqslant L^{(2)}} \bigwedge_r k_{elr}^{(2)} \bigwedge_{0 \leqslant j \leqslant J^{(2)}} \bigwedge_s p_{ejs}^{(2)}.$$

Formally, one has: $L^{(2)} = L^{(1)}$, since all new conditions 'created' from the tautology in K_{e0} may be viewed as a single (conjunction) conditions; whereas, each 'old condition', from K_e or P_e, is replaced by an 'associated tautology'. It is more useful, however, to keep track of the cardinalities of sets of 'old' / 'new' conditions, so that the forms of analogues may be monitored with adequate accuracy. In most cases of interest, e is representable through a **finite** set of LC; that is, K_e and P_e are finite sets.

For any set S put card$S =: v_S$. In general, v_S covers arbitrary (un)countable sets. For e and $i(e)$, one has: $v_{K_e}, v_{P_e}, v_{K_{i(e)}}, v_{P_{i(e)}}$ – where, for counting purposes, all tautologies are ignored, and every conjunction is 'expanded' to count all of its 'components'. Thus (when v_{R_e} and $v_{R_{i(e)}}$ are finite) the indices

$$d_K(e, i(e)) := v_{K_e} - v_{K_{i(e)}}, \text{ and}$$

$$d_P(e, i(e)) := v_{P_e} - v_{P_{i(e)}},$$

indicate the net changes in numbers of key, and, peripheral, conditions for e and $i(e)$. The same notation is used for $i_1(e), i_2(e)$, giving $d_K(i_1(e), i_2(e))$ and $d_P(i_1(e), i_2(e))$.

Interim Summary

The LC (e) are partitioned into **key** (K_e) and **peripheral** (P_e) subsets of **conditions**, which are **ordered by individual weighting**. The formation of $i(e)$ from e entails the **removal** of some conditions in R_e, the **deformation** of the rest of R_e into (part of) $R_{i(e)}$, and the **introduction** of some new condition(s) into $R_{i(e)}$. The net changes in numbers of conditions of various types may be **monitored** through **indices**. All conditions are formally defined over the (designated) set – say $(\gamma)_{v_e}$ – of **constituents** of e.

The critical choices (so far) refer only to e (identification of K_e; assignment of weights to members of K_e, P_e – with associated **ordering**). Next, the corresponding choices must be made for $i(e)$, with appropriate **comparison criteria** between e and $i(e)$. Let ${}^iK_e$ comprise all elements of $K_e \backslash K_{e0}$ that are **not** removed in forming $i(e)$ from e, and consider the restriction, say, i^*, of i to ${}^iK_e$, $i^* : {}^iK_e \to K_{i(e)}$, $k \mapsto i^*(k)$. The conditions $k, k^*(k)$ are called i**-comparable**. Similarly, p, $i^*(p)$ are i-comparable, for $p \in {}^iP_e$, etc..

The formation of analogues (of e) involves, primarily, combinations of **near-morphisms** (for some e-condition(s)) and of **mappings** (of other e-conditions) with **near-preservation** of the associated 'solution-sets'. The range of possible analogues of e corresponds to the variety of interpretations / realizations of these (transformed) conditions. The representations discussed up to now cover the internal / external properties of e (within prescribed frameworks / contexts, F/C) as conditions / relations, R_e, and mappings of R_e to $R_{i(e)}$. The admissible forms of such 'analogue mappings' are now discussed in more detail.

To produce viable criteria one must impost (minimal) sets of plausible necessary conditions for analogue-formation. Refinements may be achieved by imposing 'higher-order conditions'.

Recall that, for any e, R_e is (partially) ordered – say, by $\succ$ – with induced orders over K_e, P_e, and[264] $K_e \succ P_e$ (by assumption). Corresponding weightings of the k_m and p_n are symbolised by w, so that, in particular, $w(k_m) > w(p_n)$, etc.; which is taken as **equivalent** to $K_e \succ P_e$. The **assignment(s)** of weights over any K_e, P_e are made through 'informed agreement' – and then **recorded** permanently in $\mathcal{P}_\Sigma$, with the associated 'importance-ranking(s)'. The inherent subjectivity is regrettable, but raises no serious problems of principle, provided only that consistency is maintained. In these terms, the basic assumptions have the form[265]:

$$i(e)_I : w(k_e) >> w(P_e)$$

$$i(e)_{II} : i^*(K_e) = K_{i(e)}$$

$$i(e)_{III} : v_{K_{I(e)}} \geqslant 1$$

$$i(e)_{IV} : w(i^*(k_e)) >> w(P_{i(e)}).$$

Comments

(I) This amounts to the claim that e is essentially determined by K_e.

(II) In the formation of $i(e)$ from e, no 'new' key conditions may be created.

(III) Every analogue reproduces some key condition(s).

(IV) It is asserted that $i(e)$ is essentially determined by $K_{i(e)}$.

In general, $v_{K_e} > v_{K_{i(e)}}$, since some condition(s) from K_e may be dropped before the analogue is formed. The integer $d_K(e, i(e)) := v_{K_e} - v_{K_{i(e)}}$ may be called the K_e**-defect** of $i(e)$. The K-full analogues have defect 0, and are 'preferable to' analogues of positive K-defect, unless there are (strong) counterbalancing factors. This is natural, since the main characteristics of 'good analogues' combine to maximize imitation and environmental difference – subject to various criteria involving (at least) weights of the k_{em} and (qualitative) measures of environmental / framework differences. It should be emphasized again that **total** imitation is rarely of significant mathematical interest – since it usually amounts to **transcription**, in which transformation simultaneously changes e into 'another' (equivalent) **representation**, $i(e)$, of e. The most suggestive analogues may not only have positive K-defect, but also, only approximate / partial realization of certain mapped facets of e – where 'approximate' refers, here, to procedures over the 'standard spaces', $_SX$, namely: Γ/T/U/P/C/N, with 'auxiliary spaces', L/μ/S/F, for extra structure / probabilistic interpretations. As in E_1, these symbols stand for **closure / topological / uniform / proximity / contiguity / nearness** spaces; and for **linear / measure / statistical(-metric) / fuzzy** spaces – all, with possible additional imposed structures (as discussed in E_6).

---

[264] i.e. $k_m \succ p_n$, for all relevant m, n.

[265] Where, e.g., $w(H) := \mathcal{S}\{w(h) : h \in H\}, \mathcal{S} \equiv \Sigma$ or $\int$, etc.; and, '$A >> B$' means 'B is negligible compared with A'.

This mixture of partial realisation, standard approximation, analogy, and abstraction (taken in any order, with possible repetitions) corresponds to the operator-composition already referred to as $(\circ H)_L e$, where each of the H_j implements some admissible **approach** (to e). In the resent context, all of the **other** modes of approach are composed with 'pure analogy-formation' – to produce the most general forms of analogue relevant to Mathematics.

It should be clear by now that **abstraction** is viewed as 'approach **via** conceptual form', whereas **analogy** amounts to 'approach **via** behaviour'. The 'discrepancies' between e and $i(e) \in Ie$, somehow indicates the **quality** of the abstraction, $q(e)$, and the analogue, $i(e)$. Accordingly, the next task is to interpret these discrepancies in ways susceptible of estimation, so that different analogues of e may be compared. Since e is specified by sets of internal (c) and external (ρ) conditions, any 'valid analogue' is characterized by associated sets of **near-morphisms** : $c \mapsto i(c)$, and **near-resemblances** $\rho \mapsto i(\rho)$. The central problem for 'ranking' analogues: $e \to i(e)$ now becomes that of **gauging deviations from 'full morphism' / 'complete resemblance'**. Insofar as 'total analogy' amounts to 'equivalence', it appears that analogues, $i(e)$, may fall short of totality in the following ways:

(1) dropping of some internal e-condition(s);

(2) replacement of some morphisms by **quasi, partial, approximate ... morphisms**[266] (in senses to be specified!);

(3) introduction of **approximate realization / evaluation** for near-morphisms);

(4) dropping of some external e-condition(s);

(5) distortion / deformation of some **resemblance correspondence(s)**;

(6) approximate realization / evaluation in **resemblance correspondences**.

Comments

(a) For the 'key representation' of e:

$$e \approx \bigwedge_{1 \leqslant m \leqslant N^-} k_m^- \bigwedge_{1 \leqslant n \leqslant N^+} k_n^+,$$

the internal and external conditions are distinguished, and the overall K-defect is the sum of defects in internal and external conditions under the analogue mapping, i, where

$$i(e) \approx \bigwedge_{1 \leqslant m \leqslant N^-} i(k_m^-) \bigwedge_{1 \leqslant n \leqslant N^+} i(k_n^+).$$

In this process, some of the $i(k_j^\mp)$ are tautologies (i.e., the corresponding $k_j^\mp$ are **dropped**, in forming $i(e)$), in accordance with items **(1)** and **(4)**, above.

(b) Item **(2)** requires that various types of 'incomplete morphisms' be defined. Let e be 'internally defined' over W, with laws of combination $\#_1, \ldots, \#_J$, and let ${}^f M_j$ signify '**morphism under** f **for** $\#_j$: that is,

$$(w_1, w_2) \in W^2 \Rightarrow f(w_1, \#_j w_2) = f(w_1)^f \#_j f(w_2).$$

Further, for $S \subset W$, denote by ${}^f M_j^S$ **the restriction of** ${}^f M_j$ **to** S^2 ($\equiv S \times S$); and, by (respectively) ${}_k\varphi$, the **left**, and, **right, perturbations of** ${}^f M_j$ **by** φ **for** ${}^f \#_k$, where $D_\varphi \equiv$ dom$\varphi \subset W$, and (unless the contrary is stated) $\{w_1, w_2\} \subset$ domφ. Otherwise, φ may be arbitrary, as long as the '${}^f \#_k$-product' is well-defined. This characterises the **basic**

[266] These are **types** of near-morphisms; there may be others.

forms of perturbed morphisms. Somewhat more generally, one may take $D_\varphi \subset \xi$, which allows for some '**interactions**' of e with its environment, ξ, as well as '**internal effects**'. For the formal definition, it is useful to put:

$$_jw_{1,2} := w_1\#_jw_2;\ _jf(w)_{1,2} := f(w_1)^f\#_jf(w_2).$$

Then the perturbed morphism conditions may be written:

$$f(_jw_{1,2}) = \varphi^f\#_{kj}f(w)_{1,2} \text{ (left)};$$

$$f(_jw_{1,2}) = {_j}f(w)_{1,2}{}^f\#_k\varphi \text{ (right)},$$

where $D_\varphi \subset W$ (**basic case**), and $D_\varphi \subset \xi$ (**general case**), since W is 'part of e', and e is located within ξ, etc..

(c) The range of admissible forms of φ plainly depends on the nature of e and ξ, as well as, on f. The laws of combination, $\#_j$, correspond to e: typical examples include the laws for various 'algebraic structures', as already discussed – but there are also other possibilities, some of which are exemplified in $E_9/E_{10}/E_{11}$. Since the LC (e) are simply collections of (in)eqs. – each internal (in)eq. being mapped (in the formation of $i(e)$ by a morphism or near-morphism) – the diversity of analogues of e is considerable. The mapping of external conditions (and of all relevant peripheral conditions, for eventual refinements) adds to this diversity.

(d) Additional generality is obtained if the binary laws, $\#_j$, be replaced / augmented by m-**ary** laws (say, $_m\lambda(w)_m$) for $1 \leqslant m \leqslant m_0$. **Note that** the extended binary combinations $w_1\#_j\ldots\#_jw_q$ may be reduced to the 'pure-binary case' by putting $w_1\#_j\ldots\#_jw_{q-1}\#_jw_q = (w_1\#_j\ldots\#_jw_{q-1})\#_jw_q$ – provided only that $\#_j$ is **associative**. Otherwise, such combinations constitute special classes of q-**ary operations**, etc.. In the present context it is convenient to write $(1,\ldots,H) =: \overline{H}$, for integers $H \geqslant 2$; and then, to put: $_jw_{\bar{H}} = w_1\#_j\ldots\#_jw_H$, and

$$f[w]_{(m)} := (f(w_1),\ldots,f(w_m));$$

$$f[w]_{\{m\}} := \{f(w_1),\ldots,f(w_m)\}.$$

These, and cognate, notations facilitate the representation of intricate interrelationships within Σ . For m-ary laws, $_m\lambda$, the basic morphism condition under f becomes:

$$f(_m\lambda(w)_m) = {}^f_m\lambda(f[w]_{(m)}), m \geqslant 2$$

– which includes 'the binary case, '$m = 2$'.

General Laws of Combination

A broader decomposition of e (and of the environment, ξ) into sub objects, with various laws of combination, is introduced now.

So far, only m-element laws of combination have been considered (over the base-sets of e and $i(e)$). More generally, suppose that e is comprised of sub-objects – say, $e \equiv \{\gamma_u(e) : u \in U_e\}$, where U_e is some (un)countable index set; and, that e is **realized** through collections of conditions involving the $\gamma_u(e)$ and sets of **external** sub objects (say, G_v) of the environment, $\xi(e)$, of e. **Then:** all laws of combination of q internal, and $m - q$ external, sub objects $(q \geqslant 0)$ may be represented as $_mL(\eta)^q_m$, of **internal order** q, and **external order** $m - q$, where

$$\eta_j := \gamma_j(0 \leqslant j \leqslant q); G_{j-q}(q+1 \leqslant j \leqslant m).$$

Here, the γ_j may be elements of W, and the G_{j-q}, elements of $\xi\backslash\{e\} =: \xi'$ – but all other possibilities are also covered. In particular, the class of **external** e-**conditions** corresponds to

$\{{}_mL(\eta)^q_m : m \geqslant 2, m-1 \geqslant q \geqslant 1\}$; and, the internal e conditions, to $\{{}_mL(\eta)^m_m : m \geqslant 2\}$. The total numbers of internal / external / 'mixed' laws are denoted by $N^-_L/N^+_L/N^\mp_L$, and $N_L(e,\xi) \equiv N_L := N^-_L + N^+_L + N^\mp_L$.

The associated morphism conditions for the laws ${}_mL$ take the form (under a mapping f):

$$f({}_mL(\eta)^q_m) = {}^f_mL(f[\eta]^q_{(m)}) = {}^f_mL[\bar\eta]^q_{(m)},$$

where ${}^f_mL$ is the image of ${}_mL$ under f, and $\bar\eta_j := f(\eta_j)$. The cases of purely internal / external laws correspond to $q = m/q = 0$, respectively. Here, the individual 'points' of W, or W', are regarded as subobjects of e, or of ξ', and the other sub objects, or **constituents**, of e, or of ξ, must be specified in any context / calculation. **Properties of individual constituents**, say, γ_u, **of** e **may be included, for** $m = 1$.

The perturbed morphisms, for the ${}_mL$, under f, now become (for $m \geqslant 1$):

$$f({}_mL(\eta)^q_m) = \varphi * {}^f_mL(f[\bar\eta]^q_{(m)}) \ldots \textbf{(left)},$$

$$f({}_mL(\eta)^q_m) = {}^f_mL(f[\bar\eta]^q_{(m)}) * \varphi \ldots \textbf{(right)},$$

where the class of admissible φ is, as before, dependent (at least) on e, ξ, and f. Here, $*$ denotes any w.d. law of combination for φ and ${}^f_mL(f[\eta]^q_{(m)})$. On this basis, **the overall analogical structure associated with** e **in** ξ depends on:

(i) the **decompositions** of e, ξ' into constituents.

(ii) the **laws of combination** over e, ξ', ξ.

(iii) the range of admissible **perturbations** of 'morphisms' for the laws in **(ii)**.

(iv) mutually consistent **variations** in **(i), (ii), (iii)**.

The laws ${}_mL$ may be of any (integer) orders. Extensions to allow for (un)countable combinations – say, ${}_{v_T}L(\eta)_S$, with $v_X :=$ card$X$, and $S \subset U$ – may be defined along the same lines.

This very general scheme for analysing partial, quasi, approximate... morphisms (corresponding to domain-restriction / perturbation / approximate-evaluation / ...) yields criteria for the comparison of elements of $An(e)$ – with each other, and with e itself, to furnish interpretations of '**analogical approach to** e' (to be combined with (standard-)**approximative** and **abstract approaches to** e). The totality of approaches to e, namely, $\mathbb{A}e$, covers all of these modes of 'incomplete representation'. The representations adopted here may be characterized as: **Morphism-anatomy of analogy**. The imprecisions due to domain-restriction / perturbation / incomplete calculation broaden the spectrum of admissible analogues to include any examples of great significance for mathematical development; see, especially, E_{11}. In particular, when order structures are w.d. over certain constituents (of e, ξ'), 'the divergence from exact morphism' may be estimated – as, also, in some other cases, discussed later. The treatment of **peripheral** conditions is another aspect of the process of analogy-formation that remains to be clarified within the framework now established.

The following, mainly qualitative, notions refer to facets **(i)–(iv)**, above, for analogical structures. The idea is that **an arbitrary analogue is always improved** (unless it has reached the 'optimal form', of **transcription**) by effecting certain operations:

(a) **refinement** of the **decomposition** of e;

(b) **reduction** of **domain restriction** (for the ${}^f_mL$);

(c) **increase** in $N^-_L/N^+_L/N^\mp_L$;

(d) **increase** in the '**overall difference**' between ξ and ${}^i\xi$;

(e) **increase** in the **scope** of ${}^i\xi, i(e)$;

(f) weakening of **perturbations** of the ${}_mL$.

Of these criteria, **(a), (b), (c)** are quantitative-in-principle; **(d), (e)** are primarily qualitative, and **(f)** may be quantified for certain classes of $\{{}^i\xi, {}^i_mL, \varphi\}$. It is assumed that the operations are (mutually) independent and commutative; and that all other (admissible) analogue-improving operations satisfy the same conditions. Plainly, these assumptions are by no means obviously valid in all cases; rather, they must be 'justified in broadly typical situations'.

Some **complementary analogue criteria** may be based on those $LC(e)$ which are **not equivalent to laws of combination**. Denote such conditions by $\tilde{c}, \tilde{\rho}, \tilde{k}, \tilde{p} \ldots$; and, the 'law-based conditions', by $c, \rho, k, p, \ldots$ (as before). Then, for any $e \equiv \{\gamma_u(e) : u \in U_e\}$, one has:

$$e \approx \bigwedge_j k_j(\gamma)_{U_e} \bigwedge_m \tilde{k}_m(\gamma)_{U_e},$$

and the effect of f on any $\tilde{k}_m$ is given by:

$$f(\tilde{k}_m(\gamma)_{U_e}) = {}^f\hat{\pi} * {}^f\tilde{k}(f[\gamma]_{(U_e)}),$$

where ${}^f\hat{\pi}*$ represents (application of) the relevant perturbation(s). The decomposition of $f(e)$ becomes $f[\gamma(e)]_{\{U_e\}}$, and some of the $f(\gamma_U(e))$ may be 'formal constants' – equivalently:

$$\mathrm{dom} f \cap \{\gamma\}_{U_e} \subsetneq \{\gamma\}_{U_e},$$

because certain of the γ_u are 'dropped' in the mapping f. All of this is 'intuitively clear', but awkward to formalize!

The main result is that: **analogues of** e **are formed through 'vector mappings',** f**, producing near-morphisms of the inherent laws of combination, and (controllable) perturbations of the remaining (key)** $LC(e)$. Any **peripheral condition** retained in f may correspond to **refinements** of the analogues.[267] **NOTE that**: conditions of order 1 on individual constituents of e (or, of ξ) may be incorporated in the general scheme through 'laws' ${}_1L(\gamma)$; i.e., ${}_mL$, for $m = 1$.

From this account of the underlying processes in analogue-formation, one may formulate preliminary criteria for **the approach of** $i_1(e)$ **to** $i_2(e)$; and, **of** $i(e)$ **to** e – where i denotes the (schematic) 'imitation mapping' through which the behaviour of e is mirrored by that of $i(e)$, etc.. The conditions are given for generic mappings, f, viewed as 'projections of i' onto various 'subspaces'. These ideas may be made precise in particular cases; but their essential content should be apparent. The most elusive characteristic of analogues is the strong **context-dependence** of their '**significance**' – since crucial modes of behaviour ay be transferred between otherwise-wildly divergent environments ξ, and ${}^i\xi$. This notion of **environmental divergence** is evidently of basic importance for 'analogical analysis'.

If **transcriptions** of e – formed through transformations of $(\gamma)_{U_e}$ and sets of reversible mappings / morphisms – are viewed as **analogically equivalent to** e, then it appears that the (mathematical) '**importance**' of $i(e)$, say, imp$(i(e))$, **decreases**, as $i(e)$ approaches e. However, it is also evident that imp$(i(e))$ decreases as $i(e)$ becomes 'analogically uncorrelated with e'! Consequently, **optimal analogues** must lie somewhere between these extremes. Moreover, in a given **context**, a single (near-)morphism / (near-)faithful mapping, for a **quintessential property**, may outweigh all other considerations; and so on! Consequently, any grading scheme for analogues (of e) must take account of the relative **weights** of the conditions in $LC(e)$ – say, $w(k_m) =: w'_m; w(p_n) =: w''_n$.

From these considerations it is clear that analogical approaches to e are of primarily **qualitative** nature; and, that 'optimal analogues' are 'typically not close to transcriptions'. In spite of the (seemingly inescapable) imprecision of these statements, analogical analysis plays a crucial part in the development of mathematical concepts, and the subsequent construction of

[267] All laws of combination constitute **key** conditions. The mapping is denoted here by f, but the usual notation (for Σ) is 'i' ($\equiv$ 'imitation'), as above, and as in most other contexts.

fruitful theories. Indeed, the organization of the IB – for instance, into **environments for investigation** (see, especially, E_8) – requires a systematic classification of $\{An(e) : e \in \mathcal{D}\}$, say, where $\mathcal{D}$ denotes some specified 'mathematical domain' (e.g., conformal mapping). Further, the '$LC(e)$-representations' exhibit the essential properties of e, in various contexts / frameworks. It appears that the most suggestive **model of 'analogical approach'** should be framed in terms of **projections onto 'subspaces' associated with particular properties**. In **near-transcriptions**, all of the projections correspond to **near-morphisms / near-faithful mappings**. '**Goo analogues** of e' (almost) reproduce the seminal characteristics of e (while 'ignoring others'); and, '**adequate analogues** of e' (almost) reproduce **some** (but not all) of the seminal characteristics of e. **Some** peripheral characteristics of e may appear in good / adequate analogues – as refinements of the key properties involved. Moreover, good analogues **conserve** basic behaviour, while **enlarging scope** substantially; whereas, transcriptions **translate** full behaviour to formal distinct environments, **without** enlarging scope. Optimal analogues achieve 'an ideal balance' between these goals, as symbolised (above) in imp($i(e)$), whose precise **form** (as a function of 'behaviour', 'scope', ...) must be selected to harmonise with other 'approach indicators'.

Let $S \subset LC(e)$ and define $\Omega_S(\subset An(e))$ by $i(e) \in \Omega_S \Leftrightarrow [c \in S \Rightarrow i(c) \in LC(i(e))]$. Thus Ω_S comprises those analogues of e obtained through nontrivial mappings of the conditions in S – say, S-**comparable analogues** (of e). It should be realized that, in general, if $\Gamma_S := \{i(e) \in An(e) : S \subset LC(e)\}$, then $\Omega_S \underset{\neq}{\subset} \Gamma_S$, since some condition(s) from S may be omitted under the mappings in Γ_S, but (by definition) not for Ω_S. Further, one has: $An(e) = \bigcup_{S \subset LC(e)} \Gamma_S$, where it is assumed (for Ω_S and Γ_S) that the environments $\xi, {}^i\xi$, are specified as part of the $LC(e)$, $LC(i(e))$, respectively. Notice, also, that for every $S \subset LC(e)$ there is a minimal $T_S \equiv T \subset LC(e)$ such that $\Gamma_S \backslash \Omega_S \subset \Omega_T$, and, that (in principle) **direct comparisons of analogues may be made within any** Ω_S – through criteria already adumbrated, but not yet made precise.

Each of these comparisons concerns pairs of transformations of the same condition(s) in $LC(e)$ – either, pairs of (near-)morphisms; or else, pairs of more general mappings, intended 'to conserve certain aspects of e'.

It may appear that the apparatus introduced, to describe / monitor / compare analogues of ME specified by sets of LC, is over-elaborate. Further consideration may show, however, that general analogical analysis raises issues of great variety / complexity / subtlety – which must be confronted squarely for the effective design / operation of Σ . No reduced scheme can encompass the multifarious interrelations typical of substantial mathematics. Finally, it should be noted that the processes of **(inverse) analogical approach to** e display some parallels to those of **asymptotic convergence** – if viewed from the standpoint of 'mathematical importance' – since imp($i(e)$) is maximal (in some sense) between the extremes of 'transcription' and 'analogical independence'. For the moment (at least), this analogy will not be pursued!

Inverse Analogy

The formation of analogues of ME, e, form $LC(e)$, over ξ, has been treated in terms of laws of combination / (near-)morphisms / other property-mappings. In spite of some imprecision, a coherent theory seems to be attainable – as the material in E_{10}/E_{11} shows. By contrast, the study of **inverse**-analogues of SE/SP is fraught with difficulties. Nevertheless, the seminal importance of such processes in the creation of 'rich' mathematical theories makes even a tentative treatment of these problems worthwhile, here.

The SE/SP may be of several types, e.g.:

(i) '**constituents**' of established scientific theories;

(ii) '**parameters**' derived from various theories;

(iii) **relations** among constituents of theories;

(iv) links among various theories.

Comments

(i)′ By **constituents of theories** one means: observational / experimental **motivation**, basic **terminology**, and possible **formulations / models**.

(ii)′ Any 'skeleton scientific theory' may be extended to allow explicit calculation (broadly interpreted), and further experimental investigation. These processes generate **characteristic parameters** (taken here to include certain function(al)s, as well as 'constants') whose definitions derivations / determinations constitute SP.

(iii)′ The **fundamental laws** of any theory are (eventually) formulated as (sets of) ineqs interconnecting various 'basic constituents' / parameters. Other mathematical relations may be derived / conjectured.

(iv)′ **Intertheoretic links** are, mostly, forged from scientific analogies. For example, connections between Elementary-Particle Theories and Cosmology have been suggested in this way.

The basic task of **inverse-analogue-formation** requires that key aspects of the associated SP be **(quasi-)realized**[268] over **mathematical** structures – possibly, of novel types. For instance, several types of **algebras** are suggested by processes in **genetics**[269]. This theory – covering both non associative and non commutative algebras, generated innovative constructions / procedures which could have been overlooked in a purely mathematical development. In this case, nonstandard forms of binary combination correspond to genetic behaviour, and the algebras mirror these phenomena. More often, however, the latent mathematical content is obscured by phenomenological-data / imprecise-modelling, so it is necessary to extract the scientific content as a preliminary to possible quasi-realization: a combination of **abstraction** and **analogy**.

The **formulation of procedures** for constructing inverse-analogues of SE/SP is problematical, since the primary connection is founded on stimulus / suggestion / motivation of 'the creative imagination' – rather than on specifiable transformations / interrelations, however imprecise (as for **direct** A). The basic task is, therefore, to establish aggregations of seminal SE/SP from theories in 'the sciences' (of all types), each item (say, α) in any collection (say, A) typifying some key concept(s) / relation(s) / ... in the class of theories concerned. This parallels the LC-representation of ME; but, the setting is more informal / qualitative, since one is dealing here with observed / conjectured / empirical / ... phenomena (rarely susceptible of fully rigorous treatment), rather than with axiomatically-defined objects. Hence, one may refer to this as a **L**ocational **D**escription of α in A: say, $LD(\alpha; A)$, as compared with $LC(e; \xi)$ for ME.

The schematic representation is as follows. Every theory, $\mathcal{J}$, has descriptive **facets**, $\varphi_m(\mathcal{J})$ – say, $\mathcal{J} \approx \{\varphi_m(\mathcal{J}) : 1 \leqslant m \leqslant M_{\mathcal{J}}\}$. The φ_m may be of motivational / observational / experimental / empirical / model-based / conjectural / ... types (which are largely complementary, subject to mutual consistency). The SE/SP may pertain to any collection of $\varphi_m(\mathcal{J})$. Then, typically, one has:

$$A(\mathcal{J}) \equiv \{\alpha\}_{I_{\mathcal{J}}}, \mathcal{J} \approx \{\varphi_m(\mathcal{J}) : 1 \leqslant m \leqslant M_{\mathcal{J}}\};$$

$$\alpha \approx LD(\alpha; A); \alpha \equiv (\gamma(\alpha))_V;$$

$$\alpha :: \bigwedge_{j \in J} d_j(\gamma)_V.$$

[268]Full realization yields mathematical **models** of SP; the **quasi**-realizations correspond only weakly (if at all) with SP, but their underlying **forms** are strongly **motivated** by the SP. Thereafter, the mathematical structures evolve autonomously.

[269]A systematic account is given in LN in Biomath., #36, by A.Würz-Busekros (Springer, 1980)

Here, the $\gamma_v(\alpha)$ are **constituents** of α over $\mathcal{J}$ (strictly, $\gamma_v(\alpha;\mathcal{J})$), and the $d_j(\gamma)_V$ are the **descriptors** of α, over $\mathcal{J}$. In general, **several** theories, $\mathcal{J}_h$, of various types, may be involved – with descriptive facets $\varphi^h_m(\mathcal{J}_h)$ – and then the representations become:

$$\alpha \approx LD(\alpha; \bigcap\{A_q : q \in S \subset \bar{H}\}) \equiv LD(\alpha);$$

$$\alpha :: \bigwedge_{j \in J} d_j(\eta)_V,$$

where, now, the $\eta_v(\alpha)$ are constituents of α over $\{\mathcal{J}\}_{barH}$, and (as usual) $\bar{H} \equiv \{1,\ldots,H\}$. In these terms, one has:

$$-An(e) = \bigcup_{\substack{h \in \bar{H} \\ j \in J}} \{I^{-1}(\alpha) : \alpha \in A_j(\mathcal{J}_h)\},$$

with $I^{-1}(\alpha)$ denoting a generic inverse analogue of α – as compared with $I(e)$, over $I(\xi)$, for direct analogues of e, over ξ. The principal difference between the 'direct' and 'inverse' schemes is that, whereas e is (essentially) fully determined by $LC(e)$, the SE/SP, α, is (typically) at best, **qualitatively** – and often, only vaguely, determined through the $LD(\alpha)$. This partial indeterminacy, however, does not raise problems of principle, since the primary role of the $LD(\alpha)$ is to stimulate the invention / development of novel mathematical structures, rather than to facilitate specific calculations. It is still necessary to have a systematic representation – so that the particular sets of scientific facets 'generating' certain types of inverse analogue may be monitored within the IB. The next task is **to propose modes of generation of $I^{-1}(\alpha)$ from α.**

The underlying idea is that $I^{-1}(\alpha)$ evolves through a combination of **(i) near-morphisms** of precisely identifiable facets of α; **(ii) quasi-morphisms** – reflecting certain other facets / properties of α, without fully-defined mappings from $LD(\alpha)$; and, **(iii) aesthetic / poetic reactions** to the $LD(\alpha)$! These processes give rise to **purely mathematical developments** – of the 'new-structures' thus suggested by α.

This deceptively simple specification is, still, broadly accurate! The mappings of type **(i)** are exemplified by the laws of combination in genetics, and their associated algebras (as already cited[270]). Links of type **(ii)** may be found (for instance) in the theory of **hysterons**[271] – based originally on magnetic hysteresis phenomena. A good illustration of type-**(iii)** structures (in spite of the technicalities in detailed proofs) is furnished by Shelah's work **structure theories**[272], where the very **issues** investigated have been (at least partially) **conjured** out of puzzlement over the apparent 'resistance of certain classes of structures to efficient organization'. The most remarkable inverse analogues are 'mixtures of types **(ii)/(iii)**'; but those partially – even entirely – of type **(i)** may also be notable, as well as of technical importance (e.g., the '**genetic algebras**'). Many **examples** (mainly, of types **(i)/(ii)**) are encountered in $E_9/E_{10}/E_{11}$, where appropriate references are given.

The specification of '**inverse-analogical approaches to** e' follows the same pattern as that for direct analogical approaches – except that the $LD(\alpha)$ **need** not (and, typically, **do** not) determine / characterize α **completely**; so that, subsequent mappings – based on, or motivated by, the $LD(\alpha)$ – will establish only relative **weak** forms of 'analogical convergence'. Apart from this distinction, the frameworks for direct, and inverse, analogy are similar – and may be deployed together, without incurring inconsistencies.

---

[270] See: LN Biomath, #36, by A. Würz-Busekros

[271] See: Krasnsel'skii, MA, and AV Pokrovskii, '*Systems with Hysteresis*' (Springer, 1988)

[272] See: Shelah, S, '*Classification Theory and the Number of Nonisomorphic Models*' (North-Holland, 1978; revised ed., 1990); Hodges, W, '*What is a Structure Theory*', Bull. LMS, 19 (1987) 209–237

CHAPTER 7
Approaches to Mathematical Information

This Essay – together with E$_6$ – may be regarded as absolutely central to the 'conceptual' and 'practical' development of Σ . The notions of approximation correspond (for any item, e) to **incomplete specification**, or else, to (classes of) **modifications**. These, comparatively straightforward, prescriptions nevertheless appear to cover all cases of significance – including all 'standard approximation procedures' **and** the 'qualitative variations' associated with abstraction, analogy and inverse analogy. This set of criteria governs all admissible **approaches** to e (within associated structural frameworks).

The basic idea is that every ME, e, is determined by (its own set of) **predicate functions** (of appropriate orders / types); and, that the modes of approach are realized by combinations of imprecision in specification, or of modification, of these predicate functions.

NOTE that, although the 'apparatus' exhibited here formally covers the basic spaces, $_SX$, specified in E$_1$, the treatment of these aspects is deferred to E$_6$.

In E$_6$ the **standard modes of approximations**[273] are treated (and interrelated) in detail – with copious illustrations showing how they correspond to the scheme introduced here (as 'special cases'). **The predicate-based representation accommodates both the qualitative and the quantitative descriptions**, and so forms an adequate foundation for the investigation of **conceptual vicinities** – which embody the essence of Σ as an (O)IB. Thus, let e be **located** by some **conjunction of conditions** – say, as[274] $e :: \bigwedge_{\alpha \in S_e} c_\alpha(e)$, where S_e may be finite or (un)countable, and the $c_\alpha(e)$ are n_α**-place predicate functions** involving the ('known', or 'assumed') objects in terms of which e is defined[275]. Observe that e may be located in diverse ways – corresponding to contexts / frameworks / .... There is, initially, no question of **uniqueness** – or, even, of **minimality** – of the set $C_{S_e} := \{c_\alpha(e) : \alpha \in S_e\}$. Moreover, the conditions $c_\alpha(e)$ may have any of the forms mentioned in E$_3$, in the discussion of types of definitions – **nominal, explanatory, explicit, implicit, axiomatic, ...** For the present considerations, it is supposed that e is to be located within a prescribed GSF (say, $\mathcal{J}$), and that e is **characterised**, by the $c_\alpha(e)$, over $\mathcal{J}$, though S_e need not be 'minimal'. That is, the form and 'number' of locational conditions[276] may vary with $\mathcal{J}$; but this $\mathcal{J}$-dependence will be exhibited only when comparisons or other operations involving collections of GSF are in question[277]. Otherwise, the framework adopted will be clear from the context. On the other hand, all of the c_α, and S, are obviously dependent on e. If a family, $\{e_\lambda\}$, of ME is studied, then the LC may be indexed by λ; and so on. It is useful to write: $l(e) = (c(e))_S$, for a typical set of LC on e; and, to call $\{e' :\models l(e)\} \equiv \text{inst}(e)$ the **(set of) instances of** e.

A fuller form of $l(e)$ is given by $e :: \bigwedge_{\alpha \in S_e} c_\alpha(e; (x_{\sigma^e_\alpha})_{n_\alpha})$; or, more succinctly, by[278]

$$e :: \bigwedge_{\alpha \in S_e} {}^e c_\alpha(x_{\sigma^e_\alpha})_{n_\alpha},$$

where the σ^e_α select appropriate elements from U_t, and e occurs – explicitly or implicitly – in **some** (but not necessarily **all**) of the ${}^e c_\alpha$. On this basis, it appears that e may be **perturbed** in any combination of the following ways[279]:

(a) Some of the ${}^e c_\alpha$ may be (partially) **suppressed**;

(b) Some of the ${}^e c_\alpha$ may be **strengthened**;

---

[273]Corresponding to the spaces, $_SX$, of E$_1$

[274]This notation is used from now on – with S in place of S_e, unless ambiguities arise

[275]The simplest representation is as $e :: \bigwedge_{\alpha \in S_e} c_\alpha(e; \xi^e)$, where $\xi^e_\alpha := (x_{\alpha^e_\alpha})_{n^e_\alpha}$

[276]LC$\equiv$ LC(e). In typical cases, S_e is **finite**; but it may, in principle, be **(un)countable**

[277]Mostly, it is assumed that the set $\{c_\alpha\}$ comprises **mutually non-implicatory conditions**. This form may be obtained by selecting from any implicatory chain only the strongest condition(s)

[278]The necessity of identifying all elements of U_t in any LC requires this (or similar) cumbersome notation. Some indices may be suppressed in many contexts, but the full gamut must appear when manipulations of OT are involved

[279]Here, 'some' means 'at least one'

(c) Some of the ${}^e c_\alpha$ may be **weakened**;

(d) Some of the ${}^e c_\alpha$ may be **'almost satisfied'** (with many possible interpretations of 'almost');

(e) Some of the ${}^e c_\alpha$ may be **modified** in **non-implicatory** ways (e.g., through abstraction, analogy, or inverse analogy);

(f) Some extra condition(s) may be **added** to the ${}^e c_\alpha$.

In all cases, one refers to the **deformations** of the ${}^e c_\alpha$. Options (a), (b), (c), (f) are self-explanatory (with ${}^e c_\alpha{}'$ taken to be **weaker** / **stronger** than ${}^e c_\alpha$ IFF ${}^e c_\alpha \Rightarrow {}^e c_\alpha{}' / {}^e c_\alpha{}' \Rightarrow {}^e c_\alpha$). The scope of (d) covers virtually all calculational aspects of Σ (but it is still amenable to investigation!) Lastly, the deformations in (e) are those which may be regarded as 'irreversible' (in certain senses, to be discussed in E_7). The modified forms of e (under perturbations) are called **variants** of e. **Notice that** $l(e)$ specifies objects through sets of characterising relations (within some GSF, $\mathcal{J}$). This is **not** a form of quantitative evaluation / estimation for the 'separation' of objects of the **same type**; rather, it is concerned with monitoring **changes** in species / type under perturbations of LC. In this sense, the 'spaces ${}_S X$, of E_1 are obtained – from any 'initial space' – through a succession of changes in the defining relations. This is demonstrated in E_6, for the spaces[280] L/T/U/P/C/N/S/μ ($\equiv$ **L**inear / **T**opological / **U**niform / **P**roximity / **C**ontiguity / **N**earness / **S**tatistical(-Metric) / **M**easure Spaces). A simple illustration is furnished by the transition from a metric space to a **pseudo**metric space (when the condition: $d(p,q) = 0 \Leftrightarrow p = q$ is weakened).

In this general setting it is useful to introduce the **locational manifold** Λ_t as:

$$\Lambda_t \equiv \{l(e) : e \in \xi, l \in \mathcal{L}(e)\}.$$

Here, ξ denotes the set of all ME in M_t, and $\mathcal{L}(e)$, the set of all admissible LC for e (recall that every e may have many LC, etc.). Then Λ_t is one realisation of the set (say, ζ_t) of all w.d. **concepts** over M_t, and one may formulate criteria / problems of **stability** / **robustness** / **reversibility** / **...**, under (classes of) perturbations of LC. Some of these notions are already used[281] (implicitly) in parts of **Universal Algebra, Metamathematics**, and **Theoretical Computer Science**; here, however, they are applied in the study of (essentially arbitrary) mathematical procedures. The inclusion of inverse analogy among the allowable approaches to e (in relation to the above – and other – 'physical criteria') raises many fascinating issues – some of which are explored (mainly through examples) in E_{10}/E_{11}.

The important variations in the c_α – producing perturbations of e – are primarily **semantic**, since each of the $x_{\sigma^e_{aj}}$ is an element of U_t, the set of 'basic objects' in $\Sigma_{\ t}$. For instance, one could have: $x_{\sigma^e_{\alpha 1}} \equiv$ differential operator; .... Then, perturbations of e may be 'generated' by varying the **types** 'component ME' – for instance, to: **tri-diagonal** matrix; **strong** Markov process; **linear** OD operator; ..., with ranges of (other) possibilities in each case, etc.. The corresponding ranges of the matrix **elements**, Markov-process **parameters**, and OD-operator **coefficients**, etc., play comparatively minor roles in determining the **types** of these components – and so, of e itself. Consequently, subtly problems are encountered in attempts to define '**neighbourhoods** of e' – here referred to as conceptual vicinities of e (say, $V(e)$).

It seems appropriate to call the 'points' of $V(e)$ variants of e (as above), which may be'close to e', or (progressively) 'far from e', according to criteria to be formulated – amongst which are criteria governing abstraction, analogy and inverse analogy. For the moment, it is worthwhile to discuss the possible 'elementary perturbations', (a)-(e), in more detail, in **Comments**, C_a-C_e.

C$_a$: A condition c_k may be **ignored** (and so formally omitted from the set of LC, $l(e)$), to give $l(e) \backslash \{c_k\}$. Less drastically, c_k may be **imposed 'selectively'** (e.g., only when specified forms of (some) other conditions, $c)k$, hold).

[280] See E_1 for motivation. The notation μS for **measure**-space is used because MS stands for **metric**-space

[281] For instance, where 'theories' are treated as deductive systems (over axiom-sets); especially, in **Model Theory**.

C_b/C_c: These possibilities should be considered together, as they are, in a sense, mutually inverse (when applied to the same c_k). It is useful to consider **descending / ascending chains** of conditions, say, $C_\alpha)^{\mp m}$, where

$$c_\alpha^{\mp 1} \equiv c_\alpha \text{ and, for } m \geqslant 2,$$

$$c_\alpha^{-1} \Leftarrow c_\alpha^{-2} \cdots \Rightarrow c_\alpha^{-m} \text{ (descending to } c_\alpha\text{)},$$

$$c_\alpha^{1} \Rightarrow c_\alpha^{2} \cdots \Rightarrow c_\alpha^{m} \text{ (ascending to } c_\alpha\text{)}.$$

In this way one may consider the approaches $e^{(\pm \underline{m}_{S_e})}$ to e, defined by:

$$e^{(\pm \underline{m}_{S_e})} := \bigwedge_{\alpha \in S_e} {}^e c_\alpha^{(\pm m_\alpha)} \left(x_{\sigma_\alpha^e}\right)_{n_\alpha},$$

where $\underline{m}_{S_e} := \{m_\alpha : \alpha \in S_e\}$, and each of the m_α takes (arbitrary) positive integral values – so that, formally, at least, $e^{(\pm \underline{m}_S)}$ approaches e as $\underline{m}$ approaches $\mathbb{I}_S \equiv \{m_\alpha \equiv 1 : \alpha \in S\}$. The possible choices of the ${}^e c_\alpha^{(\pm m_\alpha)}$ are, in principle, unrestricted – provided only that, collectively, they constitute ascending / descending implicatory chains with initial / final elements given by the ${}^e c_\alpha (\equiv {}^e c_\alpha^{(\pm 1)})$.

In spite of this apparent generality, the question arises as to whether **every** representation of this form fully determines some (unique) ME. Again, if e is specified (somehow), to what extent may the ${}^e c_\alpha$ be determined? These questions require careful consideration.

C_d: Here, there is an enormous range of possible interpretations, most of which are considered in E_{10}/E_{11}. Indeed, the conditions c_α may (in principle) involve almost all ME – and hence, almost all forms of approaches. Nevertheless, some interpretations of 'almost satisfied' are so pervasive that it is worthwhile to list them now. This is done through the associated notions of approximation. Although these characterisations are by no means rigorous or exhaustive (more complete treatments are given in later Essays), they do encapsulate the criteria involved. The basic idea is that a condition is 'almost satisfied' whenever some part(s) of it are '**approximately realized**'. The criteria are listed as:
Approximation **in / with / by / through / outside / . . .**

(i) metric / norm

(ii) mean-square

(iii) measure

(iv) probability

(v) stochastic conditions

(vi) information

(vii) Baire category

(viii) Level of property p

(ix) partial preservation of information

(x) minimal response to input error

(xi) analogy / inverse analogy / abstraction

(xii) 'essential behaviour'

(xiii) dominant properties

(xiv) asymptotic forms

(xv) imposed ('unobvious') structures

(xvi) simplicial constructions

(xvii) 'simplification'

(xviii) (outside) 'small domains'

These, and other, criteria are elaborated, and diversely exemplified, in E_{10}/E_{11}. Item **(xi)** is covered in E_7; it is included under 'approximation' as an extension of the standard modes. Some of the other items may also yield extensions. The listed types have very wide application, according to the interpretations of the general properties involved. Some forms are of 'standard type', while, others require novel constructions. The relation of these variations of the conditions c_α to those of types **(b)/(c)** (above) is far from obvious; plainly, they are not mutually exclusive, but examination of detailed links is a formidable task – partially accomplished in $E_7/E_{10}/E_{11}$.

C_e: The class of **non-implicatory** deformations (of the c_α) comprises all modifications producing states c'_α, which cannot be restored to c_α through strengthening / weakening operations **alone** – though, in some cases, this restoration may be effected through a suitable 'flow' (in, or out) of **information**. This corresponds to the fact that such deformations always result in a (net) loss, or gain, of information. For a **gain**, it suffices to 'discard' the extra information (once it has been identified); for a **loss**, however, it is, typically, impossible to identify the 'missing information', thus precluding systematic restoration of c'_α to c_α. The studies of generalization / abstraction / analogy / inverse analogy, in E_7, explicate these ideas, and discuss the parallels with several scientific theories.

Perhaps **the** fundamental problem in treating deformations of LC is to define various types of (**quasi-neighbourhoods**) of e. This raises issues of great subtlety, since predicate functions may vary both syntactically and semantically. Before this matter is addressed, however, two simpler – but still significant – possibilities are considered.

Let I_e index the set of all admissible LC of e, say as $\{l_i(e) : i \in I_e\}$, and put $l_i(e) \equiv \underset{\alpha \in S_{ei}}{\wedge} {}^e c_\alpha$. For $\Gamma \subset S_{ei}$, define:

$$N^i_\Gamma := \left\{ \underset{\alpha \in S_{ei}}{\wedge} c'_\alpha \underset{\beta \in S_{ei} \backslash A}{\wedge} {}^e c_\beta : A \subset \Gamma \right\};$$

$$N_\Gamma(e) := \underset{i \in I_e}{\cup} N^i_\Gamma(e),$$

where the c'_α are ('suitable confined', but otherwise arbitrary) deformations[282] of the corresponding ${}^e c_\alpha$, and A ranges over all subsets of Γ. This specifies (crude) **quasi-neighbourhoods**[283] of e. Refinements may be obtained by stipulating the 'degrees of deformation of the c_α', in various ways – for instance, through some criteria involving the above imprecatory chains. Another class of refinements (useful, e.g., in restricting the c'_α) may be produced by requiring that the x_{σ_k} (in each of the LC, $l_i(e)$) must vary over (subsets of) associated 'spaces' – say, X_{σ_k} – with adequate structure for the introduction of (types of) **measures** and **bounded sets**. Under these conditions, one may define a **(quasi-)distance** between and $c_\alpha \in l_i(e)$ and (say) $C'_\alpha \in l_i(e')$, as follows:

$$\delta(c_\alpha, c'_\alpha; x) := \lim_{k \to \infty} \Omega_k^{-1} \mu(P_k \triangle P'_k);$$

$$X_k := \{x \in X : \| x \| \leqslant k\}, \Omega_l := \mu(X_k);$$

$$P_k := \{x \in X_k : c_\alpha(\ldots x \ldots); P'_k := \{x \in X_k : c'_\alpha(\ldots x \ldots)\};$$

$$A \triangle B := (A \backslash B) \cup (B \backslash A).$$

Here, x is any 'variable' in an n-place predicate function $c^i_\alpha(e; (x_{\sigma_\alpha ei})_{n^i_\alpha})$ from the LC, $l_i(e)$. A further extension covers the deformation of $c_\alpha(e, (x_\sigma)_{n_alpha})$ to $c'_\alpha(e', (x_{\sigma'_\alpha})_{n_\alpha})$, where

$$(x_{\sigma_\alpha})_{n_\alpha} \cap (x_{\sigma'_\alpha})_{n_\alpha} =: w_\alpha \in W;$$

$$\bar{P}_k := \{w \in X_k \cap X'_k : c_\alpha(e; (x_\sigma)_{n_\alpha})\};$$

---

[282] not necessarily implicatory

[283] or, Conceptual Vicinities (CV) – V(e), in the earlier notation.

$$\bar{P}_k' := \{w \in X_k \cap X_k' : c_\alpha'(e'; (x_{\sigma'})_{n_\alpha})\};$$
$$\bar{\delta}(c_\alpha, c_\alpha'; w) := \lim_{k\to\infty} \bar{\mu}(\bar{P}_k \triangle \bar{P}_k')/\bar{\mu}(W_k).$$

In this case, $\bar{\mu}$ is a measure on W, regarded as a cartesian product of spaces corresponding to the common elements of $(x_{\sigma_\alpha^i})_{n_\alpha^i}$, etc., for each LC, $l_i(e)$.

A different notion of '**neighbourhood**' may be obtained by allowing (subsets of) the $x_{\sigma_{\alpha k}^{ei}}$ to vary independently over certain domains within the associated spaces $X_{\sigma_{\alpha k}^{ei}}$ (for the LC $l_i(e)$, etc.). The collections of 'solutions' of these deformed LC then constitute neighbourhoods. Notice that there is no obvious form of distance function corresponding to this specification; but various possibilities could be explored. It should be reiterated that taxonomic / parsing requirements necessitate the multiplicity of indices, since the LC, $l_i(e)$, 'carries' the selection function, σ_α^i for the condition c_α^{ei}, where $\alpha \in S^i (\equiv S_e^i)$. Thus, the **full** dependence covers e, i, α and the argument, k, for σ_α^{ei}. More elegant representations may be given – by introducing suitable 'intermediate sets', etc.; but these forms do not permit precise identification of elements of U_t within any LC. Often, it is possible to ignore (and so, suppress) dependence of e and i – once the ME and its particular LC have been specified. The designations $x_{\sigma_{\alpha k}}$, etc., are relatively manageable.

This limitation, to deformations within classes of measurable spaces admitting bounded sets (in some sense(s)), is not as drastic as it may, at first, appear; for, many mathematical theories are developed broadly in this fashion (though the present **formalism** has rarely, if ever, been invoked). The interpretations of 'measurable', and of 'bounded', characterise the 'strength' of quasi-neighbourhoods defined in this way. It is found that even certain types of abstractions / analogues fall within the purview of this representation scheme.

The basic problem of defining (quasi-)distances between arbitrary predicate functions – say, $c(\xi)$, $c'(\xi)$ – seems to possess no completely general solution. The problem has been partially solved for classes of **programming languages** (and their associated theories) through a combination of techniques involving complete partial orders (CPO)[284] and **full abstraction**[285].

Although this framework (however it may be refined) seems inadequate for the syntactic / semantic complexities of 'typical mathematical activity', it is worthwhile to indicate how these ideas yield approximation theories in semantics, in classes of programming languages. It should be noted that 'mathematical language' is typified by **variability** in the imposition of **syntactical** rules, and by **flexibility** in the interpretation of expressions obtained from wfe through **semantic** rules. Consequently, **an incompletely formalized specification language** is more appropriate than any programming language in the 'representation of Mathematics' as a calculational system / IB[286]. Two species of semantics, 'operational', and 'denotational / fixed-point', are used in theoretical computer science. **Operational** semantics may be characterised by 'abstract machines', M, whose effects – through operation codes / sets of primitive instructions – on the state components of M constitute the specification(s) of M itself. The associated semantic of any language, L, is equivalent to a translation of L into an operation code. In this scheme, the meaning of a whole program in not always expressible as a function of the meanings of its components. In **denotational** semantics, it is supposed that programs take their meanings in classes of 'spaces', $\{D\}$, where, in particular, **recursive definitions** correspond to **fixed points (fp)** of operators $F : D \to D$. In some theories, these fp may be uniquely determined; in others, some canonical property is invoked (e.g., 'greatest', 'least', 'optimal'). Here, the interpretation of a program in **always** representable as some function of the interpretations of its components.

The approximation schemes may be formulated in two (essentially equivalent) ways: (a) through **lattices / CPO**; and, (b) through **systems of neighbourhoods**. In each case, the end result is a specification of **domains** (the 'spaces' within which programs take their meanings).

---

[284] See, e.g., '*Semantics of Programming Languages*', by CA Gunter (MIT Press, 1992) p 115 ff

[285] e.g., Gunter, op. cit, p 179 ff

[286] This point of view is expounded by VN Agafonov: '*From specification languages to specification knowledge-bases: The PTO approach*' (LNCD # 379 pp 1-17, Springer, 1989)

The theories corresponding to (a) and (b) may be found in several books[287]. In scheme (a), topological results (principally involving (continuous) lattices / fp theorems, and the solution of 'fp equations') are dominant. Scheme (b), emphasizing the representation of domains through neighbourhood systems, is regarded as more direct – with more explicit descriptions of 'approximation'. In both cases, languages are given semantically by mapping each (syntactic) construction to its meaning in a suitable domain. The operational semantics correspond to the classes of 'admissible constructions' for the language. Two fundamental questions arise in this context: (i) **Are the operational and denotational semantics always equivalent to each other;** and (ii) **Are the denotations of constructions necessarily unique** (in models of programming languages)? These are (respectively) the questions of **semantic equivalence**, and, of **full abstraction**. The non-triviality of these questions is assured by the existence of **counter-examples** – obtained, for (i), via[288] a 'diagonal argument', and for (ii), by showing[289] that the so-called Standard Model is **not** fully abstract. The associated concepts of approximation may be developed in considerable detail – within these structurally rigid environments. The elements of **imprecision** that pervade 'everyday mathematics' (and make it so powerful for general calculation)[290], largely invalidate these operational / denotational semantic techniques. The problem of bridging the gap between these two (rather extreme) positions is apparently unsolvable in any complete form. Some versions of 'constructive mathematics', especially that based on **Martin-Löf's Type Theory**[291], offer better prospects; but they are too far removed from 'the real mathematical world' to be much use for Σ . Only Bishop/s '**B-constructivity**', (mentioned in E_1) seems to have the potential for significant contributions to Σ . Moreover, this scheme has been formalized[292], and so may be amenable to semantic analysis. Even Bishop's criteria, however, exclude many essential facets of Σ -activity – notably, most heuristic arguments, and all 'proofs by contradiction'.

Other (partial) formalizations of Mathematics include AUTOMATH[293] (for **proof-checking-communication**); NuPRL[294] (for **proof-development**); and, ONTIC[295] (for **knowledge-representation** in Mathematics). All of these packages have already been implemented. Regular work on AUTOMATH stopped around 1990, but NuPRL and ONTIC are still under development.

These aspects of proof-checking / communication / development / representation – are obviously relevant for the design of P_Σ - of which they would form valuable 'subsystems'. Obstacles of notation / compatibility / compilation (at least!) make this aim highly problematical; but the underlying ideas / methods certainly **can** be adapted, and eventually incorporated. Whilst AUTOMATH and NuPRL are somewhat akin to 'machine languages for Σ ', the ONTIC system lies between theorem-proving packages and Σ , but with an emphasis on identifying paradigmatic **steps** in proofs – for intstance[296]: NOTE Φ / SUPPOSE Ψ / FOCUS-ON s. Again, the '**interactive** structure of Σ ' should take account of the ideas propounded by Pager (see E_0 for references) for 'computer-based mathematical communities'. As regards **automated proof / AI facilities**[297], it was remarked in E_0 that such capabilities, though undoubtable important, are severely limited by lack of flexibility / sophistication; scope for their inclusion may, however, by left without radically affecting the overall specification of Σ as a complex system. Possible formulations of 'approximation' over such frameworks range from the 'obvious' ones

[287] For (a), see, e.g., Gunter, op. cit., Section 4.5; and, for (b), the book: '*Lectures on a Mathematical Theory of Computation*' (Oxford U. Progr. Research Group PRG-19, 1981) by Dana Scott.

[288] See, Mulmuley, K, '*Full Abstraction and Semantic Equivalence*', PhD Thesis (CMU-CS-85-148), Carnegie-Mellon U, Ch. 2

[289] See, e.g., Gunter, op. cit., Sec 6.1

[290] See Agafonov, op. cit.; and Scott, op. cot, for CPO approximation

[291] See, e.g., '*Programming in Martin-Löf's type theory*', by B Nordström / K Petersson / JM Smith (Oxford UP, 1990)

[292] See, Goodman, N / Myhill, JJ, LNM #274 (Springer, 1972) pp 83-96

[293] e.g., '*Abstract AUTOMATH*', by A Rezus (Math. Centre Tract #160, Amsterdam, 1983)

[294] e.g., '*Implementing Mathematics with . . . NuPRL . . .*' (Prentice-Hall, 1986) by RL Constable et al.

[295] e.g., '*Ontic - A knowledge-representation system for Mathematics*', by DA McAllester (MIT Press, 1989)

[296] See 'ONTIC', by DA McAllester (MIT Press, 1989)

[297] Note that both NuPRL and ONTIC are of this type – but, that their basic **structures (taxonomic** and **inferential**) are of primary interest here

(in B-constructivity) to variants for models having 'near-full abstraction' (in some sense); and so on. In these circumstances, the restriction of deformations of LC to products of certain classes of 'tractable' spaces (to define 'conceptual vicinities') seems to offer the best available compromise – though there are many possible refinements / modifications of this basic scheme, depending on the contexts in which 'convergence structures for predicate functions' arise. These issues are discussed further in E_8, where proofs are treated as combinations of (suitably transformed) operational theorems.

Another form of 'closeness' for ME has already been indicated in E_4 – namely, the body of criteria corresponding to the (Boolean) **algebras of formalised theories**[298] $\mathcal{J} \equiv \{\mathcal{L}, C, A\}$, over (first-order) languages $\mathcal{L}$, with consequence relations C, and axiom set(s) A – combined with results[299] on topological / metrical properties of classes of Boolean algebras. Ostensibly, this yields characterizations of 'neighbourhood', for predict functions over specified domains – with implications for automated theorem-proving (as well as for the interactive computation in Σ). In practice, however, it is still far from clear how such routines can be implemented; so, at this stage, these possibilities should be regarded as primarily of theoretical interest – though, in certain frameworks, considerable progress can be made.

Although the interrelations among abstraction, analogy and standard modes of approximation are explored in some detail, in E_6, it is appropriate, here, to characterise these processes at the most basic level, as types of **approaches** (to any e). Thus, **abstraction** is regarded as **approximation in some essential facet(s)** – or, more succinctly , as: **approximation in essence(s)**. On the other hand, **analogy** corresponds to the **approximation of some internal / external interaction(s)** – or, briefly, to: **approximation through interaction(s)**. The notion of **behavioural approximation** is also pertinent to analogy (and seems to be substantially equivalent to interaction-approximation). A special subclass of {**abstractions**} is furnished by {**generalizations**} – viewed here as {**reversible** abstractions}. These heuristic observations are valuable aids for the formulation of more rigorous representations.

Criteria for deciding which facets of e are 'essential' may vary – and be associated with different **levels** of abstraction. The question of 'converting irreversible abstractions into generalisations' is one example of such tasks for non-implicatory deformations of LC. Another issue concerns the choice of **representation** for abstractions of e – and the associated 'stability properties'. For analogues, the above link with 'interaction' / behaviour is closely connected with (types of) **imitation** – say, **by** e' **over** (the structure) S', **of** e **over** S, as modelled by comparing the internal / external behaviour-relations, of e' in S', with a corresponding set of relations for e in S. Typically, these sets of relations have different cardinalities, and the comparisons are primarily qualitative. In these terms, analogues may range from 'total' ≡ **reproductions / transcriptions**, to arbitrarily weak – through the problem of measuring / estimating the **strength** of analogues, consistently, is a hard one, and probably has no completely satisfactory solution. For various **classes** of ME, however, covering situations of importance in 'mathematical practice', adequately precise comparisons **may** be made; while, the representations introduced **formally** apply to **all** cases. An interesting aspect of these 'abstract / analogical approaches to e' is that the **scope** of an abstraction / analogue, e', of e may encompass that of e, even though some facets of e may be missing, or 'reduced', in this 'model of e'. Here, the 'scope of e' (over a theory $\mathcal{J}$)' may be defined as $S_{c_\mathcal{J}}(e') := \{u : u \text{ is a special case of } e' \text{ in } \mathcal{J}\}$; or, more succinctly, as: $S_{c_\mathcal{J}}(e') := \{u : e' \text{ covers } u \text{ in } \mathcal{J}\}$.

These remarks illustrate how the notion of 'approaches to e' extends that of 'approximations to e'. The associated species of approach, namely **Ap, Ab, An, -An** (of **approximations, abstractions, analogues**, and **inverse analogues** – respectively) are considered in E_7, for their interrelations / algebraic (e.g., categorical) aspects. The problems associated with **approximation of ME by (families of) other ME of the same type** involve all of the basic spaces $_S X$, introduced in E_1; that is: Linear / **Topological** / **Uniform** / **Proximity** / **Contiguity** / **Nearness** / **Statistical** / **Fuzzy** / **Measure** ($\equiv \mu$) spaces – which appear to span the range of **effective calculational environments**. Theoretical issues for such structures are discusses in E_6. Their

---
[298] See Rasiowa-Sikorski, op. cit., §10
[299] See Blumenthal-Penning, op. cit.

multifarious realizations in 'mathematical activity' are diversely exemplified in E_9 / E_{10} / E_{11}. The approximative aspects of analogy / inverse-analogy are widely illustrated in E_{11}. These realizations of **approaches** (of all kinds), amounting to several thousand items, are intended as **paradigms** for the organization and use of the Σ -IB. The collection is substantial, but in no way remotely exhaustive. The main aim is to promote 'the approach point of vies' – and so to maximize effective calculability in all domains of Mathematics.

The most fundamental issue for Σ is how the qualities of OT are affected when some of the 'OT ingredients' are replaced by their approaches – of various types. These modifications of OT are discussed further in E_8, and exemplified in E_{10}/E_{11}. Here, the aim is to characterize possible 'versions of OT' obtainable in this way, and to indicate associated questions of quasi-stability / continuity.

In this context there are three broad areas in which approach-modifications affect OT: (i) **formulation**; (ii) **realization**; (iii) **application** – each of which may be further subdivided into **deterministic**, and **stochastic**, forms. To this list one may add cognate variants of all species of **diagrams** pertinent to the proofs / justifications-of-proofs / implementations.

An extensive discussion of diagrams is given in E_4, so it is appropriate, in this Essay, to consider 'generic **approaches**' to (each species of) diagrams, in terms of admissible approaches to (some of) their **components** – or, even to their **sources**. All of this reflects the stress, in Σ , on generalized approximation / t-max.-operationality – in all areas of Mathematics. The corresponding treatment of all w.d. **compound** diagrams (as defined in E_4) may then be investigated.

The **modification operators**, $\mathbb{M}^{(\lambda',\lambda'')}_{D;E*}$, are introduced, in E_4, as:

$$\mathbb{M}^{(\lambda',\lambda'')}_{D;E*} := \{M_D : D^{(\lambda')}_{E*} \to \mathcal{D}^{(\lambda'')}_{E*}\},$$

where $\mathcal{D}^{(\lambda)}_{E*} := \{D_\lambda(e) : e \in E^*\}$, and the M_D are the modification operators for individual diagrams (of species λ'). The species may (but need not) be conserved under M_D.

The basic relation, for a diagram $D^e_{\lambda,j} \equiv D^e_{\lambda,j}(\gamma)_q$, then takes the form:

$$\mathbb{M}D^e_{\lambda',j}(\gamma)_q := D^e_{\lambda'',j}((M\gamma)q),$$

where $\mathbb{M} \equiv (M)_q$, and the 'subscript' D is suppressed, for clarity. Of course, this isa partially schematic representation, since the appropriate form of $\mathbb{M}$, for transforming $D^e_{\lambda'}$, is assumed in each case. Nevertheless, the interpretation is clear – and may be rendered unambiguous, whenever specific applications are involved.

This notation, in spite of its imprecision, is very useful since it may be adapted to cover all species of diagrams relevant for Σ . **Notice that** the component set, $(\gamma)_q$, actually depends on the species, λ', and type, j, of $D^e_{\lambda',j}(\gamma)_q$; that is: $(\gamma)_q \equiv (\gamma^{(\lambda',j)})_{q(\lambda',j)}$; and, that the dependence on 'the source', e, is now shown as a superscript. All of these detailed dependences must be taken into account when the classes of admissible species / types of diagrams are listed / implemented within the OIB. In formal manipulations, however, the number of 'parameters' should be minimized, as far as possible. This general strategy is followed in all of the Essays, since the inherent complexity of the matters discussed would otherwise make the arguments unavoidably obscure.

The list of diagram species (19, in all) is given in E_4; it is repeated here for easy reference. (1) CAD representations; (2) Control-theoretic procedures; (3) geometrical / topological configurations; (4) higher-dimensional configurations (in all fields); (5) General Algebra; (6) Homological Algebra; (7) other algebraic / geometric-topological forms; (8) Algebraic Geometry; (9) Differential Topology / Geometry; (10) Real Analysis; (11) Complex Analysis; (12) Functional Analysis; (13) Lie Groups / Lie Algebras; (14) Many-Body Calculations in (Chemical) Physics; (15) High-Energy Physics; (16) Graphs / Networks (in all fields); (17) Illustrative (in all fields); (18) Pure Geometry (of all types); (19) Miscellaneous.

There are, of course, other possible species (e.g., chemical structure / reaction; optimization (in all fields)) which may be incorporated within the OIB at any time, without invalidating material already included – and this will certainly happen. The aim here is simple to indicate how **subtaxonomies** of (species of) diagrams may be cumulatively constructed. Accordingly, for each species of diagrams, '**typical components**' may be identified – along with their common variants – so that cognate forms of **modification operators** may be specified. Next, the range of **admissible approaches** to these (classes of) diagrams may be explored, in terms of the modes of approach associated with the 'basic spaces', ${}_SX$, as designated in E_1, and with the forms of analogy / inverse-analogy / abstraction considered in E_7. Once again, it is worthwhile to recall that the basic spaces are of classes **L**(inear) / **T**(opological) / **U**(niform) / **P**(roximity) / **C**(ontiguity) / **N**(earness) / **S**(tatistical) / **F**(uzzy) / μ(easure) – each of which has been studied extensively (**see** E_6 for selected references) **NOTE that 'stochastic versions'** of the basic spaces are covered by defining suitable **measures** over the deterministic forms[300] This procedure extends even to **numerical** approximation schemes[301]. Any ME defined over the stochastic spaces obtained in such ways is (by definition) a 'probabilistic variant' of the generic ME concerned.

The classification of diagrams requires preliminary lists of types (for each species), and of examples (for each type). Since an initial selection of species has been made, the next task is to assign collections of types to each species. Although this process is partially subjective, and far from exhaustive, it does lay the foundation for a more systematic taxonomy. Moreover – as for all data in Σ – refinements / additions may be made at any time in a consistent manner. With each type, j, of a diagram of species λ, for e, there is associated a list – say, $\Gamma^e_{\lambda,k}$ – of **components**[302]. In the previous notation, $\Gamma^e_{\lambda,j} = \{\gamma^{(\lambda,j)}\}_{q(\lambda,j)}$; and, $(\gamma)_q$ is the (somehow) ordered form of $\Gamma_{\lambda,j}$. Here, $1 \leqslant \lambda \leqslant 19$; $1 \leqslant j \leqslant J^e \equiv J^e(\lambda)$. The assignment of types, to produce collections, $\{D^e_{\lambda,j}(\gamma)_q\}$, is made subjectively, but adequate levels of completeness will eventually be attained, as Σ evolves. Consequently, it suffices here merely to indicate a few types for each species, as illustrations of the sorts of criteria used.

NOTE that the only requirement for ξ to be an admissible component of the diagram $D^e_{\lambda',j}(\gamma)_q$ is that ξ is, either, an element (i.e., an edge or a vertex) of some corresponding proof-graph; or else, an element of some data-diagram for e. This is so because every diagram $D^e_{\lambda',l}$ is an amalgamation of such basic (proof / data) diagrams. Consequently, the possible forms of components are essentially unrestricted – no matter how intricate / elaborate / arcane / ... they may appear. In other words, all components are viewed as single (possibly, 'compound') objects – each variously modifiable – whose classes of admissible transformations determines the possible transformations / approaches for arbitrary diagrams w.d. over Σ .

Notice, also, that many of the entries in these lists are **settings** for components (rather than individual components), furnishing contexts / environments within which (classes of) diagram-components may be specified.

The following 'topical outlines' of all the Σ -species of diagrams are intended merely as initial indications of the diagrammatic part of the OIB. An associated goal is to provide adequate foundations for the development of 'approach analyses' of diagrams in arbitrary areas of Mathematics. This Section is, therefore, unavoidably long, in spite of the brevity of the outlines. In the next Section, general properties of **approaches to** (classes of) **diagrams** are considered.

NOTE that, although copious references are given here, the great majority of citations may be found in E_{10}/E_{11}, where diverse examples of approaches are discussed.

1. CAD Diagrams

(a) Architectural drawings (plans to scale).

(b) Machine designs (detailed plans).

[300] Some of these forms are covered in studies of S-MS: See the book: '*probabilistic Metric Spaces*' by Schweizer, B / Sklar, A (ref 2.1 of E_6). See also, e.g., '*Probability on Banach Spaces*' (J Kuelbs, ed.) (Dekker, 1978) and other volumes in this (Dekker) series '*Advances in Probability and Related Topics*' (1970 –)

[301] See, e.g., '*Stochastic Finite Elements ...*', by Ghanem, RG, Spanos, PD (Springer, 1991); and, books on 'Least-Squares' / Montecarlo Methods

[302] The set of components includes all environments / elements / rules / interactions / ...

(c) Printed circuits (for VLSI).

NOTE References are unnecessary here; typical sets of components are indicated in the descriptions of such diagrams. There are many other possibilities, but the ones chosen are representative. This species of diagrams is relevant mainly to (inverse) analogy.

2. Control Diagrams

(a) General forms of block diagrams.

(b) Network diagrams in population biology.

(c) Process diagrams in Chemical Engineering.

In this case, the following references may be consulted for many examples / details / component-lists: (i) '*Feedback and Control Systems*', by JJ DiStefano, et al., (op. cit.), (Schaum, 1967) (ii) '*Theory of Automatic Control*', (A Netushil, General Ed.) (MIR, Moscow, 2nd ed., 1978) (iii) '*Digital Communications and Spread Spectrum Systems*', by RE Ziemer / RL Peterson (Macmillan, 1985) (iv) '*Dynamical System Models*', by AGJ MacFarlane (Harrap, 1970) (v) '*Network Models in Population Biology*', by ER Lewis (Springer, 1977) (vi) '*Mathematical Methods in Chemical Engineering*', by JH Seinfeld, L Lapidus, Vol. 3 (Prentice Hall, 1974) (vii) '*Introduction to System Dynamics*', by JL Shearer, et al. (Addison-Wesley, 1967), op. cit.

NOTE As for CAD diagrams, the possible range / scope is enormous here; and, again, such diagrams are relevant mainly to (inverse) analogy. In **both** species (1 and 2) the **admissible modifications** of (classes of) diagrams include: variations of **environment(s)** / **diagram-topologies** / **gross-forms-of-components** / **structure(s)-of-components**. All of these operations may be realized mathematically. **Illustrations**, covering these (and other) forms of modification, may be found in E_{10}/E_{11}, together with some of the inverse-analogical structures that they 'generate'.

3. Geometrical / Topological Configurations

(a) Triangulations of manifolds / surfaces.

(b) Finite-element decompositions of domains.

(c) Packing / Covering configurations.

References:

(a)(i) Massey, WS, '*Algebraic Topology: An Introduction*' (Springer, 1989)

(a)(ii) Maunder, CRF, '*Algebraic Topology*' (CUP, 1980)

(a)(iii) Fomenko, A, '*Visual Geometry and Topology*', (Springer, 1994), op. cit.

(a)(iv) Henle, M, '*A Combinatorial Introduction to Topology*' (Freeman, 1979)

(b)(i) Zienkiewicz, OC, '*The Finite Element Method in Engineering Science*' (McGr-H, 2nd ed., 1971)

(b)(ii) [e.g.] Brebbia, CA, ed. '*Finite Element Systems: A Handbook*' (Springer, 2nd ed., 1982)

(b)(iii) '*The Mathematics of Finite Elements and Applications*', Vols. 1-5..., (AP, 1973-1985...) JR Whiteman, ed.

(b)(iv) '*Multigrid Methods*' (W Hackbusch, U Trottenberg, eds.) (Springer, LNM #960, 1982)

(c)(i) Conway, JH, Sloane, NJA, '*Sphere Packings, Lattices and Groups*' (Springer, 2nd ed., 1993)

(c)(ii) Ziegler, GM, '*Lectures on Polytopes*' (Springer, 1995)

(c)(iii) Wells, AF, '*Structural Inorganic Chemistry*', (OUP, 4th ed., 1975)

Typical Components(-settings)

(3)(a): Environmental space
manifold / surface itself
ingredients of the triangulation
precision of triangulation orientation of triangulation

(3)(b): Environmental space
Domain of problem
ingredients of decomposition
accuracy of decomposition

(3)(c): Environmental space
Domain to be packed / covered
Shape(s) of elementary packing / covering domains
accuracy of packing / covering

4. Higher-Dimensional Configurations / Schemes

(a) Linear / Multi-Linear Algebra; Geometry

(b) Geometry of Tensor Fields

(c) Structures Involving Non-numerical Metrics

(d) Schematic Structures in Proofs

References:

(a)(i) Bloom, DM, '*Linear Algebra and Geometry*' (CUP, 1979)

(a)(ii) Sommerville, DMY, '*An Introduction to the Geometry of N Dimensions*' (Dover, 1958)

(a)(iii) Porteous, IR, '*Topological Geometry*' (Van Nostrand, 1969)

(a)(iv) Grueb, WH, '*Linear Algebra*' (Springer, 2nd ed., 1963)

(b)(i) Dodson, CTJ / Poston, T, '*Tensor Geometry*' (Pitman, 1977)

(b)(ii) Kreyszig, E, '*Differential Geometry*' (OUP, 1959); Kreyszij, E, '*Differential Geometry and Riemannian Geometry*' (U of Toronto Press, 1968)

(b)(iii) Eisenhart, LP, '*Non-Riemannian Geometry*' (AMS, 7th pr., 1972)

(b)(iv) Sharafutdinon, VA, '*Integral Geometry of Tensor Fields*' (VSP, Utrecht, 1994)

(c)(i) The briefly annotated Bibliography on '*Non-Numerical Distance Functions in Topology*', by R Schmid-Zartner / HC Reichel / B Pötscher (U of Vienna, 1984), op. cit., contains about 600 citations. Quasi-geometrical representations pertaining to higher-dimensional objects may be based on many of these 'spaces'. Diverse examples are given in E_{10}/E_{11}.

(c)(ii) Pisier, G, '*The Volume of Convex Bodies and Banach-Space Geometry*' (CUP, 1989)

Typical Components(-settings)

Environmental space(s)
Forms of ME within these frameworks
Detailed properties of 'parts' of such ME
Interrelations of these parts

(d)(i) Ziedler, E, '*Nonlinear Functional Analysis*' (Springer, 1984-1988, 5 Vols,; esp., Vols 1,2,3)

(d)(ii) Kiang, T-h, '*The Theory of Fixed Point Classes*' (Springer, 1989)

(d)(iii) Geddes, KO, et al., '*Algorithms for Computer Algebra*' (Kluwer, 1992), op. cit.

(d)(iv) Cox, D, et al., '*Ideals, Varieties, and Algorithms*' (Springer, 1992), op. cit.

(d)(v) Khilmi, GF, '*Qualitative Methods in the Many-Body Problem*' (Gordon and Breach, 1961)

(d)(vi) Oxtoby, JC, '*Measure and Category*' (Springer, 1971)

(d)(vii) Books on applications of functional analytic ('abstract space') arguments to diverse engineering / scientific / economics problems contain many proofs based on schematic structures involving diverse types of operators between suitably constructed (linear) spaces. **See**, e.g.: '*Applied Functional Analysis*', by DH Griffel (Horwood, 1981); '*Applied Functional Analysis*', by RD Milne (Pitman, 1980); '*Applied Functional Analysis*', by AV Balakrisinnan (Springer, 1976); '*Optima and Equilibria*;, by J-P Aubin (Springer, 1993); '*Modern Foundations of Systems Engineering*', by WA Porter (Macmillan, 1966) – and, Ziedler ((d)(i), above)

(d)(viii) Another area where schematic / structural diagrams are important involves **data-flow in** (general classes of) **computer algorithms. See**, e.g., '*Program Flow Analysis ...*', SS Muchnick / ND Jones, eds. (Prentice-Hall, 1981), where a wide range of flow analyses may be found.

(d)(ix) The most general class of **proof diagrams** may be envisaged as follows – for the proof / realization of $T : p \Rightarrow q$ (a deduction / construction).

Suppose that $\mathcal{J}_{\Sigma_t} \equiv (\tau)_{L_t}$ is the collection of all 'designated theories' in Σ_t (where, typically, the theories τ_i need not be logically independent); and, that the proof of $T : p \Rightarrow q$ has the form $\pi_T \equiv \pi_T\{\tau_\sigma\}_{m_\pi}$, with $\{\tau_\sigma\}_{m_\pi} \subset \mathcal{J}_{\Sigma_t}$.

Suppose, also, that the elements of $\{\tau_\sigma\}_{m_\pi}$ are depicted as 'black boxes' surrounding the 'initial data', p, of T.

Then: '**motion through**' $\pi_T\{\tau_\sigma\}_{m_\pi}$ takes the form(s) of **labelled, directed graphs** (typically, **self-intersecting**) with (say), v_i 'vertices' within the box, B_i, representing τ_i. Edges lying in any B_i correspond to 'refractions', and the other edges, to 'reflections'. This notation is in line with the scheme used in E_6, to study (inverse) analogy. Here, the 'vertices' represent (generic) OT for the theories τ_i.

Notice that each of the τ_i has its own 'internal structure' (with 'undefined objects' / axioms / deductive hierarchies), and that even quasi-metrical criteria may be defined, in some cases. In this way, not only general classes of **proofs**, but also, **Calculational Bases**, may be investigated systematically. The basic ideas about such CB are introduced in E_3. More detailed investigations of the possible 'structures' definable over the CB may be enhanced by the representations discussed here.

Classifications of theorems / proofs in terms of graph-theoretical characteristics of the corresponding proof-diagrams (relative to various sets of 'underlying theories') are likely to yield interrelations among ostensibly disparate results.

Typical Components of Proof-Diagrams

Premise components

Conclusion / Construction components

Constituent theories of the proof

Topological / Combinatorial features of proof-graphs (and corresponding compo- nents-setting)

Note Evidently, the admissible range of components in proof-graphs is virtually unlimited, since $\mathcal{M}_t$ is essentially equivalent to 'the body of established theorems (at t). The main provisos involve distinctions among fully / partially / nil-operational theorems.

5. General Algebra

The main species of structures here (as in E_4) correspond to Groups / Rings / Fields / Lattices, and Linear / Boolean / Operator / Universal / Topological Algebras – and the **basic references** are also as for E_4. Consequently, the variety of possible diagrams / components is considerable, so only the barest indications are given.

The main forms of diagrams emphasize internal structure / interrelations / embeddings / realizations of the algebraic systems listed above. The details depend on the fundamental characteristics of these systems (ultimately, on their axioms).

The most suggestive properties of diagrams, in this context, are associated with the key results for the structures / systems concerned (i.e., Groups, Rings, ..., Topological Algebras). In particular, the **proof-diagrams** for such results – and for structural embeddings, and other interrelations among collections of algebraic / analytical / ... systems – may be used to cover the whole area of General Algebra. Among the **components** are: the **base-sets** (over which structures are defined); the **axiom-sets** (of all ME / theories underlying each proof-diagram); and, **the characteristics of** all relevant **embedding procedures**, and of other types of **structural interrelations**. The admissible classes of proof-diagrams have been defined in terms of **directed / labelled / self-intersecting graphs**, located within 'environments containing families of theories', etc.. **This identification of classes of diagrams, for any mathematical field, with the set of proof-diagrams for its 'principal result', furnished a uniform process for 'generating' the central diagrammatic content of the (generally recognized) 'largescale subdomains of $\mathcal{M}_t$.**

Further, since **every** ME is regarded (in Σ) as a theorem (trivially, in the cases of 'primitive' / undefined objects, whose **existence** is simply postulated; and, of axioms – whose **forms** are postulated) the diagram(s) for any ME, e, are of two main kinds: **data** (descriptive), and, **proof** (of basic properties). On this basis, every subfield of $\mathcal{M}_t$ has (in principle, at least) some diagrammatic representation(s) / elaboration(s) – and, the deduction of any theorem, $T : p \Rightarrow q$, may be viewed as a compound transformation of the set of data / proof-diagrams, for p, into a cognate set for q.

The quasi-diagrammatic form of $\mathcal{M}_t$ is itself a **facet** of $\mathcal{M}_t$! Its **interpretation** is comprised of diverse, intricate operational compositions, so it is replete with logical interdependence and nonuniqueness. Nevertheless, there appears to be no inherent inconsistency in the development of this scheme, and it may be used to endow substantially arbitrary sets of ME with non-trivial quasi-geometrical content, of special value for heuristic investigations – and even, in some instances, in rigorous formulations. Notice that, (e.g.) commutative diagrams (in Algebra) and Feynman diagram (in Quantum Electrodynamics) are of the 'data' type; whereas, diagrams 'proving' commutativity or 'evaluating' some perturbation-expansion term, are of the 'proof' type. More precisely, the 'data diagrams' are (primarily) informative, while the 'proof diagrams' are (mainly) visual explications of methods of proof, or of evaluation. Such distinctions obtain whatever the source of a diagram may be.

IT should be realized, therefore, that the family of diagrams 'generated by a single source, e' is, in general, extensive – covering all aspects (informative / deductive / calculational) of e. The citations here, for each species of diagram, merely list a few examples / references / components. More detailed discussions are given (in context) throughout E_{10}/E_{11}. From now

on, the basic reference-list for each species of diagrams is taken to be the (corresponding) list given in E_4; additional items (if any) will be listed as necessary.
References: As in E_4 (refs. 1-10)
Typical Components(-settings) for structures s of species $\mathcal{S}$
Base sets (for realizations)
Realizations of characterizing operations
Implementation of embeddings
Internal structural relations for s
Relations with other structures (in $\mathcal{S}'$)
Relations within families, $\{s_\alpha\}$, of species $\mathcal{S}$
NOTE that the classification into diagrams of 'data' and 'proof' types may be used for **all** species of diagram (i.e., for **all** subfields of Mathematics and its applications). Hence, it is understood, for the remaining diagram-outlines here, that proof-diagrams (of arbitrary complexity / sophistication) are covered – along with all relevant data-diagrams. The chosen examples are comparatively simple, but they suffice to indicate some ways in which Mathematical information, whatever its primary interpretation, has significant quasi-graphical content.

6. Homological Algebra

There is essentially only one class of data-diagrams here – subsuming networks of spaces / objects / mappings of diverse types. The main distinctions correspond to the field(s) in which Homological Algebra is deployed; for instance:

(a) Algebraic Topology

(b) Category Theory

(c) Topos Theory

(d) Algebraic Geometry

(e) Algebra

– which are taken here as **species**.
References:

(a)(i) Hilton, PJ / Wylie, S, '*Homology Theory*' (CUP, 1960)

(a)(ii) Maunder, CRF, '*Algebraic Topology*' (Van Nostrand, 1970; CUP, 1980)

(a)(iii) Spanier, E, '*Algebraic Topology*' (McGr-H, 1966)

(a)(iv) Stückrad, J / Vogel, W, '*Buchsbaum Rings*' (Springer, 1986)

(b)(i) Freyd, P, '*Abelian Categories*' (Harper & Row, 1964)

(b)(ii) Ehresmann, C, '*Catégories et Structures*' (Dunod, 1965)

(b)(iii) Strooker, JR, '*Introduction to Categories, Homological Algebra and Sheaf Cohomology*' (CUP, 1978)

(c)(i) Johnstone, PT, '*Topos Theory*' (AP, 1977)

(c)(ii) Kock, A, '*Synthetic Differential Geometry*' (CUP, 1981)

(d)(i) Hartshorne, R, '*Algebraic Geometry*' (Springer, 1977)

(d)(ii) Stückrad / Vogel [(a)(iv)]

(e)(i) Strooker [(b)(iii)]

(e)(ii) MacLane, S, '*Homology*' (Springer, 4th pr., 1994)

(e)(iii) Hilton, PJ / Stammbach, U, '*A Course in Homological Algebra*' (Springer, 1971)

(e)(iv) Stückrad / Vogel [(a)(iv)]

Typical Components(-settings)

Spaces[303]
Mappings[304]
Objects[305]
Functors[306]
Categories[307]
Internal structural elements (of spaces)[308]
Internal structural elements (of categories)[309]

7. Other (Algebraic / Geometric) Topological Forms

(a) Algebraic curves

(b) Algebraic surfaces

(c) Surfaces in 3-manifolds

(d) Compact complex surfaces

(e) Knot Theory

(f) Topological study of polyhedra / complexes

(g) Homotopical Algebra

(h) Curves / surfaces / manifolds in Mechanics

(i) Riemann surfaces

(j) Minimal surfaces

(k) Combinatorial representations

References:

(a)(i) Brieskorn, E / Knörrer, H, *Plane Algebraic Curves*' (Birkhäuser, 1986)

(a)(ii) Seidenberg, A (ed.) *Studies in Algebraic Geometry*' (AMS, 1980)

(a)(iii) Hartshorne, R, *Algebraic Geometry*' (Springer, 1977)

(a)(iv) Griffiths, P / Harris, J, *Principles of Algebraic Geometry*' (Wiley, 1978)

(b)(i) Zariski, O, *Algebraic Surfaces*' (Springer, 1935; repr. Chelsea, 1948; suppl. ed. 1971)

(b)(ii) Hartshorne [(a)(iii)]

(b)(iii) Griffiths / Harris [(a)(iv)]

---

[303] All **species** of spaces (or, classes of **types** of spaces) are covered
[304] E.g., mono/epimorphisms; but most classes are admissible
[305] These constitute the domain / range of mappings within a category
[306] These compound transformations map the objects / morphisms of C_1 to those of C_2
[307] A category, C, comprises (a system of) objects and morphisms, suitable related
[308] A broad selection of such features is given in (e.g.) Dunford / Schwartz, '*Linear Operators*' (Vol 1; esp. ch. 4); MA Naimark, '*Normed Rings*' (Noordhoff, 1959); A Kufner / O John / S Fucik, '*Function Spaces*' (Noordhoff, 1977)
[309] Many structural elements are discussed in Strooker [(b)(iii); esp. ch. 2]

(b)(iv) Formenko, A, *Visual Geometry and Topology*' (Springer, 1994)

(c)(i) Hilton, PJ (ed), *Studies in Modern Topology*' (AMS, 1968)

(c)(ii) Fomenko [(b)(iv)]

(d)(i) Barth, W / Peters, C / Van de Ven, A, *Compact Complex Surfaces*' (Springer, 1984)

(d)(ii) Griffiths / Harris [(a)(iv)]

(d)(iii) Hartshorne [(a)(iii)]

(e)(i) Seifert, H / Threlfall, W, '*Lehrbuch der Topologie*' (Teubner, 1934; Chelsea, 1945)

(e)(ii) Kosniowski, C, '*A First Course in Algebraic Topology*' (CUP, 1980)

(f)(i) Rourke, CP / Sanderson, BJ, '*Introduction to Piecewise Linear Topology*' (Springer, 1982)

(f)(ii) Fomenko [(b)(iv)]

(f)(iii) Maunder, CRF, '*Algebraic Topology*' (CUP, 1970)

(g)(i) Baues, H-J, '*Algebraic Homotopy*' (CUP, 1988)

(g)(ii) Baues, H-J, '*Combinatorial Homotopy and 4-Dimensional Complexes*'

(g)(iii) Buonchristians, S / Rourke, CP / Sanderson, BJ, '*A geometric Approach to Homology Theory*' (CUP, 1976)

(h)(i) Fomenko [(b)(iv)]

(h)(ii) Abraham, R / Marsden, JE, '*Foundations of Mechanics*' (Benjamin / Cummings, 2nd ed., 1978)

(i)(i) Tsuji, M, '*Potential Theory in Modern Function Theory*' (Chelsea, 1975)

(i)(ii) Springer, G, '*Introduction to Riemann Surfaces*' (Addison-Wesley, 1957)

(i)(iii) Farkas, HM / Kra, I, '*Riemann Surfaces*' (Springer, 1992)

(i)(iv) Cohn, H, '*Conformal Mapping on Riemann Surfaces*' (McGr-H, 1967)

(j)(i) Fomenko, AT / Tuzhilin, AA, '*Elements of the Geometry and Topology of Minimal Surfaces in Three-Dimensional Space*' (AMS, 1991)

(j)(ii) Nitsche, JCC, '*Lectures on Minimal Surfaces*', Wol 1 (CUP, 1989)

(k)(i) Seifert / Threlfall [(e)(i)]

(k)(ii) Henle, M, '*A Combinatorial Introduction to Topology*' (Freeman, 1979)

Typical Components(-settings)

(a) Ambient Spaces
Characterization conditions
Singularity conditions
Genus
Order
Branch structure
Admissible coordinate systems
Parametric equations
Intersection criteria
Classification criteria

(b) Ambient Spaces
- Characterization conditions
- Singularity conditions
- Genus
- Admissible coordinate systems
- Parametric equations
- Classification criteria

(c) Ambient spaces
- Representations
- Decompositions
- Sections
- Piecewise-linear approximations
- Components
- Deformations
- Embedding criteria
- Classification criteria

(d) Ambient spaces
- Characterization conditions
- Singularity conditions
- Genus Graphs associated with (quadratic) forms
- Embedding conditions for curves
- Patching criteria
- Singularity conditions (resolution diagrams)
- Classification criteria

(e) Ambient space
- Equivalence conditions
- Classification criteria
- Associated groups

(f) Triangulation schemes: simplexes
- Structure of complexes
- orientation criteria
- Simplicial approximation procedures
- Deformation criteria
- Operations involving handles

(g) Chain-complex diagrams / algebras
- Cochain algebras
- Pull-back / push-out operations
- Group(kid)s of homotopies
- Function-space homotopies
- Mapping cones
- Categories of (co)chain algebras
- Telescope diagrams
- Towers of groups
- Towers of categories (as approximations)
- Towers of fibrations
- Postnikov decompositions
- Obstruction criteria

(h) Geodesic flows
- Symplectic manifolds
- AP motion in action / angle variables

'Large' / 'small' bifurcation diagrams
Chambers of mappings
Liouville tori, and associated 'surgeries'
Critical surfaces
Bifurcation graphs / 'molecules'
DPM (domain(s) of possible motion)
Graphs of integrable systems
Constant- (iso-) energy surfaces
Phase(-space) portraits
Poincaré maps
Center manifolds; stable / dense manifolds
Classification of stable bifurcation diagrams

(i) Classification criteria
Manifold representations
Covering surfaces
Singularity conditions
Genus
Period criteria (for Abelian functions)
Conformal mapping criteria
Prolongation criteria
Uniformization criteria
Triangulation procedures
Curvature criteria
Representations as compactifications of algebraic curves

(j) Minimality characterizations
Attainability conditions
Variational criteria
Criteria for representations by algebraic functions
Differential-geometric criteria
Function-theoretic criteria
Parametric representations
Local conditions
Global conditions

(k) Decomposition procedures for complexes
Identification criteria
Surface-representations
Covering spaces
Dimension criteria
Surface-classification(s)
Simplicial approximation procedures

8. Algebraic Geometry

(a) Commutative Algebra

(b) Algebraic Varieties

(c) Analytic Varieties (≡ Complex Analytic Sets)

(d) Schemes

(e) Complex Analytic Manifolds

(f) Algebraic Curves

(g) Algebraic Surfaces

(h) Analytic curves / surfaces

(i) Miscellaneous

References:

(a)(i) Zariski, O / Samuel, P, '*Commutative Algebra*', Vols 1 and 2 (Springer; originally, Van Nostrand, 1958/1960)

(a)(ii) Bourbaki, N, '*Commutative Algebra*' (transl., Springer, 1989)

(a)(iii) Atiyah, MF / Macdonald, IG, *Introduction to Commutative Algebra*' (Addison-Wesley, 1969)

(a)(iv) Rowen, LH, '*Ring Theory*' (Student Edition, AP, 1991)

(a)(v) Faith, C, '*Algebra . . .*', Vols I/II (Springer, 1973/1976)

(b)(i) Hodge, WVD / Pedoe, D, '*Methods of Algebraic Geometry*', Vols 1,2,3 (CUP; repr. 1994), esp. Vol 2

(b)(ii) Weil, A, '*Foundations of Algebraic Geometry*', (AMS, repr. 1989)

(b)(iii) Hartshorne, R [7(a)(iii)]

(b)(iv) Griffiths / Harris [7(a)(iv)]

(b)(v) Cox, D / Little, J / O'Shea, D, '*Ideals, Varieties, and Algorithms*' (Springer, 1992)

(b)(vi) Lefschetz, S, '*Algebraic Geometry*' (OUP, 1952)

(c)(i) Vitushkin, AG (ed.) '*Several Complex Variables: I*' (Springer, 1990)

(c)(ii) Range, RM, '*Holomorphic Functions and Integral Representations in Several Complex Variables*' (Springer, 1986)

(c)(iii) Lelong, P / Gruman, L, '*Entire Functions of Several Complex Variables*' (Springer, 1986)

(d)(i) Hartshorne, R [7(a)(iii)]

(d)(ii) Atuyah / Macdonald [(a)(iii)]

(d)(iii) Bosch, S / Lütkebohmert, W / Raynaud, M, '*Néron Models*' (Springer, 1990)

(d)(iv) Mumford, D (Lecture Notes) (repr. Springer, 1988 as LNM #1358)

(e)(i) Griffiths / Harris [7(a)(iv)]

(e)(ii) Hartshorne [7(a)(iii)]

(e)(iii) Yano, K, '*Differential Geometry of Complex and Almost Complex Spaces*' (Pergamon, 1965)

(e)(iv) Choquet-Bruhat, Y / DeWitt-Morette, C / Dillard-Bleick, M, '*Analysis, Manifolds and Physics*', rev. ed. (North-Holland, 1982)

(e)(v) Gindikin, SG / Khenkin, GM, (eds.), '*Several Complex Variables: IV*' (Springer, 1990)

(f)(i) Griffiths / Harris [7(a)(iv)]

(f)(ii) Hartshorne [7(a)(iii)]

(f)(iii) Brieskorn / Knörrer [7(a)(i)]

(g)(i) Zariski [7(b)(i)]

(g)(ii) Griffiths / Harris [7(a)(iv)]

(g)(iii) Hartshorne [7(a)(iii)]

(h)(i) Vitushkin [(c)(i)]

(h)(ii) Shiffman, B, '*Holomorphic curves in algebraic manifolds*' (BAMS 83 (1977) 553-568)

(h)(iii) Gindikin / Khenkin [(e)(v)]

Typical Components(-settings)

(a) – mappings (e.g., homo/iso/epi/mono/. . . morphisms)
– modules
– exact sequences (short / long / . . .)
– commutative diagrams (with specified components / . . .)
– ambient structures
– quotient structures
– tensor products
– direct products
– direct sums
– ideals
– categories
– functors
– local(ized) structures
– free structures
– global(ized) structures
– prime structures
– reducible structures
– irreducible structures
– graded structures
– spectra
– invertible structures
– singular structures
– filtered structures
– complete structures
– compact structures
– lifted structures
– finite structures
– infinite structures
– bounded structures
– valuated structures
– topological structures
– transcendental structures
– flat structures
– places
– divisors

(b) – dimension of varieties
– reducible varieties
– absolutely / relatively irreducible varieties
– 'Cayley forms' (of varieties)
– orders of varieties

- parametrizations of varieties
- multiplicative varieties
- sections of varieties
- intersection criteria
- groups of (equiv. classes) of varieties
- ideals of varieties
- projections of varieties
- differentials
- transformations of varieties
- analytic / algebraic / algebroid / . . . varieties
- varieties as topological spaces
- varieties as pre-schemes
- specializations

(c) – local representation
- local structure
- dimension
- Hausdorff dimension of zero-sets
- analytic coverings
- irreducible components
- volume estimates
- intersection indices

(d) – presheaves
- sheaves
- preschemes
- ring-topologies / ring-TS
- spectra (sheaves of rings for ring-TS. . .)
- affine schemes
- locally-affine (≡ general) schemes
- categories of schemes
- {prime ideas (of ring A)} = spec A (as a set)
- $V(a)$:={prime ideals $b \subseteq$}

(e) – sheaves of structures
- germs of holomorphic functions
- sections
- complex (analytic) spaces
- coverings of domains
- underlying TS, X
- open covers of X
- complex-analytic structures

(f) – 1-dimensional algebraic varieties
- manifolds
- projective spaces
- plane curves
- irreducible curves
- singular (multiple) points
- divisors (≡ elements of free abelian groups generated by points)
- valuations
- divisors of **functions**
- complete linear systems
- dimension of complete linear systems
- Riemann-Roch (-type) theorems
- (effective) genus

- function fields (algebraic, of transcendence degree 1)
- **models** of function fields (≡ algebraic curves)
- Jacobian varieties of nonsingular curves
- polarized varieties
- algebraic curves and abstract Riemann surfaces
- compact Riemann surfaces ↔ smooth algebraic curves

(g) – algebraic verities of dimension 2
- surfaces corresponding to algebraic functions of 2 CV
- homological structure of algebraic surfaces
- divisors / linear systems over algebraic surfaces
- resolution of surfaces (blowing-up / down)
- **non**singular surfaces
- traces
- deficiencies
- arithmetic genus
- Riemann-Roch theorems
- invariants
- plurigenera
- rational surfaces
- ruled surfaces
- linear genus
- characteristic sets
- birational transformations
- (relatively) minimal models
- exceptional divisors
- curves of first / second kinds

(h) – analytic varieties
- analytic hypersurfaces
- irreducible analytic varieties
- 2-D compact complex manifolds
- bi-meromorphic mappings
- exceptional curves
- minimal models
- holomorphic maps: {curves} → {points}
- classification(s) of analytic surfaces
- interrelation of {curves} and {chains}
- representations of chains
- normal forms for hypersurfaces

9. Differential Topology / Geometry

(a) analysis on manifolds

(b) densities

(c) curves

(d) Morse Theory (critical points)

(e) differential forms

(f) integration (of forms)

(g) degrees of maps

(h) local curve-theory

(i) global curve-theory

(j) local surface-theory

(k) global surface-theory

(l) transversality

(m) stability criteria (for maps)

(n) intersection criteria

(o) fixed-point criteria

(p) triangulation / dissection procedures

(q) cohomology criteria

(r) homotopy criteria

References All of the topics listed – as 9(a) to 9(r) – are covered, adequately for the identification of quasi-diagrammatic components, in the following expositions, where the original sources are given.

(i) Hirsch, MW, '*Differential Topology*' (Springer, 1976)

(ii) Naber, GL, '*Topological Methods in Euclidean Spaces*' (CUP, 1980)

(iii) Guillemin, V / Pollack, A, '*Differential Topology*' (Prentice-Hall, 1974)

(iv) Golubitsky, M / Guillemin, V, '*Stable Mappings and Their Singularities*' (Springer, 1973)

(v) Berger, M / Gostiaux, B, '*Differential Geometry: Manifolds, Curves and Surfaces*' (Springer, 1988)

(vi) Spivak, M, '*A Comprehensive Introduction to Differential Geometry*' (Publish of Perish, Berkeley, Ca, 1979)

(vii) Milnor, J, '*Topology from the Differentiable Viewpoint*' (U of Virginia Press, 1965)

(viii) Milnor, J, '*Morse Theory*' (Princeton, UP, 1963)

(ix) Choquet-Bruhat, Y / Dewitt-Morette, C / Dillard-Bleick, M, '*Analysis, Manifolds and Physics*' (N-Holland, 2nd Ed., 1982)

The correlations among the topics (a)–(r), and the references (i)–(ix), are multiple; hence no detailed indications are given. Moreover, each of the topics listed has facets with quasi-graphical content; and so on. Consequently, the topics detected here (and for the other species of diagrams considered) should be regarded primarily as illustrations of diagram-structure. This basic collection of diagrams, covering all of the principal areas of (pure / applied) mathematics, constitutes an initial version of 'the diagrammatic content of $\mathcal{M}_t$' – which may be designated as $\mathbb{D}_{\mathcal{M}_t}$ – which may be designated as $\mathbb{D}_{\mathcal{M}_t}$. The manifold problems encountered in any rigorous specification of $\mathbb{D}_{\mathcal{M}_t}$ do not detract significantly from its usefulness in the organisation / development of the OIB. Moreover, novel procedures, as well as 'new' types of problems may be suggested by viewing results graphically.

NOTE From now on, only collections of general references are given to cover the topics listed.

10. Real Analysis

(a) Properties of (co)domains for maps $f : U \supset A \to B \subset V$

(b) Geometrical characteristics of $f : U \supset A \to B \subset V$

(c) Forms of continuity of $f : U \supset A \to B \subset V$

(d) Forms of sequential convergence

(e) Forms of series convergence

(f) Measures (in spaces ${}_{S}X$)

(g) Integrals (over spaces ${}_{S}X$)

(h) Indefinite integrals

(i) Simplex-type representations of operations

(j) Simplex-type decompositions of domains

(k) Simplex-type generation of estimates / inequalities

(l) Higher-dimensional quasi-geometric diagrams

(m) Convex Functions

(n) Harmonic Functions

(o) Subharmonic Functions

NOTES

(i) Here (and for Complex / Functional Analysis) the spaces involved, ${}_{S}X$, correspond to the 'Basic Spaces' introduced in E_1 (and used in all subsequent Essays).

(ii) The spaces U, V are chosen from the ${}_{X}S$.

(iii) The (co)domain properties include connectivity / boundary-smoothness / exist- ence-of-singular-point / . . . .

(iv) Geometrical characteristics of the maps f include boundedness / openness.

(v) Forms of continuity include: point-wise / uniform / absolute / complete / semi / mean / stochastic-mean.

(vi) Forms of sequential convergence include: conditional / absolute / pointwise / uniform / weighted / stochastic.

(vii) Forms of series-convergence include conditional / absolute / uniform / mean / weighted / stochastic – all for various types of series (e.g., Fourier, hypergeometric, general-orthogon-al).

(viii) The measures (in spaces ${}_{X}S$) have many forms – allowing 'content' to be assigned to domains in diverse settings. These forms include: Lebesgue / Catatheodory / Radon / Haar / Hausdorff / Markov / Plancherel / Jordan.

(ix) The associated integrals (over domains in general spaces) may be approached – through the criteria / procedures developed in these Essays such approach schemes have graphic facets..

(x) The possible variants of 'indefinite integral', corresponding to admissible measures over the $_SX$, may be used to obtain estimates of ME whose representations contain such integrals.

(xi) Simplex-based definitions of the derivatives / increments of functions $f : \mathbb{R}^m \supset A \to B \subset \mathbb{R}^n$ have analogues in more general spaces. This formalism has a geometrical flavour.

(xii) Quasi-simplicial decompositions of domains (in any class of spaces) yield estimates of content, as well as allowing the representation of diverse functions in effective forms.

(xiii) Estimates / inequalities for ME defined over domains in general spaces may be derived by evaluating such ME for certain simplicial subdomains.

(xiv) Diagrams in (multi)linear algebra often depict 'n-dimensional' objects in quasi-geometrical contexts.

References

As before, a small collection of general references covers all of the topics 10(a)–10(l), with many variants / overlaps, and full references to the original sources.

(i) Royden, HL, '*Real Analysis*' (3rd ed., Macmillan, 1988)

(ii) Smith, KT, '*Primer of Modern Analysis*' (Springer, 1983)

(iii) Manhin, J, '*Analyse*' (De Boeck-Wedmael, 1992)

(iv) Halmos, PR, '*Measure Theory*' (Van Nostrand, 1950)

(v) McShane, EJ / Botts, T, '*Real Analysis*' (Van Nostrand, 1959)

(vi) Price, GB, '*Multivariable Analysis*' (Springer, 1984)

(vii) Rogers, CA, '*Hausdorff Measures*' (CUP, 1970)

(viii) Voxman, WL / Goetschel, RH, '*Advanced Calculus*' (Dekker, 1982)

(ix) Hobson, EW, '*The Theory of Functions of a Real Variable*', Vols 1,2 (3rd ed., CUP, 1927; Dover, repr., 1957)

(x) Lay, SR, '*Convex sets...*' (Wiley, 1982)

(xi) Eggleston, HG, '*Convexity*' (CUP, 1969)

(xii) Burckel, RB, '*... Classical Complex Analysis*' (Birkhauser, 1979)

(xiii) Doob, JL, '*Classical Potential Theory...*' (Springer, 1984)

11. Complex Analysis

(a) Analytical / topological properties of curves / domains

(b) Holomorphic Functions

(c) Power Series

(d) Complex Integration

(e) Families / Sequences / Series of Complex Functions

(f) Conformal Mapping

(g) Analytic Continuation

(h) General Riemann Surfaces

(i) Complex Approximation

(j) Spaces of Complex Functions

(k) Elliptic Functions

(l) Abelian Functions

(m) Meromorphic Functions

(n) Value-Distribution Theory

(o) Modular Forms / Automorphic Forms

(p) Integral Representations

(q) Subharmonic Functions

(r) Boundary Problems

(s) Extension Problems

(t) PDE Techniques

(u) Potential theoretic techniques

(v) Entire Functions

(w) Residue Theory

(x) Complex Analytic Sets / Hypersurfaces

(y) Representations in Domains of Special Types

(z) Complex Manifolds

NOTES

This general subject spans a vast range of mathematical fields, but the topics listed here are selected for their importance in the development of 'pure analysis'. Many of the topics cover (in difference ways) functions of one variable, and, of several variables. The following (minimal) comments are confined to the citation of some key examples.

(i) Such properties are important (e.g.) in the integral representation of functions, and in the solution of various boundary problems – for instance in Elasticity and in Quantum Field Theory.

(ii) The admissible representations / characterisations of 'holomorphic functions' determine corresponding 'versions' of Complex Analysis. The terms 'holomorphic' and 'analytic' are essentially synonymous.

(iii) The Taylor Series yields 'germs' of analytic functions (for analytic continuation).

(iv) This is central – for 1, of $n \geqslant 2$ complex variables. The deformation of contours introduces homotopy criteria.

(v) This includes 'normal families', and various criteria for local / almost / uniform convergence.

(vi) The Riemann Mapping Theorem, and existence of mapping onto canonical n-connected domains are the main results.

(vii) This includes the definitions of CAF and its 'branches'; of singular / branch points. The Monodromy Theorem is central.

(viii) Here, classification problems are dominant. Divisors and differentials establish links with Algebraic Geometry. Uniformization is the main goal.

(ix) This includes: approximation by specified classes of functions; uniform approximation; interpolation best approximation – all, over various types of domains.

(x) This has both function-theoretic and functional-analytic aspects. Here, the calculational aspects are stressed.

(xi) Elliptic Functions have an extensive classical theory for the two basic types, 'Weierstrass' and 'Jacobi'. Many applications are found in Mechanics, Physics and Number Theory (e.g.).

(xii) Abelian Functions are representable as quotients of suitable Theta Functions (which are themselves representable in terms of Elliptic Functions). The intricate theory of these functions is of fundamental importance in Algebraic Geometry and Number Theory.

(xiii) The distinctive character of Meromorphic Functions is reflected in Nevanlinna's fundamental theorems. 'Meromorphic continuation' is a modest generalisation of DAC. Representation criteria are exemplified by Mittag-Leffler's theorem.

(xiv) The general theory applies to both one-variable and n-variable functions – 'in principle'; but 'in practice', parts of the theory are 'essentially of one-variable character'. The 'n-variable' part is of importance for both algebraic and differential geometry.

(xv) Modular forms are (by definition) automorphic forms for the modular group SL_2 $(\mathbb{Z})$. Automorphic forms for general groups are defined through PD eigenvalue conditions and specified transformation rules. Most of the applications are in Number Theory.

(xvi) This covers, not only all of the basic 'integral theorems', but also definition of Special Functions as contour integrals with parameters, and 'Cauchy Integrals'. Derivation / solution of singular integral equations are included.

(xvii) In Complex Analysis, subharmonic functions occur chiefly in the study of H^p-spaces.

(xviii) The Riemann, and Hilbert, boundary problems require the determination of analytic functions from their boundary behaviour. The solutions depend on properties of Cauchy Integrals.

(xix) The basic problem here is to determine domains of holomorphy – for functions of several complex variables. More generally, results on 'the extension of analytic objects' are sought.

(xx) The PDE techniques involve the (generalized) Cauchy-Riemann conditions (for n CV). This yields: (α) a PDE-theory of analytic functions of n CV, and (β) a theory of 'generalized analytic functions' of 1 CV.

(xxi) These (potential theoretic) methods yield alternative representations / estimates in the study of Meromorphic Functions, Cluster Sets, Conformal Mapping Covering Surfaces and Riemann Surfaces.

(xxii) The classical theory of Entire Functions includes 'canonical factorisations' and results on zero-distributions (e.g., in sectors of $\mathbb{C}$). For functions of n CV, many of these results have analogues. The growth estimates (for 1 CV or n CV) are also of central importance.

(xxiii) The extension of Cauchy's Residue Theorem (in its most general 1-variable form) to cover functions of n CV involves concepts from Differential Geometry / Topology.

(xxiv) The theory of Analytic Sets / Hypersurfaces furnished extensions of some results in Algebraic Geometry to functions of n CV.

(xxv) Explicit evaluations / estimates may be obtained for functions of n CV over certain types of domains. These specialized cases yield (through approximation) some results for more general domains.

(xxvi) Here, notions of 'complex analytic structure' and 'manifold structure' are combined, in diverse ways, with algebraic geometric / topological techniques.

References:

(i) Newman, MHA, '*Elements of the Topology of Plane Sets of Points*' (CUP, 1951)

(ii) Ahlsors, LV, '*Complex Analysis*' (Mc Graw-Hill, 1966)

(iii) Palka, BP, '*An Introduction to Complex Function Theory*' (Springer, 1991)

(iv) Burckel, RB, '*An Introduction to Classical Complex Analysis (Vol. 1)*' (Birkhäuser, 1979)

(v) Saks, S / Zygmund, A, '*Analytic Functions*' (PWN, 1965)

(vi) Evgravov, MA, '*Analytic Functions*' (Saunders, 1966)

(vii) Rudin, W, '*Real and Complex Analysis*' (McGraw-Hill, 1987)

(viii) Jones, GA / Singerman, D, '*Complex Functions*' (CUP, 1987)

(ix) Farkas, HM / Kra, I, '*Riemann Surfaces*' (Springer, 1992)

(x) Lelong, P / Gruman, L, '*Entire Functions of Several Complex Variables*' (Springer, 1986)

(xi) Hörmander, L, '*An Introduction to Complex Analysis in Several Variables*' (Van Nostrand, 1966)

(xii) Various authors / eds., '*Several Complex Variables*' Vols 1–6 (Springer, 1990)

(xiii) Fuks, BA, '*Introduction to the Theory of Analytic Functions of Several Complex Variables*' (AMS, 1963)

(xiv) Range, RM, '*Holomorphic Functions and Integral Representations in Several Complex Variables*' (Springer, 1986)

(xv) Krantz, SG, '*Function Theory of Several Complex Variables*' (Wiley, 1982)

(xvi) Tsuji, M, '*Potential Theory in Modern Function Theory*' (Maruzen, 1959; repr. Chelsea, 1975)

(xvii) Maurin, K, '*Analysis, Part II*' (PWN, 1980)

(xviii) Miyake, T, '*Modular Forms*' (Springer, 1989)

(xix) Bruggerman, RW, '*Families of Automorphic Forms*' (Birkhäuser, 1994)

12. Functional Analysis

(a) Mappings among the Basic Spaces

(b) Bounded linear mappings

(c) Unbounded linear mappings

(d) Convexity results

(e) Duality results

(f) Extension results

(g) Spectra and resolvents

(h) Spectral representation / analysis

(i) Function spaces: general

(j) Function spaces: examples

(k) Solution of operator equations

(l) Branching analysis: Newton polygons

(m) Structural characteristics of the Basic Spaces

(n) Nonlinear mappings

(o) Topological degree

(p) Monotone mappings

(q) Implicit functions

(r) Fixed-point theory

NOTES

Here, again, the aim is merely to identify topics collectively spanning the vast territory currently included in Functional Analysis – so that quasi-graphical facets may be incorporated as admissible components of data / proof diagrams.

(i) The class of Basic Spaces (for Σ) contains, as examples, all of the types of spaces (e.g., Hilbert / Banach / Topological / Uniform / . . .) encountered in mathematical practice. The study of w.d. mappings among these spaces furnishes the raw materials for the development of this field.

(ii) Boundedness properties are especially relevant for normed linear spaces (NLS), in their divers realizations.

(iii) Unbounded linear operators have more intricate structures than those of boun- ded linear operators – though some of the fundamental theorems do have fairly direct extensions. The main applications are to differential equations, and to the Hilbert-Space formulations of Quantum Mechanics.

(iv) These results are of two types – for spaces, and for operators. The notion of locally convex TLS is important in proving geometric forms of the Hahn-Banach extension theorem – with many applications. The role of convexity in Topology may be elucidated by comparing convexity results for LS with those for LTS. The convexity properties of operators are central to the solution of positive-operator (in)eqs. – as well as in Optimization / Game-Theory / Economics. In particular, the definition of (semi)-norms, and the proof of 'separation theorems', are fundamental here.

(v) For VS there are both algebraic and topological duals – and the adequacy of these theories requires local convexity. The possible topologies on various dual spaces are significant for the study of generalized functions / integrals over such spaces.

(vi) The central results here is the Hahn-Banach Theorem, in its many versions – with diverse variants / generalizations / applications.

(vii) The resolvent set, $\rho(w)$ and spectrum, $\sigma(w)$ of an element w, are complementary (complex) sets – on which $\lambda e - w$ is (resp.) invertible / noninvertible. Here, w may be a linear operator between NLS. More generally, $\{w\}$ may be a Banach algebra. Operators have spectral representations – upon which operational calculi are based. (See (viii)). NOTE that the spectrum is partitioned into 'points', 'continuous' and 'residual' subsets.

(viii) The form of spectral representation depend on the types of operators involved (e.g., compact operators on BS, or, bound, self-adjoint operators on HS). The basic form expresses the operator as an integral of suitable 'projections'. The operational calculi are based mainly on analyticity properties of the resolvent (i.e., of $(\lambda e - w)^{-1}$) as a function of λ.

(ix) The general theories of B-algebras, and of function-algebras embody most of the results minimally dependent on detailed properties (e.g., of smoothness / growth / zero-distribution) of the 'elements' of the space. There is, however, no totally 'abstract' theory, so the distinction between this 'topic' and the next ((x)) in mainly a matter of emphasis on detailed calculation / analysis.

(x) Many types of function spaces arise in the study of PDEs, in connection with smoothness properties of solutions – when the 'coefficient functions' in the PDO are of specified classes, etc.. A range of 'embedding theorems' interrelates spaces of various 'orders' / 'species' – with cognate 'extension' results for the functions / elements. The overall classification is into: continuous / continuously-differentiable / integrable / hybrid spaces. The basic forms of these species include: $C(\Omega)/W^{k,p}(\Omega)$, i.e., $\{$continuous functions on $\Omega\}$ / Sobolev spaces / Lebesgue spaces – but there are diverse variants of each species.

(xi) These 'operator equations' include O/PDE, Integral Equations, Delay-Differential-Difference Equations, Functional-Differential Equations, and more general classes of equations over B-spaces. Most solution procedures involve 'series with parameters.

(xii) The 'branching analysis' includes the use of Newton polygons for (truncated) Puiseux series, together with Liapunor-Schmidt reduction and implicit-function theorems. The 'method of undetermined coefficients' is then used to obtain (adequate approximations to) solutions of the resulting 'branching equations'.

(xiii) This involves the interrelations among topologies / norms / neighbourhoods / ... in the designated 'Basic Spaces' – within which all 'deterministic approaches' for Σ are constructed. The stochastic approaches are then represented through suitable (probability) measures. Such structural information underlies most of the functional-analytic procedures.

(xiv) For general nonlinear mappings, there are no nontrivial 'universal principles'. All of the structure carried by (spaces of) linear mappings is lost, and the main concern is to estimate 'closeness to linearity' in various contexts. On the other hand, most interesting behaviour of operator equations obtains **only** for **non**linear equations. Nevertheless, certain classes of nonlinear mappings do exhibit quasi-universal properties – which constitute the main results of Nonlinear Functional Analysis (**see** (xv)–(xviii)).

(xv) The study of 'topological degree' of mappings – in finite-dimensional or infinite-dimensional spaces – allows estimates of the number of solutions of classes of operator equations in these spaces. The notion of 'rotation' for vector fields is closely related to that of degree; both concepts are used to obtain FP theorems. Variants include the Leray-Schauder degree, which is more widely applicable that 'standard' topological degree.

(xvi) The degree-theory of nonlinear maps allows estimates of the number of solutions of certain operator equations, but the question of **uniqueness** of solution remains open. For 'monotone operators' (on HS or BS), uniqueness problems may be solved. Some extensions to more general spaces may be achieved through the use of 'accretive operators'.

(xvii) There are two classes of Implicit Function Theorems (IFT): the 'soft IFT', where stringent conditions are imposed on the mapping and its inverse (between BS); and the 'hard IFT', where much weaker (but more intricate) conditions are required.

(xviii) These FP theorems (of several types) are basic for proving existence of the solution(s) of various classes of nonlinear operator equations – and for investigating possible branching behaviour. Other (often surprising) uses of FPT occur in many parts of Pure and Applied Mathematics.

References:

(i) Taylor, AE / LAY, DC, '*Introduction to Function Analysis*' (Wiley, 1980)

(ii) Lang, S, '*Real and Functional Analysis*' (3rd ed., Springer, 1993)

(iii) Edwards, RE, '*Functional Analysis*' (Holt, R&W, 1965; Dover, 1995)

(iv) Rudin, W, '*Functional Analysis*' (McGraw-Hill, 1973)

(v) Berberian, SK, '*Lectures in Functional Analysis and Operator Theory*' (Springer, 1974)

(vi) Brown, A / Pearcy, C, '*Introduction to Operator Theory: I*' (Springer, 1977)

(vii) Goldberg, S, '*Unbounded Linear Operators*' (McG-H, 1966)

(viii) Griffel, DH, '*Applied Functional Analysis*' (Horwood, 1981)

(ix) Balakrishnan, AV, '*Applied Functional Analysis* (Springer, 1976)

(x) Milne, RD, '*Applied Functional Analysis*'(Pitman, 1980)

(xi) Naylor, AW / Sell, GR, '*Linear Operator Theory in Engineering and Science*' (HR&W, 1971; Springer, 1982)

(xii) Deimling, K, '*Nonlinear Functional Analysis*' (Springer, 1985)

(xiii) Aubin, J-P, '*Optima and Equilibria*' (Springer, 1993)

(xiv) Zeidler, E, '*Nonlinear Functional Analysis and its Applications*' (Parts I-V, Springer, 1984/88)

(xv) Lloyd, NG, '*Degree Theory*' (CUP, 1978)

(xvi) Schwartz, JT, '*Nonlinear Functional Analysis*' (G&B, 1969)

(xvii) Rickart, CE, '*General Theory of Banach Algebras*' (Van Nostrand, 1960; Krieger, repr., 1974)

(xviii) Rickart, CE, '*Natural Function Algebras*' (Springer, 1979)

(xix) Cotlar, M / Cignoli, R, '*Introduction to Functional Analysis*' (N-Holland, 1974)

13. Lie Groups / Lie Algebras (LG/LA)

(a) LG as manifolds with extra structure

(b) L subgroups

(c) Homomorphisms

(d) Linear representations

(e) Local LG

(f) (Transitive) Actions

(g) Orbits / Stabilizers

(h) Compact LG

(i) Coset manifolds

(j) Quotient groups

(k) Semi-direct products

(l) Connectedness

(m) Simple-Connectedness

(n) Covering homomorphisms

(o) Universal covering LG

(p) Tangent algebras of LG (TA(lgebras))

(q) Vector fields on LG

(r) Differentials of homomorphisms of LG

(s) Differentials of LG actions

(t) Tangent algebras of stabilizers

(u) Path-deformations in manifolds

(v) Virtual L-subgroups

(w) Correspondence of L subgroups and sub-T-algebras

(x) 'Fundamental Theorem of Calculus' for LG/A

(y) Abelian LA

(z) Exponential maps

(a′) FP-subgroups of LG-automorphisms

(b′) TA of Aut.; Derivations:

(c′) Commutator subgroups of LG

(d′) Quasi-inverse-problems for Virtual L-subgroups

(e′) Correspondence of {mutual commutator subgroups} and {mutual commutator subalgebras of ideals}

(f′) Solvable LG

(g′) Functor from {associative algebras A} to {LA: $\mathcal{L}(A)$}

(h′) Universal enveloping algebras

(i′) Equivalence of categories {local-analytic LG} and {finite-dimensional LA}

NOTES

In this case, the topics listed are not disparate theories / procedures (as for Real / Complex / Functional Analysis), but rather, various facets of a single overall theory. Consequently, there is no need for comments here – except to stress that all of the fundamental concepts are covered in this list. All admissible proof / data diagrams associated with LG/LA may be discussed on this basis.

References:

(i) Jacobson, N, '*Lie Algebras*' (Wiley, 1962; Dover, 1979)

(ii) Sattinger, DH / Weaver, OL, '*Lie Groups and Algebras with Applications to Physics, Geometry and Mechanics*' (Springer, 1993)

(iii) Onishchik, AL, (ed.), '*Lie Groups and Lie Algebras*' (EMS Vol.20; Springer, 1993)

(iv) Richtmyer, RD, '*Principles of Advanced Mathematical Physics, Vol II*' (Springer, 1981)

(v) Warner, FW, '*Foundations of Differentiable Manifolds and Lie Groups*' (Springer (repr.) 1983)

(vi) Hermann, R, '*Lie Groups for Physicists*' (Benjamin, 1966)

(vii) Kac, VG, '*Infinite-Dimensional Lie Algebras*' (CUP, 1985)

(viii) Barut, AO / Raczka, '*Theory of Group Representations and Applications*' (PWN, 1977)

14. Many-Body Calculations in (Chemical) Physics

The intention here is simply to identify several broad areas where quasi-diagrammatic procedures are important. The references contain detailed lists of original sources.

(a) Solid State Physics

(b) Lattice Dynamics

(c) Statistical Mechanics

(d) Statistical Field Theory

(e) Chemical Reaction Theory

(f) Classical Mechanics; Celestial Mechanics

(g) Fluid Dynamics

(h) Superconductivity

(i) Superfluidity

(j) Nuclear Physics

(k) Plasma Physics

Notice that there are some overlaps among the topics here – and with 'High-Energy Physics' ((15)) and 'Graphs / Networks' ((16))

NOTES

(i) Solid-state Physics covers a multitude of 'sub-theories', for instance: Theory of Metals (including 'Fermi Surfaces'); Theory of Semiconductors; Magnetism; Crystallography; Transport Theory – along with most of the other topics ((b)-(k)) listed above.

(ii) Lattice Dynamics pertains mainly to calculations of 'collective effects' due to the motion / interaction of various types of (quasi-)particles with 'equilibrium positions' at the points of n-dimensional lattices of specified symmetry classes. The interactions may be linear of nonlinear, 'classical' or 'quantum-mechanical', 'deterministic' or 'stochastic'.

(iii) 'Statistical Mechanics' refers to the representation / calculation of 'bulk properties of matter' in terms of the (microscopic) motions of the constituent particles – primarily on the basis of Hamiltonian formulations of Dynamics (or of its quantum analogues). As for (i), most of the other topics have statistical-mechanical interpretations, but the formalism is different.

(iv) Statistical Field Theory covers both Quantum Field Theory (of several types) and the use of field-theoretic schemes in Statistical Mechanics. As usual, there are deterministic and stochastic versions. Many forms of diagrams 'arise naturally' in series expansions.

(v) In the present context, Chemical Reaction Theory refers to the (quantum-)statistical-mechanical determination / estimation of reaction products / rates. However, the elaborate **graphical notation for the synthesis / reaction of (in)organic chemical compounds** may also be included here – especially because of the rich possibilities it offers for 'reflections' (**see** E_7).

(vi) Celestial Mechanics is, of course, just one sub-theory of Classical Mechanics, but it involves many perturbation expansions – with quasi-graphical content. The study of nonlinear dynamical systems (deterministic, stochastic, discrete and continuous) includes Bifurcation-Theory / Fractals / Chaos.

(vii) Fluid Dynamics may be treated both, as a part of Continuum Mechanics and, as a part of Statistical Mechanics. Here, the PDE-based scheme is considered. Naturally, these two schemes are not independent, but the methods used are largely distinct.

(viii) Superconductivity and superfluidity are examples of many-body quantum-mechanical phenomena. Nevertheless, each has its own formalist, with many diagrammatic facets.

(ix) Nuclear Physics comprises several – largely distinct – parts: Nuclear Structure; Nuclear Reactions; Nuclear Engineering (reactor design / analysis), etc.. The potential 'diagrammatic content' of each part will be clear from the basic references given – where, as always, detailed citations may be found.

(x) Plasma Physics is concerned, mainly, with the motion of (large) assemblies of charged particles in EM fields – both, from a theoretical point of view, and for applications, notably to Astrophysics and to the fusion-generation of power.

References

(i) Ashcroft, NW / Mermin, ND, '*Solid State Physics*' (HR&W, 1976)

(ii) Ziman, JM, '*Principles of the Theory of Solids*' (CUP, 1964)

(iii) Ziman, JM, '*Electrons and Phonons*' (OUP, 1960)

(iv) Shockley, W, '*Electrons and Holes in Semiconductors*' (Van Nostrand, 1950)

(v) McKelvey, JP, '*Solid State and Semi Conductor Physics*' (Harper & Row, 1966)

(vi) Born, M / Huang, K, '*Dynamical Theory of Crystal Lattices*' (OUP, 1954)

(vii) Ludwig, W, '*Recent Developments in Lattice Theory*' (Springer Tracts in Modern Physics #43, 1967)

(viii) Fowler, RH, '*Statistical Mechanics*' (2nd ed., CUP, 1936)

(ix) Huang, K, '*Statistical Mechanics*' (Wiley, 1963)

(x) Ruelle, D, '*Statistical Mechanics*' (Benjamin, 1969)

(xi) Sinai, Ya. G (ed.), '*Statistical Mechanics*' (World Scientific, 1991)

(xii) Stumpf, H, '*Quantum Processes in Polar Semi-Conductors and Insulators*' (Vieweg, Parts 1,2, 1983)

(xiii) Itzykson, C / Drouffe, J-M, '*Statistical Field Theory*' (Vols. 1,2, CUP, 1989)

(xiv) Levine, RD / Bernstein, RB, '*Molecular Reaction Dynamics*' (OUP, 1974)

(xv) Levine, RD, '*Quantum Mechanics of Molecular Rate Processes*' (OUP, 1969)

(xvi) Purcell, KF / Kotz, JC, '*Inorganic Chemistry*' (Saunders, 1977)

(xvii) Norman, ROC, '*Principles of Organic Synthesis*' (Methuen, 1968)

(xviii) Abraham, R / Marsden, JE, '*Foundations of Mechanics*' (2nd ed., Benjamin / Cummings, 1978)

(xix) MacKay, RS / Meiss, JD, '*Hamiltonian Dynamical systems*' (Hilger, 1987)

(xx) Sinai, Ya. G (ed.) '*Dynamical Systems*' (World Scientific, 1991)

(xxi) Hagihara, '*Celestial Mechanics*' (Vols. 1,2, MIT Pr 1971/72)

(xxii) Brouwer / Clemence, '*Methods of Celestial Mechanics*' (AP, 1961)

(xxiii) Batchelor, GK, '*Fluid Dynamics*' (CUP, 1970)

(xxiv) Lamb, H, '*Hydrodynamics*' (CUP, 1879 6th ed., 1932; Dover, 1945)

(xxv) Streeter, VL, '*Handbook of Fluid Dynamics*' (McGraw-Hill, 1961)

(xxvi) Fetter, AL / Walecka, JD, '*Quantum Theory of Many-Particle Systems*' (McGraw-Hill, 1971)

(xxvii) Feshback, H, '*Theoretical Nuclear Physics: Nuclear Reactions*' (Wiley, repr. 1992)

(xxviii) Hofstadter, R (ed.) '*Nuclear and Nucleon Structure*' (Benjamin, 1963)

(xxix) Akcasu, Z / Lellouche, GS / Shotkin, LM, '*Mathematical Methods in Nuclear Reactor Dynamics*' (AP, 1971)

(xxx) Glasstone, S / Edlund, MC, '*The Elements of Nuclear Reactor Theory*' (Macmillan, 1952)

(xxxi) Brown, GE, '*Unified Theory of Nuclear Models*' (North U, 1964)

(xxxii) Satchler, GR, '*Direct Nuclear Reactions*' (OUP, 1983)

(xxxiii) Ring, P / Schuck, P, '*The Nuclear Many-Body Problem*' (Springer, 1980)

(xxxiv) Rosenbluth, MN / Sagdeev, RZ, '*H/B of Plasma Physics*, Vol. 1 (North Holland, 1983)

(xxxv) Chakraborty, B, '*Principles of Plasma Mechanics*' (2nd ed., Wiley, 1990)

15. High-Energy Physics

This branch of Physics, in the present context, involves the evaluation / estimation of various 'parameters', for the interaction of classes of Elementary articles, through several species of **Q**quantum **F**ield **T**heories. The corresponding representations / expansions require associated **R**enormalization **P**rocedures. The graphical interpretation of the terms in these expansions give rise to a wide range of **I**nteraction **D**iagrams, whose combinatorial / topological properties have been studied intensively. Accordingly, the topics listed here merely cover the most prominent schemes for such calculations.

(a) Quantum Electrodynamics (QED)

(b) Quantum Chromodynamics (QCD)

(c) Lattice Gauge Theories (LGT)

(d) Generalized Quantum Field Theories (GQFT)

(e) Properties of Interaction Diagrams

(f) String Theories

NOTES

(i) QED has many formulations, each having associated classes of diagrams. The aim is always to remove singularities, and to produce tractable expansions. Additional problems arise when the 'vacuum state' is destabilised by intense external fields. This requires new types of expansions / procedures. Asymptotic techniques (involving generalized functions) yield other forms of expansions / diagrams. Topological criteria are also important here, for all types of diagrams.

(ii) QCD evolved from the theory of quarks. Problems of expansions / renormalization parallel those for QED – but the techniques / expansions / diagrams are all qualitatively different.

(iii) LFT may be regarded as a development of the correspondences between QFT and Classical Statistical Mechanics (CSM). Both relativistic and continuum models may be treated on this basis (through appropriate limit procedures). Diffusion models, based on stochastic differential equations, play a prominent role in this theory.

(iv) Many of the expansions / renormalization / asymptotic / topological properties are common to **all** QFT models – and so characterise classes of GQFT. The rigorous development of renormalized QFT may, therefore be presented in very general forms – in which suitable types of diagrams are still of fundamental importance.

(v) The various (combinatorial / asymptotic / topological / . . .) properties of GQFT diagrams may be studied systematically. In particular, both Homotopy Theory and Homology Theory are widely used here.

(vi) String Theory actually has many forms – the current versions being types of so-called **Super**string Theories. The string diagrams are interrelated in diverse ways; they are also related to associated Feynman diagrams. Hence, they possess cognate combinatorial / asymptotic / topological properties.

References

(i) Ryder, LH, '*Quantum Field Theory*' (CUP, 1985)

(ii) Cheng, T-P / Li, LF, '*Gauge Theory of Elementary Particle Physics*' (OUP, 1984)

(iii) Ynduráin, FJ, '*The Theory of Quark and Gluon Interactions*' (Springer, 2nd ed., 1993)

(iv) Fradkin, ES / Gitman, DM / Shvartsman, SM, '*Quantum Electrodynamics*' (Springer, 1991)

(v) Schwarz, AS, '*Quantum Field Theory and Topology*' (Springer, 1993)

(vi) Smirnov, VA, '*Renormalization and Asymptotic Expansions*' (Birkhäuser,1991)

(vii) Zavialov, OI, '*Renormalized Quantum Field Theory*' (Kluwer, 1990)

(viii) Rother, HJ, '*Lattice Gauge Theories: An Introduction*' (World Scientific, 1992)

(ix) Hwa, RC / Teplitz, VL, '*Topology and Feynman Integrals*' (Benjamin, 1966)

(x) Schwarz, JH, (ed.) '*Superstrings: The First 15 Years of Superstring Theory*' (2 Vols., World Scientific, 1985)

(xi) Green, MB / Schwarz, JH / Witten, E, '*Superstring Theory*' (2 Vols., CUP, 1987)

(xii) Rivers, RJ, '*Path Integral Methods in Quantum Field Theory*' (CUP, 1987)

16. Graphs / Networks (In All Fields)

(a) Topological aspects of Graph Theory

(b) Topological aspects of Network Theory

(c) Combinatorial aspects of Graph Theory

(d) Combinatorial aspects of Network Theory

(e) Convex programming

(f) Simplicial approximation

(g) Matroid theory

(h) Machine-scheduling

(i) Homotopic routing schemes (VSLI)

(j) Surface embeddings ('representativity')

(k) Constrained cycles in graphs

(l) Diagrammatic expansions (in SM/QFT/...)

(m) Network flow problems (in OR)

(n) Electro-magnetic circuit analysis / synthesis

(o) General Systems Theory

(p) Linear Network Theory

(q) Nonlinear Network Theory

(r) Stochastic Network Theory

NOTES

(i) Of all admissible forms of graphs / networks corresponding to topics (a)–(), only generic types are taken as distinct – no matter what their fields of origin may be.

(ii) There are obvious identifications between networks and their underlying graphs, as a network problems involve flows on graphs.

(iii) Although proof-diagrams may be viewed as labelled, directed, self-intersecting graphs – and calculations, as species of flows on such graphs – the topics considered here refer primarily to 'typical' theorems / applications of graphs / networks within 'mathematical activity' in general.

(iv) Since every proof / calculation has discrete / combinatorial / graphical facets, the topics (a)-() have been selected mainly to illustrate the ubiquity of graph / network structure.

References

(i) Bollobás, B, '*Graph Theory*' (Springer, 1979)

(ii) Bollobaás, B, '*Combinatorics*' (CUP, 1986)

(iii) Colbourn, CJ, '*The Combinatorics of Network Reliability*' (OUP, 1987)

(iv) Temperley, HNV, '*Graph Theory and its Applications*' (Horwood, 1981)

(v) Kelly, FP / Williams, RJ (eds.), '*Stochastic Networks*' (Springer, 1995)

(vi) Carré, B, '*Graphs and Networks*' (Elsevier, 1979)

(vii) Dolezal, V, '*Nonlinear Networks*' (Elsevier, 1977)

(viii) Busacker, RG / Saaty, TL, '*Finite Graphs and Networks*' (McGraw-Hill, 165)

(ix) Girard, A, '*Routing and Dimensioning in Circuit-Switched Networks*; (Addison-Wsley, 1990)

(x) Lee, SC, '*Modern Switching Theory and Digital Design*' (Prentice-Hall, 1978)

(xi) Hui, JY, '*Switching and Traffic Theory for Integrated Broadband Networks*' (Kluwer, 1990)

(xii) Murota, K, '*Systems Analysis by Graphs and Matroids*' (Springer, 1987)

(xiii) Papadimitriou, CH / Steiglitz, K, '*Combinatorial Optimization*' (Prentice-Hall, 1982)

(xiv) Lewis, ER, '*Network Models in Population Biology*' (Springer, 1977)

(xv) Bachem, A / Grótschel, M / Korte, B (eds.) '*Mathematical Programming: The State of the Art*' (Springer, 1983)

(xvi) Boffey, TB, '*Graph Theory in Operations Research*' (Macmillan, 1982)

17. Illustrative Diagrams (In All Fields)

(a) Schematic diagrams

(b) CAD representations

(c) Graphics: planar diagrams

(d) Graphics: planar sections

(e) Perspective forms

(f) Holographic representations

(g) Flow-Diagrams (for programs / proofs / . . .)

(h) Flows / Interactions on diagrams

NOTES

(i) This class of diagrams ranges over all types of depiction – from those merely indicating the relative locations of (constituents of) a collection of $n(\geqslant 1)$ ME, $\{e\}_n$, within prescribed environments, through many forms of linear / planar / spatial / higher-dimensional **static** representations, to various sorts of **dynamic** representations (of flows / interactions) defined over the static forms.

(ii) The schematic diagrams do not pretend to any kind of literal accuracy; rather, they suggest underlying connections among the ME / constituents / environments. In some cases these connections may be 'upgraded' to reflect more precise / detailed / rigorous relationships – static or dynamic.

(iii) The CAD representations are, in general, both accurate and detailed. They encompass all kinds of sections / projections, and often contain numerical specifications (of components, etc.). They are, therefore, primarily relevant to Engineering and the 'Applied Sciences'. Nevertheless, their inverse reflections' may have fundamental mathematical ramifications.

(iv) All of the topics (c), (d), (e) are covered by **computer-graphics packages**; and even (f) will soon become a standard feature. Since these types of diagrams typically represent aspects of particular (classes of) objects, in various transformations, they have little direct mathematical significance, but – as for CAD – their inverse reflections may be of great importance.

(v) Flow-diagrams for computer programs have both static and dynamic aspects. The 'flow-charts' embodying algorithms were introduced almost at the inception of 'automatic computation' – and gave rise to 'algebraic flow-chart languages'. The consequent 'analysis of algorithms' is characterised – for sequential calculations – by diagrammatic representations. These forms are, however, essentially static in nature. The analysis of 'program flow' has a more dynamic character, for both 'sequential' and 'parallel' modes of computation. The extension of such notions to 'Σ -calculations' (with which all of the present Essays are ultimately concerned) incorporates a variety of flow / interaction phenomena. Thus, proof-diagrams (as discussed earlier in this Essay) embody both flows and interactions – even though such diagrams are partially schematic.

(vi) The general analysis of flows / interactions on (classes of) diagrams has the character of a theory of 'generalized lattice models' – where both the underlying lattices and the associated flows / interactions may be substantially arbitrary. There are diverse techniques for investigating such problems – and cognate realizations involving all areas of Mathematics.

References

(i) Prather, RS, '*Discrete Mathematical Structures for Computer Science*' (Houghton Mifflin, 1976)

(ii) Muchnick, SS / Jones, ND (eds.) '*Program Flow Analysis*' (Prentice-Hall, 1981)

(iii) Raynal, M / Helary, J-M, '*Synchronization and Control of Distributed Systems and Programs*' (Wiley, 1990)

(iv) Gelenbe, E, '*Multiprocessor Performance*' (Wilet, 1989)

(v) Park, CS, '*Interactive Microcomputer Graphics*' (A-W, 1985)

(vi) Jenks, RD / Sutor, RS, '*Axiom: The Scientific Computation System*'

(vii) Smith, HM, '*Principles of Holography*' (Wiley, 2nd ed., 1975)

(viii) Kim, WH / Chien, RT-W, '*Topological Analysis and Synthesis of Communication Networks*' (Columbia UP, 1962)

18. Pure Geometry

(a) Planar Euclidean Geometry

(b) Solid (3-D) Euclidean Geometry

(c) Projective Geometry

(d) Hyperbolic Geometry

(e) Elliptic Geometry

(f) Parabolic Geometry

(g) Descriptive Geometry

(h) Riemannian Geometry

(i) Non-Riemannian Geometry

(j) General Non-Euclidean Geometry

NOTES

(i) The diagrams here are (mostly) literally representative of (sections of) configurations – subject to the appropriate axiom-systems for each species of Geometry.

(ii) In some cases (e.g., (g)) metrical precision may be incorporated, though the only essential characteristic is the correct indication of 'geometrical information' (data) upon which deductions may be based.

(iii) The interpretations of geometrical figures obviously depend on the systems in which theorems are to be proven. Hence, a fixed 'visual configuration' may underlie several inequivalent collections of theorems.

(iv) All of the special types of geometries (hyperbolic / elliptic / parabolic / Riemannian / nonRiemanian . . .) have associated metrics – defined over all or part of the underlying space(s). This allows the systematic treatment of these geometries within a general theory (though detailed properties are better developed separately).

(v) Descriptive Geometry furnishes the foundation for CAD – together with Holography.

References

(i) Coxeter, HSM, '*Introduction to Geometry*' (Wiley, 2nd ed., 1969; repr., 1989)

(ii) Berger, M, '*Geometry*' (Vols. 1,11; Springer, 1987)

(iii) Berger, M, et al., '*Problems in Geometry*' (Springer, 1987)

(iv) Gray, J, '*Ideas of Space*' (OUP, 2nd ed., 1989)

(v) Richardus, P / Adler, RK, '*Map Projections*' (North-Holland, 1972)

(vi) Watts, EF / Rule, JT, '*Descriptive Geometry*' (Prentice-Hall, 1946)

(vii) Coxeter, HSM, '*Non-Euclidean Geometry*' (U Toronto P, 1965)

(viii) Coxeter, HSM, '*The Real Projective Plane*' (Springer, 3rd ed., 1993)

(ix) Busemann, H, '*Projective Geometry and Projective Metrics*' (AP, 1953)

(x) Blumenthal, LM, '*Distance Geometry*' (OUP, 1953)

(xi) Greenbarg, MJ, '*Euclidean and Non-Euclidean Geometries*' (Freeman, 2nd ed., 1980)

(xii) Smart, JR, '*Modern Geometries*' (Brooks / Cole, 1973)

(xiii) Hirschfeld, JWP, '*Projective Geometries Over Finite Fields*' (OUP, 1979)

(xiv) Sommerville, DMY, '*An Introduction to the Geometry of N Dimensions*' (Methuen, 1929; Dover, 1958)

(xv) Ziegler, GM, '*Lectures on Polytopes*' (Springer, 1995)

(xvi) Javary, A, '*Géomeétrie Descriptive*' (5th ed., Vols 1,2; Librarie Ch. Delagrave, Paris, 1899)

(xv) Yaglom, IM / Boltyanskii, VG, '*Convex Figures*' (Holt IR/W, 1961)

Approaches to Σ -Diagrams

The topical outlines for diagrams (covering a wide range of mathematical fields) offer indications of typical diagram-components. In E_4, it is remarked that the quasi-graphical content of Σ – say, Σ^G – mirrors Σ itself, since every ME has **graphical facets**. Indeed, an arbitrary ME (say, e) may be represented in terms of its 'basic facets', $\psi_k(e) \equiv {}^k e$, for $1 \leqslant k \leqslant 6$, where the $\psi_k(e)$ correspond (respectively) to the **arithmetical** / **combinatorial** / **algebraic** / **analytical** / **geometrical** / **topological** propertied of e.

It follows that, if e is located / defined within the environment ξ, then the facets, $\psi_k(e)$ may be regarded (schematically) as being realized within the 'projections', ${}^k\xi := \psi_k(\xi)$. Although the resulting reflection of Σ in Σ^G is somewhat vague, is still constitutes a form of approach – of Σ^G to Σ ! Attempts to exhibit the visual content of mathematical theories have produced some outstanding formulations, collectively spanning the main domains of Mathematics. Moreover, almost all of these treatments may now be found in books[310] – so it is quite feasible to develop a pre taxonomic skeleton for Σ^G, as a basis for fuller versions. The major task is to establish uniform notations for the representation of (quasi-) diagrams (covering essentially arbitrary topics) so that approach procedures may be systematized.

NOTE

Since the quasi-graphical facets of ME are highly nonuniform – and often subtly 'hidden' by other more dominant facets – these diagrammatic aspects are not included as 'basic facets'. They are, however, of great importance in the full characterisation of ME.

A 'general teary of approaches to Σ -diagrams' is attainable only if the (classes of) diagrams may be embedded in adequately structured environments. Consequently, the primary task is **to determine which structures may be defined over (appropriate) collections of Σ -diagrams**. Initial steps towards this aim may be taken by investigating:

(a) the admissible (types of) combinations of (classes of) Σ -diagrams;

(b) the possible specification of diagram-environments in terms of the environments of the diagram-elements;

(c) the resulting forms of diagrammatic continuity / convergence / stability;

(d) the extent to which such criteria may be applicable to Σ -**proof**-diagrams.

Some of these matters have been discussed, in a preliminary way, for both diagrams and facets, in E_4. Recall that: general Σ -diagrams involve finite, countable or uncountable collections of (substantially arbitrary) active / passive **elements** – each 'operating' within its prescribed (deterministic / stochastic) **environment**. These elements are subject to **rules** of several types: (i) rules of (relative) **location** / **orientation**; (ii) rules of (directed) **interconnection**;

[310] A sample list is given at the end of this Essay

(iii) rules of r-element **interaction** ($r \geqslant 1$). The diagram D 'as a **configuration** of elements' is denoted by C_D, while the detailed specifications of: C_D, all elements / environments, all rules for the elements, and the environment of C_D itself, together comprise the **components** of 'the diagram D'. Thus, if D(e) denotes a diagram for the ME e, then a schematic representation of D(e) may be given in the form:

$$D(e) \approx (\bar{C}, \bar{\mathbb{E}}, \rho(\bar{\mathbb{E}})), \text{where:}$$

$$\bar{C} \equiv \bar{C}_{D(e)} := (C_{D(e)}, \xi_{C_{D(e)}});$$

$$\bar{\mathbb{E}} \equiv \bar{\mathbb{E}}_{D(e)} := \{(E_\lambda, \xi_{E_\lambda}) : \lambda \in \Lambda_{D(e)};$$

$$\rho(\bar{\mathbb{E}}) \equiv \rho_{D(e)}(\bar{\mathbb{E}}) := \{\rho_S(\{E_\lambda \in S\}) : s \subset \Lambda\}.$$

Here, ξ_X denotes 'the environment of X', the E_λ are elements, the ρ_S are rules, the symbol (a,b) denotes ordered pairs, and $\{\ldots\}$ stands (as usual) for sets. The dependence on e of all parts of the representation of D(e) will be ignored in manipulations involving a single ME; but where at least two distinct ME, say, e_1, e_2, occur (nontrivially) this dependence must be shown. A slight simplification results if 'D(e)' is replaced everywhere by 'e', etc..

On this basis, **approaches to** D(e) have the (schematic) form

$$A(D(e)) \equiv [A_1, A_2, A_3](D(e))$$
$$: = (A_1\bar{C}, A_2\bar{\mathbb{E}}, (A_3\rho)(A_2\bar{\mathbb{E}})),$$

with A_1, A_2, A_3 suitably defined to produce admissible approaches to $\bar{C}, \bar{\mathbb{E}}, \rho$, respectively. Of course, this apparent simplicity masks a diversity of (typically, intricate) combinations of detailed approaches to configurations / elements / rules / environments / ... – mostly defined over the 'Basic Spaces', $_SX$; or else, over 'scientific domains' (within Engineering / Physics / Chemistry / Biology / Economics / ...). The designated species of approach include all types of (deterministic / stochastic) **approximation** – in the maximally general settings associated with spaces $_SX$– together with all w.d. forms of **abstraction / analogy / inverse analogy**.

From now on, all of the notations / definitions / procedures introduced in E_4, for diagrams / facets, are assumed without further comment. They may be applied to the detailed forms of diagrams discussed here, to produce a Σ **-Diagram Calculus (DC)** of potentially great power and generality. The fulfilment of this potential depends mainly on the development of software for efficient handling of the diverse diagrammatic operations, reflecting corresponding mathematical processes embodied within the elements / environments / rules / interactions of the diagrams. Plainly, then, the DC is intimately related to the OTC of E_4 – the main distinction lying in the diagrammatic (rather than formulaic) monitoring, in the DC, of 'the stated of mathematical expressions' in the course of (interactive) investigations.

It is only necessary to recognize the immense power of diagrammatic formalisms in Engineering / Physics / Chemistry / ... for the significance of the DC to be appreciated. Moreover, within the IT/AI contexts, the scope of such representations is greatly enlarged – along with the scale. In particular, the search software, and the facilities for 'manipulating diagrams' in various ways, allow the OTC and the DC to be **combined** effectively, so that investigations may be conducted with maximal efficiency and imagination, taking full advantage of interdisciplinary procedures / analogues / inverse-analogues 'suggested by the system' at each stage of any calculation.

The **classification** of 'approaches to diagrams', by members of collections of other diagrams, is best pursued by considering separately (i) general-approximative, (ii) analogical, and (iii) inverse-analogical, processes – each in both deterministic and stochastic forms. For cases of species (i), rigorous estimates of 'closeness' are attainable whenever all of the diagrams involved are – in suitable senses – continuous functions of their components. The possible forms of continuity (and associated convergence) are diverse – especially, where 'spaces of rules / interactions' are considered. Indeed, one may conjecture that: the 'macroscopic behaviour of diagrams' reflects the behaviour of their components. If, however, several components vary simultaneously, then the 'overall response of their diagram' may be ascertained only through detailed investigation, unless all of the (quasi-)limit operations entailed commute with each other, and so may be applied successively, in any order, without affecting the end result. This situation is, of course, common in standard mathematical contexts, but it raises qualitatively novel problems in the DC – for instance, in the possible branching behaviour of diagrams with a variety of complicated components (which may produce quasi-chaotic responses).

Indeed, the whole apparatus introduced (mainly in E_4) for the OTC may be adapted to cover classes of w.d. **functions of diagrams** – in terms of the actions of functions on the diagram-components.

A fundamental question now arises: To what extent is this formalism applicable to Σ **-proof**-diagrams? The related issue of proper interpretation of the results must also be considered. Notice that, for the OTC, 'proofs as compositions of $n(\geqslant 1)$ suitable transformed OT' are treated in some detail in E_8. Here, the emphasis is on successive transformations of **diagrams** in the course of a proof, rather than on sequences of formal expressions. Although these two species of proof-processes are closely related, the DC provides a coarse-grained description of proof-trajectories than that of the OTC, since (un)countable **class** of expressions may correspond to a single type of diagram-**component**.

Indeed, it appears that Σ **-program-flow-**diagrams constitutes **realizations** of Σ **-proof** diagrams; whereas, **OT-Composition**-proofs exhibit the **detailed I/O and successive mappings** (by OT in the composition) **at all stages** of the proof. Schematically, one has:

$$\rho(\pi^O \mid \Sigma^D) \approx \pi^P \approx r(\pi^D \mid \Sigma),$$

where $\rho(A \mid B)$ denotes the **reflection(s)** of A in B, $r(X \mid Y)$, the **realization(s)** of X over Y, and π^O, π^P, π^D (respectively) 'operational', 'program', and 'diagram', forms of 'the same proof / implementation'.[311]

There is a distinction here, for π^P, among the types of 'hardware architecture' underlying the Σ -computation-programs – including: 'sequential / pipeline / vector / array / parallel. Such distinctions are exhibited as features of the diagrams, through connectivity, graph-structure and other properties.

Recall that a Σ **-proof-diagram**, D(e) for $e : p \Rightarrow q$ has (minimally) the following constituents:

(i) a collection, $\mathcal{J}^D$, of theories, $\tau_i (i \in I^D)$, with each of the τ_i represented by a 'box', B_i, containing all of the OT in τ_i, etc.;

(ii) a labelled, directed, possibly self-intersecting (finite) graph (with initial vertex labelled as p, and terminal vertex, as q) some of whose edges may be traversed more than once (in either sense);

(iii) a corresponding set of data-labels for the vertices, with each traversal of an edge associated with 'its own data' – so that vertices are, in general, multiply-labelled (by ordered lists);

(iv) an underlying interdependence graph (exhibiting the structure of $\mathcal{J}^D$ as a set of variously interrelated elements within an overall framework).

NOTES

(i') The theories τ_i are (typically) to mutually independent – so the boxes may overlap diversely. No attempt is made to choose the τ_i either uniquely or even minimally; it suffices for each τ_i to be 'a w.d. theory'. Every interior vertex lies in at least one of the B_i.

Notation:

$\mathcal{J}$:={theories τ_i 'directly involved in $\pi^{D(e)}$'}

B_i :='box' representing (the OT in) τ_i

G^{π^D} :='basic graph' for π^D

[311] **Note that** 'implementations' may be subsumed under 'proofs' (of 'constructions', etc.)

V^{π^D} :={vertices of G^{π^D}}
E^{π^D} :={edges of G^{π^D}}
L^{v_k} :=(labels at vertex $v_k \in V^{\pi^D}$), an **ordered** list corresponding to the directed traversal of G^{π^D}.

(ii′) The boxes B_i are envisaged as mutually located in accordance with the information embodied in the $\mathcal{J}$-structure graph, $G^{\mathcal{J}}$ (as far as intersection / disjointness properties are concerned).

(iii′) From now on, the superscript D (in π^B, etc.) is omitted, unless there is danger of confusion. The integers[312]

$$v_i := \#(V^{G^\pi} \cap B_i)$$

provide a very crude – but still heuristically useful – indication of 'the **distribution of facets within the proof** π'.

Further, denote by μ_{jk} the number of edges of G^π (counted to include multiple traversals) starting in B_j and ending in B_k. The μ_{jk} give an indication of '**the frequency of theory-fluctuations in the proof** π' (for $j \neq k$); and, of '**the number of deductions**[313] **within** τ_k' (for $j = k$).

Here, **the theory τ_i is directly involved in** π. Whenever some deduction(s) in π may be viewed as OT in τ_i, appropriately formulated.

(iv′) For the 'target theorem', $e : p \Rightarrow q$, the 'initial label' ($\equiv$ 'premise data') is p (or any equivalent expression(s)). The subsequent distribution of labels depends not only on the logical nature of p, q – but also, on the **hardware architecture** (i.e., sequential / pipeline / vector / array / parallel / ...) assumed for the implementation(s) of the proof of e. The connectivity structure – and other topological / graph-theoretical properties – of G^π, reflect these choices of architecture, especially where (symbolic / numerical) evaluations / manipulations are involved. Although this complicates the estimation of the v_i and the μ_{jk}, as defined in (iii)′, no difficulties of principle seem to arise, even in the most general cases.

(v′) The graph $G^{\mathcal{J}}$ embodies all of the interrelations among the τ_i in $\mathcal{J}$. In particular, it exhibits the partial inclusions of some theories in others, the common sub theories among various sub collections of $\mathcal{J}$, and the mutual disjointness of other elements of $\mathcal{J}$. The labels in $G^{\mathcal{J}}$ serve mainly to identify structures / facets, so that the essential distinctions among the τ_i may be discerned.

(vi′) Notice that, even where there are multiple traversals of some edge (say, from B_j to B_k) the **data** will, in general, vary from one traversal to another. This corresponds to the repeated use of 'the output' from application of some OT in B_j as 'the input' for the application of a specified OT in B_k. These data assignments are incorporated in the lists at each vertex of G^π.

Approaches to Proof Diagrams

From all of the preceding remarks about the structure of proof-diagrams it follows that general approach procedures are admissible for all of the components – but, that questions of synchronization / commutation among these procedures must be considered in detail, if the convergence[314] of sequences of **diagrams** to limiting diagrams is to reflect cognate processes for associated **proofs**. This is especially relevant where pipeline / parallel architectures are

[312]Here, #X denotes (as usual) the number of elements in the set X

[313]The term 'deduction' includes both logical derivation and calculation – so that the proof π comprises a finite number of deductive steps (some possibly 'composite', in pipeline / parallel architectures, etc.). See (iv)′.

[314]The possible forms of 'convergence' reflect the structures of the OT involved – from concrete 'analytical criteria', to (e.g.) the CPOs of programming semantics.

concerned, since, in such cases, the combined effects of ‘compound operations’ on sets of r components ($r \geqslant 2$) may depend, in indicate ways, on the order in which the operations are applied.

Notice that, in E_8, the compositions of transformed OT do not include specifications of the **modes of implementation**[315] of these OT. For ‘pure deduction’ (without ‘calculation’ in the usual sense) no difficulties arise; but the limiting forms of components, under variously combined approach procedures will not, in general, be unique – or even w.d.. Consequently, the modes of all OT involved must be precisely set out before ‘states of proof-processes’ may be monitored. The problems entailed in defining ‘semantic distance’ appear to have no universally satisfactory solution; **see** E_5/E_8 for discussion of these issues. The main use of proof-diagrams lies in the (partially heuristic) identification of ‘promising lines of investigation’, and, of ‘just-solvable variants of currently intractable problems’. Moreover, diagrammatic criteria may be valuable in the exploration of the boundaries / limitations of ‘remorseless deduction / calculation’ (devoid of heuristic / imaginative content) for the solution of classes of problems – suitably specified within some branch(es) of Mathematics. Here, there may be interesting parallels with the study of ‘expressive power’ in formalized languages.

The apparent lack of any universally applicable quasi-metric, for **arbitrary** predicate functions, is not as serious a limitation as it may initially seem to be, since investigations typically proceed within frameworks / contexts. Hence, it is seldom (if ever) appropriate to compare two predicate functions which are neither known, nor conjectured, to be interrelated in **some** way(s). The (established or suspected) links then yield ‘separation estimates’ – however restricted – upon which further procedures / conjectures may be based.

Greater progress is possible within formalized (e.g., programming) languages, but these lack the flexibility and ‘scope for intuitive activity’ that are essential for creative mathematical work. By delineating the structures within which nontrivial separation-estimates **may** be made, one may combine formal manipulation and heuristic investigation fruitfully, in the OIB. The capacity to move effortlessly between these species of activity is an important design aim for Σ . It should be emphasized that problems resolvable either, by formal technique alone, or else, through ‘intuitive vision’ along, are seldom found to be of fundamental significance – in any theory. The ‘environment for investigation’ must promote the interplay of rigorous calculation and informal exploration.

Proof-diagrams are essentially ‘graphical reflections of formal proofs’, exhibiting proof-**trajectories** within $\Sigma^{\mathbb{T}}$, the set of generic OT in Σ . Since $\Sigma\,\mathbb{T}$ contains only **generic** OT, it is a finite / countable set, within which proof-trajectories are representable as planar graphs – whose detailed structures reveal basic features of the proofs involved. These graphs are obtained from the proof-**diagrams** by retaining only the constituent OT and their orders of application (allowing for suitable repeated use), and embedding the resulting graphs in $\Sigma\,\mathbb{T}$ (regarded as a point-set).

Thus: the proof-trajectories are finite, directed (planar) graphs, in which the vertices are (uniquely) labelled by generic OT, and the edges are labelled by finite sequences of ($\pm$) integers (of strictly increasing moduli) giving the step-numbers / senses for traversals of these edges, in the course of the proof. For general implementations, the proof-trajectory has a tree-structure, with congruent subgraphs corresponding to pipeline / concurrent parts of the underlying (serial / pipeline / parallel) subprocesses involved. The detailed **results** of these subprocesses appear as **vertex**-labels in the complete proof-diagram. In a sense, therefore, proof-diagrams are more detailed than proof-trajectories, but less detailed than ‘full proofs’ (where all constituent OT are stated, all intermediate calculations are given, and the overall sequence of serial / pipeline / parallel operations is rigorously specified). Insofar as **proofs** are equivalent to **programs**, these classes of descriptions correspond to program-diagrams, program-trajectories (flow-diagrams), and ‘full programs’, respectively.

As Σ evolves, various diagrammatic characteristic of classes of proofs will be identified, and used to suggest promising lines of attack on ‘continuous problems’. Moreover, quasi-stability

[315] or ‘modes’ (of OT) for short

properties of proofs under 'perturbations of their trajectories' may be investigated, to study maximal extensions of π_T to modifications, $\pi_{M(T)}$ – where $\{M(Y)\}$ covers all modifications of T provable through 'deformations of π_T'.

Associated notions of 'natural boundary for $\{M(T)\}$' arise in this context. Here, it is assumed that π_T is regarded as a function of its constituent OT, and that these OT are altered, in all admissible ways, to produce elements of $\{\pi_{M(T)}\}$. Cognate problems may be examined for the **trajectories** of π_T, $\pi_{M(T)}$, etc.. These, and allied, facilities, for studying unobvious structural aspects of proofs, greatly enhance Σ as a research environment – especially, when combined with ready access to broadly comparable proofs in all areas of Mathematics. Indeed, intelligent use of such facilities will drastically reduce the number / length of papers acceptable for publication (or, for inclusion in Σ), by eliminating (near-)duplication / triviality. Where proof-procedures are common to results in ostensibly disparate fields, **that commonality** is itself a highly significant fact – from which new / unifying theories might be developed. A remarkable example of this kind concerns the theories of so-called **white noise** (related to Brownian Motion) and of **generalized functions in infinitely many variables** – which are substantially equivalent.[316] In fact, which noise is defined as the time-derivative of Brownian Motion, and its probability distribution is a standard Gaussian measure in the space of generalized functions. It is notable that many of the basic results of these theories were obtained independently, before their near-coincidence was recognized. This case – though extreme – is not untypical over the entire range of mathematical theories. Another example (though less complete) involves the theory of diophantine approximation, and value-distribution theory – for which Roth's theorem (on rational approximation to algebraic numbers) corresponds to Nevanlinna's Second Main Theorem (on the deficiencies of meromorphic functions)[317]. The analysis of such examples in terms of proof-deformations / trajectories, etc. – using all of the facilities in P_Σ – would be most instructive. Eventually, this kind of analysis will become a routine part of proof-development within Σ , yielding (often, unsuspected) links among theories in diverse fields of mathematics and its applications.

As with many aspects of Σ , precise formalisms for diagrammatic representations / perturbations will evolve with the overall implementation of the system, but the discussion given here, together with the wide-ranging examples in E_{10}/E_{11}, should be adequate to lay the foundations for these operations. In any case, advances in hardware-design / programming may well change the form of graphics / IT packages so radically that only the theoretical foundations established in these Essays will remain essentially **un**changed; and thus is so far all implementation-aspects of the OIB. It is, of course, inevitable that the monumental task of (re)formu- lating Mathematics to take optimal advantage of the developing IT/AI/SAM capabilities of computers will be tackled primarily on a highly theoretical level – to fabricate a vast, intricate, but 'philosophically solid' superstructure, without which the crucial quality of **coherence** could never be realized. It is also daunting to observe that, even the extremely elaborate arguments developed in these Essays cover only the initial – albeit, fundamental – phase of a very longterm scheme. In view of the considerable disorganization, and explosive growth, of 'Mathematical knowledge', however, arguments for the construction of a system 'of the Σ -type' seem to be overwhelmingly strong.

[316] See, e.g., Hida, T, '*Brownian Motion*' (Springer, 1980), and Berezansky, YM / Kondratiev, YG, '*Spectral Methods in Infinite Dimensional Analysis*' (Vols. 1,2, Kluwer, 1995)

[317] See: Vojta, PA, '*Diophantine Approximations and Value Distribution Theory*' (Springer LNM #1239, 1987) for details

CHAPTER 8

Proofs as Combinations of Operational Theorems

The basic ideas of **operationally**, and of **operational theorems** (OT) are expounded in E_4 – along with some possible formulations appropriate for P_Σ. In this scheme, $\Sigma_t$is regarded as a self-consistent OIB containing a set, U_t, of (permanently ordered) 'underlying ME', u_j, with minimal definitions – a **basic dictionary**[318] – through which theorems may be expressed. Of course, as t increases, this dictionary grows larger; but it remains a collection of (mutually consistent) **terms / definitions** – deliberately excluding all conventional forms of 'theorems'. This is in line with the view of Mathematics as a **language**, allowing a range of syntactic / semantic constructions / interpretations. The major difference from most other languages lies precisely in the capacity of mathematics for **applications**, to generate 'new' information.

For every generic theorem, T, in Σ_t, there are several 'versions' – including the **explicit, task**, and **cast** forms, as well as **operational** representations (at various levels), all of which are discussed (schematically) in E_4. Here, it is useful to recall the notation, namely[319]

$$
\begin{aligned}
&{}^gT \equiv {}^gT(\theta_T) \equiv p_T(v) \Rightarrow q_T(w) \textbf{ (generic)}:\\
&{}^eT \equiv T(\theta)_h = \Gamma^T(u_\sigma)_h \textbf{ (explicit)};\\
&\begin{cases} {}^tT \equiv & {}^tT(\tau)_k; \tau_i \equiv \tau_i(\theta)_h, \\ {}^tT \equiv & {}^tT(\tau(\theta)_h)_k, \textbf{ (task)}, \end{cases}\\
&{}^cT \approx C(T) \equiv \{c(T)\}_l \textbf{ (cast)},
\end{aligned}
\tag{1}
$$

where $h \equiv h(T), k \equiv k(T), \sigma \equiv \sigma(T), l \equiv l(T), \{u_\sigma\}_h \subset U_t$, and Γ^T determines the style / formulation of T, over U_t in Σ_t.

Recall, also, that ${}^eT$ contains a technically full statement of T in which all ingredients are identifiable. For ${}^eT \equiv T(\theta)_h$, the elements, $\{\theta\}_h$ are 'local'; whereas, $\{u_\sigma\}_h$ comprises a **selection** from the 'global' set, U_t. The version ${}^\tau T$, on the other hand, lists 'the **tasks**, τ_i, accomplished through T' – the τ_i being expressed in terms of $\theta_1, \ldots, \theta_h$, as required. In ${}^cT$, only 'the (comparatively) high-level ingredients of T' are listed – separately, for the **premises**, and the **conclusions** (making ${}^cT$ and ${}^\tau T$ especially relevant in the design of '**environments for investigation**').

The most general transformations of OT considered so far (in E_4) involve compositions of (near-)type-preserving modifications of individual arguments of compound OT. As stated in E_4, extensions to apparently broader classes of OT-transfor- mations seem to be reducible to this basic form – but the existence of nontrivially distinct forms has not been precluded. Consequently, it is necessary to outline a formalism to cover all potentially admissible cases.

Two possibilities are presented as illustrations. They are both based on the representation of U_t as a union of sets of ME of distinct species:

$$U_t = \bigcup_{1 \leqslant k \leqslant N_t} U_t^k,$$

where U_t^k comprises ME of species s_k and ${}^tU_t$ is taken to be **finite** (with indexed families of 'similar ME' treated as single objects)[320].

[318]though not ordered alphabetically

[319]The suffixes, e, τ, c, should not cause confusion; they are rarely used (beyond setting contexts). Further, $(\theta)_h \equiv \theta_T$ denotes the **ingredients** of T; and, in general, $c(T) \subset \theta_T$

[320]The subscript t is dropped here, from now on.

(α) Let ${}^k\varphi : U^k \supset P \to Q \subset U^k$, for $1 \leqslant k \leqslant N$. Now define the action of ${}^k\varphi$ on the OT $T \equiv T(u_\sigma)_h$ by

$$ {}^k\varphi(u_\sigma)_h := T(u_\sigma)_h|{}^k\varphi(u_\sigma)_h^k $$
$$ (u_\sigma)_h^k := (u_\sigma)_h \cap U^k. $$

This transformation may reduce the total number of arguments in T, but the set of **distinct species** is preserved. Here, for suitable functions F, g, and sets $A, B(B \subset A)$, one puts[321] $\tilde{F}(A)|_{g(B)} := \tilde{F}(A\backslash B \cup g(B))$, the argument(s) $g(B)$, for $\tilde{F}$, replacing those in B, for F, with the original ordering in A (if any) maintained, as far as possible.

(β) Suppose, now, that ${}^k\varphi$ is replaced by ${}^{k,l}\psi$, where:

$$ {}^{k,l}\psi : U^k \supset P \to Q \subset U^l, \text{ and that} $$

$$ {}^{k,l}\psi T(u_\sigma)_h := T(u_\sigma)_h|_{{}^{k,l}\psi(u_\sigma)_h^k}. $$

Then: the total number of arguments in T may be reduced (an is (α)), but the set of distinct species is (in general) **not** preserved.

The **crucial question** (for both (α) and (β)) is whether the **transformed** OT are interpretable as **w.d. OT**.

As in E_4, various 'quasi-**closure** operations' may be considered, with the possibility of '**species transitions**'[322].

Although the **basic** definitions of $f(T(u_\sigma)_h)$, introduced in E_4, do (broadly) 'preserve OT status' – by maintaining both the total number of arguments, and the species of each argument – there are still some exceptions, and care is required to ensure consistency. The types of transformations just defined (and other, even more general, types) raise many difficult problems of classification / validity, so no definitive treatment is attempted here. Rather, some particular cases are explored – as illustrations – in $E_9/E_{10}/E_{11}$.

The main reason for introducing (almost arbitrary) **functions of OT** is that **proofs** of theorems may be regarded as equivalent to (Boolean) combinations of compositions of suitable **transformed** OT – in the sense that **every proof of** T: $p \Rightarrow q$ (with **premise(s)**, p, and **conclusion(s)**, q) **may be represented as:**

$$ \hat{\pi}p = q \vee r, $$

where $\hat{\pi}$ denotes the combination of OR and r, the 'remainder of $\hat{\pi}p$ with respect to q. Here, p, q are, respectively, n_p-place, and n_q-place, predicate functions, and all of p, c, r are wfe. **NOTE** Although it is natural to denote 'conclusion(s)' by 'c', the basic notation for (O)T in Σ is: $T : p \Rightarrow q$ (or, $T : p \Rightarrow q \vee r$, if 'extra information' is produced in the proof(s)).

The most general combinations of OT considered here have the form:

$$ \Phi\left[\bigwedge_{m\in M_k} C^{J_m}\right] \equiv \bigvee_{\substack{1\leqslant k\leqslant K\\ M_k\subset\mathbb{P}}} b_{M_k} \bigwedge_{m\in M_k} C^{J_m}, $$

where $C^{J_m} := j \in {}^oJ_m T_j(f^j(u_{\sigma^j}))_{L_j}$ is the **card**-J_m-fold composition of modified / transformed (atomic) OT, T_j, for j ranging over J_m. This may be put into a 'linear' form:

$$ \Phi\left[\bigwedge_{m\in M_k} C^{J_m}\right] = \bigvee_{\substack{1\leqslant k\leqslant K\\ M_k\subset\mathbb{P}}} b_{M_k}\xi_{M_k}, $$

321 This is more or less standard in 'multivariable calculus', apart from the ordering constraint

322 Analogues of 'phase transitions' in Statistical Mechanics.

in which $\xi_{M_k} \equiv \bigwedge_{m \in M_k} C^{J_m}$. The 'coefficients' b_{M_k} may be regarded as 'formal' unless particular interpretations are given. The conventions:

$$\left(\bigwedge_m C^{J_m}\right) e := \bigwedge_m (C^{J_m} e) ;$$

$$\left(\bigvee_q T_q\right) e := \bigvee_q (T_q e) ,$$

are also adopted – and the action of any OT T on the ME e is as specified in E$_4$. The admissible interpretations of $T_j(f^j(u_{\sigma j}))_{L_j}$ are as given in E$_4$ ('standard'), or else, as just defined in $(\alpha)/(\beta)$, above.

It should be stressed that this somewhat intricate scheme is designed to provide adequate foundations for the **O-TC (Operational-Theorem Calculus)** in the general forms envisaged in E$_1$/E$_4$, where various types of '**weakly-formalized heuristics**' are mentioned, but not developed. Recall that the basic idea is that the OT (as '**operators**') transform the ME (as **wfe**); and, a 'new theorem' is **proved** when the **premise(s)**, p, have been transformed into the **conclusion(s)** – now denoted by q – and, possibly, other ('extra') results (say, r). In symbols: $\hat{\pi}_T p = q \vee r$. The complications stem, mainly, from the requirement that all of the 'component OT' in the C^{J_m} must be **atomic** – in the sense that none of the T_j is decomposable into either of the forms $T_j' \vee T_j'', T_j' \wedge T_j''$, with both of T_j', T_j'' in admissible OT form. If, instead, the restriction to atomic T_j were dropped, then a single expression:

$$\underline{\hat{f}}({}_n T([u_\sigma])_{\underline{h}} \equiv (T_1(\hat{f}_{11}(u_{\sigma_{11}}), \ldots$$
$$\hat{f}_{1h_1}(u_{\sigma_{1h_1}})), \ldots, T_n(\hat{f}_{n1}(u_{\sigma_{n1}}), \ldots, \hat{f}_{n,h_n} u_{\sigma_{nh_n}})))$$

would suffice! This representation, however – with $T_1, \ldots, T_n$ not necessarily atomic – would obscure many of the transformational details and their interrelations. In any case, typical applications involve single compositions, C^{J_m}, acting 'locally' (in some 'segments' of proofs), to be combined in the development of complete proofs.

Recall that the OTC is concerned with the action of (substantially) arbitrary combinations of OT on collections of ME. The '**calculus**' corresponds to the (quasi-)analytical transformations; but there are also possible **algebraic** structures associated with sets of OT. If the individual (generic) OT are regarded as 'elements' of the base set, B, then various laws of combination (of **arity** $r \geqslant 2$) may be postulated, for subsets of r elements. If $r = 2$, one has **binary** combinations, and analogues of the basic arithmetical operations (and their **inverses**) may be sought. In particular (for 'the law $*$') one requires a **unit element** (say, v^*, for **neutral**), and an inverse (say, $\overset{-1}{*}T$), where

$$(\forall e) : v^* e = e; \text{ and}$$

$$(\forall T \in B) : \overset{-1}{*}T * T = v^* = T * \overset{-1}{*}T .$$

Here, T need not be atomic. The most important law of combination is binary **composition** (repeated, if necessary, etc.). This has already been denoted (as usual) by $\circ$, in the definition of C^{J_m}. One therefore requires a **composition-inverse** for (invertible) OT – it being understood that not all OT are invertible. There is, of course, nothing unusual in this situation: for instance, even ordinary matrices are both non commutative and noninvertible, in general.

It turns out (as indicated in E$_4$) that – with due care, and attention to details – **operational converses** of T yield types of compositional **inverses** of T, the unqualified term, 'inverse', corresponding to the full / complete / total converse (and weaker variants, to 'partial converses', etc). Evidently, this is, as it stands, very vague! Nevertheless, it is susceptible of rigorous formulation (with concomitant limitations in range of applicability), as indicated later in this Essay. For the moment, however, it is useful to explore various notions of quasi-convergence of sequences of wfe (with associated 'proof representations'). The problem of defining forms of

semantic (quasi-)**distance** is fraught with difficulties, and appears to have no **general** solution. Moreover, such 'distances' are of major importance in only two broad contexts: **(i) monitoring the progress of proofs**; and, **(ii) assessing the validity of conjectures** – as a prelude to their heuristic use. An equivalent characterization is that of sequences of OT converging to some limiting OT – which includes the case of **finite** sequences, corresponding to 'proofs in n steps', etc..

For Σ , there are essentially three (progressively vaguer) representations of any proof of $T : p \Rightarrow q$, namely:

$$\pi(T) : \Phi \left[\bigwedge_{m} C^{J,} \right] p = q \vee r;$$

$$'\pi(T) : \{T_j(f^j) : j \in M_k, 1 \leqslant k \leqslant K\};$$

$$''\pi(T) : \{T_j : j \in M_k, 1 \leqslant k \leqslant K\}.$$

The forms $\pi(T)$, $'\pi(T)$, $''\pi(T)$, may be called (respectively) **full, skeleton**, and **indicative** proofs of T (in terms of the T_j). Other ('intermediate') forms may be introduced later, but π, $'\pi$ and $''\pi$ suffice for most purposes. Moreover, various algebraic / analytical structures may be defined over the sets:

$$\pi\Gamma \equiv \{\pi(T) : T \in \Gamma\},$$

$$'\pi\Gamma \equiv \{'\pi(T) : T \in \Gamma\},$$

$$''\pi\Gamma \equiv \{''\pi(T) : T \in \Gamma\},$$

for 'sufficiently large / rich collections', Γ, of OT. Although many of the operations involved are somewhat formal in nature (e.g., deformation / extension / generalization / perturbation), they are still useful in formulating certain (local or global) properties of (classes of) proofs – for instance, types of '**stability**' – and in specifying types of '**averages / means**' of OT; or, over suitable sets of OT. The dependence of proofs of 'significant theorems' on particular (combinations of) 'more basic' OT may also be monitored – to produce assessments of the '**centrality**' of given sorts of results over a wide range of proofs, and to test other 'trends' involving classes of results; and so on. Such investigations would be intractable on a large scale if done 'by hand'. For a computer-based OIB, however, they are both feasible and highly relevant, since 'the flow of information' into a mathematical domain is crucial for its systematization / expansion.

Among the basic issues to be explored for the OTC are:

(a) the role(s) of approximation;

(b) the interpretation(s) of $H(T)_M$;

(c) possible analogues of Riemann Surfaces;

(d) possible analogues of 'eigen-elements'.

(a)$'$ **Approaches** (the Σ -brand of general approximations) may occur in any of the **modes**: **formulation, realization, application** (of an OT to an ME). If $\mathcal{A}$ denotes 'the approach operator', then one has (schematically):

$$[\mathcal{A}T](u_\sigma)_L \text{ (\textbf{formulation})};$$

$$T(\mathcal{A}u_\sigma)_L \text{ (\textbf{realization})};$$

$$\left.\begin{array}{l} \mathcal{A}[Te] \\ T[\mathcal{A}e] \end{array}\right\} \text{ (\textbf{application})}.$$

All of these modes may be **combined** in various ways – to encompass the full range of **deterministic approaches** in the OTC. In fact, even **stochastic** versions may be incorporated, by introducing suitable (probability) **measures** within this scheme. This process is outlined later.

Of course, the **form** of $\mathcal{A}$ is different for each mode, and for each particular case; but the logical / mathematical **structure** remains essentially unchanged. Diverse illustrations of all these representations are given in E_9, E_{10} and E_{11} (with commentaries / references).

The other remaining mode of approach relevant to the OTC involved types of 'convergence of sequences / families of wfe to limiting wfe'. This may take any of several forms defined for objects in Σ (through parametric / functional dependence of the sequences / families concerned); or else, variants of the modes of convergence for CPOs in theoretical semantics. The procedure for defining 'distances' between sets of Locational Conditions (see: E_5) may also be considered here.

(b)′ There appear to be several possible interpretations of $H(T)_m \equiv H(T_1, \dots, T_m)$ within the OTC framework – though none of them is completely general, or satisfactory. The basic requirement is that, for every admissible (ordered) set, $(T)_m$, of OT, the object $H(T)_m$ must also be an admissible OT, (unambiguously) determined from $(T)_m$. The function H may be of any class; or else, it may be defined through explicit representations (e.g., in terms of some Elementary of Special function(s)). All versions entail that '$+$', '$\times$', and 'products of OT with parameters' (or various kinds) are assigned mutually consistent interpretations. In spite of this apparent diversity of interpretations, the general representation for w.d. combinations of OT (given as $\Phi[\bigwedge_m C^{J_m}]$, above) **formally** covers all possibilities considered for Σ .

(i) For 'abstract functions', H (i.e., those having no **explicit** representation), only the **possibility** of producing some w.d. OT, $H(T)_m$, from $(T)_m$ need be established. This devolves on the Kolmogorov / Arnol'd theorem – the result that a broad class of functions $H : I^m \to \mathbb{R}(I := [0,1])$ is representable through finite sums of finite superpositions of functions of **one** variable. In symbols:

$$H(x)_m = \sum_{q=0}^{2m} g\left(\sum_{p=1}^{m} \lambda_p \phi_q(x_p)\right) \equiv \mathcal{S}(H; g, \phi),$$

where g is continuous on $[0, m]$, $\lambda_p \in [0, 1]$, and $\phi_q : I \to I$. If H **is** specified – as some combination of Elementary / Special functions; or, of series / integrals / …involving such functions, etc. – then (formal, at least), an explicit representation, say, $\mathcal{S}(H; \phi, g)$, may be assumed, as a basis for the definition of $H(T)_m$ – through the existing definitions of $f(T)$ already defined in Σ (see, e.g., E_4). It follows that $\mathcal{S}(H; g, \phi)$ may be defined for OT, **provided that**[323] 'Σ ' and '$\lambda\phi$' are w.d. in this context, to yield, simply:

$$H(T)_m = \sum_{q=0}^{2m} g\left(\sum_{p=1}^{m} \lambda_p \phi_q(T_p)\right).$$

Since theorems are just imprecatory logical statements, the most natural type of 'summation' is that used in Boolean algebra – i.e., '$+$' is replaced be '$\wedge$'. For the 'product', $\lambda\phi$, the obvious idea is to consider the 'constant function', say ${}_\lambda f$, with ${}_\lambda f(x) \equiv \lambda$ – and then, to define ${}_\lambda f(T)$ as before. Variants to cover the cases where λ is a (constant) vector / tensor / … may also be treated as in E_4. In this way, $H(T)_m$ is w.d. over a broad spectrum of functions, H.

(ii) As mentioned in E_4, **another possible interpretation of** $H(T)_m$ may be given directly, for classes of functions admitting Taylor expansions. Thus, if

$$H(x)_m = \sum_{\underline{o} \leqslant (k)_m} a_{(k)_m} x_1^{k_1} \dots x_m^{k_m},$$

[323] Here, Σ (for once) refers to **summation**!

and '$+$', '$\lambda\phi$' are as in **(i)**, then the further identification of 'x' with $\circ$ ((ordered) **composition**) yields the representation:

$$\cdot H(T)_m := \sum_{\underline{o}\leqslant(k)_m} a_{(k)_m} T_1^{(k_1)} \circ \cdots \circ T_m^{(k_M)}.$$

Here, $T^{(k)}$ denotes the $(k-1)$-fold composition of T with itself. In all subsequent discussions, the forms $f(T)$, $\cdot f(T)$ (and $H(T)_m$, $\cdot H(T)_m$) may be combined in the construction / analysis of 'operational proofs'.

(iii) Another variant of '$f(T)$', or, of '$H(T)_m$', may be obtained by identifying '$\times$' with **conjunction** ($\wedge$), while retaining the previous forms of '$+$', and '$\lambda\phi$' – and then using Taylor series for f, H (as in **(ii)**), with the extra definition

$$\left(\bigwedge_m T_m\right) e := \bigwedge_m (T_m e).$$

These forms – say, $\vdots f(T)$, $\vdots H(T)_m$ – do not seem to occur 'naturally' in the construction of proofs through successive application of OT (the basic model for Σ). Nevertheless, there may be situations where the forms f, $\cdot f$, $\vdots f$, (or, H, $\cdot H$, $\vdots H$) should be combined, in various ways, in the course of interactive investigations.

(iv) Additional variants could be formulated, but this is not attempted now. Rather, such variants are identified as they become relevant (in E_9, E_{10} and E_{11}).

(c)$'$ The seminal importance of Riemann Surfaces (RS) in complex function theory, suggests that OTC-**analogues** of the **RS of a function element** ($\equiv$ **germ** of a **C**omplete **A**nalytic **F**unction (**CAF**)) may be significant for the analytical structure of Σ . In order to proceed, one needs to identify the basic ingredients in the definition / construction of RS, and then, to associate these ingredients with objects basic for the OTC. In particular, some analogue(s) of the process of (**D**irect) **A**nalytic **C**ontinuation (**(D)AC**) along-curves / throughout-domains must be sought – with concomitant notions of 'Complete OT', etc. Although close analogy is not to be expected here, the underlying unification stemming from 'quasi-RS constructions' more than justifies this undertaking.

The adequacy of the following proposed OTC/RS analogues may be assessed only through their relative utility in formalised heuristics and other investigations. They are propounded, mainly, to exemplify the radical inventiveness essential in developing the full potential of the Σ scheme. (The next section (on 'eigen-elements') is similarly motivated).

(i) Let T be a **generic** OT: $T \equiv T(u_\sigma)_L$, with **realizations** $\{f(T) : f \equiv (f)_L \in \mathcal{J}\}$; and, suppose that one of the u_{σ_k} (say, u_{σ_1}) denotes an analytic function (of z), with RS $\mathcal{R}_{\sigma_1}$. Then $\mathcal{R}_{\sigma_1}$ is **the RS of** T **corresponding to** u_{σ_1}. Moreover, $\mathcal{R}_{\sigma_1}$ is also the RS of $f(T)$, whenever f_1 is 'regular' ($\equiv$ single-sheeted).

(ii) If T is generic, $f \equiv (f)_L$, and all u_{σ_k} are regular, but f_j has RS $\mathcal{R}_{f_j}$, then $\mathcal{R}_{f_j}$ is called the RS of $f(T)$ corresponding to f_j.

(iii) Combinations of the forms **(i), (ii)** produce more intricate RS structures for $f(T)$ – where several of the f_j, u_{σ_k} may be non regular. The range of possibilities here is considerable!

(i)* **NOTE** A more suggestive form (essentially equivalent to **(i)**) is obtained if a **set** of generic OT is indexed by the **branches** of u_{σ_1} (**not** by their **values**), as z varies. Then $T \equiv T(z)$ still has RS $\mathcal{R}_{\sigma_1}$; but now, the **branches of** T are given by $T^h := T(u^h_{\sigma_!}, \ldots, u^h_{\sigma_L})$, where the $u^h_{\sigma_1} \equiv u^h_{\sigma_1}(z)$ are the branches of u_{σ_1}. The **precise** relation of **(i)*** to **(i)** may be more subtle than this.

(iv) As a final variation (for now!), on this quasi-RS theme, consider the modification of **(ii)** corresponding to families, $\{f_{\lambda(z)} : z \in \Omega\}$ of ('vector') transformations. Here, one may write $T^{\lambda}(z) \equiv T(f_{\lambda(z)}(u_{\sigma})_L)$, with 'branches', $T^{\lambda^h} \equiv t^{\lambda^h}(z) := t(f_{\lambda^h(z)}(u_{\sigma})_L)$, where h indexes the branches of λ. As usual, care is required to ensure that all of the objects involved are w.d.. More general formulations of quasi-RS associated with OT (simple / multiple / families / ...) are indicated in the context of **examples**, in $E_9/E_{10}/E_{11}$.

(d)′ Since the OT are regarded as **operators**, transforming wfe, it is pertinent (and irresistible!) to seek **analogues (for the OT) of the eigen-elements of (non)linear operators** in Algebra / Analysis / Topology.... Moreover, for each possible analogue, problems arise in **obtaining approaches to the associated eigen-elements**. As for the quasi-RS analogues, preliminary definitions are given, of the proposed analogues, for future practical assessment in examples – where some other variants may also be found (**see E_9/E_{10}/E_{11}**) The **generic OT** is (as before) denoted by:

$$^{g}T \equiv {}^{g}T(\theta_T) \equiv p_T(v) \Rightarrow q_T(w)$$

The elements ε are just wfe. The possible **action** of T on ε has been specified in E_4. On this basis, the most obvious analogues of eigen-elements for T may be characterized as:

(i) $T\varepsilon = \lambda\varepsilon$ (if $\lambda\varepsilon$ is w.d.);

(ii) $T\varepsilon = \varepsilon \vee r$ (for suitably restricted r);

(iii) $T\varepsilon = \varepsilon \wedge s$ (for suitable restricted s).

Less obviously, one could require that:

(iv) $T\varepsilon = \varepsilon_1$ (with 'ε_1 suitably close to ε');

(v) $T\varepsilon = \varepsilon_2$ (with 'ε_2 qualitatively similar to ε').

Other possibilities are considered later.

Although '$\lambda\varepsilon$' may be defined, formally, to 'make sense' for all combinations λ, ε (and this **is** done in certain OTC developments), the form **(i)** is relevant, mainly, when $\varepsilon, \lambda\varepsilon$ belong to the same **vector space** (however that space may arise in some investigation). Even with this overall restriction, the form **(i)** appears to furnish some extension(s) of the standard forms – for matrices / differential operators / integral operators /..., since λ and ε may be of arbitrary types, provided only that $\lambda\varepsilon$ is w.d.. However, the crucial limitation in such extensions lies in the difficulty of **determining** pairs, λ, ε, of eigen-elements, from T **alone**. Consequently, there are two classes of extensions – those for which some (quasi-)algorithm may be formulated (from which **approaches** to eigen-elements may be obtained); and those for which only the **existence** of eigen-elements may be established. All other cases are purely formal, and have no OTC interpretations.

As a preliminary to more precise descriptions of the forms **(i)–(v)**, the representation of possible actions of OT on wfe is recalled (and slightly modified). Here, a typical wfe is denoted by ε, with **cast**, $c(\varepsilon)$, of (atomic) ME; while, $c(T)$ still denotes 'the cast of T' (as in E_4). Once again, one has:

$$T(\theta_T) = \bigvee_{1 \leqslant j \leqslant 3} b_j \alpha_j(T(\theta_T));$$

$$T(\theta_T)\varepsilon = \bigcup_{1 \leqslant j \leqslant 3} b_j \alpha_j(T(\theta_T))\varepsilon,$$

where the b_j may be 0 or 1, independently, and $\alpha_1, \alpha_2, \alpha_3$ denote (as in E_4) the **informative, substitutive**, and **constructive** components of T. **NOTE** that, similar representations correspond to $T(u_{\sigma})_L\varepsilon$ – and are equivalent to those for $T(\theta_T)$ in all that follows. The basic plan is to regard ε as a **B**(oolean)-**expression** in the members of $c(\varepsilon)$: say, as $\varepsilon \equiv \beta(e^{\varepsilon})_k$, and then to consider $T\varepsilon \equiv T\beta(e^{\varepsilon})_k$, to obtain:

$$T\beta(e^{\varepsilon})_k = \left[\bigcup_{1\leqslant j\leqslant 4} b_j\alpha_j(T(\theta_T))\right]\beta(e^{\varepsilon})_k.$$

Since the arguments e_i^{ε} occur within a B-expression, this action may be 'expanded' as:

$$T\beta(e^{\varepsilon})_k = \beta\left(\bigcup_{1\leqslant j\leqslant 3} [b_j\alpha_j(T(\theta_T))](e^{\varepsilon})_k\right) \equiv \beta(T((e^{\varepsilon})_k)).$$

The detailed form of $T((e^{\varepsilon})_k)$, expressed through the α_j, obviously depends on that of T; but, each of the α_j yields a characteristic **type** of subexpression, namely:

$$\alpha_1(T)e = e \cup \tilde{T};$$

$$\alpha_2(T)e = e_T^* \cup R_T(e);$$

$$\alpha_3(T)e = \gamma_T(v_e, w_e).$$

Here, $\tilde{T}$ is the purely **informative** (**non**operational) part of T; **for calculational purposes, $\tilde{T}$ is discarded**; e_T^* is obtained from e through T-based **substitution(s)** – with 'remainder' R_T (if any); and, $\gamma_T(v_e, w_e)$ comprises the term(s) **constructed** from e-realizations of v, w, through the T-prescriptions in $\alpha_3(T)$.

On this basis, $T\varepsilon$ has an essentially unambiguous interpretation, provided that T is fully specified, and so the conditions **(i)**/**(ii)**/**(iii)** may be investigated. Although no systematic **general** procedures are to be expected, adequate techniques for certain classes of OT may be developed to mirror results for matrices, differential / integral operators, etc., in case **(i)**. Additional variants of **(i)** that are **not** reducible to standard algebraic / analytical / topological/... routines, require quasi-constructive methods – in which the separate images of each $e_i^{\bar{\varepsilon}}$ under $\alpha_1, \alpha_2, \alpha_3$ are suitably combined to produce $\bar{\lambda}\bar{\varepsilon}$ from $\bar{\varepsilon}$, with $\bar{\lambda}, \bar{\varepsilon}$ determined (in this sense) by T, and '$lam\bar{b}da\bar{\varepsilon}$' is w.d.. There may be some parallels with continuous / residual / point spectra of operators.

The other forms: **(ii)** $T\varepsilon = \varepsilon \vee r$, and **(iii)** $T\varepsilon = \varepsilon \wedge s$, may be treated within the same framework. Once again, appropriate combinations of types $\alpha_1, \alpha_2, \alpha_3$ are sought, to produce pairs $\bar{\lambda}, \bar{\varepsilon}$, satisfying the conditions **(ii)**, or **(iii)** – in nontrivial ways. Also as before, no general procedure seems to be attainable, but, for each T, eigen-elements (of types **(ii)** or **(iii)**) may be sought by quasi-constructive methods, in which (e.g.) 'imbalances' are rectified at each stage, until the required forms are obtained. Thus, in all cases, $\bar{\varepsilon}$ (and $\bar{\lambda}$) depends on T, and the basic structure of T may suggest plausible initial choices of $\bar{\varepsilon}$; and so on. These variants of eigen-conditions appear to be 'natural'. Their 'status' would be greatly enhanced if corresponding 'existence theorems' could be established – for instance, to show that certain classes of OT do possess particular types of eigen-elements (even if no algorithms are known for determining these eigen-elements). This aspect is explored later.

The types **(iv), (v)** involve primarily qualitative criteria. Some progress may be made through the notions of **neighbourhood** of an (atomic) ME; and, of **analogical closeness** of a pair of (atomic) ME. There are essentially three cases: **(a)** wfe with the same **form**, β, but different **casts**; **(b)** wfe with **different casts and different forms**; **(c)** wfe with **different forms, but the same casts**. (**NOTE**: these cases are not completely independent; they are adopted here because of their typicality.) The central problem is to obtain criteria for **(a), (b), (c)**, in terms of conditions on the (pairs of) individual ME – for which some definitions have already been proposed (see E_5).

In E_5, two types of **neighbourhoods** are considered – each based on LC for ME. In special cases – for instance, where the basic spaces, ${}_SX$(see E_1) are involved – these types coincide (substantially) with the standard forms. They may, however, by used in far more general contexts; in particular, the LC may be replaced (formally) by collections of B-expressions, as required here. Although these forms virtually ignore explicit **semantic** content, they are still useful here. A preliminary discussion of 'semantic distance' is given in E_4.

The previous definitions are repeated now, for convenience. Let the set $L(e) := \{l_j(e) : j \in I_e\}$ comprise all admissible LC of e (in Σ_t), and put $l_j(e) \equiv \bigwedge_{\alpha \in S_{ej}} {}^e v_\alpha$. For $\Gamma \subset S_{ej}$, define

$$N^j_\Gamma(e) := \{\bigwedge_{\alpha \in A} c'_\alpha \bigwedge_{\beta \in S_{ej} \backslash A} {}^e c_\beta : A \subset \Gamma\},$$

$$N_\Gamma(e) := \bigcup_{j \in I_e} N^j_\Gamma(e).$$

This remains somewhat vague, unless the permissible ranges of variation of the individual c'_α are specified. This requirement, in turn, may be (partially) met, **either**, by stipulating that each c'_α be **confined** to a certain **set** (say, $c'_\alpha \in \Phi_\alpha$); **or else**, by introducing the quasi-**distance** function, δ (also defined in E_5), where:

$$\delta(c_\alpha, c'_\alpha; x) := \lim_{k \to \infty} \Omega_k^{-1} \mu(P_k \Delta P'_k),$$

$$x_k := \{x \in X : \| x \| \leqslant k\}, \Omega_k := \mu(X_k),$$

$$P_k := \{x \in X_k : c_\alpha(\ldots x \ldots)\}, P'_k := \{x \in X_k : c'_\alpha(\ldots x \ldots)\}.$$

Here, X is a 'space' with 'adequate' (quasi-)metrical structure', etc.. The extension to allow for variations of $(x)_m$, or of $(x_\sigma)_m$, within c_α, with $x_{\sigma_m} \in X_{\sigma_M}$, etc., was also proposed in E_5, to give a modified form, $\bar{\delta}$, of δ.

So far, no assumptions have been made on the admissible classes of wfe; indeed, $\varepsilon, \varepsilon', \varepsilon_1, \varepsilon_2$ are essentially unrestricted. In 'mathematical practice', however, these wfe are, typically, comparatively restricted in range – for instance, 'sets', 'vectors', 'matrices', 'linear operators', 'Markov chains', or various 'formulae' containing such objects. Moreover, in OT, **predicate-functions** (of specified orders) are of fundamental importance in the representation of 'premises' and 'conclusions'.

It will become clear (on reflection) that, of the cases **(a), (b), (c)**, above, only **(a)** is immediately applicable here. The range of admissible deviations in **(b)**, or in **(c)**, must be highly constrained, for two structurally distinct B-expressions to be (nontrivially) mutually close. In **case (a)**, combinations of the definitions of $N_\Gamma(e)$, and of $\bar{\delta}(c_\alpha, c'_\alpha; (x_\sigma)_m)$ may be used to characterize closeness, for wfe $\varepsilon, \varepsilon_1$. In cases **(b), (c)**, extra conditions must hold, in order that ε and ε_1 may be 'sufficiently close' (for some types of eigen-elements to be so defined). Examples are considered later where such relationships hold.

The notion of '**qualitative similarity**' envisaged for ε and ε_2, in the condition **(v)**: $T\varepsilon = \varepsilon_2$, amounts to preservation of the **B-structure** of ε, and of the **type** of each element in $c(\varepsilon)$ – so that (for instance) **matrix → matrix, Markov chain → Markov chain, set → set**, ..., but only the **detailed form** of these ME is modified (e.g., **symmetric** matrix → **non**symmetric matrix). Plainly, this is a very weak kind of similarity; it may be strengthened by restricting the ranges of variation of the elements of $c(\varepsilon)$, within their basic types. This prescription furnishes a basis (for given T) to determine pairs $\varepsilon, \varepsilon_2$; in general, there will be many of these pairs – even **families** of pairs – since the requirements are weak. Nevertheless, all of these types of quasi-eigen conditions may be used to organise the IB, especially in the development of (optimal) environments for investigation – and in the formal treatment of operational proofs within Σ .

All of the material in this Essay, so far, is prerequisite for any systematic discussion of Σ **-proofs** as **compositions of OT**. On a basic level, every Σ -proof of T is a realization of the 'conversion' of one B-expressions (the **premise(s)** of T) into another B-expressions (the **conclusion(s)** of T), through the successive application of approximately modified / transformed generic OT. In this process, some generic OT may occur several (even, (un)countably many!) times, in various instantiations. From a strictly **constructivist** point of view, only **finite** compositions are admissible – but one of the fundamental aims of Σ is to embody a pragmatic approach to constructivity, by encompassing **a hierarchy of levels of constructive procedures**, maximising calculational facilities in **all** mathematical domains. Consequently, finery criteria are relaxed (minimally, if possible) where full algorithms are not available.

Since each member of the cast of any theorem within an OT-composition must be independently transformable, and generic OT may be repeated, it follows that a typical (finite) operational proof of $T : p \Rightarrow q$ has three possible representations:

(a) $(T_\sigma\circ)_L p = q \vee r$ (**generic / indicative**);

(b) $(f_{\phi_\sigma}(T_\sigma)\circ)_L p = q \vee r$ (**schematic / skeleton**);

(c) $(f_{\phi_\sigma}(T_\sigma(u_\sigma)_{h_\sigma})\circ)_L p = q \vee r$ (**detailed / full**).

Here (by convention – see E_1/E_4) the outer subscript, L, 'acts on all of the inner occurrences of σ (to produce, in turn, $\sigma_1, \ldots, \sigma_L$). The ordered list $(\sigma)_L \subset \mathbb{N}^+$ may contain arbitrary repetitions, and the same is true of $(\phi_\sigma)_L$. The associated transformations $f_{\phi_{\sigma_k}}$ are selected from the underlying collection of admissible transformations in Σ_t.

This (unavoidably prolix) notation reflects the facts that **(i)** each OT in the proof has its own cast; that **(ii)** each cast is transformed in some particular way; and, that **(iii)** every member of any cast is transformed by a corresponding function-component. The definition of $f_{\phi_{\sigma_k}}(T_{\sigma_k}(u_{\sigma_k})_{h_{\sigma_k}})$, as in E_4, gives:

$$T(f_{\phi_{\sigma_k 1}}(u_{\sigma_k 1}), \ldots, f_{\phi_{\sigma_k h_{\sigma_k}}}(u_{\sigma_k h_{\sigma_k}})).$$

In spite of the apparent complexity, these representations are direct consequences of the simplest form, '$f(T) := T(f)$'. For (un)countable OT-compositions (which may occur, for instance, when parametric families of OT are involved), the above representations must be modified. This is done later.

NOTE Both of $(\phi_\sigma)_L, (u_{\sigma_k})_{h_{\sigma_k}}$ are 'generated by T_{σ_k}' – in the sense that they yield some w.d. realization(s) of T_{σ_k}, etc.. It is unnecessary to display full dependence on the T_{σ_k} (and, ultimately, on the 'target theorem', T) except where there is danger of ambiguity. It is supposed that the sets of underlying objects, u_m, and of generic OT, T_n, are permanently ordered as $\Sigma_t$expands – so that selections, $(T_\sigma)_L$, etc., are unambiguously defined, for **generic** OT, together with the **casts** of these OT. Most of these considerations would be intuitively satisfied / irrelevant for calculations performed **without** an IB. For Σ , by contrast, they are absolutely crucial, in the efficient organization of the OIB.

The expanded forms of **(a), (b), (c)**, are[324]:

(a)$'$ $(T_{\sigma_L} \circ \cdots \circ T_{\sigma_1})p = q \vee r$;

(b)$'$ $f_{\phi_{\sigma_L}}(T_{\sigma_L}) \circ \cdots \circ f_{\phi_{\sigma_1}}(T_{\sigma_1})p = q \vee r$;

(c)$'$ $f_{\phi_{\sigma_L}}(T_{\sigma_L}(u_{\sigma_L})_{h_{\sigma_L}}) \circ \cdots \circ f_{\phi_{\sigma_1}}(T_{\sigma_1}(u_{\sigma_1})_{h_{\sigma_1}})p = q \vee r$

As in E_4, the 'residual information', r, produced in the proof of $T : p \Rightarrow q$, may be used to invoke other OT (say, T') whose premises (say, p') satisfy the condition: $q \vee r \Rightarrow p'$ where $T' : p' \Rightarrow q'$. All known conditions of this kind – with their associated OT – are recorded as part of the overall taxonomy of Σ_t.

Thus there are four 'standard forms of OT' within Σ_t(as well as a variety of **descriptive** forms) to be located in the IB, namely: the minimal rigorous **statement** (with no indication of other OT used in the proofs); and the forms **(a), (b), (c)** – giving progressively more information about (operational) proofs. The descriptive forms are both explanatory and 'inter relational' (establishing links among ostensibly disparate mathematical domains). Since the OT are (logically) simply pairs of (possibly 'compound') predicate functions – connected by '$\Rightarrow$' (implication) – the OT are elements of various algebraic structures (notably, Boolean algebras). Moreover, every **logical theory** possesses associated algebraic structures, and all of the corresponding

[324]Although $(T_\sigma)_L := (T_{\sigma_1}, \ldots, T_{\sigma_L})$, the **convention** here is that $(T_\sigma\circ)_L := T_{\sigma_L} \circ \cdots \circ T_{\sigma_1}$

results may be applied in the manipulation of sets of OT. The ramifications of this scheme are explored, in detail, in the book: '*The Mathematics of Meta-Mathematics*', by H. Rasiowa / R. Sikorski (PWN, 1963), to which reference has already been made.

Several other possibilities are created by the representations **(a), (b), (c)**, in the OIB context. In particular, suppose that the IB_t contains the (**maximal**) set of OT:

$$\{T_h : p \Rightarrow q_h : h \in S_p\} := \mathcal{J}_p.$$

Then the compound theorem ${}_{S_p}T : \bigwedge_{h \in S_p} (p \Rightarrow q_h)$ is called a **forward multiple OT** of (finite) **order** $|S_p|$. Similarly, a **backward multiple OT** of order $-|S_p|$ is defined as: ${}_{=S_p}T : \bigwedge_{h \in S_p} (q_h \Rightarrow p)$. These are special forms of **compound OT**, as defined in E_4. In general, S_p need not be maximal in Σ_t. The possibility that S_p may be an **infinite**, (un)countable set is considered later. All of the q_h, and p, are assumed to be either, **axioms**, or else, **already proven** over Σ_t. These types of compound OT play important roles, by focusing on the **scope of** p in Σ – where

$$\text{sc}_\Omega(p) := \{v : p \to v \text{ in } \Omega\}$$

– and, on the **range** of premises q_h which, separately, imply p. In the analysis of proof in an OIB, the 'information flow', into / out-of some 'problem domain' may suggest the 'shape of a proof'; after which, progressively more specific versions may be produced, quasi-iteratively – until (in favourable cases) a fully rigorous proof is obtained. The manner in which such heuristic processes evolve cannot be formalized (since, otherwise this would itself constitute a formal proof); but many suggestive criteria 'emerge' along the way. These ideas are diversely illustrated in $\text{E}_9/\text{E}_{10}/\text{E}_{11}$.

Another facet of the analysis of operational proof concerns **perturbations, stability, deformations, and proof-paths** (all in various sense). These, and cognate, notions stem naturally from the study of **proofs as multi-step mappings within sets of predicate functions** (taking premises to conclusions). The **realizations** of such mappings by OT-compositions may be (almost arbitrarily) modified through changes in some **generic** OF / transformations of the ('fixed') constituent OT. Indeed, in many important cases, these modifications could involve virtually all of the established fields of Mathematics. These intricate processes furnish excellent illustrations of **conceptual unification** produced by information-flow into restricted domains – through the extreme adaptability of generic mathematical procedures. In other words: All the (mathematical) world's a stage, and one OT in its time play (infinitely) many parts!

Apart from their intrinsic elegance, these proof representations offer diverse possibilities for (heuristic) exploration of hypotheses, and of 'exotic constructions'. In particular, the effects of (the above) controlled changes in 'the components of proofs' on both 'conservation of validity', and 'variation in strength', may be monitored systematically within Σ_t. Further, the ranges of applicability of classes of (general) techniques may be correlated with the types of realization of these techniques. Although such tasks may be denigrated, if viewed as primary (human) research activities, their value as a basis for innovative / imaginative work is considerable within Σ – especially, because the OIB search software – properly designed – can accomplish its aims exhaustively. **In short**: the study of proofs as **boolean combinations of compositions of transformed generic OT** (regarded as **functions** of these OT) raises novel areas of investigation, likely to be of fundamental importance in mathematical research, as governed by interactive systems of the Σ -type.

Among the topics of immediate relevance to this (O-T-composition) formulation of operational proof are:

(i) General analysis of forward / backward multi-OT; and, of the (Σ_t-maximal) collection (say, $\pi_t(T)$) of distinct proofs of any $T : p \Rightarrow q$.

(ii) OT as computer programs, with (formal) **inversion** procedures.

(iii) Relations between **(ii)** and Martin-Löf's 'Type Theory'.

(iv) Carnap's 'L-provability' (vs 'Derivation'), with associated (Hausdorff) TS structures over {L-provable statements}.

(v) Possible representations involving POLYADIC algebras (of which Lindenbaum-Tarski algebras are special cases).

(vi) Forms of **proof-stability** under perturbations of the components in OT compositions.

(vii) **Variations** of proofs under classes of perturbations.

(viii) Proof-preserving **deformations: quasi-homotopy** criteria.

Although some of these topics (and others discussed later) are primarily of academic / aesthetic significance, **now**, it is difficult to predict their medium / longterm roles. Indeed, even in the comparatively **short**-term phases of Σ -development, and of these formalisms could assume fundamental importance – for instance, on account of some improved / new software facilities; and so on. In the history of Mathematics over the Twentieth Century, there are many instances of such sudden rises to stardom! Notable examples include (finite) Group Theory; Graph Theory; Matrix Theory, and Mathematical Logic (as a foundation of Computer Science).

(ix) Another important field of investigation for operational proofs is **the classification of** (operational) **converses** – in terms of **'standard' / approximate / weak /...** versions, their interrelations, and their admissible uses in OT compositions. This is an intrinsically **mathematical** topic (in contrast with topics **(i)–(viii)**, which have a partially **meta**-mathematical flavour). Mostly, this distinction is not pertinent for Σ , since the interplay of descriptive levels plays a key part in the development of investigations. The main difference lies in **methodology** (specific theorems require detailed proofs – often of a highly technical nature. On the other hand, questions connected with topics **(i)–(viii)** tend to be discussed within broader frameworks.). The principal **aim**, here, is to establish general formalisms for the representation / treatment of all these topics. The extent of the analysis varies markedly according to the propensity for systematic calculation.

For instance, the analysis of **converses** (of various types) is conducted on both detailed and descriptive levels – in line with the types of questions considered. The associated Σ -software reflects these fluctuations in level – but 'leaves room for eventual equalisation'.

(x) If an OT-composition proof in the (above) form **(c)**$'$ is displayed in successive '**lines**' (or '**rows**') of symbols; say, as:

$$\pi_T[(T_\sigma(u_k))_l] = \vec{\rho_L} \equiv (\rho)'_l,$$

the column vector with elements $\rho_1, \ldots, \rho_L$, **then**:

$$\rho_{kH} := f_{\phi_{\sigma_k}}(T_{\sigma_k}(u_{\sigma_k})_{h_{\sigma_k}})\rho_k, 1 \leqslant k \leqslant L-1;$$

$$\rho_1 := f_{\phi_{\sigma_1}}(T_{\sigma_1}(u_{\sigma_1})_{h_{\sigma_1}})p;$$

$$\rho_L = q \vee r.$$

In formal language – including the (enhanced) **first-order predicate calculus** (which is adequate to express virtually all investigations[325] covered by 'standard mathematical activity'), the study of (formal) **proofs via unification** is well established; indeed, it is the basis of automatic theorem-proving. In this context, various **bounds** are known for **proofs of minimal size** (given the existence of a **proof of** k **lines** for a **sequent of size** m). In this approach, a **proof** is a (specified) **rooted tree**, whose **vertices** are labelled by **segments**. The **size** of the proof is the sum of the sizes of its sequence. The **number of proof lines** is the number of vertices. The **size of a formula / semiterm** is the

[325] See, e.g., Enderton, HB, '*A Mathematical Introduction to Logic*' (AP, 1972) §2.0

number of (atomic) symbols in it; and the **size of a sequent** is the sum of the sizes of its constituent formulae / semiterms. Finally (!), the **depth**, dp(t), **of a semiterm**, t, is the length of the longest path in the tree-representation of t. Here, **terms** contain only **free** variables, but **semiterms** contain both free **and bound** variables. In 'pure' proof theory, only the (typographical) **forms** of elements are considered – **not** their **interpretations**. Relaxation of this condition yields various levels of **semantic** systems[326]. The limitations of **purely syntactic** procedures have been exemplified, not only in the inadequacy of Hilbert's 'Consistency Program', but also, in the uninteresting / trivial results obtained so far from automated theorem-proving packages[327] – though, additional AI / heuristic devices may improve this situation somewhat. Nevertheless, these facilities are accommodated within P_Σ both, to cover routing 'book-keeping tasks' – and, to allow for possible enhancements of their scope / power, through improvements in software, or otherwise. The decisive extra ingredient in Σ -investigations lies in **the interactive participation of users**, who may 'react to the output', at any stage of the process. This enables users to develop proofs quasi-heuristically, until they assume forms susceptible of (retrospectively!) rigorous expression. It is just this combination of subconscious / intuitive / pictorial /... interpretation of 'the output generated through application of OT to wfe', with the motivated selection of 'appropriate OT' from the comprehensive IB, that produces radical increases in the scope / power of investigation – with a concomitant increase in the role of serendipity!

This problem, of **choosing OT pertinent to the operational proof of** T, is, of course, highly **non**-algorithmic, in most cases. If this were not so, Mathematics would be reduced to a cranking process! The assumption of **random** choices, however, would produce a Monkey / Shakespeare situation. The resolution of this issue lies, rather, in the development of **environments for investigation** of specified **problems** – in which the proofs of theorems may be included. Since the IB is very large, so is the associated CB (Calculational Base). The possible structures of such CB are discussed briefly in E_4. The **overall** CB is composed of **sub**-CB, corresponding to the various subdomains comprising Mathematics within Σ_t. These subdomains, in turn, are characterized by their own skeleton / partial / full **subtaxonomies**. The term **pre-**taxonomies covers certain combinations of these substructures: see E_3.

The process of taxonomy-construction may now be described (schematically) as follows.

(i) Selection of all general theories $\mathcal{J}_m$ collectively covering Mathematics over Σ_t: say, as

$$\bigcup_{1\leqslant m\leqslant L} \mathcal{J}^m = \mathcal{M}_t.$$

(ii) Decomposition of each $\mathcal{J}_m$ into sub theories: say, as

$$\mathcal{J}^m \equiv \bigcup_{1\leqslant i\leqslant K_m} \mathcal{J}_i^m.$$

(iii) Construction of 'presubtaxonomies', say,

$$s_i^m(\mathcal{J}_i^m).$$

(iv) Extension of the s_i^m (via '**interpolation**') to produce subtaxonomies – say, $\overline{s_i^m}$ – by procedures to be specified.

NOTE that steps **(iii)/(iv)** yield presubtaxonomies / subtaxonomies $s^m, \overline{/s^m}$, for the basic theories $\mathcal{J}_{\Updownarrow}$. The procedure is essentially independent of the 'topics' treated in the $\mathcal{J}_m$. An adequate choice of the collection $\{\mathcal{J}_m\}$ may be based on **the set of 'main headings' for**

[326]See, e.g., Beth, EW, '*The Foundations of Mathematics*' (North-Holland, 1968) Chs. 11, 12

[327]This judgement is, merit ably, subjective – but well-founded!

M(athematical) R(eviews). This MR list has been greatly refined (since 1940) and 'practically optimized'. Moreover, it is familiar to researchers, internationally. The additional subclassifications used for MR (also regularly refined) have stood the test of time, and so may be taken as a **basis** for the pre-subtaxonomies. It should be realized, though, that the MR lists **alone** are far from furnishing (even weakly adequate) taxonomic skeletons, in the sense of E_3. Rather, these skeletons, and their successive refinements, must be developed through the contributions of panels of experts, with regular maintenance, etc.. It is important to recognize that all refinements of subject classifications may be effected without disrupting / invalidating the existing frameworks – subject only to suitable **consistency criteria**.

So far, it has been assumed that the basic theories $\{\mathcal{J}_m\}$ are **mutually non-interacting**. Each (pre-)subtaxonomy is composed of '**primary constituents**; (with their axioms / definitions / basic interrelations), and '**principal aims**'. From such 'initial data' follow '**ket results on the attainment of these aims**' – together with the **techniques for deriving the key results**. Next, come all **established variants of the sub theory concerned** (with associated techniques). Other components of the (pre-)subtaxonomy include '**internal results**' (for proving the above 'key results' – of the original sub theory / variants); and so on! The taxonomies for theories $\mathcal{J}_m$, are constructed, from those for sub theories, in stages. **First**, by simple **aggregation; second**, by '**pairwise interaction**', and **at stage** k, by interactions involving k **subtheories of** $\mathcal{J}_m$. Next, one considers n**-fold interactions among** (different) **main theories** – say, $\mathcal{J}_{j_1}, \ldots, \mathcal{J}_{j_n}$. Formally, this is equivalent to the procedure for interactions among subtheories of any $\mathcal{J}_m$, so it suffices to consider this case.

Suppose that the s_{mi} have been specified (for each $\mathcal{J}_k$), and denote by $\rho(s_i^m | s_{j_1}^{p_1} \ldots s_{j_n}^{p_n})$ the collections of results derived from the s_i^m through information from $s_{j_1}^{p_1} \ldots s_{j_n}^{p_n}$ – where, for $1 \leqslant n \leqslant L-1$, $(p)_n \subset \bar{L} \equiv \{1, \ldots, L\}, j_n \in \overline{K_{p_n}}$, and $m \notin (p)_n$. Then one obtains the extended collections, $\overline{s_i^m}$ in the forms:

$$\overline{s_i^m} := \bigcup_{1 \leqslant n \leqslant L-1} \rho(s_i^m | s_{j_1}^{p} \ldots s_{j_n}^{p_n}) \cup s_i^m.$$

In spite of superficial complication, this representation – in which $1 \leqslant m \leqslant L$ and $1 \leqslant i \leqslant K_m$ – just combines the 'internal information' for $\mathcal{J}_{j_r}^{p_r}$. Plainly, this furnishes a systematic scheme for longterm development of the underlying 'overall taxonomy' of Σ_t– **provided that** associated classification procedures may be specified (to produce genera / species / types / realizations / examples /...). Since $\mathcal{J}^m$ is maintained / augmented regularly (as new results are proven, etc.), one may consider $\mathcal{J}^m(t)$, for each m in $\bar{L}$, at various times, $(t)_q$, and the taxonomic construction process may be **iterated**. The original 'global decomposition of $\mathcal{M}_t$ into $\bigcup_{1 \leqslant m \leqslant L} \mathcal{J}^m$ is permanent; but the possibility is left open for **extra** theories, $\mathcal{J}^{L+h}, h \geqslant 1$, to be added to $\mathcal{M}_{t'}$. These various (sub)theories are only rarely **strictly** disjoint. Rather, they are 'predominantly independent'. Examples of recent origin could include: 'Soliton Theory', 'Chaos Theory', and 'Catastrophe Theory'; each is embedded within broader, established mathematical domains, but still has a distinctive body of formulations / techniques / applications. Many other examples are cited in E_9/E_{10}/E_{11}.

Of course, this general process of taxonomy-construction is far from uniquely determined. Subjective elements in the choice / decomposition of the $\mathcal{J}^m$, and in the formation of the $\rho(s_i^m | s_{j_1}^{p_1} \ldots s_{j_n}^{p_n})$ are unavoidable; but all that matters is the accumulation of sound, nontrivial, adequately classified mathematical information, subject to 'practical consistency criteria'. The formalism / notation just introduced may be further refined, **subsets** of s_i^m and the $s_{j_r}^{p_r}$ are admitted as 'arguments' in the 'interaction sets', $\rho(\ldots | \ldots)$. In particular, s_i^m may be replaced by subsets with 'very few elements' (even, by 'singletons'). In this way, the effects of various (sub-)theories on central results in another throw may be monitored and analysed. Once again, it should be emphasized that **all of these steps may be iterated without invalidating the (currently) existing classification(s)** – since only a form of **interpolation** is involved. This 'conservation of IB validity under updating procedures' is of fundamental importance for the feasibility of $\Sigma_t$as an expanding, computer-based OIB, potentially adequate to cover arbitrary mathematical theories / applications, quasi-exhaustively. Many of the possible indicators of 'centrality' / 'power' / 'interconnection' / 'coherence' /... of (sets of) OT – which would have (at best) academic / aesthetic significance without IB facilities – assume crucial importance for Σ , where indiscriminate application of OT produces rubbish. Moreover, these measures of centrality / coherence / stability /... raise fascinating questions about the nature of powerful / ubiquitous (as contrasted with merely adequate / consistent) theories – both within Mathematics, and through reflections / refractions from the Sciences. These matters are discussed (primarily) in E_{11}. Although this is far from being a prescription for generating 'important results', such criteria – intelligently interpreted – undoubtedly broaden the purview of Σ -investigations – and so enlarge (rather than reduce) the roles of speculation / contemplation / insight / imagination. This broadening includes the option of pursuing many (classes of) **enumerative problems** in depth, to obtain exact or asymptotic results; or else, to provide evidence for well-founded **conjectures**. Many parts of Number Theory, Combinatorics, Graph Theory, Linear Algebra, Coding Theory and Design Theory (to take only the most obvious examples) would be profoundly affected in this way. The secondary ('feedback') influence on other – ostensibly, **non**-enumerative – investigations would also be considerable.

The topics **(i)–(ix)**, listed above, are now discussed briefly.

(i) Multiple OT

These are special forms of **compound** OT (as defined in E_4). For the **forward** / **backward multiple OT** of (finite) **order** $v_p / -v_p$, the notation[328]:

$$_{S_p}T : \bigwedge_{h\in S_p} (p \Rightarrow q_h); v_p := |S_p|;$$

$$_{-S_p}T : \bigwedge_{h\in S_p} (q_h \Rightarrow p),$$

has already been introduced. More generally, one may put:

$$_{\Omega_p}T : \bigwedge_{q\in\Omega_p} (p \Rightarrow q);$$

$$_{-\Omega_p}T : \bigwedge_{q\in\Omega_p} (q \Rightarrow p),$$

of orders $|\Omega| / -|\Omega|$, allowing for **infinite** (un)countable sets Ω. Although the distinction between finite and infinite multiple OT is important (especially, in implementation / realization over P_Σ), a **uniform** notation is also valuable; so the symbols, $_\Omega T, _{-\Omega}T$ – with Ω finite or (un)countable – is adopted as 'the standard form', with '$_{\pm\Omega}T$' covering both types. The relevance of the $_{\pm\Omega_p}T$ lies in **information flow** (from, or to, p), as a measure of the **scope** / **centrality** of p. The cases where Ω is infinite usually correspond to (sequential / continuum) **parametrizations** of **families** of (closely interrelated) theorems. As usual, the exceptions to this 'rule' are of particular interest, as illustrated in E_9/E_{10}/E_{11}. Formal manipulations over sets of multiple OT (as compound OT with 'simple structures') may be codified for applications – as part of the OTC. These, and cognate, enhancements of the OTC allow Σ -investigations to be conducted within highly stimulating environments – in the style of weakly formalized **heuristics**, with maximal opportunities for the discovery of felicitous links among ostensibly disparate domains.

The problem of selecting **pertinent** OT, at any stage of a proof of $T^* : p \Rightarrow q$, may be approached, at the most basic level, as follows. Suppose that a full proof has been represented as:

$$f_{\phi_\sigma}(T_\sigma(u_\sigma)_{h_\sigma}\circ)_L p = q \vee r,$$

[328]The subscript p (on S and Ω) may be suppressed it there is no danger of ambiguity. The symbols $|S|, |\Omega|$, denote card$S$, cardΩ; $\#S, \#\Omega$ are also used.

and that a suitable collection, $(T_\sigma)_L$ is sought. Although there is no infallible procedure for 'determining the T_{σ_k}', the range of 'sensible choices' may be drastically reduced by rewriting this representation 'iteratively' as:

$$q^{(o)} := p, q^{(L)} := q \vee r,$$

$$q^{(k+1)} := f_{\phi_{\sigma_k}}(T_{\sigma_k}(u_{\sigma_k})_{h_{\sigma_k}})q^{(k)},$$

for $k = 0, 1, \ldots, L-1$,and then considering 'quasi-measures of relevance' involving the content, $c(T_{\sigma_k})$, at each stage. Denote by $T^{(j)}$ the theorem $q^{(j)} \Rightarrow q \vee r^{(j)}$, where the $r^{(j)}$ simply correspond to the 'extra information' produced. (**NOTE** If $T : p \Rightarrow q$, then $r^{(o)} = \phi$.)

For any theorem T, define

$$\gamma\{T\} := \{T' : \bar{c}(T') \cap \bar{c}(T) \neq \phi\}$$

$$\gamma^k\{T\} :=' gamma\{\gamma^{k-1}\{T\}\}, k = 2, 3, \ldots.$$

Here, $\bar{c}(T)$ is the Σ -equivalence class of the cast, $c(T)$, etc.. Evidently, this may be strengthened in various ways – especially, by requiring (in the overall design of Σ) that $c(T)$ embodies only the **essential** case of T (in a sense that is intuitively clear, but virtually impossible to formalize!).

If T^* is prescribed, then the user may peruse $\gamma\{T^*\}, \gamma^2\{T^*\}, \ldots$, and then select certain OT that seem 'promising'. Exploratory application of 'suggestive transforms' of these OT produces 'new prescribed theorems' – to which similar operations may be applied; and so on. A fundamental question arises, in this context: Is it possible for an OT to be 'significantly relevant for T^*', and yet to lie outside **all** $\gamma^k\{T^*\}$?

Without a long (and tedious) discussion of 'all possible cases', no definitive answer can be attempted. On 'intuitive grounds', however, it appears that only types of (weak) analogues / reflections / refractions could, possibly, play such roles – for instance, by suggesting otherwise-unsuspected frameworks for interpretations of T^*. If this appreciation of the question is correct, then the (apparently inherent) limitations of automated theorem-proving packages become evident – and the potentially **unlimited** power of Σ (through interactive combinations of the OIB with **users**) may be understood.

(ii) OT as (Computer) Programs

From certain points of view, **constructive mathematics 'is** (equivalent to)' **programming** (in some 'suitable language(s)')[329]. This is not so surprising if one considers the SAM packages, where many basic mathematical procedures are implemented in LISP, or in C – as underlying languages. The broader claim – that **all** of constructive mathematics may be regarded as 'the development and application of programs' – is not (so far) as all-embracing as it may at first appear, since it amounts to a **specification** of 'constructiveness'. Many of the OT in Σ do not (strictly) satisfy this criterion, which seems more relevant to a theorem-proving package than to an interactive OIB. Nevertheless, and result so represented **is** necessarily **operational** in 'the Σ -sense'. Moreover, this approach may be used to identify the computational con tent in 'standard proofs'[330], and hence, to exhibit the (often 'hidden') **non**algorithmic parts of such proofs, so that these parts may be modified, eventually – either through the use of new techniques, or else, through the introduction of approach representations, to obtain variants of the original theorem(s) concerned. This it the (pragmatic) view adopted for Σ ; the potential scope / power of theorems is of primary importance. The quest for fully operational versions is maintained but theorems are incorporated within the IB even if they are purely descriptive / informative; after which they are gradually converted into (maximally) operational forms.

[329]See: Constable, RL, '*Mathematics as Programming*', LNCS #164, pp116–128

[330]See: Constable, R / Murthy, C, '*Finding Computational Content in Classical Proofs*', pp 341–362 in: Huet, G / Plotkin, K (eds.) '*Logical Frameworks*' (Cambridge UP, 1991)

When is **is** possible to develop 'program versions' of OT, the **inverses** – in the sense of the O-T Calculus – may be regarded as inverse **programs** for which there are already some results in the literature of Theoretical CS[331]. For consistency, this approach must be equivalent to the use of operational **converses** (discussed later in this Essay). Since more 'interesting theorems' are **not** fully operational, however, the program versions are somewhat restrictive for Σ – though they do constitute a subclass of OT with efficient implementations (and so should be identified). Other implications of 'proofs as programs' are better treated as part of the next item.

(iii) Martin-Löf's Type Theory

Here, **propositions** are identified with **sets** – where each set comprises all **proofs** of the proposition concerned (so that false propositions correspond to ϕ). The first-order predicate calculus is then **interpreted** within this system. On this basis, it may be shown that the evaluation of a well-typed program always terminates – a form of exact calculability. The main results in Type Theory generate (finite) program constructions – and hence, associated **proofs**. For instance, if the proposition $(\forall x \in A)(\exists y \in B)P(x, y)$ is viewed as a **specification**, and if $f : A \to B$ is constructed, with $f(a) = b$, where $P(a, b)$, then f is equivalent to a program satisfying the specification. It is claimed that even B-constructive[332] proofs are representable in this way (as programs)[333]; so the scope is (increasingly) broad, since the purview of B-constructivity is still being enlarged – and no absolute limits to this enlargement have been established[334]. In spire of this trend, the effective use of OT in 's depends mainly on responses to wfe in **standard** notation (rather than in various programming logics, etc.). If adequate **compilers** are developed, then the 'program representation' may be useful in raising the efficiency of fully constructive procedures. Meanwhile, most implementations will be formulated within enhanced versions of existing SAM packages.

Another aspect of type theory is its role in the proof-development system Nuprl[335], in which a 'uniform notation for discourse' is proposed – along with facilities for 'creating an encyclopaedia of results in this notation'. Definitions are introduced as 'templates' (for new notation). **Proofs** are formulated as **sequence**: $x_1 : T_1, \ldots, x_n T_n >> S[\text{ext}s]$, where the hypothesis list (premises) is separated from the consequence(s) (conclusion) by '>>'; the x_i are (free) variables, and the T_i are terms. The relevance of Nuprl for Σ lies principally in the areas of proof-**checking**. The goal of **uniformity** (of notation) runs counter to the Σ aim of 'free-wheeling mathematical speculation' promoted by evocative representation (including the graphical / diagrammatic / pictorial). It is possible, though, that the more formal scheme will be valuable for the eventual development of formal proofs of conjectures generated through 'informed speculation'.

Recall that the **type theory**, on which Nuprl is based, is distinct from **set theory** (but admits **interpretations** of it)[336]. A more directly mathematical interactive theorem prover / verifier (based on **Z**ermelo-**F**rankel set theory, and using **forward**-chaining inference techniques) is ONTIC[337], which synthesises ideas from AI and theorem-proving packages. An important feature is the use of **focus objects** – to obviate 'combinatorial explosions' in search software. Further, the system contains a 'library of lemmas', whose relevance to focus objects may be assessed (to minimise 'superfluous information'). It is claimed that 'ONTIC integrates **constraint-propagation**' (inference bounded – in running-time / number-of-steps – by the size of a finite constraint network / graph), **congruence-closure** (a variant of 'unification'), **focus(s)ed binding** (a form of 'inheritance mechanism'), and **automatic universal generalisation** (**via** proofs

[331] See, e.g., Gries, D, '*The Science of Programming*' (Springer, 1981) Ch. 21

[332] i.e. constructive in Bishop's sense (see E_2) and (for full details) the books (B), (B/B) listed there.

[333] This use of Type Theory (in Constructive Mathematics) is treated in: Nordstrom, B, et al., '*Programming in Martin-Löf's 'Type Theory*'', (OUP, 1990)

[334] From the book (B/B), indications of the situation up to 1985 may be obtained. (See, especially, the Bibliography)

[335] See: Constable, RL, et al., '*Implementing Mathematics with the Nuprl Proof Development System*' (Prentice-Hall, 1986) for a detailed account of 'Version-1'.

[336] See Constable, op. cit., Sec. 2.6.

[337] See McAllester, DA, '*ONTIC: A Knowledge Representation System for Mathematics*' (MIT Press, 1989)

of results for generic objects of type τ – with **no extra assumptions**), to produce 'a single labelling process on a fixed graph structure'. Another comparison may be made with AUTOMATH[338] (based on a typed lambda calculus – closely related to constructive logics). Once again, direct usefulness for Σ is confined to proof-**checking**, since the AUTOMATH format is opaque and unsuggestive, as well as being unwieldy[339]. The ultimate aim, of course, is to synthesise from (the successors of) these systems, powerful / transparent facilities for proof-development / checking – with adequate compilation into the top-level P_Σ language.

(iv) Carnap's L-Systems / Axiomatics

In some respects this may be regarded as an anticipation (essentially 'theoretical') of **Bourbaki's** (continuing) development of (Pure) Mathematics through diverse but interrelated **structures**, with associated **Axiom Systems**. For Σ , the 'ideal OT' is fully constructive, but **partially operational** forms (even, **descriptive** forms) are covered by the OTC, where the aim is to promote 'mathematical activity' – rather than to 'validate calculations'. Fortunately, both approaches (Axiomatic, and Constructive) may be accommodated within the Σ framework, as expounded in these Essays, without inconsistency or contradiction. The Bourbaki scheme is briefly discussed in E_1, as a possible superstructure for Σ . The underlying **logical** justification – with representations of formal languages / predicate functions / propositions / proofs / confirmations /..., and copious **examples** – may be found in the book: '*Introduction to Symbolic Logic and its Applications*', by R Carnap (English translation, Dover, 1958). Although there are now many expositions of these matters[340], Carnap's combination of clarity, cogency and coverage, is still unsurpassed, at least for the purposes of Σ . In particular, the role of 'logic' in the formation of **axiomatic theories** is treated in detail.

Carnap's presentation is characterized by 'underlying languages' (A, B, C) in terms of which all systems of axioms are specified. These languages differ somewhat in power and scope, but also in succinctness. Church's λ-(Lambda) operator occurs in B, C, but not in A. Various special notation is used to abbreviate long / complex expressions; e.g., $Ixy \equiv x = y, J_k x_1 \ldots x_k \equiv$ 'all of $x_1 \ldots x_k$ are distinct' (which may be expressed **via** '$\neq$'). Many basic concepts are formulated over the logical frameworks (A, B, C). The notation 'L', for language, 'L-true', for 'true within L', etc., is used systematically to study 'the structure of n-place relations / predicate functions' (involving a hierarchy of 'types' – to avoid the set-theoretical antimonies). The overall scheme is adequate to accommodate the Σ -transactions encountered in proof-development. It is essential that the foundational apparatus does not obscure the content of the OT (which is, in many cases, unavoidably intricate). The Carnap formalism seems to offer about the optimal mixture of rigour and expressiveness – though it is, of course, essentially equivalent to the other treatments of predicate calculus covering the same ground. The overriding aim is to facilitate the determination of axiomatic systems (AS), over suitable languages (L), for (arbitrary) domains of Mathematics and its applications. This process raises problems , if strictly constructive procedures are required; but, in the context of Σ , no such restrictions are imposed. Several examples are discussed by Carnap, from Geometry, Physics, and Biology. Another source of AS in Geometry and Physics is the Symposium Proceedings, '*The Axiomatic Method*', edited by L Henkin / P Suppes / A Tarski (North-Holland, 1959). This approach to the study of mathematical theories is relevant primarily to structural questions; but the possible **deformations of theories**' through perturbations of their AS, may also be investigated **via** this model. Problems of proof-**stability** / proof-**variation** / proof-**deformation** – for individual **theorems** – are important for Σ ; these matters constitutes items **(vi)/(vii)/(viii)**, here.

---

[338] See, e.g., Rezus, A, '*Abstract Automath*' (Tract No. 160, Math. Centrum, Amsterdam, 1983) and references given there.

[339] The AUTOMATH version of E. Landon's '*Foundations of Analysis*' (1976) was issued in five volumes!

[340] For instance: Kneebone, GT, '*Mathematical Logic and the Foundation of Mathematics*' (Van Nostrand, 1965); Suppes, P, '*Introduction to Logic*' (Van Nostrand, 1957); Beth, EW, '*The Foundations of Mathematics*' (North Holland, 1968); Hermes, H, '*Introduction to Mathematical Logic*' (Springer, 1973); Kleene, SC, '*Mathematical Logic*' (Wiley, 1967); Enderton, HB, '*A Mathematical Introduction to Logic*'.

(v) Polyadic Algebras

These algebras may be characterized as **Boolean algebras with operators** (which are abstract analogues of **substitution**). More precisely, they correspond to **endomorphisms of a Lindenbaum-Tarski algebra** (the algebra of wff of a first-order logical theory). An extensive discussion of L-T algebras is given in H Rasiowa / R Sikorski, '*The Mathematics of Meta-Mathematics*' (op. cit.), where many concepts basic to Σ are formalized over the first-order predicate calculus – which (see Enderton, HB, '*A Mathematical Introduction to Logic*' (op. cit.)) is adequate for all investigations conducted in 'general mathematical activity'. This body of results therefore underpins a fundamental facet of the OTC, namely, the systematic modification of OT through substitution, to obtain representations of operational proofs (on which investigations of proof stability / variation / deformation /... may be based).

(vi) Proof-Stability

There are several possible notions of stability associated with the general representation (given above)

$$\pi(T) = underseta \leqslant k \leqslant L \bigvee_{M_k \subset \mathbb{P}} b_{M_k} \xi_{M_k} \equiv \Phi_T(\xi_M)_K$$

$$\xi_{M_k} := \bigwedge_{m \in M_k} C^{J_m}$$

$$C^{J_m} := j \in {}^\circ J_m T_j(f^j(u_{\sigma^j}))_{L_j},$$

with the dependence of all entities of T being understood. Here, the 'target theorem' is $T : p \Rightarrow q$, and the interpretation requires that $\pi(T)p = q \vee r$, where r denotes the 'extra information' generated from p (if any). Thus, formally, $\pi(T)$ is an operator-valued function of all T_h such that $h \in \bigcup_{1 \leqslant k \leqslant K} M_k \equiv M_K$. This form allows for arbitrarily complicated premises / conclusions – in practice, $K, |M_k|, |J_m|$ are 'small' (often 1 or 2) and the proofs have simple / transparent structures.

For Σ , however, as a computer-based OIB, the length / complexity of operational proofs (OP) is not a significant factor (given the capacity / speed of modern systems); and, some results of fundamental importance may be provable – at least, initially – only be exploiting the full IT/SAM/AI resources available, though subsequent proofs could become progressively shorter! This situation is fairly typical in 'standard modes of research'; except that many procedures beyond the capacities of unaided human investigators are routing options for Σ , though these (comparatively) brute-force methods should be used only after more conceptual / elegant approaches have been explored.

It is in this context that variants of 'stability' are to be considered; either, as reflections of types of stability already established, or else, as 'new types' suggested by the OT-composition scheme. Of course, not all forms of stability have obvious 'OT analogues', but – as for definitions of $f(T)$ over general class of functions, $\{f\}$ – the stability of $\pi(T)$ may, in some cases, be specified in terms of comparable properties of some constituent(s) of $\pi(T)$. This gives rise to an abundance of possibilities, ranging widely over Mathematics (in all fields of application); many examples are given in $E_9/E_{10}/E_{11}$. The following (preliminary) definitions refer directly to the basic OT-Composition representations of proofs.

Let the substitution operation s be defined by:

$$s(A|B)\Psi(A) := \Psi(B),$$

for equipollent, ordered sets, A, B. Notice that A and B need not be disjoin; indeed, one may have all possibilities, from $A = B$ to $A \cap B = \phi$. Moreover, if $A \equiv (a)_n, B \equiv (b)_n$, then the substitution operation may also be written as:

$$s((a)_n|(b)_n)\Psi(a)_n = \Psi(b)_n.$$

In particular, for OT-proof-representations, $\Phi(\xi_M)_K$ is transformed into $\Phi(\xi'_M)_K$, say, where each of the ξ'_{M_h} is obtained from ξ_{M_k} by replacing every T_j in an OT-Composition by T'_j.

For any wff, e, denote by $[e]$ the maximal proposition contained in e, and, by $c(e)$, the wff $e\backslash[e]$. Let $T : p \Rightarrow q$, and $\Phi(\xi_M)_K p = q \vee r$. Consider the set

$$\{q'\}_{\Phi,p} \equiv \{[\Phi(\xi'_M)_K p'] : p' \in \mathcal{N}(p)\},$$

where $\mathcal{N}(p)$ demotes some 'neighbourhood of p'. Then a basic form of stability requires that $[q'] \equiv [\Phi(\xi'_M)_K p']$ 'remains sufficiently close to $q\wedge$'. The definition of 'neighbourhood' given in E_5 – in the form

$$\bar{\delta}(c_\alpha, c'_\alpha; w) := \lim_{k\to\infty} \bar{\mu}(\bar{P}_k \Delta \bar{P}_k{}' / \bar{\mu}(W_k)$$

– may be used here to cover a broad spectrum of cases. The basic stability condition thus becomes:$\{q'\}_{\Phi,p} \subset \mathcal{N}'(q)$, which is just a sort of **continuity**. The underlying property is that: small enough variations of **stable** operational proofs (of T) yield operational proofs of OT 'close to T', and, 'of the same type as T'. By contrast, an **unstable** operational proof (of T) has neighbourhoods containing **no** proof(s) of any T' near T; or else, containing some proof(s) of OT of different type (from that of T).

Of course, these notions are speculative and somewhat vague, but they do play a role in brands of 'weakly formalized heuristics' characteristic of Σ . Another useful variant of proof-stability is best formulated in terms of classes of 'proof-variations'. **NOTE** that both proof-stability and proof-variations are considered as consequences of **perturbations** of the constituent OT.

(vii) Proof-Variations

The principal aim here is to study variations of Φ_T under perturbations of its constituent OT (contained in compositions within the ξ_{M_h}). Rather than seeking quasi-continuity (as for 'stability'), one explores the propensity of Φ_T to generate OP, $\Phi_{T'}$, of theorems 'emanating from T', through changes of procedures at various stages of Φ_T. These procedures / stages correspond to the application of particle OT (say, T_j) in Φ_T. The admissible perturbations of such T_j encompass **all** Σ **-approaches to** T_j (i.e., all standard forms of approximation', and all w.d. abstractions / analogues). Although this may appear to be of purely theoretical interest, the nature of the Σ IB makes the exploration of 'potential scope' of established proof-schemes a matter of fundamental importance. Indeed, on this basic, valuable indications of 'centrality of results' may be obtained, along with 'tactical information' for incorporation into heuristics routines. More formal (algebraic / analytical / logical /. . .) structures over sets of OP may be developed, to formalize these exploratory procedures; but the intention here is simply to identify proof-variation as a significant topic.

(viii) Proof-Deformation

This topic is concerned with classes of OP producing **'the same target theorem'** (say, $T : p \Rightarrow q$). If $\mathcal{P}$ denotes a suitably defined space of propositional / predicate functions, then (schematically) every OP corresponds to some **path** in $\mathcal{P}$; and every op **of** T, to a path from p to q in $\mathcal{P}$. In fact these paths are (typically) **polygonal**, with internal vertices at the 'intermediate propositions' through which p is transformed to derive q. **Equivalently**, every proof-path joins **theorems** in **chain** cumulatively proving T; namely as: $\{q^{(k)} \Rightarrow q^{(k+1)} : q^{(o)} \equiv p, q^{(n)} \equiv q, 0 \leqslant k \leqslant n-1\}$. Now, there are several ways of specifying algebras associated with formalized theories; and, of specifying topologies / geometrical properties (including **arcs** joining pairs of points). If this is viable for the present discussion, then **families** of paths from p to q may be introduced – with cognate forms of (quasi-)homotopy. Once again, the utility of such constructions cannot be predicted, but the resulting formalism is likely to raise novel problems, in unfamiliar contexts – which is (almost!) a prescription for generating 'interesting mathematics'.

The most pertinent sources here are: the book by Rasiowa-Sikorski (op. cit) and the papers: '*Boolean Geometry*': I (LM Blumenthal) / II (LM Blumenthal / CJ Penning)[341].

NOTE that: a variant of proof-deformation (as discussed here) may be called proof-**perturbation** – where the constituent OT in an OP are modified (in prescribed classes of ways) and the corresponding set of different OP/OT is studied. Questions of **stability** arise naturally for such processes. This is considered briefly in **(vii)**, above; and, in a variety of contexts, in other Essays.

(ix) Converse OT

The OTC is modelled (distantly) on the Heaviside Operational Calculus (OC) – in the sense that algebraic / analytical structures are **imposed**, on classes of OT, **before** the adequate **realization** of these structures[342]. For the OC, problems of representation / convergence / uniqueness / version /... were eventually solved; initially, through the Laplace transform[343]; more generally, through the 'convolution quotient ring', introduced by Mikusinski[344]. The corresponding issues in the OTC, however, cannot be resolved so conclusively, since they involve **all** of Mathematics, as a universal calculational scheme. Consequently, justification of the OTC is unlikely to be achievable 'globally' – rather, through gradually augmented procedures over collections of mathematical **sub**domains. The (longterm) possibility of unifying all of these procedures (by means of some form of 'conceptual continuation') has been raised in earlier Essays – and this is discussed further (mainly, in examples) in $E_9/_{10}/E_{11}$. Nevertheless, even the 'local development' of the OTC within Σ is of considerable value, both aesthetically, and for use in heuristics. Perhaps the most fundamental property of the OTC is the existence of **invertible** OT – in terms of which the processes of Algebra / Analysis / Topology /... may be employed in Σ -investigations.

The basic prescription[345] for operational inverses of OT is that they correspond to **converses** – at suitable levels. That is: a **strict inverse** is given by a **strict converse**; a **weak inverse**, by a **weak converse**; a **partial inverse**, by a **partial converse**; and **approach inverse**, by an **approach converse**; and so on, with possible combinations of these and other types of converses.

Let $T : p \Rightarrow q$ be any generic OT, with converse, κT (say). Each variant of T has an associated converse. The Σ -variants of T include: $\mathcal{A}T$ (approach), AT (standard approximation), WT (weak), PT (partial), GT (generalized), ST (strong). Thus, each of $\mathcal{A}$, A, W, P, G, S, ... may be viewed as a schematic operator, which generates from T the corresponding class of variants of T. In Logic, if κT exists, then it is **unique**, and represented as: $\kappa T : q \Rightarrow p$. In practice, however, modified forms of converses are of great importance. These modified forms are obtained by replacing p and/or q by $m(p)$ and/or $M(q)$, where $m(p), M(q)$ are admissible variants of p, q, respectively. This yields possible **quasi-converses** of $T : p \Rightarrow q$ as $m\kappa T : q \Rightarrow m(p), MkT : M(q) \Rightarrow p, Mm\kappa T : M(q) \Rightarrow m(p)$. The form κT is called **strict**; certain special forms (defined later) are called **full**; and the other forms, **modified** (which, formally, covers **all** types).

Recall that every OT has, in Σ , three representations – referred to as

$$^{e}T \textbf{ (explicit) } \equiv T(\theta)_h \equiv \Gamma^T(u_\sigma)_h;$$

$$^{t}T \textbf{ (task) } \equiv {}^{t}T(\tau)_k \equiv {}^{t}T(\tau(\theta)_h)_k;$$

$$^{c}T \textbf{ (cast) } \equiv C(T) \equiv \{c(T)\}_l$$

341 Rend. Circ. Mat. Palermo Ser. 2 (1952) 343–361; (1961) 175–192

342 Heaviside's methods were criticized as unsound, even though he 'solved' previously intractable problems.

343 The most stimulating treatment is given by Van der Pol, B / Bremmer, H, '*Operational Calculus Based on the Two-Sided Laplace Integral*' (Cambridge UP, 2nd Ed, 1955)

344 A lucid account may be found in: Berg, L, '*Introduction to the Operational Calculus*' (North-Holland, 1967), where the one-sided Laplace transform is also (briefly) treated.

345 See Elvey (1980): '*Symbolic Computation and Constructive Mathematics*' (U. Waterloo Research Report CS-80-49, pp249, Section 14.4)

– as well as the **generic** form,

$$^gT \equiv {^gT}(\theta_T) \equiv p_T(v) \Rightarrow q_T(w).$$

Here, all of h, k, l, σ are (formally) dependent of T, and help to characterize T, in terms of its constituents, $(u_\sigma)_h$, tasks, $(\tau)_k$, and cast, $\{c(T)\}_l$. Thus, the most significant variants of T are formulated by modifying the elements in these representations. Although **all** of the types AT, WT, PT, GT, ST (and others not mentioned yet) are formally covered by AT alone – because of the quasi-universal interpretations of 'approach' within Σ – the emphases are very different, among these classes of modified OT (as in 'normal mathematical activity'). In the present context, 'AT' refers primarily to standard-**approximative** versions of T (rather than those obtained through abstraction / analogy / inverse-analogy). Under this restriction, the class {AT} no longer subsumes all of the other classes of modified OT, and genuine combinations of modifications may be interpreted consistently. The most important task is to specify all types of **converses** of T; in other words, to define $\kappa T, A\kappa T, WA\kappa T, GAW\kappa T$, etc. – each corresponding to some kind of operational inverse.

It is convenient to use an alternative notation for (generic) OT, in the discussion of converses. If $T : p \Rightarrow q$, the $[T] \equiv p, \langle T\rangle \equiv q$, so that, equivalently: $[T] \Rightarrow \langle T\rangle$. This is readily adaptable to the most general types of theorems / converses, since (e.g.) $AWT : [AWT] \Rightarrow \langle AWT\rangle, \kappa AWT : [\kappa AWT] \Rightarrow [AWT]$. **Notice that:** for a **strict** converse of T, one requires that $[\kappa T] = \langle T\rangle$; whereas, for a **full** converse, it suffices that, **either**, $[kT] \Rightarrow \langle T\rangle$, **or else**, $\langle T\rangle \Rightarrow [kT]$, where, now[346], κT denotes strict converses, and kT, full converses. Of course, it may happen that $[kT]$ and $\langle T\rangle$ are logically independent; in which case the pair of OT, $[kT] \Rightarrow [T], \langle T\rangle \Rightarrow [T]$, constitute (part of) a **multiple** OT (as considered earlier). If $[kT]$ and $\langle T\rangle$ are only **partially** independent – in a sense to be discussed later – then forms of '**partial converses**' may be defined. The other types of converses mentioned so far may be defined as soon as the variants AT, WT, PT, GT, ST, ... have been specified.

Variants of T

The overall procedure, for all variants, is based on the standard representations, $^eT/{^\tau T}/{^cT}$, mentioned above – each composed of collections of elements (ME from $\mathcal{U}_t$, for ^eT; tasks accomplished, for $^\tau T$; primary ingredients, for ^cT). Each element may be approximated / weakened / partially-realized / generalized / strengthened / abstracted / analogised – or left unchanged! If all modifications of some OT (in a specified representation) are of **the same type**, then a **pure variant** is obtained; otherwise, the variant is of **mixed type** (the 'mixture' being depicted by the (finite) sequence of letters chosen from {A, W, P, G, S, ...}, with preservation of order, but no restriction on repetition). Most of the applications involve pure variants or 'low-order mixtures', but the full potential of Σ can be realized only if arbitrary mixtures of types are admitted.

Even this range of variants does not cover all situations, **unless** the **modes** of combination (of various elements) are **themselves** designated as elements – albeit of qualitatively different sorts. Therefore, this assumption is made from now on. Notice that the classes $\{^eT\}, \{^\tau T\}, \{^cT\}$, are not disjoint, since (e.g.) explicit representations will, in general, contain some elements more characteristic of the other forms of representation. With these preliminaries, a basic classification may be developed for the strict / full converses of all w.d. variants of any OT, and for the other modified forms.

To systematise this process, it is useful to combine the generic and local-explicit forms of T, to obtain:

$$T : [T](\xi)_m \Rightarrow \langle T\rangle(\eta)_n, \text{ with converse}$$

$$\kappa T : \langle T\rangle(\eta)_n \Rightarrow [T](\xi)_m.$$

[346] and henceforth

Here, $(\xi)_m, (\eta)_n$ denote the sets of constituents selected from $\mathcal{U}_t$, etc.. Now introduce, for each finite subset S of $\mathbb{P} \equiv \{1, 2, 3, \dots\}$, the 'modification set', say, w^S, defined as $w^S := \{w^k \in \Omega : k \in S\}$, where Ω denotes the set of all Σ -admissible modification operators, and, for $S \subset \bar{j} \equiv (1, \dots, j), w^S(\lambda)_j \equiv (\chi)_j$, where

$$\chi_k := \begin{cases} w^k \lambda_k & , l \in S \\ \lambda_k & , k \in \bar{j} \backslash S. \end{cases}$$

The choice of the w^k from Ω is unrestricted; repetitions are allowed. The form of w^S in any application depends on the nature of the ingredients in $[T]$ and $\langle T \rangle$. In other words, w^S effects prescribed types of modifications of the ingredients λ_k, for $k \in S$ – thus producing the modified (quasi) converses $m\kappa T, M\kappa T, Mm\kappa T$, already mentioned. In these terms, w^S encompasses realizations of $m, M, Mm, \dots$; and m, M, may be regarded as 'global operators', expressed through the 'local operators', w^k, etc.. **Notice that** the operators m, M, κ **do not commute**, in general, since (e.g.) $m\kappa T$ is distinct from $\kappa m T$, typically – if both OT are even w.d..

This formalism is flexible enough to accommodate all possible variants of T; and hence, also, the converses of these variants, whenever they are w.d.. The associated modification operators may be 'indexed' by the types of variant produced –

$$m_{\mathcal{A}}, m_A, m_W, m_P, m_G, m_S, \dots \text{ for } p \equiv [T];$$

$$M_{\mathcal{A}}, M_A, M_W, M_P, M_G, M_S, \dots \text{ for } q \equiv \langle t \rangle.$$

The 'mixed modifications' may be multiply-indexed (e.g., $m_{A,W}, M_{PSA}$), with realizations involving the w^S, etc.. All of this may be 'intuitively obvious', but it cannot be incorporated into the OIB structure without a systematic notation.

It is necessary now to indicate how the 'operators' m_Q, M_Q may be defined, for finite ordered sets, Q, of symbols chased, possibly with repetition(s), from $\{\mathcal{A}, A, W, P, G, S, \dots\}$. Since (by convention) m_Q acts on $[T]$, and M_Q, on $\langle T \rangle$, one may write, unambiguously, $m_Q T, M_Q, T, M_{Q_1} m_{Q_2} T (\equiv m_{Q_2} M_{Q_1} T)$, as appropriate in any context. Observe that, whatever the form of Q may be, all of these modifications of T are themselves 'potential OT', and so may be further modified – say, by $m_{Q'}, M_{Q'}$, etc.. The only restriction on this process is that the 'output' **is**, in each case, a w.d. OT. In particular, if T **is** as (strict) converse – say, $T \equiv \kappa T'$ – then all of the OT $m_Q T, M_Q T, M_Q m_Q T$ (provided only that they are w.d.) may be regarded as quasi-converses of T', of the type(s) associated with Q, Q_1, Q_2; and so on! In this way, the whole range of admissible types of OT-**inverses** may be 'constructed'. **NOTE that** such inverses cannot be obtained through the representation of $f(t)e$ and $T(\dots f \dots)e$ (with $f(x) \equiv 1/x$, etc.), which correspond primarily to frameworks for manipulating **transformations** of OT – and so may be called **transformational** inverses, rather than **operational** inverses, as discussed here.

In specifying the operators $m_{Q_1}, M_{Q_2}, M_{Q_2} m_{Q_1}, \dots$, it suffices to consider the case where $|Q_1| = 1 - |Q_2|$, since the global operators are just compositions of local operators; and, as m_Q, M_Q have identical forms, then it is enough to treat just M_Q, for $Q = \{\mathcal{A}\}/\{A\}/\{W\}/\{P\}/\{G\}$ $/\{S\}/\dots$. This, in turn, entails giving 'working interpretations', for the terms 'approach' / approximate / weak / partial / generalized / strong / ..., within Σ . Several of these terms have already been discussed in detail; whilst the central concepts involved overlap, there are distinguishing 'essential characteristics' adequate to allow nontrivially different representations of rate different types of modifications / variants.

The classification of operators, M_Q, may be based on the fundamental properties of each type of variant covered – for instance, as follows[347].

(i)′ $\{A\}$ involves all forms of (standard) **approximation** of ME / elements (including modes of combination, etc.). Global modification operators that are purely approximative – say

[347] Since $\mathcal{A}$ subsumes all of the other types of modification, as well as covering aspects of other types (abstraction / analogy), the 'case $Q = \{\mathcal{A}\}$' is treated last.

w^S, with all w^k approximative $k \in S$ – correspond to various modes of approximation discussed in E_5/E_6 (especially). The resulting OT may be approximately-**formulated** /approximately-**implemented**, through the treatment of subsets $\xi^S/\eta^{S'}$ of $(\xi)_m/(\eta)_n$, where $\lambda^S := (\lambda_t : t \in S)$, etc., and the selection of approximative operators in w^S determines the variant of T concerned. The underlying **framework(s)** for approximation are associated with the w^k, here.

(ii)′ The range of '**weak variants**' of ME in Σ is broad; but the characteristic properties mostly involve species of 'weak convergence' within some approximation environment. Again, standard variants always yield forms through relaxation / weakening of some condition(s) – hence the name. Consequently, weak OT have wider applicability than the standard OT; and, more generally, weak ME are obtained by weakening some defining condition(s) for the corresponding standard ME, as discussed in E_5/E_7. For **theorems**, T, the possible range of weak forms is determined by the variety / combination of weakened ingredients in $[T]/\langle t\rangle$. **Note that** the changes between weak, standard, and strong versions (of any T) are **not** necessarily **reversible**; this point is considered further in **item (iv)**. **NOTE also, that** the distinction between 'weak' and 'partial' versions of OT (and of ME in general) is somewhat blurred; the reversible weak versions result from **implicatory** changes in conditions; other weak forms are **not** reversible (through condition-modification) to obtain standard forms.

(iii)′ **Partial realization** of any ME is associated with **(a)** incomplete specification; **(b)** non-implicatory deformation of the LC[348]. For instance, in a partial converse of T (say, $P\kappa T$), one has $\langle P\kappa T\rangle \subsetneq [T]$, where $\langle p\kappa T\rangle$ cannot be 'strengthened to $\langle \kappa T\rangle$' (since this would be an implicatory deformation). On this basis, the partial forms of ME are linked, primarily, with abstractions / analogues, which 'complement A in $\mathcal{A}$'. In other words, 'Approach' subsumes Approximation / Abstraction / Analogy – as envisaged from the outset, when the 'Approach scheme' was introduced (in E_1). Moreover, in partial forms of OT (or of other ME) some of the LC may be dropped – also a non-implicatory operation! – and new LC may be added.

(iv)′ **Generalizations** are characterized in E_7 as 'reversible abstractions'. In the present context, they are generated by **implicatory** deformations, whose inverses produce **specializations**. In this process, the **strength** of statements may be **conserved**, while their **scope** is **extended / reduced**; alternatively, **both** the scope **and** strength may be altered. The claimed 'reversibility' of the process refers to the logical interconnection of any ME with its generalizations / specializations. These remarks are pertinent to arbitrary ME (through their LC representations); and hence, to OT with modified ingredients. Diverse examples are given in $E_9/E_{10}/E_{11}$.

(v)′ For **strong** versions of OT, the presence of **non**-implicatory deformations[349] is, perhaps, typical; so, a relatively small proportion of strong OT may be changed **reversible** back to standard forms. This behaviour (true, in some degree, also for **weak** OT (**item (ii)**)) complicates the representation of W, S, in the formation of WT, ST from T – in terms of operations on some ingredient(s) of T. When 'new' conditions have to be 'interpolated' – or 'old' conditions removed – from LC, elements of abstraction / analogy enter the process – and so must be incorporated within the definitions / specifications of W, S, as 'operators'. These matters are discussed in E_7.

(vi)′ The notion of **approach** to (arbitrary) ME – as a 'universally applicable extension of approximation' – is fundamental for Σ . The associated 'operators', $M_{\{mathcalA\}}$, must, therefore, be compounded of operators for the formation of standard approximations (of all kinds), abstractions, and (inverse-)analogues – all of which have been discussed in

[348]Locational Conditions: see E_5
[349]relative to the standard form(s)

earlier Essays (especially, E_5/E_7). Although no **general** representation of $M_{\{\mathcal{A}\}}$ (or, of any other M_Q) is attainable, the overall formalism is of great value in the systematic study of Σ as an OIB. Moreover, in particular cases, the M_Q may be given definite representations, as many illustrations in $E_9/E_{10}/E_{11}$ will indicate.

It follows, from all of these considerations, that the 'operators' $\mathcal{A}, A, W, P, G, S, \ldots$ may be represented only as (usually somewhat complicated) combinations of deformations of the LC specifying the ingredients of the OT in question. Nevertheless, classes of OT will admit modification operators of 'broadly the same structure'; so a degree of systematization is attainable. The cumulative formation of this collection of (classes of) modification operators will furnish a basis for the systematic design of 'environments for investigation' within Σ .

NOTE The generation of OP through (the action of) compositions of transformed generic OT on wfe (premises) may be combined with approach criteria to extend the concept of proof to heuristics. This leads to various types of **quasi-proof** (of $T : p \Rightarrow q$), including:

(i) quasi-proofs using 'probability logic';

(ii) quasi-proofs as rigorous proofs of weak forms of q from p;

(iii) quasi-proofs as rigorous proofs of q from strengthened forms of p;

(iv) quasi-proofs using some probabilistic estimate(s);

(v) quasi-proofs based on simplifying assumptions in the underlying model(s);

(vi) quasi-proofs with acknowledged gaps;

(vii) quasi-proofs over Probabilistic Spaces;

(viii) quasi-proofs over fuzzy structures.

There are other interesting possibilities, but this list illustrates the basic ideas. Here, the **scope** (of a CB) and **centrality** of an ME, e, are pertinent.

References:

Hintikka, J, et al. (eds.), '*Aspects of Inductive Logic*' (N-H, 1966), esp.: Suppes, P (pp 49–65), and Scott, D, et al. (pp 219–264);

Wand, H, '*A Survey of Mathematical Logic*' (N-H, 1963);

Schweizer, B, et al., '*Probabilistic Metric Spaces*' (Elsevier, 1982);

Jeffreys, H, '*Scientific Inference*' (CUP, 1957);

Jeffreys, H, '*Theory of Probability*' (3rd ed., OUP, 1961) (For **fuzzy structures**, see the References in E_6);

Wallay, P, '*Statistical Reasoning with Imprecise Probabilities*' (C and H, 1991).

CHAPTER 9
An Approach View of Mathematics

The notion of '**approach** combines all forms of '**standard approximation**' with **abstraction** and **(inverse-)analogy**. The various processes covered as 'species of approach' are discussed (mainly) in E_5/E_6/E_7. Problems of interpretation / consistence, and calculational procedures, dominate these discussions, since the primary aim is **to maximize calculability** in **all** domains of Mathematics and its applications. Here, it is argued that: every ME is representable, **either, as** (part of) **some approach-scheme(s) / process(es) / procedures, or else, in terms of the results(s) of some approach routine(s). In short**: every ME is approachable by other ME, in some way(s) – as discussed later. The artificiality of this uniform mode of representation, in certain cases is vastly outweighed by its unifying power, reflecting the 'operational viewpoint of Σ '.

In particular, **novel mathematical questions are realised** over the **possibility / attainability / speed / efficiency / optimality / scope / stability of the approach processes** for diverse classes of ME, where such questions have seldom (if ever) been investigated[350]. The motivation for this view of ME lies in '**the generic incompleteness of calculations**' – in the sense that, **typically**, calculations may be performed only **partially** – and so constitute (cases of) **approximation procedures**, of various types. In this context, **exact / complete calculations** correspond to '**exceptional elements**'. Here, as always in Σ , 'calculations' encompass computations / manipulations / transformations / deductions / constructions /.... Among the more direct **sources** of approximation (and, more generally, of **approach**) are: **definitions; limitation conditions;** realizations of **quasi-varieties** (through (non)constructive schemes, in each case). Further – and in some respects most importantly – there is the specification of concrete / abstract **boundaries**, within various mathematical domains, where each form of boundary characterizes some aspect(s) of approach. In E_2, the fundamental SAM algorithms are discussed, and it is argued that a very small 'core' of algebraic / analytical /... techniques 'essentially spans M_t' – in the sense that virtually all of the (ostensibly more complex) ME may be obtained to arbitrary levels of precision / completeness through 'reductions' / limit-representations, in terms of objects defined over the core. Although no conclusive **proof** of this claim can be given, it can be substantially justified on the basis of detailed **synopses** (regarded as 'skeleton pre-taxonomies').

The corresponding task in the present context – to justify the 'approach-proced- ure / approach-output representation' of arbitrary ME – is of a similar nature to that considered in E_2: no absolute resolution is possible, but very persuasive arguments may be based on the treatment of diverse (and apparently 'typical') examples. A heterogenous collection of (nearly 3,000) items pertinent to the developments in this Essay (E_9), as well as in E_{10}/E_{11}, has been compiled over many years (and will be maintained / augmented[351] as a basic Σ -resource). Since the overriding objective is to establish a universal framework / methodology for the productive organization / use of operational mathematical information, the relatively small size of this 'initial collection' is not a significant obstacle to the cumulative construction of Σ , to a point where **all** 'new information' is formulated in terms of 'approach criteria', and the underlying collection of examples contains (at least) **millions** of items! Indeed, it was precisely to take maximal advantage of the (increasingly powerful) DBMS / IT / AI / SAM /... facilities, to organize, unify / deploy Mathematical Knowledge, that the entire 'Σ scheme' was envisaged. Without the unifying power of such a scheme, Mathematics will become progressively fragmented and 'globally disconnected' – with a catastrophic (and probably irrevocable) loss of effectiveness and scope. Of course, no matter how it is organised (within reason!) Mathemat-

[350] Preliminary comments on these issues are made later in this Essay.

[351] Indeed, this collection may be almost doubled (even now) through the citation of results from papers / books in virtually all areas of Mathematics and its applications. These extra references are given with the relevant discussions (rather than being added to the basic list).

ics will remain immensely powerful; but immeasurably less so that it **can** be if approach-based criteria are adopted is fundamental. In short: these Essays furnish a preliminary model for the interpretation / formulation / development of **Mathematics as a computer-based universal calculational scheme**.

The challenge is to establish an interactive framework within which arbitrarily sophisticated mathematical information may be accommodated systematically, in maximally operational forms. This is a task of such enormity and complexity that all of the previous Essays are concerned, primarily, with fundamental questions of design / interpretation / structure /... rather than the multifarious aspects of 'mathematical activity' for which this research environment is intended. In the following Essays (E_{10}/E_{11}), therefore, the emphasis is on **examples** – of the most diverse kinds – selected as paradigms[352]. The eventual aim is to develop 'a quasi-basis for M_t'; that is, a collection of elements from which essentially arbitrary ME in M_t may be produced / approached 'quasi-algorithmically'. This requirement is, plainly, related to those considered in E_2, since the aim there is to demonstrate that specifiably enhanced SAM packages are adequate to cover all of the domains of 'modern mathematics' – even though the key SAM algorithms may appear to be of very limited scope.

In E_2, the structures of the principal subdivisions of Mathematics (Algebra, Real / Complex Analysis, Topology, Geometry,...) are examined through extensive citations from existing synopses / contents-lists. It is found that (as claimed) systematic reductions to 'the simplest substructures' (which are potentially within the scope of specifiably enhanced SAM routines) are possible in principle. Moreover, approaches to such reductions may be constructed – and implemented over P_Σ. There will, however, be a substantial time-lag between the initial discovery / proof of results and the corresponding ('final') Σ -versions. Indeed, the processes of proposing conjectures / hypotheses, seeking initial proofs (of any sort), and producing Σ -versions of these proofs (at various levels of operationality) are largely distinct facets of research.

The main aim in E_{10}/E_{11} is to present a varied collection of results, ranging widely over the standard mathematical sub domains, and formulated in terms of approach concepts / criteria. The intention is for the results selected to exhibit a certain degree of 'typicality'; but there is no claim to any form of exhaustiveness or completeness – or even, of 'denseness'. Moreover, the discussion of (inverse) analogues is almost totally deferred to E_{10} and (mainly) E_{11}, though the important notion of 'location / neighbourhood in conceptual space' (considered formally in some of the earlier Essays) is developed here in some examples.

At this point, an obvious objection to the whole scheme adopted in E_9/E_{10}/E_{11} must be confronted – namely, that the initial selections, S_0, of terms / procedures / paradigms / examples / illustrations /... is too subjective / limited to 'produce' (asymptotically) M. In this simple form, the objection is substantially valid, but the intended relation of Σ_t to M is somewhat more subtle than this. If, schematically, Σ_s denotes 'the state of Σ generated by S', then, since $\Sigma_{\ S}$ may itself be regarded as a set,[353] one may consider the **sequence** $\{S_k\}$, where $S_k := \Sigma_{S_{k-1}}$ and S_0 corresponds to the 'initial data'. Although 'the generation process' (based on systematic use of approach criteria in the formulation / proof of 'new results', and the introduction of 'new concepts') is far from **uniquely** defined, it **is well**-defined – at the level of 'mathematical practice'. In these terms, the **central claim** is that: **forms of S_0 may be specified such that: for any w.d. ME, e, there is a minimal S_k explicitly containing** e[354].

Here, 'e is **explicitly contained** in S_k' means that e is wd over S_k, and that the **interpretation(s)** of e, its links with other ME, etc., are covered within S_k. Another way of putting this condition is that the principal **theories** involving e are treated in S_k – and, of course, in all S_l with $l > k$. **In short**, as k increases, S_k explicitly contains increasingly large subsets of M, and any prescribed e is ultimately explicitly contained in **some** S_k. In this sense, 'S_k **becomes closer to** M', though no form of convergence is involved, yet, since M is regarded (platonically) as 'the edifice of Mathematics, awaiting full discovery', no obvious measure-based criterion seems feasible; nevertheless it is plain that S_l is closer to M than S_k is, if $l > k$, simply because

[352] The present Essay (E_9) concentrates on methodological issues.
[353] This is 'intuitively obvious'; full proof is straightforward but tedious!
[354] Equivalently, 'there is a minimal k, with e explicitly contained in S_k'.

$S_k \subset S_l \subset M$!

Recall that, if $\mathcal{A}$ denotes the collection of all admissible approach procedures, then 'formation of the $\mathcal{A}$-closure of S_{k-1}' constitutes an essential part of the generation of S_k from S_{k-1}. In this process, of course, only elements of $\mathcal{A} \cap S_{k-1}$ are involved; but ultimately, all elements of $\mathcal{A}$ are covered. If $\mathcal{A} \subset S_0$, then every procedure in $\mathcal{A}$ is used at all stages in the sequence $\{S_k\}$. Although this scheme for the 'production of $\{S_k\}$ from S_0' is precise in principle, it is only partially constructive (in any sense), and so must be viewed pragmatically – as a framework for 'Σ -transactions', each of which may be rigorously specified over some appropriate domain(s).

It has already been observed that calculations are 'typically incomplete' – and so correspond to results of certain approach processes; whence, **every ME that is** (somehow) **calculable has an approach-based representation**. If it can be demonstrated that all other ME are w.d. in terms of appropriate (possible, formal) calculational procedures, then the association of **all** ME with approach schemes will be established. The overriding aim, in Σ , **to maximize calculability**, in **all** contexts, dictates that a global 'approach-based representation' must be adopted – even though it may be unnatural / inelegant, in some situations. Variants / alternatives may be introduced wherever this enhances calculability. Evidently, all ME admit many (possibly, inequivalent) representations. For Σ -calculations, the only essential requirement is that **mutual consistency** be maintained. Mostly, this is covered by checking routines in the SAM packages underlying P_Σ; extra criteria may be imposed as necessary.

Perhaps the most fruitful 'approach-view of M_t' is obtained by associating with each set of conditions, $LC(e)$, the 'neighbourhoods' comprising (subsets of) classes of (non)implicatory deformations. This appears to cover all admissible 'standard approximations', specializations, generalizations, abstractions, variants, and (with due attention to detail!) (inverse) analogues. In other words, all forms of ME considered in these Essays are 'derivable' – formally, at least – from this 'LC representation'. Such a scheme is discussed in $E_5/E_6/E_7$ as a basis for the treatment of 'generalized approximation of mathematical information'. Here, however, the emphasis is on the effective **description / interpretation** of properties, qualitative or quantitative, of 'arbitrary ME'. So far, the formation of **stochastic variants of ME** has not been treated directly. Although these variants are important – especially, in applications – they are covered through analysis over**measure spaces** ($_S\mu$, in the list of 'basic spaces' for Σ). In fact, all 'stochastic versions' of any generic ME may be obtained by intro ducting suitable probability distributions in to the relevant LC. This process is widely illustrated in E_{10}/E_{11}. **NOTE that** the use of probability in statistical-metric spaces also yields certain types of stochastic ME – and exemplifies the 'LC-based procedure' just mentioned.

It is now natural to consider the '**span**' of e under specified sets of deformations of $LC(e)$. In particular, one may introduce the 'complete ME' generated by $LC(e)$ under the class of **all** admissible (non)implicatory deformations – with associated forms of 'conceptual continuations'. (This is paralleled by the **scope** of a collection of OT[355] (or, of elements of the CB[356]), as discussed in E_3/E_4). Plainly, one should not expect the CME to furnish a full imitation of the 'CAF' in Analytic Function Theory. Nevertheless, the formalism seems apt for the investigation of qualitative changes in ME-produced by modifications of underlying axioms, etc. – since **the essential characteristics of any ME are largely determined by the nature of its 'neighbourhoods' in the 'conceptual space' whose 'points' are equivalence classes of LC**.

Thus: M_t is depicted as an (un)countable set of disjoint equivalence classes of the $LC(e)$, each comprising all possible 'representations' of some ME, e, in locational form. These representations may be interpreted as 'quasi-states of e' – in the sense the **only ME in mutually compatible states may be admissibly combined within** Σ .

In these scheme it is desirable that elements of $V(e)$ – the set of **variants** of e – are (relatively) 'close to e', since they emanate from e, through modifications of the $LC(e)$. This requirement involved semantic, syntactic and metrical criteria – none of which is easily specified, let along satisfied! The intuitive meaning seems clear enough: the interpretation of e is to be

[355] Operational Theorem(s)
[356] Calculational Base(s)

changed 'incrementally' – but not, in general, implicatorily. Oneway of achieving this is for individual conditions in the (generic) $LC(e)$ to be treated separately (as far as possible), so that the principal effects of each modification may be identified / classified. Although it is simplest to start with the **generic** ME, e, and then to form its variants, $v(e) \in V(e)$, consistency of this process implies that $V(e(e))$ coincides with $V(e)$. For pedagogical reasons, however, it is always preferable to introduce 'new concepts' in the most basic forms compatible with general development of the subject(s) involved. On this basis, the 'origin' of any $v(e)$ should be 'optimally recognisable', as e. The parallel with musical themes / variations is apposite here: all variations must be (at the very least) suggestive of the basic theme, being obtained from it **via** changes of rhythm / key / structure (e.g., by 'inversion' / fugal construction /. . .); but the analogy is very limited, since there is no question of exhaustive treatment, and 'retrieval of the theme' from arbitrary variations in highly problematical! The crucial property of **mathematical** variants (compared with other types) is that they may be **precisely specified** – for instance, through incremental modifications of LC – and hence used to retrieve (partially) the ME from which they were obtained. The main obstacles to **complete** retrievability lie in **stochastic** modifications[357], where information may be lost, as well as in various types of 'averaging procedures[358].

In spite of such elements of indeterminacy, this picture of M_t as the collection of all ME – each represented by an equivalence class of LC, '**surrounded by variants**', and subject to multifarious 'interactions' with other ME[359] – seems essentially sound, in a qualitative / heuristic way. Rigorous forms of standard approximation, along with other, exactly realizable , forms of approach, are readily accommodated within this framework. The 'interactions' are to be interpreted somewhat figuratively, including: the **effects of OT on other OT / the effects of T on wfe / the formation of theories** from aggregations of ME. Many reflections from Chemical Physics arise here – for instance: n**-body interactions** $(n \geqslant 1)$ / **valency / cohesion / bonding / isomerism** – the cite only the most obvious examples.

The fundamental role of **approach criteria** in this overall representation of M_t stems from **the apprehension of ME in terms of near-variants / standard approximants, over a unified conceptual framework**. Detailed structural / calculational aspects of M_t as a quasi-space may be developed (albeit, very heterogeneously), but of far greater importance is the resulting **synthesis** of 'general mathematical activity'. Moreover, the **simulation** of this activity, through interactive computation, constitutes the primary purpose of Σ as a 'system', and the LC-based environment is ideally suited to this task.

In this context, an **approach-vicinity of** e comprises 'all ME sufficiently close to e', with respect to (some of) the Σ -approach criteria – including all types of 'standard approximation' and all forms based on abstraction / analogy / inverse-analogy. For the standard procedures, established tests of precision / optimality / speed /. . . may be adopted directly. The remaining (non-quantitative) modes of approach admit only partially quantitative estimates – from the quasi-metrical to the essentially qualitative. This wise range of potentially attainable 'locational accuracy' complicates the assessment of approach, but also diversifies the possibilities for such assessments. For the primarily non-quantitative cases, concerns with precision / optimality / speed are largely replaced by discussions of **the basic characteristics of qualitative approach**. For 'mixed cases', these two views may be combined.

Some quasi-metrical elements may be introduced though the measures specified in E_5, where the separation of individual LC, c_α, from possible variants / deformations, c'_α, is discussed. In this way, a broad class of notions of 'qualitative approach' may be (partially) quantified. In cases where the separation of certain conditions – denoted generally as $\bar{c}_\beta, \bar{c}'_\beta$ – cannot be quasi-metrically estimated, a formal separation (say, $s(\bar{c}_\beta, \bar{c}'_\beta)$) may be assigned – to be estimated, somehow, whenever this may be possible. Although this is far from a complete solution of 'the separation problem', it does allow maximal progress to be made, together with identification of all pairs of 'poorly-differentiated conditions'. It seems likely that eventual estimates of

[357] E.g., through stochastic LC

[358] E.g., when various 'mean values' are introduced

[359] Say, $I_q[e]_k$ for type-q interactions among $e_1, \ldots, e_k$

the $s(\bar{c}_\beta, \bar{c}'_\beta)$ will involve both non-numerical distance functions[360] and extensions of the Basic Frameworks[361], Σ $(X \equiv \Gamma/T/U/P/C/N)$.

It follows that, in the 'approach-representation of M_t', every ME is set in the neighbourhoods of its near-variants / deformations; and that, to some degree, the **stability** (under deformation) of its essential characteristics may be investigated.

The most obvious source of **in**stability is the change of certain individual LC to 'states' from which the original LC 'cannot admissibly be regained'. As usual, new questions are raised (to be studied in examples), but the aim is to investigate 'the structure of (in)stability', and so to render the basic concepts more accessible to nontrivial definition. Accordingly, the nature of: 'ME-dependence on LC'; of 'typical changes'; and, of irretrievable LC-properties must be considered in very general settings.

For dependence on individual LC, the main concern is with **continuous** versus **discontinuous** behaviour (of e). An obvious example is furnished by **the number of real zeros of a complex polynomial** under variations of any coefficient. A **typical change** here corresponds to random selection from some neighbourhood about the 'initial value'; or else, to variation along an arc within the neighbourhood. The property of 'possessing k **real** zeros' is **not retrievable algorithmically** – unlike that of 'possessing k zeros, real or complex', which is **conserved**. A comparable, but not equivalent, example involves **properties of the spectra of matrices**, when the elements are modified in prescribed ways. Further instances may be found for **Sturm-Liouville problems**, and, for **Fredholm integral equations**[362]. Examples of very different types may be based on '**perturbations**' (finite) **graphs**[363]; and so on.

Another class of instabilities corresponds to **random** perturbations of certain LC – to 'produce' objects that are 'very hard', or even 'impossible' to specify / construct, yet '**dense**' in the generic class of objects concerned[364]. The principal innovation embodied in Σ is that the notions of stability are applied to **arbitrary** ME, through formulations over a diverse range of approach frameworks – for which the attainable levels of precision also vary widely. Although the formulations of this type may appear to be little more than schematic, their underlying structure(s) still reflect the characteristics of conceptual vicinities – and so may be refined, as relevant results become available. Indeed, this process of cumulative refinement is itself fundamental for Σ .

In the discussion of **ME-dependence on sets of** $m(\geqslant 1)$ **LC**; of **typical perturbations of LC**; and, of **irretrievable conditions**, it is useful to recall the representation for wfe, OT and ME, introduced (respectively) in E_3, E_4 and E_5. To distinguish these types of objects here, the wfe may be denoted by φ ('formula(e)'); the OT, by T; and the ME, by e. Then the basic forms become:

(a) $\varphi(\theta)_h \equiv' \Gamma^\varphi(U^\varphi)$,

(b) $T(\theta)_k \equiv' \Gamma^T(u_\sigma)_k$,

(c) $e :: \bigwedge_{\alpha \in S_e} c_\alpha(e; (x_{\sigma^e_\alpha})_{n_\alpha})$.

Of course, **(c)** formally subsumes **(a)** and **(b)**[365]. Here, in **(a)**, the wfe, φ, is expressed alternatively in terms of 'local' entities, $(\theta)_h$, or of entities selected from the underlain list, U_t. In **(b)**, the OT, T, is represented both locally, and, over U_t. The ME, e, in **(c)** is determined / located by the conjunction of conditions $\bigwedge_\alpha c_\alpha(e; X^e_\alpha)$, with $X^e_\alpha \equiv (x_{\sigma^e_\alpha})_{n_\alpha} \subset U_t$; a 'local form' for e – say, $e(\theta)_l$ – may also be given, the θ_i being distributed over the c_α in various combinations,

[360] See the Annotated Bibliography... by HC Reichel, et al. (op. cit.)

[361] See, especially, E_6

[362] Related ideas are treated in connection with '**ill-posed problems**'. See, e.g., Tikhonov, AN / Arsenin, VY, '*Solution of Ill-Posed Problems*' (Wiley, 1977); Bukushinsky, A / Concharsky, A, '*Ill-Posed Problems ...*' (Kluwer, 1994)*

[363] See, e.g., Bull LMS, 22 (1990) 209–216 (P. Rowlinson), and references there.

[364] See, e.g., Kahane, J-P, '*Some Random Series of Functions*' (Heath, 1968*; 2nd ed. CUP, 1980)

[365] Nevertheless, each representation has its 'main context(s)', where it is particularly appropriate.

as appropriate to characterize e. The symbol $\equiv'$ denotes 'qualified identity' – in the sense that $\Gamma^{\varphi}, \Gamma^{T}$, incorporate some syntactic information and a specification, which are only **implicit** in $\varphi(\theta)_h, T(\theta)_k$. Any of the representations **(a), (b), (c)**, may be modified in certain ways, without essentially changing φ, T or e; for more restricted classes of modifications, **absolute** invariance of γ, T, e may be established. The relevant forms of stability devolve on the preservation of seminal properties under changes producing **in**essential variations of ME. Nevertheless, a wide spectrum of behaviour may be covered by this scheme, since the 'inessential facets of e' (for its characterization) may still be of great interest from other points of view.

In general, the $LC(e)$ are variously **coupled**, so their individual effects on the properties of e are hard to monitor. There are two extreme cases:

(i) mutually isolated $LC(e)$, and

(ii) inseparable $LC(e)$ – with intermediate forms:

- **(i)′ almost isolated** $LC(e)$, and
- **(ii)′ partially separable** $LC(e)$.

There are several possible interpretations of **(i)′** / **(ii)′** – mainly corresponding to the relative numbers of isolated / weakly-coupled / . . . / strongly-coupled conditions within the overall $LC(e)$; but, also, to the comparative **dominance** of certain conditions in determining the basic properties of e. The 'conjunction form' of the $LC(e)$ is by no means unique, but this does not affect the fundamental classification of (subsets of) conditions into **types** (isolated / weakly-couple / strongly-coupled / . . . / dominant). Thus, each set of $LC(e)$ has the detailed representation:

$$LC(e) \equiv\subset \alpha \in S \bigwedge c_{\alpha}(e; X_{\alpha}^{e}) =: \bigwedge_{1 \leqslant j \leqslant J^e} \bigwedge_{k \in S_j^e} c_{jk}(e; X_{jk}^{e}),$$

where $S^e = \bigcup_{1 \leqslant j \leqslant J^e} S_j^e$, and the subsets, S_j^e are **not**, in general, mutually disjoint – each being associated with one of the designated types of condition-coupling / condition-dominance.

Accordingly, the subsets of $LC(e)$ may be classified (broadly) as: **isolated, weakly-coupled, strongly-coupled, dominant, heterogeneous. unexceptional** – so that $J_e = 6$, and the associated subsets of $LC(e)$ are labelled (respectively) as $C_k^e, 1 \leqslant k \leqslant 6$. More refined (quasi-)partitions of S^e may be given covering extra interrelations within subsets of $LC(e)$, but this raises no new issues of principle. Therefore it suffices to discuss the typical behaviour associated with each of the C_j^e in the decomposition[366]

$$LC(e) = \bigwedge_{j \in \bar{J}^e} \bigwedge_{k \in S_j^e} c_{jk}(e; X_{jk}^{e}) = \bigwedge_{j \in \bar{J}^e} \hat{C}_j^e$$

where $\hat{C}_j^e \bigwedge \{c_{jk} : k \in S_j^e\} \equiv \bigwedge \{c : c \in C_j^e\}$. When there is no chance of confusion, $LC(e)$ is denoted by $\mathcal{L}(e)$ – especially in complicated manipulations involving the LC of several distinct objects, $e_1, \ldots, e_N$. Since the S_j^e need not be mutually disjoint, the C_j^e may have some common element(s) – so that (to take the simplest possibility) c_1 may be isolated from c_2, but strongly coupled to c_3. The complete enumeration of possibilities for specified $LC(e)$ will rarely be realized 'in practice'; but the Σ -formalism must still allow for this. Indeed, in 'typical LC representations', the decompositions are comparatively simple, with several of the C_j empty, and most of them disjoint from each other. Moreover, some of the interconnections are found to be 'effectively negligible for the characterization of e' – and so may be ignored. The chosen forms of the $LC(e)$ within Σ must reflect such simplifying tendencies, which render the overall system 'cumulatively manageable'.

The most basic descriptions of the sets C_j are as follows:

[366] In E_5, the notation $e :: \bigwedge_{\alpha} c_{\alpha}(\ldots)$ was introduced mainly to cover 'nonstandard approximation'. Here, a more detailed study of associated types of **approach** is attempted.

C_1: The elements of C_1 may be varied **independently** of each other – but not, of the other C_j.

C_2: Variation of some element(s) of C_2 produces only '**small changes**' in the rest – but necessarily in the C_j.

C_3: Variation of some of the elements of C_3 produces '**fundamental changes**' in the rest – but not necessarily in the other C_j.

C_4: Elements of C_4 are '**quasi-stable**' under variations of elements outside C_4 – but not, in general, of other elements in C_4.

C_5: Elements of C_5 are common to at least 2 of $C_1,\ldots,C_4$.

C_6: $C_6 := \mathcal{L}(e)\setminus \bigcup_{1\leqslant j\leqslant 5} C_j$

For C_2, C_3 (respectively) the terms 'small changes', 'fundamental changes', 'quasi-stable' must be adequately defined, before significant progress may be made. The interpretations of C_1, C_5, C_6 are self-evident. The somewhat convoluted phrasing of C_1–C_4 is necessary because the C_j are not, in general, disjoint. In other words, **the defining properties of the C_j are 'internal'**. A full specification of 'the structure of $\mathcal{L}(e)$' must account for 'interactions **among** the C_j' – as well as, **within** each of the C_j. Plainly, the pattern of such interactions over $\mathcal{L}(e)$ varies diversely with e, and no general representations can be found. Indeed, 'ostensibly close ME', say, e', e'', may correspond to 'very different types of behaviour', for $\mathcal{L}(e'), \mathcal{L}(e'')$. In spite of this extreme variability, 'basic trends' may be identified, in terms of which the set of (all) ME may be partitioned into subsets of elements having broadly similar LC decompositions' – in the sense of internal / external interactions. Of course, these subsets will be formed **cumulatively**, as Σ_t evolves; but, at any value, $t*$, there will be a well-defined distribution of ME into such subsets – as one facet of the organization of the IB. The 'broad similarity of pairs of elements' within a subset is associated with qualitative types of interactions among various collections of conditions from each set of LC. These types have already been designated as: isolated / weakly-coupled / strongly - coupled / dominant / heterogeneous / unexceptional.

In this context, C_2 is weakly-coupled when, under certain classes of modifications of some of its elements, the rest **retain their essential characters**. By contrast, certain classes of modifications of some elements of C_3 produce **changes in the essential characters** of the rest. For C_4, 'most changes' in elements outside C_4 leave C_4 'invariant'. This is a slight advance on the previous tentative forms – but still far from adequate! The next task is to define 'essential character of an ME' – and thence, 'retention / change of essential character'. The term 'invariant' here refers to the **forms** of conditions $c_\alpha \in C_4$, rather than their detailed realizations. The specification of 'essential character' for ME depends on the establishment of a detailed **taxonomy** of Σ_t; possible ways of achieving such precise classifications have breed discussed in other Essays[367]. On this basis, **categories** of ME may be formed, producing cognate partitions of (subsets of) Σ_t. Although this may appear subjective, it is still **definite** – provided that the assignment of ME to categories is **permanent** (new elements being added, if necessary, as t increases). In general, membership of some category requires that an ME exhibit several distinct (qualitative) facets – which collectively determine its overall character. Of these facets, the **primary** ones (identified as part of the taxonomy) fix the **essential** character / category.

In these terms the sets $C_1,\ldots,C_4$ may be represented as follows. Let $M(S) \equiv \{m(s) : m \in M\}$ denote the class of admissible modifications of the set S, and put:

$$LC(e) := \bigwedge_{j\neq i} \hat{C}^e_j \bigwedge m(C^e_i), \text{ for } 1 \leqslant i \leqslant 3,$$

where $\bigwedge S := \bigwedge\{s : s \in S\}$. Then:

$$^m_1LC(e) \approx e$$

[367] See, especially, E_1, E_8

$$\text{cat}({}^m_2LC(e)) = \text{cat}(e)$$

$$\text{cat}({}^m_3LC(e)) \neq \text{cat}(e).$$

Further, if ${}^m_4LC(e) := \bigwedge_{1 \leqslant i \leqslant 3} m(C_i^e) \wedge C_4^e$, then

$$\text{cat}({}^m_4LC(e)) = \text{cat}(e).$$

Here, cat(x) denotes the taxonomic category of x. The sets C_k^e, of conditions c_α from $LC(e)$, are **characterized** by the above criteria – at least, qualitatively. In other words: change of essential character is identified in Σ with change of taxonomic category deformations of subsets of $LC(E)$ leaving cat(e) unchanged are regarded as fundamentally different from those producing a change in cat(e). In particular, category-preserving deformations are mostly reversible; whereas, changes in cat(e) text to be irreversible – either, because information is lost, or else, on account of other 'obstacles'. For various subclasses of ME, all of these criteria may be formulated **quantitatively**. Examples include: (non)singular perturbation expansions[368]; continuation procedures[369]; bifurcation phenomena[370]; singularity theory[371]; and, even, collections of approach environments! In such cases, rigorous (if not always exhaustive) analyses may be given, other classes of ME may be treated in combinations of rigorous and heuristic ways, according to the nature of the structure(s) in which they are 'embedded'. The formal identification of e with its sets of LC is essential for the systematic development of Σ . The (still increasing) capacity / speed of computers makes this permanent identification feasible, even in the most intricate search software. The task of furnishing appropriate sets of LC for each ME remains a **mathematical** one – though, for certain ME, the existing system may be indispensable in generating and checking suitable conditions. Here, the issue of what is 'valid mathematics' may arise with increasing frequency. The case of the Four-Colour Problem /? Theorem exemplifies many of the difficulties involved[372]. Eventually, when 'Σ -transactions' are established as 'standard modes of mathematical activity', most of these difficulties will disappear – mainly through systematic checking procedures. At any time, typical OT may be either 'initially proven' (unchecked), or 'finally proven' (fully checked); say, i.p., f.p., respectively[373]. The upgrading from i.p. to f.p. may take from μ-seconds to centuries! This raises no problems of principle, provided that the 'correct status' is ascribed to every OT at all times. Plainly, any proof containing at least one i/p/ OT is itself of i.p. status. A proof in the form of an OT-composition of n 'factors', k of which are f.p., may be assigned the 'confirmation index' k/n though there are, of course, several other possible estimates of confirmation[374]. **Notice that** this is quite different from 'proofs subject to (various) hypotheses', of which 'proofs on RH' are among the commonest – RH denoting the Riemann Hypothesis, that all non-trivial zeros of the Riemann Zeta Function lie on the 'critical line', $\{z : \text{Re}z = 1/2\}$[375]. Although neither of these forms of incomplete proof is conclusive, it is possible, within Σ , to use them consistently and fruitfully – 'new results' being suitable labelled to exhibit dependence on such initial / conditional proofs. Any result obtained on this provisional basis is open to possible 'conclusive derivation' through the use of other (f.p.) OT. Once a result is proven (somehow) alternative – often, much shorter – derivations tend to be discovered. In this process both original proof and its successors may

[368]e.g.: Giacaglia, GEO, '*Perturbation Methods in Nonlinear Systems*', (Springer, 1972)*; Smith, DR, '*Singular Perturbation Theory*' (CUP, 1985)*.

[369]e.g.: Allgower, EL / Georg, K, '*Numerical Continuation Methods*' (Springer, 1990)*.

[370]Golubitsky, M / Stewart, I / Schaeffer, DG, '*Singularities and Groups in Bifurcation Theory*', Vols I/II (Springer, 1985/1988)*.

[371]e.g.: Golubitsky, M / Guillemin, V, '*Stable Mappings and Their Singularities*' (Springer, 1973)*; various editors: '*Dynamical Systems*' Vols I–VII (Springer, 1988–1994)*⁻.

[372]See, e.g., Tymoczko, T, '*The Four-Colour Problem and its Philosophical Significance*', J. Philosophy LXXVI (2) 1979, 57–83; Ibid. LXXVII (12) 803–820 (M Detelfsen / M Luker).

[373]The term 'conclusively proven' (c.p.), equivalent to f.p., is also used from now on.

[374]e.g. there may be other forms of partial proof involved; each must be identified, and the constitution of the 'mixture' taken into account, in all manipulations / investigations.

[375]Many papers on analytic number theory contain conditional proofs ('on RH'). See, e.g., Revs. in Number Theory (WJ Le Veque, ed.) Vol 4, Sec. 30 (AMS, 1974) for several examples.

span the range of characteristics: heuristic / initial / conditional / conclusive – in various combinations. In the most interesting cases, both the lengths and the underlying theories of the proofs vary widely. Thus, a theorem may (eventually) have algebraic, analytical geometrical, topological, probabilistic – even quantum-mechanical! – proofs, of varying status / length / sophistication / complexity. Many illustrations of such 'multiple proofs' are given in E_{10}/E_{11}: the role of **analogy** is central in the development of proofs over diverse structures. The elimination of all 'non-f.p. OT' from a compositional OT-proof does not necessarily increase the strength of the consequent result, though the **status** of the result **is** raised.

The formidable problems encountered in formulating general notions of 'approximate proof' – for instance, the effective definition of (semantic) 'separation of predicate functions', with cognate forms of convergence – are discussed in E_5, where it is concluded that such a program is viable only over 'subclasses of sufficiently rich, quasi-metrical structure'. Nevertheless, there are many theories within which such criteria are valuable – in heuristic, and even rigorous, calculations. Indeed, a blend of these regimes characterizes **formalized heuristics** (FH) – the exploration of various hypotheses, guided by 'feedback' from existing results, and buttressed by rigorous estimates / OT; and this constitutes a principal motif of Σ .

In 'the approach view of M_t' (the main topic of E_9) **every** ME is **approachable** by (sequences / families of) other ME. This idea is viable, since 'approach' covers 'standard approximation', 'abstraction', and 'analogy' (direct or inverse). Even ME that are precisely specified may be considered as approaches to (similar, but) more general / abstract ME; or, to other analogous ME. Within most mathematical contexts, however, **constraints, invariants, exact realizations,...** are still well-defined, and may be used unambiguously / consistently. Moreover, the scope of calculation is greatly enlarged through the introduction of approach criteria – so that, ME of arbitrary complexity / sophistication (typically, calculable only in relatively trivial cases) may be represented, and (at least) approached, even in very general settings. The detailed **development** of such approaches, especially, over ostensibly disparate frameworks, makes heavy demands on ingenuity and technique. Frequently, the necessary results are well known in certain fields – but their relevance to broader syntheses has not been appreciated. One of the primary contributions of Σ is to promote these syntheses as a fundamental facet of the evolution of M_t.

From the diverse discussions of E_1–E_9, a picture of Σ emerges – as an interactive environment for the recording / organization / manipulation / development / application of Mathematical Knowledge, with cognate feedback at all stages of this formal cycle. In view of the profundity and intractability of the issues involved, these discussions are inescapably intricate, but, once the key ideas propounded here have been absorbed into 'mathematical practice', it should be possible to produce a comparatively succinct 'Σ -Manual'. The wide-ranging examples of 'approach phenomena' in E_{10}/E_{11} are intended to establish paradigms for the operational-formulation / global-synthesis of Mathematical information. The realization of Σ within a central computer facility – as an evolutionary process – may be based on the model(s) developed in these Essays.

In the remainder of this Essay, an attempt is made to develop a **minimal classification scheme** adequate to organize the topics considered in E_{10}/E_{11}[376]. Each topic has some primary interpretation(s), involving operational formulation or approach criteria – as well as various secondary aspects. The types of operationally, or of approach (introduced, and treated at length, in E_1–E_8) may be used to classify entries in E_{10}/E_{11}. An **alternative classification** may be based on the collection of **synoptical views of** M_t **'through a conceptual prism'** – corresponding to a spectrum of fundamental criteria motivated by Mathematics itself, and all of the (biological / chemical / engineering / physical / economic / social) Sciences. Each of these 'views' constitutes a **map** of M_t – and the acronym **MAP** (for Mathematical Analogies (in) Perspective) seems apt here. The **MAP** scheme is outlined later in this Essay.

The following brief remarks refer to 'the qualitative treatment of approach processes'. As mentioned in the introduction to this Essay, many novel mathematical questions arise naturally

[376] and, all of the related topics introduced as Σ evolves.

in this context – the principal novelty stemming from the application of approach criteria to essentially **arbitrary** classes of ME, rather than to restricted classes for which such criteria are standard[377]. The aim here is to state, in general terms, what is meant by the possibility / attainability / speed / efficiency / optimality / scope / stability of **approach to** $e(\in D_1)$ **through a process** P**, over the domain** D_2. Illustrations of these ideas are given in E_{10}/E_{11}. **Note that** e itself **need not** be in D_2 – e.g., when e is a 'point of adherence' of a set; but also in far more general situations[378].

(1) Possibility of Approach

This covers **existence** criteria, regardless of specification of procedures, constructive or formal. Although this is common for 'operator equations' of all types, in Σ , the existence of approaches to **arbitrary** ME is considered. In some cases, ME of the type sought may be **dense** (in some sense) in spite of there being no indication of any approach procedure[379]. Such ME are viewed as **isolated**[380], and may not be used in the formulation of calculational schemes. The class of all Σ_t-isolated ME is 'dynamic'. There are both inward and outward flows of elements, as Σ_t evolves. This class (say, $\mathcal{C}_I$) is of great importance, since it embodies 'snapshots' of the **constructive deficiency** of Σ_t. In the discussion of conceptual categories and their associated classes of LC, a basic distinction is drawn between 'initial' and 'full' proofs (especially, those in OT-compositional forms). The study of $\mathcal{C}_I$ addresses comparable issues from a different angle. The 'structure of $\mathcal{C}_I$' (in various senses) is worthy of investigation.

(2) Attainability of Approach

Here, it is **the existence of an approach scheme for** e (rather than of e itself) that is in question. Plainly, it could happen that **both of** e and some approach scheme(s) for e were elements of $\mathcal{C}_I$ (as introduced in **(1)**, above); in other words, that the existence of e, and of at least one approach scheme for e, were established **indirectly** – with no indication of the form of scheme involved[381]. Then the task of **developing** such schemes for e would be 'worthwhile' (since it would be – in principle – feasible). In this sense, **attainability** is equivalent to **feasibility**; but not, in general, to **viability**! Once again, the main interest of these distinctions is revealed in applications to 'arbitrary ME', where issues of this kind have seldom even been raised.

(3) Speed of Approach

In standard approximation of all types, speed of approach essentially determines viability of computation (even for modern computer systems)[382]. For the more general species of approach, similar considerations obtain – though precise estimates are then exceptional, since the underlying spaces typically lack quasi-metrical structures. Nevertheless, qualitative analogues of 'speed' may often be introduced, to assess the 'accessibility' of classes of ME for

[377] See, e.g., Stetter, HJ, '*Analysis of Discretizations Methods*' (Springer, 1973); Blum, EK, '*Numerical Analysis and Computation*' (A-W,1972)

[378] Even the '**Cauchy convergence criterion**' allows 'limit ME' to differ in type from 'sequence ME'

[379] A good example is furnished by the (typical) result of REAC Paley / A Zygmund (Proc. Camb. Phil. Soc. 28 (1932) 266–272): Given $\{a_n\} \subset \mathbb{R}, \sum_{\mathbb{N}} a_n^2$ divergent, the series $\sum_{\mathbb{N}} \pm a_n \cos nt$ fails to be a Fourier-Lebesgue series for almost all $\{\pm\}$ sequences; yet no procedure was known (at least, in 1968) for the construction of such a 'bad series'. (See Kahane, op. cit. for discussion of this phenomenon).

[380] More precisely, as '**calculation ally isolated**'.

[381] A 'standard' example involves sufficient conditions for the existence of asymptotic formulae / expansions for integrals with parameters; see, e.g., JP Rosser, pp371–387, in '*On Numerical Approximation*' (JS Langer, ed.), U Wisconsin Press, 1959. **Another example** is furnished by representations of (classes of) functions of n variables in terms of functions of fewer variables. **See** Golomb, M, ibid., pp 275–326.

[382] In large-scale projects (e.g., for high-energy physics / celestial mechanics /. . .) the expected running-time / cost must be considered.

which standard approximation schemes are inappropriate[383]. In this way, the minimum number of iterations – or, of 'terms' – required to ensure the prescribed closeness of approach may be estimated. Intuitively, one concludes that: the faster the approach process, the fewer iterations / term /... are required – and this is readily verified in most 'standard approximation schemes'[384].

(4) Efficiency of Approach

The emphasis here is on the effective use of calculational resources (for standard forms of approximation)[385]: the fastest mode of approach need not always be the most efficient. For 'non-standard' approach schemes, the basic aims are similar, but the criteria are necessarily more qualitative. These ideas are crucial in the design of 'environments for investigation' of arbitrarily complex problems. In this context, 'measures of information' and 'information flows' (to / from neighbourhoods of ME) are highly pertinent. Indeed, one way of classifying (types of) approaches to e (specified partially / fully in some domain) is in terms of the information locally gained / lost in the approach process(es).

(5) Optimality of Approach

The purview of 'Optimal Approximation' is broad, and there are many treatments of various standard forms[386]. As for the other aspects of approach discussed here, the principal task is to adapt / extend the existing 'standard-approximation optimality criteria' to cover the non standard species of approach – with the inevitable reduction of precision entailed, balanced by a gain in applicability. Many problems of optimal approximation arise naturally in **signal processing** – including aspects of **detection, filtering**, and **control** – where many (interrelated) optimality conditions must be satisfied[387]. **NOTE that** speed and efficiency are among common optimality requirements and, that, typically, the achievable forms of optimality combine 'partial optimality' for several 'competing' properties.

(6) Scope of Approach

Here, $e(\in D_1)$ is approached within D_2 ($\neq D_1$, in general) and, at the simplest level, the scope is 'estimated' as $D_1 \cup D_2$. More subtle forms depend on the detailed structures of the domains / approach-processes involved. In any case, it seems clear that enlargement of scope (at best) leavers the 'precision of approach' unchanged; mostly, the precision is reduced. The interrelations among scope (of a procedure), centrality (of a result), and the processes of[388] 'conceptual continuation', are widely illustrated in E_{10}/E_{11}.

(7) Stability of Approach

This property is concerned, mainly, with 'local equivalence of approaches to e' – in the sense that: all admissible approach processes (for e) produce essentially equivalent results, 'within

[383] See, e.g., '*Computers and Intractability*' by MR Garey / DS Johnson (Freeman, 1979) for examples of qualitative analysis of algorithms.

[384] See, e.g., '*Numerical Recipes*', by WH Press et al. (CUP, 1989); '*The Art of Computer Programming Vols 1,2,3*', by DE Knuth (A-W, 1969) 1981/1973).

[385] One variant – for **linear** approximation, in Banach / Hilbert / Measure-Space settings – is treated in detail by A Sard: '*Linear Approximation*' (Math. Surveys, No. 9, AMS, 1963), esp., Ch.9. **See also** Sard, A, pp191–207 in Langer, JS (ed.), op. cit.

[386] For linear approximation subject to **non**linear constraints, an extensive review is given by M Golomb / HF Weinberger, pp117–190, in '*On Numerical Approximation*', RE Langer, ed., U. Wisconsin Press, 1959. Other schemes are considered by Sard, op. cit., Ch. 10, for 'minimal response to error'.

[387] See, e.g., '*Signal Processing*' (IMA Volumes, Nos. 22 (ed. Grünbaum, et al.) / 23 (ed. Auslander, et al.) Springer 1990).

[388] This idea is discussed in: Elvey, JSN, '*Varieties of Approximation*', pp77–94 of 'Artificial Intelligence in Mathematics' (J Johnson, et al., eds., OUP, 1994).

sufficiently small neighbourhoods of e'. Thus 'stability' is, to some extent, a property of e itself: all modes of approach passing arbitrarily near e must (in fact) **converge** to e (in suitably generalized sense), a species of '**attraction**' (as understood for 'dynamical systems'). A related property is that of **uniqueness** of approach: an approach process can have at most one value / result (when it is applied to fixed, specified 'data').

Outline of the MAP Scheme

In the MAP scheme, M_t is regarded as an amalgamation of diverse manifestations of (qualitatively disparate) basic concepts, say, $c_q, \ldots, c_N$. The 'prism' splits M_t into **conceptual views**, $v_j(M_t)$, each embodying all possible realizations of c_j in M_t. It is supposed that every ME, e, in M_t lies in at least one of the $v_j(M_t)$, so that $M_t \subset \bigcup_{j=1}^{N} v_j(M_t)$. Evidently this corresponds to the decomposition of 'white light' into its spectral components. Although the **choice** of the c_j is obviously subjective, their total number is essentially unrestricted; so there is no limitation on the range / scope of views selected. For initial development, however, **some** choice must be made. The concepts listed below have very strong claims for inclusion; others may be added at any time; some may be combined, or even deleted; and so on! The flexibility of the OIB allows such operations to be performed consistently.

The concepts selected are predominantly of mathematical origin, but there are still many 'reflections from scientific theories' – and the proportion of such concepts will increase as Σ evolves. The procedure is the same for all concepts involved: they are **defined abstractly / formally** (perhaps in several species) along with their **negation-forms**. After that, the entire formalism treated in E_1–E_8 may be deployed, as appropriate, to produce versions of the $v_j(M_t)$ regularly. The possible '**transmission**' of properties / behaviour /..., within or among structures, is of particular interest.

NOTE that: the 'negation forms' of all MAP concepts (e.g., **dis**continuity, **non**lin- earity, **in**stability) are also treated here.

BASIC CONCEPTS (C_j)

1. CONTINUITY

2. DIFFERENTIABILITY

3. ANALYTICITY

4. BOUNDEDNESS

5. UNIFORMITY

6. COMMUTATIVITY / ASSOCIATIVITY / DISTRIBUTIVITY

7. EMBEDDABILITY

8. LINEARITY

9. INEQUALITY

10. CONSTRUCTIBILITY

11. DENSENESS

12. DENSITY / DISTRIBUTION / MEASURABILITY

13. TRANSCENDENTALITY

14. INVARIANCE

15. APPROXIMABILITY

16. PERIODICITY

17. INTEGRABILITY

18. STABILITY

CONNECTED DUALITY

20. CONNECTEDNESS

21. ACCESSIBILITY

22. APPLICABILITY

23. DEPENDENCE

24. EXTREMALITY

25. INDETERMINACY

26. REPRODUCIBILITY

27. REPRESENTABILITY

28. REPLACEABILITY

29. COMPLEXITY

30. SINGULARITY

31. UNIVERSALITY

32. ADDITIVITY / MULTIPLICATIVITY

33. CONVERGENCE

34. COMPACTNESS

35. INVERTIBILITY

36. ITERATION / RECURSION

37. METRIZABILITY / NORMABILITY

38. DISTINGUISHABILITY

39. DECOMPOSABILITY

40. PROPAGATION

41. DEFORMABILITY

42. (RE)CONSTRUCTABILITY / RETRIEVABILITY

43. EFFICIENCY

44. SEPARABILITY / INTERACTION

45. ORDERING / ORIENTATION

46. INTERFACIAL PHENOMENA / BOUNDARIES

47. INTERFERENCE

48. ASSOCIATION / ASSOCIATIVITY

49. REDUCIBILITY

50. CONSTRAINT

51. FINITENESS / COUNTABILITY

52. CONVEXITY

53. INFECTION / CONTAGION

54. EXTENDABILITY

55. COMPLETENESS / COMPLETION

56. POSITIVITY

57. CHARACTERIZABILITY

58. COMPATIBILITY / COMPARABILITY

59. ERGODICITY

60. BRANCHING PHENOMENA

61. LOCALIZABILITY

62. RESPONSE

63. CALCULABILITY

The material contained in E_{10}/E_{11} may be classified through all of the approach criteria already developed, together with the categories associated with the list of basic concepts (and its extensions, etc.). The main task here is to provide sufficiently general definitions / characterizations of the c_j, from which the $v_j(M_t)$ may be cumulatively produced. Since the aim is to encompass all viable forms of each c_j, it appears that combinations of qualitative attributes – with as much precision as this allows – must be sought. This, in turn, dictates that primarily **descriptive** ('non-symbolic') versions are required – their **interpretations** in all admissible frameworks covering various regimes of rigour / technicality / complexity / abstraction / sophistication /..., according to the nature of the concept considered. The following definitions / characterizations are therefore somewhat exploratory / heuristic; but they may be refined / modified, 'experimentally', until stable forms are obtained!

1. **The ME $e \equiv e(\xi)$, defined within structure S, over framework F, is CONTINUOUS at $\alpha \in D \subset S$ IFF when ξ approaches α (in D) in any w.d. sense, $e(\xi)$ approaches $e(\alpha)$, in some corresponding sense.**

2. **The ME $e \equiv e(\xi)$, defined within S, over F, is DIFFERENTIABLE at $\alpha \in D \subset S$ IFF e is continuous at α, and e is linearly approachable at α** (in some w.d. sense(s)).

3. **The ME $e \equiv e(\xi)$, defined within S, over F, is ANALYTIC at $\alpha \in D \subset S$ IFF e is differentiable at α, and the derivative of e at α is independent of the way in which ξ approaches α within D.**

4. **Let $e \equiv e(\xi)$ on $D \subset S$, over F, and let $\psi : e(D) \to S'$. Then: e is ψ-BOUNDED on D IFF $\psi[e(D)]$ is S'-bounded,** where S' admits bounds in any of the measure / metric / norm / order / geometrical / topological /. . . senses.

   Here, it is supposed that the standard forms of boundedness are already w.d.! In this way appropriate types of boundedness may be associated with arbitrary ME.

5. **The ME $e \equiv e(\lambda)_m, \lambda_h \in \Lambda_h \subset S_h$ (over F_h) behaves UNIFORMLY in $(\lambda_\sigma)_j \subset (\lambda)_m, j < m$ IFF a uniform topology is definable over $\underset{1 \leqslant i \leqslant j}{X} \Lambda_{\sigma_i}$.**

6. **Let p, q, r be arbitrary elements of a structure S (over framework F) admitting binary laws of combination l_1, l_2. Then: S is:**

   COMMUTATIVE IFF $l_i(p, q) = l_i(q, p)$;

   ASSOCIATIVE IFF $l_i(p, l_i(q, r)) = l_i(l_i(p, q), r)$;

   DISTRIBUTIVE (l_1 over l_2) IFF $l_1(p, l_2(q, r)) = l_2(;_1 (p, q), l_1(p, r))$.

   NOTE that, a general form of quasi-associativity (over $(e)_k$, with distinct laws $(l)_{k-1}$) **may be defined recursively by:**

$$(l(e))_1 := e_1;$$

$$(l(e))_k := l_{k-1}((l(e))_{k-1}, e_k), \text{ for } k \geqslant 2;$$

$$V_q(l(e))_k$$

   independent of q, where $V_q(l(e))_k$ denotes the q-th mode of evaluation of $(l(e))_k$ with the relative positions of the e_i, e_j fixed.

7. **For categories of spaces $C \equiv \{S_\lambda : \lambda \in \Lambda\}$, the structure-preserving mappings $m_{\lambda'\lambda''} : S_{\lambda'} \to S_{\lambda''}$ determine EMBEDDINGS of $S_{\lambda'}$ in $S_{\lambda''}$.** If the $m_{\lambda'\lambda''}$ are merely **continuous**, then they determine **IMMERSIONS of $S_{\lambda'}$ in $S_{\lambda''}$.**

8. **Let $e \equiv e(\xi)$ be defined on $D \subset S$ (a TVS). Then: e is LINEAR over D IFF**

$$e(q_1 * \xi_1 \oplus q_2 * \xi_2) = q_1 * e(\xi_1) \oplus q_2 * e(\xi_2),$$

   where q_1, q_2 are scalars in S, and, as usual, $*, \oplus$ represent the two laws of combination for S. If, instead, $e \equiv e(\xi)_n$, then e is **MULTILINEAR IFF e is linear for each ξ_i separately.**

9. **Every INEQUALITY in an ordered set $(O, <_O)$, for wfe defined over countable subsets of a set, $W \equiv \{w\}$, of 'symbols', has the general form:**

$$\alpha^m(f^m(w_\sigma)_m) <_O \beta^n(g^n(w_\varphi)_n),$$

for $w_{\sigma_i} \in D_{\sigma_i}, w_{\varphi_j} \in D_{\varphi_j}$, **where f^m, g^n, determine wfe (over W) which are mapped into O by α^m, β^m, respectively**[389].

10. **The ME e is CONSTRUCTIBLE within S, over F IFF there is an algorithm, A, implementable within S, such that $e \in A[S]$.**

11. **Let S be any space / structure admitting a closure operation, say, $S \supset X \mapsto \bar{X}$, and let $B \subset S$. Then: A is DENSE in B** (within S) **IFF $A \subset S \wedge \bar{A} = B$. NOTE that A need not be a subset of B.** Although this appears to be a very specialised formulation, the possible scope is wide, since both S and the closure operation may be of any (mutually) admissible forms. Moreover, approach versions, where $A \subset S \wedge \bar{A} \approx B$ may also be considered.

12. **A MEASURE on a space S is a functional on 2^S that is countably additive over disjoint subsets, and finite on bounded subsets. A DENSITY on S is a nonnegative functional on 2^S, of total mass 1. A property $p \equiv p(\xi)$ is DISTRIBUTED over S with density $\rho \equiv \rho(\xi)$ IFF 'the total extent of p over $A \subset S$' equals $\mathcal{J}_A(\rho * p)$,** for suitable forms of '**integral**', $\mathcal{J}_A$, and '**product**', $\rho * p$.

13. **The ME e is quasi-TRANSCENDENTAL over the structure κ IFF: No condition of the form $\underset{\lambda\in\Lambda}{S}\, \alpha_\lambda * \xi^{[\lambda]} - \theta_\kappa$ is satisfied by ξ, unless** $\alpha_\lambda = \theta_\kappa \forall\lambda \in \Lambda$. Here, all $\alpha_\lambda \in \kappa, \xi^{[\lambda]}$ is any 'quasi-power' of ξ (of order λ), and $*$ is any w.d. law of combination for α_λ and $\xi^{[\lambda]}$. The operation $\underset{\lambda\in\Lambda}{S}$ covers all w.d. combinations of 'summation' and 'integration' (of all possible types). When Λ is a finite subset of $\mathbb{N}$ and $\xi^{[\lambda]}$ is the standard power, ξ^λ, the basic form of transcendence is obtained.

14. **The ME e is φ-S-M-INVARIANT for the process P IFF**

$$\varphi(e)\ \text{wd}S \Rightarrow P * \varphi(e)\ \text{wd}M(S).$$

Here, S is some **set**, $\varphi(e)$ is some **facet** of e, and $*$ denotes the **mode of action** of P on $\varphi(e)$.

NOTE that: e is **strongly** invariant when (in addition) $P * \varphi(e) \approx \varphi(e)$ – where '$\approx$' denotes any wd class of **morphisms**.

[389] The (total / partial) order relation, $<_O$, is arbitrary – as are the structures of the wfe and the mappings into $(O, <_O)$. Since 'inequality' is 'generic', and equality, 'exceptional', this MAP is of particular importance.

15. **The ME e is APPROACHABLE, within S, over F, through process P, with tolerance τ and speed σ IFF**

$$\forall s \in S, \exists n \equiv n(s) : k > n \Rightarrow P^{\langle k \rangle} s \in N_\tau(e); \text{ and}$$

$$\sigma \geqslant n^{-1}.$$

Here, $P^{\langle k \rangle}$ denotes 'an application of order k, of P'. This may be 'P to k terms', the composition of k copies of P, etc.; and $N_\tau(e)$ covers all admissible forms of 'approach neighbourhood'.

NOTE that aspects of **measurability** may be viewed in this way.

16. For any set S admitting an **addition operation** (say, $+'$), **the ME** $e \equiv e(s)$**, defined over** S **has PERIOD** λ **IFF**

$$[s \in S \wedge s +' \lambda \in S] \Rightarrow e(s +' \lambda) = e(s).$$

Here, $e(S)$ may be a subset of **any** w.d. structure. If, instead, the condition

$$e(s +' m_1\lambda_1 +' \cdots +' m_k\lambda_k) = e(s)$$

holds for all combinations of integers $m_1, \ldots, m_k$ – with all quotients λ_p/λ_q suitable restricted, and $s +' m_1\lambda_1 +' \cdots +' m_k\lambda_k$ and s, in S, then e is k**-fold-periodic**. The ultimate direct extension is obtained as: $e \equiv e(s)$ is ε-λ**-periodic** over S IFF

$$\forall s \in S, \exists \lambda(\varepsilon) \ni \lambda(\varepsilon) + s \in S \wedge e(s + \lambda(\varepsilon)) \in \mathcal{N}_\varepsilon(e(s)),$$

where, again, $\mathcal{N}_\varepsilon$ covers all possible 'approach-based neighbourhoods'.

In the standard versions, **Banach-space-valued, weakly-almost-periodic functions on topological groups** seem to cover all known cases.

17. **The concept of INTEGRABILITY** involves **the inversion of (classes of) differentiation operations.** This covers systems of (non)linear differential equations within all admissible spaces, etc.., as well as all forms of (in)definite **integration** – with respect to various types of **measures**. One possible definition, of wide applicability is: e **is integrable (within structure** S**, over framework** $\mathcal{F}$**) with integral** $I \equiv \{I\}_L$ **IFF** e **is expressible as a combination of wfe involving derivatives of (some of) the** I_k – where 'derivative' is interpreted in any consistent way.

This allows for very general realizations within the basic Σ -spaces, ${}_S X$(as structures), over general algebraic / topological /. . . frameworks. There is even scope for (nonstandard) approach-based interpretations. For types of **definite** integration, integrability is broadly equivalent to **measurability** – a species of approximability / **approachability**.

18. The broadest **quantitative version of STABILITY, for** $e \equiv e(\eta)$**, within** S**, over** $\mathcal{F}$, runs: $e \equiv e(\eta)$ **is** P_1-P_2**-stable** IFF

$$\forall \varepsilon > 0 \exists \delta(\varepsilon) : [P_1 e](P_2 \eta) \in \mathcal{N}_\varepsilon(e(\eta)), \text{ if}$$

$$\max\{\|P_1\|, \|P_2\|\} < \delta(\varepsilon).$$

This formulation covers all situations where the terms are consistently interpretable. Here, P_1 perturbs the **structure** of e, and P_2, the **argument(s)** of e; $\mathcal{N}_\varepsilon$ is a **neighbourhood** of 'size ε' (in some sense); and, the condition '$\max\{\|P_1\|, \|P_2\|\} < \delta(\varepsilon)$' is, potentially, interpretable over **any ordered set** (according to the definitions of $\|P_1\|, \|P_2\|$, etc.). Further, e **may have several 'components'**, with associated arguments – so that '$e(\eta)$' is an amalgam; and the basic definition must be formulated to reflect this.

NOTE. The possibility of (fully) **approach-based definitions** is covered here, for 'preservation of conceptual type / species'.

19. **The ME e, e', are mutually DUAL for the theories $\mathcal{J}, \mathcal{J}'$ IFF a correspondence**

$$\mathcal{J}'[\ldots e' \ldots] \approx \mathcal{J}[\ldots e \ldots]$$

is valued. More generally, one has:

$$\mathcal{J}'[\ldots (e')_m \ldots] \approx \mathcal{J}[\ldots (e)_m \ldots 0, \text{ for } m \geqslant 1,$$

where '$\approx$' denotes some form of 'similarity' – from **identity / isomorphism**, through all w.d. graduations, to **weak forms of approach**. In this duality relation, of $(e')_m$ to $(e)_m$, the individual dualities between e'_i and e_i, for $1 \leqslant i \leqslant m$, are combined. The transformation of $\mathcal{J}$ into $\mathcal{J}'$ is obtained by interchanging e_i and e'_i, for each i. In the most general situation, continuous ordered **families** of objects – say $(e)_\Lambda, (e')_\Lambda$ – may be interchanged: $e_\lambda \leftrightarrow e'_\lambda, \lambda \in \Lambda$.

20. The notion of **CONNECTEDNESS** is specialized, but fundamental. **An ME, e, is connected when it is not decomposable into m ($\geqslant 2$) disjoint ME** – say, $e_1, \ldots, e_m$. Thus, if e is **dis**connected, then $e \approx \{e\}_m$, where the e_k are 'mutually separated', in various possible ways. The standard form (for TS) require that:

$$e \neq e_1 \cup e_2 \wedge e_1 \cap \bar{e}_2 = \phi = \bar{e}_1 \cap e_2,$$

which may be extended to 'approach formulations' if '**closure**' is w.d.. Possibilities for '**near / almost /. . . -connected ME**' may be based on admissible criteria for **negligibility**.

21. **The ME e_2 is ACCESSIBLE from e_1, within the structure S, over the framework $\mathcal{F}$, when there is at least one process for 'converting e_1 into e_2'** – possible, subject to extra constraints. Of equal, if not greater, importance is the notion of **INaccessibility**, since 'interesting phenomena' often result from this property. **NOTE that**, if e_1, e_2, are points in a 'space' S, then any suitable mapping $\varphi : S \to S$ suffices. More generally, 'transformations' for which $e_1 \mapsto e_2$ are required; or, alternatively, 'processes' changing the 'state' e_1 into e_2. The most interesting phenomena correspond to INaccessibility: e_2 is **inaccessible** from e_1, within S, over $\mathcal{F}$, when there is **no** admissible process taking e_1 to e_2. Notice that n-connectedness ($n \geqslant 2$) 'induces' inaccessibility.

NOTE that the 'process' involved must be realizable within S, over $\mathcal{F}$.

22. The central aim in Σ is to make arbitrary mathematical procedures maximally **applicable** – mainly through the introduction of approach-based versions of ME (universally viewed

as OT, even where definitions are involved). Since all OT have the same **logical** form – namely, $T : p \Rightarrow q$ – it follows that: T **is APPLICABLE to** e **IFF** E **(as wfe) contains 'some realization of** p'. More generally, the '**apparatus** A' (possibly comprised of several OT in various combinations) **is applicable to** e **(in context** C**, within the structure** S**, over the framework** $\mathcal{F}$**) IFF** e **contains some realization(s) of** A (for $C/S/\mathcal{F}$). **NOTES. (i)** This allows for modification of axioms / adjunction of new axioms, etc. **(ii)** The possible **approaches** to applicability may be explored, systematically or heuristically. **(iii)** Here, $\underline{\text{dom}A}$ denoted the domain of the operator A.

23. An essential facet of mathematics lies in the detailed investigation of **logical / functional (IN)DEPENDENCE** – among prescribed (classes of) ME. This covers all notions of **correlation**, as well as **separation / decomposition / decoupling** of arbitrarily defined 'systems' (of direct of 'reflected' mathematical origin). In particular, **sets of mutually independent ME** could be used to determine **quasi-coordinates** (in various contexts). An '(in)dependence MAP of M_t' includes all of these notions – along with the whole range of I(mplicit) F(unction) T(heorems).

24. **The concept of EXTREMALITY covers both local and global aspects of 'max / min / sup / inf'**, as well as **the study of boundary points of** (arbitrary classes of) **sets**. This involves **calculus (of variations (in-the-large))**, and schemes for '**nondifferentiable ME**'. **NOTE that** ME may have diverse 'extremal **attributes**' without being 'extremal-as-a-whole'. **Hence** the notion may be associated with virtually **all** ME. The study of **optimality**, in any form, may be included here, along with problems involving **boundaries** (of all types).

25. **The notion of INDETERMINACY has many facets: intrinsic non-uniquen- ess / intrinsic unspecifiability / incalculability / undecidability / essential indefiniteness /. . . .** The **degree** of indeterminacy of e (within S, over F), say, $d_i(e; S, F)$, and the possibility of **approach to** e **by** (sequences / families of) **determinate ME**, are of particular interest. In certain cases, e may be effectively isolated (i.e. **inaccessible**) from determinate ME – hence, **un**approach- able. One **aim** is, therefore, to characterize (un)approachable ME (in the above sense); another, to develop procedures for estimating $d_i(e; S, F)$ within suitable environments.

26. Although **exact** reproduction is of limited importance, **approaches** to reproduction of **form** – say, **quasi-REPRODUCIBILITY** – are fundamental for Σ . **The attribute** α**, of** e**, is** P**-reproducible (within S, over F) IFF** $P\alpha(e) \approx \alpha(e)$**.** For **exact reproduction of** e, one has: $Pe = e$ (i.e., e in invariant / fixed /. . . under P), so that $\alpha \equiv 1$ (identity) and $\approx$ coincides with $=$ (equality). **More generally, one may have only**: $P\alpha(e) \approx [f_P(\alpha)][g_P(e)]$ – a very weak form of reproduction, which, nevertheless, includes **complex multiplication**. The function(al)s f_P, g_P, depend on the 'process' P (at least). The scope of this representation is very broad. The (exact) reproduction of kernels in HS, and within families of probability distribution functions, furnish obvious illustrations.

27. The concept of **representation** is central in M_t, since most investigations are conducted in terms of 'apposite forms of ME' – for tractability / effectiveness. Thus: e **is REPRESENTABLE, by $\rho(e)$, within the structure S, over the framework F, IFF $\rho(e)$ furnishes a realization of** e, in this context; that is, **there exists a model within which the behaviour of e is realized by $\rho(e)$**. The possible models are characterized by **interpretations** of systems of **axioms / basic equations / conditions.**

28. **The ME e_1 is REPLACEABLE by e_2 in the wfe ε, within S, over F, for context C, when $\varepsilon[\ldots e_2 \ldots]$ is C-indistinguishable from $\varepsilon[\ldots e_1 \ldots]$. More generally**, one requires that $\varepsilon[\ldots e_2 \ldots]$ **approaches** $\varepsilon[\ldots e_1 \ldots]$, **in this 'environment'** – the mode / degree of approach fixing the type of replaceability. This formulation covers **analogues**, as well as manipulation / calculation.

29. **The COMPLEXITY, $c(e; E)$, of the ME e, over environment E, is an estimate of the number of 'elementary operations' required to construct / produce / realize / implement / apply /...** e **in various contexts**. Consequently, there are several **types** of complexity associated with e – depending on the form of e, the classes of admissible 'elementary operations', and the specified limitations on their use. **Typically:** $c(e; E) = O(\gamma_e(p)_k)$, **where γ_e is some function of the parameter(s), $(p)_k$, characterizing e and E.**

30. **The ME e_1 is SINGULAR for e_2 in environment E IFF 'the behaviour of e_2 is uncontrollable near e_1'. Typically**, this means that **at least one property**, say, $\pi(e_2)$, **holds, arbitrarily-nonuniformly, in all sufficiently small deleted neighbourhoods of** e_1 – but **not at** e_1.

    In this formulation, e_1 may range over some subset of any of the admissible Σ -spaces, ${}_SX$ (in suitable realizations) for which both e_1 and e_2 are w.d.. All of the well-known cases (e.g., for analytic functions, operators, general transformations) are covered here. **NOTE:** 'CV singularities' are associated with 'infinities' / multivaluedness, but such behaviour is not generic for other singular ME.

31. **The ME e is UNIVERSAL for the process P (on structure S, over framework F)** IFF $P_s = (p_s \circ e)s$, **for every (compatible) subobject, s, of S.**

    One version / interpretation of this condition may be stated algebraically as: 'P **may be factored through** e **on** S'. Fundamental examples occur in Category Theory and in Algebraic Topology.

32. **The most general definitions of ADDITIVITY involve arbitrary forms of 'addition' – with sub/super-additivity defined in terms of the relevant** (partial) **ordering**. This covers functions on (nOn)abelian groups, Boolean functions, etc.. **Among the obvious types of addition are:** $+, \oplus, \bigcup, \vee$.

    NOTE that: non-numerical distance-functions (of considerable diversity) may be used to exemplify sub/super-additivity; and, that **approaches to (sub/super-)additivity** are readily introduced, in this formulation – with diverse, concomitant applications.

33. **Notions of CONVERGENCE involving metrizable / norm able structures (including summability / stochastic forms) are ultimately reducible to 'standard convergence over subsets of** $\mathbb{R}^1$ (possibly, **via** convergence over subsets of $\mathbb{C}^k$ of $\mathbb{R}^k$, for $k \geqslant 2$). For **non-metrizable / norm able structures, variants of Moore-Smith ('net') convergence may be introduced**. On this basis, the **convergence of arbitrary families of ME** may be covered.

34. The notion of **COMPACTNESS**, in Topology (though not in general) is an extension of **finiteness**. In this (Heine-Borel) form, '**open covers**' occur; hence, **generalizations devolve on diverse realizations of 'open-ness'**. The other, **topological / metrical, formulation** involves **convergent subsequences** – which may be interpreted for any admissible type of convergence, of sub**families. Completely continuous** (= compact) **mappings** may also be treated in this way – and '**approach versions**' of all types of Σ -compactness are easily formulated, on this basis.

35. In the Σ -scheme, **every** ME, e, is (at least, **formally**) also an OT – applicable to arbitrary wfe ε (though only trivially so, unless 'the valencies of e and ε match'). Consequently, on this basis, e **is INVERTIBLE (within** S**, over** F**)** IFF e **has an PT-converse** – which nullifies the effect of e on (any matching) ε. With (sufficiently careful) natural interpretations, this definition covers all of the obvious cases (operators / transformations / processes /...) while leaving scope for more exotic examples, and, for 'approach versions'.

36. **A general ITERATION process, for a family** $\{e_\lambda : \lambda \in \Lambda\}$ **of ME with values in some ordered structure(s), has the form:**

$$e_{\lambda_1} = \varphi_{\lambda_1} \circ \psi_{\lambda_1}(e_{\lambda_0});$$

$$e_{\lambda_k} = \varphi_{\lambda_k} \circ \psi_{\lambda_k}(e_{\lambda_{k-1}}), \text{ for } k = 2, 3, \ldots.$$

Here the φ_{λ_i} are 'instances' of some 'generic operations', φ; but the ψ_{λ_j} are essentially unrestricted. If all of the φ_{λ_i} coincide, and all of the ψ_{λ_j} are 'identity operators' then a (purely) **recursive** process is obtained.

NOTE that both the φ_{λ_i} and the ψ_{λ_j} may be algorithms (of any mutually consistent types), so that a wide range of **calculational** (including **deductive**) procedures is covered here.

37. The notion of **METRIZABILITY** covers **all forms of 'distance functions' into ordered structures**, and – for TVS – all possible **norms. NOTE that:** only a TVS may be normable – whereas, a wider class of (e.g., 'regular-Hausdorff 2nd-countable') TS is metrizable. **Distance within certain stochastic structures** may be handled through variants of '**statistical**' (≡ **probabilistic**) **MS**, as well as by more specialized techniques.

38. The basic notion for Σ is that of p-**DISTINGUISHABILITY within structure** E**, over framework** F, where the elements s_1, s_2, of S, are p-**distinguishable** (for the **property** p) IFF **exactly one of** s_1, s_2 **has** p. Very general forms (for **categories of spaces**) occur in Topology, with '$p \equiv$ **homeomorphism / diffeomorphism**'. For **algebraic** structures,

other types of morphisms may be taken for p. In Analysis, the most obvious example is given by the Stone-Weierstrass theorem, involving 'algebras that separate points'. However, the general definition covers **arbitrary** properties / procedures /... p, in all areas of mathematics.

39. **The ME e, defined within environment** E, may be **DECOMPOSED** – say, as $e == (e)_l$ – IFF e **may be exactly 'reassembled' from** $(e)_l$**, with 'zero net flux of any attribute'**.

    This is contrasted with general **separation**, where there is a net gain / loss in certain attributes (e.g., in potential energy, for collections of particles in Chemical Physics). The e_i are assumed to be 'of the same type as e'; but 'approach versions' allow for controlled type-differences, as well as for various forms of **in**exactness of representation of e by $(e)_l$.

40. A general form of **PROPAGATION of** e**, within** S**, over** F may be defined by the condition

$$[X \cup Y \subset S \wedge e(X)] \Rightarrow e(Y),$$

where all of the possibilities **(i)** $X \subset Y \subset S$; **(ii)** X **disjoint from** Y; **(iii)** $X \cap Y \neq \phi$, are allowed, and $e(X)$ means 'e exists / is w.d. / operates /... over X', etc..

NOTE that, according to context, $e(Y)$ may obtain 'logically', as a **deduction** from $e(X)$; or else, only 'ultimately' (but certainly), as in physicochemical processes. The precise mechanism(s) / mode(s) of propagation are not relevant, here – except that, for Σ , 'the purely **tactile** transmission of e from a point source in S', is treated separately (as 'infection / contagion').

41. **The ME e_1 is DEFORMABLE into e_2, within S, over F**, when: **there exists a continuous, invertible transformation $T : S \to S$ such that $Te_1 = e_2$, where T is independent of the choice of e_1, e_2.**

    NOTES

    (i) e_1, e_2 may be elements of S; or else, 'values of $e \equiv e(s)$', if e is parametrized by $s \in S$.

    (ii) Approach-forms have Te_1 'close to e_2', in any of the Σ -modes of approach.

    (iii) If the **invertibility** condition is omitted, then **plastic**, and **hysteretic**, transformations may be covered.

42. Whereas **decomposition** posits 'flux-free reassembly' of an ME e from 'subobjects' $e_1, \ldots,$ e_l, **RECONSTRUCTION / RETRIEVAL of e from the e_i** is, in general, **I'm**perfect – and **the imperfections are to be identified / estimated** (within S, over F). Among the most important cases are: reconstruction / retrieval from **(i) several partial representations** of e; **(ii)** representations in which there is a **net loss of information**; **(iii) data on models of** e – producing classes of '**inverse problems**' .

    The **extent of reassembly / restoration** may be estimated through approach criteria (corresponding to S, F).

43. The concept of **EFFICIENCY** is relevant mainly to **the implementation of algorithms**. Since every OT embodies 'operational realizations', possible formulations of efficiency are important for Σ . Many versions appear in the CS / Automata-Theory / Complexity / Approximation / Optimization literature.

44. The concept of **SEPARABILITY, of the ME** e**, within** S**, over** F**,** is an extension of **decomposability** – the crucial difference being that **the separation of** e (into $\{e\}_l$, say) **involves**, in general, **net flux of certain attributes of** e **(or, of the** e_i**)**. One paradigm here may be based on physico-chemical phenomena (where there is **energy**-flux). For classes of PDE, **exact** separation ($\equiv$ decomposition) is possible; whereas, other classes of (systems of) operator equations may be separated only by introducing some extra condition(s), precluding full restoration of the original equation(s). (The **K-S equations** in Statistical Mechanics furnish a nontrivial example). The term 'energy' is used here for a typical attribute that need not be conserved in separation processes.

45. **An ME** e **is ORDERABLE within** S**, over** F, when **an order relation is definable on** e**. Orientation** is equivalent to direction-preserving triangulation (for surfaces / manifolds). Hence, orientation may be viewed as a type of ordering (of the points of the boundary, etc.). The standard definition of an 'order-relation' reflexive / antisymmetric / transitive) seems to cover all cases, corresponding to admissible forms of "$<$' (the order relation).

46. **An** $(\alpha)_k$**-INTERFACE for** $e \equiv e(s)$**, within** S**, over** F is a **quasi-boundary (say,** B**) across which the attributes** α_i **change qualitatively / discontinuously.**

NOTES

(i) e itself may be one of the $\alpha_i(e)$.

(ii) Other $\alpha_j(e)$ may be arbitrary functions of e; and so on.

(iii) The quasi-boundary B may be (partially) **specified**; otherwise, it is just 'located relative to S, F'. In many cases, B is a system of **curves** / **surfaces**.

The '**interface MAP of** M_t' therefore covers all aspects of generalized-bound- ary behaviour of ME – in all admissible approach formulations.

47. **The ME** e_1, e_2 **INTERFERE with each other when: either of them – in a simultaneous realization of both – differs from its sol realization(s)** – all, within S, over F.

NOTE that: the question of mutual **compatibility** (of $e_1, \dots, e_l$ within S, over F) may be treated, in part, through interference criteria.

48. **The term ASSOCIATION, for a collection,** $\{e\}_n$**, of ME,** may be regarded as: **an estimate of quasi-resistance to separation (into sub collections)**; and so, as: **a measure of linkage among the** e_i. Various forms of **dependence** are included. Obvious examples are furnished by: **correlation** (in statistical mechanics); **functional dependence** (of all types); **statistical association** (of random variables).

The separation of any associated collection $\{e\}_n$ into sub-collections entails a **net** (typically, **irretrievable**) **flux** in some attribute(s), α. For instance, exact representations of α within S for $\{e\}_n$ become approximate, for any partition of $\{e\}_n$.

49. The distinctive quality of **quasi-REDUCIBILITY, for the ME** e, in the schematic form: $e \approx [e]_l$, where '$\approx$' has many interpretations, is that **all of** e **and the** e_i **are of the same basic type.** Thus **investigation / evaluation** of e is broadly equivalent to the corresponding task(s) for the (typically, '**simpler**') ME $e_1, \ldots, e_l$. A slightly **more explicit view** of the reduction of e – within S, over F – may be obtained by setting:

$$e = {}^e\Gamma(e)_l \oplus {}^eR,$$

where ${}^e\Gamma$ **is some specific combination of (some of) the** e_i**, and** ${}^eR$ **is the difference of** e **and** ${}^e\Gamma(e)_l$ – however this may be represented.

NOTE that obvious forms of $\oplus$ include variants of 'standard addition', 'direct summation', logical disjunction, and set union.

50. **The ME** $e \equiv e(\xi)$ **is CONSTRAINED (within** S**, over** F**) by the conditions** $C \equiv C(\xi; e)$ IFF $\{\xi \in S : C(\xi, e)\} \neq S$. That is: e is formally w.d. within S but admissible interpretations of e require that the condition(s) $C(\xi, e)$ hold throughout $S_1 \subseteq S$. If $C(\xi, e)$ comprises **explicit** (in)equations for e then the constraint is **direct**; otherwise (e.g., if there is at least one **implicit** (in)equation), the constraint is **indirect**. Perhaps the archetypal 'constraint view of M_t' is furnished by '**perturbation of axiomatic structures**' – where the axioms may be viewed as constraints, etc..

51. **The attribute** $\varphi(e)$ **of the ME** e **(defined within** S**, over** F**) is** μ**-FINITE** IFF $\varphi(e)$ **is of finite** μ**-measure** (in any w.d. sense). **If all** admissible attributes of e are μ-finite (i.e., each $\varphi(e)$ is μ_φ-finite) **then** e **is finite** (without further qualification). Here, μ may take its values in any ordered structure – though $\mathbb{R}^+$-valued measures are the most important.

    Typically, e has a mixture of **discrete** and **continuum** attributes, located in arbitrary realizations of the Σ -spaces ${}_SX$ – and corresponding to diverse 'counting procedures' / boundedness-criteria.

52. **The geometrical definitions of convexity for the ME** e **(within** S**, over** F**)** may be formulated over any OTVS, as:

$$e \equiv e(s) \text{ **is CONVEX within** } S \subset V_{\leqslant}$$

    IFF, $\forall p, q \in K^+, \forall \xi, \eta \in S$,

$$e\left(\frac{p}{p+q}\xi + \frac{q}{p+q}\eta\right) \leqslant \frac{p}{p+q}e(\xi) + \frac{q}{p+q}e(\eta),$$

    where $V_{\leqslant}$ **is an OTVS with ground field** K, K^+ **is the positive cone of** K**,** $p, q \in K^+$ **are arbitrary, and '**leq**' is the order-relation for** K **(and so, of** $V_{\leqslant}$**).**

53. The – somewhat distasteful! – notions of **INFECTION / CONTAGION** are, nevertheless, important for Σ . **Infection** is usually **localised** – possibly at a countable set of points – and may **spread**, either, through **propagation**, or else, through **contact**.

Both of these phenomena are exemplified by **germs of analytic functions**, and, **the Principle of DAC**, respectively.

54. **The concept of EXTENSION** is fundamental for Σ . The general form is:

The ME e_1 is extendable to e_2 under condition(s) C (within S, over F) IFF **there is a finitary procedure, say, P, producing e_2 from e_1, while maintaining C.**

55. The notion of **COMPLETENESS** is ubiquitous in Σ (and has many forms). A general formulation is: **The ME e is α-complete (within S, over F) when α is maximally realized for e.**

Here, α is some w.d. attribute of e – for MS, $\alpha \equiv$ 'convergence of Cauchy sequences' (which may be extended to quasi-Cauchy convergence, in any space where this is w.d.). Other specifications of α yield **'sup/inf-completen- ess', edge-completeness (for graphs), axiom completeness (of theories);** and so on.

56. **The notion of POSITIVITY is w.d. for all posets / ordered TVS** (at least). More generally, the attribute $\alpha(e)$ may be positive (in the above sense) even if e itself is not so – in which case e **is α-positive.**

NOTE This may appear narrow / limited; but, in the broadest interpretations, it covers a wide range of topics.

Thus:

$e \equiv e(s)$ is **α-positive (within S, over F)** IFF $\alpha(e) > \theta$ – **for the order relation on $\alpha(e(s))$, with 'zero-element' θ.**

57. **The ME e is CHARACTERIZABLE through properties $\{p\}_l$, within S, over F** IFF

$$e \in \{x : \bigwedge_{1 \leqslant i \leqslant l} p_i\},$$

and all properties of e are derivable from $\{p\}_l$.

The essence of **characterization** lies in **minimality – of l, and of the 'stren- gths' of the p_i;** and also, **in diverse realizations of e** in such forms.

Simple **examples** include the characterization of TS (as 'T_k', for $k = 1, 2, 3, 3\frac{1}{2}, 4, \ldots$); and, of probability distributions (e.g., 'normal', 'Poisson').

58. **Let $V_T(e)$ denote the set of valid statements about the ME e, within a theory T.** Then: **e_1 and e_2 are COMPATIBLE for T** IFF **$V_{T_2}(e_1) = V_{T_1}(e_2)$, where $T_i := T \cup V_T(e_i)$,** for $i = 1, 2$.

Hence: theories remain consistent under the adjunction of statements compatible with the relevant axioms.

NOTE that this formulation is more general than it may initially appear, since arbitrary calculations over any theory proceed through sequences of operations that maintain consistency. For instance, all **interfacial** conditions are of this kind.

59. **Let S be any structure admitting a quasi-measure, μ, and let the ME e be defined within S, over F.** Then: e **is μ-ERGODIC** IFF

$$[D \subset S \Rightarrow \mu(e(D)) = \mu(D) \wedge e(D) = D] \Rightarrow \mu(D) = 0.$$

Here, the 'action of e over $D \subset S$' is arbitrary. 'Standard ergodicity' corresponds to w.d. (rather than quasi-measures μ on S. The 'approach versions' may have $\mu(D) \approx 0$ (or even, $\mu(D) \approx \theta$, if μ takes values in some space(s) more general than $\mathbb{R}$), with μ 'quasi' and $\approx$ interpreted in any admissible way.

60. **For the ME $e \equiv e(s)$, defined within S, over F, the BRANCHING of e into $\{e\}_k$, as s passes through s_0 (say), in some w.d. sense(s),** may also be **(partially) reversed, with $\{e\}_k$ coalescing into $\{e'\}_j, 1 \leqslant j \leqslant k-1$.**

Here, all of e, e_p, e'_q, are assumed to be (broadly) 'of the same **type**'. The possible 'approach versions' of branching devolve on interpretations of: $e \approx \{e\}_k \approx' \{e'\}_j$, and the scope is vast, since all of e, S, F may (in principle) be substantially arbitrary – covering, e.g., chemical reactions, and abstract theories / subtheories! The case '$j = 1, e'_1 \neq e$' is of particular interest.

61. **The attribute α of the ME $e \equiv e(s)$ defined within S, over F, ant taking values in (μ, M), is $'mu$-LOCALIZED to $N \subset S$** IFF

$$\mu[\alpha(e)|_S] = \mu[\alpha(e)|_N].$$

NOTE that N may be any subset of S – of any relative 'size' or connectivity – and that (μ, M) is any w.d. measure space.

As usual, the replacement of '$=$' by '$\approx$' (in any admissible 'approach interpretation') and the essentially unrestricted forms of α, e, S, F, endow this formulation with very wide scope.

62. **For any ME e, denote by $\alpha(e)$, an attribute of e, and by $\tau(e)$, the type of e. Define:**

$$\Gamma_{21} \equiv \Gamma_{21}(e_2, e_1) := \{\alpha(e_2) : \tau(\alpha(e_2)) = \tau(e_1)\}.$$

Then: **the RESPONSE of e_1 to e_2** may be defined by

$$R_S(e_1; e_2) : e_1 \Big|_{\bigcup_{\alpha \in \Gamma_{21}} S(\alpha)} \sim e_1 \Big|_S,$$

where $S(\alpha) \equiv$ 'S subject to α'.

NOTES.

(i) If $\tau(e_2) = \tau(e_1)$ then e_1, e_2 are directly comparable, so the 'perturbed structure becomes $S(e_2)$.

(ii) In the definition of $R_S(e_1, e_2)$, '$=$' may be repaved by any admissible approach, '$\approx$', etc..

63. There are several types of CALCULABILITY to be considered.

(i) **Weak calculability** amounts to **exact** evaluability, **in principle.** This is the constructive equivalent of 'existence' in 'standard proofs'.

(ii) Σ_t**-calculability** is equivalent to **exact evaluability** over Σ_t.

(iii) Other forms of calculability correspond to various **frameworks / constraints.**

NOTE that, although Σ -approaches may be defined for (essentially) arbitrary ME, only weakly calculable ME admit (potentially) **convergent** sequences / families of approaches (in some sense(s)).

APPENDIX 1
Some Quasi-Operational Schemes

In initial discussions of ideas underlying the development of Σ , the primary aim was to exhibit diverse areas of Pure / Applied / Applicable Mathematics 'potentially realizable' over suitable enhanced SAM packages[390]. In spite of the improvements in speed / efficiency / graphics-facilities, the **format, scope**, and, above all, the **milieux**, have barely changed since 1980 – though a few of the **topics** have, gradually, been (partially) implemented – mainly as 'library routines', in the leading systems. **The approach / OT framework**, however, has remained unrecognized since its introduction in 1994[391]. Consequently, it is important to include (now, 'from the Σ -perspective') sketches of these earlier formulations. No attempt is made to update the basic references – beyond remarking that the rigorous development of 'B-constructive Mathematics' (already exemplified by the book, '*Constructive Analysis*', by E Bishop / D Bridges, Springer, 1985) is fundamental for the establishment of 'Σ -style(s) of proof' in the OIB – but some **additional references** are given, to extend the original terrain covered, and to identify relatively up-to-date Bibliographies.

Ultimately, **every generic ME** in Σ_t will be t**-max-operational** – other forms of 'the same ME' being available as contexts require them. The intention in SCCM[392] is to identify nontrivial, 'research-level' SAM calculations that are potentially (near-)fully implementable. The issue of **conventional mathematical I/O** is (in spite of LaTeX, etc.!) anything but a matter of mere Luddite reaction. Indeed, it is universally acknowledged that Mathematics progresses largely through hunches / heuristics, rather than chains of impeccable deductions from axioms – the retrospective versions! So far, computers have been used mainly for **number**-crunching and (since the late 1960s) **symbol**-crunching – the performance of prohibitively large **computations**, for which all relevant **procedures** are (algorithmically) **known**. The paradigms for Σ are, by contrast, **investigations**: attempts to perform **calculations** (covering **estimates / computations / constructions / deductions**). The following outlines (adapted from SCCM) should be viewed in this light – and, as indications of quasi-operational modes of representation of arbitrary mathematical theories, in Σ -calculations[393].

In SCCM, the term **symbolic investigation** (SI) is used to cover the development of operational forms of (arbitrary) ME. This includes the deployment of effective / (B-)constructive representations / procedures – with emphasis, always, on **maximal calculability** (over the various Σ -frameworks discussed in other Essays). The corresponding operational **proofs** are, in a sense, 'dual' to the **results** embodied in the OT. That is: the **technique(s)** of proof of some OT, T_1, may have only tenuous connections with the **sub theories** in which T_1 plays a central role. Attempts to 'harmonize' these sub theories, $\mathcal{J}(T_1)$, with the OP, $\pi(T_1)$, to produce 'logically homogeneous realizations', raise many novel – and recondite – problems, which are worthy of systematic study. Degrees of (mis-)matching between $\mathcal{J}(T_1)$ and $\pi(T_1)$ may be defined, and estimated (within various frameworks). In the preliminary stages of exploration of a conjectured theorem, such considerations may seem aesthetic / philosophical. For Σ , however, as an environment for formalized heuristics, the **structure** of (O)P is of fundamental importance. As always in Σ , the pursuit of these issues generates new material for the (O)IB!

The main intention here is to extract from SCCM the briefest descriptions of some subtheories (of Pure / Applied Mathematics) where quasi-algorithmic processes are readily identified – with indications of the (collections of) (in)equations involved. Plainly, this furnishes only a

[390] See pp127–247 of '*Symbolic Computation and Constructive Mathematics*' (CS-80-49, Dept. of CS, U. of Waterloo, 1980, by JSN Elvey), where a broad selection of topics is considered.

[391] Actually, these ideas have been explored by the present author since the mid-1970s, but this outline (pp77–94: '*Varieties of Approximation*', in '*Artificial Intelligence in Mathematics*', J Johnson, et al., eds., OUP, 1994) is the first published account.

[392] **S**ymbolic **C**omputation and **C**onstructive **M**athematics ($\equiv$ Elvey (1980)).

[393] As stressed in E_0, this approach-based scheme essentially **characterizes** Σ as a research environment. **Notation:** $\underline{\text{cal}}(e)$, for the ME e.

minute sample of the eventual possibilities, but the idea is **to demonstrate the longterm feasibility of this aspect of Σ .** The distinction between computation / manipulation and 'deduction procedures' is not always transparent, so it is maintained only where clarity would otherwise be compromised. In any case, no **formal** distinction is made from an 'operational point of view'. Somewhat more details are given in SCCM, where references to full treatments may also be found. The primary aim here is to **state** some broadly representative problems, with adequate basic references.

Before the examples are sketched it is necessary to explain the sense(s) in which (partial) **solutions** may be obtained in SI. In problems where all of the (algebraic / analytical / combinatorial /...) procedures in w.d. algorithms may be accomplished **exactly**, the resulting solution is also **exact**. Although such schemes may be 'conceptually straight-forward', their **realizations** could still be highly nontrivial. (Consider, for instance, the solutions of **(i)** the **2-D Ising Model** in zero magnetic field[394]; **(ii)** solutions of **Einstein's field equations** in GR[395]; **(iii)** exact computations in the computer-based proof of **4CC**[396]; **(iv)** various steps in the classification / representation of **finite simple groups**[397].)

Next, one has problems where 'most' steps in the solution are exactly implementable, but the rest (e.g., truncation of power-series, determination of integrals, use of infinite matrices) involve limit operations / asymptotics[398]. Lastly, there is a vast spectrum of SI where – to vary ing degrees – essential steps can be tackled only be invoking diverse approximation procedures, for many of which rigorous (symbolic) error estimates are obtainable only through intricate analyses[399]. Most of the following examples are of this type – exactly solvable forms in each case being useful, mainly, to assess general schemes.

1. Implicit Functions [of n (Complex) Variables]

This covers several types of results, with varying SI potential. The basic formulation, in all cases, involves a relation, $F(z,w) = 0$, over prescribed domains – say, $\mathcal{D}', \mathcal{D}''$ – which (subject to suitable restrictions, on $F, \mathcal{D}', \mathcal{D}''$) determines functions $W \equiv W(z), Z \equiv Z(w)$ such that the conditions **(a)** $F(z, W(z)) = 0$, **(b)** $F(Z(w), w) = 0$, hold identically – over $\mathcal{D}', \mathcal{D}''$, respectively.

(i) In the standard case, $\mathcal{D}', \mathcal{D}''$ are intervals in $\mathbb{R}^1$; see, e.g., Apostol, TM, '*Mathematical Analysis*', 2nd ed., A-W, 1974, Th. 13.7.

(ii) A fairly general version, with $\mathcal{D}', \mathcal{D}''$ subsets of (distinct) BS is given in (e.g.) Field, MJ, '*Differential Calculus and its Applications*', Van Nostrand, 1976, Th. 3.3.1.

(iii) Versions with weaker conditions on F (and vastly harder proofs) may be found in Schwartz, JT, '*Nonlinear Functional Analysis*' (G&B, 1969), CH.II; and, in: Hamilton, RS, BAMS 7 (1982) 65–222 - both, for Fréchet-space formulations, with applications to PDE, etc..

The SI content stemming from **(i) – (iii)** depends strongly on the detailed properties of $F, \mathcal{D}', \mathcal{D}''$, but **iterative procedures** are often essential for the (approximate) determination of W, Z, etc.. These results extend (for **(i)**) to allow $z \in \mathbb{R}^m, w \in \mathbb{R}^n$, with $\mathcal{D}' \subset \mathbb{R}^m, \mathcal{D}'' \subset \mathbb{R}^n$, though this is formally included in case **(ii)**. Further, the study of **branching phenomena** for nonlinear operator equations involves several specialized types of implicit function routines (discussed later in this Essay).

(iv) If F is a **polynomial** in z, w, over $\mathbb{C}$, the W, Z are (mutually inverse) **algebraic functions** – with finite numbers of **branches**, each analytic over the associated **Riemann surface** (say, $\mathcal{S}_F$) essentially determined by the nature / distribution of the **critical points** of F.

[394] See, e.g., Huang, K, '*Statistical Mechanics*' (Wiley, 1963), Ch. 17.
[395] See, e.g., Kramer, D, et al., '*Exact Solutions of Einstein's Field Equations*' (CUP, 1980)
[396] See, e.g., Saaty, TL / Kainen, P, '*The Four-Colour Problem...*' (Dover, 1987)
[397] See, e.g., Gorenstein, D, '*Finite Groups*' (Harper & Row, 1968)
[398] e.g., for solution of DE; or, for summability techniques
[399] Such procedures extend calculability – a basic aim in Σ .

(v) If F is a **polynomial** in z (or, w), with 'coefficients **meromorphic** in w (or, z)', then Z and W are **algebroidal functions.**

For **(iv)**, see, e.g., Bliss, GA, '*Algebraic Functions*', Dover, 1966 (orig., AMS, 1933); and, for **(v)**, and, Saks, S / Zygmund, A, '*Analytic Functions*', (PWN, 1965) Ch. VI. Evidently, **(v)** covers both the extension of single-valued to multi-valued analytic functions; and, of algebraic, to transcendental, functions.

For types **(i), (ii), (iii)**, iterative procedures may be developed – but the details depend closely on the forms of $F, \mathcal{D}', \mathcal{D}''$ (as, also, for the systematic study of branching behaviour). Hence, there seem to be general SI schemes here. More incisive results may be obtained for types **(v), (vi)**. Thus, for **(v)**, three basic problems are:

P$_1$: to develop effective methods for the **identification** / (approximate) **determination of critical points of** F – and thence, to give a concise **description of** (the structure of) **the associated Riemann surface.**

P$_2$: to produce effective representations for the branch(es) of Z, W.

P$_3$: to apply the results of P$_1$, P$_2$ in significant investigations within diverse parts of (pure / applied) mathematics.

Since an algebroidal function (as defined here) may be equivalently regarded as an n-valued analytic function (over $\mathcal{D}'$ or $\mathcal{D}''$ – except for finite sets of poles or algebraic branch points), problems P$'_j$ analogous to P$_j$ are relevant – along with basic questions originally posed for entire functions (including: maximum principles, Picard's theorem, asymptotic directions / values). Some of the Nevanlinna theory of meromorphic functions may also be extended. The corresponding problems, say, P$'_4$, P$'_5$,..., are amenable to SI. Some examples will be indicated later – as 'effective techniques in Function Theory'.

Basic References:

P$_1$: Saks / Zygmund (op. cit.); Forsyth, AR, '*Theory of Functions*' (3rd ed., CUP, 1918 / Dover, 1965); Bliss, GA, '*Algebraic Functions*' (AMS, 1933 / Dover, 1966); Jones, GA / Singerman, D, '*Complex Functions...*' (CUP, 1987); Petrovič, M, '*Iterative Methods for ... Polynomial zeros*' (Springer, 1989)

P$_2$: Kung, HT / Traub, JF, JACM 25 (1978) 245–260; Bliss, GA (op. cit.); Farkas, HM, et al., '*Riemann Surfaces*' (2nd ed., Springer, 1992)

P$_3$: Examples include: (asymptotic) expansions of various special functions (see, e.g., Olver, FWJ, '*...Asymptotics and Special Functions*', AP, 1974) also, treatments involving generating functions (say, W) satisfying algebraic identities (e.g., $f \equiv F(W, z, t) \equiv (1 - 2tz + z^2)W^2 - 1 = 0$, for Legendre polynomials).

NOTE that: the **reciprocal, Nth-root**, and **reversion** are all representable in terms of algebraic (or, algebroidal) functions. Moreover, extensions to **Puixeux series** (especially, in Algebraic Geometry), and to allow **other types of 'ground fields'**, may also be established.

2. Effective Procedures in Function Theory

This covers a multitude of subtheories. It suffices, here, to indicate a few disparate examples.

(i) 'Classical approximation problems'.

(ii) Various estimates for 'schlicht functions'.

(iii) Representations of analytic functions.

(iv) Conformal mapping.

(v) Potential-theoretic procedures.

(vi) Miscellaneous 'applications'.

(vii) Generalized Analytic Functions.

Example (i) deals mainly with the (uniform) approximation of analytic / meromorphic functions (over specified domains / boundaries) by sequences of 'simpler' functions – polynomials, rational functions, series of polynomials (via partial sums) – subject to prescribed 'closeness criteria'. Many subtle / intricate results are involved.

B(asic) R(eference(s)): Walsh, JL, '*Interpolation and Approximation by Rational Functions in the Complex Domain*' (AMS, 1956); Smirnov, VI / Lebedev, NA, '*Functions of a Complex Variable: Constructive Theory*' (transl., Iliffe, 1968); Davis, PJ, '*Interpolation and Approximation*' (Blaisdell, 1963).

Example (ii) Many of the estimates were derived primarily in attempts to prove the 'Bieberbach Conjecture' – whose original general proof was functional-analytic in character (though function-theoretic versions were eventually distilled from it). Nevertheless, there are many other problems concerning univalent ($\equiv$ schlicht) functions, requiring precise estimates. Such results are paradigmatic of the simplest types of 'operational information' (OI) in Σ .

BR: LNM #478 (G Schober); LNM #646 (O Tammi); LNM #913 (O Tammi), Springer, 1975; 1978; 1982; Schaeffer, AC / Spencer, DC, '*Coefficient Regions for Schlicht Functions*' (AMS, 1950); de Branges, L, Acta Math. 154 (1985) 137–152; Henrici, P, '*Applied and Computational Complex Analysis* (Vols. 1,2,3: Wiley, 1974, 1977, 1986); Goluzin, GM, '*Geometric theory of functions of a complex variable*' (AMS transl. #26, 1969).

Example (iii) One broad class of representations of analytic functions of n variables ($n \geqslant 1$) is furnished by the **Cauchy Integrals**. The case '$n = 1$' is, however, qualitatively different from the cases '$n \geqslant 2$, in spite of the validity of a general formula. Call these "*simple*', and "*compound*', Cauchy integrals, respectively. Several classes of 'densities' in these representations may be considered. Many variants of simple CI occur 'naturally' in (e.g.) Elasticity / Aerodynamics / Scattering Theory. Compound CI are of importance in '*Q*uantum '*F*ield '*T*heory – in both cases, mainly in the solution of associated integral equations. Such representations may be used to determine functions having prescribed forms of discontinuities across (parts of) specified 'boundaries' in $\mathbb{C}^m$ (mostly, finite unions of 'arcs', or their equivalents in $\mathbb{C}^m, m \geqslant 2$).

More general representations (originating in problems of analytic continuation) cover H_p – the Hardy class of functions, f, regular in the open unit disc, D, of $\mathbb{C}$, and satisfying the BC: $\int_0^{2\pi} |f(re^{i\theta})|^p d\theta = O(1), (r \to 1^-)$.

Here, the basic result it that: f may be 'recovered' over D^0 from its BV, $r(z)$, over **any set,** E**, of positive** (1-dimensional) **measure**. More precisely:

$$\models (\mathcal{H}[h_\lambda]\gamma)(z) \twoheadrightarrow f(z) \qquad (z \in D^0, \lambda \to \infty, p \geqslant 1),$$

where, if

$$q^{\pm}(w,z) \equiv q^{\pm}(e^{i\theta}, z) := (e^{i\theta} \pm z)^{\pm 1},$$

$$h_\lambda(z) := \exp\{-\frac{1}{4\pi}\log(1+\lambda)\int q^+ q^- d\theta\},$$

then:

$$\mathcal{H}[h_\lambda]\gamma := \lambda h_\lambda + \frac{1}{2\pi i}\int_E q^- \bar{h}_\lambda r dw.$$

Moreover,

$$\|\mathcal{H}[h_\lambda]\gamma - f\|_p \to 0 \qquad (p \neq \infty, \lambda \to \infty).$$

This general result was proved via functional-analytic techniques involving Toeplitz operators. Earlier, suggestive results include:

(a) $\{f \in H_p, p \geqslant 1, f|_E \equiv 0 \quad (E \subset \delta D, \mu(E) > 0)\} \Rightarrow f \equiv 0$ over D^0;

(b) $\{f$ analytic $(|z| < 1), \lim' f|_E \equiv 0 \quad (\lim' :=$ nontangential limit, $\mu(E) > 0,$ $E \subset \delta D)\} \Rightarrow f \equiv 0$ over D^0.

In view of the uniformity of convergence, asymptotic approximations to $f(z)$ may be sought. In particular, both the 'Riemann Hypothesis' and 'Goldback's Conjecture' may be formulated within this framework.

BR: Henrici (op. cit.), Vol. 3, Ch. 14; Range, RM, '*Holomorphic Functions and Integral Representations in Several Complex Variables*' (Springer, 1986); Duren, P, '*Theory of H_p-Spaces*' (AP, 1970); Patil, DJ, BAMS 18 (1972) 617–620; Young, LC, in Ney, P / Port, S, eds., '*Advances in Probability*', Vol. 3, 00166–176 (Dekker, 1974).

Example (iv) There are essentially two classes of procedures for **conformal mapping:** for **simply**, and **multiply**, connected domains, respectively – though some methods may be used in both cases. There are **eight** distinct types of '**canonical domains**', say, D'_k, namely: concentric-circular-slits annulus / disc / plane; radial-slits annulus / disc / plane; exterior of several discs; parallel-slits plane. The corresponding mapping functions may be represented in terms of **Bergman kernel functions**, K, constructed from systems of orthogonal functions – over the domain or its boundary. For dimly-connected domains D, put

$$\bigwedge_S := K_S(q,s)/K_S(q,q), (S \equiv D', \delta D'), \text{ and}$$

$$I_{q,z}(\bigwedge_S) := \int_q^z \bigwedge_S(q,t)dt.$$

Then the following theorem may be proved (via **variational principles**):

$$\models f(z) = I_{q,z}(\bigwedge_{\delta D'}^2); g(z) = I_{q,z}(\bigwedge_{D'})$$

are mapping functions for D'.

For multiply-connected domains, the corresponding forms are:

$$\psi'_j(w) = \chi_j(w) + \int_S \Gamma_j(s)K_S(w,s)d\bar{s},$$

where the χ_j, Γ_j depend suitably on parameters characterizing the D'_j. These, too, may be derived from variational principles. Other schemes involve (non)linear integral equations of various types, with iterative solution-procedures.

BR: Nehari, Z, '*Conformal Mapping*' (McGraw-Hill, 1952 / Dover, 1975); Gaier, D, '*Konstructive Methoden der Konforme Abbildung*' (Springer, 1964); Henrici (op. cit.), Vol. 3, Chs. 16, 17. See MPCPS 2007 (D Crowdy) for n-connected domains.

Example (v) The use of potential-theoretic techniques in function theory is diverse – from the solution of the Dirichlet / Nenmann problems (over domains in $\mathbb{R}^2$) to the Nevanlinna theory of meromorphic functions and the properties of Riemann surfaces. The fundamental connections between harmonic and analytic functions furnish a basis for such methods – especially in the study of conformal mappings of n-connected domains ($n \geqslant 1$), for which the 'kernel-function scheme' may also be used. However, the potential-theoretic framework allows sharper results to be obtained about the structure of the mappings (e.g., boundary behaviour).

The techniques developed for the standard BVP of Potential Theory, and their extensions, may be 'transferred' to cognate problems involving analytic functions. In particular, the **Poisson-integral-representation**:

$$u(z) \equiv u(re^{i\theta}) = \frac{1}{2\pi}\int_{-\pi}^{\pi} P(r,\theta,\varphi)d\varphi, 0 \leqslant r < 1,$$

$$P(r,\theta,\varphi) := (1-r^2)(1-2r\cos(\varphi-\theta)+r^2)^{-1},$$

is of basic importance – as are the various maximum principles, and the theory of subharmonic functions. The possible scope for SI routines lies mainly in the multifarious inequalities and constructions derivable from detailed examination of potential-theoretic procedures over planar domains.

BR: Tsuji, M, '*Potential Theory in Modern Function Theory*' (Maruzen, 1959 / Chelsea, 1975); Henrici, op. cit., Vol 3, Ch. 15; Bergman, S / Schiffer, M, '*Kernel Functions and Elliptic Differential Equations in Mathematical Physics*' (AP, 1953).

The book by Tsuji covers the whole range of function-theoretic problems, with copious detailed results, including many sharp estimates; Henrici emphasises numerical routines, but treats the background results quasi-constructively.

Example (vi) The applications of effective function-theoretic procedures are wide-ranging – among the obvious ones are conformal mapping, integral representations (of solutions of (o)DE, etc.), asymptotic representations of various classes of functions. All of these procedures have wide application in (e.g.) Elasticity, Aerodynamics, Fluid Mechanics and 'Electromagnetics', to cite the most obvious instances. In particular, the 'CV formulations' in Fracture Mechanics, alone, are subtly and diverse. Moreover, Plasma Physics, Acoustics and Quantum Field Theory also admit a variety of nontrivial 'CV applications'; and so on. Saddle-point / steepest descent, Wiener-Hopf, and other specialized techniques for asymptotics all involve CV results in essential ways.

BR: Milne-Thomson, LM, '*Theoretical Aerodynamics*' (Macmillan, 1966, 4th ed., Dover, 1973); Milne-Thomson, LM, '*Theoretical Hydrodynamics*' (3rd ed., Macmillan, 1955); Crighton, DG et al., '*Modern Methods in Analytical Acoustics* ' (Springer, 1992); Muskhelishvili, NI, '*Singular Integral Equations*' (Noordhoff, 1953); Henrici, op. cit., Vols 1,2,3; Carrier, GF / Krook, M / Pearson, CE, '*Functions of a Complex Variable...*', (McGraw-Hill, 1966); Pearson, CE (ed.), '*Handbook of Applied Mathematics...*' (2nd ed., V. Nostrand, 1983); Jones, DS, '*Theoretical Electromagnetism*; (Pergamon, 1964); Bleistein, N / Handelsman, RA, '*Asymptotic Expansions of Integrals*' (Dover, 2nd ed., 1986); Sih, GC (et al.) ed(s): '*Series on Mechanics of Fracture...*', Vols 1–7 (Noordhoff, 1973–1981); Newton, RG, '*The Complex j-Plane*' (Benjamin, 1964); Barton, G, '*Dispersion Techniques in Field Theory*' (Benjamin, 1965); Takahashi, T, '*Mathematics* *of Automatic Control*' (H/R/W, inc., 1966); Guillemin, EA, '*The Mathematics of Circuit Analysis*' (Wiley, 1949); (etc.)^N!

Example (vii) There are several theories of '**generalized analytic functions**' (**g.a.f**), but all of them are based on the notion of replacing the Cauchy-Riemann equations by other sets of (first-order) PDE. The corresponding theories are due, mainly, to Bers / Nirenberg, and to Vekua, namely: **pseudo-analytic functions (ps. a. f.)** and **g.a.f.**. More modern developments are covered by Wen / Begehr – for the BVP of (systems of) (non)linear elliptic PDE – and, by Gilbert / Buchanana (where (generalized) **hyperanalytic functions (g. hp. a. f.)**, and **functions over Clifford algebras**, are emphasized).

Within these formalisms, generalized Riemann-Hilbert (R-H) problems, and others, may be formulated and – to carrying degrees – effectively solved. Conversion into (systems of) integral equations may sometimes be effected, through the use of Pompieu operators. Although most of the representations are unpleasantly intricate, it appears that operational schemes may be fashioned from them. The range of applications includes: Elasticity, Fluid Mechanics, Electromagnetics and other domains of Continuum-Mechanics / Theoretical-Physics.

BR: Bers, L, *Theory of Pseudoanalytic Functions*', (Courant Institute Notes, NY, 1953); Vekua, IN, *Generalized Analytic Functions*' (transl. Pergamon, 1962); Wen, GC / Begehr, HGW, *Boundary Value Problems for Elliptic Equations and Systems*' (Pitman, 1990).

3. Summability (and analogous) Procedures

These procedures cover both **series** and **integrals** which are (in some way(s)) not convergent. Thus, the scope of summability is broad – extending classical results and furnishing various asymptotic results in both pure and applied mathematics. A convenient framework for many of these methods is furnished by **matrix-transformations of sequences** of various types; in particular, the series may be treated in this way (**via** partial sums, etc.). Notions of **summability for classically divergent integrals** may also be defined – both for integrals over unbounded domains, and for integrals whose integrands have various types of singularities in the domain of integration. NOTE that these interpretations are mainly distinct from Cauchy principal values, finite parts, etc..

The detailed **summability schemes**, for Σ , S, are diverse and intricate – but still readily accommodated within the OIB, within CB, etc.. Although the fundamental forms of summability were envisaged (and partially justified) by Euler and Cauchy, the general theory has evolved gradually, and is still far from complete, with new criteria and extensions to more general structures (e.g., **topological groups**). However, two (mutually quasiconverse) types of results are of basic importance here: **abelian theorems**, and **tauberian theorems**. Within Σ , these furnish **OT** and their operational **quasi-inverses**.

BR: Prullage, DL, Math. Zeit 96 (1967) 259–278; ibid 103 (1968) 129–138; Ninham, BW, Numer. Math. 8 (1966) 444–457; Cooke, RG, '*Infinite MAtrices and Sequence Spaces*' (Dover, 1966); Hobson, EW, '*The Theory of Functions of a Real Variable...*' (2nd ed., CUP, 1926 / Dover, 1957); Knopp, K, '*Theory and Application of Infinite Series*' (Blackie, 1928: transl. 2nd ed.); Widder, CV, '*The Laplace Transform*' (Princeton, 1941); Moore, CN, '*Summable Series and Convergence Factors*' (AMS, 1938 / Dover, 1966); Peyerminhoff, A, '*Lectures on Summability*' (Springer LNM #107, 1969); Bosenquet, LS, PLMS (3) Vol III (1953) 267–304; Wilansky, A, '*Summability Through Functional Analysis*' (N. Holland, 1984); Malgrange, B, Exposit. Math. 13 (1995) 163–222; Zygmond, A, '*Trigonometric Series*' (2nd ed., 2 Vols., CUP, 1959); Pitt, HR, '*Tauberian Theorems*' (OUP, 1958); Bureau, F, '*Divergent Integrals and PDE*', pp143–202 in Second Symp. Applied Math. (FE Grubbs, et al., eds.) Wiley / Interscience, 1955; Wiener, N, '*Generalized Harmonic Analysis / Tauberian Theorems*', (repr. by MIT Press, 1964); Lyttkens,

S. Math. Scand. 35 (1974) 61–104.

4. Differential Calculus in General Spaces

Several types of 'differentiation', with associated calculi, have been defined – for BS, LTS, and so on. The most well-known of these are **Gâteaux, variational**, and **Fréchet**, derivatives; but there are many other variants. Among the most important **fields of application** are: **Statistical Mechanics, Quantum Field Theory**, and **Optimization** (especially, in **Control Theory** and **Mathematical Economics**). The outline in SCCM identifies the key characteristics of several species of derivative, along with their (dis)advantages – some of which are far from obvious consequences of the basic definitions! In spite of this, the development of associated schemes for the OIB would be comparatively straightforward, since the rules for application / calculation of the underlying operators (subject to precisely specified limitations) may be readily distilled. Therefore, it is enough, here, to list some key references for the definitions / theories / applications / comparisons of the theories mentioned.

BR: Nashed, MZ, '*The role of differentials...*', in '*Nonlinear Functional Analysis*' (MZ Nashed, ed., AP, 1971); MZ Nashed (ed.) '*Generalized Inverses and Applications*' (AP, 1976 – with an annotated bibliography of over 250 pages!); Frölicher, A / Bucher, W, '*Calculus in Vector Spaces without Norm*' (LNM #30, Springer, 1966); Yamamuro, S, '*Differential Calculus in Topological Linear Spaces*' (LNM #374, Springer, 1974); Averbukh, VI / Smolyanov, OG, '*The Theory of Differentiation in Linear Topological Spaces, I;II*' (Russian Math. Surveys 22 (1967) 201–258; 67–113); Ver Eecke, P, '*Fondements du Calcul Différentiel*' (esp., the bibliography); Volterra, V, '*Theory of Functionals...*' (Blackie, 1929 / Dover, 1959); Croxton, CA, '*Liquid State Physics...*' (CUP, 1974); Visconti, A, '*Quantum Field Theory, Vol 1*' (Pergamon, 1969); Intriligator, M, '*Mathematical Optimization and Economic Theory*; (P-H, 1971); Pearson, CE, (ed.), '*Handbook of Applied Mathematics*' (end ed., V. Nostrand, 1983); Gabasov, R / Kirillova, F, '*The Qualitative Theory of Optimal Processes*' (Dekker, 1976)

5. Branching Analysis for Operator Equations

This is a vast subject, covering matrices, linear integral operators, nonlinear integral operators, and much more. The analysis of parametrized families of matrices[400], and of linear integral operators[401], are comparatively tractable (but far from trivial). The study of singularities is discussed separately (as a framework for Catastrophe Theory, and more generally). Here, the emphasis is on nonlinear (systems of) integrodifferential equations over BS – for which a comprehensive approximation / perturbation scheme has been highly developed. Even the summary of this theory in SCCM (based on the comprehensive treatise by Vainberg / Trenogin) is intricate and somewhat lengthy; so, the aim here is just to identify the main SI routines involved[402]

It turns out that **all** of the following classes of systems may be studied within a **single framework**:

1. implicit functions
2. linear and nonlinear-integral-equations
3. linear and nonlinear-integrodifferential-equations
4. singular-integral-equations
5. periodic-solutions-of-DE, perturbation-theory

[400] See, e.g., papers by VI Arnold in the BR.
[401] See, e.g., books on linear integral equations in the BR.
[402] NOTATION (in this subsection): $P_w/P_s/TS \equiv$ power / Puiseux / Taylor series.

6. linear and nonlinear-BS-mappings (of 'Fredholm / Noether' types).

This is achieved by combining the Lyapunov / Schmidt reduction of the branching problem (from infinite to finite dimension) with extensions of the 'Newton Polygon construction' for the systematic determination of fractional P_wS corresponding to the branches of algebraic / algebroidal functions (of the parameter(s), $(\lambda)_n$, involved). The 'cases with $n = 1$' are the commonest, yielding Puiseux series, in fractional powers of $\lambda_1 \equiv \lambda$.

The basic problem (for all types of operator equations) takes the form: find all 'small solutions' of the equation(s)

$$\text{(a)} \qquad \Sigma H_{ik} x^i \lambda^k = 0,$$

as fractional PS, $x_j(\lambda)$, where the sum is over nonnegative integers i, k, with $i + k \geqslant 1$, the x_j continuous near $\lambda = 0$, and all $x_j(0) = 0$. Here, it is convenient to put $F_{ik} =: H_{ik}/(i!k!), F_{10} =: -B$. All of the F_{ik} are viewed as Fréchet derivatives (of order i in x, k in λ) at $\lambda = 0$, and B is a linear operator, etc.. Such quasi-TS may be found (e.g.) in the book by Hille / Phillips. Then (a) becomes:

$$\text{(a}'\text{)} \qquad Bx = F_{01}\lambda + \sum_{i+k \geqslant 2} x^i \lambda^k.$$

It is proven in V/T that (a′) is L/S transformable into the equivalent system (for $i = i, \ldots, r$):

$$\text{(b)} \qquad \Sigma_1 L^{(i)}_{(k)_r 0} \langle \xi^k \rangle_r + \Sigma_2 L^{(i)}_{(j)_r k} \langle \xi^j \rangle_r \lambda^k = 0,$$

where $(k)_r := (k_1, \ldots, k_r), \langle \xi^k \rangle_r := \xi_1^{k_1} \ldots \xi_r^{k_r}$, etc., $k_1 + \cdots + k_r \geqslant 2$ in Σ_1, and $j_a + \cdots + j_r \geqslant 0$ in Σ_2. Formally (at least), the system (b) determines, quasi-recursively, small solutions $\xi_m^{(i)} \equiv \xi_m^{(i)}(\lambda)$ as P_sS – through the elaborate use of extended N/P constructions.

SI routines for the formation of branching systems for all types of operator equations covered may be developed. Although such systems are typically complicated and somewhat intractable, they do offer excellent possibilities for SAM, at various levels. The usefulness of this whole scheme obviously hinges on the quality of the (many!) approximate systems involved – whose branching behaviour must mirror that of the underlying system, (a'). The analysis in V/T (and other BR) shows that such techniques are, in principle, tenable.

BR: Vainberg, MM / Trenogin, VA, '*Theory of Branching of Solutions of Nonlinear Equations*' (Noordhoff, 1974); Krasnosellkii, MA, et al., '*Approximate Solution of Operator Equations*' (Noordhoff, 1972); Golubitsky, M, et al., '*Singularities and Groups in Bifurcation Theory, Vols I, II* (Springer, 1985; 1988); Hille, E / Phillips, R, '*Functional Analysis and Semigroups*' (2nd ed., AMS, 1957); Pimbley, G, '*Eigenfunction Branches of Nonlinear Operators and their Bifurcations*' (LNM #104, Springer, 1969); Bartsch, T, '*Topological Methods for Variational Problems*' (Springer, 1993).

6. Calculations in Celestial Mechanics

This item is motivated by the remarkable 4-volume presentation of **theoretical** results, covering virtually all aspects of this field in near-exhaustive detail, by Y Hagihara: Vol I, '*Dynamical Principles and Transformation Theory*' (pp689); Vol II: '*Perturbation Theory*' (Parts 1,2; pp504, 921); Vol III: '*Differential Equations in Celestial Mechanics*' (Parts 1,2; pp1–504, 505–1160); Vol IV '*Periodic and Quasiperiodic Solutions*; (Parts 1,2; pp1–552, 553–1243) [I: MIT, 1970, II (1), MIT, 1971; II (2), MIT (1972); III, IV Japanese Society for the Promotion of Science, 1974; 1975].

This gargantuan work merits close study! Its high relevance for Σ is incontestable.

NOTE. See Zentralblat 242 #7003; 305 #700125; 344 #70007; 355 #70009.

7. Function-Theoretic Methods in Elasticity

This material is widely known. Its great importance in engineering design, alone, justifies attempts to produce SI routines. The formalism covers both (an)isotropic and (non)simply connected bodies – effectively planar, though some of the procedures have partial extensions to spatial systems. The primary methods here are due to Muskhelishvili / Sherman / Lekhnitski. All of these schemes are fully expounded in the BR; their suitability for effective SI will be apparent – though, as always, the development of operational versions raises significant problems.

BR: Muskhelishvili, NI, '*Some Basic Problems of the Mathematical Theory of Elasticity*' (3rd ed., transl., Noordhoff, 1953); Lekhnitskii, SG, '*Theory of Elasticity of an Anisotropic Elastic Body*' (Holden Day, 1963); Lekhnitskii, SG, '*Anisotropic Plates*' (G&B, 1968); Kalandiya, AI, '*Mathematical Methods of Two-Dimensional Elasticity*' (MIT, 1975).

8. Singular Integral Equations

These equations occur in many applications, so key procedures from their basic theory offer 'natural' SI routines for (generalized) BVP. As in Item 7, much of this material is standard, so it suffices to cite some BR where full details may be found. Once again, the challenge of producing operational forms remains. It should also be remarked that the numerical solution of these equations raises recondite problems (over convergence estimates, etc.) many of which are also pertinent to the SI routines. Such results are less familiar: this is an active research area.

BR: Muskhelishvili, NI, '*Singular Integral Equations*' (2nd ed., transl., Noordhoff, 1946); Gakhov, FD, '*Boundary Value Problems*' (transl., Pergamon, 1966);
Pogorzelski, W, '*Integral Equations..., Vol 1*', (transl., Pergamon, 1966); Prössdorf, S / Silbermann, B, '*Numerical Analysis for Integral and Related Operator Equations*' (Birkhäuser, 1991); Wenland, WL, '*Elliptic Systems in the Plane*' (Pitman, 1979).

9. Ill-Posed Problems

Operator (in)equations whose (quasi-)solutions are not continuous functions of the I/B/...data may still be effectively solve (relative to specified topologies, etc.) provided that 'suitable interpretations' are given. The original designation, of 'well-posed problems', is due to Hadamard, for the PDE of Physics. It appears, however, that ill-posed problems span a substantial part of (pure, and applied) mathematics, so it is essential to devise w.d. procedures for their resolution. Among ill-posed problems of common occurrence are: **(i) 'inversion' of $\int Kf = u$; (ii) differentiation of 'partially known functions'; (iii) evaluation of Fourier series; (iv) the planar 'Cauchy Problem' with perturbed I/BC; (v) AC from an arc into a domain; (vi) inverse gravimetry problems; (vii) some basic problems in Linear Algebra**, and in **Optimization**. Practical applications of the various 'regularizing schemes' for the (iterative) solution of such 'awkward problems' include calculations in: **image-processing, tomography, seismology, acoustics**. Formal procedures with rigorous error / convergence criteria may be derived.

BR: Colton, D, Pitman Research Notes #4 (1976); #6 (1976); Tikhonov, AN / Arsenin, VY, '*Solution of Ill-Posed PRoblems*' (Wiley, 1977); Bakushinsky, A / Goncharsky, A, '*Ill-Posed Problems: Theory and Applications*' (Kluwer, 1994) [with a Bibliography of 142 items]. Knops, RJ (ed.), LNM #316 (Springer, 1972); Kaplin, A, et al., '*Stable Methods for Ill-Posed Variational Problems*' (Akad. Verlag, 1994).

10. Galois-Field Procedures for Switching Circuits

The main problem here is the synthesis of complex circuits with prescribed characteristics (for instance, to control traffic flow). This use of prime-power-order fields may be modifiable to handle aspects of microchip design, etc.. Although there are now diverse procedures for the design / analysis of various types of automata / robots, the Galois-scheme is highlighted because of the near-exhaustive, symbolic representations. In the BR, more recent sources for other approaches to such problems are expounded.

BR: Moisil, Gr. C, '*The Algebraic Theory of Switching Circuits*' (Pergamon, 1968); Korte, B, et al. (eds.) '*Paths, Flows and VLSI-Layout*' (Springer, 1990); Murgai, R, et al., '*Logic Synthesis for Field-Programmable Gate Arrays*' (Kluwer, 1995); Sherwani, N, '*Algorithms for VLSI Physical Design Automation*' (Kluwer, 1993).

11. Problems in Robotics

These problems are primarily kinematical / dynamical in nature. Once again, it suffices to cite a few key references exhibiting the potential for SI routines.

BR: Yoshikawa, T, '*Foundations of Robotics...*' (MIT, 1990); Magnus, K (ed.) '*Dynamics of Multi-Body Systems*' (Springer, 1977); Bianchi, G, et al. (eds.) '*Dynamics of Multibody Systems*' (Springer, 1985); Vukobratović, M, et al. (eds.): '*Scientific Fundamentals of Robotics*', Vols. 1–6 (Springer, 1982–1986); Duffy, J, '*Analysis of Mechanisms and Robot Manipulators*' (Arnold, 1980).

12. Problems in Control Theory

This, too, is an enormous subject, ranging from the most delicate (functional-)analytic investigations to packages for (e.g.) optimal chemical processing. The tools for studying control problems are correspondingly diverse: algebraic / analytical / graph-theoretic / diagrammatic / numerical /.... Moreover, systems may be linear / nonlinear / deterministic / stochastic, etc.. **Criteria for stability** are of basic importance.

BR: Merriam III, CW, '*Automated Design of Control Systems*' (G&B, 1974); Netushil, A (ed.), '*Theory of Automatic Control*' (Mir, 1978); Nagrath, IJ / Gopal, M, '*Control Systems Engineering*' (2nd ed., Wiley, 1982); MacFarlane, AGJ, '*Dynamical System Models*' (Harrap, 1970); Fuhrmann, RA (ed.) '*Mathematical Theory of Networks and Systems*' (Proc., Springer, 1984); DiStefano, JJ, et al., '*Feedback and Control Systems*' (Schaum, 1976); Atherton, DP, '*Nonlinear Control Engineering*; (V. Nostrand, 1975); Bromberg, PV, '*Matrix Analysis of Discontinuous Control Systems*; (Macdonald, London, 1969).

13. Problems in Circuit Theory

Two fundamental problems dominate Circuit Theory: the **analysis**, and **synthesis**, of classes of circuits (with specified types of components). Although there is much common ground with Control Theory, there are also distinctive types of problems – for instance: **reception, transmission, resonance, amplification, interconnection, matching**. Further, the components may behave **nonlinearly**, and circuits may be affected by '**random noise**'; etc. etc.! Consequently, a very broad spectrum of mathematical techniques may be found in this field, and the SI potential is high – as the following (brief) section of **BR** attests.

BR: Atabekov, GI, '*Linear Network Theory*' (Pergamon, 1965); Ferris, CD, '*Linear Network Theory*' (Merrill, 1962); Messerle, HK, '*Dynamic Circuit Theory*' (Pergamon, 1965); Paul, CR,

'*Analysis of Multiconductor Transmission Lines*' (Wiley, 1994); Willson, AN (ed.), '*Nonlinear Networks: Theory and Analysis*' (IEE Press, 1974); Haring, DR, '*Sequential Circuit Synthesis...*' (MIT, 1966); Chen, W-K, '*Broadband Matching...*' (World Scientific, 1988); Boite, R / Dewilde, P (eds.), '*Circuit Theory and Design*; (Delft UP / N-H, 1981).

14. Analysis of Nonlinear (O)DE

The obvious applications of these techniques (to problems in Electronics, Vibrations, etc.) with their implications for 'chaotic behaviour', have produced a vast literature, formally covered by 'Dynamical Systems Theory', in its mathematical aspects. Although most of the detailed criteria involved are intricate, the scope for SI is great – as perusal of the listed BR with demonstrate.

BR: Minorsky, N, '*Nonlinear Oscillations*' (V. Nostrand, 1962); Kaliski, S, et al., '*Vibrations and Waves, Part A...*' (Elsevier, 1992); Rouche, N, et al., '*Ordinary Differential Equations*' (Pitman, 1980); Sansone, G / Conti, R, '*Nonlinear Differential Equations*' (Pergamon, 1964); Reissig, R, et al., '*Nonlinear Differential Equations of Higher Order*' (Noordhoff, 1974); Dincă, F, et al., '*Nonlinear and Random Vibrations*' (AP, 1973); Thompson, JMT, et al., '*Nonlinear Dynamics and Chaos...*' (Wiley, 1986); Vejvoda, O, et al., '*Partial Differential Equations: Periodic Solutions*' (Nijhoff, 1982) [with a 29 page Bibliography]; Mitropolskii, Y, '*Problems of the Asymptotic Theory of Non-Stationary Vibrations*' (Israel Progr. Sci. Transl., 1965); Cesari, L, '*Asymptotic Behaviour and Stability Problems in Ordinary Differential Equations*' (3rd ed., Springer, 1971); Pliss, V, '*Nonlocal Problems in the Theory of Oscillations*' (AP, 1966); Guckenheimer, J, et al., '*Nonlinear Oscillations, Dynamical Systems, and Bifurcations of Vector Fields*' (Springer, 1983); Morsden, JE, et al., '*The Hopf Bifurcation...*' (Springer, 1976).

NOTE: Many of the sharpest results on boundedness / periodicity / convergence of solutions of nonlinear ODE are given only in papers (mostly listed in the Bibliographies of the books in the BR).

15. Basic Calculations in Algebraic Topology

Only the barest indication is given, here, of three species of calculation of combinatorial / algebraic type. Many more procedures could be developed – covering other branches (Geometric, Differential) of Topology; but, as usual in this topical list, the crucial point is to show that there is significant SI content. The main classes of problems are:

(i) decomposition / assembly procedures for triangulable spaces;

(ii) implementation of the constructions embodied in the 'weak', and 'strong', 'simplicial approximation theorems';

(iii) calculation schemes for some groups of basic topological importance in 'nontrivial cases';

(iv) routines for handling classes of (co)homological diagrams;

(v) axiomatic (co)homology calculations: CW complexes.

In SCCM, items **(i)–(iii)** are considered and the key steps are identified; item **(iv)** is partially covered in E_4/E_5, in the extensive discussion of 'diagrams'. No problems of principle seem to be raised by any of these procedures – which are indicative of cognate schemes in other parts of (Pure) Mathematics. Within Σ , these are all 'natural routines', where flexible I/O and graphics facilities may be combined with arbitrary forms of 'approximation' to extend the realm of 'calculability'. Among the obvious 'applications' in this area are algorithms in Graph Theory – which, in turn, appear in (electrical) network theory, etc.. The BR given here adequately reflect several interesting possibilities.

BR: Maunder, CRF, '*Algebraic Topology*' (CUP, repr., 1980); Kondo, K (ed.), '*RAAG Memoirs*', Vols 1 (1958), 2 (1962), 3 (1968) (Dept. of Engineering, U. Tokyo); Kim, WH, et al., '*Topological Analysis and Synthesis of Communications Networks*' (Columbia UP, NY, 1962); Magnus, W, et al., '*Combinatorial Group Theory...*' (Interscience, 1966 / Dover, 1976); Hu, S-T, '*Elements of General Topology*' (Holden Day, 1964); Hu, S-T, '*Homology Theory*' (Holden Day, 1966).

16. Calculations in Differential Geometry

This topic is already partially covered by packages for routines in General Relativity; but there are many other fields of application, e.g., **Twistor Theory, Linear and NonLinear Elastic Shell Theory, Rotating (Electrical) Machines, multivariate (statistical) calculation**, and other topics to be found in the **RAAG Memoirs** (See also the BR#15). Moreover, there are (continuing) researches in 'pure differential geometry / topology'.

As usual, the main point for Σ is that there are already substantial bodies of quasi-algorithmic routines that have been regarded as intractable 'by hand' (except in simple cases), but may now be seen as a powerful resources in SAM / Σ frameworks.

BR: Kreyszig, E, '*Introduction to Differential and Riemannian Geometry*' (U. Toronto P., 1968); Klingenberg, W, '*Riemannian Geometry*' (de Gruyter, 1982); Eisenhart, LP, '*Non-Riemannian Geometry*' (AMS, 1972); RAAG Memoirs, op. cit. (BR#15); Penrose, R, et al., '*Spinors and Space-Time*', Vol 1 (CUP, 1984); Farrell, RH, '*Multivariate Calculation*' (Springer, 1985); Gallavolti, G, '*The Elements of Mechanics*' (Springer, 1983); Rutten, HS, '*Theory and Design of Shells on the Basis of Asymptotic Analysis*' (Rutten & Kruisman, Holland, 1973);; Michael, AD, '*Matrix and Tensor Calculus...*' (Wiley, 1947); Thomas, TY, '*Plastic Flow and Fracture in Solids*' (AP, 1961); Kunin, IA, '*Elastic Media with Microstructure, I; II*' (Springer, 1982; 1983); Brulin, O, et al. (eds.), '*Mechanics of Micropolar Media*' (World Scientific, 1982); Kröner, E, (ed.), '*Mechanics of Generalized Continua*' (Springer, 1968); Nabarro, FRN (ed.), '*Dislocations in Solids, Vol 1*' (N-H, 1979); Fichera, G (ed.), '*Trends in Applications of Pure Mathematics to Mechanics*' (Pitman, 1976); Hanyga, A, '*Mathematical Theory of Nonlinear Elasticity*' (PWN, 1985); Berger, M, et al., '*Differential Geometry: Manifolds, Curves and Surfaces*' (Springer, 1988); Bott, R, et al., '*Differential Forms in Algebraic Topology*' (Springer, 1982); Choquet-Bruhat, Y, et al., '*Analysis Manifolds and Physics*' (N-H, rev. ed., 1982); Kramer, D, et al., '*Exact Solutions of Einstein's Field Equations*' (CUP, 1980).

17. Calculations with Generalized Functions

There are two partially distinct species of 'generalized functions' – characterized by **(a)** linear functionals on various spaces of test functions; **(b)** rings of 'convolution quotients'. Of these schemes, **(a)** has been developed in great detail in both 'theory' and 'applications', while **(b)**, though less developed, has also been widely applied.

For SI the main classes of routines involve **(i)** manipulation of established 'explicit formulae'; and **(ii)** general formal manipulations. For **(i)**, a large collection of results involving particular GF may be deployed as part of a CB for 'concrete function(al)s'. The material of type **(ii)** includes FE/DE procedures for PDE (e.g., analysis over Sobloev spaces). The procedures **(b)** are applicable in some situations where those in **(a)** fail. The approximations introduced here (for **(a)** or **(b)**) reside, not in the operational relations, but in their realizations involving sequences / integrals /... with Elementary / Special functions.

The main large-scale routines are for handling Fourier integrals and solutions to (systems of) ODE with singular I/BC, using matrix / Laplace-transform schemes.

BR: Gel'fand, IM, et al., '*Generalized Functions, Vols 1–5*' (AP, transl., 1964–1968) [esp., Vol 1, for basic operations / examples]; Jones, DS, '*The Theory of Generalized Functions*' (2nd ed., CUP, 1982); Berg, L, '*Introduction to the Operational Calculus*' (N-H, 1967); Mikhusinski,

J, '*Operational Calculus*' (Pergamon, 1959); Zemanian, AH, '*Distribution Theory and Transform Analysis*' (McGr-H, 1965 / Dover, 1987); Zemanian, AH, '*Generalized Integral Transformations*' (Interscience, 1968 / Dover, 1987); Erdélyi, A, '*Operational Calculus and Generalized Functions*' (HRW, 1962); Liverman, TPG, '*Generalized Functions and Direct Operational Methods*' (1964).

18. Procedures in Asymptotics

In almost all calculational schemes special techniques are required for cases where the domain(s) of calculation become unbounded, or where some element(s) become 'singular' in various senses. Thus the purview of Asymptotics is vast – covering parts of every established branch of (pure / applied) Mathematics, with many specialized methods 'tuned to particular applications'. In spite of this, there is a core of essential procedures (for summation / integration / singular-evaluation /...) which may be 'distilled' for use in all appropriate situations. As usual, it suffices to cite some key sources for such techniques. The longterm aim is to produce a classified collection of asymptotic techniques in all fields, in operational forms (some basic methods having several – not necessarily equivalent – 'embodiments' in diverse (analytical / algebraic / probabilistic /...) fields.)

BR: Olver, FWJ, '*Intro. to Asymptotics and Special Functions*' (AP, 1974); Sirovich, L, '*Techniques of Asymptotic Analysis*' (Springer, 1971); Copson, ET, '*Asymptotic Expansions*' (CUP, 1965); Erdélyi, A, '*Asymptotic Expansions*' (Dover, 1956); de Bruijn, NG, '*Asymptotic Methods in Analysis*' (N-H, 1958 / Dover, 1981); Murray, JD, '*Asymptotic Analysis*' (OUP, 1974); Bleistein, N, et al., '*Asymptotic Expansion of Integrals*' (2nd ed., Dover, 1986); Wilcox, CH (ed.) '*Asymptotic Solutions of Differential Equations...*' (Wiley, 1964); Batchelder, PM, '*Intro. to Linear Difference Equations*' (Harvard UP, 1927 / Dover, 1967); Ramm, AG, '*Theory and Applications of Some New Classes of Integral Equations*' (Springer, 1980); Majina, H, '*Asymptotic Analysis for Integrable Connections...*' (Springer, 1984); Carleson, L, et al., '*Complex Dynamics*' (Springer, 1993); Mishchenko, AS, et al., '*Lagrangian Manifolds and the Maslov Operator*' (Springer, 1990); Mitropol'skii, Y, '*Problems of the Asymptotic Theory of Nonstationary Vibrations*' (Israel Progr. Sci. Transl., 1965); Molchanov, IS, '*Limit Theorems for Unions of Random Closed Sets*' (Springer, 1993); Kalashnikov, VV, et al., (eds.) '*Stability Problems for Stochastic Models*' (Springer, 1989); Pfanzagl, J, '*Asymptotic Expansions for General Statistical Models*' (Springer, 1985); Borokov, AA, '*Asymptotic Methods in Queueing Theory*' (Wiley, 1984); Jeffreys, H, '*Asymptotic Approximations*' (OUP, 1962).

19. Operational Schemes for Partial Differential Equations

The comparative simplicity of solution routines for (systems of) ODE does not obtain for PDE. Indeed, the only known procedures for sets of PDE (or, **P**seudo DE) are of daunting complexity, and may yield only asymptotic – rather than complete – solutions. Nevertheless, these procedures (essentially, of two types) are of great importance, since they cover fundamental problems in diverse areas of (pure and applied) Mathematics.

The procedures of 'Type 1' are due mainly to Ehrenpreis / Hörmander/ Palamdov / Trèves; those of 'Type 2', initially to Maslov. The Type-1 forms are essentially sums of quasi-exponential terms – for sets of PDE with constant coefficients. The Type-2 Schemes cover **PDE with variable coefficients, differential-difference equations,** and **pseudo-differential equations** – all of which may be **non**-linear. This is based on elaborate extensions of the original Heaviside method and its subsequent rigorous formulations, **transforming** the sets of ODE into equivalent **algebraic** problems (for the PDE, equivalent **integral equations** are obtained).

The Ehrenpreis approach extends the domain of Fourier transforms over $\mathbb{C}^n$ through the concept of 'analytically uniform spaces'. Another scheme, due to Palamodov, is founded on homological algebraic representations. By contrast, Maslov's method originated in problems of Mathematical Physics (Plasma Physics / Solid-State-Physics / Nuclear Physics / Electron-Optics /...). It is based on a calculus for 'linear, noncommutative operators over algebras with μ-structures'. Typically, equations of all species may (eventually!) be transformed into Volterra / Fredholm integral equations – for which efficient (asymptotic) approximation procedures have been developed.

Despite the intricate form of both Type-1 and Type-2 representations, many concrete criteria / routines may be 'extracted' – in line with the universal priority in Σ : to maximize calculability in all contexts. An indication is given in SCCM of the key principles; but only by consulting the BR is it possible to assess these techniques.

BR: Ehrenpreis, L, '*Fourier Analysis in Several Complex Variables*' (Wiley, 1970); Berenstein, CA, et al., '*Analytically Uniform Spaces...*' (Springer, 1972); Hörmand- er, L, '*Linear Partial Differential Operators*' (Springer, 1963); Palamodov, VP, '*Linear Differential Operators...*' (Springer, 1970); Maslov, VP, '*Operational Methods*' (Mir, 1976); Mishchenko, AS, et al., '*Lagrangian Manifolds and the Maslov Operator*' (Springer, 1990); Pommaret, JF, '*Differential Galois Theory*' (G&B, 1983); Trèves, F, '*Linear Partial Differential Equations...*' (G&B, 1967).

20. Topological Approximation of Fixed Points of Mappings

These techniques stem from Brouwer's proof (1912) of his fixed-point (fp) theorem for mappings between finite-dimensional spaces. Until comparatively recently, only the Banach fp theorem was used in numerical routines – even though the premises for Brouwer's theorem are weaker, and many other fp theorems have been proved since these original results were obtained. Among more recent results are the fp theorems of: Kakutani; Lefschetz; Poincaré-Birkhoff; Leroy-Schauder; Tihanov; Kleene.

Apart from the (standard) uses of fp theorems to prove the existence of solutions to various types of algebraic / difference / differential / integral equations, there are diverse applications in (e.g) Game Theory, Mathematical Economics, Optimization, Theoretical Computer Science, and Control Theory. Consequently, constructive procedures for the approximation of fp – under various sets of premises – are of considerable importance.

The basic procedure involves nested sequences of simplexes (with diameter $\to 0$) each constructed to contain the sought fp. Other schemes depend on homotopic transformations of the original fp problem (into 'equivalent, simpler' ones). A further variant of these ideas yields the so-called homotopy-continuation procedures. NOTE that the SI content devolves mainly on the use of such algorithms for n-parameter problems – where issues of estimation / convergence, etc., may be considered. Although this topic emphasizes topological methods, the iterative procedures (over BS) typical of Banach-type fp theorems may also be mentioned here (but see also Topic 4).

BR: Zeidler, E, '*Nonlinear Functional Analysis, Vol. 1*' (Springer, 1986); Smart, DR, '*Fixed Point Theorems*' (CUP, 1974); Istrătescu, VI, '*Fixed Point Theory*' (Reidel, 1981) [with a 40-page Bibliography]; Allgower, EL, et al., '*Numerical Continuation Methods*' (Springer, 1990); Todd, MJ, '*The Computation of Fixed Points...*' (Springer, 1976); Border, KC, '*Fixed Point Theorems...*' (CUP, 1985); Ellickson, B, '*Competitive Equilibrium...*' (CUP, 1993); Gunter, CA, '*Semantics of Programming Languages...*' (MIT, 1992); Krasnosel'skii, MA, '*Translation Along Trajectories of Differential Equations*' (AMS, Transl., 1968); Krasnosel'skii, MA, et al., '*Geometrical Methods of Nonlinear Analysis*' (Springer, 1984); Wang, Z, et al., '*Algebraic Systems of Equations...*' (Kluwer, 1994).

21. Calculations in Singularity Theory

The emphasis here is on three areas:

(a) determinacy / unfolding calculations;

(b) critical points of functions on manifolds;

(c) integer-valued characteristics of vector fields.

For **(a)**, Catastrophe Theory is the motivation; for **(b)**, 'Morse Theory' and its applications; and, for **(c)**, nonlinear operator equations – but each item has a far wider purview. Brief indications are given in SCCM; but see the 'asterisked BR' for minimal accounts of the key results of each type.

BR: *Poston, T, et al., '*Taylor Expansions and Catastrophes*' (Pitman, 1976); *Poston, T, et al., '*Catastrophe Theory...*' (Pitman, 1978); Gilmore, R, '*Catastrophe Theory...*' (Wiley, 1981); Morse, M, '*The Calculus of Variations in the Large*' (AMS, 1934); Morse, M, '*Topological Methods in the Theory of Functions...*' (Princeton, 1946); Morse, M, et al., '*Critical Point Theory in Global Analysis...*' (AP, 1969); Krasnosel'skii, MA, '*Topological Methods in the Theory of Non-Linear Integral Equations*' (Pergamon, 1964); *Krasnosel'skii, MA, et al., '*Plane Vector Fields*' (Iliffe, 1966); *Krasnosel'skii, MA, et al., '*Geometrical Methods of Nonlinear Analysis*' (Springer, 1984); Hartman, P, '*Ordinary Differential Equations*' (Wiley, 1964); Rouche, N, et al., '*Ordinary Differential Equations...*' (Pitman, 1980); *Milnor, J, '*Morse Theory*' (Princeton, 1963); Klingenberg, W, '*Riemannian Geometry*' (de Gruyter, 1982); Bartsch, T, '*Topological Methods for Variational Problems...*' (Springer, 1993).

22. Banach-Space Calculations: Bases

Many procedures for finite-dimensional spaces become undefined / intractable for infinite-dimensional spaces. Partial exceptions to this behaviour are furnished by BS with bases – and by BS with generalized bases (of various kinds). The bases are countable sets of elements for which norm-convergence results of the form: $\|x - \sum_1^m \alpha_k x_k\| \to 0$ as $m \to \infty$ hold.

Criteria for **basicity** of specified sequences, and **'transmission** of basis properties' (under various operations) are of special importance. Some of the results may be **extended** to classes of TVS other than BS (e.g., LCTVS). This has implications for certain problems involving 'best-approximation behaviour'. Although the **bases** are defined to be **countable**, the **generalized bases** may have **any cardinality**. For applications to Physics / Engineering, the most important results deal with **quasi-eigenfunction expansions** for various differential / integral /... operators – in particular, with expansions over **sets of Elementary / Special functions**, for which completeness / basis tests are required. In this connection, the distinction between basicity and **totality** (characterized by the condition $\|x - \sum_1^m a_k(m) x_k\| \to 0$ as $m \to \infty$) is fundamental – and emanates from the **Weierstrass Approximation Theorem**, for (uniform) approximation by sequences of (trigonometrical) polynomials. The primary SI content here is of two sorts: **(i)** collections of sequences whose completeness / basicity / totality properties (relative to specified (normed) spaces) are established – together with relevant parts of CB to handle these sequences t-max-operationally; **(ii)** general results on bases and their extended forms – with emphasis on OT formulation wherever possible. Applications dance from structural criteria for BS/HS, through topics in best-approximation theory, to very practical problems in signal-analysis image-processing.

BR: Marti, J, '*Introduction to the Theory of Bases*' (Springer, 1969); Singer, I, '*Bases in Banach Spaces, Vols 1;2*' (Springer, 1970; 1981); Higgins, JR, '*Completeness and Basis Properties of Sets of Special Functions*' (CUP, 1977); Boas, RP, Jr., et al., '*Polynomial Expansions of Analytic Functions*' (Springer, 1958); Singer, I, '*Best Approximation in Normed Linear Spaces...*' (Springer, 1970).

NOTE that Singer, Vol 2, op. cit., contains a diverse collection of **generalizations** of base / basis (as originally defined) of both countable and uncountable types, with many sharp results

/ interrelations. The aim of producing 'Vol. 3', including results for concrete BS, seems to have been abandoned.

23. Calculations in Statistical Mechanics / Quantum Field Theory

These (intricately linked) subjects span a substantial part of Theoretical Physics – call them SM / QFT, here. Each admits both lattice ($\equiv$ discrete) and continuum models – of various basic phenomena. Whereas calculations in General Relativity motivated much of the initial development of SAM packages (see, e.g., Kramer, D, et al., op. cit., Topic 16), virtually no SM / QFT SI routines seem to have been developed – perhaps, because the structure of GR (embodied in Einstein's field equations) is comparatively simple to explore, using fairly straightforward differential-geometric procedures. Nevertheless, several fundamental problems (say, P_k) may be identified (mainly, in SM, here) for which SI routines could be implemented.

For simplicity, only 'classical' (as opposed to 'quantum') SM is considered, for a system composed of N 'particles' moving 'within / over' a container, Ω, of volume / size $|\Omega|$, with the motion governed by a Hamiltonian, $\mathcal{H} \equiv \mathcal{H}(\alpha) \equiv \mathcal{H}(p,q) \equiv \mathcal{H}((p)_N,(q)_N)$, where α denotes a generic point in the phase space, Δ, of the system, and p_j, q_k are (generalized) momentum, and position, coordinates. In these terms, the 'thermodynamic potentials' are given by:

$$|\Omega|^{-1} \log Z(\beta, N, \Omega) \qquad \text{(canonical);}$$

$$|\Omega|^{-1} \log \Xi(\beta, z, \Omega) \qquad \text{(grand canonical),}$$

where

$$Z := \frac{1}{N!} \int_{\Delta_N} e^{-\beta\mathcal{H}} d\alpha,$$

$$\Xi := 1 + \sum_{N \geqslant 1} \frac{z^N}{N!} Q_N,$$

$$Q_N := \int_{\Omega^N} e^{-\beta V(q)_N} (dq)_N,$$

$(k\beta)^{-1}$ is the (absolute) temperature, and $V(q)_N$ is the potential energy – typically of the form

$$V(q)_N = \sum_{1 \leqslant i < j \leqslant N} \psi(q_i, q_j).$$

The so-called thermodynamic potentials (from which observable behaviour of the system may be determined) are defined as:

$$\zeta := |\Omega|^{-1} \log Z, \text{ and}$$

$$\xi := |\Omega|^{-1} \log \Xi.$$

Within this framework (say, $\mathcal{F}^*$), the following basic problems may be formulated.

P$_1$ Prove that the established thermodynamic properties are derivable from $\mathcal{F}^*$.

P$_2$ Determine possible equations of state (functional relations interconnecting pressure, volume, temperature, etc.) from $\mathcal{F}^*$ – for a system, $\mathcal{S}$, in overall equilibrium.

P$_3$ Prove that $\mathcal{S}$ tends towards an equilibrium state, if $\mathcal{S}$ remains isolated for long enough.

P$_4$ Account, mathematically, for the phenomena of phase transitions ($\equiv$ changes of macroscopic state) in $\mathcal{S}$ – and find necessary / sufficient conditions for the occurrence of such transitions.

P$_5$ Characterize the possible equilibrium states of $\mathcal{S}$.

P_6 Obtain adequate mathematical characterizations of the various states of matter (solid, liquid, dense gas, dilute gas, plasma, etc.).

P_7 Give a detailed mathematical description of the critical region(s) of S (i.e., of the neighbourhoods in 'thermodynamic space' of transition points) and determine / estimate the associated critical exponents (governing singular behaviour, etc.).

NOTES

(a) Ω may be a geometrical lattice in $\mathbb{R}^d$, $(d = 1, 2, 3)$ – or even in $\mathbb{R}^m, m \in \mathbb{P}$, for some models.

(b) Most of the formalism – and of the problems P_k – have cognate formulations for Quantum SM, and for QFT.

(c) The so-(mis)-called **second quantization** schemes are fundamental in QSM.

The problems P_k require a wide range of powerful techniques – even for non-trivial investigation. For general models, most of the P_k are (at best) only partially solved – the fullest solutions covering a diversity of lattice models, in CSM / QSM / QFT. Even for general continuum models, however, many profound results have been obtained, and, the mathematical schemes developed for all (lattice / continuum) classes of models span vast conceptual domains.

In spite of the wide scope of the P_j, the dominant processes involved (at the 'first level') are few: **(i)** simulation; **(ii)** cellular automata models; **(iii)** limit processes (mainly, for $|\Omega| \to \infty$). An alternative framework, where infinitely large systems are treated from the outset, involves procedures over C^*-algebras. Possible SI routines include:

(α) asymptotic representations of large, but finite systems (of all types)

(β) miscellaneous 'second-quantization calculations'

(γ) exact calculations of various lattice-model properties

(δ) 'Renormalization-group' calculations

(ε) Phase-transition tests / analyses.

It should be emphasized that 'many-body problems' in all domains (e.g., lattice dynamics) may also be covered (in several aspects) by these techniques.

BR: Ruelle, D, '*Statistical Mechanics...*' (Benjamin, 1969); Baxter, RJ, '*Exactly Solved Models in Statistical Mechanics*' (AP, 1982); Domb, C, et al., (eds.), '*Phase Transitions and Critical Phenomena*' Vols 1–10$^+$ (AP, 1972–1987$^+$); Barber, M, (Renormalization Group) Phys. Reports 29 (1977); Lieb, EH, et al., (eds.) '*Mathematical Physics in One Dimension*' (AP, 1966); Doolen, GD, et al., (eds.), '*Lattice Gas Methods for Partial Differential Equations*' (A-W, 1990); Berezin, FA, '*The Method of Second Quantization*' (AP, 1966); Fetter, AL, et al., '*Quantum Theory of Many-Particle Systems*' (McGr-H, 1971); Itzykson, C, et al., '*Statistical Field Theory*' Vols 1,2 (CUP, 1989); Ring, P, et al., '*The Nuclear Many-Body Problem*' (Springer, 1980); Sinai, Ya. G (ed.), '*Mathematical Problems of Statistical Mechanics*; (World Scientific, 1991); Li, X, et al., (eds.), '*Lattice Gauge Theory Using Parallel Processors*' (G&B, 1987); Dewitt, C, et al., (eds.), '*The Many Body Problem*' (Wiley, 1959); Ludwig, W, '*Recent Developments in Lattice Theory*' (Springer, 1967); Israel, RB, '*Convexity in the Theory of Lattice Gases*' (Princeton UP, 1979).

24. Group-Theoretic Calculations

Here, one may consider established sets of routines for **(a)** High-Energy Physics; **(b)** Solid-State Physics; **(c)** Crystallography; **(d)** Chemistry (etc.!). Most of these routines are already (close-to) implementable for Σ ; the rest will be, after suitable (re)formulation. Here, only the briefest list of BR need be given.

BR: Hamermesh, M, '*Group Theory...*' (A-W, 1962); Bishop, DM, '*Group Theory and Chemistry*' (OUP, 1973); Harris, DC, et al., '*Symmetry and Spectroscopy*' (OUP, 1978); King, RB, (ed.), '*Chemical Applications of Topology and Graph Theory*' (Elsevier, 1983); Ynduráin, FJ, '*The Theory of Quark and Gluon Interactions*' (2nd ed., Springer, 1993); Feld, BT, '*Models of Elementary Particles*' (Blaisdell, 1969); Streitwolf, H-W, '*Group Theory in Solid-State Physics*' (Macdonald, 1971); Bradley, CJ, et al., '*The Mathematical Theory of Symmetry in Solids*' (OUP, 1972); Slanina, Z, '*Contemporary Theory of Chemical Isomerism*' (Reidel, 1986); Jones, H, '*The Theory of Brillouin Zones...*' (N-H, 1960); Ballhausen, CJ, '*Intro. to Ligand Field Theory*' (McGr.-H, 1962).

25. Finite-Element / Boundary-Element Routines

Although these procedures for solving classes of operator equations – especially, PDE – are primarily numerical, there are many analytical / algebraic sub procedures involved. Moreover, in problems where extra 'parameters' occur, even the numerical phases have symbolic aspects. Again, the FE functions – corresponding to FE dissections of the PDE domain – are, typically, 'polynomial'; but, in some situations (e.g., for domains with (intricately) curved boundaries) **rational** FE functions are appropriate, raising some nontrivial issues in **Algebraic Geometry** (into sections / singularities /...).

The Boundary-Element method (where the boundary, rather than the domain, is dissected) also requires various analytical / algebraic operations (possibly involving extra parameters). In both the FEM and the BEM, (quasi-)variational problems 'generating' the sought solutions are derived. The underlying algorithms are formulated over Sobolev spaces (of integer / fractional orders) for proofs of existence / uniqueness / (speed of) convergence / singularities. **Notice that** the implementation of FEM/BEM routines may yield sets of (coupled) **difference equations** – for which a range of sophisticated solution-procedures may be used. Indeed, these are examples of quasi-quadrature schemes for classes of operator equations of all kinds. Here, too, many of the constituent subprocedures are of analytical / algebraic types.

It is evident, therefore, that the application of FD/FE/BE techniques to non-trivial problems may be greatly enlarged in scope / efficiency through the deployment of suitable SAM routines. Some attempts in this direction were made long ago; but much remains to be done – for instance, in dealing with operations over Sobolev spaces through a systematic SAM package.

NOTE. Infinite EM Schemes have also been developed (See, e.g., Numer. Math. 39, 39–50 (Han, H))

BR: Strang, G, et al., '*An Analysis of the Finite Element Method*' (P-H, 1973); Oden, JT, et al., '*Finite Elements, Vol IV: Mathematical Aspects*' (P-H, 1983); Sloan, IH, '*Error Analysis of Boundary Integral Methods*', pp287–339 in Acta Numerica (1992) (CUP, 1992); Marchuk, GI, '*Methods of Numerical Mathematics*' (2nd ed., Springer, 1982); Marchuk, GI, et al., '*Difference Methods and their Extrapolations*' (Springer, 1983); Yserentant, H, '*Old and New Convergence Proofs for Multigrid Methods*', pp285–326 in Acta Numerica (1993) (CUP, 1993); Ghanem, RG, et al., '*Stochastic Finite Elements...*' (Springer, 1991); Wachpress, E, '*A Rational Basis for Finite Elements*' (AP, 1975); Fairweather, G, Finite Element Methods' (Dekker, 1978); Nazarov, SA, et al., '*Elliptic Problems in Domains with Piecewise Smooth Boundaries*' (de Gruyter, 1994); Godunov, SK, et al., '*Theory of Difference Schemes...*' (N-H, 1964).

26. Solution of Stochastic / Functional Operator Equations

This covers a vast terrain, with diverse techniques, so the aim here is just to demonstrate that nontrivial SI routines may be developed. There are essentially four broad classes of such equations: **(i) deterministic linear; (ii) deterministic non-linear; (iii) stochastic linear; (iv) stochastic nonlinear**. Within each of these classes there may be arbitrary combinations of algebraic / difference / differential / integral / BS/GS types of equations – where 'BS' covers general forms of operator equations over BS, and equations over even more general spaces (GS) are also envisaged. Cases **(i), (ii)** are partially covered in Topic 5 (for equations with some extra parameter(s)); Topic 17 (for ODE); Topic 19 (for PDE), as well as the schemes in Topic 9 (for ill-posed problems) and Topic 14 (for qualitative analysis of nonlinear (O)DE); and so on.

Nevertheless, it is useful to consider various types of approximation procedures broadly applicable to all (or 'most') of these types of equations. Moreover, schemes for classes of operator **in**equations are also of great importance in applications.

For the stochastic versions the indeterminacy may enter in any combination of several ways, including: **(a)** random I/BC; **(b)** random 'coefficients' (which may be functions); **(c)** random 'forcing functions' (for non autonomous equations). Here, the additional techniques developed for random ODE, and PDE – and, for linear, or **non**linear DE – are fundamentally similar; but quite distinct from all procedures for deterministic equations. It appears that (for approximation) the key tasks are: **(α)** evaluation of stochastic integrals; and **(β)** solution of stochastic ODE, of various types.

The associated (non)deterministic **in**equations have been studied intensively, on account of their wide application in (e.g.): Noise Theory; Control Theory; Optimization; Viscoelasticity; Nuclear Physics; Plasma Physics; Free-Boundary Problems; Plasticity; Antenna Design; Nonlinear Oscillations; (etc.)^N!

Techniques for the (quasi-)constructive solution of all these types of (non)deterministic (in)equations are now well established and may be found in many treatises and current papers. [In Section 15.21 of SCCM, an indication is given of how SI may be used for the basic problems.] One further aspect, of great potential for SI is the Itô Calculus, which has been widely applied in Engineering, the Sciences, and even (largely disastrously!) in Stock-Market prediction.

The following BR cover most of the theoretical / applied procedures, adequately.

BR: Patterson, WM, '*Iterative... Solution of a Linear Operator Equation in Hilbert Space*' (Springer, 1974); Saaty, TL, '*Modern Nonlinear Equations*' (McGr-H,1967); Kantorovich, LV, et al., '*Functional Analysis*' (2nd ed., Pergamon, 1982); Ostrowski, AM, '*Solution of Equations in... Banach Spaces*' (AP, 1974); Lattes, R, et al., '*The Method of Quasi-Reversibility...*' (Elsevier, 1969); Cameron, RH, '*A Simpson's Rule for... Wiener Integrals...*' TAMS 58 (1945) 184–219; Cameron, RH, et al., '*Transformation of Wiener Integrals...*' BAMS 51 (1945) 73–90; Montroll, EW, Commun. Pure Appl. Math. 5 (1952) 415–453; Feynman, RP, et al., '*Quantum Mechanics and Path Integrals*' (McGr-H, 1965); Edwards, SF, et al., pp I61–I75 in Frisch, HL, et al. (eds.), '*The Equilibrium Theory of Classical Fluids*' / ≡ J. Math. Phys. 3 (1962) 778–792; Gel'fand, IM, et al., J. Math. Phys. 1 (1960) 48–69; Young, LC, in (i) Ney, P (ed.), '*Advances in Probability, Vol 1*' (dekker, 1970); and in (ii) Ney, P, et al. (eds.), '*Advances in Probability, Vol 2*' (Dekker, 1974); Jazwinski, A, '*Stochastic Processes and Filtering Theory*' (AP, 1970); Wong, E, et al., (i) Int. J. Eng. Sci. 3 (1965) 213–229; (ii) Ann. Math. Stat. 36 (1965) 1560–1565; (iii) Report No. 65-5 AF-A FOSR-139–64 (UC Berkeley, 1965); McShane, E, Proc. Sixth Berkeley Symp. Prob. Stat. Vol 3, 263–294; Ladde, GS, et al., '*Random Differential Inequalities*' (AP, 1981); Duvant, G, et al., '*Inequalities in Mechanics and Physics*' (Springer, transl., 1976); Cottle, RW, et al., (eds.), '*Variational Inequalities...*' (Wiley, 1980); Laksmikantham, V, et al., '*Differential and Integral Inequalities...*' Vols 1,2 (AP, 1969); Prodi, G (ed.), '*Problems in Nonlinear Analysis*' Proc. (CIME Ediz. Cremonese, Roma, 1971); Varian, HR, (ed.), '*Economic and Financial Modeling with Mathematica*' (TELOS / Springer, 1993); Hourison, SD, et al., (eds.), '*Mathematical Models in Finance*', Phil. Trans. Roy. Soc. London 347A (15/6/94) pp449–598; Elworthy, KD, et al. (eds.) '*Stochastic Analysis...*' (Pitman, 1989).

27. Group Representation Calculations

This topic is partially covered in #24, but the scope for SI in generating / analysing the diverse forms of representation for (in)finite groups – including Lie groups and 'discontinuous groups' (mainly associated with classes of automorphic functions) – is so broad that collections of specialized routines are required. There is, of course, some overlap with procedures for High-Energy / Solid-State Physics, Quantum Chemistry, etc., but the techniques considered here are more orientated towards Algebra / Analysis than towards scientific applications. Although many of the general methods are already quasi-constructive, the problems of converting them to operational forms are daunting – but definitely solvable in principle.

BR: Curtis, CW, et al., '*Representation Theory of Finite Groups...*' (Interscience, 1962); Curtis, CW, et al., '*Methods of Representation Theory*' Vols 1;2 (Wiley, 1981; 1987); Barut, AO, et al., '*Theory of Group Representations...*' (PWN, 1977); Dornhoff, L, '*Group Representation Theory*', parts A; B (Dekker, 1971; 1972); Carter, RW, '*Simple Groups of Lie Type*' (Wiley, 1972 repr. 1989); Carter, RW '*Finite Groups of Lie Type*' (Wiley, 1985, repr. 1993); O'Meara, OT, '*Symplectic Groups*' (AMS, 1978); Johnson, DL, '*Topics in the Theory of Group Presentations*' (CUP, 1980); Bruggerman, RW, '*Families of Automorphic Forms*' (Birkhäuser, 1994); Donkin, S, '*Rational Representations of Algebraic Groups*' (Springer, 1985); Auslander, M, et al. (eds.), '*Representations of Algebras*' (Springer, 1982); Webb, P (ed.), '*Representations of Algebras* (Springer, 1982); Webb, P (ed.), '*Representations of Algebras*' (CUP, 1985); Humphreys, JE, '*Linear Algebraic Groups*' (Springer, 1975); Kac, VG, '*Infinite Dimensional Lie Algebras*' (2nd ed., CUP, 1985); Magnus, W, et al., '*Combinatorial Group Theory...*' (2nd rev. ed., Dover, 1976); Jacobson, N, '*Lie Algebras*' (Interscience, 1962; Dover, 1979).

28. Some Calculations in Perturbation Theory

From the SI perspective there are 2 classes of routines here: **(i)** finite-dimensional spaces and **(ii)** infinite-dimensional spaces. For case **(i)** there is an essentially complete theory, already in substantially operational form, treating transformations representable by n-parameter families of matrix operators. For this framework the topics include: singularities of eigenvalues; determination of Neumann series for resultants; manipulation with projection operators; differentiability / analyticity properties of eigenvalues / eigenvectors. Some of these topics have cognate sub theories in case **(ii)** – but there are extra (e.g., convergence) complications. Applications to solutions of the Schrödinger equation (in various models) yield additional problems, and motivation. The underlying theme is, of course, approximation – for specified types of (linear) operators (e.g., Fredholm / Noetherian / Toeplitz). Some of these matters re also relevant to Topic 5 (Branching Analysis for Operator Equations). Although the theory has abstract formulations, most of the applications (and the continuing research) deal with classes of (integro-)differential operators – especially in quantum-mechanical Scattering Theory, where linear, self-adjoint mappings between HS furnish the basic examples. The bifurcation problems covered in Topic 5 involve **non**linear operators – as do diverse problems currently investigated.

BR: Kato, T, '*Perturbation Theory for Linear Operators*' (Springer, 1966; 2nd ed., 1976); Knowles, IW, et al. (eds.), '*Differential Equations*' (N-H, 1984); Riesz, F, et al., '*Functional Analysis*; (Dover, 1990 / orig., Unger, 1955); Demuth, M, (ed.), '*Operator Calculus and Spectral Theory*' (Birkhäuser, 1992); Edmunds, DE, et al., '*Spectral Theory and Differential Operators*' (OUP, 1987); Martin, M, et al., '*Lectures on Hypernormal Operators*' (Birkhäuser, 1989); Friedrichs, KO, '*Perturbation of Spectra in HS*' (AMS, 1965); Berezansky, YM, et al., '*Spectral Methods in Infinite Dimensional Analysis*', Vols 1,2 (Kluwer, 1995).

Another (more specialized) part of Perturbation Theory covers the approximate solution of (mainly) (O)DE with 'small parameters'. This is less abstract than the general theory for operators in BS/HS/...; but its great importance in many applications (e.g., in Hydrodynamics, Vibration Theory, Control Theory) justifies a separate listing here – though the most general frameworks cover **all** types of operators / perturbations.

BR: Bellman, RW, '*Perturbation Techniques...*' (HRW, 1964); Smith, DR, '*Singular Perturba-*

tion Theory' (CUP, 1985) [with a substantial Bibliography]; Hille, E, et al., '*Functional Analysis and Semigroups*' (2nd ed., AMS, 1957); Yosida, K, '*Functional Analysis*' (4th ed., Springer, 1974); Titchmarsh, EC, '*Eigenfunction Expansions...*' Parts I; II (OUP, 1946; 1958); Levitan, BM, et al., '*Intro. to Spectral Theory*' (AMS, 1975); Zauderer, E, '*Partial Differential Equations...*' (Wiley, 1983).

29. Integral Operators (and Some Extensions)

This is a very broad field, with natural division into: **(i)** One-dimensional / regular problems; **(ii)** n-dimensional / regular problems; **(iii)** one-dimensional / singular problems; **(iv)** n-dimensional / singular problems (for $n \geqslant 2$, integer).

NOTE that fractional-integral operators may also be considered – as partial extensions of this classification. The areas of application are also varied. There are both (partially) formal, and (quasi) constructive procedures (some of which yield numerical routines). Moreover, systems of coupled integral (in)equations / representations give rise to a variety of challenging problems – most of which are still actively investigated. (Some forms of (non)linear (P)DE may also be treated by these techniques – as can some classes of pseudo DE.)

BR: Baker, CTH, '*The Numerical Treatment of Integral Equations*' (OUP, 1977); Prössdorf, S, et al., '*Numerical Analysis for Integral Equations*' (Birkhäuser, 1991); Mikhlin, SG, et al., '*Singular Integral Operators*' (Springer, 1986); Pugachev, VS, '*Theory of Random Functions*' (Pergamon, 1965); Schetzen, M, '*Volterra & Wiener Theories of Nonlinear Systems*' (Wiley, 1980); Krasnosel'skii, MA, et al., '*Approximate Solution of Operator Equations*' (Noordhoff, 1972); Krasnosel'skii, et al., '*Integral Operators in Spaces of Summable Functions*' (Sijtoff, 1976); Krasnosel'skii, MA, '*Positive Solutions of Operator Equations*' (Noordhoff, 1964); Krasnosel'skii, MA, '*Topological Methods in the Solution of Nonlinear Integral Equations*' (Pergamon, 1964); Krasnosel'skii, MA, et al., '*Geometric Methods of Nonlinear Analysis*' (Springer, 1984); Roach, GF, et al, '*Fractional Calculus*' (Pitman, 1987); Sneddon, IN ,'*Mixed Boundary-Value Problems in Potential Theory*' (N-H, 1966); McBride, AC, '*... Fractional Integration for GFs*', I (SIAM J. Math. Anal. 6 (1975) 583–599); II (Proc. RS Edin. 77A (1977) 335–349); McBride, AC, '*Solution of Hypergeometric Integral Equations...*' (Proc. Edin. Math. Soc. (1974) 265–285); Cooke, JC, '*The Solution of Triple Integral Equations in Operational Form*' (QJ Mech. Appl. Math. XVIII (1965) 57–72).

30. Calculations in Abstract Harmonic Analysis

As usual, the aim is merely to show that there are obvious SI routines to 'open the Σ -account in this area', by formulating operational versions of some basic procedures – notably, the construction of Haar measures and their associated integrals. Many other, comparatively straightforward procedures may be found involving harmonic analysis on (e.g.) Lie groups. Above all, the existing effective techniques are already SI-implementable; so, even a substantial collection of such results would be valuable, as a precursor to the more intricate theorems. **NOTE that** some nontrivial examples have been discussed within the 'B-constructivity paradigm' – which has continued to provide preliminary formulations for operational versions of diverse mathematical schemes since the publication of 'Constructive Analysis', by E Bishop / D Bridges (which has a substantial Bibliography).

BR: Hewitt, E, et al., '*Abstract Harmonic Analysis, Vols I, II*' (Springer, 1963; 2nd ed, 1979; 1970); Gilbert, JE, et al., '*Clifford Algebras... in Harmonic Analysis*' (CUP, 1991); Warner, G, '*Harmonic Analysis on Semisimple Lie Groups, Vols 1; II*' (Springer, 1972; 1972); Varadarajan, VS, '*Intro. to Harmonic Analysis on Semi-Simple Lie Groups*' (CUP, 1989); Bishop, E, et al., '*Constructive Analysis*' (Springer ,1985); Gel'fand, IM, et al., '*Generalized Functions, Vol.*

4: *Applications of Harmonic Analysis*' (AP, 1964); Gel'fand, et al., '*Representation Theory...*' (Saunders, 1969); Rudin, W, '*Fourier Analysis on Groups*' (Wiley, 1962, repr., 1990).

31. Calculations in Computational Geometry

Typical procedures here are: **(i)** construction of convex hulls; **(ii)** determination of intersections of w.e. sets; **(iii)** solution of 'closest-point' problems; **(iv)** search routines. Although the motivation stems from sets in $\mathbb{R}^k, k = 1, 2, 3$ – where applications include: applied statistics, computer graphics, CAD, computer vision, VLSI design, linear programming, and mathematical morphology – the procedures are w.d. for all integers k, and even for subsets of more general spaces. Indeed, for subsets of BS, there are possible algorithms in Game Theory, Statistical Mechanics and Mathematical Economics – to only only obvious examples.

The key observation is that most of the applications involve real-valued functions on collections of (finite) sets in some space(s) – the basic aim being efficient implementation, through complexity analyses. The list of BR covers the most important mathematical problems and a selection of applications. NOTE that (linear) transformations over various spaces play a basic role – so algorithms from (generalized) Linear Algebra are often invoked.

BR: Brown, KQ, '*Geometric Transforms...*' (Report CMU-CS-80-101, Carnegie-Mellon Univ., 1979); Serra, J, '*Image Analysis...*' (AP, 1982); Bentley, JL, et al., (Report CMU-CS-80-109); Bryant, RE (Report CMU-CS-87-106, 1987); Sheffler, TJ, (Report CMU-CS-87-123, 1987); Kanade, T, (Report CMU-CS-79-153, 1979); Sleator, D, et al., (Report CMU-CS-88-108, 1988); Dwyer, RA, (Report CMU-CS-88-132, 1988); Ziegler, GM, '*Lectures on Polytopes*' (Springer, 1995); Preparata, F, et al., '*Computational Geometry...*' (Springer, 1985); Sherwani, N, '*Algorithms for VLSI...*' (Kluwer, 1993); Wechsler, H, '*Computational Vision*' (AP, 1990); Pratt, WK, '*Digital Image Processing*' (Wilet, 1978); Gonzalez, RC, et al., '*Digital Image Processing*' (2nd ed., A-W, 1987); Ying-Lie O, et al., (eds.),'*Shape in Picture...*' (Springer, 1994); Wolff, LB, et al., (eds.), '*Shape Recovery*' (Jones and Bartlett, 1992); Holmes, RB, '*Geometric Functional Analysis...*' (Springer, 1976).

32. Probability Theory and Statistics

The scope for SI activity here is almost unlimited, since the subject is replete with explicit formulae, exact / asymptotic representations and intricate applications (within pure and applied mathematics). Is suffices, therefore, to list a random (!) selection of BR.

BR: Stuart, A, et al., '*Kendall's Advanced Theory of Statistics, Vols 1; 2*' (Arnold, 1994; 1991); Feller, W, '*... Probability Theory*', Vols I; II (Wiley,3rd ed., 1968; 1966, 2nd ed., 1971); Rényi, A, '*Probability Theory*; (N-H, 1970); Johnson, NL, et al., '*Distributions in Statistics*', 4 Vols., (Wiley, 1969; 1970; 1970; 1972); Anderson, TW, '*... Time Series*' (Wiley, 1971); Wilks, SS, '*Mathematical Statistics*' (Wiley, 1962); Rohatgi, VK, '*Intro. to Probability Theory and Mathematical Statistics*' (Wiley, 1976); Moran, PAP, '*... Probability Theory*' (OUP, 1968); Asmussen, S, '*Applied Probability and Queues*' (Wiley, 1987); Ephreinides, A, et al., (eds.) '*Random Processes...*' (DH&R, 1973); Rogers, GS, '*Matrix Derivatives*' (Dekker, 1980); Preisendorfer, RW, '*Radiative Transfer on Discrete Spaces*' (Pergamon, 1965); Penrose, O, '*Foundations of Statistical Mechanics*' (Pergamon, 1970); Bharucha-Reid, AT, '*Probabilistic Methods...*' Vols. 1; 2; 3 (AP, 1968; 1970; 1973); Saaty, TL, '*Elements of Queueing Theory*' (McGr-H, 1961 / Dover, 1984).

33. Effective p-adic Analysis

The use of p-adic schemes in SAM is exemplified by 'Hensel-type methods', but the scope of the p-adic techniques is far wider – covering virtually all domains of Analysis, and parts of Algebra and other fields. Moreover, many of the results are finitely (and so automatically effective). **NOTE that** this is essentially part of 'analysis over nonarchimedean fields' – with diverse implications for fundamental results in Number Theory and AlgebraicGeometry. Some idea of the SI possibilities may be obtained from the Bibliographies of (most) items in the BR.

BR: Weiss, E, '*Algebraic Number Theory*' (McGr-H, 1963); Zariski, O, et al., '*Commutative Algebra*' Vol II (V Nostrand, 1960, repr. Springer); Borevich, ZI, et al., '*Number Theory*' (AP, 1966); Van Rooij, ACM, '*Nonarchimedean Functional Analysis*' (Dekker, 1978); Dwork, B, '*Lectures on p-adic Differential Equations*' (Springer, 1982); Koblitz, N, '*p-adic Numbers, p-adic Analysis...*' (Springer, 1977); Baldassarri, F, et al., (eds.), '*p-adic Analysis*' (Springer, 1990); Kani, E, '*Potential Theory on Curves*' (pp475–543 in: De Konick, JM, et al. (eds.), '*Théorie des nombres / Number Theory*' (de Gruyter, 1989)); Koblitz, N, '*p-adic Analysis: a short course...*' (CUP, 1980).

34. Calculations in Complexity Theory

This is, mainly, concerned with the assessment / analysis of algorithms – in all domains – so it is of great importance for Σ -implementation, and as a source of results for the OIB! Thus the range of topics covered here essentially spans M_t; but the frameworks / schemes involved may be characterized comparatively succinctly – and extended to furnish fundamental CB for estimating the complexities of (essentially arbitrary) OT. The following brief list of BR suffices here.

BR: Traub, JF, (ed.) (a) '*Complexity of Numerical Algorithms*'; (b) '*Analytic Complexity*'; (c) '*Algorithms and Complexity...*' (AP, 1973; 1976; 1977); Traub, et al., '*...Optimal Algorithms*' (AP, 1980); Garey, MR, et al., '*Computers and Intractability*' (Freeman, 1979); Knuth, DR, '*The Art of Computer Programming*' Vols 1; 2; 3 (AW, 1968; 2nd ed., 1981; 1973); Watanabe, O (ed.), '*Kolmogorov...and Computational Complexity* ' (Springer, 1992); Steele, JM, et al., (Panel) '*Probability and Algorithms*' (Nat. Acad. Press, Washington DC, 1992); Ambos-Spies, K, et al., (eds.), '*Complexity Theory...*' (CUP, 1993)I Borodin, A, et al., '*The...Complexity of Algebraic...Problems*' (American Elsevier, 1975); Sedgenrick, R, '*Algorithms*' (A-W, 1988).

APPENDIX 2
Diverse Examples of Approach

The basic idea in this Essay is to use the classes (introduced in E_7): $\mathbb{D}, \mathcal{S}, \mathcal{G}, \mathcal{A}$, $\mathcal{L},^{\pm} \mathcal{J}$ – for Determinations, Specializations, Generalizations, Approximations (in all w.d. standard forms/spaces), Abstractions, Analogues, Inverse-Analogues– to classify the entries in the 'SiMA[403] / MaAT[404]-List.' For each 'approach of ϵ to e' the following (minimal) information is given:

(i) the (broad) **subtheories/environments** in which e, ϵ are 'based';

(ii) the **forms of approach** ('pure' or 'mixed') from $\mathcal{M} \equiv \{\mathcal{S}, \mathcal{G}, \mathcal{A}, \mathcal{L},^{\pm} \mathcal{J}\}$, of ϵ to e;

(iii) descriptions of the **modes/realizations of approach**, of ϵ to e.

The m-th item on the S/M-List is denoted by i_m – with associated attributes, $p[i_m], \langle i_m \rangle, \ldots$, to be specified later.

Although such classification will rarely be exhaustive, their dominant characteristics may be 'near-definitive' – fundamental, invariant under the evolution of Σ_t, and unambiguous. of course, some of the interpretations to be given here may be challenged on various grounds, but mostly they will be generally acceptable. In any case, changes may be made at any time without affecting the consistency/validity of the overall OIB.

Ultimately, these 'approach-outline' may generate **pre-algorithms**[405] – within suitable mathematical frameworks – including details of all 'inherent limitations to approach', of which obvious examples are furnished by Roth's theorem[406] (on the approximation of algebraic numbers by rationals) and numerous results on the complexity of procedures[407]. Moreover, the (t-max-) operational versions of all relevant theorems will be cited (if references exist for them); otherwise, conjectured quasi-operational forms may be stated – as future research problems!

The treatment of the material in this Essay is deliberately terse, since the number of items considered is large, and their scope is immense. An indication of the size/complexity of a near-systematic scheme – even for a narrowly-defined mathematical domain – is provided by the '**Handbook of Nonparametric Statistics**', by J.E Walsh [408], (mentioned in other Essays). This voluminous work, totalling over 2,000 pages was written when computers were primitive with virtually no IT/AI capabilities. Consequently, the need for copious cross-referencing, alone, added greatly to the size of the books. Nevertheless the irreducible core of 'pre=operational information' presented is impressive but also illustrative of the urgent need for a system like Σ , where a computer based version of Walsh's Handbook could be accommodated in a highly compressed, but far more efficient, form.

The choice of topics for the initial SiMA/MaAt-List (S/M-List) requires some explanation. The guiding aim is to exhibit characterize, in E_10, diverse forms of '**approach phenomena**' in **all areas** of mathematics and its applications – with a more extensive discussion of selected **analogues**, in E_11. The challenge so presented is to **to create a conceptual framework** rich enough to accommodate the immense range of results (potentially) involved. It is suggested that this challenge has been met, to a high degree, by the body of formalisms/procedures developed in $E_0 - -E_9$ – where arbitrary ME are represented in terms of admissible approach schemes/LC [409].

---

[403] Studies in Mathematical Analogy

[404] Mathematics as Approximation Theory

[405] i.e. quasi-effective procedures (potentially) convertible into **algorithms**

[406] See, e.g., LeVeque, W.J., 'Topics in Number Theory' (A-W, 1956, Vol II)

[407] See, e.g., Garey, M.R./Johnson, AAS. 'Computers and Intractability' (Freeman, 1979); Ambos-Spies, K.,et al. (eds.) 'Complexity Theory' (CUP, 1993)

[408] Vols I,II, III (Van Nostrand, 1962, 1965, 1968)

[409] Locational Condition(s)

The detailed **selection** of references for the S/M-List combines widespread search and chance discovery, to produce a heterogeneous collection of books and papers (acquired over many years), and another list of citations to publication not directly accessible.

It has already been observed that such limitations of objectivity/size (in the **initial** S/M -List) are substantially illusory – since the evolution $\Sigma_t$from Σ_{t_0} may be compared (in this respect) with the determination of a 'complete analytic function' from any of its 'germs', though the processes involved for Σ are, of course, far more elaborate! Therefore, the emphasis is broadly on finding interesting/nontrivial examples, many of which may be unfamiliar, and should also prove stimulation. The overall **scope** of the list is, evidently, wide, but it has no pretence to 'completeness'; rather, it is just a basis for development – with **approach/structural-transfer** as the underlying themes.

Experience in developing an **IB for solid mechanics** (especially, Fracture Mechanics)[410] shows that a combination of **taxonomies, tagging,** and quasi-universal **problem-representation(s)** is indispensable. The tagging covers detailed annotations of research papers, so that the **types/techniques/solutions** of a wide range of **problems** may be studied systematically. The extreme scope/complexity of this partial IB-even for a single mathematical domain – underlines the gargantuan challenge presented by Σ ; yet it also demonstrates some of the basic techniques required for the construction of the Σ - OIB (even though OT, as such, are not used). Above all, it is clear that **the Σ -superstructure**, designed to promote the efficient organization of the entire compass of Mathematical Knowledge, **must be highly elaborate**. The intricate constructions and complicated arguments deployed in these Essays are, therefore, unavoidable. The present task – of exhibiting a large, heterogeneous collection of 'mathematical phenomena' within an approach framework – is best tackled by establishing a 'standard template layout', on which the essential characteristics and their various interconnections may be displayed. A minimal list of these characteristics is as follows.

(1) FACETS (from $\{\varphi_1, \ldots \varphi_6\}$);

(2) PRIMARY SUBTHEORIES/ENVIRONMENTS (Mathematical);

(3) PRIMARY SUBTHEORIES/ENVIRONMENTS (Scientific);

(4) SPECIES OF SPACE(S) (from $\{L, \Gamma, T, U, C, P, N, S, \mu\}$);

(5) APPROACH-TYPE(S) $(\mathcal{D}, \mathcal{S}, \mathcal{G}, \mathcal{A}, \mathcal{L},^{\pm} \mathcal{J}$);

(6) OPERATIONAL LEVEL(S) (from { **nil, partial, full**}

(7) FORMULATION(S)

Let the entries, for species (1)-(7), be denoted (respectively) by $\Phi_m, H_m, \Gamma_m, B_m, A_m, L_m, F_m$ – for the item i_m.

This scheme may be **codified** by introducing notation for **the primary mathematical/scientific subtheories,** and for the collection of **admissible formulations** — so that every item in the SiMA/MaAT - list [411] has some cipher -like representation(s) – referred to, in the Solid Mechanics IB, as the **tagging(s)** of the item concerned [412]. On this basis the S/M-list may be **scanned**, through the Σ -search -software, along with the body of operational information of all species/types constituting the OIB. [413] The next task is, therefore, to devise suitable notation for (a)**primary mathematical subtheories;** (b)**primary scientific subtheories**; and

[410]This project was sponsored by the **R**oyal **A**ircraft (now, **D**efence **R**esearch) **E**stablishment, Farnborough, UK– mainly at the universities of Nottingham (1987-1991) and Oxford (1991-1994). extensive reports, and a preliminary software package, under the heading: Information Base for Mathematical Fracture Mechanics (refs: No. 2024 /085 / XR / STR, and, CB / RAE / 9 / 4 / 2057 / 141 / MA) are retained at DRA, Farnborough. Eventually, this package may be fully implemented/published.

[411]abbreviated to S/M - list

[412]The operational levels may be denoted as n, p, f

[413]The **full** tagging must be based on detailed **taxonomies** of all the (sub) theories in the mathematical/ physical/ chemical/ Engineering/ Biological/ Economic/ in Sciences – as well as in Mathematics itself!

(c) **formulations** (in terms of characteristic, (1)-(6)) of the 'mathematical phenomena' in the S/M-list. Many items may admit several nontrivially distinct formulations; other, apparently only one (though new formulations may eventually be found, unless 'uniqueness criteria' are established).

(a');(b') The designation of 'primary subtheories' – mathematical, or scientific – is inevitably a partially subjective process. Nevertheless, the **cores** of all such designations will be substantially identical; it is only for the finer points that possible alternatives need to be discussed. in any case, the **structure** of the Σ - OIB is largely independent of the detailed choices that are made. It suffices, therefore, to indicate how the cores of $\{$ mathematical/scientific subtheories $\}$ may be specified. As with many aspects of Σ , there is no requirement of uniqueness or completeness – only consistency and 't-operational-adequacy' are essential.

Consequently, the MR primary heading will cover the core of $\{$ primary mathematical subtheories $\}$; and the comparable lists for Review/Abstract journals in Physics, Chemistry, Engineering, Economics, Social Sciences, Biological Sciences, Medical Sciences... will together cover the core of $\{$ primary scientific subtheories$\}$, to be used, mainly, in the investigation of **inverse** analogues. All of these lists are compiled through comprehensive studies of the fields involved – to produce the 'skeleton pretaxonomies' discussed in earlier Essays (especially, in E_1), which are embodied in comparatively simple numerical coding schemes, permitting almost unlimited refinement/extension, to treat the various fields in detail as they are developed. The system proposed here corresponds to using the first and second levels of the MR lists (and comparable Science Reviews), where **Level**-1 contains the broad headings (e.g., Number Theory; Field Theory; Algebraic Geometry; Group Theory) and **Level-2** covers the established subdivisions of the topics in Level-1 (e.g., Geometry of Numbers, Additive Theory, Probabilistic Theory; Finite Fields, Field Extensions, Topological Fields; Local Theory, Schemes, Curves, Surfaces; Permutation Groups, Representation Theory, Algebraic Groups).

Since the MR, and Scientific labelling schemes are close to optimal for their intended purposes, they may be adopted virtually unchanged for Σ , except for the harmonization of notations, to avoid inconsistencies / clashes / ambiguities. This might be achieved by assigning a prefix to each cipher, indicating which of the subjects (Mathematics/Physics/Chemistry/...) may be regarded as the principal source of the item concerned. Sometimes, several subjects appear to contribute significantly to certain items. In such cases, appropriate multiple prefixes may be used. **Notice that** no distinction should be drawn between 'pure' and 'applied' mathematics , since the results / techniques / constructions / ... are independent of the mode(s) in which they are employed. On the other hand, the **interpretation(s)** of any 'mathematical situation' should be made purely on the basis of 'observed collective behaviour of its constituent ME' – regardless of their 'standard individual interpretations'. Thus general structures associated originally, with particular mathematical fields may be identified (or defined) within seemingly unrelated domains. A good example of this phenomenon is furnished by the widespread use of differential - geometric techniques in Statistics. See, for instance, **Farrel, R.H, 'Multi-variate Calculation'** (Springer, 1985), where several other techniques previously unfamiliar in statistical analysis are also deployed -among them, measure theory, topological groups and multivariate ('correlation') inequalities. Other aspects of geometric methods in statistics are covered in: '**Differential Geometry and Statistics**', by **M.K. Murray and J.W. Rice** (C & H, 1993).This notion [414] of 'transferred structures' is diversely illustrated in the S/M-list. Although such ideas have been ubiquitous in mathematics over the past century (and have been used in some specialized areas) their systematics **general** recognition/exploitation remains to be accomplished – and constitutes one of the basic aims of Σ . Indeed, these structural transfers form the fundamental ingredients in many of the most interesting (inverse-) analogues considered, permitting, 'standard approximation procedures' to be introduced in a wide range of (apparently disparate) settings. In these terms, a **formulation** of the item i_m in the S/M-list utilizes the information under headings (1)-(6) to produce a succinct **representation** of i_m. The existence of multiple formulations of some i_m stems from the nonuniqueness of the above information – for instance,

[414] This, term is used in the author's paper, 'Varieties of Approximation' (pp77-94 of: **Artificial Intelligence in Mathematics**', IMA Proc. JH Johnson, et al., eds, OUP, 1994

where there are several, alternative, 'admissible versions'. In the full tagging of i_m, all versions are to be included– in arbitrary order, unless there are definite 'precedence constraints'.

(c') It is useful to view the **interpreted** full tagging of i_m as the **profile** of i_m – say, $p[i_m] \equiv p_m$. The collection [415] $\mathcal{P}_t := \{p[i_m] : 1 \leqslant m \leqslant J_t\}$ is then, in an obvious sense, a partial profile of $\mathcal{M}_t$! Of course, any $p[i_m]$ involves in its specification a subset (say, S_{i_m}) of $\mathcal{M}_t$. The evident circularity – including self-reference for recursive definition – raises no problems, provided that consistency is maintained. This remark applies to many specifications/ routines encountered in the development of Σ . The formal deduction of Mathematics from Set Theory is not at issue here; rather, it is effective/longterm organization of 'mathematical knowledge', regardless of its origin. The associated taxonomies are composed of many interdependent elements, and comprehensiveness outweighs elegance and minimality – though brevity (preserving clarity) is another important aim. **Note that** the item-profiles are essentially elaboration/extensions of the item-approach -outlines mentioned earlier. The main distinction lies in the degree of detail involved. Both forms are useful in the systematic study of approach procedures in all areas of mathematics and its applications – and the formal notation/coding the two forms reflects this situation. The **admissible formulations** of i_m may be regarded as **realizations of the approach-scheme(s)** for i_m as specified by the other entries in the item-profile. Symbolically, this may be represented as:

$$p_m =: (\pi_m)_6 = ((\pi_m)_5, \rho[(\pi_m)_5]),$$

where $(\pi_m)_6$ denotes the ordered list

$$(\Phi_m, H_m, \Gamma_m, B_m, A_m : L_m, F_m)$$

,

$$A_m : L_m := \{A_{mj}(L_m j) : 1 \leqslant j \leqslant r_m\},$$

$A_{mj}(L_{mj})$ stands for 'A_{mj} implemented at level L_{mj} (selected from $\{n, p, f\}$)' and $\rho[W] :=$ 'the set of admissible realizations of W'. Although this is purely schematic, it is still useful for organization/manipulation of item-profiles, and even for symbolic operations on such profiles! In this context it should be emphasized that the scale of the OIB, and the versatility of IT/AI software, make it crucial that mathematical data may be displayed /manipulated **multifariously**. When this synthesis is achieved, the boundaries between schematic and explicit 'information-regimes' largely disappear – and the concept of **information - flux through problem-domains** (as discussed in earlier Essays) becomes central for Σ . The **analysis** of **how** information modifies problem-domains may be modelled by systems of (in)equations, possibly raising some qualitatively new types of problems; and so on!

This notation – $p_m \equiv (\pi_m)_6$ for the (full) profile of the item i_m in the S/M-List, is used from now on; but, before examples can be treated explicitly, the nature of π_{m6}, in terms of the elements of $(\pi_m)_5$, must be clarified. Fundamentally, this is a matter of **modelling**, since an arbitrary mathematical phenomenon, β, may be realized over any framework, $\mathcal{F}$, within which the set of defining characteristics of β(say, $C_\beta \equiv \{c_\beta\}$) is **reproducible**. More generally, $\beta$is called Q -**realizable** IFF the subset Q of C_β is reproducible (over $\mathcal{F}$). Other types of **partial/approach/imperfect/ ...realization of** β are introduced as they arise (in examples), yielding cognate types of **formulation**. In particular, **full** realizations of βcorrespond to (restricted) **transcriptions**. If full/partial/approach/imperfect/ ...realization is denoted symbolically as $1/2/3/4/\ldots$ – realization, and Q_k is the subset of k-realizable elements of C_β, then , in the most general case, one may have $(Q)_l$-**realization** of β, where the subsets Q_k of C_β need not be disjoint. Here, the value of $l \equiv l_t$ depends on the number of distinct types of realizability recognized in Σ_t. The set, C_β, of defining characteristics of β, is typically 'small', but the notation allows C_β and the Q_k to be of all mutually consistent forms. These distinctions

[415]The S/M-list will eventually become a major resource, containing millions of items submitted by users. The initial form given in these Essays exemplifies the styles of 'mathematical phenomena' considered – with $J_{t_0} \approx 3000$

among types of realization/formulation would have a relatively minor impact on research outside the IB-setting; but, written in the Σ -environment, such shades of implementation are of major importance, and give rise to some seminal problems concerning 'hybrid formulations'.

Thus, with the, notational conventions introduced in E_1, the formulation(s) of i_m may be written as

$$\pi_{m6} = F_m(\pi_m)_5 = \rho[(\pi_m)_5]\langle i_m\rangle$$

To recapitulate : $p[i_m]$ denotes the **profile** of i_m, with admissible **formulation(s)** $\langle i_m\rangle \equiv \pi_{m6}$, defined in terms of the other profile-entries, listed as

$$(\Phi_m, H_m, \Gamma_m, B_m, A_m : L_m) := (\pi_m)_5.$$

Formally, π_{m6} is a **set-valued function**; for each assignment of attributes of i_m– embodied in some interpretation of $(\pi_m)_5$ – there is a corresponding realization of i_m, and the **set** of all such realizations of $i_m$is denoted by $\rho[(\pi_m)_5]$.

NOTE that, where any item in the S/M-List is 'naturally indexed as a **family**': $i_m \equiv \{i_m(\lambda) : \lambda \in \Lambda\}$, the profile notation may still be used but with $\pi_{mk} \equiv \pi_{mk}(\lambda)$, etc ...

It is evident from the intricacy of this notation, that the construction of the profile-list, $\mathcal{P}_t$, is a considerable task, since each item must be suitably located, characterized and formulated, and the main interrelations among the 'item-components' are to be indicated succinctly. Inevitably, therefore, most of the entries will be subject to regular updating/modification, as $\Sigma_t$evolves. The key observation, however, is that **such 'maintenance' embodies the essence of Σ -activity**, since it reflects **the course of mathematical research**. The requirement of accommodating extensions/variants/... of all ME within a uniform scheme makes heavy demands on the flexibility/universality of the formalism – so the '**template**' specified here is probably at least as transparent and tractable as any adequate alternative design would be.

So far, item-profiles have been regarded as **linear arrays** of components chosen from 'admissible classes of objects'. This is convenient for many IB operations (e.g., parsing, modification, substitution) but it is unsuitable for exhibiting **interrelations** among components. For this purpose, the **proof-diagrams** introduced in E_5 may be **adapted**, as follows. Recall that, in Σ , every ME is (formally) treated as an OT – so that, ever S/M-item corresponds to some function of OT **formulating** the result(s) involved. Thus, one has the (formal) representation(s): $i \approx \alpha(T_\sigma)_q$. where all of i, α, σ, q are indexed by m, T_{σ_J} is an OT selected from U_t, for $1 \leqslant j \leqslant q$, and $\alpha(T_\sigma)_q$ is constructed from compositions/Boolean-combinations of the relevant OT. In particular, part of $\alpha(T_\sigma)_q$ coincides with $\langle i\rangle$, the set of formulations of i, and each of the T_{σ_j} has some associated **proof-diagram**. Consequently, the proof-diagram(s) for $\langle i\rangle$ may be 'fabricated' from those of the T_{σ_k} – and, in this way, **interrelations** among the ME in $\langle i\rangle$ may be exhibited. For the remaining ME (i.e., those ME in $p[i]\backslash\langle i\rangle$) there are also associated proof diagrams, allowing additional interrelations (among subtheories, etc.) to be indicated. All of these features of profiles are processed through the IB graphics software. The overall profile-representation combines calculational and locational information. An extension of the format associates with each item $p[i]$ in $\mathcal{P}_t$ the sets $c_1\{i\}c_2\{i\}/c_3\{i\}$ of other profiles having strong/medium/weak connections with $p[i]$. **NOTE that** these 'connection-sets' are compiled as part of the maintenance of the OIB, with brief commentaries on the nature/justification of the various links involved. Ultimately, some of this information may be generated by AI routines– at least, in preliminary forms. The sets

$${}^V p[i] := \{p[i]\} \cup \{c_j\{i\} : j \in V \subset \bar{3} \equiv \{1, 2, 3\}\}$$

give very rough forms of '**vicinities** of $p[i]$' within $\mathcal{P}_t$ – subjective, partial, imprecise..., but **useful** in the organization of the OIB, especially for 'environments for investigation'.

Since the profiles, $p[m]$, have basic representations as ordered lists of wfe containing OT (at specified operational levels), **the OTC formalism is applicable over subsets of $\mathcal{P}_t$**. This makes it possible to produce '**explicatory versions of proofs**', in which the underlying motivation and the (faltering) evolution of the derivations/constructions are emphasized. Such versions offer vital supplements to the clinically elegant, but typically unenlightening, journal presentations. The OIB facilities make it feasible – even tractable – to include **several** versions of every nontrivial result (and of all definitions). This diversity of representations greatly increases the potential for mathematical discovery. Indeed, DBMS capabilities must be pushed to their limits to attain the performance standards demanded by Σ , and profound issues in the design of very large interactive systems arise here[416]. The comparison of interrelated subtheories in the same profile may be handled through the IB graphics package – with similarities and differences being displayed (perhaps, in different fonts).[417] Such information is embodied in the modified/amalgamated proof -diagrams – to which a range of 'admissible operations' may be applied. The sets ${}^V\mathcal{P}_t := \{{}^Vp[i] : p[i] \in \mathcal{P}_t\}$ contain the 'essential data' about the items in the S/M-List, and so constitute a fundamental part of the Σ -information superstructure. The denseness, precision, and heterogeneity of 'typical mathematical information' necessitate the development of radical/novel organizational schemes – both for efficient retrieval and for transformation/modification. Synthesis on this scale – and of such complex information – is now possible, with current IT resources. The **construction** of this synthesis **for** $\mathcal{M}_t$ is also **essential**, no matter how imperfect the early versions may be. The enormity of the challenge so presented becomes apparent when it is recognized that these Essays furnish only a (coherent, comprehensive but **near-minimal**) **foundation** for this longterm undertaking! In this light, the '**approach** apparatus', the emphasis on **operational** theorems, and the OTC - **formalism**, together characterize **milieux** within which Mathematics may be progressively developed as **a universal theory of approach procedures** (over specified classes of elements). In view of the fact that 'approach' covers all forms of 'standard approximation' – together with, abstraction, analogy and inverse analogy – it is **plausible** that Mathematics may be accommodated in this framework. The difficulties – and these are multifarious – are encountered when **rigorous justifications**, and the corresponding **overall schemes**. are to provided! It would be fruitless to pursue such an ambitious project as Σ without confronting – and adequately resolving the fundamental taxonomic/operational/calculational/ philosophical issues involved. This attempt to accomplish these aims – in a collection of essays, rather than in a formal treatise – is intended to establish a somewhat detailed **plan**, and a **style** for largescale, longterm development. In any unconventional scheme of the magnitude of Σ , there are bound to be infelicities of expression, (near-) repetitions, obscurities – and downright mistakes. Subsequent versions may (and should!) be elegant, minimal, pellucid and logically unimpeachable – but such standards are seldom attained in initial expositions of original/unfamiliar ideas. At all events, the diverse approach - outlines/profiles given in E_{10} – and the indications of (inverse-) analogues in E_{11} – amount to a substantial collection of examples/ explications of the intricately interwoven ideas/representations/operation /proofs/ constructions/... introduced and developed in $E_1 - E_9$ – on which the later (collective) phases of the Σ -project may be founded using the general **methodology/tool-kit** that has evolved during the initial, innovative, phase. With these preliminaries, it is possible to produce **basic taggings** for the items in the S/M-List, by using the 'template format' – augmented to include some **Basic Reference(s)** (BR) and **Capsule Description(s)** (CD), for each item. The BR do not, in general, correspond to the origins of the items, and no attempt is made to cover recent developments; rather, it is the **approach - characteristics** that are of central concern. Some of the more elusive aspects of **mathematical analogy** are considered in greater detail in $E_1 1$.

The writing of CD presents formidable challenges, since (mostly) they cannot be compiled from Abstracts/Reviews, because of the emphasis on 'approach-behaviour', in Σ . Moreover, extreme **brevity** is essential (as there are some 3,000 items), but each CD must still 'make sense' on its own – with obvious prerequisites/background-material. The collection of CD must constitute an 'acorn', from which the 'Σ -oak' may grow. The soil/environment, and a broad description of the oak, form the content of $E_1 - E_9$; in E_{10}/E_{11}, one variety of acorn is specified!

In the notation already introduced for 'templates of the S/M items', the entries in the aug-

[416]See, e.g., Adv. Computer Res. Vol 3 (1986)

[417]The use of colour-codes would be lost on at least one mathematician of my acquaintance!

mented S/M-List (for the item i_m) have the form:

$$T_m \rightarrow \mu_m/R_m/D_m/\Phi_m/H_m/\Gamma_m/B_m/A_m/L_m/F_m,$$

where : $T_m \rightarrow \mu_m, R_m, D_m$ denote (respectively) **T**itle → (dominant) **MAP**(s), Basic Reference(s), and Capsule Description(s), for i_m. As before, the terms $\Phi_m, \ldots, F_m$ are (generically) labelled as $(1), \ldots (7)$ in the i_m- entry. The subtheories also constitute 'environments', here; the associated sets –H_m, Γ_m – are based on the classifications in various Reviews/Abstracts publications; the precise specification of H_m, Γ_m, in Σ is assumed to be given (the detailed listings will form part of Σ_{t_0}. Consequently, only a broad **selection** of items is tagged here – as a fairly typical sample from the whole S/M-List.

This selection illustrates both the unity and the diversity of mathematics. The full S/M-List is intended to become 'cumulatively dense in $\mathcal{M}_t$, as $t \rightarrow \infty$'! The aim is to promote an 'approach attitude' to mathematical information (on a global scale) and to establish 'the operational representation in the formulation/proof of theorems'. The possible losses in elegance/brevity/clarity maybe offset by gains in effectiveness/scope of applications. At all events, the conceptually transparent/elegant versions may be retained as complementary information, since theorems should be presented in several forms within a comprehensive IB.

For a **final taxonomic representation** of i_m, the symbol m itself may be labelled – along with all of the other entries through the introduction of a 'topical notation' – say, as : $\Delta_1, \ldots, \Delta_{11}$, where the correspondence may be displayed in tabular form:

Δ_m^1	Δ_m^2	Δ_m^3	Δ_m^4	Δ_m^5	Δ_m^6	Δ_m^7	Δ_m^8	Δ_m^9	Δ_m^{10}	Δ_m^{11}
m	$T_m \rightarrow \mu_m$	R_m	D_m	Φ_m	H_m	Γ_m	B_m	A_m	L_m	F_m

and the Δ_m^j constitute **descriptors**. Thus, the m-th item in the S/M list has 11 descriptors – of which the first is the symbol for 'the m-th topic', and the last is the set of its formulations over Σ . For purposes of parsing, searching, transforming... information in the OIB, such uniform representations are essential – and are used for all of the 'fundamental data-sets' in Σ .

NOTE. The partial descriptors μ_m may be represented as **lists** of numbers (in increasing order) of the MAP(s) primarily illustrated by i_m. The 'separator' ('→', here) allows separate **parsing** of T_m and μ_m, etc. ; and modification, as 'new interconnections of MAPs' are discovered.

Thus: the S/M-List is essentially equivalent to

$$\mathcal{D}_t \equiv ((\Delta_m)^1 1 : 1 \leqslant m \leqslant J_t)$$

, the (ordered) list of 'fully-interpreted profile - taggings'. It has been observed that the OTC-operations may be applied to the elements of $\mathcal{D}_t$; the resulting 'modified S/M-Lists' correspond to the (designated) 'generic list'. This raises no problems of principle, but these modifications must be monitored within the OIB, to determine precisely the form(s) of 'i_m' involved in any particular calculation. All of this is in line with the 'dynamic nature of Σ ', where most 'elements' are approachable within (suitably defined) classes of other elements. An important longterm aim for Σ is to render all such approach - interrelationships accessible to users of the system, in the course of investigations. This will enhance the scope and power of mathematical research profoundly.

The tagging may be based on the (revised) 1980 Mathematics Subject Classification (1985)[418], even though major changes have been made (1990, and later). The methodology is independent of the detailed forms of (mathematical/scientific) classifications used. In any case, the cognate listings for all of 'The Sciences' are barely hinted at – for (inverse) analogies – here, and in E_{11}, since a proper treatment of these aspects must await the implementation of Σ_{t_0}, and the overwhelming emphasis in these Essays is on **mathematical** structures and their interrelations. Therefore only the principal scientific subfields (at most) are specified, so that the scope of analogues may be assessed. even for mathematical properties, attention is confined (mainly) to levels 1/3 of MSC-85.

The designation of mathematical information /ME as **nil/ partially/ fully - operational** requires further explanation now, since this is a key ingredient in the tagging of items. The

[418]The author happens to have this version (denoted, from now on, as MSC-85).

simplest characterization is thatL
e **is nil-operational** means that e is purely descriptive;
e **is fully operational** means that all procedures/processes entailed by e may be effectively realized/applied over all appropriate classes of ME (for which e is an admissible operator);
e **is partially operational** in all other cases.

Plainly, this is a very crude specification and, as Σ evolves, it will be refined greatly, to allow for detailed comparisons of 'operationality', etc.. Some possible extensions of this kind have been outlined (especially, in E_4), but many subtle issues remain to be addressed before an adequate scheme is developed.

In forming a broadly representative collection of profiles of S/M - items, the principal guiding criteria adopted are as follows.

(i) Nontriviality

(ii) Existing level(s) of rigour

(iii) Scope/interest

(iv) Unfamiliarity / Novelty

(v) Diversity

(vi) Heuristic Potential

(vii) Unification

Although the assessment of such attributes is unavoidably subjective, their **centrality** will be generally acknowledged. Again, the comparative smallness of the number of profiles considered is balanced by giving **a complete list of the item-titles/key-references** identified, so far, as embodying phenomena of essential relevance for Σ . These qualities of relevance '**generate**' (as aims):

(a) the exhibition of ME in t- **max – operational forms** (as varied as possible);

(b) the exhibition of ME in all admissible **approach-contexts**;

(c) the estimation of **'degrees of operationality/approach'**;

(d) the identification of (primarily) **approximative items** (for E_{10}), and, of (primarily) **analogical/ inverse-analogical items** (for E_{11}).

In order to achieve these aims, a 'standard format' must be established, and deployed uniformly. The eventual incorporation of taxonomies will allow S/M-List to be **scanned**, through the use of sophisticated search-software, on the basis of a comprehensive tagging scheme. For the (general-) **approximative** material in E_{10}, aims (a), (b), (c) are substantially achievable – with explicit estimates, in many cases. By contrast the (inverse-) **analogical** items considered in E_{11} are far less susceptible of precise analysis, since the interrelations are mainly qualitative in character. The overall classification must devolve on a 'small collection' of properties/features; for instance, as follows.

size (finite/infinite, bounded/unbounded);

cardinality (finite/countable/uncountable, discrete/continuum);

functional form(s) (linear/nonlinear, continuous/discontinuous,
differentiable/non-differentiable, . . .);

algebraic properties ((non)commutative, (non)associative, (non)distributive, . . .);

algebraic structure (nil/group/ring/field/. . .);

representation (combinatorial/ algebraic/ analytical/ topological/ geometrical/ . . .);

analytical structure (space/manifold/ . . .);

topological structure (homotopy/(co)homology/. . .);

stochastic structure (deterministic/measure-space/. . .)

calculational structure (semi-numerical/algebraic/analytical/topological/ combinatorial/. . .)

All of these '**basic properties**' must be labelled; after which, '**derived properties**' (obtained from admissible operations on w.d. combinations of the basic properties) may also be suitably to reflect their forms of derivation; and so on! On this basis, classes of (inverse-) analogues may be specified for appropriate collections of ME, with cognate analyses of validity/strength/. . . and 'heuristic implications' – broadly interpreted. Before the selection from the S/M-List is indicated it is worthwhile to comment briefly on the main selection-criteria.

(i)' Non triviality

This is a matter of depth,substance, subtlety and so on in relation to the underlying (sub)theories for each item considered. Although quantitative assessments may be made (e.g., based on the lengths, complexities of definitions, derivations, constructions/. . . within these theories) the intuitive judgements of list-compilers are probably more reliable (at least in the phases of Σ covered by these Essays).

(ii)' Existing Levels of Rigour

For the (general-) **approximative** aspects of approach, the most stimulating results correspond to **highly-developed/sophisticated (sub)theories**, with rich mathematical structures. For **analogues,** and **abstractions**, the distinctions are somewhat less marked: 'simple ME' may yield important analogues/abstractions. The situation for **inverse** analogues, however, is quite different. Highly-developed **scientific** theories often produce **direct mathematical** analogues; but the most significant **inverse** analogues tend to stem from **vaguely-formulated/partially-empirical theories** whose fundamental concepts are ostensibly observational or experimental. Once again, **the key distinction** between mathematical models and inverse analogues must be drawn.

(iii)' Scope/Interest

By 'scope; one understands 'capacity for (in)direct influence' – within some framework(s); assessments of this quality are largely retrospective, but general indicators may also be identified. In particular, an ME may be definable, in various forms, over several, apparently disparate, (sub)theories; then 'the essential notion embodied by the ME' is recognized as 'possessing wide scope'. Equally, an ME admitting only a single type of definition may, nevertheless, suggest manifold developments in different contexts: another sign of broad scope. Other forms of scope will be exhibited in many of the S/M items.

The attribute of 'interest' is essentially equivalent to **stimulus** (e.g., through suggestiveness/provocation/ quasi-paradox). As with most of the properties considered here, it cannot be defined unambiguously (still less, quantified); but its importance in framing selection -criteria is undeniable.

(iv)' Unfamiliarity/Novelty

Although every S/M-item will be potentially familiar to some nonempty collection of readers[419] – at least, as a basic **result** – many of the **interpretations/ elaborations/ (inverse)analogues/ ...** envisaged here may well be 'new', in published form. Moreover, the enormous range of topics covered virtually guarantees that 'most' readers will be unaware of 'many' results in the list. It is hoped that the selection of comparatively unobvious/ uncommon examples will enhance the pedagogical value of these Essays by prompting readers to formulate their own examples – to be 'developed along Σ -lines.

(v)' Diversity

The introduction of $\mathcal{M}_t$, over Σ_t, as a)very large-scale) deductive/calculational system – cumulatively asymptotic to $\mathcal{M}_t^* \equiv$ 'Mathematics at time t' – entails that **all** parts of ('pure' and 'applied') mathematics are to be covered. Consequently, a wide range of examples is essential here. As a basic criterion, it could be required that all of the main MR categories are included; further, all of the principal **sub**categories (as Σ evolves), etc..

(vi)' Heuristic Potential

One of the central aims of Σ is to promote the use of 'quasi-formalized heuristics', in all areas of mathematics. The OIB framework is especially opposite to such activity, since its facilitates the combination of procedures from loosely interrelated fields – on a speculative basis – in the preliminary stages (at least) of investigations. The identification of 'ME of high heuristic potential' is, plainly, subjective, but not entirely so; indeed, the processes of 'discovery' are now widely studied[420] as a branch of the 'Theory of Knowledge', involving philosophy, psychology and AI. Nevertheless, the intuitive grasp of heuristic potential in mathematics is comparable with the (strong!) chess-player's recognition of a few 'promising lines' among the enormous collection of admissible continuations.

(vii)' Unification

The concept of **approach**, subsuming approximation (of all standard types), abstraction and (inverse analogy) is central in Σ , as attested by $E_1 - E_9$. The potential calculability of arbitrary ME – within an approach framework – **unifies** mathematics remarkably, without blurring the boundaries between precise calculation, rigorous estimation, and qualitative assessment. **The most stimulating results are compounded of heuristic assumptions, exploratory approaches, and (mutually) convergent sequence/ families of upper/lower bounds – over 'spaces admitting suitable order relations'**. Such spaces may be multifarious species/types – for which the underlying solution-scheme is universal. Many of the S/M-items are chosen to emphasize this underlying pattern. **NOTE that** the upper/lower-bound limits need to coincide – so, both equations and **in**equations are covered here.

Paths to 'Mathematical Progress'

Before the S/M -examples are discussed it is appropriate to consider (in general terms) the substantially distinct species of 'paths to mathematical discovery' that may be identified within the evolution of mathematics and its applications. In spite of the vastness of this canvass, a surprisingly small number of 'basic discovery-paths' may be characterized – potentially adequate to accommodate arbitrarily chosen results from 'the literature'. This is, perhaps, less remarkable than it may initially appear since the ideas involved furnish only possible 'plans of attack' –

[419] in particular, to the author(s) of the reference(s) cited

[420] See, e.g., '**Scientific Discovery**', by P.Langley/H-Simon/G-L Bradshaw/ JM lytkow (MIT Press, 1987), where many references are given.

without offering clues as to their successful realization in mathematical practice! Some of these 'research strategies' have been diversely explicated – especially in pedagogical contexts; but the emphasis here is, rather on (a) **technically intractable results**, (b) **results extending the frontier of established fields,** and (c) **results fundamental in the creation of radically new mathematical fields.**

There is, of course, some overlap among the classes (a), (b), (c); but the specifications given are useful in producing the item-profiles – as well as, for taxonomic purposes. in any case, classifications based on rigorous **partitions** of data (into strictly disjoint subsets) do not accurately reflect the many-faceted **interplay** among the ingredients of mathematical proofs. This is one of the principal limitations of the automated theorem-proving packages developed up to now. Attempts, in AI schemes, to mitigate these faults have been (at best) only partially successful. The interactive/OIB structure of Σ , establishing a symbiotic relationship between the user and the system, furnishes an optimal environment for partner ship: the user's imaginative scope is unfettered – and may even be enlarged through the supply of (often, unsuspected, but) highly suggestive results, and the capacity to apply them to the data 'generated through interactive calculation'. Although the selection of these results is governed by the cumulative state of Σ_t(through 'environments for investigation' – **see** E_8), judgements about the relevance/form/aesthetics/..., of all wfe obtained, remain absolutely the user's prerogative: nothing is lost, but much may be gained.

Strategies for Class (a)

The **intractability** of calculations may take several forms, for instance:

a1: problems of **definition/interpretation**

a2: problems of **size**, for **arithmetical computations**

a3: uncontrolled **numbers/complexities of terms** in (ostensibly) evaluable wfe

a4: lack of procedures for (some) **local operations**

a5: lack of procedures for (some) **global operations**

a6: problems of **interpretation, representations and estimation** for (some) **partial evaluations**

The associated **basic strategies** may be outlined as follows

S$_{a1}$: Formation of all approach structures compatible with the given one – to explore possible variants of the intractable calculation, and to seek alternative realizations.

S$_{a2}$: Conversion of all indicated arithmetical operations to conform with the format for the Σ -processor, P_Σ.

S$_{a3}$: Use of the basic simplification/reduction/ transformation facilities in P_Σ to obtain optimal forms of individual terms for further use of (more advanced) P_Σ – functions.

S$_{a4}$: Identification of all models/structures over which the original (local) calculations are w.d. – with concomitant (partial) evaluation schemes

S$_{a5}$: Modification (if necessary) of the selection 'local-model calculations' to produce realizations as compositions of suitably transformed OT over Σ .

S$_{a6}$:
S$_{a7}$: } Similar to S$_{a4}$/S$_{a5}$ – but now, for **global** calculations (e.g., summation, integration, 'over-aging', global (analytic) continuation.

Strategies for Class (b)

Here, the 'established fields' (typically, $\mathcal{F}$ are essentially subtheories of accepted theories, $\mathcal{J}$, whose (consistent) axiom - sets have been specified – possibly, in several (equivalent) forms. Extensions of the frontier of $\mathcal{J}$ often result from cognate extensions for $\mathcal{J}$, but there are exceptions to this pattern.

It is supposed that $\mathcal{J} =: \mathcal{J}_\Phi$ (and so, also, $\mathcal{F}_\Phi$) is 'basically realized over some framework Φ – to which other frameworks, say Φ', may be equivalent ($\approx$), or else (weakly) **in**equivalent ($\not\approx$). Similarly, other theories $\mathcal{J}^*_\Phi$, may be equivalent, or (weakly) **in**equivalent to $\mathcal{J}_\Phi$ (with corresponding conditions for $\mathcal{F}^*_\Phi$, and $\mathcal{F}_\Phi$), etc... In the present context, the terms 'near-equivalent' and 'weakly-inequivalent' are regarded as synonymous. Distinct frameworks may (but need not) be equivalent. The fundamental domain of validity of $\mathcal{J}_\Phi$ is denoted by $\Delta(\mathcal{J}_\Phi)$, etc ...

S_{b1}: **Equivalent realization** over equivalent frameworks, $\Phi'(\approx \Phi)$

S_{b2}: **Equivalent realization** over **in**equivalent frameworks, $\Phi'(\not\approx \Phi)$

S_{b3}: **Removal** of some axiom(s) of $\mathcal{J}_\Phi$

S_{b4}: **Introduction** of some new axioms into $\mathcal{J}_\Phi$

S_{b5}: **Perturbation** of some axiom(s) of $\mathcal{J}_\Phi$

S_{b6}: **Weakly-equivalent realization** over $\Phi' \not\approx \Phi$

S_{b7}: **Weakly-equivalent realization** over $\Phi' \not\approx \Phi$

S_{b8}: **Characterization** of (sub)theories $\not\approx \mathcal{J}_\Phi/\mathcal{F}_\Phi$

NOTES on the S$_{bi}$

(b1)' This is a form of transcription – often of importance when calculations in $\mathcal{J}_\Phi$ seen to lack context/motivation, while the (equivalent) corresponding calculations in $\mathcal{J}_{\Phi'}$ are transparent/tractable. This amounts to a 'quasi-trivial extension of frontier' for $\mathcal{F}_\Phi$. The term 'virtual extension' may also be appropriate.

(b2)' Here, the extension is certainly 'real' – but **basic**; the results are unchanged, but their range of applicability is enlarged.

(b3)' This is one type of 'elementary generalization'. The theories $\oplus'$ obtained by dropping some axiom(s) of $\mathcal{J}_\Phi$ (at time t) may, over the period $[t, t_1]$, have 'no interesting realizations' – apart from specializations to variants of $\mathcal{J}_\Phi$! Even so, this process does yield)at least, formal) extensions of $\mathcal{J}_\Phi$ – with the potential for important development.

(b4)' Since this is one form of **specialization**, the only possible extensions devolve on the formulation/proof of **sharper results** (including **constructions and estimates**). This may be viewed as 'interior extension' of the frontier of $\mathcal{J}_\Phi$

(b5)' Questions of quasi-**stability** arise for this mode of extension, e.g.: if some axiom(s) were modified 'within small neighbourhoods of their $\mathcal{J}_\Phi$ forms' – to produce $\mathcal{J}^*_{\Phi'}$ (say) – would $\mathcal{J}_\Phi$ necessarily be derivable from $\mathcal{J}^*_{\Phi'}$? If so, $\mathcal{J}_\Phi$is **axiom-stable**. More generally, $\mathcal{J}_\Phi$is **strongly stable** IFF $\mathcal{J}_\Phi$is derivable from **all** theories $\mathcal{J}_\Phi{}^*$ obtained under 'sufficiently small perturbations of some ingredient(s) of $\mathcal{J}_{\Phi'}$ – and $\mathcal{J}_\Phi$is **weakly stable** IFF this condition holds only for **some** (not all) $\mathcal{J}^*_\Phi$

(b6)' One defines $\mathcal{J}^*_{\Phi'}$ to be **weakly-equivalent** to $\mathcal{J}_\Phi$IFF **some** approach-path from $\mathcal{J}^*_\Phi$, to $\mathcal{J}_\Phi$ **exists** (and, **constructively - weakly - equivalent**, when the approach-path maybe **constructed** – over the relevant environment). Here, $\Phi' \approx \Phi$.

(b7)' This is similar to (b6)', but with $\Phi' \not\approx \Phi$.

(b8)' The **characterization** of $\mathcal{J}^*_{\Phi'}$ with $\mathcal{J}^*_\Phi \not\approx \mathcal{J}_\Phi$ is important for the study of 'limitation/impossibility results'.

NOTE that weak equivalence is also, formally, a type of **in**equivalence from which equivalence may be restored, along **some** approach-path(s). It is therefore useful to introduce the notion of 'strong **in**equivalence', from which **no** combination of approach procedures can restore equivalence. In such cases there must be 'generic differences' between $\mathcal{J}^*_\Phi$, and $\mathcal{J}_\Phi$.

Strategies for Class (c)

The creation/development of a 'radically new mathematical field' is, evidently, a rare event! Consequently, no remotely effective 'prescriptions' are to be expected here. It is, rather, a matter of designing OIB environments to enhance the exploration of 'unusual mathematical ideas' – against a rich background of accepted theories, with detailed operational content. This basic aim may be broadly pursued by outlining a selection of radical innovations/discoveries, and considering the nature of the break-through in each case. From this process, some vague (but not vacuous!) principles may be distilled. **Notice that** there is no conflict here between 'discovery' and 'creation'; that is an issue for the Philosophy of Mathematics. As always the Σ -view is primarily heuristic, with rigour and operationality as distant goals.

Even at this level, some distinctions (e.g., over motivation/scope) should be drawn. The principal attributes of such novel researches include:

(i) the recognition of new patterns in familiar structures;[421]

(ii) the introduction of new schemes of calculation in technically sophisticated structures;[422]

(iii) the development of calculational procedures in structures 'apparently resistant to quantitative analysis';[423]

(iv) the characterization of substantially new areas of study – and, **the formulation/use of techniques** for solving some of the associated '**key problems**'.[424]

Many examples of types (i)-(iv) may be found in the S/M-List; the few cited here are 'chosen at random' to indicate the basic ideas involved. Although only a small proportion of these examples embodies 'fundamentally **revolutionary** extensions of Mathematics', the cumulative effect of such research is to **to enlarge both the scope and the depth of viable calculation** – to degrees not attainable through strategies of types $(a)/(b)$. It is worth repeating that these broadly- defined species of 'research strategies' do not determine **strictly** disjoint classes – but, that the significant overlaps are marginal/small. For purposes of OIB - design, this scheme is very convenient, since it allows 'new results' to be listed either, in the standard MR format, or else (exceptionally) as 'of type(s) $(a)/(b)/(c)$;. Moreover, 'transitions' from 'standard' to 'exceptional' may be accommodated within this set-up – in the sense that some new discovery may transform the status of certain structures, etc., by radically changing the range of admissible/feasible calculations. it could be claimed that **the introduction and regular improvement of SAM packages** has produced quasi-transitions in several mathematical fields with potentially high symbolic-computational content. Indeed without (e.g.) **Axiom**, the conceptual scheme of Σ could only be formal/ futuristic – though **the 'approach orientation', and its associated mathematical frameworks, would still be highly significant, in promoting the (re)formulation of theorems in operational modes.**

[421] E.g., 'Feigenbaum universality' for iterated maps; 'elementary catastrophes

[422] E.g., 'inverse-scattering procedures for nonlinear waves/PDE's; 're-normalization schemes' for phase transitions/critical phenomena.

[423] E.g., 'Communication-complexity' (circuit-depth analysis)

[424] E.g., 'Classification-theory of models' (Set Theory

It should be emphasized that, although there are (so far) 63 MAP **types**, most of the S/M-items have been selected to exemplify/extend some particular MAP(s) – often, just one MAP. The degrees of relevance of any item to MAPs other than the one(s) of principal interest will vary from 'almost nil' to 'high', but such possible links are not mentioned here. Rather, it is hoped that these peripheral interconnections will be elaborated by future users of Σ , to increase the range/power of the OIB as a research environment. This situation will be maintained, as the S/M-list is enlarged, even when new MAP types are introduced. consequently, the sets μ_m (for the descriptor $\Delta^2_m \equiv T_m \to \mu_m$) will be 'typically small' – often, singletons – illustrating a fundamental aspect of the MAP(s) concerned (e.g., the possibility of realizing (partially) operational form(s)).

Style of the CD

For parsing/reference purposes, some uniformity must be imposed on the forms of the CD, but - in view of the extreme heterogeneity of the S/M-topics – only broad guidelines are feasible. Although brevity is essential, independent comprehensibility is also of great importance. Hence, the CD should be 'mildly' discursive, while encapsulating the fundamental results for which items were selected – with extra references for more general related results. (the basic references are listed separately in the S/M-profiles). A degree of mathematical literacy/maturity is assumed. The 'prerequisites for understanding any S/M item' may be indicated[425], so that users may study each topic suitably prepared! Since the levels of abstruseness/profundity vary widely among the S/M topics, the lengths/complexities of the corresponding CD reflect this variation.

The following sample CD (say, γ_k) correspond to the items i_k on the S/M List.

1. The **Schubert calculus** covers **enumeration problems** for **geometrical configurations** in projective n-space, P^n (especially, for $n = 2, 3$). This exemplifies CALCULABILITY in some nontrivial settings. The extensions – to **topological/algebraic intersection theory**, for varieties/manifolds – are of fundamental importance in (effective) **algebraic geometry**.[426] The possible SAM routines greatly enlarge the scope of such calculations.

The primary technique is **specialization** (of configurations), combined with the **principle of conservation of the number of solutions** (over classes of problems 'continuously equivalent to a prescribed specialized problem). The basic space here is the '**Grassmann manifold**', $G_{d,n}$, of dimension $(d+1)(n-d)$, whose points represent classes of d-planes in P^n.

The following results are known (from algebraic topology)[427]:

(i) The cohomology groups $H^i(G'_{d,n}\mathbb{Z}$ vanish, unless $0 \leqslant i \leqslant 2(d+1)(n-d)$.

(ii) $H^*(G_{d,n};\mathbb{Z} := \bigoplus_i H^i(G_{d,n};\mathbb{Z})$ is an oriented, graded ring under the 'cup product' – so that

(iii) $H^{2(d+1)(n-d)}(G_{d,n};\mathbb{Z}) \overset{\varphi}{\approx} \mathbb{Z}$ (with $\varphi(u) \equiv$ deg(u)).

(iv) There is a surjection, θ, from H^* to a set V of subvarieties of $G_{d,n}$, defined by finite systems of polynomial equations.

(v) The cohomology classes ('**Schubert cycles**') Γ_Ω, of the '**Schuberts subvarieties**', $\Omega(A_0, \ldots, A_d)$, defined by: $\Omega(A_0, \ldots, A_d) := \{L : L$ is a $d-$ plane $\wedge$ dim$(L \cap A_i) \geqslant i\}$ are **fully characterized** by the $a_i \equiv$ dim(A_i).

(vi) The $\Gamma_{\Omega(A_0,\ldots,A_d}$ are **unchanged** under **continuous variation** of the $\Omega(A_0, \ldots, A_d)$. Hence:

---

425 This idea was suggested by Pager (op. cit.) for the communication of proofs among users.

426 The structure outlined here is given in the survey by S.L Kleiman/D.Laksov (Amer.Math.Monthly 79(1972)1061-1082)

427 See, e.g., Hodge / Pedoe Vols. 1,2; Greenberg / Harper; Griffiths / Harris; Spivak (Vol V)

(vii) All enumeration - problems whose configurations are continuously interchangeable have the same number of solutions (counted according to multiplicities, etc). This is a rigorous version of the '**conservation principle**', on which most of the classical calculations were based.

2. Between-ness/Separation Groups

(i) Linear order corresponds to **directed lines; between-ness,** to **UNdirected lines; separation** (of point pairs), to **undirected circles**; and, **cyclic order**, to **directed circles**.

(ii) The aim is to generalize the (linearly) ordered groups to group structures for the other types of ORDERABILITY, through **axiomatizations** of (α) a **ternary between-ness relation**, $B[a,b,c]$, over a base-set L - subject to the **group conditions**: $(\forall g, h \in L)\ B[a,b,c] \Rightarrow B[gah, gbh, gch]$; ($\beta$) a **quaternary separation relation**, $S[a,b,c,d]$, over L - subject to the **group condition**: $(\forall g, h \in L) S[a,b,c,d] \Rightarrow S[gah, gbh, gch, gdh]$.

(iii) B/S-relations may be **induced** by linear orders, $<$, in ways determined by admissible axiomatizations of B, S, and of **transitivity relations**. Some S-relations induce B-relations; and conversely.

(iv) Diverse extensions of these ideas and cognate developments for
semi-groups, **rings**, **fields**, and **lattices**
are surveyed in the monograph: '**Partially Ordered Algebraic Systems**', by **L.Finchs** (pp229, Pergamon, 1963) For Semigroups, between-ness and order are studied in some detail by **Gilder**, J. PCPS 61 (1965)13-28

3. Constructive (D)AC

This may be formulated as a **matrix - iterative procedure** (with matrix M, say) for the continuation of 'analytic elements' $K(z) \approx \{a_i(z)\} \equiv \underline{a}(z) : \mathbb{C}^p \supset W \to B$, from z_0 to z_l along a parametrized path. Here, B may be a complex Banach space – or, in principle, any class of space for which suitable convergence criteria w.d. .Although this basic ('Weierstrass') procedure cannot be realized in terms of limits of rational functions of the coefficients in $K(z_0)$ – to obtain $K(z_l)$ – a **constructive** scheme **may** be (nontrivially!) obtained through the sequences $\{\underline{a}^{(q)}_{n\gamma^q}\}$, where $\underline{a}^{(q)}_{n\gamma^q} = M_{n\gamma^q, n\gamma^{q-1}} \underline{a}^{(q-1)}_{n\gamma^{q-1}}$, for $1 \leqslant q \leqslant l, 0 < \gamma \leqslant$, such that, for $0 < \gamma < \gamma_0$, $\lim_{\substack{n\to\infty \\ n\gamma^l \in \mathbb{P}}} \|\underline{a}^{(l)}_{n\gamma^l} - \underline{a}_{n\gamma^l}(z_l)\|_B = 0$, with $\mathbb{P} \equiv$ the set of positive integers, γ_0 fixed. **NOTE that** the proof would appear to hold, with only minor modifications, both for $p = 2, 3, \ldots$, and for $B \equiv$ complex BS. Further generalizations would require more radical changes in the basic proof.

4(a)(i) Positive-definite (pd) functions occur in many branches of mathematics – for instance, in Probability Theory, Operator Theory, Moment Problems, Complex Function Theory, Embedding Problems, Integral Equations, Group Representation Theory. The definitions cover locally compact groups, locally convex TVS, and classes of semigroups (where pd functions provide generalizations of Laplace/Fourier Transforms).

(ii) For a (semi)group G, the function

$$f : G \to \mathbb{C} \in P(G) \text{ IFF } (\forall n \in \mathbb{P}, \forall (x)_n \subset G, \forall (c)_n \subset \mathbb{C}) \sum_{i,j=1}^{n} f(x_i^{-1} x_j) c_i \bar{C}_j \geqslant 0$$

(iii) If $\bar{f} \equiv$ complex conjugate of f, $\tilde{f}(x) := f(x^{-1}), p_1, p_2 \geqslant 0$ then, for $f_1, f_2 \in P(G)$ all of: $\bar{f}_1, \tilde{f}_1, \mathbb{R}f_1, |f_1|^2, f_1 f_2, p_1 f_1 + p_2 f_2$ are in $P(G)$

(iv) $P(G)$ is closed under (pointwise) **net** - convergence

(v) With each $f \in P(G)$ there is associated a **H**ilbert **S**pace, $H(f)$, having several basic group - representation properties

(vi) Fourier-Stieltjes transforms over $M^+(\Gamma)$ – the set of positive measures on the Character Group, Γ, of G- are **continuous -** pd; and **conversely**. ('Bochner's Theorem')

(vii) **If** $f \in P(G)$ **then** $M \equiv (m_{ij})$ **is a pd matrix** where $m_{ij} := f(x_i^{-1} x_j)$; but the sufficient

(/necessary) conditions for M to be pd cannot be directly extended from 'the standard Linear Algebra cases'.
(viii) The class of **hermitian functions** ($f(x^{-1}) \equiv f(x)$) such that $AA * f \in P(G)$, for A in a special subclass of the algebra, A, of **shift operators**, comprises the **definitizable functions** on G.

4(b)(i) The notion of **approximate factorization** in a commutative (e.g. Banach) algebras, A, devolves on the existence of (bounded) **approximate** (left/right) **identities** in A – for an A-module, X – defined as (bounded) nets, $\{e(\lambda)\}_{\lambda \in \Lambda}$, such that $e(\lambda)x \xrightarrow{\lambda} x$ $(x \in X)$
Notation: $e \simeq {}^{\prime}{}_{X;A}$
(ii) Theorem: $\{e \simeq 1_{X;A}, z \in X, \delta > 0\} \Rightarrow (\exists y \in X)(\exists a \in A) : z = ay \wedge \|z - y\| \leqslant \delta$
(iii) The proof exhibits a in the (constructible form):

$$a = \sum_{k=1}^{\infty} \gamma(1-\gamma)^{k-1} e(\lambda_k)$$

for a sequence $\{\lambda_k\} \subset \Lambda$. By taking **partial sums** (say, of order M) one obtains elements a_M such that $z \simeq a_M y$ Variants of this scheme, for other types of algebras/structures, are readily envisaged.

5. The notions of (i) **exchangeability**, (ii) **indiscernability**, (iii) **rearrangement-invariance** may be defined very generally – for functions, $f(x)_k$, with arguments selected from a sequence, $\{x_m\}$. For (i), $f(x_m)_k = f(x)_k$ $(\forall k, \forall (x_m)_k$ with the x_{m_i} distinct);
for(ii), $f(x_m)_k = f(x)_k$ $(\forall k, \forall \nearrow (m)_k)$;
for (iii), $f((\epsilon x_m))_k = f(x)_k, (\epsilon_i^2 = 1$, with the x_{m_i} distinct).
NOTE. These terms are not standardized; in probability literature, a basic form of **permutability** is specified by $f(\sigma(x)_k) = f(x)_k$, with $\sigma(x)_k$ any permutation of $(x)_k$.

An important application involves '**almost-sure limit theorems**' for sequences of **i**ndependent, **i**dentically-**d**istributed **r**andom **v**ariables – say, $\{X_m\}$, for which measures $\Pi \in P(\mathbb{R})$ are w.d.

If $\mathcal{P}(\mathbb{R}) \equiv$ space of probability measures on $\mathbb{R}$, and, for $\Pi \in \mathcal{P}(\mathbb{R})$,

$$\alpha(\Pi) := \begin{cases} +\infty, & \text{if } \langle |x| \rangle_\Pi = +\infty \\ \langle x \rangle_\Pi, & \textit{otherwise} \end{cases}$$

$$\beta(\Pi) := \langle (x - \alpha(\Pi))^2 \rangle_\Pi$$

then **statutes**[428] may be defined as measurable subsets of $P(\mathbb{R} \times \mathbb{R}^\infty$ such that $(\#)(\pi \in P(\mathbb{R}) \wedge \{X_m\} \in i.i.d.(\pi)) \Rightarrow (\pi, \{X_m(\omega)\}) \in A$, a.s.. In these terms, it may be shown that: **If** $\{X_m\}$ is a 'concentrated sequence' of v.v.[429] defined on the probability space (Ω, F, P) **then** there exists a measurable function $\mu : (\Omega, F, P) \to \mathcal{P}(\mathbb{R})$, and a subsequence, $\{x^\mu_{m_i}\} \subset \{X_m\}$, all of whose subsequences, $\{X_{n_j}\} \subset \{X^\mu_{m_i}\}$, satisfy $(\mu(\omega), \{x_{n_j}\}) \in A$ a.s. The proof entails the specification of $\{X_{n_j}\}$ 'suitably close to a **permutable** sequence, $\{Z_j\}$'

6. The development of 'routines' for the **differentiation of algebraic functions** is essential for the implementation of procedures in Algebraic Geometry. A basic discussion of this problem is given by Van der Waerden ('**Modern Algebra**', 2nd ed. (transl.), Ungar, 1949/1950, Vol I §65), for algebraic functions of one generic variable, over a field k. Fairly simple modifications of that proof yield schemes covering **differentiation of rational functions of m algebraic functions in n variables, over k.**

The essential difficulty stems from the **multivaluedness** of algebraic functions; the definition(s) formulated for **generic** sets of variables must be suitably modified for **specialization** (to particular points). The formulae for generic variables are based on the **minimum polynomial** (say, F) defining $Z(X)_n$ through the condition $F(Z, (X)_n) = 0$.

---

[428] This term is used by DJ Aldons (BAMS 83, 121-3, 1977).

[429] The usual term is **tight**: for a metric space, M, the measure μ is tight IFF $\forall \epsilon > 0 \exists$ compact $K \subset M : \mu(M/K) < \epsilon$. For Borel probability measures, of for sequences of d.f., corresponding conditions hold. Many related topics are discussed in 'weak convergence' and empirical processes (Springer, 1996, pp 508) by A.v.d Vaart/JA Wellner

The corresponding **representation** is $z_{x_i} := F_{x_i}/F_z$, is separable (as a polynomial in z). Here, F is indeterminate up to a nonzero factor independent of z. If, now, $z \equiv R(z(x)_m)_n$, where R is **rational**, then the generalization of this process yields:

$$z_{x_i} = \sum_{1 \leqslant h \leqslant m} R_{z_h} z_{h_{x_i}}$$

with cognate forms of the implicit function theorem. The notion of a **specialization** (over the ground field k) of $((x)_n, (z)_m)$ to $((x')_m, (z')_n)$ is **defined** to be **equivalent** to the **admissibility** of the z'_i as **values** of the z_i for the **arguments** x'_j. If $(z(x)_n)_m$ is a given collection of algebraic functions, and $\{F(X,Z)\}_r$ generates the **the prime ideal** in $k[X,Z]$ with generic zero (x,z), then **the derivative** $z_{i_{x_j}}$ **is w.d. at** $(x',z') \equiv ((x')_n, (z')_m)$, **provided that the Jacobian matrix** $(\frac{\partial F_i}{\partial Z_j})$ **has rank** m. All further manipulations for the study of algebraic varieties may be developed on this basis.

8. The basic notions of ERGODICITY stem from **phase-space distributions** over constant-energy surfaces in Statistical Mechanics. Cognate properties for **finite-order Markov chains** are well known. **Extensions to classes of countable chains may be made** – in terms of **infinite stochastic matrices**, viewed as **linear transformations** of the BS, m, of bounded sequences (x_i) with $\|x\| := \sup_i |x_i|$; or, equivalently, of the space of stochastic matrices, under multiplication by a fixed stochastic matrix.

For any set X let $F \equiv (f_\lambda : X \to X, \lambda \in \Lambda)$;
$F_t := (of_i)_t$ ($t \in \mathbb{P}$, repetitions allowed, $(i)_t \subset \mathbb{P}$);
$F_{k,n} := \underset{k \leqslant i \leqslant k+n-1}{0} f_i$ (for $\Lambda = \mathbb{P}$);
$l_{t,n}(F \cup G) :=$ Max. no. of F-blocks of length n in $(F \cup G)_n$, for $F, G : X \to X$;
$\delta(A) := \sup \sup_{i_1, i_2 \{n'\} \subset \mathbb{P}} \sum_{j \in \{n'\}} (a_{i_1 j} - a_{i_2 j})$;
$= \sup_{i_1, i_2} \sum^+ (a_{i_1 j} - a_{i_2 j}), \sum^+ \equiv \sum_{\{a_{i_1 j} \geqslant a_{i_2 j}\}}$
$\gamma(A) := \inf_{i_1, i_2} \Sigma \min (a_{i_1 j}, a_{i_2 j})$
A stochastic matrix is defined to be **stable** iff all of its rows are identical.

Here, F, G are (mostly) sets of (stochastic) matrices. F is **weakly uniformly ergodic** iff $\forall \epsilon > 0 \exists t_0(\epsilon) : \forall F_t : t \geqslant 0, \delta(F_t) \leqslant \epsilon$. F is **weakly uniformly stable** relative to G iff $\forall \epsilon > 0 \exists l_0(\epsilon) \exists t \geqslant ' : \delta((F \cup G)_n) \leqslant \epsilon \forall (F \cup G)_n : l_t((F \cup G)_n) \geqslant l_0$.

If $A_k : m \to m, k \in \mathbb{P}$ satisfies $\|A_k \ldots A_1 - S\| \overset{k}{\to} 0$, for some stable matrix S then the sequence (A_k) is **strongly ergodic**; and (A_k) is **strongly stable** relative to (B_k) iff there is a stable matrix $S : \|A_k B_k \ldots A_1 B_1 - S\| \overset{k}{\to} 0$. $\models 1 - 2\gamma(A) \leqslant \delta(A) \leqslant 1 - \gamma(A); \models \delta(AB) \leqslant \delta(A)\delta(B)$ for **conditions for** (asymptotic) **stability** – may be formulated in terms of $\rho : Z \to \mathbb{R}^+$ satisfying: $(i) \exists (t \geqslant 1, x_0 \in X) : \forall F_t\ \rho(F_t(x_0)) \leqslant \alpha \rho(x_0)$ and $(ii)\max\ (\rho(f(x_0), \rho(g(x_0)) \leqslant \rho(x_0)$. For (i)(ii) $\to \rho((F \cup G)_n(x_0)) \leqslant \alpha^{l_0} \rho(x_0)$ whenever $l_t((F \cup G)_n) \geqslant l_0$.

More general results and many references may be found in: **Senata, E., 'Non-negative Matrices and Markov Chains'**(Springer (2nd ed.), 1980).

8. The general theory of Stochastic processes is developed for classes of CA ($\equiv$ **countably** - additive) set functions – i.e., **measures**. In some contexts there are advantages in using FA ($\equiv$ **finitely** additive) set functions – to obtain modifications of the standard theory. A comparison of the theories so obtained show that FA theories yield intrinsically **dis**continuous processes (in w.d. senses), in some cases. The **discrepancy**, $\delta(m) := \sup_\epsilon \lim_{n \to \infty} m(E_n)$, for $\epsilon := (E_n), E_k \supseteq E_{k+1}, \cap_n E_n = \phi$ gauges the degree of **non** - CA behaviour of (E_n). Thus, if $(E_n) \subset X$, then the non vanishing of $\Delta(m)$ may reflect certain 'inadequacies of X'.

Forms of (almost-) denseness play a basic role here: for $\zeta \subset \mathbb{P}(\mathcal{H}), X \subset \mathcal{H}$is ζ-**dense iff** $\phi \neq c \in \zeta \Rightarrow X \cap C \neq \phi$; and, if $(\mathcal{H}, \mathcal{F}, \mu)$ is a measure space (with $\mathcal{J}_0$ any 'field' generally iff $(E \in \mathcal{F}_0 \wedge \mu(E) > 0) \Rightarrow E \cap X \neq \phi$. (**NOTE that** this generalizes the topological definition of 'dense'). Further, X is defined to be **thick** (for μ) **iff** $\mu^*(X) = \mu(\mathcal{H})$, where μ^* denotes outer measure for μ.

In a specified sense, a class of FA set functions may be regarded as '**contractions**' of CA set functions.

9. Some classes of (O)DE may be solved by (local) 'factorization'/decomposition of the associated differential operator(s). For the ODE $y''+r(x,m)y+\lambda y=0$, equivalent representations may be found as:

$$\begin{cases} H^-_{m+1}H^+_m y &= p^2(\lambda - L_{m+1}(x))y \\ H^+_{m-1}H^-_m y &= p^2(\lambda - L_m(x))y \end{cases}$$

twice-differentiable and $H^{\pm}_n(x) := k(x, n+\frac{1}{2}\pm\frac{1}{2}) \mp (p(x)\frac{d}{dx}+\alpha(x))$, where $\alpha(x) := -\frac{1}{2}p'(x)$. These forms are early (Revs Mod Phys. 23(1951)21-68) examples of 'ladder-operators schemes', introduced in Quantum Mechanics, and useful in the study of special functions. By taking $k(x,m) \equiv k_0(x) + mk_1(x)$, one may derive a Ricatti equation for k_1 and an associated linear, first-order ODE for k_0. Solutions at the points, x^*, where $(\lambda - L_{m+1}(x))(\lambda - L_m(x)) = 0$, with λ specified in the basic ODE, are readily obtained. Raising/lowering operators $(y(m) \to y(m\pm1))$ may be constructed, and more general forms of the $H^{\pm}$ maybe given (extending the scope of this method).

Connections between such 'factorization schemes' and the theory of representations of Lie groups are discussed in the book: **'Lie Theory and Special Functions'** by W.Miller, Jr (AP, 1968, Ch7).

10. This is an extension to **semi-**groups of the ideas/procedures in $\#2$. A between-ness semigroup admits a tertiary relation, 'b lies between a and c', that is invariant under the semi group operation. Here, such semigroups are **characterized** through criteria generalizing those for groups.

11. For polynomials f, of degree n, with leading coefficient in $\mathbb{P}$ it is known that $f(x) = (g(x)^k)\forall sl\ x \in \mathbb{P} \Rightarrow f \equiv g^k$, for some polynomial g over $\mathbb{Z}$. The condition may be weakened to require only that: $x \to \infty$ through a non-thin sequence of integers in $\mathbb{P}$, where (x_j) is **thin** IFF $\exists M \in \mathbb{P}\exists\alpha > 0 : x_{j+M} - x_j > x_j^{\alpha}$.

A further extension characterizes polynomials f 'close to k0th powers'. Let $\bar{x} := \{1, 2, \ldots X\}$, $S' := \mathbb{P}\backslash S$, then: $\models$ **If** $S := (x_i)$, and $\inf_{n\in\mathbb{P}}|f(x) - n^k| = 0(x)$, as $x \to \infty$ in S, where $\#(S' \cap \bar{X}) = 0(X^{\frac{1}{k}})$ **then** $f(x) \equiv (g(x))^k + A$, for some polynomial g over $\mathbb{Z}$ and some constant $A \in \mathbb{P}$.

The (indirect/nonconstructive) **proof** depends crucially on '**Hilbert's Irreducibility Theorem**' (irreducibility of polynomials $p(x)_m$ over K may be preserved , for $p(x)_h$, $1 \leqslant h < m$, under suitable interpolation in $x_{h+1}, \ldots, x_m$).

12. Let X be a metrizable TS, f a continuous self-map of X such that

(i) $\exists\xi X : \forall x \in X, f^n(x) \to \xi = f(\xi)$

(ii) There exists an open neighbourhood $U(\xi) \in \tau(X) : \forall V(\xi) \in \tau(X) exists m_V : n \geqslant m_V \Rightarrow f^n(U) \subset V$

Then: $\forall\lambda \in (0,1)\exists$ some metric ρ_λ on X, satisfying: $(\forall x, y \in X)\rho(f(x), f(y)) \leqslant \lambda\rho_\lambda(x,y)$ Moreover, if X has a compatible complete metric, then ρ_λ is complete.

This is one of several (partial) **converses of the Banach fp Theorem.**

NOTE Probabilistic analogues of the Banach fp Theorem are mentioned in MR$46\#4592$ and MR $81f:54029$. Partial converses of these variants would be of interest for Σ .

13. For functions defined implicitly – e.g., via PDE – it is often necessary to consider $\lim_{\epsilon\to0^+}\lim_{t\to\infty} F(\underline{x}, t, \epsilon)$, for $x \in \mathbb{R}^n, t \in R, 0 < \epsilon < 1$. The 'more tractable case' $\lim_{t\to\infty}\lim_{\epsilon\to0^+}$ often corresponds to initial linearization, in which the crucial (physical) behaviour is lost. Possibilities for combining $\lim_{\epsilon\to0^+}$ into a single ('Kaplun') limit process are important in 'singular perturbation theory', where, for $\delta \equiv \delta(\epsilon), \underline{x}_\delta := \delta^{-1}\underline{x}$, $\lim_\delta F(\underline{x}, t, \epsilon) := G(\lim_{\epsilon\to0^+} x_\delta, t, 0^+)$, with $F(\underline{x}, t, \epsilon) =: G(x_\delta, t, \epsilon)$.

The **order** (equivalence) **classes** – for continuous functions $\mathbb{R}^+ \supset A \to B \subset \mathbb{R}$ – may be defined by:

$$\text{ord}\gamma := \{\theta : \theta = O(\gamma), \text{as } \epsilon \to 0^+\}$$

The set $\{\text{ord}\gamma\}$ becomes a TS if

(i) $S \subset \{\text{ord}\gamma\}$ is **convex** and $(\forall \text{ord}\theta \in S)$

(ii) $\exists \gamma, \delta$ with $\text{ord}\gamma, \text{ord}\delta \in S$: $\text{ord}\gamma < \text{ord}\theta < \text{ord}\delta$. Here, $\text{ord}\gamma_1 < \text{ord}\gamma_2$ IFF $\gamma_1 = o(\gamma_2)$ as $\epsilon \to 0^+$ i.e.: (i) specifies the topology $\tau(\{\text{ord}\gamma\})$

NOTE. S is **convex** IFF $(S \ni \text{ord}\gamma_1 < \text{ord}\theta < \text{ord}\gamma_2 \in S) \Rightarrow \text{ord}\theta \in S$; and if also $\exists \text{ord}\gamma_1, \text{ord}\gamma_2 \in S : \text{ord}\theta \in S \Rightarrow \text{ord}\gamma_1 \leqslant \text{ord}\theta \leqslant \text{ord}\gamma_2$, then S is a closed interval. (open intervals have '$<$' instead of '$\leqslant$' in this definition.) With these preliminaries, and the **definition**:

$H(\underline{x}, t, \epsilon)$ is a uniform $\Delta(\epsilon)$- approximation to $F(\underline{x}, \epsilon, t)$ over the convex set $S \subset \{\text{ord}\gamma\}$ IFF $\text{ord}\gamma \in S \Rightarrow \lim_\gamma \|(\frac{1}{\Delta})(F - H)\| = 0$, the following 'Extension Theorem' holds. **Theorem If** $H(\underline{x}, t, \epsilon)$ is a uniform $\Delta(\epsilon)$-approximation to $F(\underline{x}, t, \epsilon)$ over the closed interval, S_0, **then** this property **extends** to some **open** $S \supset \mathrm{S}_0$

Many comparable results, covering diverse situations – and copious references – may be found in the books: '**Introduction to singular Perturbations**', by **RE O'Malley, Jr** (AO,1974), and, '**Singular Perturbation Theory'**, by **D.R.Smith**(CUP, 1985).

14. The simplest form of distance on a probability space of 'events' is suggested by the inequality:

$$(\#) P(A \Delta C) \leqslant P(A \Delta B) + P(B \Delta C)$$

Here, it is observed that 'algebras of events' are isomorphic to 'algebras of sets' (which are extended – if necessary – to σ - algebras). $P(X)$ denotes the probability of X, and $P(X \Delta Y) := P(X - Y) + P(Y - X)$, where $X - Y := X\bar{Y}$, with '+' (for events/sets) denoting 'union', 'juxtaposition', intersection, and $\bar{Y}$, the event/set complementary to Y. (**NOTE:** $X \Delta Y$ is often called the **symmetric difference** of the sets X, Y.)

Plainly, $(\#)$ furnishes a 'triangle inequality' for a **metric**, $d(X, Y) := P(X \Delta Y)$. It may be shown that the assignment

$$d^*(X, Y) := \begin{cases} P(X \Delta Y)/P(X + Y), \text{ if } P(X + Y) > 0 \\ 0 \qquad\qquad\qquad\qquad \text{Otherwise} \end{cases}$$

also yields an admissible metric. Such metrics are important in '**Boolean Geometry**'; see, e.g. Blumenthal LM, Rend. Circ mat Palermo 1(1952)343-460; Blumenthal/Penningibid 10(1961) 175–192. Also: 'On a certain distance of sets and the corresponding distance of functions' by E.Marczewski/H. Steinhaus (Collo .Math. VI (1958)319-327

15. If the general theory of **Dynamical Systems** is viewed as a branch of Topology/Ergodic Theory, then the question arises as to the most appropriate 'carriers' of such systems – in particular for so-called **local** dynamical systems. For global systems, various classes of TS are adopted; but it is claimed that some **local** systems (corresponding, e.g., to 'non-prolongable (O)DE') are better specified over the **induced TS** generated by **uniformities** or **proximities**. In view of the central role played in Σ by the 'basic spaces' ${}_SX$, and the approaches realizable over these spaces, cognate forms of **information-flux** could be developed within this abstract framework. As usual in Σ , the quasi-metamathematical aspects of this proposal need not be feared!

The book cited as 'basic references', here ('**Dynamical Systems in the Plane**'), by **O.Hajek**, AP, 1968) contains many fundamental formulations and procedures.

16. The notion of (maximally) extending the domain of applicability of an ME from its 'canonical setting' is ubiquitous in mathematics; but it is seldom treated as a process–say **conceptual continuation** – susceptible of formal analysis. The obvious model here is '(D)AC from a PS germ', in Function Theory. The crucial property, inviting generalization, is that IF 'two versions' of 'the same ME' have a **common domain of validity,** where **they coincide,** THEN **each version furnishes some CC of the other** to the relevant super domain(s). This is unambiguously interpretable for 'analytic elements'; but, cognate schemes for (substantially) arbitrary ME demand extremely careful formulation.

The basic ideas of CC, and a few examples, are discussed in the paper: '**Varieties of Approximation**', by **JSN Elvey** (pp 77-94 in '**Artificial Intelligence in Mathematics**', J.Johnson, S. McKee and A.Veller, eds., OUP, 1994. Other examples are identified as they occur in these **C**apsule **D**escriptions.

23. **Converses** – of various degrees of completeness – play an important role in Σ , since they furnish operational **in**verses of OT (suitably interpreted!) in the OT-composition representation of proofs. A comprehensive collection of 'important theorems', with associated 'standard'/'strong'/'weak'/'quasi'/...converses will provide a powerful calculational resource. It appears that no significant collection of this sort has been produced up to now – though the developers of **B**(ishop)-Constructive version of theorems in diverse areas do seem to acknowledge the urgency of this task.

For **Rouché's theorem** (of indirect use in many **stability analyses**[430] it is necessary to strengthen the usual 'CV version' before a viable converse can be obtained. The standard form has $|f(z) - g(z)| < |g(z)|$, on ∂D, for f, g, analytic inside $W \supset \bar{D}$. This may be strengthened to (i): $|f(z) - g(z)| < |f(z)| + g|(z)|$ on ∂D, but this form still admits no proper converse.[431]

Instead, the further strengthening to (ii): $\exists$ **finite Blaschke products** α, β – – **of the same order**[432] – **such that** $|\alpha f + \beta g| < |f| + |g|$ **on** ∂D, allows the proof of a full converse. Partial converses (for the standard version, or for condition (i)) are also useful in providing cognate types of OT - inverse for the construction of operational proofs.

The proof of the converse result here depends on several subtle properties on the extension/approximation of (conformal) mappings.[433]

Note. Some generalizations of Rouché is theorem to domains in $\mathbb{C}^n, n \geqslant 2$, may be found in JLMS (2)20(1979) 259-272 (NG Lloyd).

24. The setting here is the uniform rational approximation of functions analytic in $S^o \subset \mathbb{C}$ and continuous in S. Characterization of the classes of domains over which such forms of approximation exist, involves estimation of analytic capacities, domain densities, and of distances from the space $\mathbb{Q}(S)$ of rational functions over S, to $A(S)$, the algebra of functions analytic over S^o and continuous over S. The underlying scheme is surveyed in Ruso-Math. Surveys 22(1967) 139-200 (transl.) by A.G. Vitushkin. The key results are extended to vector=valued analytic functions. Problems where S is of infinite connectivity, and for pointwise-bounded approximation, raise additional questions as to attainability/effectiveness.

26. The notions of **characteristic exponents,** and **invariant manifolds** occur in the qualitative theory of differential equations – and its extensions to **Dynamical Systems Theory**. There are several definitions – essentially equivalent, but not obviously so! For characteristic (also called '**Lyanpunov**') exponents, three possible specifications are as follows.

(a) The (non unique (mod $2\pi k$)) complex numbers α in the representation $\lambda = e^{\alpha T}$ of the eigenvalues of the **monodromy (matrix) operator** for T-periodic solutions of ODE

(b) The (non unique (mod $2\pi k$)) complex numbers $log\lambda_i$, where the $\lambda_i, 1 \leqslant i \leqslant m$ are eigenvalues of $\Lambda_x := \lim_{n\to\infty} (T_x^{n*} T_x^n)^{\frac{1}{2n}} \in \mathbb{M}_m$. Here, M is a manifold; p, an **ergodic** ($\equiv$ indecomposable) measure on M;

$$T : M \to \mathbb{M}_m \equiv \{\text{square matrices of order } m\};$$

$$f : M \to M, \text{ a \textbf{measure-preserving} mapping}$$

and ρ satisfies: $\int \rho(dx) \max(0, \log\|T(x)\| < \infty$. Finally, $T_x^n : \underset{n-1 \geqslant k \geqslant 0}{\circ} T(f^k(x))$, with $f^0(x) \equiv x$, and T_x^{n*} is the **adjoint** of T_x^n.

---

[430]This use is mainly for studying pole-distributions of transfer functions; but many zero-distributions are studied in other applications

[431]An example is given to show this.

[432]See, e.g. Rudin,W.,'Real and Complex Analysis (3rd ed., McG-H,1987, Ch9. 17, 18, for Blaschke products; also P.Koosis, Intro. to H^p Spaces', (CUP, 1980)

[433]The crucial properties are: continuity at the boundary, and uniform approximation by sequences of quotients of finite Blaschke products; references are given in the BR (AM 89(5) 302-305, 1982.

(c) For a diffeomorphism f of a compact manifold M, x is regular IFF $\exists(\lambda(x))_m$, strictly increasing in $\mathbb{R}$, with associated decompositions $T_x M =: \underset{1 \leqslant i \leqslant m}{\oplus} E_i(x)$, such that $\forall(0 \neq u \in E_i(x), 1 \leqslant i \leqslant m)\ n^{-1}\log\|(D_x f^n)u\| \to \lambda_i(x)$ – the Lyapunov exponents of f at x. (Here, $D_x g \equiv$ the differential of g at x)

NOTE There are still other ways of defining characteristic exponents, but the ones just given should suffice for illustration. Since each λ_i is an eigen**value**, it is paired with some eigen**vector(s)**. This allows the introduction of '**partial dimensions**'

Invariant manifolds may be specified more directly; they are simply subsets of the relevant phase space(s) where prescribed dynamical functions (e.e., the Hamiltonian) assume constant values. This may be extended to cover **finite collections** of dynamical functions – in terms of the corresponding **inverse functions** of cartesian products of mappings.

27. **W**eak **L**aws of **L**arge **N**umbers cover the convergence **in probability** of sequences $B_n^{-1}(S_n - A_n)$ to the degenerate ($\equiv$ null) r.v.. Here, S_n is the sum of $n\ rv$ (possibly, i.i.d.); for convergence $a.e.$, **S**trong laws are involved.In theorems of the **C**entral **L**imit type, the limit distribution is **non** degenerate (**normal**, in the basic case), Moreover, the r.v. in S_n may be BS - valued, etc.. The investigation of **rates** of convergence for such classes of problems is highly technical. The **asymptotic** behaviour of these approximations is best understood for **stable** limit distributions – where a r.v. Y over (Ω, A, μ), with $d.f.$, G, and $c.f.$, g is called **stable** IFF $(\forall a_1, a_2 > 0)(\exists\ a > 0, b \in \mathbb{R}) : g(a_1 t).g(a_2 t) = \exp\{itb\}g(at)$ – a **reproducibility property**.

NOTE. WLLN may be formulated in terms of a **metric** function, p, defined by:

$$V := \{r.v. \in (\Omega, A, u),\ \text{with finite expectations}\ \}$$

$$\rho(X) := E\{\frac{|X|}{1+|X|}\},\ \text{for}\ X \in V (E\{\} \equiv\ \text{expectation})$$

Then $\models d(X,Y) := \rho(X-Y)$ yields a MS, (V, d) such that $d(W_n, W) \to 0$ IFF $W_n \xrightarrow{p} W$. Further,

$$\models \rho(X+Y) \leqslant \rho(X) + \rho(Y)$$
$$\rho(\sigma X) \leqslant\ \max(|\sigma|, 1)\rho(X)$$

Here, r.v. that are a.s. equal are **identified.**

Typical results in this area give estimates of the distances between the d.f. of a standardized sum of i.r.v and a stable d.f. (e.g., Gaussian). Some fundamental results may be found in **Feller W.**, '...**Probability Theory** ...'vol II (Wiley, 1966) Ch XVI, for instance:

Let G_n denote the (normalized) n -fold convolution of the (one-dimensional) d.f. G (with c.f. g): $G_n(x) := G^{n*}(x\sigma/\sqrt{n})$. **Then**:

$\models G_n(x) - \Phi(x) - (m_3/6\sigma^3\sqrt{n})\tilde{\Phi}(x) = 0(1/\sqrt{n})$,

where

$\Phi(x) := (1/\sqrt{2\pi})\exp(-x^2/2); \tilde{\Phi}(x) := (1-x^2)\Phi(x); m_3 := \int x^3 G(dx);$
$\sigma := \int x G(dx)$

NOTES (i) Minor Modifications are required if G is an **arithmetic** ($\equiv$ '**lattice**' **distribution** – i.e., one concentrated on $\{0\} \cup \{\pm q\lambda, q \in \mathbb{P}\}$

(ii) It is assumed that m_3 is $w.d.(M < \infty)$.

Many **variants/extensions** of such results, covering diverse stochastic processes, may be found in (e.g.): **Christol,** G./ **Wolf, W.**, '**Convergence Theorems with a Stable Limit Law**' (Akad. Verlag, 1992); **Jabod, J./Shiryaev, A.N**, '**Limit Theorems for Stochastic Processes**' (Springer 1987), both of which have substantial **Bibliographies**; Rachev, ST, '*Probability Metrics...*' (Wiley, 1991).

28. The span of '**G**eometric **F**unction **T**heory' is very broad, including mapping properties of (simply/multiply – connected) domains, approximation problems (for functions of one class by those from other classes), inequalities for series -coefficients in classes of n-valent functions ($n \geqslant 1$). For **meromorphic functions**, more detailed investigations may be based on the **Nevanlinna representations** – and their extensions to functions of several CV. In particular,

rates of growth (along specified families of curves), and the study of **covering surfaces**, are of basic interest here. Further extensions involve quasi-conformal mappings. More generally, the links between geometrical/topological characteristics of domains in $\mathbb{C}^n (n \geqslant 1)$ and the properties of various species of functions $w.d.$ over these domains constitute the major part of GFT. For Σ , cognate (analogues of) properties for species of functions over subdomains of the Basic Spaces, $_SX$, introduced in E_1, are of primary concern here. The treatise: 'Potential Theory in Modern Function Theory', by M.TSUJI (Chelsea, 1975) has much on GFT. Apart from the BR (Goluuzin: ' Geometric Theory of Functions of a a CV'(AMS, 1969)), the books by **WK Hayman** (**Meromorphic Functions**, OUP, 1964) and **G. Julia** ('Principes Geometriques D'Analyse' - vols 1,2, Gauthies - Villars 1930, 1932) are useful references. ALSO: **Sansone/Gerretsen**: 'Lectures on ... Fns of CV' vol II (Wolters Noordhoff, 1969).

29. The **numerical stability** of (matrix) **eigenvalue** and **least-squares procedures** under (small) perturbations of **data** may be assessed in terms of **condition-numbers**.

$$\text{Let } C(\varphi, a_0) := \lim_{\delta \to 0^+} \sup_{\|\Delta a\| \leqslant \delta} \frac{\|{}^0\Delta x\|_X}{\|{}^0\Delta a\|_A}$$

Here, the procedure corresponds to: $\varphi\ A \to X; x = \varphi(a); x_0 + \Delta x = \varphi(a_0 + \Delta a)$, for the NLS X, A with associated norms with ${}^0\Delta x := x_0^{-1}\Delta x, {}^0\Delta a := m\ a_0^{-1}\Delta a$.
$\models \|{}^0\Delta x\| \leqslant c(\varphi, a_0)\|{}^0\Delta a\|$, 'in 1st -order approximation'.

Further, **if** φ is F(réchet)-differentiable then: $\Delta x = \varphi'(a_0)o\Delta a + o(\Delta a)$, as $\|\Delta a\| \to 0$, so that: $c(\varphi, a_0) = \|\varphi'(a_0)\|\|a_0\|/\|x_0\|$.

The $e-v$ problem for $M \in \mathbb{R}^{n^2}$ with typical $e.v\ \lambda_1 \in \mathbb{C}$ has $\varphi : \mathbb{R}^{n^2} \to \mathbb{C}$, in a neighbourhood of M, where $\lambda_1 = \varphi(M)$ and $\models \varphi'(M)o\Delta M = v_1^T \Delta M u_1/(v_1^T u_1)$, with v_1^T, u_1 the eigen-vectors associated with λ_1

In **the least squares problem**, a matrix $M \in \mathbb{R}^{mn} (m > n)$ and an m-vector $b \in \mathbb{R}^m$ are prescribed; the task is to determine $x \in \mathbb{R}^n$ such that

$$\|b - Mx\|_2 = \min_{y \in \mathbb{R}^n} \|b - My\|_2$$

For this problem, one may show that:

$$c(\varphi, M) = \|M\|_F \|M^+\|_2[(\|M^+\|_2\|r\|_2/\|x\|_2)^2 + 1]^{\frac{1}{2}}$$

where

$$\|M\|_F := (\sum_{i=1}^{m} \sum_{j=1}^{n} \|m_{ij}\|^2)^{\frac{1}{2}}$$

M^+ is the pseudo-inverse of M, $\gamma := b - Mx = (I - MM^+)b$. For Σ , notions of solution-stability are of fundamental importance. A more general scheme devolve on the **sensitivity analysis** of function(al)s occurring in various applications – especially in all areas of Engineering Science; see, e.g., $LNM\#1086$ (V.Komkov, ed.) for several – mainly, review-articles.

30. Diverse examples of **interface problems** are of fundamental importance in Σ , for two main (interrelated) reasons, namely: (a) that they are central for **continuous schemes**, in which the effective domains of definition of ME are (maximally) extended; (b) that – partly through schemes (a) – they facilitate **the removal of artificial barriers** between ostensibly disparate ME, and **the identification/characterization of inherent/nonremovable barriers.**

Continuation problems for PDE, where solutions are extended across the interface(s) between subdomains, furnish basic illustrations of such phenomena. For polygonal domains triangulated from the (interior) origin by 'rays', '$\theta = \theta'_j, 1 \leqslant j \leqslant J$, various types of **interface conditions**, as paths cross rays, may be imposed. If the solutions are known to exhibit singular behaviour, then numerical procedures require refinement near the singular points. Denote by $u \equiv u(r, \theta)$ the (class of) solutions considered.

One form of refinement generalizes the FEM to an **I(nfinite)EM**. For domains $\Omega \subset \mathbb{R}^2$, with polar coordinates (γ, θ), **quasi-eigenfunction expansions** of solutions of **variational problems** equivalent to the original PDE may be determined through associated (sets of) **e.v problems** corresponding to the interface conditions. if the polygon Ω is triangulated by rays, and a **sequence**, $\{\Omega_k\}$, of interior polygons similar to Ω produces **layers** between neighbouring polygons, then a **Sobelev-Space framework** may be used to formulate, and analyse the approximation procedures 1generated by $\{\Omega_k\}$'. If Ω has m vertices, and y_k is the m-vector of values of u at the vertices of Ω_k, then **the y_k satisfy a countable set of linear algebraic equations** with remarkable **decomposition properties** from which **rates of convergence** to the (singular) solution may be estimated.

One significant **application** involves the calculation of **S**tress-**I**ntensity **F**actors in (linear, elastic) **Fracture Mechanics.**

31. The **bifurcation phenomena** exhibited by the solutions of nonlinear operator equations may – with additional restrictions – be mirrored by the solutions of corresponding discrete approximations. For evolution equations, steady sates may evolve into periodic orbits, under variation of some parameter(s): the Hopf bifurcation.

The equation $u_t + F(\lambda, u) = 0$, with $F : \mathbb{R} \times V \to V'$ (the dual of V) may be investigated near an arbitrary point (λ_0, μ_0), for functions F of the form: $F(\lambda, v) =: Av + G(\lambda, v)$, where $A : V \to V'$ is a coercive isomorphism of class $\zeta^p (p \geqslant 3)$, and, $G : \mathbb{R} \times V \to V'$ and $G|_{\mathcal{X}} : \mathbb{R} \times \mathcal{X} \to \mathcal{X}$ are also $\zeta^p (p \geqslant 3)$, $\mathcal{X}$ being a suitable closure of an associated space of 2π -periodic functions, with $D^p G$ bounded over bounded subsets of $\mathbb{R}x\mathcal{X}$.

The key approximation have is to the **inverse**, $\mathcal{J}$, of $\omega_0 \partial_s + A$, by a family of functions $\mathcal{J}_h : \mathcal{X}' \to L^2(0, 2\pi; V_k) \cap \mathcal{X}$, where the V_n are finite-dimensional subspaces of V, and $h \to 0^+$, subject to the condition $f \in \mathcal{X}' \to \|(\mathcal{J} - \mathcal{J}_h)f\| \to 0^+ \ (h \to 0^+)$.

This reduction to a family of finite dimensional problems, $\{P_h\}$, simplifies the bifurcation analysis for computation – through the details are more complicated than for 'the continuous problem', P_c. The eventual result is that: every branch of solutions of P_c has (as $h \to 0^+$) a unique branch of P_h 'arbitrarily close to it' – in a precise sense – with the rate of convergence proportional to $\|(\mathcal{J} - \mathcal{J}_h)\varphi\|_{\mathcal{X}'}$ for specified φ, independent of h.

For Σ , the significance of this example lies primarily in **the quasi-stability of bifurcation behaviour** under classes of perturbations of the underlying operator equations – an important property which suggests possible **extensions** to more abstract frameworks /ME.

38. The approximation of **shape**, for smooth/regular objects, may be approached by the methods of Differential Geometry and (algebraic) Topology. The extension of such methods to **irregular** objects (e.g., fractals) requires cognate extensions of the techniques of Algebraic Topology to (essentially) **arbitrary** spaces. The original version of **G**eometric **S**hape **T**heory (due to K. Borsuk) introduced the homotopy category of finite polyhedra (say, A) and the homotopy category of compact MS (say, B) together with a functor, $K : A \to B$. Another aspect here involves the **extension of invariants** from well-understood models to less familiar contexts. The associated notions of **approximation** are somewhat elusive, devolving on **categorical versions of GST**, where the topological properties of models in some category, C_1, are suitably mapped in another category , C_2, to allow the approximation of objects in C_2 corresponding to these models. The formalism covers both 'abstract ME' and 'concrete ME' encountered in The Sciences. **See: 'Shape Theory: Categorical Methods of approximation',** by **J-M. Cordier/T.Porter**(Ellis Horwood, 1989)

40. In an obvious sense, TS are generalizations of MS, but the designation **Generalized MS** is reserved for 'classes of spaces close to metrizable spaces' – with additional **quasi-stability properties** under specified topological operations. Moreover, some of the accepted GMS are also GC(ompact)S, etc.. A common procedure for obtaining GMS is '**to perturb sets of NSS metrization conditions** in various ways, so that primary characteristics of MS' remain valid. The resulting class of GMS is diverse, comprising subclasses corresponding to each choice of (generic) perturbations. Although no exhaustive classification has been made on this basis, the most important cases have been studied extensively. **See: Ch10** (by G. Gruenhage) of **HBS(et-theoretic)T** (ed. K.Kunen/J.E.Vaughan), Elsevier, 1984, and references cited there.

NOTE. For Σ , variants of metrizability are fundamental, since they allow quasi-metrical criteria to be extended to subclasses of the other Basic Spaces introduced in E_1.

43. **The study of 'geometry' within constructive formalisms** is problematical, since many of the continuity properties central to classical/standard treatments are lacking. For the

B-constructivity introduced by Bishop[434], and currently being developed in virtually all areas of Mathematics[435], the geometrical limitations are broadly acceptable for Σ . The more extreme systems, based on $\mathbb{N}, \mathbb{Q}$, and $\{$ constructive real numbers$\} \equiv CRN$, proposed by Markov [436]m and defined over 'the constructive continuum' (say, D), lead to various counter-intuitive results – e.g., the non-intersection of suitable deformations of the diagonals of a square! Such apparent anomalies may be removed by extending D to 'richer sets' – with cognate continuous extensions of the 'curves', etc., to forms exhibiting (most of) the intuitively expected behaviour. Such extension schemes exemplify subtle types of **approach**, which are likely to be important also in ostensibly **non**-geometrical settings.

45. Notions of **stability** are pre-eminent for Σ , as they embody quasi-permanence of ME within fluctuating environments. Consequently, there are many types of stability – roughly equivalent in the broadest interpretation(s), but extremely diverse in levels of intricacy and technicality. In Probability Theory, (a) the 'tendency of distributions to normality', and (b) the 'hereditary normality' of i.i.d.r.v. whose sum is normally distributed (Cramér's Theorem) are especially notable. In fact, (a) covers theorems of 'the **Central Limit type**', while (b), proved **via characteristic functions**, is a nontrivial **converse** (of the result that the sum of normally distributed r.v. is itself normal).

Since the d.f. of the sum of r.v. is the **convolution** of the d.f. of the summons, the stability of Cramér's theorem is formulated in terms of sequences of convolutions – say, $\{F_n\} \equiv \{F_n^{(1)} * F_n^{(2)}\}$. If ρ_1, ρ_2 are specified **metrics** over the space of $\mathcal{D}$ of all d.f., and N comprises all **normal** d.f., then:

Cramér's Theorem is **stable** (for prescribed metrics ρ_1, ρ_2) IFF

$$\delta_n := \max_{j=1,2, G\in\mathcal{N}} \inf \rho_2(F_n^{(j)}, G) \to 0 (n \to \infty)$$

whenever $F_n = F_n^{(1)} * F_n^{(2)}$, and $\epsilon_n := \rho_1(F_n, \Phi) \to 0 \ (n \to \infty)$, where Φ is the normal distribution with zero mean and unit variance. The **quantitative** stability, here, refers to estimates of the form.

$\delta_n \leqslant \delta(\epsilon_n) \to 0 \ (\epsilon_n \to 0)$. Significant progress has been made in the special (but important) case where $F_n^{(1)} = F_n^{(2)}$, and $\rho_1 = \rho = \rho_2$, for

$$\rho(F, G) := \sup_x |F(x) - G(x)|, \rho(F, \Phi) := \epsilon < 1$$

. If F_1 has median 0 then $\models \rho(F_1, \Phi_{0,\frac{1}{2}}) \leqslant C\epsilon^{\frac{1}{36}} \log(\frac{1}{\epsilon}$, where $F = F^{(1)} * F^{(2)} = F^{(1)} * F^{(1)}$ is the d.f. of a sum $X = X_1 + X_2$ of i.i.d.r.v, and C is an absolute constant.

Hence: if $\epsilon_n := \rho(F_n, \Phi)$, as above then $\delta(\epsilon_n) = C\epsilon_n^{\frac{1}{36}} \log(\frac{1}{\epsilon_n}) \to 0 \ (\epsilon_n \to 0^+)$ gives an estimate for quantitative stability.

46. The problem of '**d**imensional **r**eduction' (e.g., from $\mathbb{C}^n$ to C^{n-1}; hence, inductively, to $\mathbb{C} \equiv \mathbb{C}^1$) is fundamental for constructive aspects of Σ , where it may be combined with arbitrary approach procedures. Such reduction criteria may be derived from **sufficiency conditions** for subsets $S \subset \mathbb{C}^n$, as follows. Denote by $<,>$ the scalar product on $\mathbb{C}^n$ (as a LVS) and put $H_{\bar{D}_k}(u) := \sup_{Z\in\bar{D}_k} \mathcal{R} < u, z >$, for $\bar{D}_k(\subset \mathbb{C}^n)$ convex and $\mathcal{R}w \equiv$ real part of w. Then $H_{\bar{D}_k}$ is **subadditive/positive-homgeneous**(of order 1): the **support functions of** $\bar{D}_k$, say $H_{\bar{D}_k} \equiv H_{\bar{D}_k}(\bar{\lambda})_n$. If $P_D \equiv$ the set of entire functions of λ such that $\varphi \in P_D \Leftrightarrow |\varphi(\lambda)| \leqslant c_\varphi \exp\{H_{k(\varphi)}(\lambda)\}$, with $\lambda \in \mathbb{C}^n$, then $P_k(\subset P_D) := \{\varphi \in P_D : \|\varphi\|_{k,S} < \infty\}$ yields a sequence of (bounded) subspaces determining an **inductive limit topology** (say, μ_S) in P_D – where $\|\varphi\|_{k,S} := \sup_{\lambda\in S} |\varphi(\lambda)| \exp\{-H_k(\lambda)\}$, and $\mathcal{S}$ is **weakly sufficient** IFF $\mu_S = \mu_{\mathbb{C}^n}$.

A procedure for constructing minimal w.s. subsets of $\mathbb{C}^n$ for (i) an n-ball, and (ii) a tubular domain (as families of 'analytic $(n-1)$ - planes') may be formulated, and used to represent

[434] Bishop, E., 'Foundations of Constructive Analysis' (McGraw-Hill, 1967)
[435] Bishop/Bridges, 'Constructive Analysis' (Springer, 1985)
[436] See, e.g., 'Foundations of Constructive Mathematics' (Springer, 1985)

functions in $H(D) \equiv \{f$ analytic D with the topology of uniform convergence on compact sets $\}$ as (generalized) Dirichlet series. Such results extend those originally found for 'classical cases' $(n = 1)$.

47. The idea of **infinitely divisible** distributions devolves on the representation of a r.v., ξ, as a sum of n r.v. ξ_i. If this is possible $\forall n \in \mathbb{N}$, then the corresponding d.f., F, is the n-fold convolution of the d.f.,F_i, of the ξ_i. The F_i may have any (distinct) forms, provided only that $F = F_1 * \cdots * F_n$. There is a correspondence between infinitely divisible distributions and processes with independent increments – and semigroups play a basic role in the analysis of these ME. The controlled approximation of d.f. for sums of i.d.r.v. by infinitely divisible d.f. may be considered on the Lévy metric:

$$L(G,H) := \inf\{\epsilon : G(x-\epsilon) - \epsilon \leqslant H(x) \leqslant G(x+\epsilon) + \epsilon\},$$

where $G, H \in \mathcal{F} \equiv \{1-D$ distributions$\}$. The set $\mathcal{G} \equiv \{$infinitely divisible $1-D$ distributions in$\mathcal{F}\}$ is also introduced, together with

$$\rho(G,H) := \sup_x |G(x) - H(x)|, \Phi_\sigma (\equiv \Phi_{0,\sigma})$$

$$E_a \approx \delta(a), E \equiv \delta(0)$$

for the G.F. δ, and products/powers of measures represented by convolutions. For $\epsilon > 0$, define Γ by $\Gamma(\epsilon) := \{G \in \mathcal{F} : L(G,E) \leqslant \epsilon\}$.

The **accuracy of approximation** of $F \in \mathcal{F}$ by $G \in \mathcal{G}$ is measured by the function $\varphi \equiv \varphi(\epsilon) := \sup_n \sup_{F_i \in \Gamma(epsilon} \inf_{D \in \mathcal{G}} L(F_1 * \cdots * F_n, D)$. **Note that** the F_i are all L-close to E; if the F_i were unrestricted in $\mathcal{F}$ then some choice would yield $L(\tilde{F}_1 * \cdots * \tilde{F}_n, D) = 0$, since $D \in \mathcal{G}$. However, since F is not infinitely divisible, such a choice cannot be made to cover all values of n, etc. The form of φ corresponds to a 'worst-case estimate', involving a class of F_i for which rigorous results may be obtained. The main estimates are summarised by the inequalities:

$$0 < \epsilon \leqslant 1 \Rightarrow c\epsilon \leqslant \varphi(\epsilon) \leqslant c\epsilon(1 + |\log\epsilon|)^3$$

where the upper bound improves on previous bounds: $\varphi(\epsilon) = O(\epsilon^q), q = \frac{1}{5}, \frac{1}{3}, \frac{1}{2}$. The restriction to $\Gamma(\epsilon)$ produces relatively concentrated d.f.. The proofs of such results are intricate but their main importance lies in partial characterizations of species of d.f..

48. The **'method of fictitious domains'** replaces a specified I/BVP over a domain Ω_1 by a 'nearby problem' over a simpler domain Ω_2. Call these problems $\mathcal{P}_1, \mathcal{P}_2$, with solutions σ_1, σ_2. It turns out that $\Omega_2 =: \Omega_2' \cup \Omega_2''$, where $\sigma_2|_{\Omega_2''} \approx 0$; in this sense, Ω_2'' may be viewed as fictitious. ore precisely, $\mathcal{P}_2$ involves some parameter(s) ϵ, and $\sigma_2|_{\Omega_2''} \equiv 0(1)$ as $\epsilon \to 0$. The method was originally used for elliptic (no IC), but it may be adapted to hyperbolic/parabolic PDE. In typical cases, $\Omega_2' = \Omega_1$, so that Ω_2 **is an extension of** Ω_1 (to a parallelepiped, for elliptic PDE). A rigorous treatment of this case is given in: **'Methods of Numerical Mathematics'**,by **G.I. Marchuk** (Springer, 2nd ed., 1980) sec. 2.7. Here, **hydrodynamics problems** (for viscous/ ideal fluids) **over doubly-connected domains** are replaced by **'neighbouring problems' over extended, simply-connected domains.** The **formalism** combines **BEM/FEM** techniques with techniques from (singular) **perturbation theory**. For Σ , the 'closeness criteria' for pairs of (I)/BV problems are of special relevance, since they allow 'the quasi-metrization of spaces of problems', etc . . . .

49. The **Edge-of-the-Wedge Theorem** is an extension to function of n CV of a method of (D)AC underlying the Schwarz Reflection Principle is a basic Function Theory. Denote by $\mathbb{C}^{\pm}$ the upper/lower half -planes in $\mathbb{C}$ Let $D^{\pm}$ be open in $\mathbb{C}^{\pm}$; $F^{\pm}$ analytic in $D^{\pm}$; $F^{\pm}(z) \overset{\text{unif}}{\to} F^+(x) = F^-(x)$, continuous for $x \in (a,b) \subset D^+ \cap D^-$. **Then**: $F^{\pm}$ are analytic on (a,b), and are elements of **the same** CAF.

Various forms of AC for functions in radial, tubular domains, $T^{\pm}$ (with conical bases, $V^{\pm}$), enlarge the domain of analyticity from $T^{\pm}$ (for $F^{\pm}$) to $T := H(T^*) \cup H(T^-) \cup N$, where $H(W) \equiv \bigcap\{\text{dom} f : f \in \mathcal{A}(W)\}$, and $\mathcal{N}$ is a w.d. neighbourhood in $\mathbb{R}^n \subset \mathbb{C}^n$. The original conditions on $F^{\pm}$ are of two types

(a) $|F^{\pm}(z)| \leqslant Be^{\epsilon|y|}|y|^{-\rho_{\pm}}(1+|z|)^{m_{\pm}}, (z \in T^{\pm}; \epsilon > 0)$

(b) For $y \to 0$ in $V^{\pm}, \phi \in C^{\infty}(\mathbb{R}^n)$, supp$\phi \subset G$(open), lim$\langle F^{\pm}, \phi\rangle$ exist (finitely/ uniformly) and coincide. Here $\langle , \rangle$ denotes inner product, and the set N_G(above) $:= \bigcup_{\xi \in G} \{z : |z-\xi| < \theta \text{ dist}(\xi, \partial G)\}$, for some w.d. $\theta(V^{\pm}) > 0$.

Note that (b) is GF-related; and, that one obvious extension of 'E-WT' involves the replacement of GF by **hyperfunctions:see** Dokl. **21** (transl.) 21- (Zarinov, V.V.). Indeed, the **Sato hyperfunctions** may be defined as equivalence classes of functions analytic in some product of punctured neighbourhoods of $\mathbb{R}^1$ in $\mathbb{C}^n$, with $F_1 \sim F_2$ IFF $F_1 - F_2$ is AC to the full neighbourhood.

Another class of extensions of E-WT cover functions, F, analytic in **tuboids** of $\mathbb{C}^n$ (defined via fibering maps) with profiles Λ and projections Ω (say), with the BV of F given by Schwartz distributions: **Let** $T^{\pm}$ by fiberwise-connected tuboids with profiles $\Lambda^{\pm}$ over Ω, where $x \in \Omega \Rightarrow \Lambda_x^+ = -\Lambda_x^-$, and let $F^{\pm} \in A_{\mathcal{D}'}(T^{\pm})$, bv$F^+ =$ bv F^- on Ω. Then $\exists N_\Omega, \mathbb{C}^n \supset N_\Omega \supset T^{\pm} \exists F \in A(N_\Omega) : F|_{T^{\pm}} = F^{\pm}$.

NOTE. This ('**microlocal**') formulation [437] may be further extended to allow bv $F^{\pm}$ to be Sato hyperfunctions.[438]

Finally, an **algebraic-topological version** of E-WT may be obtained for functions analytic in tuboids, and 'locally slowly increasing as $y \to 0$'. The key result, here is that a certain sequence of (bv-related) linear spaces (and associated maps) is **exacts**. This subsumes the (above) micro local formulation and several other results analogues to E-WT.

51. The construction of a (**pseudo**)**metric** d_H**, on the space of optimality criteria for machine design** furnishes a striking example of quantitative analysis in a primarily qualitative area. Naturally, this level of precision is attained at the cost of adopting comparatively simple machine models governed by 'criterion vectors' $(\Phi(\alpha)_n)_n$. An 'idea; machine/vector, Φ^*, is postulated , and the pseudometric is constructed 'to within distances from Φ^*', under 'translational invariance of d_H, and $\models$ this is a condition weaker, than 'independence wrt (Pareto) preference'm expressed **via** $1-$dim^2 '**value functions**', V_j. If $\Phi := \underset{1 \leqslant j \leqslant n}{X} \Phi_j$, then n-dimensional value functions, V_Φ are definable on (the TS) (Φ, τ), and $:\models \exists V_\Phi \Leftrightarrow (\Phi, \tau)$ is ps-metrizable. Further, if $(\alpha)_n \in D$, then $\Phi(D) \subset \Phi$, and $V_\Phi(\Phi), V_\Phi(\Phi(D))$ are w.d.; and the set $V_\Phi(\Phi)$ is a Hausdorff TVS, provided that the ps-m is translationally invariant. **NOTE that** the (continuous, monotonic) value functions, V_j (on $[0,1]$) are readily constructed – hence, so if d_H. Lastly, $\models$ when such a pseudometric d_H exists on $V_\Phi(\Phi)$,there is a **metric** ρ_H such that $x \in \Phi \Rightarrow d_H(\Phi^*, x) = \rho_H(\Phi^*, x)$.

Approximation criteria for d_H may be given.

52. A basic characterization of a class of **optimal quadrature schemes** may be given in terms of **exponential polynomials (EP) deviating least from** 0 (in the integral mean) **over the quadrature interval.** The EP are represented as elements of **kernels**, M_r, of **OD O(perators)** $P_r(d_t)$, where – for a prescribed sequence $\{\lambda_j\} \subset \mathbb{R}$ - one sets $P_k(z) := \prod_{1 \leqslant j \leqslant k} (z - \lambda_j)$, so that $M_r = M_r(\lambda)_r$.

The associated '**best-approximation-in the mean problem**' has the form: **determine** $\inf_{g \in M_r} \{\int_a^b |y(t) - g(t)|dt\} \equiv F_r$, **for a specified element** $y \in M_{r+1} \backslash M_r$. For the analysis, $[a,b]$ may (WLOG) be transformed to $[O, \Delta]$. The main result then becomes:

If $\lambda_{r+1} \geqslant 0$ **then** F_r **is strictly convex in** Δ. The **optimality criteria** are formulated for quadrature schemes

$$I(x, \mathcal{Q}_p, T) := \sum_{1 \leqslant i \leqslant n} Q_i(\partial_t)x|_{t=t_i}$$

[437] A review of Microlocal Analysis is given in EMS Vol 33, PP 7-147 (Yu. V. Egorov); for higher microbial analysis see, e.e. LNM #1555 (O.Liess).(resp.Springer 1993, 1993)

[438] For Hyperfunctions, see, e.g., LNM # 126 (Schapira (French)), and LNP # 39 (resp: Springer, 1970, 1975) The statements of theorems require only minor changes from the basic E-WT.

for $x \in K, Q_i \in \mathcal{Q}_p$ (polynomials of degree p) $T \equiv (t)_n$ (nodes), with x periodic and 'adequately differentiable'. The periodicity is governed by extra constraints on the P_r (and hence, also, on the M_r).

The principal result here is that: **if** $\{\lambda_j\} \subset \mathbb{R}, K = \hat{C}^r(\mathbb{R}) \cap \mathcal{P}_1, \{\rho = r-1, r \in \mathbb{P}$ **or** $\rho = r-2, r \in 2\mathbb{P}\}$. **ess sup**$|P_r(\partial_t)x| \leqslant 1$ **then the (unique) optimal quadrature scheme** (up to rigid translation of T) **has equally -spaced nodes over** $[0,1]$

NOTE. Here, $\hat{C}^p(I) \equiv \{$ functions on I with absolutely continuous derivatives of order $j, 0 \leqslant j \leqslant p\}$ $\mathcal{P}_1 \equiv \{1-$periodic functions $\}$, and (as usual)

$$\text{ess sup}_\mu\, g := \inf[b : \mu\{t : g(t) > b\} = 0]$$

Optimality results for more general classes of integrands may be found (e.g.) in the books by **V.I.Krylov:'Approximate Calculation of Integrals**' (tranl., Macmillan, 1962); **A.Sard: 'Linear Approximation**' (AMS,1963), and by **E.K. Blum, 'Numerical Analysis and Computation – Theory and Practice** (A.-W, 1972).

The 'uniform spacing condition' applies to many diverse problems – for instance, minimal potential-energy states in (1-D) crystals. This should be reflected in suitably - defined mores of approach, for all problems admitting this type of optimal solution.

57. The Weierstrass Approximation Theorem covers uniform approximation, by polynomials, of real-valued, continuous functions on bounded, closed intervals; but the **rate and accuracy** of approximation (for various classes of functions) are not considered. In particular, results for Lipschitz classes, $\text{Lip}_\alpha I$, of functions f over $I \subset \mathbb{R}$ satisfying $x, y \in I \Rightarrow |f(x) - f(y)| \leqslant |x-y|^\alpha$ are of basic importance in Analysis.

For the class $\text{Lip}_1[0,1]$, rather precise estimates may be obtained – by constructing suitable approximating polynomials. The proofs of such estimates are mostly quite intricate – and not readily modifiable to treat ostensibly similar problems (e.g., for Lip_β instead of Lip_α, etc.), though the underlying **schemes** of the proofs do exhibit some patterns. Here, the basic result is that:

$$f \in \text{Lip}_1[0,1] \Rightarrow \forall n \in \mathbb{P} \exists P_n :$$

$$|f(x) - P_n(x)| \leqslant \frac{\pi}{2n}\sqrt{1-x^2} + O(\frac{|x|}{n^2})$$

which yields the simpler error term $\frac{\pi}{2(n-1)}(1 + O(\frac{1}{n})).\forall n \in \mathbb{P}$, it is also shown that the error term $[\pi?(2n+2)]\sqrt{1-x^2}$ is **not** attainable for some $f_n \in \text{Lip}_1[0,1]$ – by exhibiting an appropriate f_n.

Cognate results, for Lip_α and some other function classes , are discussed in 'Constructive Approximation', by RA De Vare/GG Lorentz (Springer, 1993).

58. The elements x, y of a group G are **conjugate** $(c(x,y))$IFF$\exists a \in G : y = a^{-1}xa; G$ is **free** IFF the only products of generators and/or their inverses that equal 1 are of form $b^{-1}b = bb^{-1}$. Let a group G, a set of groups $\{G_i\} \equiv D$, and a set $H \equiv \{h_{ij}\}$ of homomorphisms of G onto G_i, be specified. Then: **D approximates** G **in conjugacy**, via H (NOTATION: $A_c(D, G; H)$) IFF $[x, y \in G \wedge \sim c(x,y)] \Rightarrow \exists h_{ij} \in H :\sim c(h_{ij}x, h_{ij}y)$. Since this is a very general concept of approximation, concrete (counter) examples are valuable. Here, the set of groups $M_{pq}(a,b)$ generated by a, b, where $a^p = 1 = b^q$ is studied. For $p, q = 0, 2, 3, 4, 6$ (excluding $p = 0 = q$), $\models \exists F_2$ (free, of rank 2): $\sim A_c(M_{pq}, F_2)$; whereas (e.g.) $\models A_c(M_{3,2}, F_2)$. The definition of F_2 depends on various properties of its generators (say, x, y). The result: $\sim A_c$ (PSL$(2,p)$) for $p = 4k+1$ **prime**, also holds. The underlying method involves **Nielsen transformations**, $\mathcal{V}$, defined as follows. Let G be free, $H(\subset G)$ be finitely generated by $\gamma_H \in \Gamma_H \supset \bigwedge_H \equiv \{$ free sets of generators of $H\}$. Then $\mathcal{V} : \Gamma_H \to \bigwedge_H, \gamma_H \mapsto v(\gamma_H) \equiv \lambda_H$.

The broader interpretations/implications of this form of approach should be explored.

59. A curve L in $\mathbb{C}_z$ is **quasiconformal** (qc) IFF $(\forall z_1, z_2 \in L \exists c > 0 :$ $\text{diam}_L z_1 z_2 \leqslant c|z_1 - z_2| < \infty$. For bounded domains $G_k, \zeta\bar{G}_k =: \Omega_k, \partial\bar{G}_k$ q.c, $k = 1, 2)$ qc mappings $\Phi_k : \mathbb{C}_z \overset{\text{onto}}{\to} \mathbb{C}_z, \Psi_k := \Phi_k^{-1}$ and $\varphi := \Psi_2 \circ \Phi_1 =: \tau^{-1}$ are constructed (via extension procedures, etc.). For such q.c. maps, the so-called **D-property** holds: $\max_{K_r}\Delta_z f / \min_{K_r}\Delta_z f \leqslant D$, for $z \in S_f$, where $K_r := \{\zeta : |\zeta - z| = r\}, \Delta_z f := |f(\zeta) - f(z)|$, and the set S_f is w.d.. On this basis, the following **key estimates** may be proven.

(1) If $\Gamma_{\Omega_1}(\zeta)_3$ is the **family of separating curves** (of ζ_1, ζ_2 from $\zeta_3(\equiv z), \infty$ for $\bar{\Omega}_1$) with **modulus** ($\equiv$ reciprocal of **extremal length**) $m_\Omega(\zeta)_3$ **then**

$$|\frac{z-\zeta_2}{z-\zeta_1}| \leqslant B_1 \exp\{2\pi m_{\Omega_1}(\zeta)_3\} \leqslant B_2|\frac{z-\zeta_2}{\zeta_1}|^A$$

, $z \in L_1 \equiv \partial\bar{\Omega}_1, \zeta_k \in \bar{\Omega}_1, |z-\zeta_1| \leqslant B|z-\zeta_2|, B, B_1, B_2$ are independent of z, ζ_1, ζ_2, and A depends only on $\partial\Omega_1$.

(2) For $\tau := \varphi(z), \tau_k := \varphi(\zeta_k)$

$$|\frac{\tau-\tau_2}{\tau-\tau_1}| \leqslant B_3|\frac{z-\zeta_2}{z-\zeta_1}|$$

where B_3 is independent of τ, τ_k, z, ζ_k, and A depends only on $\partial\bar{\Omega}_1$.

(3) With G_1, G_2, as above (with the Ω_k now 1 - connected) and φ mapping Ω_1 conformally and univalently onto Ω_2, further conditions may be formulated such that $\forall n \in \mathbb{P}, f \in A(\bar{G}_1)\exists P_n$ (polynomials of degrees $\leqslant n$):

$$|f(z) - P_n(z)| \leqslant B_4\omega[\lambda_n(\varphi(z))]$$

where B_4 is independent of z, ω **modulus of continuity** of $f \circ \varphi^{-1}$ on ∂G_2, and $\{\lambda_n\}$ is defined in terms of **level curves** of ∂G_2.

This result is based on an integral representation $f \in A(\bar{G}_1)$, namely

$$f(z) = -\frac{1}{\bar{k}}\iint_{\Omega_1}\frac{((f\circ y))(\zeta)}{(\zeta-z)^2}y_{\bar{\zeta}}d\sigma_\zeta, \zeta \equiv \zeta + i\eta$$

Here, $y(\zeta) := \Psi_1\{[\phi_1\bar{(}\zeta)]^{-1}\}$ is **the q.c. reflection in** $L_1 \equiv \partial\bar{G}_1, y_{\bar{\zeta}} := \frac{1}{2}(y_\xi + iy_\eta)$ and σ_ζ is standard planar measure.

60. Species of **quasi-universal representation** for classes of entire functions are subsumed by the following result:

(1) Let the **T**aylor **S**eries, $\sum a_k z^k$, of the entire function f satisfy the conditions (i) $a_n \neq 0$; (ii) $\{|a_{k-1}/a_k|\}$ is nondecreasing, Then: $\exists\{\lambda_k\} \subset \mathbb{C}$ (with all terms distinct), $\lambda_k \to \infty$, such that $F(z) = \sum_{k\in\mathbb{N}} d_k^F f(\lambda_k z)\forall z \in C_R \equiv \{z : |z| <\}, 0 < R < \infty$. (**NOTE that** the d_k^F are not constructively determined, here).

(2) If, now: (i) $a_k \neq 0$ (ii) $\{|a_{k-1}/a_{k+1}|\} \nearrow$ (iii) $\limsup_{k\to\infty}|a_{k-1}/a_{k+1}| \equiv \Delta < 1$, then $\{f|a_{k-1}z/a_k|\}$ forms a basis for $F \in A_R \equiv \{g$ analytic in $C_R\}$ for $\Delta < R \leqslant 1$, with regular convergence in C_R and coefficients d_k given by:

$$d_k = \omega_L(H_k; F)/L'Hx_k), \text{ where } : L'(z) \equiv \partial L/\partial z$$

$$\mathcal{H}_k := |a_{k-1}/a_k|; F(z) =: \sum b_k z^k$$

$$L(z) := \Pi(1 - z/\mathcal{H}_k) =: \sum c_k z^k$$

$$\omega_L(z; F) := \sum(b_k/a_k)\sum_{m\geqslant k+1} c_m z^{m-k-1}$$

(3) If, instead, $\{\mathcal{H}_k\} \nearrow, a_k \neq 0$, $\limsup_{k\to\infty}\mathcal{H}_k/\mathcal{H}_{k+1} = 0, \{\lambda_k\} := \{\mathcal{H}_m\} \cup \{0\}$, then : $\{f(\lambda_k z)\}$ forms a basis for A_R, ‘ $< R \leqslant \infty$, and $d_k = \omega_L(\lambda_k; F)/(L'(\lambda_k)$

The essentiality of these sets of conditions may be demonstrated.

61. For subsets K (of a MS) admitting some T (Chebychev)**system** $\psi)_n \subset C(K)$, generating the subspace P_n of 'polynomials', the **best approximation operator**, $\pi_n : C(K) \to P_n, f \mapsto \pi_n(f)$ produces the best approximation from P_n to $f \in C(K)$. (One-sided/directional) **derivatives** of π_n at f may be defined – together with **G**âteaux derivatives – under various sets of conditions. The **existence** of such derivatives 'at f (along φ)', of order $m \geqslant 1$, may be characterized comparatively simply, as follows.

(1) Let $A(f) := \{x \in K : |f(x) - \pi_n(f)(x)| = \|f - \pi_n(f)\|_\infty\}, f, \varphi \in C(K), f \notin P_n$. **Then** $\exists!$ $(\alpha \in \mathbb{R}, p_n(f,\varphi) \in P_n) : x \in A(f) \Rightarrow [\varphi(x) - p_n(x)]$ sgn $[f(x) - \pi_n(f)(x)] \leqslant \alpha$.

(2) Let $f, \varphi, (\psi)_n, p_n(f,\varphi)$ be as just specified (if $f \notin P_n$), and $p_n(f,\varphi) := \pi_n(\varphi)$ (if $f \in P_n$). **Then:** $\pi_n(f + t\varphi) = \pi_n(f) + tp_n(f,\varphi) + o(t), t \to o^+$. **Hence:** π_n is once-differentiable along φ IFF $p_n(f,\varphi) = -p_n(f,-\varphi)$.

NOTE: by definition, π_n is G- **differentiable** at f IFF π_n is differentiable along arbitrary φ, at f.

The operator π_n is m**- times-differentiable** at f along $\varphi, m \geqslant 1$,
IFF: $(\#)\exists(D(f,\varphi))_n \subset P_n : \pi_n(f + t\varphi) = \pi_n(f) + \sum_{1 \leqslant i \leqslant m} t^i D_i(f,\varphi) + o(t^m), t \to 0$.

For **one-sided** differentiability, '(D_n)' is replaced by $(D^\pm)_n$, and '$t \to 0$', by $t \to 0^\pm$. A typical result, here, depends on certain '**smoothness properties**' of $f, \varphi, (\psi)_n$. Let $m(\geqslant 2) \in \mathbb{P}, K \equiv [a,b] \subset \mathbb{R}$, or else $K \equiv$ circumference $[0, 2\pi], (\psi)_n \equiv$ a T-system on $K, f, \varphi, \psi_i \in C^m(K), (\partial/\partial x)^2[f(x) - \pi_n(f)(x)] \neq 0$, for $x \in A(f)$; **then** $(\#)$ holds for some $(D^+)_n \subset P_n$, with $t \to 0^+$

NOTE that: 'm^+ - differentiability (of π_n) at f along φ' **is equivalent to** 'm^- - differentiability (of π_n) at f along - φ'; and, that $\models D_i^-(f,\varphi) = (-1)^i D_i^+(f,\varphi)$, for $1 \leqslant i \leqslant m$.

62. Denote by X either of the spaces $C(T_2), L(T_2)$ of **functions over the 2-D torus; and, by** $E_{n,m}(f)_X$ **the best approximation of** f **(in the metric of** X**) by trigonometrical polynomials of order** n **in** x **and** m **in** y. $E_{n,\infty}(f)_X, E_{\infty,m}(f)_X$ correspond to trig. polynomials in x (resp., y) with 'coefficients' depending on y (resp., x). **The estimates:** $E_{n.m}(f)_X \leqslant A\log\{2+ \min(n,m)\} \times [E_{n,\infty}(f)_X + E_{\infty,m}(f)_X], A$ an absolute constant, are classical. **Stronger results** – for the special sequence $\{E_{n,n}(f)_X\}$ may be formulated succinctly in terms of **sequences** $\{\epsilon_n^1\} \searrow 0, \{\epsilon_n^2\} = o(1/\log n), \{\omega_n\} \nearrow \infty$, and $\bigwedge_n(f)_X := E_{n,\infty}(f)_X + E_{\infty,n}(f)_X$, as follows.

(1) $\exists f \in X : \limsup_{n\to\infty} E_{n,n}(f)_X / \bigwedge_n(f)_X > 0$

(2) Given $\{\epsilon_n^1\}\exists f^{\{\epsilon_n^1\}} : E_{n,n}(f)_X \leqslant \epsilon_n^1$ and (1) holds.

(3) Given $\{\epsilon_n^2\}\exists f^{\{\epsilon_n^2\}} : E_{n,n}(f)_X \geqslant \epsilon_n^2$ and (1) holds.

(4) Given $\{\epsilon_n^1\}\exists g^{\{\epsilon'_n\}} : 0 < limsup_{n\to\infty}(\epsilon_n^1)^{-1} E_{n,n}(f)_X < \infty$ and (1) holds for g

(5) If $\epsilon_n^1 \leqslant c\epsilon_{2n}^1, \omega_n = 0(\log n), \epsilon_n^1 \leqslant E_{n,n}(f)_X \leqslant \omega_n \epsilon_n^1$ then $E_{n,n}(f)_X \leqslant c'\omega_n \bigwedge_n(f)_X$

(6) If $\epsilon_n^1 \leqslant c\epsilon_{2n}^1, \omega_m - \omega_n \leqslant c''\log(m/n), m > n$ and $\exists\alpha > 0 : n^\alpha \epsilon_n^1 = o(1)$, then $\exists f \in X :$

$$c_1\epsilon_n^1 \leqslant E_{n,n}(f)_X \leqslant c_2\omega_n\epsilon_n^1$$

All n-conditions hold for $n \in \mathbb{N}$

63. Uniform estimates of the rate of convergence in the **C**entral **L**imit **T**heorem for **i**dept, **n**on-**i**denticaly **d**istributed r.v. with values in H (a real, separable **H**ilbert **S**pace) may be obtained for (substantially) arbitrary covariance operators. This formulation covers many 'special cases'. **All of the results involve estimates of the function**

$$\Delta_n(a) : \Delta_n(a) := \sup_{t \geqslant 0} |Pr\{B_m^{-1} \sum_1^m X_i \in S_t(a)\} - \mu(S_t(a))|,$$

where μ is the Gaussian distribution with mean 0 and covariance operator T; the X_i have non degenerate distribution F_i, with mean 0 and covariance operator $\sigma_i^2 T$; $S_t(a)$ is the 'sphere' of radius t about a in $\mathcal{H}$; $B_m^2 := \sum_1^m \sigma_i^2$; and, $|\ldots|$ denotes the norm $|x| := (x,x)^{\frac{1}{2}}$ of the i.p. (,) in $\mathcal{H}$.

If $\mathcal{V}_{ir} := \int_H |x|^r |F_i - \mu_i|(dx)$; $\bigwedge_r := \sum_{i=1}^m \mathcal{V}_{ir}$; $\mathcal{V}_r := n^{-1} \bigwedge_r$, then these estimates (under various extra conditions) have the forms:

(1) $\Delta_n(a) \leqslant c_1(T,r)(1+|a|) \bigwedge_r^{\frac{1}{r}} B_m$

(2) $\Delta_n(a) \leqslant c_1'(T,r)(1+|a|)[\mathcal{V}_r n^{(2-r)/2}]^{[1/r]}$

Here (WLOG), $\{\sigma_i\} \searrow, \sigma_i > 0$; the μ_i are d.f. for i. r.v. Y_i (independent of $(X)_m$), and $\mu_i = \mu(0, \sigma_i^2 T)$. **NOTE** that the X_i may be **non**-identically distributed ; and, that the Y_i may be 'conveniently chosen' to facilitate the proofs.

64. In the **quasi-metrical theory of finite groups**, various types of **length function** may be defined – in particular, the p-length, $l_p(G)$, and the **derived length**, $d_p(G)$, of the finite, p-solvable group G. The result: $l_p(G) \leqslant d_p(G)$, for p prime, $p \geqslant 3$, is well known: P.Hall/G.Higman, PLMS (3)6(1956)1-42 – where bounds for $l_p(G)$ in terms of the structure of the sylow p-subgroup of G are obtained (and used to obtain cognate bounds on $d_p(G)$). The 'extra case' ($p = 2$) requires different methods from those for '$p \geqslant 3$'; but the result is the same: $l_2(G) \leqslant d_2(G)$, where G is now a sylow 2-subgroup of a finite solvable group. The subject is rich in intricate finitary procedures and highly specialized constructions. Many of the results are proved by induction (on the order(s) of subgroups, etc.); others make use of 'central series', sets of commutators, invariants, etc.. (The theory of group representations – (in terms of characters/modules/blocks/…) has its own typical techniques, some of which apply, also, to infinite groups …).

65. The term **quasi-power basis** refers to families/sequences of (analytic functions spanning various function spaces. Typically, it has the form

$\{beta_n z)\}, \{\beta_n\} \subset \mathbb{C}, f_n(z) := \sum_{k \geqslant 0} a_k^{(n)} z^k \in \mathcal{A}_{R_n}$ and $\inf(R_n/|\beta_n|) \equiv r_0 > 0$ and hence, the 'qp-basis property' holds (if at all) only in those A_r with $r \leqslant r_0$. **Sufficient conditions** for $\{z^n f_n(\beta_n z)\}$ to be a qp-basis in A_r may be established under certain conditions on the $\{a_k^{(n)}\}$ and $\{\beta_n\}$: see I.I.Ibragimov, et al., **Sov. Math. Dokl. 15(1974)**169-173. **Here**, a simpler prescription is formulated, as follows. **Let** $\Phi(z) := \sum_{n \geqslant 0} a_n z^n (a_n \neq 0, |a_n|^{\frac{1}{n}} \to 1); \Pi_m(t, (\lambda)_m) \equiv \Pi_m^{(\lambda)}(t) := \prod_{1 \leqslant k \leqslant m} (t - \lambda_k)^{-1} \; (|\lambda_k| \leqslant 1); F_n(z) := \frac{1}{2\pi i} \int_{|t|=R_1} \Phi(tz)\Pi_n(t)dt$, where $|z| \leqslant r \in (0,1), R_1 \in (1, 1/r); f_n(z) := \frac{1}{2\pi i} \int_{|t|=R_1} (1-tz)^{-1} \Pi_n^{(\lambda)}(t)dt$.

Then: (1) $\models T : f_m \mapsto Tf_m = F_m$ **is an automorphism of** $A_R(R > 0)$. (2) $\{f_n\}$ **(and hence,** $\{F_n\}$**)** is a qp-basis $\Leftrightarrow |\pi_n^{(D)}(s)|^{\frac{1}{n}} \to |z|$,uniformly for $|z| > 1$.

Connections among these (and other) qp-basis constructions are worthy of investigation.

66. Certain properties of BS, X, have probabilistic interpretations. The **Banach-Saks Property** (BSP) is that: **every bounded sequence** $\{x_n\} \subset X$ **has a subsequence** $\{x_m'\} \subset \{x_n\}$ **that is** (C,1) - **convergent-in-norm to** $x \in X$ – i.e., $\|M^{-1} \sum_{1 \leqslant m \leqslant M} x_m' - x\| \to 0 (M \to \infty)$. The (C,1) - summability corresponds to the 'summation matrix' $S \equiv (s_{pq})$, where $s_{pq} := p^{-1}$step$(1 - p^{-1}q)$. If the x_m' are r.v., and prob$\{\sum_{q=1}^{Q} \|s_{pq} x_q' - x\| \overset{Q}{\to} 0\} = 1$, then $\{x_m'\}$ is a.s. - S **summable (in norm) to** x. (This definition applies to any summation - matrix S).

Ramifications of the BSP are considered in **Partington, JR** MPCPS 82 (1997) 369-374. Summability of random sequences is treated in **Kahane, J-P., 'Some Random Series of Functions** (Heath, 1968; 2nd ed., 1985); a.s.-convergence (in Analysis) is discussed, e.g., in: **'Stochastic Convergence'**, by **E. Lukacs** (Heath, 1968). BS-aspects of summability are treated (e.g.) in: '**Summability Through Functional Analysis**', by **A. Wilansky**(North-Holland, 1984). The interplay of Geometry, (Functional) Analysis, and Probability Theory is reflected in the **study of stochastic processes on general spaces**

72(ii) **Probabilistic Number Theory** (PrbNT) is concerned primarily with the distributions of values of various classes of **Arithmetic Functions** (AF). Such functions, say, f, are called **additive** IFF $f(ab) = f(a) + f(b)$, whenever the positive integers a, b are coprime. The **asymptotic distributions** of AF (in specified senses) are of particular interest – for instance, as forms of the **W**eak/**S**trong **L**aws of **L**arge **N**umbers (W/S LLN). The convergence involved may be weak ($\overset{\omega}{\to}$), strong ($\overset{s}{\to}$), or 'to be specified' ($\to$). Results in this area are typically prolix (since several 'side-conditions' may be imposed.) Nevertheless, **quasi-tauberian theorems**- relating asymptotic behaviour of (individual) AF to stochastic behaviour of associated classes of AF – may be established, in some cases. A wide range of tools from Measure Theory, Functional Analysis, Combinatorial Theory, Probability Theory, ... has been deployed in PrbNT – through (complex) Function Theory is not so common here as in **An**alytic NT. The further extension to (say) the AF in **Al**gebraic NT may be considered by combining the basic methods of PrbNT with those in: '**Abstract Analytical NT**' $\equiv (K)$ by **J. Knopfmacher** (N-H, 1975) and, '**Probabilities on Algebraic Structures**', by **U. Grenander** (Wiley, 1963) $\equiv$ (G). Indeed, a surprisingly broad collection of 'admissible structures' may be used here.

In (K), the aim is to extend to 'arithmetical semigroups', S, the function-theoretic methods used in 'classical Analytic NT'. Here, the basic structures are free, commutative semigroups (with unit) generated by a countable set, B, of free generators, and admitting a (real-valued) multiplicative norm, $\|$, such that $|a| > 1$ on B, and $N(x) \equiv \#\{a \in S : |a| \leqslant x \in \mathbb{R}\} < \infty$. Analogues of D(irich-let) - convolution, and of basic AF may be found, and **G**enerating **F**unctions for enumeration problems may be formulated for S (covering, e.g., the categories of finite abelian groups, semi-simple finite rings compact Lie groups, and symmetric Riemannian manifolds). Analogues of some key theorems in classical NT may be obtained for those S where $N(x) = Ax^{\theta} + O(x^{\eta})$ $(A > 0, \eta < \theta, x \to \infty)$ [see: MR54 (1977) #7404 and BAMS 83 (9/77)1021-7, for useful reviews of (K)]. In (G), the emphasis is on limit laws for n-fold iterations of suitable binary combinations of i.(i.)d.r.v., as $n \to \infty$ – for all algebraic structures S over which such problems are w.d.. The aim is to produce analogues of the 'main theorems' of 'standard Probability Theory'. Topological spaces/ groups, l.c.groups, nuclear spaces are among the admissible structures over which limiting distributions – and other topics – may be investigated. [See MR 34 (1967) # 6810 for a substantial review with several references.]

NOTE. The use of species of 'harmonic analysis' over S is central to (G) and so partially determines the range of S for which nontrivial results are provable.

The asymptotic distributions of concern in the present item involve the so-called frequencies, $v_x(f, \alpha; z, \beta) := x^{-1}\#\{n \leqslant x : f(n) - \alpha(x) \leqslant z\beta(x)\}$, where α, β, are positive-valued functions defined for $x \geqslant 2$, and β (but not $\alpha or f$) must satisfy certain growth conditions. For the (quasi-) WLLN, it suffices that $\beta(x) \to \infty (x \to \infty)$ and $\sup_{\omega \leqslant x^2} \beta(\omega) \leqslant C\beta(x) (x \geqslant 1, C$ constant). If, instead, one has $y > 0 \Rightarrow \beta(x^y) \sim \beta(x)$, as $x \to \infty$, and $\sup_{\omega \leqslant x} \beta(\omega) \leqslant C\beta(x)$ then a WLLN result holds for any (positive/...) α; but a type of SLLN may be proven under extra conditions on α.

72(ii) There is an extensive theory of best approximation in NLS (by elements of subspaces): See **Singer, I., 'Best Approximation in NLS by Elements of Linear Sub Spaces**' (Springer, 1970). For the particular case of **Von Neumann Algebras**, the **characterization** and **uniqueness** of **b**est **a**pproximations may be studied in terms of **central projections** and **Chebyshev sub spaces**, $V \subset M$ (a v.N. algebra), where V **is a chebyshev subspace of a BS X** IFF **every** $\xi \in X$ **has a unique b.a. by elements in** V. Denote by $\gamma(M)$ the **centre** of M.

Theorem 1. If $x \in M$, then $\mathbb{C}x$ is Chebyshev for M IFF $\exists$ a projection $p \in \gamma(M)$: px is left-invertible in pM, and $(1-p)x$ is right-invertible in $(1-p)M$.

This covers $1 - D$ chebyshev subspaces. In general, no $n - D$ chebyshev subspaces can exist, since:

Theorem 2. Let N be a $k - D$ $*$-subalgebra of M $(k > 1)$. Then N is not Chebyshev for M. These criteria, though not profound, are relatively simple in form, and so useful in applications.

NOTE. A **von Neumann** (or, W^*-) **Algebra**, M, is a $*$-subalgebra of the set, $B(H)$, of bounded linear operators on a (separable) HS, H, such that M is closed under the weak operator topology (see: **Berberian**, S.K., '**Lectures in Functional Analysis and Operator Theory**' (springer, 1974) p173, for this topology). The '*-subalgebra property' is equivalent to the condition(s): $\{A, B \in M \wedge \lambda, \mu \in \mathbb{C}\} \Rightarrow \{\lambda A + \mu B, A^*, B^*, AB \in M\}$ ($*$ denoting **adjoints**).

(Alternative definitions may be given for strong/uniform operator topologies: see $pp172/168$ of Berberian, op.cit).

The **centre** of an algebra, A, comprises all elements commuting with every element of A.

The v.N. algebras offer one 'natural' analogue (over infinite-dimension sets) of the finite-dimensional **matrix algebras.**

73. The question of the extent to which 'linearity' determines the structure of **L**inear **A**lgebra may be explored by constructing 'models of L.A. type' involving **non**linear operations. here, '**monotone sets/gauges**' (defined over $P_n := \{x \equiv (\xi)_n : \xi_j \geqslant 0, 1 \leqslant j \leqslant n\}$) model the usual point-line duality, and multivalued **monotone convex/concave processes** T, possessing adjoints, T^*, inverses T^{-1} – and admitting $S + T, ST, \lambda T$ as w.d. operations ($\lambda \in \mathbb{C}$) – mirror many properties of matrices. The corresponding modelling of 'quasi-**e**igen **v**alues' is more problematical. Nevertheless, by calling $\lambda(>0)$ a **sub** - e.v. of T IFF $\exists x > 0 : \lambda x \in Tx$, one may characterize the set of all sub-e.v. of T (or, of T^*) as follows. For T concave (and so T^* convex) $\bigwedge(T) = (0, \bar{\lambda}]; \Lambda(T^*) = [\underline{\lambda}, \infty)$, where (for $y- \;>> 0$ IFF $y - z \in P_n^0)\bar{\lambda} := \inf_{x^*>>0} \sup_{x>>0} F(x, x^*)$ and

$$\underline{\lambda} := \sup_{x>>0} \inf_{x^*>>0} F(x, x^*), F(x, x^*) := \frac{\langle Tx, x^*\rangle}{\langle x, x^*\rangle}$$

. Here, $F(u, v)$ is the **Kuhn-Tucker functions:** $F(x, y^*) := \sup\{\langle y, y^*\rangle : y \in Tx\}$ and $\langle , \rangle$ is the usual i.p. in $\mathbb{R}^n$. Analogues of the Perron-Frobenius Theorem follow from this. Applications to (Econometric) optimization problems may be made.

88. A general axiomatization of sets/ functions (originally due to Von Neumann) may be adapted to cover both **fuzzy sets** (whose characteristic functions have ranges in posets, L- e.g., in $[0, 1]$ – rather than in $\{0, 1\}$) and **multi-sets** (which may have arbitrarily repeated members – e,g, the collection of zeros of a polynomial.). This is important for Σ , since many of the analogues considered may be formulated within such variants of 'standard sets'. In the v.N axiomatization there are three (nondisjoint) types of objects: **argument, functions**, and **arguments-functions** (which may play either role). The key 'new' axiom here is, in fact, derivable from the '**limitation-of-size principle**' discussed in **Fraenkel, A.A., Bar-Hillel, Y., Levy, A.**, '**Foundations of Set Theory**' (N-H, 2nd ed., 1973) (esp., section 7.4). on this basis, the following definitions may be given:

- An **L-fuzzy class/set** is any function/arg-function taking only values in L
- A **multi-class/set** is any function/arg-function taking only cardinal-number values

A neutral term – for instance, collection, or, assembly – could be used to include all w.d. class/set - like ME, with extra conditions for each of the special cases.

89. In Algebraic Geometry (and SAM integration) **Puiseux series**, typically of the form

$$\mathcal{S}_{d,r} \equiv \sum_{i>r} a_i X^{i/d} \; (i, r \in \mathbb{Z}, d \in \mathbb{P}, a_i \in F)$$

, play an important part. The field $\mathcal{S} \equiv \bigcup_{n\in\mathbb{P}} F\{X^{\frac{1}{n}}\}$ of all Puiseux series is **algebraically closed** (a.c.), contains $F[X]$ (the **polynomial ring**) and also furnishes an a.c. extension of **the field of formal PS.** A natural generalization may be formulated, in which analogues $\mathcal{S}_m$, of $\mathcal{S}$, for $m(\geqslant 2)$ indeterminates, are obtained – namely **a.c. fields containing the** m**-variable polynomial ring**, $F[X])m$. This may be achieved by defining a **field-family**, $\mathcal{F}_\Gamma$ (over F, with Γ a totally ordered abelian group) comprising subsets of Γ, where:

(i) $\mathcal{F}_\Gamma$ (as a set) generates Γ,

(ii) $A \in \mathcal{F} \Rightarrow A$ is w.-o rel Γ

(iii) $\mathcal{F}$ is closed under union and set-inclusion;

(iv) $\Gamma + \mathcal{F} \subset \mathcal{F}$; $(\vee)$ if $\Gamma^+ \equiv$ the positive part of Γ and $\langle A \rangle \equiv$ the additive semigroup generated by A, **then** $(A \in \mathcal{F}, A \subset \Gamma^+) \Rightarrow \langle A \rangle \in \mathcal{F}$.

On this basis, one has: $\models F^{\Gamma}(\mathcal{F}) := \{\varphi : \Gamma \to F,\ \text{supp}\varphi \in \mathcal{F}\}$ **has a unique prolongation of its natural valuation to any of its extensions.** $\models (F$ a.c.(char. 0), $\Delta \equiv$ **divisible hull of** $\Gamma)$ $\Rightarrow F^{\Delta}(\mathcal{F}_{\Delta})$ is a.c. Further, if $W(\Gamma) := \{W \subset \Gamma : W$ is w.o $\}$ then $\models W(\Gamma)$ is a field family – and so are $\mathcal{P}_1 := \{\frac{1}{n}W : W \in W(\Gamma), n \in \mathbb{P}\}$, and $\mathcal{P} := \mathcal{P}_1 \cap A(A := \{D \subset \Delta : \langle \Gamma \cup D \rangle$ is $\Gamma -$ f.g.$\})$ Then: $\models W(\Gamma) \subset H \in \{\mathcal{F}.\} \Rightarrow \mathcal{P} \subset H$; and $\models F^{\Delta}(\mathcal{P}$ is an a.c. extension of $F^{\Gamma}(W(\Gamma))$. For Δ, see, e.g., C.Faith, 'Algebra vol.1 (Springer, 1973).

NOTES. (a) $\Gamma \equiv \mathbb{Z}, \Delta = \mathbb{Q}, F \equiv \mathbb{C}$ yields S. (b) $\Gamma = \mathbb{Z}x \ldots x\mathbb{Z} \equiv \mathbb{Z}^m$ (with lex. order) yields S_m. (c) If char $(F) = p \neq 0$, then the situation is more complicated, with few general results.

90. In the study of **derivatives** (of all orders) of functions $F : \mathbb{R} \supset A \to B \subset \mathbb{R}$, there are many variants of 'the derivation process'. In particular, the **upper/lower Peano derivates** may be defined by:

$$uF_k(x) := \limsup_{h \to 0} \frac{k!}{h^k}[F(x+h) - F(x) - \sum_{i=1}^{k-1} \frac{h^i}{i!} F_i(x)]$$

in the case where $A \equiv [a, b]$ and $x + h \in [a, b]$ in the lim sup process; the definition of $lF_k(x)$ is obtained if 'lim inf' replaces 'lim sup'. In this inductive scheme uF_1 and lF_1 are taken as the **standard** upper/lower derivates, lim sup/lim inf $h^{-1}[F(x+h) - F(x)]$. When $uF_k(x) = lF_k(x) =: F_k(x)$, the **ordinary** $k^{\text{th}}P$**–derivative** is obtained (with corresponding one-sided forms for $h > 0/h < 0$, etc.) FInally, if lim sup/lim inf are replaced by lim sup ap/liminf ap, then **approximate** k-th P-derivatives may be determined. Here, the 'ap versions', i.e., approximate upper/lower limits (of F at x_0) may be defined as (respectively):

$$\underline{bd}[y : x_0 \in D\{x : F(x) > y\}], \bar{bd}[y : x_0 \in D\{x : F(x) < y\}]$$

where DS denotes the set of **points of dispersion** of the set S. In general , the relations:

$$\liminf_{x \to x_0} F(x) \leqslant \liminf_{x \to x_0} \text{ap} F(x) \leqslant \limsup_{x \to x_0} \text{ap} F(x) \leqslant \limsup_{x \to x_0} F(x)$$

are valid. The 'ap derivatives' are denoted by $F_{(k)}(x)$, etc., and the standard derivatives by $F^{(k)}(x)$, etc..

Here, the **the main results** are as follows:

(α) For $n \geqslant 2$, let $F_{(n-1)}(x)$ exist $(< \infty)$, and be bounded on one side, throughout $[a, b]$; then $F^{(n-1)}(x)$ exists and equals $F_{(n-1)}(x)$ throughout $[a, b]$.

(β) For $n \geqslant 2$, let $F_{(n-1)}(x)$ exist $(< \infty)$ over $[a, b]$. **If** $uF_{(n)}(x) \geqslant 0$, **a**lmost **e**verywhere, and $> -\infty$, **n**early **e**verywhere (both, over $[a, b]$), then $F_{(n-1)} \nearrow$ continuously, over $[a, b]$.

NOTES (i) The term 'a.e.' is standard, but 'n.e.' is defined in Potential Theory (for potentials and capacities). For $E \subset \mathbb{R}^m$, points of dispersion are points at which the strong upper/lower derivates of the measure function of E both varnish. Here the measure function of $E : X \mapsto |E.X|$ ($||\equiv$ outer measure, $\cdot \equiv$ intersection) is usually denoted by $L_E : \mathbb{R}^m \to \mathbb{R}$; it is additive and absolutely continuous.

94. The study of periodic (P) solutions of (non)linear systems of (O)DE has many technological applications (e.g., in Mechanical/Civil Engineering, Electronics), and several solution procedures are extensively covered in the literature. For almost -periodic (AP) (O)DE, by contrast, far less is known, and general methods are less tractable than those for periodic (O)DE. The most general problem in this area is concerned with **the characterization of systems of operation equations admitting AP solutions** – and with the development of cognate solution procedures. **Comparisons** among systems possessing P, AP, and non - P solutions are also of importance, since they reflect the special properties of AP functions, in

analytic/algebraic contexts. **NOTE that**, even linear systems with AP-matrix representations exhibit a rich variety of behaviour. Here, the basic system has the form $Lx(t) = f(t; x(t))$, where $(i) L \equiv D_t^m + \sum_{k=1}^{m} A_k(t) D_t^{m-k}$ for $D_t \equiv \frac{d}{dt}, A_k, f$ are AP in t (f uniformly AP for x in compact subsets of $\mathbb{R}^m$), and L is **regular** – i.e., for each bounded, continuous $f : \mathbb{R} \to \mathbb{R}^m$ the equation (ii) $Lx = f$ has a unique bounded solution, say, Tf, on $\mathbb{R}$, such that Tf is AP if f is AP. It turns out that Tf is representable in the form (iii): $(Tf)(t) = \int_{\mathbb{R}} G(t,s) f(s) ds$, with a suitable **Green's function**, G (since L induces an **exponential dichotomy**); and that **the existence of AP solutions of** $Lx(t) = f(t, x(t))$ **is equivalent to that of fixed points for** F, where $F(x)(t) := f(t, x(t))$.

These f.p. may be investigated in terms of results for positive monotone operators – typified by $f \mapsto \int Gf$, under suitable conditions on G (via L), which allows for some of the analysis to be extended from linear to (some) **non**linear AP operators. Other aspects of the theory of AP oscillations are covered in the books: 'APDE', by AM Fink (Springer LNM#377, 1974, pp336 with 548 references!) and 'AP Functions and DE', by BM Levitan/ VV Zhikov (CUP, 1982, with 132 references); both of these books refer to the book by Kransnosel'skii/Burd Kolosov considered here. In Levitan/Zhikov, BS-valued AP functions and associated AP operator - DE are covered. For Σ , the possible extensions to ${}_SX$- valued AP functions (with ${}_SX$ a 'basic space', as in E_1) are of primary interest. Connections with Generalized Harmonic Analysis (in the sense of Wiener: see, e.g., 'Statistical Theory of Communication', by Y.W. Lee, Wiley, 1960) may also be made.

98. Let G be a graph with vertex-set $V(G)$ and edge-set $E(G)$. Two basic **reconstruction problems** (say $\mathcal{P}_V, \mathcal{P}_E$) require the determination (up to isomorphism) of G from the isomorphism sets of $G\backslash\xi, G\backslash\lambda$, for $\xi \in V(G), \lambda \in E(G)$, respectively. If G is finite, simple, and $|V(G)| \geqslant 3, |E(G)| \geqslant 4$, then one possible solution may be formulated through the conjectures $C(\mathcal{P}_V), C(\mathcal{P}_E)$, as follows.

$C(\mathcal{P}_V)$: If ϕ maps $V(G)$ bijectively to $V(H)$ and $G\backslash\xi \cong H\backslash\phi(\lambda)$, for $\xi \in V(G)$, then $G \cong H$

$C(\mathcal{P}_E)$: If ϕ maps $E(G)$ bijectively to $E(H)$ and $G\backslash\lambda \cong H\backslash\phi(\lambda)$, for $\lambda \in E(G)$, then $G \cong H$.

$\models C(\mathcal{P}_V)$ for **trees**; but this is a very limited solution. Partial solutions of $C(\mathcal{P}_E)$ include: $\models C(\mathcal{P}_E)$ for $|V(G)| = n, |E(G)| > n\text{log}n/\text{log}2$. The case where G, H are (in)finite, simple, and $|V(G)| \geqslant 3, |V(H)| \geqslant 3$ is covered by another conjecture; say, C^*: under the premise of $C(\mathcal{P}_V)$, each of G, H is isomorphic to a subgraph of the other. This is equivalent to $C(\mathcal{P}_V)$ if $|V(G)| < \infty$.

More generally, one might seek reconstruction schemes with other types of 'deletion-specifications'.

NOTES.(i) The Brownian Motion is used in the proofs to establish relevant properties of surfaces/manifolds; it is not studied here for its intrinsic interest. (ii) The definition of BM over various sets depends on the structure of the sets– for instance, atlases on manifolds. In such cases, Bj is defined quasi-re-cursively, by local extension, etc. (iii) For the 'tail σ-algebra' of a BM see Sec. 2X. 11 of Doob, J.L., 'Classical Potential Theory and its Probabilistic Counterpart' (Springer, 1984) (iv)$K - q - c$ mappings are treated in (e.g.): **Caraman, P., 'n-Dimensional Quasiconformal Mappings'** (transl., Abacus Press, Tunbridge Wells, Kent, UK, 1974).

104. Classes of problems in **L**inear **C**ircuit **T**heory may be treated by techniques of **non-Euclidean functional analysis**. In particular, non-Euclidean ($n-e$) analogues of **B**eurling-**L**ax-**H**almos (invariance/representation) theory over variants of H^p-spaces; and, of **N**evanlinna-**P**ick (interpolation) theory. In this scheme, the **Poincaré metric** plays a basic part. The underlying procedures (for the analysis of interconnected n-parts, power consumption, gain, etc.) involve combinations of **complex bilinear transformations** – in very general settings – associated **symmetric spaces** and **Lie groups**, and **rational approximation to elements of** $\mathcal{B}H^\infty (\equiv \{$uniformly bounded elements of $H^2(B)\}$, where $B \equiv$ the unit ball in $\mathbb{C}$). Most of this framework is covered (i.a.) in: **Foias, C./Frazho A., 'The Commutant Lifting Approach to Interpolation Problems'** (Birkhauser, 1990, pp623) many of the LCT topics occur in the study of **broadband matching** (of impedance, in power-transfer between circuits). These techniques

may be applied to diverse problems of filtering/stability/amplification/...in active/passive circuits. A broad selection of matching problems is treated in: **Chen, W-K. 'Broadband Matching ...'** (world Scientific, 2nd ed, 1988).

106. The original **P**aley-**W**iener theorem is concerned with the boundary behaviour of classes of analytic functions, namely, entire functions of exponential type. A slight generalization of the P-W theorem states that: a slowly increasing function f can be extended to an entire function of exponential type $\leqslant b$ IFF the support of its F.T., $\mathcal{F}f$ (as an elements of $\mathcal{S}'$) is contained in the set $\{x : |x_k| \leqslant b_k, 1 \leqslant k \leqslant n\}$. Here, $z \equiv (x+iy)_n \in \mathbb{C}^n$. For $n = 1$, the basic P-W theorem takes the form: Let E^σ denote the class of entire functions of exponential type σ, and, $S^\sigma \equiv \{f \in L^2 : \text{supp} f \subset [-\sigma, \sigma]\}$. Then: $\models \mathcal{F}S^\sigma = L^2(\mathbb{R}) \cap E^\sigma$. The theorem has applications in Control/Circuit Theory (e.g., in network realization criteria). In the **n**on-euclidean version, the n-e FT (Say, $\tilde{f} \equiv \mathcal{F}^{ne}f), \tilde{f}(\mu, \beta) := \int \exp\{(\frac{n-1}{2} - i\mu)\langle z, \beta\rangle\} f(z)dV$ is introduced, where $z := \{x, y\}, x \in \mathbb{R}^{n-1}, y > 0, ds^2 := y^{-2}(dx^2 + dy^2)$, in the n-dimensional real hyperbolic space, H_n, with volume surface elements dV (resp., dS), and $n-e$ distance $\langle z, \beta\rangle$ from $z_0 \equiv \{\underline{0}, 1\}$ to the horosphere through z and tangent at $\beta \in H_n$

The corresponding extension to $n.e.$ Radon transforms may be effected **via** Cauchy sequences of C^∞ - functions with compact support – together with a **Pl**ancherel-type integral identity – in the form: $\hat{f}(r, \beta) := \int_{\xi(r,\beta)} f(z)dS$, where $\xi(r, \beta)$ is the horosphere tangent to the x-plane at $(\beta, 0)$, with radius r: $\xi(r, \beta) : |x - \beta|^2 + |y - r|^2 = r^2$. The Pl-identity equates quasi-averages of $|f|^2$ and $|J_n\hat{f}|^2$ through associated integrals. Here, J_n is a DO (n odd) and a composite DO/**Con**(volution)O (n even). If, now, $\mathcal{C} : |x - x_0|^2 + |y - y_0|^2 \leqslant c^2 (y > 0, |y_0| < c)$ is a spherical cap in $\mathcal{H}_n$ then:
$\models f \in L^2(\mathcal{C}) \wedge \{J_n\hat{f} = 0$ on almost all horospheres contained in $\mathcal{C}\} \Rightarrow f = 0$ a.e. in $\mathcal{C}$ This is the local P-W Theorem for Radon transforms. The proof is intricate (involving 'partial FT', Volterra integral equations, and several manipulations for the evaluation of integrals.

107. The formal structure of Calculus may be given a purely algebraic representation if the underlying set, $\mathbb{R}$, is replaced by $\mathbb{R}^*$ – the **hyperreal** number system. Let $\mathcal{R}^P$ denote **the ring of sequences of real numbers**. This is not a field (or even an integral domain), but the set $\mathbb{R}^P/M$ (where M is a suitable **maximal** ideal of $\mathbb{R}^P$) **is** a field, say, $\mathbb{R}^*$. It turns out that, if $F := \{a \in \mathbb{R}^P : \text{supp}(a) < \infty\}$, then $\models F$ is a proper ideal of $\mathbb{R}^P$, and M (above) may be taken to be the maximal ideal containing F. The maximality of M guarantees that $\mathbb{R}^*$ **is** a field; moreover, card $\mathbb{R}^* =$ card $\mathbb{R}(= 2^{\psi_0})$, and the order relation $<$ on $\mathbb{R}$ ay be extended to $\mathbb{R}^*$ – to produce a total ordering.

If $\mathbb{R}^+ \equiv$ {positive real numbers } then one defines:
$a \in \mathbb{R}^*$ is **infinite** IFF $|a| > \mathbb{R}^+$
$a \in \mathbb{R}^*$ is **infinitesimal** IFF $|a| < \mathbb{R}^+$
$a \in \mathbb{R}^*$ is **finite** IFF a is not infinite

For $\mathbb{R}^+$, only $+\infty$ is infinite, and only 0 is infinitesimal; but in $\mathbb{R}^* \models \omega \in \mathbb{R} : (\forall r \in \mathbb{R})\omega > [r] \in \mathbb{R}^*$ – and hence, $1/\omega$ **is an infinitesimal** $\neq 0$. **This allows the basic operations of Calculus to be represented, and proven, purely algebraically.**

NOTE: this includes the differentiation formula for composite functions (awkward to prove by 'ϵ, δ methods').

This is essentially the basis of the far more extensive **Nonstandard Analysis**: and (with some further 'apparatus') of Synthetic Differential Geometry'. W.S. Hatcher's paper (Amer Math Monthly 89(1982)362-370, on which this item is based, summarises the 'hyperreal scheme;. The corresponding development of 'elementary calculus' is covered by HJ Keisler: 'Elementary Calculus: An Infinitesimal Approach' (2nd ed., pp979, PWS Publ., 1986). More advanced developments may be found in (e.g.) 'Nonstandard Analysis', by A.Robinson (N-H, 1996), and in ' Foundations of Infinitesimal Stochastic Analysis', by KD Stroyan/JM Bayod (N-H, 1987). **NOTE that,** although it has been shown that every result provable by nonstandard procedures is (in principle) provable by 'standard procedures' (see: 'The Strength of Nonstandard Analysis' by C.W. Henson) the 'infinitesimal method' may provide insights and suggest further investigations, that are overlooked (or masked) in the usual frameworks.

The book: 'Synthetic Differential Geometry', by A.Kock (CUP, 1981) combines Nonstandard

Analysis with Topos Theory, with cognate results from Algebraic and Differential Topology.

110. Generalized Radon transforms furnish basic examples of reconstruction/retrieval (of functions, from various collections of 'partial representations'). From a slightly different perspective these are inverse problems (with zero loss of information). For Σ , such classes of problems are embedded within the class of '**partial-retrieval ME procedures**' where (typically) information is lost, and only incomplete reconstruction is possible. This includes not only, many types of 'transforms', but also, mappings with stochastic components. Here, there is associated with each function on some 'geometric space', (some of) its integrals over a suitably specified collection of 'geometrical objects'. **The aim** is to reconstruct the function as fully as the conditions allow. This is the **G**eneralized RT Problem (for **inversion).** There are several quasi-formal inversion techniques (depending on the formulation of the RT). A unified representation (for odd/even n) is:

$$f(\underline{x}) = C_n \Delta_{\underline{x}}^{(n-1)/2} \int\limits_{|\underline{\xi}|=1} \overset{v}{f}(\underline{\xi}.\underline{x}, \underline{\xi}) d\underline{\xi}$$

, where $C_n := 2^{-n}(\pi i)^{1-n}, \Delta_{\underline{x}}$ is the euclidean $n-$space Laplacian (interpreted in terms of fractional derivatives for n even) and $\overset{v}{f}$ denotes the standard RT of f.

Another inversion procedure involves the FT and its inverse, but this if of limited use in evaluation. The unification is achieved by introducing the operator $\mathcal{R}^t$, where : $\mathcal{R}^t\psi(\underline{x}) := \int\limits_{|\xi|=1} \psi(\underline{\psi}.\underline{x}, \underline{\xi}) d\underline{\xi}$. A variety of formal relationships may be derived among these, and associated operators. More effective (quasi-numerical) inversion schemes have been devised for applications in tomography, geophysics, nondestructive testing, etc.. Extensive development/examples of the GRT (mainly, for full reconstruction) are covered in: **Deans, S.R¿, 'The RT and some of its Applications' (pp289, Wiley, 1983); P.C.Sabatier (ed.), 'Basic Methods of Tomography and Inverse Problems'** (pp 671, Hilger, 1987). The challenge for Σ is to formulate diverse analogues of the GRT over the Basic-spaces/convergence structures introduced in E_1 – with 'integration' replaced by w.d. determinations of 'mean values' over appropriate collections of 'subobjects'. Here, stochastic components may be incorporated.

111. One of the underlying features of Σ is the emphasis on t-maximal operationality (pragmatically defined). In E_1, types of constructivity are discussed, and it is found that 'B-constructivity', based on the system introduced by **E.Bishop** in his book: '**Foundations of Constructive Analysis** (McGraw-Hill, 1967) and since widely elaborated – see, e.g., '**Constructive Analysis**' by **E.Bishop and DS Bridges** (Springer, 1985), and references there – provides a broadly applicable foundation for most of the implied developments envisaged in the evolution of Σ as an OIB.

In this connection it is important to compile 'a lexicon of basic B-constructive formulations' – to be used as 'standard' in the stylistic design of proofs. Indeed, such standardization enhances comprehensibility and soundness, because the fundamental components of proofs meet the essential constructive criteria, and so may (usually) be constructively combined within more intricate structures – to produce 'new entries' for the lexicon .... **NOTE**, however, that pragmatism is a crucial element in the Σ scheme in view of the vastness of scope and diversity of concepts.

123. The domain $D(\subset \mathbb{C})$ is **l**inearly **a**ccessible IFF the complement, $\mathbb{C}\backslash D$ is representable as a union of half-lines. Here, it is also assumed that $D \neq \mathbb{C}$, so that D is the image of $D_1 \equiv \{z : |z| < 1\}$ under an analytic, univalent mapping, say f. It has been shown that the classes, say l.a. (of such f), and c.c., of 'close-to-convex functions', coincide. The aim is **to find analytical conditions equivalent to the geometrical characterization of l.a.**. The motivation for the definition(s) of 'l.a.' lies mainly in the theory of **conformal mapping**. **NOTES**(i) f is c.c. relative to h in D_1 IFF h is convex, f is regular in D_1, and $z \in D_1 \Rightarrow \mathcal{R}\{f'(z)/h'(z)\} > 0(\geqslant 0$ in some formulations). (ii) This is quintessentially a part of 'Geometric Function Theory'. Many of the criteria devolve on variations in the arguments of certain quotients of analytic functions (over associated regions/paths/...). (iii) Some of the basic properties may be expressed as

inequalities involving combinations of arguments. The corresponding **proofs have a highly geometrical flavour**, with types of **quasi-convexity** playing important roles – along with certain **maximum principles**, and **constructions**. (iv) Variants of the results on l.a./c.c., 'slightly strengthened' would imply the Bierbach Conjecture (now a Theorem). (v) Notions of **starlikeness** (of order α, etc.) are also important in this area.

126. The **chordal metric** corresponds to the distance a function for pairs of points on the Riemann Sphere, W, along the chord between the points. For points w_1, w_2 in the w-plane, this distance is given by $\sigma(w_1, w_2) = |w_1 - w_2|[(1 + |w_1|^2)(1 + |w_2|^2)]^{-\frac{1}{2}}$. This metric is used when ∞ is included in the domains involved – since it gives $\sigma(w_1, \infty) = (1 + |z_1|^2)^{-\frac{1}{2}}$, and also allows a basis for the **compactified plane** to be taken as $\{Q(z_0; r) : r > 0\}$ for finite z_0, and $\{\zeta Q(0; r) : r > 0\}$, where $Q(a; r)$, with complement $\zeta Q(a; r)$ denotes the circle with center a and radius r.

Here, the main concern is with **cluster sets,** C, of classes of measurable functions, where, if ζ denotes one of a chord at S, a Stoltz angle at ζ, or D(for $\zeta \in D$, the open unit disc), and $S_r := S \cap Q(z; r)$, then $w^* \in C(f, \zeta, S)$ IFF $\Gamma_r(f) \equiv S_r \cap f^{-1}(\{w \in WL\sigma(w, w^*) < \epsilon\}) \neq \phi(\forall \epsilon > 0, \forall r > 0)$. The corresponding **essential** cluster sets are specified by: $w^* \in C_e(f, \zeta, S)$ IFF $\limsup\limits_{r \to 0^+} m(\Gamma_r(f)/m(S_r)) > 0$, where m is the Lebesgue measure appropriate to S. This apparatus is used to show that, for '**normal**' functions f($\equiv$ functions uniformly continuous in the chordal metric on W) the set $C_e(f, \zeta, S)$ is **connected** – a geometrical/topological function-theoretic property.

127. **Frames** form a class of complete **lattices** satisfying the distribution law: $a \wedge \bigvee\limits_{p \in A} b_p = \bigvee\limits_{p \in A} (a \wedge b_p)$, for each element a and each family $\{b_p : p \in A\}$ of elements, of the lattice (say, L, with least, and greatest elements $0_L, 1_L$). If, in addition card $A \leqslant \alpha$, L becomes an α-**frame**. An α-map, ϕ, between α-frames L, M, satisfies: $\phi(0_L) = O_M, \phi(1_L) = 1_M, \phi(a \wedge b) = \phi(a) \wedge \phi(b)$, and, $\phi(\bigvee(a_p : p \in A) = \bigvee(\phi(a_p) : p \in A)$ if card $A \leqslant \alpha$. Frames generalize the class of topologies (on a TS) from which many of the key topological results may be proven (without resorting to arguments involving individual points of the TS).

The **covering dimension** was studied first for **separable MS**, extended to **normal spaces** ($\equiv$ TS where two disjoint closed sets always have disjoint neighbourhoods). Here, **a dimension theory for a σ-frames** is developed. In particular, normal α-frames are defined, and spaces of prime dual ideals of α-frames are introduced. Certain basic results of dimension theory may be extended to α-frames, e.g., :(i)$A \subset X \wedge A$ closed $\Rightarrow$ dim$A \leqslant$ dim$X$; (ii) X normal $\wedge A_m$ closed in X with dim$A_m \leqslant m(m \in \mathbb{P}) \Rightarrow$ dim$X \leqslant n$; (iii) several conditions equivalent to 'dim$X \leqslant n$' may be formulated in terms of continuity/open covers/sequences of (disjoint) open sets.

128. **Slowly oscillating** (SO) **functions** are basic in the analysis of general **tauberian conditions** for **summability procedures** (for series/integrals). Let $T := \{\phi : \mathbb{R}_K \to \mathbb{R} : \phi \in \mathcal{A}, \phi(x), \phi^{-1}(y) \to \infty$, as $x, y \to \infty\}$, where $\mathbb{R}_K := \{t \in \mathbb{R} : t > K\}, A \equiv \{f : f$ abs. cts $\}$ and $T \equiv \{$ tauberian functions $\}$. In these terms, $\{f : f$ is SO rel. to $\phi \in T\}$ is defined by: SO(ϕ) $:= \{f : \mathbb{R} \to \mathbb{R} : \forall \epsilon > 0 \exists N_\epsilon, [s > x > N_\epsilon \wedge \phi(s) - \phi(x) < 1/N_\epsilon] \Rightarrow |f(s) - f(x)| < \epsilon\}$, and: $\models f \in$ SO (ϕ) IFF $\forall \epsilon > 0 \exists N_\epsilon\ \exists g \in \mathcal{A} : x > N_\epsilon \Rightarrow g'/\phi \in L^\infty \wedge |f(x) - g(x)| < \epsilon$. Several other criteria equivalent to '$f \in$ SO (ϕ)' may be obtained.

If $\langle Q, \rho \rangle$ is **directed**, and $f : Q \to \mathbb{R}$ then: $\lim_q f(q) = r$ means that: $\forall \epsilon > 0 \exists q_\epsilon : q \rho q_\epsilon \Rightarrow |f(q) - r| < \epsilon$. On this basis $\lim_q \int\limits_J K_q f$ may be estimated for classes of functions K_q, f – from which summability procedures mapping SO into itself may be characterized. Further, the **Wiener-Pitt tauberian theorem** ($\{g \in L^1, \hat{g} \neq 0$ on $\mathbb{R}, f \in L^\infty \cap$ SO ($\mathbb{I}$), $\lim_x (g * f) = H \int g \Rightarrow \lim_x f = H\}$) **may be extended** to the R.S FT (from the Lebesgue FT, $\hat{g}$) and to $f \in$ SO(ϕ) for $\phi \in T$. For Σ , possible analogues of such results within maximally general quasi-calculus frameworks must be sought.

129. **Hilbert's inequality** (i): $|\sum\limits_{r \neq s} (r - s)^{-1} \times u_r \bar{u}_s| \leqslant 2\pi \sum\limits_r |u_r|^2$ improved (by Schur) to replace '2π by 'π' (best-possible) may be generalized in various ways – for instance (ii) to replace $(r - s)^{-1}$ by cosec$\pi(x_r - x_s)$, with the x_t distinct (mod 1); or, (iii) to replace $(r - s)^{-1}$

by $\lambda_r - \lambda_s)^{-1}$, with the λ_t distinct. Here, $1 \leqslant r, s \leqslant R; x_t, \lambda_t$ are real; and, the u_t may be complex. **The original proof** of (i) is based on **the identity:** $\int_{-\pi}^{\pi} tQ_R(t)dt = 2\pi(S - T)$, where: $Q_R(t) := \sum_{r=1}^{R}(-1)^r[a_r\text{cos}rt - b_r\text{sin}rt]$;

$$S := \sum_{r=1}^{R}\sum_{s=1}^{R}(r + s^{-1}a_r b_s; T := \sum_{r=1}^{R}\sum_{s=1}^{'R}(r - s)^{-1}a_r b_s$$

For (ii), (iii), the basic results take the forms:

$$\sum_r \sum_s{}' u_r \bar{u}_s \text{cosec}\pi(x_r - x_s) \leqslant \delta_1^{-1} \sum_r |u_r|^2 \text{ (ii)'};$$

$$\sum_r \sum_s{}' u_r \bar{u}_s (\lambda_r - \lambda_s)^{-1} \leqslant \pi\delta_2^{-1} \sum_r |u_r|^2$$

where (here and above) Σ ' omites all terms with $r = s$, and
$\delta_1 := \min_{r,s} + \{x_r - x_s\}$
$\delta_2 := \min_{r,s} + \{\lambda_r - \lambda_s\}$,
$\{y\}$:= distance from y to nearest integer (= dist $(y, \mathbb{Z})$), $\min_+ f$:= least positive value of f. **Several refinements** may be derived from these basic versions. **Applications** in **Function Theory** (e.g., **AP functions**), and **Analytic Number Theory**, have been pursued (e.g., for '**the large sieve**').

130. Let Ω be a MS, (i)$\tau : \mathcal{P}\Omega \supset \zeta \to \mathbb{R}$, (ii) $\phi \in \zeta$; (iii)$\tau(\phi) = 0$; (iv) $c \in \zeta \Rightarrow 0 \leqslant \tau(C) \leqslant \infty$. Then τ is a **pre-measure** – from which (v) **auxiliary measures** ω_δ^τ $(\delta > 0)$ and thence (vi) a **metric measure**, $\omega^\tau := \lim_{\delta \to 0^+} \omega_\delta^\tau$ may be constructed – both via 'method III', and improvement on the previous methods I, II! This construction (PLMS (3)26 (1973)521-546) yields **cartesian-product measures** more general (and better behaved) than those of Caratheodory type'. Here, $\omega_\delta^\tau := \inf \sum_{i=1}^{\infty} \omega_i \tau(C_i)$, with 'inf' over al ordered pairs (C_i, ω_i), where $C_i \in \zeta$, diam$(C_i) < \delta$. If Ω is **complete** and **separable**, and A has the form $\bigcup_{\{i_k\} \subset \mathbb{P}} \bigcap_{r \in \mathbb{P}} F(i)_r$, where all $F(i)_r$ are **closed** then A is called an **analytic set** (not the universal terminology; e.g., **Souslin kernel ...**). Certain basic theorems originally proved for Caratheody measures may be reformulated for 'method III measures' (say, ν), including (vii) $\nu(\bigcup_{n \in \mathbb{P}} E_n) = \sup_{n \in \mathbb{P}} \nu(E_n)$, for any **increasing** sequence of sets, $\{E_n\}$; (viii) $\nu(A) = \sup\{\nu(K) : K \subset A, K$ **compact** $\}$. **NOTE that** { analytic sets} $\supset$ {Borel sets } The difficulties in such proofs lie mainly in demonstrating that sets defined/obtained in various ways **are** of specific types (analytic, Borel, . . .). The overall aim is to effect these constructions from pre-measures.

131. For the series $f(x) = \zeta + (z - \zeta) + \sum_{n \geqslant m+1} a_n(z - \zeta)^n, a_{m+1} \neq 0, m \geqslant 1$ **the formal identity** $f_\lambda \circ f(z) = f \circ f_\lambda(z), [f_\lambda(z) := f(z)|_{a_n = a_n(\lambda)}, a_{m+1}(\lambda) := \lambda a_{m+1}, a_n(\lambda)$ w.d. polynomials, $n \geqslant m + 2]$, is satisfiable $\forall\lambda \in \mathbb{C}$. Since $f_k(z)$ coincides with the k-th **iterate** of $f(z) \equiv f_1(z)$, if $k \in \mathbb{P}$, the series f_λ is called the (**complex**) **fractional iterate of** f (of **order** λ). Further, if $f(\zeta) = \zeta, f'(\zeta) = 1$, then ζ is a fp ($\neq \infty$) of f (of **multiplier 1**)

Further $\models S_f \equiv \{\lambda$: series for f_λ has radius of cdce $> 0\} = \mathbb{C}$ or else $[= L_1$ or $= L_2]$ where L_h is a (discrete) lattice of dimension h. If $S_f = \mathbb{C}$ then f is called **embeddable** various **non**embeddability conditions may be derived for classes of functions $\{f\}$, in terms of analyticity/natural-boundary criteria – and for meromorphic functions in $\mathbb{C}\backslash E$, where E (comprising essential singularities) is countable. it may be shown also that L_2 (above) must, in fact, be **empty** – even though the general proof admits L_2 as a consistent possibility.

132. The **embedding of cones** (esp., of cones of functions) into CS may adversely affect some useful properties valid for the cone. If some **distance function** is given/defined on the

cone C, then a natural embedding of C into its second dual cone, C^{**}, may preserve certain properties (though not necessarily, distance). If C generate V, and ρ is **positive and sublinear** on C, the ρ-**topology** at $x_0 \in C$ has **base** $\{U(\epsilon; x_0) : \epsilon > 0\}$; and **countable** base $\{U(1/n; x_0; n \in \mathbb{P}\}$ where $U(\epsilon, x_0) := \{y : y \geqslant x_0, \rho(y - x_0) < \epsilon\}$, for '$xgey \Leftrightarrow x - y \in C'$. Here, ρ is **increasing** IFF $0 \leqslant x \leqslant y \Rightarrow \rho(x) \leqslant \rho(y)$ – not always true of positive, sublinear ρ in this topology. Several conditions on ρ equivalent to 'ρ is increasing' (on C) may be derived – and some of these conditions may be extended to V. For $\tilde{C}$ generating $\tilde{V}$, define P by $P(C, \tilde{C}) := \{f : C \to \tilde{C} : f$ cts, linear, $f(\lambda x) = \lambda f(x), \lambda > 0\}$ If $\tilde{\rho}$ on $\tilde{C}$ corresponds to ρ on $C, T : C \to \tilde{C}$ is linear an $T(\lambda x) = \lambda T(x), \lambda > 0$, and S is given by $S_{\rho,\tilde{\rho}}(T) := \sup\limits_{x \neq 0} \frac{\tilde{\rho}(T(x))}{\rho(x)}$ then $\models S$ is positive-sublinear; increasing (if $\tilde{\rho}$ is); and $P(C, \tilde{C})$ is closed if $\tilde{C}$ is closed an $\tilde{\rho}$ is continuous. These, and associated, conditions allow the **extension of the** ρ **-topology to l.c. topologies on** V. If V is also a NLS then criteria for the embedding of C in C^{**} to be **surjective** and **isometric** may be derived.

133. The so-called **automatic continuity** of mappings between certain classes of **operator algebras** – and the cases where **dis**continuity occurs – motivate an investigation of '**near-continuity**', for various modifications of the mappings involved. For commutative complex algebras, A, with identity, e, the extensions $A\alpha := A[x]/(\alpha(x))$, for monic polynomials α, are of central interest. The **basic result**, for a Banach algebra, B, all of whose higher point derivations are continuous, is that: **any homomorphism**, $v : B \to A\alpha$, **is necessarily continuous**.

Here, if B is **any** Banach algebra allowing a homomorphism into $A\alpha$, conditions are found for a type of near-continuity – together with extra conditions ensuring full continuity. Essentially: if ν^* is the adjoint of ν and (i) ν^* is finite-to-one outside a nowhere-dense set; (ii)ν^* preserves isolated points, for $\nu : B \overset{\text{homo}}{\to} A\alpha$ then: $\exists$ closed ideals I_1, I_2, in B, and a finite-dimensional ideal $J = u(A\alpha)$, with $u^2 = u \in A$, such that: (iii) $\nu|_{I_1}$ is continuous; (iv) I_2 has at most finitely many maximal regular ideals; (v) $\nu(I_2) \subset J$; (vi) $B = I_1 \oplus I_2$

Consequently: if (i), (ii) hold, then each of the following conditions suffices for continuity of v:

(vii) $\Phi_{A\alpha} \equiv$ domv^* has no isolated points;

(viii) Rad (I_2) is finite dimensional;

(ix) B is semi-simple;

(x) ν is one-to-one.

Finally, even if ν is an isomorphism, both (i) and (ii) are essential for full continuity.

134. **Amalgams**, (E, ω_ρ), are generalizations of **sequence spaces**. Here, ω_p is a **p.o. VS** of real sequences over $\mathbb{Z}$, endowed with a **Riesz norm**, and $E := \prod\limits_{n \in \mathbb{Z}} E_n$ (E_n NLS, $x = (x_n) \in E$, addition and scalar multiplication defined coordinatewise), with $\tilde{x} := (\|x_n\|) \in \omega_p$. In many instances, properties shared by ω_p and by all E_n are also possessed by (E, ω_p); e.g., **completeness separability**. Moreover: $\models \omega_\rho$ complete $\Rightarrow (E, \omega_\rho)$ is a BS $\Leftrightarrow$ all E_n are BS. The **duals** of some amalgams may be determined explicitly. For instance, $\models (C_0, \omega_\rho)^* = \{\Phi : \Phi(f) := \int f d\mu, f \in (C_0, \omega_\rho)\}$ and $\|\Phi\| = \inf\{\rho^*((|\mu - a_n\delta_n|[n, n+1) + |a_{n+1}| : \rho^*(|a|) < \infty\}$ where δ_n is the δ - distribution concentrated at n, ρ^* is the norm on w_ρ^*m and the sequence $(\max\{|f(x)| : n \leqslant x \leqslant n+1\}) \in \omega_\rho, f \in C_0(\mathbb{Z})$. Next **multipliers,** $\phi \equiv (\phi_\lambda) \in \Pi\{L(X_\lambda, Y_\lambda) : \lambda \in \Lambda\}$, from $A \subset \Pi\{X_\lambda : \lambda \in \Lambda\}$ to $B \subset \Pi\{Y_\lambda : \lambda \in \Lambda\}$, with all X_λ, Y_λ NLS are specified by: $\phi \in [A, B]$ IFF $\phi(x) \equiv (\phi_\lambda(x_\lambda)) \in B$ whenever $x \equiv (x_\lambda) \in A$. Then, e.g.: $\models [(E, \omega_{\rho_1}), (F, \omega_{\rho_2})] = (\Pi\mathcal{L}(E_n, F_n), [\omega_{\rho_1}, \omega_{\rho_2}])$. A variant of the **Riesz-Thorin theorem** (for operators into amalgams), and the continuity of the translation operator, τ_t (where $(\tau_t f)(x) := f(x-t)$) are derivable within this framework. **F.T. spaces of amalgams** may be identified.

135. For entire functions f of exponential type (represented by PS), the properties of the sequences of zeros of $f^{(k)}, k \geqslant 0, f^{(0)} \equiv f$, together with the order of f, determine many aspects of 'general analytic behaviour'. If, in addition, the sequence of coefficients of the PS for

f exhibits w.d. gaps, then several detailed estimates may be obtained. The basic parameters are:

(a) order: $\rho := \limsup_{r\to\infty} \log \log M(r)/ \log r$

(b) type: $\tau := \limsup_{r\to\infty} r^{-\rho} \log M(r)$

where $M(r) := \max_{|Z|=r} |f(z)|$;

(c) central index, $\nu \equiv \nu(r)$ of the PS for f

(d) $\delta(f) := \liminf_{r\to\infty} r^{-1}\nu(r)$

(e) $\gamma(f) := \limsup_{r\to\infty} r^{-1}\nu(r)$

(f) $t(f) := \liminf_{r\to\infty} r^{-1} \log M(r)$

(g) $T(f) := \limsup_{r\to\infty} r^{-1} \log M(r)$

When $T(f) < \infty$, f has **exponential type** $T(f) \equiv T$

NOTE: for properties of $\nu(r)$, see, e.g., Pitman Research Notes in Mathematics, No.104: 'Analytic Functions - Growth Aspects', by O P Juneja/GP Kapoor (Pitman, 1985), esp., pp 29-33. Further, if $\rho = 1$, then $\models \delta \leqslant t \leqslant T \leqslant \gamma \leqslant eT$; and, if $\rho < 1$, then $T = 0$.

In this framework, the following (sample) results may be proven.

(i) If $\{\alpha_n\} \subset \mathbb{C}, f^{(n)}(\alpha_n) = 0, \limsup_{n\to\infty} n^{-1}\sum_1^n|\alpha_h| = k, T(f) < (ek)^{-1}$, then $f(z) \equiv 0$.

(ii) If $\{x_k\} \nearrow$ in $\mathbb{R}^+$ is defined by $\sum_{p\geqslant 1} x_k^{kp}(kp)! = 1$, and $f(\neq$ polynomial) has a PS about 0 with radius of cgce R, and gaps $\geqslant k-1$ then $\models Rx_k \leqslant \mathit{limsup}_{n\to\infty}\, n\rho_n$, where ρ_n is the radius of univalence of $f^{(n)}$ about 0.

(iii) $\delta^{-1}x_k \leqslant \limsup_{n\to\infty} \rho_n$

(iv) $\lim_{k\to\infty} [(k!)^{\frac{1}{k}} - x_k] = 0$

(v) If $N(n) := \#\{$ ran $\nu_n(1,\ldots,n)\}$ and lim sup(lim inf)$n^{-1}N(n) = d(= c)$, then lim sup(lim inf) $(n+1)^{-1}[\rho_0 + \cdots + \rho_n] \geqslant dx_k/\gamma(\geqslant cx_k/\gamma)$, for $n \to \infty$, in all cases.

From these estimates, sufficient conditions for classes of entire functions to be polynomials (or, to vanish identically) may be derived.

137. For (positively oriented) Jordan curves, γ, the space $\zeta(\gamma)$ of functions continuous on γ is the uniform limit of spaces of polynomials in $z^{\pm 1}$, for $z \in \mathbb{C}$. If $l(\gamma) < \infty$, then **all** powers of $z^{\pm 1}$ are required to span $\zeta(\gamma)$ – since, for any $s \in \mathbb{Z}$, and any polynomial $P = \sum_{m\in K\subset\mathbb{Z}, s\notin K} c_m z^m$, one has $|2\pi i| = |\int_\gamma (z^s - P)dz| \leqslant \|z^s - P\| l(\gamma)$; and hence, dist$(z^s, \overline{\text{span}}\{z^m : m \neq s\}) > 0$. If $l(\gamma) = \infty$, then $\models$ **at least one power z^r may be omitted. More generally, curves γ may be constructed for which exactly n powers**, or else, **any finite number of powers, may be omitted.** Even this choice may be extended: simple Jordan curves γ^* may be constructed such that: if $\{p_k\} \subset \mathbb{P}$ **has positive density, then**: $\sigma^*(\{p_k\}) \equiv \{z^m : m \in \mathbb{N} \cup \{-p_k\}\}$ spans $\zeta(\gamma^*)$; and hence, $\models \zeta(\gamma^*)$ **must admit some spanning set** (s) say, $\sigma^*(\{p'_k\})$, **of density** 0 **in** $\mathbb{Z}$. Denote by $z^{\sigma *}$ the set $\{z^k : k \in \sigma^*\}$, and by $\mu_\gamma(z^{\sigma*})$ a complex Borel measure on γ orthogonal to all elements of $z^{\sigma *}$; then $\exists! g \in K(D)$: (i) $\mu \in WBV(dg)$); (ii) $g(z) = \sum b_j z^{q_j}, \{q_j\} := z^{\mathbb{Z}} \backslash z^{\sigma *}$,

near $z = 0$; here, γ bounds D, externally. Conditions on γ for $\sigma^*(\{p_k\})$ to span $\zeta(\gamma)$ may be derived.

138. '**Vector ODE**' of the form $(*)\frac{dz}{dt} = f(t,z), z \in \mathbb{C}^n; f, \omega$ - **periodic in** t, continuous: $\mathbb{R} \times \mathbb{C}^n \to \mathbb{C}^n$, **holomorphic in** z, may be studied for the number of periodic solutions (within specified z -domains). Some of the results may be **extended** to ODE over **complex manifolds**, M - especially, compact manifolds (e.g., surfaces). If M is a **Hodge manifold** (of complex dimension m) then there is a (bi-) holomorphic embedding in the **complex projective space** $P^m(\mathbb{C})$, by **Kodaira's Theorem** (see, e.g. 'Principles of Algebraic Geometry' by **P.Griffiths/J.Harris** (Wiley, 1978; n.e.p/b, 1994)Ch1, sec.4). For $\Gamma \equiv \{\omega$ o[periodic vector fields, holomorphic on $M\}$; then it turns out that, in local coordinates $(z)_m$ on M, with $f \in \Gamma_2(*)$ becomes $(**)\frac{dz}{dt} = g(t,z)$, where g is essentially determined by f. The aim is to **estimate the numbers of k-periodic solutions of** $(*)$, for $k \in \mathbb{P}$ – and thence, for $(**)$, with g suitably related to f. **NOTE that**, with the topology of uniform convergence on compact subsets of $\mathbb{R} \times M, \Gamma$ is a BS. The **orbits**($\equiv$ trajectories) of $(*)$, $(**)$ of period k, correspond to fp of the mapping, say, $\Phi : M \times \Gamma \to M$. Variants of Rouché's theorem yield estimates of the numbers/multiplicities of periodic solutions, as g varies in Γ. Several topological properties (e.g., open-ness connectedness) of special subsets of Γ may be established – to obtain fairly detailed characterizations of orbit structures for $(**)$

141. The **discrepancies** of analytical inequalities may be studied fairly generally. Properties analogous to convexity/concavity, for functions f whose domains comprise **finite collections of elements** (possibly, with repetitions) are exemplified by the **Hlawka function**, H_f, of f, where: $H_f(X,Y,Z) := f(X \cup Y \cup Z) - f(Y \cup Z) - f(Z \cup X) - f(X \cup Y) + f(X) + f(Y) + f(Z)$, with $\underline{\text{ran}}(f) \subset \mathbb{R}$ when $\underline{\text{dom}}(f) \subset \mathcal{P}(\mathbb{R}^+)$/ For the '**standard inequalities**', definitions of '$f(X \cup Y \cup Z)$' corresponds to particular results, and f is called 'H-**positive**' 'H-**negative**', when $H_f(X,Y,Z)$ is 'always $\geqslant 0/ \leqslant 0$'. **NOTE that**, here, '$A \cup B$' denotes the collection of all elements in A, B, regardless of repetition. **Examples**(i) $X, Y, Z = (x,a), (y,b), (z,c)$; $\models f_1(X \cup Y \cup Z) \equiv (a+b+c)x^{a/(a+b+c)}y^{b/(a+b+c)}z^{c/(a+b+c)}$ is H-positive. (ii) $\models f_2(X \cup Y \cup Z) := ax + by + cz - f_1(x,y,z)$ is H-negative. (iii) $\models f(X \cup Y \cup Z) := (a+b+c) \times \phi(\frac{ax+by+cz}{a+b+c})$, with ϕ twice-differentiable - convex.concave, is H-positive/H-negative.

The **AM/GM**, **Holder, Minkowski,** and other, discrepancies may be covered within this framework. Superadditivity $(f(A \cup B) \geqslant f(A) \cup f(B)$ may also be treated in this formalism – from which classes of specific inequalities (mainly, for the associated discrepancies) are derivable. Some of these results extend to NLS.

142. For **NLS** $E, F, d : E \supset C \to F$ is K-**lip**schitzian IFF $x, x' \in C \Rightarrow \|f(x) - f(x')\| \leqslant K\|x - x'\|$. If E, F are HS then $\models f$ has a K-Lip. extension $\bar{f}$ to E – but **not** for arbitrary (or even for complete) NLS; but, for certain classes $\{C\}$, αK-Lip extensions **may** be obtained, with $\bar{f}$ explicitly determined. In fact, $\models \{E$, **real**; C, **bounded, closed, convex; diam**$C = \rho; C \supset B_\delta(O); g(x) := x\Lambda_C(x)\} \Rightarrow g$ is $\rho\delta^{-1}$-Lip on E. (Here, $\Lambda_C(x) := \chi_C(x) + (1 - \chi_C(x))\chi_D(x), D := [0,x] \cap \partial C$). Next, for E, F, real NLS, C, ρ, B_δ, f as above, and $f : C \to F$ K–Lip $\models \bar{f} \equiv f \circ g : E \to F$ is $K\rho\delta^{-1}$ Lip (on E). The proofs of such results have a geometrical representation (for points on various segments in the NLS) which also makes plausible many of the norm-inequalities involved. The most basic of all extension theorems for NLS is the **H**ahn **B**anach Theorem, where preservation of the given (semi-) norm is proven. The main conditions for 'controlled Lipschitz extension' have obvious application to the extension of contraction mappings – which are basic for existence proofs for many types of (algebraic/differential/integral/operator/...) equations.

143. '**Analysis in valued fields**' includes p-adic **analysis** – of major importance in Algebraic Number Theory. The problem of constructing **Haar-type measures**, for function spaces over $\mathbb{Z}_p$, with range in $\mathbb{Q}_p$, is intractable (since

$$\models \text{Hom}_{\mathbb{Q}_p}(\zeta(\mathbb{Z}_p, \mathbb{Q}_p), \mathbb{Q}_p)^{\mathbb{Z}_p} \neq 0);$$

but progress may be made for complete, valued $\mathbb{Z}_p$-algebras of continuous functions, $f : \mathbb{Z}_p \to \mathbb{Z}_p$, with the usual pointwise operations.

Notation: $\mathbb{Z}_p \equiv$ ring of p-adic integers; $\mathbb{Q}_p \equiv$ field of p-adic numbers; $v_p(p) : \mathbb{Q}_p \to \mathbb{Z}_p \cup \{\infty\} \equiv$ p-adic valuation of $\mathbb{Q}_p$ (with $\rho_p(p) = 1$, for all primes p).

For $\zeta(\mathbb{Z}_p, \mathbb{Q}_p)$, define v by: $v(f) := \inf_{x \in \mathbb{Z}_p} v_p(f(x))$. The aim is to obtain a subalgebra, say, $\mathcal{A}_0 \subset \mathcal{A} \subset \zeta(\mathbb{Z}_p)$, where $\mathcal{A} \equiv \{$ **uniformly differentiable functions** $f : \mathbb{Z}_p \to \mathbb{Z}_p \wedge f' : \mathbb{Z}_p \to \mathbb{Z}_p; V\}$, with $V(f) := \min(v(f), R_f), R_f := \inf_{x \neq y \in \mathbb{Z}_p} v_p(\frac{f(x)-f(y)}{x-y})$, $\mathcal{A}_0 \equiv \{f \in \mathcal{A} : f' = 0\}$, such that $\mathrm{Hom}_{\mathbb{Z}_p}(\mathcal{A}, C_{p^\infty}) \cong \mathbb{Z}_p$ (as $\mathbb{Z}_p$-modules)l $C_{p^\infty} := \mathbb{Q}_p/\mathbb{Z}_p$. Hence $\exists! J : \mathcal{A} \to C_{p^\infty}, J \not\equiv 0$, **continuous, linear, nondegenerate** (i.e., $f \in \mathcal{A} \wedge (\forall g \in \mathcal{A}) J(fg) = 0 \Rightarrow f = 0$). For $f \in \mathcal{A}, n \geqslant 0$, put $b_n = p^{-n} \sum_{0 \leqslant i \leqslant p^n - 1} f(i)$; then $\models \{b_n\}$ is Cauchy in $\Omega_p :=$ algebraic closure of $\mathbb{Q}_p$, and so $I_0(f) := \lim_{n \to \infty} b_n \equiv$ the I_0-integral of f. Then $\models J := I_0 \circ \mu$ ($\mu \equiv$ reduction mod$\mathbb{Z}_p$) has the required properties. (See: JLMS(2)7 (1974) 681-693: C.F.WOODCOCK, for further details).

144. **Wedges**, W, in (l.c.) TVS, $V(x, y, \in W, \lambda \geqslant 0 \Rightarrow x + y, \lambda x \in W)$ induce **orderings** – on V, and on function spaces over subsets of V. If K is a **compact, convex subset** of $V, A(K) \equiv \{f : K \to \mathbb{R}, f$ **continuous, affine**$\}$, $W \subset A(K), W \supset \{$**constant functions** $\}, P(K) \equiv \{$ weakly compact probability measures on $K\}$, then (if V is **Hausdorff**) results on **the approximation of convex/concave/l.s.c-W-decreasing functions** may be established through basic propositions in **Choquet Theory/Banach-Lattice Theory/Riesz Spaces.** Further, an **extension** of the **Riesz decomposition property** may be obtained, and shown to be **dual** to a certain **filtering condition** – and an associated **separation theorem** may be generalized, with a **norm-preserving function-extension procedure.** This decomposition ensures **compatibility** between the W-induced ordering and the underlying convexity (and it is **equivalent** to the filtering condition on W). Another consequences of this scheme is an existence proof for **barycenters** (say; $b(\mu) \in K$) of measures $\mu \in P(K)$, with additional forms of decomposition properties. **NOTES**(i) For **c**ompact **c**onvex sets K, the subset F is a **face**. IFF $\{F$ **is convex** $\wedge F$ **carries every positive measure** μ**, of unit mass, such that** $b(\mu) \in F\}$. The **facial structure** of c.c. sets may be studied within this framework. (ii) Choquet's Theorem may be viewed as an example of the **balayage** defined by a convex cone of continuous functions on a compact set.

145. The so-called **R**iemann **c**omplete integral surpasses the Lebesgue integral for scope an power; indeed, the Rc (or 'gauge') integral is both simpler to define and broader in applicability. For extensions to (ordered) TVS, however, direct generalization is inadequate to produce the strongest convergence theorems (e.g., 'dominated convergence'). If complete (regular) vector lattices, K, and integrands $f : I \to K$ are considered, then adequate extensions **may** be obtained. The constructions are intricate, and depend on the definition of suitable 'upper' (sup) and 'lower' (inf) integrals – as in the classical Riemann-Darboux form. Here, the max length of intervals (in an 'approximating sum') becomes a function – the **gauge**. $\models$ gauge-integrability of f and $|f|$ implies Lebesgue-integrability. The underlying structure is a **division space** - typically, $(T, \mathcal{I}, A)$, where T is a specified set, $\mathcal{I}$, the family of intervals in T (**disjoint** unless their intersection contains a member of $\mathcal{I}$); and $A \equiv \{S : S$ **is a family of ordered pairs** $(I, x), I \in \mathcal{I}, x \in T\}$. Further, if $J \in \mathcal{I}, J \subseteq I$ then J **divides** I IFF J coincides with some element in a (finite) division of I (say, $\{I_j : 1 \leqslant j \leqslant n\}$). There are various conditions involving $T, \mathcal{I}, A$, designed to promote definitions of upper/lower sums approximating types of (definite) integrals. In particular, the 'N-variational integrals' are w.d., and (under certain sets of conditions). '$N - v$' is equivalent to 'R-c'. Moreover, a version of the 'monotone convergence theorem' holds, for regular lattices, K. **NOTE:** the original papers are in J/PLMS (1955-1961); see also: LNM # 14

146. **Computable algebra** has assumed extra significance (beyond the purely academic) through the development of SAM packages – with Σ as a prospect! The ((**non)constructive extensions** of B_a (the countable, atomless **B**oolean **A**lgebra) are important in various set-theoretic algorithms relevant to the foundations of constructive mathematics. Here, the links with **recursive function theory** may be demonstrated, by exhibiting effective procedures for the underlying computations – which combine **algebraic** and **topological** properties. Operations preserving the **c**onstructive **e**xtensions of B_a may be characterized as certain classes of

unions of chains of c.e. of B_a - whereasL $\models \exists$ infinite chains of c.e. of B_a whose union is **non** constructive. The interplay of **computability, constructiveness**, and **recursiveness** is evident here (where B_a is identified with the collection of **copen sets of the Cantor** (discontinuum) **set**, S). Notions of **completeness, regularity, topological closure, strong constructivity, decidability, maximality,** ... are deployed to produce nonconstructive extensions of B_a and **algorithms** for constructive extensions of B_a.

147. The **ultimate boundedness** of solutions of (sets of) (O)DE is fundamental for many applications – for instance in Control Theory, and in the study of nonlinear oscillations. Other properties of (non)autonomous equations include (**almost-)periodicity**, and forms of **stability/convergence**. Several of these phenomena are exemplified in the paper: **IMA J. Applied Math.**(1983)31, 235-252, by **JSN Elvey**, where a range of general theorems, applied to a simple non linear ODE, yields use full 'practical criteria'. The most important class of equations (second-order, nonautonomous (O)DE) are covered (e.g.) in '**Nonlinear ODE**' by **G.Sansone/R.Conti** (pp 592, Pergamon, 1964). Higher-order ODE (especially, order 53, 4) are treated in: '**Nonlinear DE of Higher order**', by **R.Reissig, G.Sansone and R.Conti** (pp669, Noordhoff, 1974). A broad spectrum of vibration problems in Engineering/Physics is discussed in: **Studies in Applied Mechanics**, vol.30A ((**S.Kaliski, et al., eds**) Elsevier, pp488, 1992). Here, in $\#147$, a class of 4th-order, periodically forced, ODE with specified forms of dependence of 'coefficient functions' on lower derivatives, is analysed for detailed boundedness criteria – derived from modified **Liapunov stability fucntions.** The resulting conditions are (typically) complicated and unwieldy, but essentially **effective**. For Σ , the manipulations involved are tractable; but a prime aim, in this area, should be to obtain criteria of greater transparency and elegance.

149. The **correspondence** $D : [M, M] \to \mathbb{Z}$, with continuous **self-maps**, $f : M \to M$ of 1-**connected, finite,** m**-dimensional Poincaré complexes,** M**, with oriented fundamental class** $[M]$, and **induced homomorphism** f_*, defines the **degree** of f by : $f_*(u) = D(f)u$ $(\forall u \in [M])$. Hence, possible **characterizations** of M by $D(M) := \{D(f) : f \in [M, M]\}$ may be investigated – for instance: **when do the conditions** (i) $D(M) = \mathbb{Z}$; (ii) $-1 \in D(M)$; (iii) $\mathbb{Z} \cap D(M) = \{0\}$ **hold**? These questions are considered for special types of M (e.g., **manifold, bundle**) with 'few cells'. **NOTE.** A **Poincaré complex** (oriented, of dimension m) is a finite, connected CW-complex for which $\exists [X] \in H_m(X, \mathbb{Z})$, such that, for the 'cap product', $\cap$, one has: $[X]\cap : \hat{H}^r(X, \mathbb{Z}[\pi, X] \cong \hat{H}_{n-r}(X, \mathbb{Z}[\pi, X], r \geqslant 0$. (**The notation, etc,** is explained in **Baues, H.J., 'Combinatorial Homotopy and 4-Dimensional Complexes'** (pp 380, de Gruyter, 1991, §7). For **CW-complexes, see, e.g., Massey, W.S., 'Singular Homology Theory'**(pp 265, Springer, 1980, Ch.IV) and, **Maunder, C.R.F.,'Algebraic Topology'**(pp375, CUP, repr., 1980, ch.8). The complexes M constructed to answer questions (i)-(iii) are obtained by attaching various j-cells, k-cells, etc, to spheres S^l, by means of **characteristic maps** – say $f^k : E^k \to \bar{e}^k$ (where $f^k : U^k \underset{homeo}{\overset{onto}{\to}} e^k \wedge f^k(S^{k-1}) \subset S^k$).

148. A **strip** in $\mathbb{C}$ has the form $S := \{w \equiv u + iv : |u| < \infty, \varphi_-(u) < v < \varphi_+(u)\}$, with $\varphi_\pm$ continuous. If , for $u_2 > u_1, u_1 \to +\infty$, the 'slopes'

$$\frac{\varphi_\mp(u_2) - \varphi_\mp(u_1)}{u_2 - u_1} \to 0,$$

then $S \equiv$ an **L-strip of boundary inclination** 0 **at** $u = +\infty$. If, further, $S_{||} := \{z \equiv x + iy : |z| < \infty, |y| < \frac{\pi}{2}\}$ and $w = W(z) \equiv U + iV$ maps $S_{||}$ conformally onto S, with $U(z) \to +\infty$ as $x \to +\infty$ then (i) $\models$ arg$W'(z) \twoheadrightarrow 0$ in $S_{||}$ as $x \to +\infty$; (ii) $\models [\varphi_+(U) - \varphi_-(U)]^{-1}|W'(z)| \twoheadrightarrow \frac{1}{\pi}$ in $\{|y| \leqslant \alpha < \frac{\pi}{2}\}$, as $x \to +\infty. If$'Z(w)' is **inverse** to '$W(z)$' then, for $u + iv \in S$, (iii) $\models Z(w) \sim \pi \int_{u_0}^{u} \frac{dt}{\varphi_+(t) - \varphi_-(t)}$ as $u \to +\infty$, where '$\twoheadrightarrow$' signifies **uniform convergence**, and '**arg**' is 'suitably chosen'. These estimates are relevant to the **boundary behaviour of conformal maps**, and to **growth estimates for analytic functions**. Several proofs of these results may be given, exemplifying **special types of domain-convergence** (as in the **Caratheodory Kernel Theorem** - see, e.g., '**Conformal Representation**' (CT $\neq$ 28, CUP, 1932) by **C.Caratheodory**(§§ 120-123)),**and the structures of RS admitting prescribed classes of function** – e.g., H^p -, or **Orlicz** functions (see, e.g. LNM # 1027, by M.Hasumi, Springer, 1983). Connections with '**extremal length**' criteria are also used in the derivations (see: BAMS 80(1974)587-606, by **B.Rodin**, '**The method of Extremal Length**', for an extensive bibliography).

152. If $u \equiv g \circ f : U \to V \to G$, with U, V, open subsets of BS E, F, and f, g, smooth, then it is useful (for estimates/evaluations) to have formulae for the higher derivatives of $u : U \to G$ (also a BS). When $E = F = G = \mathbb{R}$, one has the classical result: $(a) D_x^n u = \sum_{S_l} \prod_{i \leqslant i \leqslant l} D_y^p g \frac{n!}{q_i!} \frac{(D_x^i f)^{q_i}}{i!}$, where $S_l \equiv S(l, (q)_l)$ comprises all sets, $\{l, (q)_l\}$, of positive integers, such that $\Sigma i q_i = n, \Sigma q_i = p$, for $1 \leqslant i \leqslant l$ (Bruno, F. de, Q.J. Math 1 (1856), and later variants ...

The general 'BS case' requires more abstraction, but is widely applicable; the 'finite-dimensional case' is 'intermediate' in this hierarchy (and includes 'surface transformations'). With the usual notation ('multi-indices') for PDE, the BS version may be formulated analogously to (a) – except that, for evaluation, the derivatives must be viewed as (multi) linear maps. In this context, a 'symmetric dot product' ($\odot$), and a corresponding form of 'Leibuiz' theorem', may be obtained – in terms of which $\models (b) D_x^k u(x) = \sum_{1 \leqslant q \leqslant k} \sum_{m, + \cdots + m_q = k} a) k(m)_q D_y^q g(y) \times \bigodot_{1 \leqslant ii \leqslant q} D^{m_i} f(x)$, where $a_{k+1}(k+1) = 1 (k = 0, 1, 2, \dots), a_{k+1}((m)_j, k+1-q) = \binom{k}{q} a_k(m)_j$, and $1 \leqslant k \leqslant n$ (for C^n maps f, g – and so, u). A more conceptual derivation (also covering the finite/mito-dimensional cases separately) is given in #152, with ostensibly different formulae! Proofs of equivalence for the various representations require some care. **NOTE** See Field, M.J., 'Differential Calculus and its Applications' (Von Nostrand, 1976) p164, for (b) [incorrectly formulated D^k]

153. A (**pseudo**) metric on G, or on G/H, for a **topological group** G, with **subgroup** H, is $l/r/$**bivariant** IFF $d(ax, ay)/d(xa, ya) = d(x, y) \forall x, y, a \in G$, and $\models$ the **topology** of G may be specified in terms o f continuous $\{l$- **invariant pseudometrics** $\} \equiv P$. In particular, $\models \{x \in G : p(x, e) < \epsilon, p \in P, \epsilon \in \mathbb{R}^+\}$ is a **basis at** e (the identity). Further $\models$ **top** (G/H) is **specifiable by continuous pseudometrics**, say, p^* – where $p^*(xH, yH) :=$ **inf**$\{p(a, b) : a \in xH, y \in bH; p, r$-**variant**$\}$ Next: if G is a MS and H is **closed**, then $\models$ G/H is **metrizable**; and, if H is **compact** then **top** (G/H) is specifiable by l-invariant pseudometrics (rather than merely continuous ones0. Some metrizability criteria for G may be formulated in terms of properties of H, G/H; e.g., if G is a T_0 groups and both of H, G/H have countable dense subsets, and countable open bases at every point, then G itself is metrizable and has a countable open basis for all open sets. There are many variants of this, involving special conditions on H, G/H (e.g., H may be normal); extension/separation criteria may also be established within this framework – with invariance of the underlying (pseudo) metric(s) as the unifying theme. **NOTE** From a general point of view, the basic **Theory of Topological Groups** creates a fusion of **Topology** and **Algebra**. The inclusion of (invariant) integration and harmonic analysis extends this fusion to **Analysis**. The books by **Pontryagin 'Topological Groups'** (transl, Princeton UP, 1939; 1946), and **Hewitt/Ross, Abstract Harmonic Analysis'** Vol1 (Springer, 1963) provide a conspectus of all this material.

158. **Variants of the Banach** fp **Theorem** stem from definitions of **quasi-contraction**, for classes of mappings of (M)S. For a MS Z, with $X, Y \in \mathcal{P}(Z)$, and $f : X \to Y$, the condition: $\exists (\epsilon > 0, L \geqslant 0) : (p, q \in X, d(p, q) < \epsilon) \Rightarrow d(f(p), f(q)) \leqslant L d(p, q)$, defines (ϵ, L)-**Lipschitzian mappings**, f.

$\models$ If $f : Z \to Z$ is cts and $f|_E \in (\epsilon, L)$- Lip, with $\bar{E} = X$, then $f \in (\epsilon, L)$-Lip.

$\models$ **If** Z is **convex** and $f \in (\epsilon, L)$-Lip then $(\forall \omega > 0) f \in (\omega, L)$-Lip. If $\omega < 1$ then f is (ϵ, ω)-uniformly contractive. If $W \subset \mathcal{P}(Z^2), (p, q) \in W, d(p, q) < \epsilon L \Rightarrow d(f(p), f(q)) \leqslant l d(p, q)$, for some $l \in [0, 1/L), L \geqslant 1, f \in (\epsilon, L)$-Lip, **then** f is (ϵ, L, l) - predominantly contractive, say, $f \in (\epsilon, L, l)$- Pred. On this basis, sufficient conditions on $f : Z \to Z$ for $f \circ f \equiv f^2$ to be uniformly contractive may be proven – and shown to be 'sharp', for the quasi-contraction parameters. Under these primary conditions (as specified in the basic theorems) it may be shown that: **for a complete,** η**-chainable MS,** Z**, every** $f : Z \to Z$ **has a unique** fp; which yields a further fp result for classes of **Fréchet-differentiable self-mappings of** BS. NOTES. (i) Such results are useful for existence/uniqueness theorems for diverse types of (non)linear operator (in)equations ; (ii) many of the conditions pertain to the geometry of MS/BS (iii) For a wide range of fp theorems see, e.g., VI Istsatescu 'Fixed Point Theory' (pp 466, Reidel, 1981); DR smart, 'FP Theorems' (CUP, 1974, pp93); Border, KC, 'FP Theorems with Application to Economies and Game Theory' (CUP, pp129, 1985). For the general BS background relevant here, see: 1An introduction to Functional Analysis' by M.Cotlar/R.Cignoli (North-Holland, pp584, 1974).

159. One of the key SAM algorithms is the GCD algorithm for n-variable polynomials ($n \geqslant 1$), with cognate results for rational functions – all, over $\mathbb{R}$ or $\mathbb{C}$. Versions for fields of algebraic functions are important in more sophisticated contexts. The 'intermediate forms' involve algebraic **number** fields, which have been treated extensively in the SAM **factorization routines** (see, e.g., **Geddes**, K., et al., 'Algorithms for Computer Algebra', (Kluiver, 1992; pp 585); esp., Ch. 8), where the aim is to generate explicit factorizations. In Algebraic Geometry, algebraic functions play a basic role, with divisors, ideals, valuations,. . . furnishing the main formalism. Thus there is a little prospect of effective procedures in this framework, but criteria for such function fields to be euclidean (in terms of valuations/norms) may be derived. The analogies with more concrete versions require clarification.

161. The study of Green's functions, and associated functions, for classes of harmonic potentials has strong geometrical aspects. In such cases, maximum principles may be derived and properties of equipotential surfaces may be investigated. For scientific applications, only $\mathbb{R}^2$ or $\mathbb{R}^3$ should be considered, but many results hold for (subdomains of) $\mathbb{R}^m$. Here, a linear combination, $U(P) + V(Q) - 2W(R) =: E(P, Q; R)$- with U,V,W 3-dimensionally harmonic, $U(P) \equiv V(Q)$, and R dividing the segment PQ as $b : a$– is shown to be (wide-sense) maximal, et (P_0, Q_0, R_0), only if: $\exists A_i, \underline{r}_i \equiv \underline{r}/A_i, 1 \leqslant i \leqslant 3 : U(\underline{r}/A_1) = V(\underline{r}/A_2) = W(\underline{r}/A_3)$, for $\underline{r}/A$ 'near P_0'. On this basis. by taking $U \equiv V \equiv W = g$ (the Green's function of a finite, convex, 3-dimensional domain, D, with a single pole at $0 \in D$) it may be shown that all equipotentials, of g within D are convex. It seems probable that more general techniques could produce diverse results of this kind within a unified framework. **NOTE**. Under these conditions, $\models g$ **is a convex functions**, if D is a convex domain.

162. In the study of summability methods for classes of sequences/series, various results for infinite matrices are basic. The sequence $\{s_n\}$ is A-limited IFF $\exists\sigma : \lim_{n\to\infty} \sum_{k=0}^{\infty} a_{n,k}s_k = \sigma$; and $C \equiv (c_{n,k}) \supseteq A$ IFF A-lim $s_n = \sigma \Rightarrow C$-lim $s_n = \sigma$. Given any $C \equiv (c_{n,k})$, it is of interest to seek conditions on A sufficient (or, nss) for C-lim s_n to exist whenever A-lim s_n exists (though, in general, with distinct values). Mostly A is lower-triangular with nonzero diagonal elements, and $\lim_{n\to\infty} a_{n,k} = 0$. An additional condition has a 'MVT form': $|\sum_{k=0}^{n} a_{m,k}s_k| \leqslant K|\sum_{k=0}^{n'} a_{n',k}s_k|$, for $\{s_k\} \in S, 0 \leqslant n' \leqslant n \leqslant m$, where S is a specified class, and n', K are fixed. More general MV conditions (corresponding to specified summability procedures) may be treated in a unified way – **see**, e.g., **Peyerimhoff, A., 'Lectures on Summability'** (LNM # 107, Springer, 1969), esp, Sec.5, and Sec. 6, for applications to **comparison theorems, summability factors**, and **tauberian theorems. NOTE.** For Σ , classical summability with its diverse extensions (e.g., to topological groups) is fundamental, since it establishes notions of **generalized convergence**, progressively, in hierarchies of structures/frameworks. Cognate criteria for the basic Spaces, ${}_SX$, introduced in E_1, form a unifying theme throughout these Essays.

163. Let $(a)_n \prec (b)_n$ mean that (as in 'HLP'): $a_i \searrow, b_j \searrow, \sum_1^k a_i \leqslant \sum_1^k b_j (1 \leqslant k \leqslant n-1)$ and $\sum_1^n a_i = \sum_1^n b_j$ (for $(a)_n, (b)_n \subset \mathbb{R}$). Then

(i) $\models$ For $(\omega)_n, (\lambda)_n \subset \mathbb{C}\exists M \in \mathcal{M}_n$ with $(\omega)_n$ as eigenvalues, and $(\lambda)_n$ as diagonal elements (both, as ordered sets) $\Leftrightarrow \sum_1^n \omega_i = \sum_1^n \lambda_j$

(ii) $\models$ For $(\omega)_n, (\lambda)_n \subset \mathbb{R}\exists M \in \mathcal{M}_n, M$ real/symmetric, spec$(M) = (\omega)_n$, diag$(M) = (\lambda)_n \Leftrightarrow (\lambda)_n \prec (\omega)_n$

(iii) $\models$ If $(\omega)_n \subset \mathbb{C}, (\lambda)_n \subset \mathbb{R}$ then $(Re\omega)_n < (\lambda)_n \Leftrightarrow \exists M \in \mathcal{M}_n$: spec$(M) = (\omega)_n \wedge$ spec $(\frac{1}{2}M + \frac{1}{2}M^*) = (\lambda)_n$.

Here, M^* is the transposed conjugate of M, and M_n denotes the set of all square matrices of order n.

NOTE These results solve corresponding **inverse problems** for matrices – many other examples may be found in the literature. More generally, **quasi**-inverse problems, in all parts of Mathematics, play a crucial role in Σ – and diverse instances will be found in the S/M List. The underlying question is always the same: **To what extent do certain properties characterize some primary class(s) of ME?**

164. In the formation of **algebras**, or of other structures, over prescribed '**base-sets/spaces**' suitable forms of field operations must be sought. In the S/M List there are copious examples of such prescriptions, covering virtually all areas of Mathematics – each example furnishing a basis for systematic algebraic/analytical study of the structure(s) concerned. in particular, it is of interest to compare **basic indices/parameters** for the 'product' with their values for the 'factors'. In the case of **entire functions** (e.f.) the **Hadamard (composition-) product** of F_1, F_2 is defined by $F(z) := \frac{1}{2\pi i}\int_c F_1(u)F_2(z/u)du = \frac{1}{2\pi i}\int_{C'} F_2(u)F_1(z/u)du$ For entire ('integral') functions, the order, type, and **growth indicator** (in direction $\theta)\rho, h, h(\theta)$, resp., may be compared among F, F_1, and F_2 – for specified classes of e.f., and directions of max. growth, θ^k, may be sought. The singularities (say, z') of F are found to lie in the set $\{z_1'z_2'\}$, of products of singularities of F_1, F_2. Further, if F_1 is a 'generalized exponential' and F, any e.f., then $\models \theta^*(F) \in \{\theta^*(F_1) + \theta^*(F_2)\}$. Some results for e.f. may be derived from results for PS with finite radii of cgce, but stronger versions may be proven through generalized LT. : $f(z) = \int_\theta^\infty e^{-t}t^{\sigma-1}F(t^\sigma/z)z^{-1}dt, F(Z) = \frac{1}{2\pi i}\int_{-\infty}^{(0)} e^\zeta f(\zeta^\sigma/z)z^{-1}d\zeta$, with $\sigma = \rho^{-1}$, here.

165. The **directions of max growth** for classes of entire functions (e.f.), F, are related to the **location of singularities** of the **generalized LT**, f, of F – where: $f(z) = \int_0^\infty e^{-t}t^{\sigma-1} \times F(t^\sigma/z)z^{-1}$dt; $F(z) = \frac{1}{2\pi i}\int_{-\infty}^{(0)} e^\zeta f(\zeta^\sigma/z)z^{-1}d\zeta, \sigma = \rho^{-1}$, and ρ is the order of F. In fact: (i) $\models$ **The indicator**, $h \equiv h(\theta)$ of $F \equiv F(z)$ **is max for** arg$z = \theta^* \Leftrightarrow$ arg$z = -\theta^*$ **is a singular direction for** f; and

(ii) $\models$ **If** F **has order** $\rho(\neq 0)$ **and type** h, **then, for** arg$z = \theta^*$, all $\epsilon > 0$ **and some** $\epsilon' \geqslant \epsilon$:

$$r^{-\rho}\log|F(re^{i\theta})| > h - \epsilon \wedge r^{-\rho}\log M(r) > h - \epsilon'$$

$(M(r) :=$ max$\{|F(z)| : |z| = r\})$, **over a set** $\mathcal{S}$ (of **values of** r) **of positive upper density (at least).**

(iii) Several **inequalities** may be derived to cover cases where the direction(s) of max growth must be (a) **unique** (b) **finite in number;** (c) **all contained in** $\{|z| \leqslant \alpha\}$/

(iv) Corresponding results for specified location/isolation-properties of singularities (of f); or, of max-growth directions (of F) may also be formulated as inequalities. **NOTES**(r) Most of these inequalities amount to **lower bounds** on $|\log F(z)| \equiv |\log F(re^{i\theta})|$ (or, for $F(z) \equiv \sum_0^\infty a_n z^n, f(z) = \sum_0^\infty a_n\Gamma(n\sigma+\sigma)z^{-n-1}$) on $\log|a_n\Gamma(n\sigma+\sigma)|$.

166. Various criteria for **the comparison of topologies,** $\mathcal{J}$, over the same base-set, X (or else, over sets X_1, X_2, etc.), as well as the formation of other topologies from some specified one(s), may be found in books on General Topology (e.g., J.L. Kelley, 'GT' (van Nostrand, 1955; pp 298); **A. Csaszar, 'GT'** (Hilger, 1978, pp 488); **K. Kurratowski, Topology**' (vol 1, AP, 1966; pp 560). Nevertheless, **a systematic view** of such interrelations (of great importance for Σ) seems to be available only in journal articles (e.g., #166). From $(X, \mathcal{J}_1), (X, \mathcal{J}_2)$, the g.l.b, l.u.b. topologies $\mathcal{J}_1 \cap \mathcal{J}_2, \mathcal{J}_1 \cup \mathcal{J}_2$ may be defined on X. The basic separation/regularity conditions are, in general, not conserved in the formation of 'new' topologies, and closure operations for $\mathcal{J}_2$ may be applied to sets in $(X, \mathcal{J}_2)$, etc.. Here, the fundamental relation is called **coupling**

('c'): $\mathcal{J}_1$ **is coupled to** $\mathcal{J}_2 \Leftrightarrow \forall x \in X$ **the** $\mathcal{J}_1$ **– closure of any** $\mathcal{J}_2$ **- nbhd of** x **is a** $\mathcal{J}_1$ **- nbhd of** x– which **is equivalent to:** $G \in \mathcal{J}_1 \Rightarrow cl_1 G \subseteq cl_2 G$. Coupling conditions may be found for topological groups $(X, \mathcal{J}_1), (X, J_2)$, and posets of topologies (on X) may be defined, with $\mathcal{J}_1 \leqslant \mathcal{J}_2 \Leftrightarrow \mathcal{J}_1 \subset \mathcal{J}_2 \wedge \mathcal{J}_1 \subseteq \mathcal{J}_2$ Minimality/Maximality (wr.t. '$\leqslant$') for (e.g.) **hausdorff** topologies may be characterized, with cognate results when $\mathcal{J}_1$ is **consistent** with $\mathcal{J}_2$ (i.e., $\forall x_1, x_2 \in X, x_1 \neq x_2 \exists N_{\mathcal{J}_1}(x_1), N_{\mathcal{J}_2}(x_2)$, **disjoint**). **Applications** of the comparison/coupling criteria to T-groups/ TVS include results on equicontinuity, and on relations between the Baire category of T-groups (say G) and the continuity of maps onto l.c. separable groups. See ,e.g., **Edwards, R.E.,** 'Functional Analysis ...' (HR8 W, 1965; pp 781) for many comparable results.

167. **Compounding** operations (over collections of p convex bodies in $\mathbb{R}^m, p \geqslant 1, m \geqslant 2)$ lead to comparison inequalities for the volumes of bodies so generated. The process of Steiner-Symmetrization (in specified subspaces) may be applied to the compound bodies, to generate further bounds. For closed, bounded 0 - symmetric convex bodies K_i, **the convex hull of the set** $\{\langle x \rangle^p \in \mathbb{R}^{\binom{m}{p}} : x^i \in K_i, 1 \leqslant i \leqslant p \leqslant m-1\}$ is denoted by $\kappa \equiv \langle K \rangle^p$, **the compound** of $\{K\}_p$. Here, $\langle x \rangle^p = (X)_{\binom{m}{p}}, X_q := \det(x^{iq}_{kl}), 1 \leqslant k, l \leqslant p$, for any chosen ordering, $1 \leqslant i^q_1 < \cdots < i^q_p \leqslant m, 1 \leqslant q \leqslant \binom{m}{p}$, of $\{i\}_p$. Denote by $S_r(K)$ the **Steiner symmetrization** of K in $\Pi_r \equiv \{(x)_m \in \mathbb{R}^m : x_r = 0\}$, obtained from K (viewed as a continuum of segments normal to Π_r) by translating each segment of K to be bisected by Π_r. Then :$\models$(i) $\exists c : V(\mathcal{S}_r K) \leqslant cV(K)$. If $f_K : \mathbb{R}^m \to \mathbb{R}^+, f_K(x) := \inf\{\lambda > 0 : x \in \lambda K\}$ then $\models f_K(x) > 0 (x \neq 0), f_K(0) = 0; f_K(\alpha x) = \alpha f_K(x), \alpha > 0; f_K(x+y) \leqslant f_K(x) f_K(y)$. The numbers $\lambda_i(K), \lambda_1 \leqslant \cdots \leqslant \lambda_n, \lambda_i := \inf\{\lambda > 0 : \dim(\lambda K \cap L_{\mathbb{Z}}) \geqslant i\}, L_{\mathbb{Z}} \equiv \{(x)_m \in \mathbb{Z}^m\}$, are called **successive minima** of K; f is called the **distance function of** K. Within this framework,

$$\models \text{ (ii) } \exists c' : V(\kappa) > c' \prod_{1 \leqslant i \leqslant p} [V(K_i)]^\tau, \tau := p^{-1}\binom{m}{p};$$

$$\models \text{ (iii) } \exists c'' : \prod_{1 \leqslant j \leqslant m} \lambda_j(\langle K \rangle^p) \leqslant c'' [\prod_{l=1}^{p} \prod_{j=1}^{m} \lambda_j(K_l)]^\tau.$$

NOTE. Extensive coverage of such estimates (and many references) may be found in: **Lekkerkerker, C.G., 'Geometry of Numbers'** (Wolters-Noordhoff/N-Holland, pp 510, 1969).

169. Although 'Vector Analysis', in its standard version, requires certain smoothness conditions on the domains and functions involved, some of the key results may be proven under significantly weaker conditions – or, generalized to 'n-dimensional spaces'. For **Green's Theorem**, the usual differentiability assumptions may be relaxed on $E \subset D \cup C$, where $C \equiv \delta D$ and E is closed, with **logarithmic capacity** 0m provided that the relevant integrands are (i) **Lebesgue-integrable over** D; (ii) **differentiable over** $D \backslash E$, and (iii) **Continuous over** $C \cup D$.

This weakening of the conditions is important for applications in Potential Theory. (In the basic version , the integrands must be continuously differentiable over $C \cup D$).

NOTE: logarithmic capacity is defined (e.g.) in **Doob, J.K., 'Classical Potential Theory and its Probablilistic Counterpart'** (Springer, 1984; pp 846), p:252. **The proof** involves (a) **'generalized Laplacian'**, Δ_1, where $\Delta_1 F(X) := \lim_{t \to 0} t^{-2}(M_{\Gamma_t} F - F), X \equiv (x_1, x_2); \Gamma - t \equiv$ circle, center (x_1, x_2), radius t; $M_{\Gamma_t} F \equiv \frac{1}{2\pi} \int_{\Gamma_t} F \, d\theta$; (b) **'spherical methods of summation'** of multiple FS, (c) properties of **S-H functions**, and of various types of sequential convergence. The extension to euclidean n-space is direct, by corresponding results for manifolds require extra conditions.

170. The basic properties of **sequence spaces**, σ, are representable in terms of conditions on classes of **infinite matrices**, A, acting on elements, x, of σ; etc. Then: $\{u : \forall x \in \sigma \exists ux\} =: \sigma^*$ is the **dual** of σ, and σ is called **perfect** IFF $\sigma^{**} = \sigma$. Further, if $x^{**} = \alpha$, then $\{A : A\alpha \subseteq \alpha\} \equiv \Sigma(\alpha)$ is a **ring**. Let, now, $\mathcal{F} \equiv \{A \in \delta : \delta$ **is closed under** $+, . \wedge \delta$ **contains all scalar infinite matrices**$\}$.

Problem: $?\exists | \, | : \mathcal{F} \to \mathbb{R}_+, |cA| = |c||A|, |I| = 1, |A+B| \leqslant |A|+|B|, |A.B| \leqslant |A||B|, |a_{ij}| \leqslant |A|$ (all (i,j)). **The map** $| \, |$ defines a **bound** on $\mathcal{F}$ (now taken to be a **complex normed alge-**

bra), for which $\models$ **every bound is (also) a norm**. The **converse** is **not** generally true (since '$|a_{ij}| \leqslant |A|$' need not hold). The space of sequences each of which has only finitely many elements $\neq 0$ is denoted by ϕ; if $|x| := \sum_i |x_i|$ (for $x_i \in \mathbb{R}$), then the **space** $\{x, \sum_i |x_i|\} =: \sigma_1$; and, $\{x, \max_i |x_i|\} =: \sigma_\infty$. Various sets of n/s conditions for $\mathcal{F}$ to admit a bound (or, for $\Sigma(\alpha)$ to admit a norm) may be formulated: e.g. (a)$\models$ **If** α **is perfect then** $\exists \| \|$ **on** $\Sigma(\alpha)$ IFF α **is a Kothe-Banach space**; (b) $\models$ **If** α **is perfect then** $\exists | \, \|$ on $\Sigma(\alpha)$ IFF α **is a K-B space** $\wedge \sigma_1 \subseteq \alpha \subseteq \sigma_\infty$; whereas, **if** $F_r(r > 0) \equiv \{\{x_k\} : \sum_{k \in \mathbb{P}} k^r |x_k|$ **is cgt**$\}$, then $\models \Sigma(F_r)$ **may be represented as a normed algebra admitting no bound**. NOTE. The terminology and basic results for sequence spaces may be found in the books: **Infinite Matrices and Sequence Spaces**', and '**Linear Operators'** (both published by Macmillan, 1950, 1953) by **R.G. Cooke** more recent references: '**Sequence Spaces**', by W.H. **Ruckle** (Pitman, 1981), '**Theorie der Limitierungsverfahren**', by **K.Zekk/W.Beekmans**(Springer, 1970) and, 'Summability through Functional Analysis', by A.Wilansky (N-H, 1984).

171. Properties of the (combined) **sets of zeros/α-values** of a **whole family of entire functions** – as opposed distributions for individual functions – are of interest in 'Probabilistic Function Theory'. If $g \equiv g(z) :\equiv \sum_{n \in \mathbb{N}} a_n Z^n$ is a (non polynomial) entire function of (in)finite order, the **family** $\mathcal{F}_g \equiv \{f(z,t) : 0 \leqslant t < 1, f(z,t) := \sum_{n \in \mathbb{N}} r_n(t) a_n z^n\}$ may be studied in detail, for $\{r_n\} \equiv$ **The Rademacher functions** $[r_n(t+1) = r_n(t) = r_0(2^n t), n \in \mathbb{P}; r_0(t) := 1 (0 \leqslant t < \frac{1}{2}), = -1(\frac{1}{2} \leqslant t < 1)]$. Whereas most results in 'PFT' pertain to **single** sets of α-values, or to '**most sets**' (in some sense(s)), certain **quasi-global** results for the totality (say, $\mathcal{S}(\mathcal{F})$ of any specified family, $\mathcal{F}$) may also be obtained. Here, $\mathcal{S}(\mathcal{F}_g)$ is considered, and it may be shown that: (i)$\mathcal{S}(\mathcal{F}_g)$ is **uncountable**; (ii) $\mathcal{S}(\mathcal{F})$ is **perfect**; (iii) $\mathcal{S}(\mathcal{F}_g)$ has **fractional dimension 0.**

For(i), it may be shown (by **Picard's Theorem**) that $\mathcal{S}(\mathcal{F}_g)$ has at least one zero (in fact, countably many) whence, a '**Rouchés Theorem argument**' shows that every $z_0 \in \mathcal{S}$ is surrounded by an **un**countable 'cluster' of other elements of $\mathcal{S}$. **For** (ii), it must be proven that $\mathcal{S}$ is both **dense-in-itself** and **closed**, (see, e.g., **Newmann, MHA, 'Topology of Plane Sets of Points** (CUP, 2nd ed., 1951)pp 35ff). **The proof** is a combination of **construction** and **reductio**. **For**(iii), it may be shown that $\mathcal{S}(\mathcal{F}_g)$ is a countable union of sets S_R comprising all points of $\mathcal{S}$ having modulus $< R-1$; and, that each of the S_R is itself of fractional ($\equiv$ hausdorff) dimension 0.

NOTE. The probabilistic approach to such problems was initiated in papers by JE Littlewood/AC Offord (esp., Ann. Math 49 (1948) 885–952.) (See also: Collected Papers of JE Littlewood, OUP, 1982, Vol. II). For hausdorff dimension, see 'Hausdorff Measures'. by CA Rogers (CUP, 1970).

172. The derivation of bounds on the e.v. of (classes of) matrices is of great importance in diverse parts of Algebraic Geometry, Probability Theory, ODE, Vibration Theory, etc... some of these bounds apply to general (square) matrices (e.g., Gerschgorin's Theorem and its extensions); whereas, others cover only special types of matrices – e.g., stochastic matrices. In $\#163$, conditions are given for the existence of types of matrices with prescribed e.v. and diagonal elements (in a specified order). For non-negative matrices, such a prescription (with diagonal elements χ_i and e.v. λ_k, both in any order) implies the basic inequality:

(a) $\sum_{1 \leqslant i < j \leqslant n} \chi_i \chi_j \geqslant \sum_{1 \leqslant i < j \leqslant n} \lambda_i \lambda_j$, which since $\sum_1^n \chi_i = \sum_1^n \lambda_i$, so that $\sum_2^n \chi_k = \sum_1^n \lambda_i - \chi_1$ – – yields a quadratic inequality for χ_1 (and so for any diagonal element), giving:

(b) $\lambda_1 \geqslant \chi_1 \geqslant \max\{0, n^{-1} \sum_1^n \lambda_i - - \sqrt{[(n-1) \sum_{1 \leqslant i < j \leqslant n} (\lambda_i - \lambda_j)^2]}\}$, where λ_1 denotes the maximal (simple) e.v. for a nonreducible nonnegative matrix.

173. The original **Theory of Almost-Periodic** (A.P) **Functions** is based directly on a single **quasi-periodicity condition** on **translation numbers**

$\tau \in E\{\epsilon, f(x)\}$, where $\tau \in E \Leftrightarrow \sup_{|x| < \infty} |\Delta_\tau f(x)| \leqslant \epsilon, \Delta_\tau f(x) := f(x+\tau) - f(x)$. Then f is **u.a.p** $\Leftrightarrow f$ **is cts** $\wedge E\{\epsilon, f(x)\}$ is **relatively dense** (in $\mathbb{R}$), i.e. $\exists l > 0 : I \subset \mathbb{R} \wedge |I| = l \Rightarrow$

$I \cap E \neq \phi$. An **equivalent characterization** of {u.a.p} is as: **the class of limit functions of u.cgt 'trigonometric polynomials'**, $\sum_{n \in \mathbb{P}} p_n^{(k)} A_n e^{i \Lambda_n x}$, with $0 \leqslant p_n^{(k)} \leqslant l, n > n_k \Rightarrow p_n^{(k)} \equiv 0$. One class of generalized (g) a.p. functions need not be cts, but satisfies certain 'integral-mean conditions'; another variant in {g.a.p } corresponds to extended forms of convergence (of the sequences of trig-polynomials). This in turn involves (e.g.) L^p-**spaces**,etc.. Further criteria (based on **types of 'total variation'**) give rise to {**R**iesz-a.p. functions }. The definition is: $f \in \{R^p$ -a.p.$\} \Leftrightarrow$ (i) ($\forall$ finite $I \subset \mathbb{R}) f \in R^p(I)$; (ii) $E_{R^p}\{\epsilon, f(x)\}$ is r.d. (in $\mathbb{R}$), where $f \in R^p(I) \Leftrightarrow f \in \int\{L^p(I)\}$. The spaces of a.p. functions are closed under the actions of u-cts functions and uniform limits (over $\mathbb{R}\backslash\{\pm\infty\}$) of sequences. The class of **analytic a.p. functions** replaces $\Delta_\tau f(x)$ by $\Delta_{i\tau} f(z) := f(z+i\tau) - f(z)$; etc.. Another extension, to u.a.p. classes of groups, is important for the study of a.p. functions on (Banach) algebras, and for **group representations Applications** of {a.p.} to (non)**linear (O)DE** are discussed in - '**Almost-periodic functions and differential equations**', by **BM Levitan/VV Zhikov**(CUP, 1982, transl).

174. **Symbolic logics with** n (finite), **or countably many, truth values** were introduced by **Lukasiewicz** (1922, 1930, resp.) and subsequently developed by **Post, Tarski, Lewis, etc.**– exemplified by **Post Algebras** (analogous to Boolean Algebras) for n-valued logics; see, e.g., **Rosenbloom, P.C., 'The Elements of Mathematical Logic'**(Dover, 1950), Sec4., and, **Rosser, J.B./Turquette, A.,'Many-Valued Logics'** (N-H, 1952). The logics are called **complete** IFF **every valid formula is provable**. Notions of **degree(s)**, C(L), **of completeness** for n-logics/χ_0-logics may be defined for **formalizations of the corresponding propositional and predicate calculi**. For the Lukasiewicz m-logic, $L_m, \models$ (i) $C(L_m) \leqslant d(m-1)+1$ (on certain extra conditions), here, if $\forall (P)_n \subset S_{L_m}$,where d is the usual divisor function.

(ii) If $1 \leqslant i \leqslant n-1 \sim [\{P\}_i \models P_{i+1}] \wedge n \leqslant N_{\max}$ then $C(L_m) := N_{\max} + 1$. For L_{χ_0}, both **cardinal** and **ordinal degrees of completeness** may be defined more abstractly: **see A.Tarski, Monats. f Math und Phys. 37 (1930) 361-404** for original forms, and **M. Tokarz, Studia Logica** $\overline{XXX}$ (1972)53 - 58 (ordinal)

If L_{χ_0} is **weakly complete**, has **substitution Modus Ponens** as its **only** primitive procedures, then $\models$ **Every formalization of** L_{χ_0} **satisfies**:

(iii) $C_{\text{card}}(L_{\chi_0}) = \chi_0; C_{\text{ord}}(L_{\chi_0}) = \omega$. In these cases, a nss condition for (e.g., generalized implication) **functions,** φ**, to be definable via 'Lukasiewicz primitives'** is that: **a finite set of linear forms,** λ_j**(over** $\mathbb{Z}$**), in the prop** l **variables of** φ (say, $(x)_\eta$) **correspond 1 - to - 1 with the truth values** x_j. Several **quasi-diophantine conditions** (on sets of truth values) and the'Heine-Borel Theorem' are also used to prove (iii)

NOTE. The structures of these proof is **very** obscure **schematic explication** may yield nontrivial insights!

175. Many Tauberian theorems (for series/integrals/...) involve limits as $z \to z_0 \in \mathbb{C}$, with extra restrictions on the class of paths from z to z_0. In certain important cases, some restriction(s) may be modified/relaxed without affecting the result(s), thus yielding generalizations/extensions of the basic theorems. The key result governs the limits of analytic functions on the boundary of their analyticity domains, D (s.c., with accessible boundary point(s), b, analytic mappings $\{f_n : D \to D\}$, h (bounded)). Sufficient conditions for $h(f_n(z)) \to \lambda(n \to \infty)$ whenever $z \in D_1 \subset D, b \in D_1, z_n \to b, h(z_n) \to \lambda(n \to \infty)$.

This (somewhat opaque) scheme may be used with $D \equiv$ (i) a closed sector of $\{z : |z| =\}$; (ii) $D \equiv$ the half-strip $\{R(z) > 0, a < \mathcal{I}(z) < b, a < b\}$; (iii) $D \equiv \{z : 0 \leqslant |z| < \infty, |\arg z| < \alpha\}$ (unbounded, open sector) in generalizations of Abel, and Borel, summability criteria – to integrals as well as series (Abel); and, for limits through sparser sets of points' (Borel). An important intermediate step in such applications establishes that bounded analytic function (of some class) do have limits at boundary points; after which, further conditions may be formulated to derive (variants of) associated Tauberian theorems.

176. **Fatoú's Theorem** is concerned with **the boundary limits of harmonic functions in UHP for non tangential approaches to the boundary.** $\models$ **Such functions,** H**, admit the Poisson-Stieltjes representation** $H(\xi, \eta) = K\eta + \int \eta[(t-\zeta)^2 + \eta^2]^{-1} dg(t)$, with $g(t) := \int_0^t (1+u^2) dA(u), A \in BV[\mathbb{R}\backslash\{\pm\infty\}]$ IFF H **is the difference of two harmonic functions** – and

in that case $\models$ (i) **if** $g'(x)$ **exists then** $H(\xi,\eta) \to \pi g'(x)$ as $(\xi,\eta) \to (x,0)$, **non tangentially**. (i.e., on a path between two rays with common endpoint $(x,0)$). $\models$ (ii) **The mere existence of** lim$H(\xi,\eta)$, as $(\xi,\eta) \to (x,0) \not\Rightarrow \exists g'(x)$; whereas, $\models$ (iii) $[g \nearrow \wedge H(\xi,\eta) \to A(x)$, as $(\xi,\eta) \to (x,0)$ **along each of two rays**$] \Rightarrow \exists g'(x)(= A(x)/\pi)$. An **extension** of (iii), for $(\xi,\eta) \to (0,0)$ along $\{s\cos\alpha, s\sin\alpha\}$ concerns $H_\alpha(s) := \int s\sin\alpha[t^2 - 2ts\cos\alpha + s^2]^{-1}dg(t)$, as $s \to o^+$, **subject to:** $(a) g \nearrow, g(0) = 0; (b)\Theta_{\delta,\lambda}(s) := s^\delta H_\lambda(s); (c) F_\delta^\pm(t) := sgn(t)t^{\delta-1}g(\pm t); (d)\{\lambda = \alpha,\beta; 0 < \alpha < \beta < \pi; |\delta| < 1\}; (e)\Theta_{\delta,\lambda}(s) \to C_\lambda(s \to 0^+); (f) M(\alpha,\beta,\gamma,\delta) := [C_\alpha \text{Sin}\delta(\beta-\gamma) + C_\beta \sin\delta(\gamma - \alpha)]\text{cosec}\delta(\beta-\alpha)$. Then
$\models (iv)\Theta_{\delta,\lambda}(s) \to M(\alpha,\beta,\gamma,\delta)(s \to 0^+; 0 < \gamma < \pi)$;
$\models (v) F_\delta^+(t) \to [\pi(1-\delta)]^{-1}M(\alpha,\beta,0,\delta), (t \to 0^+)$;
$\models (vi) F_\delta^- \to -[\pi(1-\delta)]^{-1}M(\alpha,\beta,0,\delta), (t \to 0^+)$.

These results are derived from a combination of **Tauberian and Abelian theorems, conformal mapping** (of an infinite sectors to get g^* from g), **contour integration**, and **the relation**: $\{g\text{cts at}a^{\frac{1}{2}} \wedge g = 0(|t|^\alpha)$ for $t = o(1), 2 > \alpha > -1\} \Rightarrow 2\int_0^{a^{\frac{1}{2}}} t dg(t) = \int_0^a dg^*(t)$. The results (iv) - (vi) give a **partial converse of (i)** in a generalized/extended form.

177. The relation between **entire functions**, of finite order ρ, and (partial sums of) their Taylor Series is basic for the proofs of **representations of entire functions of finite order (for 'most values of $|z|$') in terms of the partial sum of their TS** – in the forms:

(i) $f(z) = \sum_{N-k}^{N+k} a_n z^n\{1 + \theta(z)\}, |\theta(z)| = 0(N^{-\gamma}\log^{-\frac{1}{2}} N)$ for z
where $|f(z)| > M(|z|)N^{-\beta}, \beta \geqslant 0$, if $k = [\alpha^{-1}(1 + 2(\beta+\gamma))]^{\frac{1}{2}}N^{\frac{1}{2}}\log^{\frac{1}{2}} N$,
$M(r) := \max\{|f(z)| : |z| = \tau\}$.

(ii) $f(ze^\tau) = e^{N\tau}f(z)\{1 + \omega(\tau)\}$, for z, where $|f(z)| > M(|z|)N^{-\beta}, \beta > 0, |\omega(\tau)| = o(1/\psi(N))$, for $\tau = o[(N^{\frac{1}{2}}\log N\psi(N))^{-1}]$. Here, ψ must satisfy:
$(\psi(N)N^{-\gamma}\log^{-\frac{1}{2}} N = o(1)$. Further,

$\models$ **(iii) The 'density' of omitted values** of $|z|$ is at most $\alpha\rho$. These results are derived from **the estimates**

(iv) $|\varphi(z,z)| \leqslant \delta_1(N)\mu(|z|)\frac{3N}{k\alpha}e^{-k^2\alpha/2N}(\alpha p < 1, \delta_1(N) \to 1, k \leqslant N^{\frac{2}{3}-\epsilon})$

(v) $|\varphi(ze^\tau, ze| \leqslant \delta_2(N)\mu(|z|)\frac{3N}{k\alpha}e^{k|\sigma| - k^2\alpha/2N}(\alpha p < 1, \delta_2(N) \to 1, \sigma \equiv \mathcal{R}(\tau), k \leqslant N^{\frac{2}{3}-\epsilon}, N \equiv N(r), 1 > (2/\alpha k)(N + k + 1)|\sigma| \to 0)$, where ρ is fixed, $\mu \equiv$ max. term, and $N \to \infty$; and $\varphi(\zeta, w) := f(\zeta) - \sum_{N-k}^{N+k} a_n w^n$. Initially, (iv),(v) are proven for $F_\alpha \equiv F_\alpha(|z|) := F_\alpha(r) = \sum_{n=0}^{\infty} n^{-n\alpha}r^n$; then extended to general f. **A third estimate** (on $|\sum_{N-k}^{N+k} a_n z^n(e^{n\tau} - e^{N\tau})|$) is also used.

NOTE. Such results exemplify subtly **controlled approximation** under intricately restricted conditions.

178. From the **basic inequality**: $\max_i|\lambda_i| \leqslant \max_i \sum_{1\leqslant j\leqslant n} |a_{ij}|$, for the e.v. of a complex matrix $A \equiv (a_{ij})$ of order n, many non trivial extensions/variants are obtainable by taking for A **compound matrices** of various types. WLOG, let $R_i \equiv \sum_l |a_{ij}|$ and $|\lambda_i|$ be $\searrow$. Then $\models (i) \prod_{1\leqslant i\leqslant k}|\lambda_i| \leqslant \prod_{1\leqslant i\leqslant k} R_i, 1 \leqslant k \leqslant n$ – with '=' (for **irreducible** A) IFF (for $1 \leqslant k \leqslant r \leqslant n)|\lambda_1| = \cdots = |\lambda_r| = R_1 = \cdots = R_n(\Rightarrow |\det A| = \prod_{1\leqslant i\leqslant n} R_i)$. On this basis (with use of similarity transformations) several **inequalities** may be derived for the **determinants of matrices with (principal) diagonal dominance** (i.e., $h_i := 2|a_{ii}| - \sum_{1\leqslant j\leqslant n} |a_{ij}| \times (x_j/x_i) > 0, 1 \leqslant i \leqslant n$; or, $h := X^{-1}MXe > 0$, for M such that $(|a_{ij}|) = (|m_{ij}|), m_{ii} \geqslant 0, m_{ij} \leqslant o(i \neq j))$. Next, $\models (ii)$

If $s_i := (1 - h_i/|a_{ii}|), S_1 := \max_i s_i, S := \sum_i s_i$ then $h > 0 \Rightarrow e^S \prod_{i=1}^{n} |a_{ii}|(1 - s_i) \leqslant |\det A| \leqslant e^{-S} \prod_{i=1}^{n} |a_{ii}| \prod_{i=1}^{n} (1 - s_i)^{-1}$. **A simpler result** (for A diag-dominant) is: $\models (iii)$ If $t_i := \sum_{j=i+1}^{n} |a_{ij}|$ then $\prod_{i=1}^{n} (|a_{ii}| - t_i) \leqslant |\det A| \leqslant \prod_{i=1}^{n} (|a_{ii}| + t_i)$

NOTE several (intricately interrelated) inequalities involving $|\det A|$ may be derived – with widely varying strengths/régimes. **See**, e.g. : **Marcus, M./ Minc, H. 'A survey of Matrix Theory and Matrix Inequalities'** (P,W and S, 1964); **Horn, R.A./Johnson:(a)'Matrix Analysis** (CUP, pp561,1985); (b) **'Topics in Matrix Analysis'** (CUP, pp 607, 1991).

179. Let A, B, be **infinite (summability) matrices**, and let A_1, B_2 be obtained from A, B (resp.) by taking 2 copies of all rows of A (except the first), (resp., by taking two copies of **all** rows of B). Then **define:** $(A * B)_{nk} := (i) \sum_{1 \leqslant i \leqslant k} a_{n_i} b_{nk-1}$, and $A \underline{*} B := A_1 * B_2$. Further, **let** ${}_\lambda M$ **denote the matrix for** (C, λ) **summability**. On this basis: $\models$ (ii) ${}_r M = {}_1 M^{r*}(r-$ fold); $\models (iii) \forall (\rho, \pi, r > 0), \# \equiv *$ or $\underline{*}, {}_\pi M \#_\rho M \supset_r M$ **only if** $r \leqslant \min(\pi + 1, \rho + 1, \pi + \rho)$ – with **sufficiency**, also if $\mathbf{min}(\pi, \rho) \in \mathbb{P}$. Moreover, $\models (iv)_1 M \underline{*}_1 M \equiv {}_2 M$ (**equivalence**); whereas $\models (\nu) A \equiv B \not\Rightarrow A * C \equiv B * C$, in general.

NOTES. (a) $\models (vi)$ **If** ${}_r M \equiv ({}_r m_{ij})$ **then** ${}_r m_{nk} = \prod_{l=1}^{r-1} (k + l) \times [(n+1)^r (r-1)!]^{-1}$ – so that the proofs of (ii) - (v) involve various basic combinatorial identities. (b) Comparable schemes may be considered for other types of summability for which 'canonical transformation matrices' are known. (c) In this way, an '**OC of summability procedures**' may be established (as a part of the Σ 'OTC'). (d) The operations $*, \underline{*}$ are **quasi-convolutions**.

180. One **effective procedure** for establishing **the boundedness of solutions** of (non)-linear (O)DE is to exhibit a function $V \equiv V(x, y) : S \to \mathbb{R}^+$ such that V **is bounded** IFF $|x|, |y|$ **are bounded, as** $t \to +\infty$, where **the trajectories** $(x(t), y(t))$ are determined by the specified equation(s). For the class of ODE given by $\ddot{x} + \phi(x, \dot{x})\dot{x} + h(x) = e(t), \models (i)$ **the conditions:** $\phi(x, \dot{x}) \geqslant 0$ (for all $x, \dot{x}$), $H(x) := \int_0^x h(u)du > 0$ (for all $x \neq 0$), $H(x) \to \infty$ **with** $|x|$, and, $\int_{\mathbb{R}} |e(t)| < \infty$, **are sufficient for** $|x(t)| < C_1, |\dot{x}(t)| < C_2$ (**as** $t \to +\infty$) $\models (ii) V(x, y) := \sqrt{y^2 + 2H(x)} \Rightarrow V\dot{V} = -y^2\phi + ye \leqslant ye$ (since $y^2\phi \geqslant 0$) $\leqslant (|y| + \sqrt{2H})e \Rightarrow \dot{V} < \sqrt{2}|e|$ (since, by '**Cauchy's Ineq.**', $\sum_1^n a_i b_i \leqslant \sqrt{\{\sum_i a_i^2 \sum_i b_i^2\}}, |y| + \sqrt{2}\sqrt{H} \leqslant V$).

For ϕ **linear in** $\dot{x}$, '$\phi(x, \dot{x}) \geqslant 0$' cannot hold $\forall (x, \dot{x})$; but **the substitution**

$$a(x) := \exp(\int_0^x g(u)du), b(x) := \int_0^x a(u) f(u) du$$

yields the system $\dot{x} = (y - b(x))/a(x), \dot{y} = -a(x)(h(x) - e(t))$ – for which $\models$ (iii) $H^*(x) := \int_0^x a^2(u) h(u) du (> 0) W(x, y) := \sqrt{y^2 + 2H^*(x)}$, allows the argument in (ii) to be extended to cover this case.

NOTE The functions V, W, exemplify Liapunov Functions, widely used in Stability Theory.

181. The relations between entire functions of finite order p and (partial sums of) their Taylor Series are outlined in #177. **Here,** extensions to cover **general entire functions** and **further results on central indices,** and on **representation of derivatives** (both, for functions of order ρ) – are obtained. Once again, **the key results hold only outside 'excluded sets of values of** $|z|$**'. The estimates in #177** may be replaced by others, independent of ρ, namely:

$$\models (i) |\varphi(z, z)| = 0(k^{-1} \chi(N) \exp\{-\frac{1}{2} k^2 \lambda(N)/\chi(N)\} \mu_f(|z|))]$$

$$[\forall s.l.r; N \equiv N(r, f); k = o(N)]$$

$$\models (ii) |\varphi(ze^\tau, ze^\tau)| = 0(k^{-1} e^{N\sigma} \chi(N) \exp\{-\frac{1}{2} k^2 \lambda(N)/\chi(N)\}) \times \mu_f(|z|)$$

$$[\forall s.lr; N \equiv N(r, f); R(\tau)| = |\sigma| = 0((N^{\frac{1}{2}} \log N)^{-1}); k = o(N)].$$

Notation: for any analytic function $g : N(r, g) \equiv$ **central index** $\equiv$ rank of max term on $\{|z| = r\}M(r, g) \equiv \max\{|g(z)| : |z| = r\}; \mu(r, g) \equiv$ **max. term** (in TS of g for $\{z : |z| \leqslant r\}$); $\lambda \equiv \lambda(t) > 0, \lambda(t) \to 1(t \to \infty)$. The **auxiliary functions** $F_\alpha \equiv F_\alpha(z) := \sum_{\mathbb{N}} z^n n^{-n\alpha}; \mathcal{F}(z) := \sum_2^\infty \exp(n/\log n) z^n$ are used for initial proofs – which may be extended to general cases. On this basis (from some extension of results in #177) $\models (iii) N(r, f) \leqslant N(r, f^{(q)}) \leqslant N(r, f) + 2q\lambda(N(r, f))\log^2(N(r, f))(\forall$ s.l.$N \equiv N(r, f)$, **and with** $\lim\sup_{N\to\infty} N^{-1} q \log N \leqslant 1 - \epsilon$), for some $\epsilon > 0$, and 'allowable values of $|z|$'. **Analogues of the representations (i), (ii) in #177 may be derived for** $f^{(q)}(ze^\tau) \equiv q$**-th derivative of** f.

182. **The fundamental process** in proofs of Cauchy's Theorem and Green's Theorem is that of **approximation of the path of integration by some 'simpler path(s)'. Modes of path-approximation** may be specified to yield proofs of these theorems **under comparatively weak conditions**. In particular:

$\models (i)$ **Cauchy's Theorem** may be proven for **rectifiable (not nec. Simple) cycles,** C, with f **analytic inside** C and f **cts inside and on** C; and,

$\models (ii)$ **Green's Theorem** may be proven with C **a closed chain of rectifiable paths** (in $\mathbb{C}$) f, **cts on the union**$K(C)$ of these paths and of $O'(C)$, the open set the bound; and f_x **existing in** $O'(C)$ (**except on a 'suitably small subset'**) where, for $p \in O'(C), u(p, c)f_x$ **is summable** $u(p, C) \equiv$ **order** of p wrt C – now called the **winding number**). **The approximation** is based on **nets**, H (of **closed squares,** with **mesh**η). The sequence $\{H_k\}$ is **regular**, if: $\eta \to 0$ and each square of H_k is contained in some square of H_{k+1}. The **basic theorem** now takes the form:

$\models (iii)$ [C : **cycle of rectifiable paths**; $\{H_k\}$: **regular**; f, g: **cts on** $K(C) \cup O'(C)$; $\{C_n\}$: **sequence of 'sums'** of specially constructed **cycles** 'over $\{H_k\}$']$\Rightarrow \int_{C'_n} \omega_n \to \int_C \omega (\omega := f dx + g dy)$.

NOTES.(a) The details are messy, but the **key point** is that: **proofs under 'weak' premises entail intricate forms of approximation of the boundary,** C, **by sequences of 'simpler curves'**. (b) Similar considerations apply to all theorems involving line-integrals (or to integrals over domains of higher dimensions, etc.).

183. For a linear algebra with identity and submultiplicative norm, $|\ |$, let $A \equiv A(t) : \mathbb{R} \supset K \to \mathbb{R}$ be differentiable and invertible. Then
$\models$(i) $\limsup_{t\to t_0} \left|\frac{|A_{-1}|^{-1} - |A_0^{-1}|^{-1}}{t - t_0}\right| \leqslant A'_0$, where $A_0 \equiv A(t_0), A'_0 \equiv \frac{dA}{dt}(t_0)$. Since the matrix norm $\|A\| := \sum_{L_{ij}}^{\frac{1}{2}} |a_{ij}|^2$ is differentiable $\models (ii) |\frac{d}{dt}(\|A^{-1}\|^{-1}) \leqslant \|\frac{d}{dt}A\|$. If $A(t) \in \mathcal{HS}$ then
$\models (iii) \frac{d}{dt}|A^{-1}|^{-1} = \mathcal{R}\langle A^{-1}, A^{-1}A'A^{-1}\rangle |A^{-1}|^{-3}$ (e.g., for matrices, $A(t)$). More generally, if $|A|_{\mathcal{HS}} := \sup_{\xi\neq 0} \frac{|A\xi|}{|\xi|}$, and $\{\xi : |\xi| = 1\}$ is ordered via $\xi_\alpha \leqslant \beta_\beta \Leftrightarrow |A\zeta_\alpha| \leqslant |A\zeta_\beta|$, then $\models (iv) D^+(|A|^2) = \limsup_\alpha 2\mathcal{R}\langle A\zeta_\alpha, A'\zeta_\alpha\rangle$, for the right/left derivatives, D^+/D^-. Here, $\langle,\rangle \equiv$ i.p.

184. The **A-G mean** of $x_1, x_2 (> 0)$ is obtained by **indefinite iteration** of the transformation $x_1 \hookrightarrow \frac{1}{2}(x_1 + x_2), x_2 \hookrightarrow (x_1 x_2)^{\frac{1}{2}}$; its **value** is: $M(z_1, x_2) = \frac{\pi}{2}[\int_0^{\frac{\pi}{2}} d\lambda (x_1^2 \cos^2\lambda + y^2 \sin^2\lambda)^{-\frac{1}{2}}]^{-1}$.

For $x_i > 0, 1 \leqslant i \leqslant n$, the process $x_1 \hookrightarrow (x_1 + x_n)/2, x_n \hookrightarrow (x_1 x_n)^{\frac{1}{2}}$ is iterated (after reordering to make $(x)_n \nearrow$ at each stage). Then $\models (i) x_i^k \to M(x)_n$ as $k \to \infty, 1 \leqslant i \leqslant n$. If, now the process $x_1 \to f(x_1, x_n), x_n \mapsto g(x_1, x_n)$ is iterated (with reordering) then $\models (ii)$ **for 'suitable classes,** M_q, **of** f, g', **convergent** $[f, g]-$ **processes** are obtained. **The conditions on** f, g **include** $(*)$ $[f(x, y) = f(y, x); f(x, x) = x; f \nearrow$ in x, y separately]; (a)$f(x^-, y) > x$, (b)$f(x, y^+) < y$ (for $0 < x < y$). **Denote by** M_1 **the class** $\{f : (*) \wedge (a)\}$**, by** M_2**, the class** $\{f : (*) \wedge (b)\}$ and by $M_0, M_1 \cap M_2$. Then $\models (iii) x \leqslant y \Rightarrow x \leqslant f(x, y) \leqslant y \wedge f \in M_0, x < y \Rightarrow x < f(x, y) < y$. $\models (iv)\{[f_m, g_m]\}$, with $[f_m, g_m]$ acting on the system S_{m-1}, determine a unique mean, provided $\{[f_m, g_m]\}$ satisfy certain (uniformity) conditions; $\models (v)$ **the simple means**, $f(x,y) :\equiv F(x,y) := \phi^{-1}(\frac{1}{2}\phi(x) + \frac{1}{2}\phi(y))$, for ϕ cts, $\nearrow$, determine $F(S) := \phi^{-1}(n^{-1} \sum_{1 \leqslant i \leqslant n} \phi(x_i))$. Several (e.g. L/R - triangular) **variants** of the $[f, g]$- process may be defined with concomitant **inequalities** (e.g., $S < T \Rightarrow [F, G](S) \leqslant [F, G](T); [F, G] \leqslant_N [F', G']$). Further, $\models (\nu i)$ (e.g.) $f, g \in M_0 \Rightarrow \liminf_n x_n \leqslant \liminf_n [F, G]_n \leqslant \limsup_n [F, G]_n \leqslant \limsup_n x_n$. **NOTE.** Here, $[F, G](S)$ is obtained from iteration of $[f, g]$; or else, from successive use of $[f_m, g_m]$, etc.. Hence: **compound/reciprocal/homogeneous means** may be defined (for S).

185. **The condition:** $a_0 > ... > a_n > 0 \Rightarrow$ all zeros, ζ, of $f_n \equiv f_n(z) := \sum_{0 \leqslant i \leqslant n} a_i z^i$ have $|\zeta| > 1.\{\models |z| \leqslant 1 \Rightarrow |(1-z)f(z)| \geqslant a_0 - |a_n z^{n+1} + \sum_{0 \leqslant j \leqslant u-1} (a_j - a_{j+1}) z^{j+1}| > a_0 - [a_0 - a_1 + a_1 - a_2 + ... + a_{n-1} - a_n + a_n] = 0\}$. **This result may be generalized/extended** to (finite sections of) **PS over** $\mathbb{C} \models (i)[\forall r \in \mathbb{N}$ arg$(a_{r-1} - a_r) = \alpha + \theta_r(|\theta_r| \leqslant \cos^{-1}(\cos^r\theta), 0 \leqslant \theta < \frac{\pi}{2} \wedge \lim_{r \to \infty} a_r = \rho \exp[i(\alpha + \beta)], \rho \geqslant 0, |\beta| \leqslant \frac{\pi}{2}] \Rightarrow f(z) \equiv \sum_{r \in \mathbb{N}} a_r z^r = 0$ **only if** $|z| \geqslant \cos\theta$. $\models (ii)$ **If, further,** $a_0 - \alpha_1$, and **at least one of** $a_1 - a_2, a_2 - a_3 \neq 0$ **then** $f(z) = 0$ **only if** $|z| > \cos\theta$. $\models (iii)$ **Under the premises of** (i), with **at least two of the** $a_{r-1} - a_r \neq 0, f(z) = 0 \Rightarrow |z| > \cos\theta$. **except, possibly, for countably many values of** θ.$\models (iv)[0 < a_r \searrow \wedge \lim_r a_r \neq 0] \Rightarrow f$ **has no zero on** $\{|z| = 1\}$. $\models (v)[0 < a_r \searrow \wedge \lim_r a_r = 0] \Rightarrow$ **every zero of** f **on** $\{|z| = 1\}$ **is also a zero of** g, **where** $g(z) := \sum_{0 \leqslant i \leqslant l-1} z^i, l :=$ **gcd** $\{t \in \mathbb{P} : a_t < a_{t-1}\}$.

NOTE. Zero-distributions of (sequences/families of) **analytic/entire functions** are of fundamental importance in all areas of mathematics (e.g., '**RH**', **in Number Theory; stability criteria, in Control Theory; Network Analysis/Synthesis; Algebraic Geometry**). **See, e.g. Marden, M., 'Geometry of Polynomials'** (AMS, 1966); **Levin, B.,'Distribution of Zeros of Entire Functions'** (transl., AMS, 1964); and Burckel, R.B., **'An Introduction to Classical Complex Analysis**' Vol 1 (Birk-hauser, 1979), where there is an extensive Bibliography.

APPENDIX 3
Analogue Representation

In E_{10}, the standard-approximative[439] facets of approach are emphasized (with schematic operator, $\mathcal{A}$). The full scope of approach – covering standard approximation, abstraction, analogy, and, inverse analogy (with schematic operator, $\mathbb{A}$) is introduced in E_1. Although it is tedious to maintain these distinctions among the possible forms of approach – and 'the context' usually suffices – this Essay concentrates on (figuratively) the subset, $\mathbb{A}M_t \backslash =: AM_t$, of M_t, whose elements are primarily (inverse-)analogical in character[440].

Whereas the main aim for $\mathcal{A}M_t$ is to estimate the **separation**[441], $\sigma(\mathcal{A}e, e)$, of $\mathcal{A}e$ from $e \in M_t$, the major interest for AM_t lies in the (inverse-)analogue **mechanism(s)** exhibited by Ae – since there are (typically) only (partially) qualitative representations of $\sigma(Ae, e)$, which precludes precise evaluation / estimation. However, bounds on 'the quantitative part' of $\sigma(Ae, e)$ may still be obtained, and, as examples show, such results may be of crucial importance in Σ - Analysis[442].

In each case it must be demonstrated that the process(es) / construction(s) /. . . deployed, in the realization of Ae, really do produce (inverse) analogues. This scheme includes both deterministic and stochastic constituents – provided only that adequate algorithms have been implemented. A good illustration is furnished by the book '*Stochastic Finite Elements: A Spectral Approach*', by RG Ghanem / PD Spanos (Springer, 1991), where the **FEM**(ethos), originally purely deterministic, is extended to include '**systems with random parameters**'. So far, only $\mathcal{A}$ is involved, but: 'quasi-(stochastic) FEM over triangulable TS' **would** have A-content!

If $(*)$ denotes the basic FEM for domains in $\mathbb{R}^d (d = 1, 2, 3)$ with rectilinear decompositions, then $(*)$ may be extended in several ways, for instance: **(i)** to domains in $\mathbb{R}^n (n \in \mathbb{P})$[443]; **(ii)** to admit **curvi**linear decompositions[444]; **(iii)** to allow **mesh-refinement** near singularities[445]; **(iv)** to cover 'random systems'[446] and, **(v)** to subdomains to **triangulable TS** with suitable modifications of the problem for various operations to be w.d.[447].

(Recall that the TS X is **triangulable** IFF X is homeomorphic to a polyhedron – comprised of simplicial complexes). Plainly, the modifications in **(v)** will not, in general, be 'fully reversible', since basic characteristics of $(*)$ may be irretrievably lost, in this process of 'conceptual enlargement'[448].

The specification of (inverse) analogues of e amounts to the prescription of **correspondences** between the set of 'designated **features**, φ_j, of e' and the set of (inverse) analogues of these features – to take account of (some of) the known internal / external **interactions** involving e, as discussed in E_7.

NOTE. From now on, in this Essay, the terms 'analogue', 'analogy', 'analogical',. . . , will cover both direct and inverse relationships (unless there is a danger of ambiguity). For symbolic representations, the notation in E_7 distinguishes between these types of analogue – as $+\alpha(e) \equiv \alpha(e), -\alpha(s)$, etc..

So, further distinctions are drawn between **inner**(I) and **outer**(O) features – where e is **located** within some **environment**, ξ. Then e **embodies** its **inner** features / **interrelations**;

[439] i.e., those facets involving the Basic Spaces of E_1, together with LCS, μS.

[440] See E_7 for a discussion of abstractions / analogy 'as' approximation.

[441] This is a generalized (possible, non-numerical) 'distance': see, e.g., Reichel, H-C, et al., '*Non-numerical Distance Functions in Topology. . .*' (op. cit.)

[442] e.g., in extending the scope of '**calculability**'.

[443] See, e.g., '*Topics in Numerical Analysis III*' (JJH Miller, ed., AP, 1977, pp278–280).

[444] See, e.g., '*The Finite Element Method in Engineering Science*' (McGr-H, 1971) Ch. 8.

[445] See, e.g., LNM #960 (Springer, 1982): '*Multigrid Methods*' (W Hackbusch, et al., eds.).

[446] See, e.g., Ghanem / Spanos, op.cit..

[447] No attempt is made here to formulate such a scheme, but it appears to raise no problems of principle.

[448] The terms: **aspect, view, facet, characteristic, feature. . .** are all used in these Essays – with **quasi-local interpretations**. Some **global harmonisation** may be necessary – eventually!

and, e **interacts** (in various ways) with its **surroundings / environment**, through its **outer** features.

Accordingly, the task is to specify the sets – say, $\varphi_O(e), \varphi_I(e)$ – of designated outer / inner features of e, and then to define mappings – say, h_O^A, h_I^A – of subsets of $\varphi_O(e), \varphi_I(e)$ onto corresponding sets $\varphi_O(Ae), \varphi_I(Ae)$. Hence, **card**$\varphi_O(Ae) \leqslant$ **card**$\varphi_O(e)$, **card**$\varphi_I(Ae) \leqslant$ **card**$\varphi_I(e)$, since some features of e may lack counterparts in Ae. Indeed, **strict** inequalities obtain, 'in practice'. Unless there are 'systematic correspondences', expressible symbolically, the mappings $h_O^A, h)I^A$ are to be defined point wise. The sets $\varphi_O(e), \varphi_I(e)$ are, 'typically', finite, of 'small cardinality'; but the formalism extends to arbitrary (non-)denumerable feature-sets.

It follows that: the process of analogue-formation, for e, has several stages[449].

(a) 'Intuitive recognition';

(b) (partial) identification of $\varphi_O(e), \varphi_I(e)$;

(c) (informal) association of Ae with e, **via** preliminary forms of h_O^A, h_I^A;

(d) adjustment / modification of the pairings in **(c)** to satisfy (as far as possible) all relevant 'interaction conditions';

(e) detailed interpretation – designed to harmonise with **(a)**;

(f) formulation of (retrospectively!) rigorous versions of **(a)–(e)**.

Of these stage, **(a)** and **(e)** are the most important for Σ . Of course, **(a)** entails (at least) 'latent appreciation of **(b), (c)**' – though the details / subtleties may be largely hidden in **(a)**.

One way of systematising the emergence / generation of analogues is developed in E_{10}. Recall that each S/M-item has 11 **descriptors**, as follows[450].

Δ_1: item LABEL/NUMBER [L]

Δ_2: Title / Dominant MAP(s) [M]

Δ_3: Basic REFERENCES [R]

Δ_4: (CAPSULE) DESCRIPTION(s) [(C)D]

Δ_5: Facets / VIEWS [φ/V]

Δ_6: Mathematical ENVIRONMENT(s) [E_M]

Δ_7: Scientific ENVIRONMENT(s) [E_S]

Δ_8: Species of SPACE(s) [S]

Δ_9: Type(s) of APPROACH [T_A]

Δ_{10}: OPERATIONAL LEVEL(s) [Λ_O]

Δ_{11}: Model(s) / FORMULATION(s) [F]

The term '**mechanism**' is chosen to describe the **structure(s)** of Ae – in the sense of the **modus operandi** of the analogue(s), as exhibited through the (inner / outer) interactions for e – which may be patterned, initially, on the n-body interactions ($n \in \mathbb{P}$) of Statistical-Mechanics / Chemical-Physics[451]. The detailed **forms** of these interactions, however, must reflect the wealth of nuances routinely found in high-level mathematical activity – as diversely illustrated in

[449] More detailed specifications of 'analogues' are discussed in E_7.

[450] See, E_{10}, Part II for the motivation / details of this representation.

[451] See E_{10} Part I, Section 23 for basic ideas / references. This is a 'fine-grained description'; a coarser-grained scheme (based on 'engineering devices') is mentioned later. There are, of course, other (families of) schemes **variously analogous** to the ones given here!

E_{10} (and in the other topics in the S/M-List); and, in scientific theories / phenomena (for inverse analogues). A varied selection of CD is given in E_{10}, each CD containing extreme condensations of the main result(s) for the topic(s) considered. The CD with label N exemplifies (primarily) the MAPs in the collection $M(N)$, whose descriptions are listed (e.g., CONVERGENCE, METRIZABILITY,...). The CD may be read (or printed) by the user. The facets possibly involved in (L) have been listed (in E_5) as: arithmetical / combinatorial / algebraic / analytical / geometrical / topological (and all w.d. combinations of these 'basic facets').

For the environments, E_M, E_S, any mutually consistent combination of sub theories (as defined, e.g., by MR, or by the various scientific review journals) may be selected – though (initially, at least) only the dominant sub theories are listed, so that appropriate interpretations may be given. The precise forms of $E_M(N)$, $E_S(N)$ are not uniquely determined; rather, they are 'prescribed' within Σ (and may be modified in any consistent way). Again, the choice of space(s) stresses the 'primary' (or, 'most convenient') one(s); but, since the Basic Spaces are multifariously interrelated[452], care must be exercised!

NOTE. For the vast collection of special forms of the Basic Spaces (e.g., types of **function-spaces**) thorough classifications are essential; a substantial start in this task is provided by the scheme in '*Function Spaces*', by A Kufner, et al. (Noordhoff, 1977), op. cit..

The **scope** of $\mathbb{A}$ within Σ (as a combination of 'general approximation', abstraction, analogy, and inverse analogy) characterized Σ as a research environment of great fertility – rich enough to accommodate, and promote, investigations in all currently recognised areas of Mathematics and its applications.

Some possible levels (say, $L_{\text{op}}(e)$) of operationally for e are considered in E_4, with range from 'nil' to 'full' – and 't-max' as the best attainable over Σ_t. Plainly, the $L_{\text{op}}(e)$ are weakly increasing in t, since improved techniques may be discovered. In the 'Δ-representation', the intention is to indicate 'the overall operational level' of each S/M item, rather than to give detailed assessments. Where more refined analyses seem appropriate, the OIB may be modified as necessary.

Lastly, the **formulations**, $F(N)$, are designed to 'demonstrate' the abstraction / analogue mechanisms – in transparently structural ways, showing 'how' the elements in Ae 'combine' to produce the analogue(s). This entails the development of **quasi-diagrammatic representations**, exhibiting the various interactions, figuratively – or, in some cases, explicitly[453].

NOTE. A limited, but suggestive, comparison may be made with **compendia of engineering mechanisms** (see, e.g., '*Mechanisms in Modern Engineering Design*', by II Artobolevsky (5 Vols., MIR, Moscow, 1979); '*Mechanisms, Linkages, and Mechanical Controls*', NP Chironis (ed.) (McGraw-Hill, 1979)). These collections, though mainly 'mechanical', also contain hydraulic, pneumatic, and electromagnetic, devices[454]. In such compendia, specified 'actions' are realized through ingenious 'constructions'. If the required behaviour may be produced in several ways, then all of the 'solutions' (for a fixed specification) are mutually analogous – since their 'outputs' may differ arbitrarily, over other 'régimes', but (essentially) coincide for the applications considered.

For direct analogues of e, imitation of one 'mathematical system' by another must be demonstrated – for prescribed sets, $B_e \equiv \{b_{e\lambda} : \lambda \in \Lambda_e\}$, of **behavioural properties**, $b_{e\lambda}$. Possible formalisms / representations of classes of (inverse) analogues are considered in E_7, where it is emphasized that analogues may 'vary', from 'the vaguest resemblance', to transcription / simulation. The aim is **to construct the mechanism(s)**, say, $\mu(N)$, **for item** N of the S/M-List. If item N is primarily standard-**approximative** (and so governed by $\mathcal{A}$) then $\mu(N)$ essentially depicts the associated approximation scheme(s). Otherwise (for combinations of $\mathcal{A}$ and A; or, for A alone) $\mu(N)$ characterizes (schematically) the approximation / analogue processes

[452]See E_6 for an account of the fundamental interrelations among the Basic Spaces – with references for full treatments.

[453]The diverse forms of **control-theoretic diagrams** amy be cited as prototypes, here. See E_5 for remarks / references on Control Diagrams – among many other classes of diagrams.

[454]Physico / chemico / biological devices are also important (in Industry and in Nature); there is a great scope for inverse analogues here!

exemplified in (N).

In this Essay, the **selection**[455] $\mu[N_s]_{\{J\}}$ is developed, to illustrate typical mechanisms, for a diverse range of sources. For **direct** analogues, the **(partial) transfer of structures** – say, from $\langle S_1, \xi_1 \rangle$ to $\langle S_2, \xi_2 \rangle$ – is pervasive. Here, the $\langle S_i, \xi_i \rangle$ are **mathematical** structures (S) on environments (ξ). If, instead, S_1 is a structure over a **scientifically-based** environment, ξ_1, and S_2, ξ_2 are either 'vaguely transferred' from; or else, merely 'suggested by' S_1, ξ_1, then an **inverse** analogue is involved. These ideas are discussed in E_7, where **scientific reflections / refractions** are introduced, in the study of inverse analogues.

The analogues considered in E_{11} devolve on (partial) realizations of the fundamental concepts, c_j, in the MAP List[456] (as discussed in E_9); either, as direct mathematical variants of the c_j, or else, as – often, somewhat 'strange' – **versions** of the c_j motivated by phenomena rooted in The Sciences. In this way, both direct, and inverse, analogues are covered, and the overall scope may be broadened even further by modifying / extending the MAP List, as Σ evolved. Meanwhile, it is convenient to reproduce the list from E_9, for reference in the illustrations to be developed here.

NOTE. Outline definitions of $\{c_j\}_{m_j}$ ($\equiv$ variants of c_j) are assumed in this Essay. Recall that the '**negation forms**' of the c_j (e.g., **dis**continuity, **non**linearity, **in**stability) are also to be considered.

Notation: $\sim c_j$ denoted **the negation(s) of** c_j. For every variant of c_j there are cognate variants of $\sim c_j$ – say, $\{\sim c_j\}_{l_j}$. Although all **strict** negations of c_j must be (logically) **equivalent**, their variants need not be so. Indeed, these **in**equivalences furnish characterizations within $\{\sim c_j\}_{l_j}$ – as, initially, for $\{c_j\}_{m_j}$.

BASIC CONCEPTS $\equiv (C)_N$

1. CONTINUITY
2. DIFFERENTIABILITY
3. ANALYTICITY
4. BOUNDEDNESS
5. UNIFORMITY
6. COMMUTATIVITY / ASSOCIATIVITY / DISTRIBUTIVITY
7. EMBEDDABILITY
8. LINEARITY
9. INEQUALITY
10. CONSTRUCTIBILITY
11. DENSENESS
12. DENSITY / DISTRIBUTION / MEASURABILITY
13. TRANSCENDENTALITY
14. INVARIANCE
15. APPROXIMABILITY
16. PERIODICITY

[455] Recall the notation: $f[x]_{\{n\}} := \{f(x_1), \ldots, f(x_n)\}$. Here, the $N_{s_k}, 1 \leqslant k \leqslant J$, correspond to J items from the S/M-List.

[456] represented as $(c)_N$ (with $N = 63$, so far).

17. INTEGRABILITY
18. STABILITY
19. DUALITY
20. CONNECTEDNESS
21. ACCESSIBILITY
22. APPLICABILITY
23. DEPENDENCE
24. EXTREMALITY
25. INDETERMINACY
26. REPRODUCIBILITY
27. REPRESENTABILITY
28. REPLACEABILITY
29. COMPLEXITY
30. SINGULARITY
31. UNIVERSALITY
32. ADDITIVITY / MULTIPLICATIVITY
33. CONVERGENCE
34. COMPACTNESS
35. INVERTIBILITY
36. ITERATION / RECURSION
37. METRIZABILITY / NORMABILITY
38. DISTINGUISHABILITY
39. DECOMPOSABILITY
40. PROPAGATION
41. DEFORMABILITY
42. (RE)CONSTRUCTABILITY / RETRIEVABILITY
43. EFFICIENCY
44. SEPARABILITY / INTERACTION
45. ORDERING / ORIENTATION
46. INTERFACIAL PHENOMENA / BOUNDARIES
47. INTERFERENCE
48. ASSOCIATION / ASSOCIATIVITY
49. REDUCIBILITY

50. CONSTRAINT
51. FINITENESS / COUNTABILITY
52. CONVEXITY
53. INFECTION / CONTAGION
54. EXTENDABILITY
55. COMPLETENESS / COMPLETION
56. POSITIVITY
57. CHARACTERIZABILITY
58. COMPATIBILITY / COMPARABILITY
59. ERGODICITY
60. BRANCHING PHENOMENA
61. LOCALIZABILITY
62. RESPONSE
63. CALCULABILITY

Recall that each of the c_j 'generates' some **view(s)**, $v_j(M_t)$, of M_t. Plainly, the $v_j(M_t)$ are highly polarised (by their very construction); but various 'superpositions' of collections – say[457], $V_S(M_t) := \bigcup\{v_J(M_t) : j \in S \subset \bar{N}\}$ – yield increasingly rich representations of M_t, as $\{v\}_N$, and S, are enlarged. Eventually, the inclusion, $M_t \subset V_{\bar{N}}(M_t)$ will obtain – at least, asymptotically (**via** $(c)_N$) as $N, t \to \infty$. Of course, the $v_j(M_t)$ are 'generically incomplete', but this does not affect the utility of such figurative representations, as environments for formalize heuristics, from which rigorous proofs may (sometimes) be developed. Indeed, this is the essence of the 'Σ -philosophy' for the fruitful organization / development of Mathematical Knowledge.

In the S/M-List, every item, (L), exemplifies primarily the MAPs M_k with $k \in I_L \subset \bar{N}$. Typically, $\#I_L$ is 'small' –here, $\#I_L \leqslant 2$, in many examples, since the topics are 'designed' to illustrate a particular MAP[458]. If there are m interpretations of 'roughly comparable dominance', then $\#I_L = m$. The notation is flexible, however, and subsequent increases in $\#I_L$ (as 'new analogues' are identified) may be incorporated without affecting the underlying scheme. Accordingly, **the aim is to represent** (L) **as some element(s) of** $\bigcap_{j \in I_L} v_k(M_t)$, by considering the relevant v_j in turn. Hence, the **basic task** amounts to the **demonstration** that (L) **exemplifies** $v_j(M_t)$, for $j \in I_L$ – over as many nontrivially distinct 'environments' as possible.

With each c_j, as set of 'c_j-basic ME' – say, $\{^j m\}_{k_j}$ – is associated[459]. These sets of ME are, of course, not uniquely determined; rather, their specification constitutes part of 'the state of Σ_t'. Modifications / extensions of the $\{^j m\}_{k_j}$ would, in general, alter the nature / distribution of '**MAP-based (inverse-)analogues**', but the fundamental structure would remain unaffected (as for the $\#I_L$, above). The **level** / **quality of analogues** of c_j may be measured (roughly) by the extent to which the $^j m_i$ are **reproducible** (in the variants concerned)[460]. **Notation:** $\{^j m\}_{k_j} \equiv \beta_j \equiv \beta_j(t)$.

A good illustration is furnished by c_2: DIFFERENTIABILITY. Here, the variants all stem from the 'standard form(s)', for $f : \mathbb{R} \supset A \to B \subset \mathbb{R}$ – from which differentials, (higher) derivatives,

---

[457] Here (as before) $\bar{N} := \{1, \ldots, N\}$.

[458] This is, of course, partly subjective, but the aim is for 'high resolution'. Indeed, the ideal forms have $I_L = 1$ (on 'intuitive criteria').

[459] The j-dependence may be suppressed, unless $p(\geqslant 2)$ of the e_j are treated together, or compared.

[460] i.e., by the extent to which 'behaviour' is (broadly) imitated.

derivatives of composite functions, MVT, Taylor series, Inverse / Implicit FT,...
may be obtained – and should be elements of $\{^2m\}_{k_2}$. The full specification of $\{^2m\}_{k_2} \equiv \beta_2$ – as for arbitrary c_j – will be 'readable' by users of Σ , and may be augmented as new results pertinent to some β_j are discovered.

Variants of c_2 have been diversely investigated – including several forms over BS, and others, over TVS. In all but essentially trivial cases, at least one member of β_2 is either 'lost', or else, weakened / 'bifurcated' in complicated ways[461]. Each 'version' of c_2 constitutes a direct analogue – to be assessed in relation to the degree(s) of reproduction of (internal / external) facets of β_2. Although this is a comparatively simple example, the underlying criteria are quite general, and may be applied to any direct analogue.

For **inverse** (of 'hybrid') analogues, the criteria are rather different, since combinations of reflections / refractions may be involved. See E_7 for the formalism – and for some comments on the structure of inverse analogues. The main interest (in all cases) lies in the mathematical formalisms / systems /... ultimately emerging from 'interactions' among 'constituents' initially w.d., either, in some 'established mathematical domain(s)', or else, in some scientific field(s). The 'initial data', D_1, generating such inverse analogues, $-\alpha$, therefore (typically) comprise combinations of ME/SE/SP. If $D_1 \cup D_2 \subset \{\text{ME}\} \equiv M$, then the analogue[462] is 'mathematical'. Let $\mathcal{S} \equiv \{SE/SP\}$, and suppose that the process of inverse-analogue formation involves transformations (in some sense(S)) from D_1 to D_l 'via', successively, $D_1 \ldots D_{l-1}$ ($l \geqslant 3$). If $l = 3$, then: **either**, a mathematical **reflection** of $D_1 \subset M$ in $D_2 \subset \mathcal{S}$ is obtained; **or else**, a mathematical **refraction** in M from $D_1 \subset \mathcal{S}$. More generally, $(D)_l$ is partitioned as $(D)_l =: D^M \cup D^{\mathcal{S}}$, where (always) $D_l \in D^M$, for $\pm\alpha(e)$ – but the further distribution of the D_j between D^M and $D^{\mathcal{S}}$ determines the detailed character of $\pm\alpha$, through the sequence of reflections and refractions involved.

NOTES. (1) If the $D_i, 1 \leqslant i \leqslant l$, are depicted as 'boxes', then $\pm\alpha$ may be represented as some directed graph(s) with vertices in $D_1, \ldots, D_l$ (in that order). **(2)** Although D^M and $D^{\mathcal{S}}$ are disjoint, their (respective) **elements** (viewed as **sets**) need not be so – since distinct facets of certain (mathematical, or scientific) 'theories' may figure in the 'formation of $-\alpha$'. **(3)** If the **direct** analogues are denoted[463] by $+\alpha$, then they also have quasi-graph-representations, $G(+\alpha)$, with $D_1 \equiv e$, corresponding to cases where $(D)_l \subset M$. **(4)** There are nontrivial links between the $G(\pm\alpha) \equiv$ **graphs of analogues** and the **proof-diagrams** discussed in E_5. **(5)** There are also links with **program-flow diagrams**, in theoretical CS[464].

Let $\pm\alpha \equiv \pm\alpha(e) \in \pm\text{An}(e)$. Denote by $\mathcal{J}_M, \mathcal{J}_S$ (reps.) the sets of all w.d. 'mathematical', and 'scientific', sub theories in Σ_t; and, by $\mathcal{J}_{\pm\alpha} =: \mathcal{J}_{\pm\alpha M} \cup \mathcal{J}_{\pm\alpha\mathcal{S}}$, the **constituents** of $\pm\alpha$ (i.e., the sub theories 'directly involved in the construction of $\pm\alpha$'). Then the **vertices** of $G(\pm\alpha)$ may be labelled by the elements of $\mathcal{J}_{\pm\alpha}$. The associated set of **edges** – with numbers of oriented traversals (possibly, individually weighted[465]) then furnishes a description of $G(\pm\alpha)$. Thus, the complete form of $G(\pm\alpha(e))$ may be found only after $\mu(\pm\alpha(e))$ has been fully elucidated, in terms of 'interactions within $\mathcal{J}_{\pm\alpha}$'. Let $\mathcal{J}_e$ denote the constituents of e.

Within this framework the perception, and development, of analogues may be studied systematically, by exhibiting the interrelations among constituents of $\pm\alpha(e)$. In this way a growing accumulation of standardised analogues will be produced, as Σ_t evolves. For such analogues, α, both the graphs, $G(\alpha)$, and the mechanisms, $\mu(\alpha)$, will be documented – and **may be further analysed / modified / extended by applying 'OTC operations'.** This raises many intriguing

[461] See, e.g., LNM#30 (A Frölicher / W Bucher); LNM#374 (S Yamamuro), Springer, 1966, 1974; Russian Math. Surveys 22 (1967) 291–258; 23 (1968) 67–113 – both by VI Averbukh / OG Smolyamov. For an outline aimed at SAM-potential, see Sec. 15.6 of Elvey, JSN, '*Symbolic Computation and Constructive Mathematics*' (Res. Rep. CS-80-49, U. of Waterloo Dept. of CS, 1980). For LCTS, see: Graff, RA, TAMS 293 (1986) 485–509.

[462] direct, or inverse

[463] Thus $\pm\alpha$ may be used to show that both direct and inverse analogues are considered; so the symbols $+\alpha, -\alpha$ must be distinguished. Mostly, however, the nature of 'α' is clear from the context.

[464] See, e.g., '*Program Flow Analysis*', Muchnick, SS / Jones, CD (eds.) (P-H, 1981).

[465] More precisely: 'labelled' by the **data** for the application(s) of the PT concerned at each stage of the analogue process(es).

issues, for instance: **(i)** types of **stability**[466] of α; **(ii)** '**transplantation** of α' to a different environment[467]; **(iii)** '**Continuation** of α' into a 'maximal environment'[468]; and, **(iv)** investigation of the '**centrality** of concepts' under analogical (e.g., inter-theory) transformations[469].

As for proof-diagrams, the D_i may be specified through subclassifications in MR, and in scientific review journals. Such specifications may be refined at later stages without invalidating the 'earlier versions'. This is an aspect of DBMS development. Indeed, the paper, '*Updating logical databases*', by R Fagin, et al., (pp1–18 in: Advances in Computing Research, Vol. 3, ed., PC Kanellakis, JAI Press, 1986, op. cit.) treats **databases** as **collections of theories**. Although the detailed 'roles' or sub theories D_i in the formation of (inverse) analogues are (typically) hard to ascertain precisely, the identification of 'primary ingredients' (among the D_i) '**suggesting** the analogue α', is usually attainable. Thus $\mathcal{J}_{\alpha(E)}$ is comparatively easy to construct; but $\mu(\alpha)$ is, in many cases, far more elusive.

NOTE. For the **proof** (or, **program-flow**) **diagrams**, the equivalent of $\mu(\alpha)$ – say, the **proof-mechanism** (or, **program-mechanism**) – is **embodied** in the proof (or, program) as a **chain of deductions** (or, an **algorithm**). These species of diagrams therefore admit detailed representations (at various levels), in line with the formulation in E_7; or, with elaborations of flow-charts (of algorithms) for which **special languages** may be designed[470]. A quasi-inverse scheme (to those for proof / program / analogue diagrams) is furnished by (e.g.) the **block-diagram algebra** for control systems[471], where the system-diagrams constitute 'initial data' – from which algebraic structures are inferred. In the terminology of E_7, this algebra involves diagrams of **data**-type (rather than **proof**-type).

The **main questions about** Σ **-analogues** are therefore as follows[472]:

(a) What **forms**, $\{b(e)\}_J$, of behaviour of $e[\mathcal{J}_e]$, are quasi-reproduced in $\pm\alpha(e)[\mathcal{J}_{\pm\alpha(e)}]$?

(b) **How** are the $b_j(e)$ **related** to the MAP items?

(c) **How** is $\mathcal{J}_{\pm\alpha(e)}$ **related** to $\mathcal{J}_e$?

(d) **How** are the $b_j(e)$ **quasi-reproduced**?

NOTES. **(a)** concerns the **scope** of $\pm\alpha(e)$; **(b)**, the MAPs (separately) **exemplified** by $\pm\alpha(e)$; **(c)**, the **interrelations** among mathematical and scientific subtheories; and **(d)**, the **mechanisms** of reproduction.

Possible criteria for the estimation of **levels / degrees of reproduction** may be formulated over certain classes of analogues – according to the 'structural richness' of the environments involved. Some discussion of this problem may be found in other Essays (especially, in E_7).

It is convenient to visualise M and S as '**contiguous media**', with a formal **interface**[473]. Direct analogues, $+\alpha(e)$, evolve – through 'appropriate interactions' – within M; whereas, inverse analogues (of any type) must involve some element(s) in each of M, S – through patterns of reflections / refractions. The (internal) constitution, and structure, of analogues $\pm\alpha(e)$ is partially embodied in the associated graphs, $G(\pm\alpha(e))$; but the detailed **mechanisms(s)**, $\mu(\pm\alpha(e))$, may be explicated only in terms of the technicalities of the subtheories, D_j. Consequently, the primary aim, here, is to present studies of a representative collection of direct / inverse MA, $\{\pm\alpha\}_H$, say, where $\pm\alpha_h$ now stand for $\pm\alpha_h(e_h)$, and $h \in \bar{H}$. In each case, $g(\pm\alpha_h)$ and $\mu(\pm\alpha_h)$

[466] **Stability** is a fundamental property for Σ ; see E_8 for 'proof' forms.

[467] This is a sort of **reproducibility**.

[468] See Elvey, JSN, in '*AI in Mathematics*', J Johnson et al., eds. (OUP, 1994), where this is viewed as '**conceptual continuation**'.

[469] In particular, to what extent is 'centrality' **conserved**?

[470] See, e.g., Prather, RE, '*Discrete Mathematical Structures for Computer Science*' (Houghton Mifflin, 1975) op. cit.

[471] See, e.g., DiStefano, JJ, et al., '*... Feedback and Control Systems*' (Schaum / McGraw-Hill, 2nd ed., 1976) Ch. 7.

[472] Investigation of these questions amounts (ultimately) to the study of (quasi-)**morphisms**, within the most diverse frameworks.

[473] The association of quasi-physico-chemical properties with these 'media', governing the 'effectiveness of combination of collections, $\{D\}_k$', etc. etc.!, is irresistibly intriguing....

are determined, as far as possible, to establish a general method for handling analogues as 'active elements' of the OIB, arbitrarily transformable via OTC procedures – to generate an **OAC**[474].

For each example to be analysed here, the 'constituent subtheories' (once identified) may be 'locally labelled'; but, ultimately, a list, say, $(D^*(t))_L$, permanently ordered, and regularly maintained / extended, will form a key feature of Σ . **Notice that** a subtheory is viewed as 'any coherent body (however small) of results naturally embedded in some established theory'. As before, MR, or some set of science review journals, may be cited for the underlying list of 'established theories'. It follows that a prerequisite for the formation of $(D^*)_L$ is the construction of an adequate **taxonomy** – as envisaged in other Essays (see E_1, E_3, and, especially, E_8)[475].

The permissibility of local labelling for $(D)_l$, in the formation of $\pm\alpha(e)$, makes it possible to specify $\mathcal{J}_e$ and $\mathcal{J}_{\pm\alpha(e)}$ 'element-wise' – by introducing suitably defined subtheories 'as they arise in the construction(s)'. This latitude simplifies the process(es), while allowing the global numbering (from $(D^*)_L$) to be substituted in the 'official Σ -version(s)'. In this scheme, the subtheories D_k need not be 'minimal' (in any sense) in preliminary formulations, but a reduction to 'quasi-minimal subtheories' based on the relevant sub taxonomies, is a natural aim in the OIB design. In fact, a sequence of realizations of $\pm\alpha(e)$, with successive refinements of constituent subtheories, yields maximal insight into the structure of the analogue(s) concerned[476].

In this Essay, the collections, $(D)_l$, corresponding to the chosen S/M-items, are adequate, but (mostly) far from minimal. Plainly, this suffices for illustrative purposes – while inviting readers / users with specialized knowledge of particular S/M-items to refine / extend the associated representations of $\mu(\pm\alpha(e)), G(\pm\alpha(e))$. This is in line with the intended mode of evolution of Σ as a whole – where 'new ME' are accepted only provisionally, until they 'pass through' the **Universal Editorial Filter** (as envisaged in E_3). Even if some ME, e, is subsequently generalized / extended / optimized /... the 'improved form(s)' are added to the OIB to produce (say) $(e)^{(k_t)} \equiv (e^{(1)}, \ldots, e^{(k_t)})$, where $e^{(1)} \equiv e$, and all of the $e^{(r)}$ have suitable permanent labels – e.g., in the 'underlying set', $\mathcal{U}$: see, E_1/E_4/E_8.

This framework for the study of (inverse) analogues suffices for the discussion of diverse examples – from which quasi-axiomatic versions will emerge. For direct analogues, systematic procedures (involving the construction of appropriate quasimorphisms) may be developed. Inverse (or, hybrid) analogues, however, raise recondite problems, for general schemes – which will evolve through characteristic Σ -activity: **formalized heuristics**.

The overall **context** of such explorations is that of **mathematical / scientific theories** – viewed as 'structured object', evolving through (arbitrarily complicated) self / mutual **interactions**, to be imitated in various ways in the associated classes of analogues. In the study of the foundations of (meta-)mathematics / science, the concept of **formalized theories** is fundamental. Formalized **mathematical** theories are treated in detail in **Rasiowa / Sikorski** (op. cit.). The **scientific** theories are obtained as augmented mathematical theories – the scientific content being expressed through sets of **fundamental laws**. The entails that the physical / chemical / biological / economic /... **environments** (with associated nomenclatures) are consistently representable over the relevant mathematical structures – with 'proper interpretations'. These matters are discussed extensively in journals on the philosophy of science, and in diverse symposia[477].

The emphasis in Σ is on **approach**, so the formalized theories do not furnish the most appropriate setting. It is more productive to introduce quasi-**heuristic representations** of the form[478]: $\mathcal{J} \equiv \langle\{S\}; \{W_j\}; \Omega; \alpha; \Lambda\rangle$. Here, the **theory** $\mathcal{J}$ is represented as: **the underlying space(s)**, $\{S\}$; a family $\{W_\gamma\} \subset \mathcal{P}(\{S\})$, where the elements of W_γ admit **laws of combination** C_γ (of orders $2, \ldots, n_\gamma$); a family, $\Omega \equiv \{w_\mu\}$ of **operators** acting in prescribed ways on spec-

[474]**O**perational **A**nalogue **C**alculus.

[475]Recall (from E_1) that there are both 'abstract' (algebraic / analytical) and 'concrete' (numerical / statistical) approaches to taxonomic design / implementation. For Σ , the key property is the analogue of 'resolving power' in optics.

[476]The quintessential properties of the original ME, e, are also highlighted in this process.

[477]For instance, '*Philosophy of Science*' (USA); J. Symbolic Logic; British J. Phil. Science; Synthese – and diverse citations in these journals.

[478]There are many variants, but the essential characteristics are exemplified here.

ified classes of objects built over $[\{S\};\{W_\gamma\}]$; a set, α, of **axioms** for the basic mathematical structure of $\mathcal{J}$; and a collection, Λ, of **scientific laws** for $\mathcal{J}$. In purely mathematical theories, $\Lambda = \phi$ (the empty set). This does not, in general, mirror the distinction between Pure, and Applied / Applicable, **Mathematics** – for instance, Graph Theory, or Combinatorics, may be given 'pure representations' and yet will usually appear in diverse applications within 'The Sciences'. In other words, such ostensibly mathematical theories typically constitute the backgrounds for various models of scientific phenomena. Another major example is furnished by Probability / Statistical Theory; and so on.

NOTES

(i) The axioms α may (but need not) involve Ω.

(ii) Possible 'mechanisms of self interaction within $\mathcal{J}$' include action of Ω on α – in all w.d. forms – to modify α (regardless of whether the original axioms α depend on Ω).

(iii) For interactions of order n – say, for $i(\mathcal{J}_\sigma)_n$ – there are many intricate patterns of mutual interaction, as diversely exemplified in the S-M List. In particular, OT from $\mathcal{J}_{\sigma_k}$ may act on objects of $\mathcal{J}_{\sigma_j}$ in all w.d. / consistent ways[479]. The primary aim is the quasi-reproduction of ME in any $\mathcal{J}_m$ over 'interesting combinations', ay, $\Gamma(\mathcal{J}_\varphi)_n$ – to produce stimulating (rather than merely 'valid') new results / structures. Of course, the basic / routine results must also be systematically recorded in the OIB, to underpin the more exotic explorations!

(iv) 'Mostly', for $\Gamma(\mathcal{J}_\varphi)_n$, n is 'small' – the value in specific cases depends on how 'subtheories' are specified. It may be important to study decomposition properties of theories, in terms of 'essentially indecomposable subtheories'. In this representation, it appears that, 'typically', an ME is defined over a collection of subtheories together constituting some mathematical subdomain in (say) the main MR classification: the 'broad view', or high-level description.

(v) In the list of basic concepts c_j (reproduced, in this Essay, from E_9) DECOMPOSABILITY and SEPARABILITY are distinguished[480]. Similar criteria arise naturally in the study of the constituent subtheories of compound theories. The basic distinction is between**exact** reassembly of $\mathcal{J}$ from the $\mathcal{J}_{\varphi_l}$ (DECOMPOSITION), and **quasi**-reassembly – involving interactions / net-flux (SEPARATION). In particular, the individual / collective **effects** of the $\mathcal{J}_{\varphi_l}$ on $\mathcal{J} =' \Gamma(\mathcal{J}_\varphi)_n$ may be considered[481], and the possibilities for exact / approximate / approach '**splitting**' of $\mathcal{J}$ into essentially autonomous (but, typically, compound) subtheories may be explored. In this context, all of the OTC operations may be used (subject to local / global admissibility). This will blur (if not eliminate) the boundaries between theories and metatheories. It is convenient to denote by $M(\mathcal{J})$ the class of metatheories for $\mathcal{J}$; and, by $S_\Lambda(\mathcal{J})$, the class of scientific theories constructible from $\mathcal{J}$ – by 'extensions' / adjunctions (of sets, $\lambda \in \Lambda$, of basic laws) /.... All of the relevant operations must be consistently interpretable over $\mathcal{J}$, to produce **models** of the $s_\lambda(\mathcal{J}) \in S_\Lambda(\mathcal{J})$.[482]. Let ${}_\eta\mathcal{J}^* := (\mathcal{J}_\eta, \lambda_\eta)$; ${}_\eta s^* := s({}_\eta\mathcal{J}^*)$. Then the objects of primary interest here are of the forms **(i)** $e(\mathcal{J}')$, for any ME, e, w.d. over $\mathcal{J}'$, and **(ii)** $\{\alpha[\varphi_\eta s^*]\}(\mathcal{J}'')$, for every analogue, α, of the facet $\varphi_\eta s^*$, w.d. over $\mathcal{J}''$. Thus, both of α and $\mathcal{J}''$ are defined **on the basis of** $\varphi_\eta s^*$; indeed, α **is motivated**[483] **by** $\varphi_\eta s^*$ (in various senses) and the admissible forms of $\mathcal{J}''$ (if any) are constructed / specified to allow optimal **realizations** of $\alpha[\varphi_\eta s^*]$. For some selections[484] of ${}_\eta s^*, \varphi$, both of the $\alpha[\varphi_\eta s^*]$ and $\mathcal{J}''$ may be **effectively determined**; typi-

[479]'$j = k$' corresponds to self-interactions.

[480]See ##39, 44 of the preliminary specification in E_9

[481]Here, $='$ signifies that all of exact / approximate / approach realizations are covered.

[482]Thus, $S_\Lambda(\mathcal{J}) = \{s_\lambda(\mathcal{J}) : \lambda \in \Lambda\}$. In practice, the viable forms of Λ vary with $\mathcal{J}$, so the strict notation has $\Lambda \equiv \Lambda(\mathcal{J})$.

[483]This is the essence of **analogy** (rather than **modelling**): free association comes into play – subject to mild constraints!

[484]See, e.g., many group-theoretic developments via High-Energy Physics

cally, however, elements of **imprecision**[485] must be accommodated – to obtain **mutually consistent structures** (possibly, of novel types)[486]. This is because 'natural SE' over $_{\eta}s^{*}$ ofter 'suggest' facets of φ whose realization over mathematical SF is problematical. Consequently, the rigorous analysis of inverse analogues presents formidable challenges.

A wide-ranging discussion of calculational procedures in Physics may be found in the symposium: '*Structure and Approximation in Physical Theories*' (Härtkamper / Schmidt, eds., Plenum, 1981), op. cit.. Most of the issues discussed appear to be relevant to **scientific** theories in general[487] – and hence, also, to the examination of **inverse SA** (over appropriate **mathematical** environments). In particular, various **embeddings** / (approximative-)**reductions** of sequences of **theories** in associated **limit theories** involve **uniform** (topological) structures[488]. Again, the role of **empirical** notions – in the formulation **and** application of theories – may be studied. Moreover, all of these schemes are also applicable[489] to the mutual **approaches** among collections of **mathematical** theories.

The general frameworks outlines so far in this Essay will allow the development of quasi-**templates** for the skeleton representations of essentially arbitrary analogues – direct or inverse. Accordingly, the following sections are concerned with fundamental issues in the design of such templates[490].

The ABC (MFM) Templates

As a preliminary to the development of S/M-templates for (inverse) analogues, it is useful to study (in some detail) the scheme devised for the **A**nnotated **B**ibliography / **C**ommentary (for **M**athematical **F**racture **M**echanics). This sophisticated package evolved over nearly ten years (under the auspices of the (then) **R**oyal **A**ircraft **E**stablishment, Farnborough, UK) to a high level of mathematical completeness, with specialized search software[491]. Since the Σ -project predates this 'ABC' by many years – and leopards tend not to change their spots – it is hardly surprising that the information structure of the ABS is pertinent to the (O)IB design of Σ !

The ABC was implemented over a WINDOWS environment – initially, to classify solution-procedures for crack problems[492]. It is, however, plainly applicable (as a scheme) to any class of scientific problems governed by systems of operator (in)equations. Particular domains are identified in the associated **taxonomies**, from which **tagging** assignments (for information retrieval) may be produced. Even more: these principles extend to **all branches of (Pure / Applied / Applicable) Mathematics[493], and may be extended operationally.**

It is remarkable that – i spite of its near-universal validity – this prescription for problem / solution classification is (mostly) **effective**, over all scientific / mathematical fields – a consequence of the concentration on **absolute fundamentals**. Exploratory designs of templates for (inverse) analogues may evolve from the 'ABC versions', provided that suitable links may be forged between the classes $\mathcal{P}$, of **Problems**, and $\pm$An, of **Analogues**. The first task is to show how the 'ABC model' may be generalized (in stages) to cover arbitrary (in)equationally-specified theories in Science / Mathematics. This is, of course, a basic aim in Σ , motivated

---

[485]Consider, e.g., notions of **friction / dissipation** in Algebra / Analysis /. . . .

[486]For instance: **genetic algebras** (LNB #36), or **hysterons** (Krasnosel'skii, et al., Springer, 1989).

[487]No characteristics unique to Physics are identified – only 'structures'.

[488]The Cauchy criterion for convergence does not involve the 'limit object' – which may be of a 'different species' . . . .

[489]**Mutatis mutandis**

[490]Comparable templates for **Mathematical Fracture Mechanics** are mentioned in E_5. The 'Σ -case' is, ostensibly, far more complex!

[491]This project: '*Information Base for MFM*' (1985–1994) was based, successively, at RMCS, Shrivenham, Nottingham University, and Oxford University.

[492]The implementation was developed, from a basic mathematical plan, in collaboration with **Active Information Systems** (John Mehers).

[493]This is, again, 'obvious' – from the axiomatic / structural perspective of (e.g.) Bourbaki. The notion that **arbitrary** ME may (and should!) be viewed in this light – with (inverse) analogy 'thrown in' – motivates, and guides, the entire Σ -scheme.

here by engineering considerations. The crucial extension – to general approach phenomena – forms the principal content of this Essay.

The ABC scheme is based on the **MF-C (Master Flow-Chart)** characterising (in)equationally-formulated crack problems / solutions[494]. This deceptively simple chart emerged as the Nth iterate (N large!) of a process for extracting the essential ingredients of such characterizations. For Solid Mechanics (not just fracture phenomena) it has proven adequate for a DBMS package – on the basis of refined taxonomies, allowing detailed tagging of the research literature for efficient information retrieval[495]. All of this should be recognisable as 'the canonical Σ -method', of locating ME in conceptual space[496].

The **schematic chart** has the form:

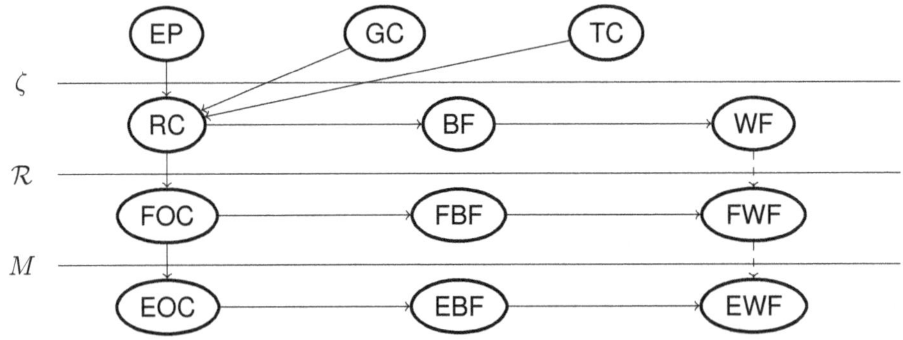

Here, **EP** $\equiv$ **Engineering Properties, GC** $\equiv$ **Geometrical Conditions, TC** $\equiv$ **Topological Conditions, RC** $\equiv$ **Required Conditions, BF** $\equiv$ **Basic Functions, WF** $\equiv$ **Wanted Functions, FOC** $\equiv$ **Formal Operator Conditions, FBF** $\equiv$ **Formal Basic Functions, FWF** $\equiv$ **Formal Wanted Functions, EOC** $\equiv$ **Effective Operator Conditions, EBF** $\equiv$ **Effective Basic Functions**, and, **EWF** $\equiv$ **Effective Wanted Functions**'

The '**filters**', $\zeta \equiv$ **Coordinate Systems**, $\mathcal{R} \equiv$ **Representations**, and $M \equiv$ **Mathematical Transformations** are defined implicitly by their effects on the 'data passing through them'[497]. **For a problem** $\mathcal{P}$, MC-F$(\mathcal{P}) =: \mathbb{M}(\mathcal{P})$.

The **EP** cover material behaviour; **GC**, the shapes, etc.; **TC** the connectivities, etc.; **RC**, the I/BC and extra constraints; **BF**, the 'potentials', etc.; ζ, the appropriate frameworks; **R**, representations of potentials, etc. (e.g., **via** integral transforms); M, all techniques deployed to obtain t-max-effective forms of all 'formal ME'; and **EOC** / **EBF** / **EWF**, the resulting 'effective ME'. The **C**capsule **D**escription(s), and **their** flowcharts, are constructed in terms of the entries in the Master Flowchart, as illustrated by the following ABC (MFM) examples[498].

3. Examples of CD / Annotations

For illustrative purposes, one CD, one 'short annotation, and one 'long annotation, are reproduced here – together with the explanation of CD-form, and comments (from R3).

These should be read in relation to the Master Flow-Chart – which is relevant, not only for CDs, but also (less directly) for all annotations.

[494] See: JSN Elvey, Interim Report (RAE, Farnborough, February, 1989)

[495] See: MFM (Tax) System: User Manual ©Active Info Systems (J. Mehers) for details of search software, etc..

[496] This is a variant of the notion of 'approach'.

[497] Fuller descriptions are given for each problem, as necessary.

[498] Taken from '*Interim Report*' (RAE, Farnborough, November, 1989) by JSN Elvey. **NOTE that** $\mathcal{ST}$ denotes **S**summary **T**itle (to enhance **search** routines).

1. Capsule Descriptions (CDs): The Basic Idea

1.1 The vast size and scope of the ABC make it desirable to produce alternatives to detailed annotations for some papers, in the current I project. (However, such annotations will form an important feature of any marketed version for which there is longterm sponsorship). These CDs are also important in their own right, as 'condensed structural forms.

1.2 Consequently, in many cases, the following scheme will be adopted. The 'master flow-chart will be given the ('standard') **abbreviated form** (cf., Figs. 3, 4 in Report (2/89))

$$\mathcal{P} = \mathrm{ECT}\zeta(\mathrm{R} \to \mathrm{B}) \to \mathrm{W}\mathcal{R}(\mathrm{FO} \to \mathrm{FB}) \to \mathrm{FW}\mathcal{M}(\mathrm{EO} \to \mathrm{EB}) \to \mathrm{EW}$$

in which the filters are denoted (now) by $\zeta, \mathcal{R}, \mathcal{M}$, (and 'motion through the filters is understood).

1.3 For additional compression, each 'component' in $\mathcal{P}$, may be labelled, so that (with '$(\cdots \to \ldots)$' replaced by ';', and, '; by ':') one has:

$$\mathcal{P} = \mathrm{a\,b\,c}\,\zeta\,\mathrm{d; e : f}\,\mathcal{R}\,\mathrm{g\,; h\,; i}\,\mathcal{M}\,\mathrm{j\,; k : l}$$

1.4 The associated capsule description takes the form of consecutive specifications of the 'components', a,...,l, together with $\xi, \mathcal{R}, M$. The notation, 'Nx' $(N - 1, 2, \ldots; x - a, \ldots, l)$, will be used for 'component x of item N in the ABC'; this may be extended to include '$N\mathcal{F}$', where $\mathcal{F}$ is a filter in item N. This is a 'local' labelling.

1.5 The **basic abbreviations** listed on pp5–6 of the (February, 1989) Report will be labelled as '$i, \ldots, xxv$' (and so on, as new abbreviations are introduced). There should be no conflict with the 'local' labelling, since '$a, \ldots, l$' will be used only within capsule descriptions – even though some components may have two distinct labels (e.g., '$d \equiv \mathrm{RC} \equiv ix$').

1.6 **Summary titles** may be given for each capsule description (since the titles of papers rarely reflect their content adequately); these extra 'titles' will be denoted by $\mathcal{SJ}$.

1.7 In some cases, one has:

$$\mathcal{P} = \mathcal{P}_1 \cup \cdots \cup \mathcal{P}_n$$

where the $\mathcal{P}_k$ are pairwise (almost-)disjoint. This corresponds to papers where two or more essentially distinct calculations are presented.

1.8 In other cases, no 'calculation' (as such) is attempted – the aim being to prove **existence**, or **uniqueness**, or, **classifications**, of solutions, etc.. For such papers, modified species of capsule descriptions will be given – indicating, broadly, the general **framework**, the **aim(s)**, and the overall strategy followed.

1.9 Detailed specifications (of **constitutive properties, equations of equilibrium**, or **motion**, etc.) may be indicated via the **reference-lists** of the papers – and otherwise, through citation of certain **general sources** (listed, once and for all, with suitably abbreviated titles of labels) to be used for **all** items in the ABC. (Thus: all capsule descriptions will include reference lists). The labels, 'a,...,l' will be used only in references to (parts of) CDs. Within each CD, the abbreviations 'EP,...,EWF' will be retained.

1.10 Although capsule descriptions do not provide the detail given in some annotations, they indicate the **scope** of papers, and the **range of techniques** deployed, along with the '**sequence of operations**'. This is sufficient to allow 'users' to assess **(i)** the **relevance** of a paper, **(ii)** the **depth and scope of the results**. Moreover, with this information, a detailed study of the paper will be facilitated considerably.

1.11 Even within the 'CD format' levels of detail will vary, so that a 'long CD' could cover several pages – but, only when the underlying concepts and methods are judged to have outstanding depth and variety. This may be exhibited, primarily, through representations of the 'filters', $\zeta, \mathcal{R}, M$. Where a paper is written sketchily, this will be reflected in the outline – there is no intention of giving extra details.

NOTE that: partial descriptions of the **filters** may be given at several stages of a CD, as they become relevant. In particular, M may comprise a mixture of **analytical, numerical** and **graphical** components.

So far, the CDs constructed have involved only a selection of the 'species' and 'calculational types' introduced in the last Report (February 1989), although several of the other entities listed there (under 'BASIC ABBREVIATIONS') appear **indirectly**. The level of detail will vary among CDs, in particular, **there is scope for expansion of the specifications of 'filters'** (in terms of elements of the set, BMO, of 'available basic mathematical operations'). This possibility will be explored for selected CDs through the use of a 'trial collection of BMOs'. Although the CDs are relatively short (compared with most annotations), their **construction** often requires more time than is needed for the corresponding annotation. This seems to be mainly because the CD is (essentially) a **'parallel program scheme' for the solution**; whereas, **the standard annotation is 'sequential' in form** (and more discursive). It follows that (as mentioned before) a **combination** of the two types of commentary / description is desirable (for the most intricate papers, at least).

2. Capsule Descriptions: Examples

In general, the CDs will be deliberately succinct – each one corresponding to the 'charts' in Figs 3 and 4 (of the Report (2/89)), and intended **to be expansible to a comprehensible calculational 'flow-chart',**, with minimal extra material. (See Section 4 for examples). Of course, the **detailed** form of any calculation considered can be followed only when the CD is studied together with the original paper; but this will be done after the CD **alone** has indicated that the paper is relevant (and potentially comprehensible) for the calculations of interest to the 'user'. **Here** (but **not** in the ABC, except in its Introduction), **comments are made on the CD examples**, in order to clarify the procedures, and conventions, used to construct the CDs.

CD – 682 CHATTERJEE, AK, and KNOPOFF, L,
'*Spontaneous growth of an in-plane shear crack*',
Int. J. Solids Structures 20 (1984) 963–978

$\mathcal{SJ}$**:** Bilateral unsteady crack growth (under 'critical SIF' criterion) of two-dimensional crack subjected to in-plane shear, with dynamical frictional stresses on the fracture surfaces. Comparison of results from 'iterated scaling approximations' (refs. [4,5]) with exact results for uniform self-similar crack-growth via functionally invariant solutions of the wave-equation (e.g., ref [14]) for displacements u, w (in plane strain').

EP: homogeneous, isotropic linearly elastic – wit P-, S-, and Rayleigh, wave-speeds α, β, v_R (respectively).

GC: $\mathbb{R}^3$, with **crack-site** (at time $t > 0$):

$$\{l(t) < x < l_+(t), y \in \mathbb{R}, z = 0\}.$$

Crack motion along $0x$; crack initiation over $0y$, at $t = 0$.

TC: —

ζ: no preliminary transformations; subsequent transformations naturally incorporated in RC.

RC: (on deviations of all parameters from their values for $t < 0$):
$\sigma_{xz}(x,t) = -f(x,t)$, for $z = 0, y \in \mathbb{R}, l_-(t) < x < l_+(t)$;
$w(x,z,t):$ continuous across $\{z = 0\}$;
$u(x,z,t):$ continuous outside crack;
coupled wave-like equations for u, w (Eqs [(2.2)]).
Constitutive equations (Eqs [(2.3)]); all stresses and displacements $\to 0$ as $z \to \infty$; all stresses have '$r^{-\frac{1}{2}}$-crack-border-singularity'.

BF: $\sigma_2(x,t) : -\sigma_{xz}(x,z,t)_{z=0}; [u](x,z,t)_{z=0} \equiv [u(x,t)]$,
where,
$[u](x,t) : -u(x,0^+,t) - u(x,0^-,t)$.

WF: modified SIFs $g_{+,+}^{(m)}, g_{+,-}^{(m)}, g_{-,+}^{(m)}, g_{-,-}^{(m)}$ (Eqs [(3.1/2)]);
crack-font: $\{x - l_\pm(t), y \in \mathbb{R}\}$
Here, $F_\pm(x,t) =: (1/\pi) g_{\pm,+}^{(m)}[x - l_+(t)]^{-\frac{1}{2}} + O(1)$,
$\sigma_2(x,t) =: (1/\pi) g_+[x - l_+(t)]^{-\frac{1}{2}} + O(1)$,
as $x \to l_+(t)^+$, with analogous definitions for $g_{\pm,-}^{(m)}$ and $g-$.

FOC: (on modified shear-stress ($F_\pm$) and modified displacement discontinuity across $0xy(G_\pm)$):

(i) $F_+(x,t) := -[1 - A_+\sigma^{(1)} - \mathcal{S}_B\sigma^{(1)}]\sigma_2(x,t)$

(ii) $G_+(x,t) := -\frac{\mu}{4}[(1-m^2)(1 + \mathcal{S}_{2/B}u^{(2)})][u]$

(iii) $\sigma_2(x,t) = [1 + \mathcal{S}_{1/B}F_+^{(1)}]F_+(x,t)$

(iv) $[u](x,t) = \frac{4}{\mu(1-m^2)}[1 - A_+G_+^{(2)} - \mathcal{S}_B G_+^{(2)}]G_+(x,t)$

with four extra eqs. obtained VIA: '$+ \leftrightarrow -$' and '$^{(1)} \leftrightarrow ^{(2)}$' [for F_-, G_-, etc.].
Here,

(v) $\mathcal{S}_C h := \frac{1}{2\pi} \int_{\alpha^{-1}}^{\beta^{-1}} C(\lambda)h(\lambda)d\lambda$

(vi) $f^{(j)}f := \delta_t \int_0^{t/s} f[x + (-1)^j\eta, t + (-1)^j\eta s]d\eta$

(vii) $A_+(v_R) := (v_R^{-1} - \alpha^{-1})^{\frac{1}{2}}(v_R^{-1} - \beta^{-1})^{\frac{1}{2}}/\mathcal{S}(-v_R^{-1});$
$B(\lambda) := \{1/\mathcal{S}(-\lambda)\}(\lambda - \alpha^{-1})^{\frac{1}{2}}(\beta^{-1} - \lambda)^{\frac{1}{2}}(v_R^{-1} - \lambda)^{-\frac{1}{2}}$

(viii) $S(\lambda) := \exp\{-\frac{1}{\pi}\int\limits_{\alpha^{-1}}^{\beta^{-1}} \tan^{-1}[\varphi(\alpha,\beta,k)]\}\frac{dk}{k+s};$

(ix) $\varphi(\alpha,\beta,k) := -\frac{4k^2(k^2-\alpha^{-2})^{\frac{1}{2}}(\beta^{-2}-k^2)^{\frac{1}{2}}}{(2k^2-\alpha^{-2})^2};$

(x) $\{g\}(x) := g(x+i0) + g(x-i0)$, for any function such that both limits exist.
There are analogous definitions (and inversions) for $F_-(x,t), G_-(x,t)$. It is found (ref [6]) that:

(xi) $\alpha\bar{F}_\pm(q,p)(p^2 - \alpha^2 q^2)^{-\frac{1}{2}} + G_\pm(q,p) = 0$

FBF: F_+, F_-, G_+, G_-

FWF: $l_+(t), l_-(t);\ \ g_{a,b}^{(m)}\ \ (a,b = \pm)$ as defined in Eqs [(3-1/3)], for $g_{a,b}^{(m)}$, and Eq [(4.3)], for an initial estimate of $l_+(t)$ [with analogues for $l_-(t)$].

NOTE that the $g_{a,b}^{(m)}$ may be related to $g\pm$ VIA FOC (see Eqs. [(3.3)]).

$\mathcal{M}$: **(a)** approximation: **(i)** since, $G_\pm(x,t) = 0$, if $x^{\pm 1} > [l_\pm(t)]^{\pm 1}$, FOC (xi) and $\mathcal{M}$ (i) allow the problem of in-plane shear to be reduced to that of anti-plane shear (see ref [6, Eqs [(3.2)$_2$, (3.5), (3.10)$_2$], and Eqs [(2.7), (2.10)], here). The values of $\sigma_2(x,t)$ off the crack, as found VIA approximation in refs [4,5] are now used to calculate $F_\pm(x,t)$ approximately.

(b) comparison: the formal solution (for uniform, self-similar crack-growth) is obtained VIA **(i)** inversion of the double Laplace-transforms in FOC (xi); **(ii)** transformation to characteristic coordinates: $\xi := \alpha t + x, \eta := \alpha t - x$; **(iii)** conversion of the inverse transform of FOC (ix) to an integral equation of 'Abel type' – to determine $F_\pm(x,t)$ off the crack from the (known) values on the crack. See Eqs [(2.5/8)], and ref [6, Eqs [(3.1), (3.16/26)]]. This yields an expression for g_+ (Eq [(3.9)]), and a formal 'equation of motion' for $l\pm(t)$ (Eq [(4.3)]); but this is effectively solvable only if some form for $\sigma_2(x,t)$ off the crack is assumed (see Eqs [(3.10)]). This 'expressions for g_+ in terms of l_+' is in fact a 'criterion for crack motion along $0x^+$'; an analogous criterion for 'motion along $0x^-$' may be given.

NOTE: reference to the book: '*Operational calculus for functions of two variables, and its applications*', by Ditkin and Prudnikov (transl., Pergamon Press, 1961) may be useful for understanding the manipulations involving double Laplace transforms.

EOC: (a) for approximation via refs [4,5], Eqs [(2.5/6), (3.1/4)], with σ_2 now assumed to have the distribution of Eqs [(3.10a,c)], with the corresponding modified stresses, F_-, F_+, given by Eqs [(3.10b,d)], respectively. **This is relatively tractable** (compared with (n));

(b) for the **exact solution** (unsteady growth of (semi-in)finite crack), ref [6, Eqs [(3.10/11), (3.25/31)]. This is **exact**, but complicated, and not evaluable in closed form.

EBF: **(a)** as in Eqs [(2.6)] with σ_2 taken as in Eqs [(3.10a,c)]

(b) as in ref [6, Eqs [(3.25/6)]]

EWF: **(a)** as in Eqs [(3.1/4)] and Eqs [(3.9/10)], for the modified SIFs; and, Eqs [(4.1/5)]ff, for the 'crack-tip loci', $l_{\pm}(t)$ (via iteration).

(b) as in ref [6, Eqs [(3.27/34)]], for the (standard) SIFs; the case of in-plane, uniform, bilateral crack propagation is solved exactly in the Appendix (cf., ref [14]) – and this solution is also used to assess the approximation procedures

Conclusion

This is an intricate paper where several powerful techniques are used . However, the arguments are hard to follow in detail, since constant references are made to related papers (some by Kostrov, others, by the present authors). Although the primary motivation for this work lies in geophysical applications (where the 'dynamical friction effects' are thought to occur), the generality of the procedures – and their extension to cover **initially-finite cracks** – should be recognised. In particular, **the possible relevance of modified SIFs in other applications** is worthy of study.

3.2 CD – 682 CHATTERJEE, AK, and KNOPOFF, L,
'*Spontaneous growth of an in-plane shear crack*',
Int. J. Solids Structures 20 (1984) 963–978

$\mathcal{SJ}$**:** This is an intricate paper, and the given title offers virtually no indication of what is involved. The $\mathcal{SJ}$ is intended to provide a minimal description.

EP: Not only the static material properties, but also the relevant wave-speeds, are required; no equations are necessary.

GC: The crack configuration is time-dependent; the initial values, $l_{+}(0), l_{-}(0)$, are specified, but $l^{\pm}(t)$ must be **found** from suitable 'equations of motion' (see: EWF).

TC: As before, the topological configuration plays no part in the calculations (though it **is** important in more general situations).

ζ**:** All primary equations are expressed in terms of rectangular cartesian coordinates (though a moving coordinate frame, and a set of P-wave characteristic coordinates, are introduced later). **The structure of this paper is such that all transformation are incorporated in [FOC and]** M**.**

RC: Here, again, the full equations need not be given; verbal descriptions, e.g., '$w(x, z, t)$: **continuous across** $(z = 0)$', are adequate.

BF: In this case, just **two** BFs suffice; but their effective **determination** involves intricate manipulations.

WF: The '**modified SIFs**' (allowing for 'dynamical friction effects' on the fracture surfaces) are to be determined – along with the 'coefficients of cohesion', and the **crack-font set**, $(x - l \pm (t), y \in \mathbb{R})$.

FOC: The formal manipulations here are complicated, but their essential structure can be expressed succinctly through a partial reformulation. Although a full understanding requires concentrated study of the original paper, the main procedures are effectively summarised here.

FBF: The nature of the FOC makes in necessary to determine **four** functions $(F_\pm, G_\pm)$, from which the original BF may be retrieved through relations included in the FOC. Strictly, the FBF should be the (formal) results of these further manipulations, but the overall strategy seems clearer as presented in the CD.

FWF: Owing to the complexity of these calculations, no direct formal expressions are available for any of these functions; but equations determining them **implicitly** are identified in the CD.

$\mathcal{M}$**:** The mathematical procedures through which the various approximations are obtained are hard to encapsulate; the best approach is purely descriptive, with **adequate references** for details. Even with this format, the specification is intricate, but the essential structure is exhibited.

EOC: Again, some reference to other papers seems unavoidable (within the CD format), but the basic relations are identified as the EOC for this problem.

EBF: The (partially indirect) prescription(s) for obtaining these functions are indicated; no (quasi-)closed-form solutions are available for the main problem.

EWF: Here, also, the best approach is to **identify the equation(s) where various WF are determined as far as possible**; no explicit 'formulae' are derived.

CD - FLOW CHARTS

In the Report (2/89), (calculational) flow-charts, $F_P \in \mathcal{F}$, for problems $P \in \mathcal{P}$, were introduced – in relation to Figs. 3 and 4. It follows that, if the CDs are constructed along these lines, then each CD should yield a typical flow-chart – displaying (and slightly amplifying) the basic features of the underlying calculation(s). This may be achieved by converting the 'generic charts' of Figs. 3 and 4 into specific charts, VIA the replacement of entries in Fig. 4 by the particular entries within the CD concerned. In this sense, a CD-flow-chart is essentially a (slightly amplified) diagrammatic version of the CD itself – with one additional feature: the relative oder – and possible interdependence – of 'component processes', may be displayed through appropriately places arrows (in an accompanying 'skeleton diagram'). Such properties depend critically upon the precise nature of these component processes – and so cannot be specified on the 'generic chart'. Of course the CD-flow-charts are not unique, but all admissible charts are 'equivalent' (in 'an intuitively obvious sense', whose formal specification would be tedious without being instructive!).

Flow Chart for CD-682

1.

Homogeneous, isotropic, linearly elastic material, with $P-, S-$, and Rayleigh, wave-speeds, α, β, v_R respectively.

2.

In-plane, shear crack propagation along the x-axis (in Cartesian System $0xyz$) with crack-initiation at $t = 0$, along $0y$, and 'crack-front', $(x = l_\pm(t), y \in \mathbb{R}, z = 0)$. No 'topological complications'. This is a highly nonlinear dynamical problem with complicated 'feedback' VIA the 'crack-edge locus histories'.

3.

All conditions initially formulated within 'the $[xyz, t]$-framework'. Methods introduced for semi-infinite cracks are used here for 'point sources'.

4.

All the field components are 'measured from their values in the pre-stressed state (at $t < 0$)'. The z-displacement, $w(x, z, t)$, is continuous across the plane $0xy$. The x-displacement, $u(x, z, t)$, is continuous outside the crack domain. All displacements $\to 0$ as $z \to \infty$; all stresses $\to 0$ as $z \to \infty$. The stress component, $\sigma_{xz}(x, z, t)_{z=0} \equiv \sigma_2(x, t)$, has an '$r^{-\frac{1}{2}}$-crack-border singularity', and is prescribed over the crack-site (as (x, t), modified by wave-motion to $f(x, t)$: $\sigma_2(x, t) = -f(x, t)$), for $\{l_-(t) < x < l_+(t), y \in \mathbb{R}, z = 0\}$. Coupled wave-like equations of motion (Eqs. [(2.2)]) for u, w.

Constitutive equations (Eqs. [(2.3)]).
5.

The solution may be obtained from $\sigma_2 \equiv \sigma_2(x,t)$, and
$[u] \equiv [u](x,z,t)_{z=0} =: u(x,0^+,t) - u(x,0^-,t)$.
Thus, σ_2 and $[u]$ are the basic functions.

6.

The aim is to determine the modified SIFs $g_{a,b,}^{(m)}(a=\pm, b=\pm)$ and the crack-front locus $\{x - l_\pm(t)\}$; see Eqs [(3.1/2)]. These are the WF. The $g_{a,b,}^{(m)}$ are defined in terms of 'dynamically modified stresses', $F_\pm(x,t)$, where $F_\pm$ has the form $F_\pm = (1-\Gamma_\pm)\sigma_2$ (see FOC (i) and its analogue; and the rest of FOC, for details) VIA:
$F_+(x,t) =: (1/\pi)g_+^{(m),}{}_+[x-l_+(t)]^{-\frac{1}{2}} + 0(1)$,
$F_1(x,t) =: (1/\pi)g_-^{(m),}{}_+[x-l_+(t)]^{-\frac{1}{2}}{}_0(1)$,
$\sigma_2(x,t) =: (1/\pi)g_+[x-l_+(t)]^{-\frac{1}{2}} + 0(1)$,
as $x \to l_+(t)^+$,
with analogous definitions for
$g_{+,-,}^{(m)}, g_{-,-,}^{(m)}$, and g_-, as $x \to l_-(t)^-$.

7.

The basic functions are constructible from $F+_-, F_-$ VIA known relations (as in FOC (i) – (xi)) if σ_2 is known off the crack.

8.

Consequently, the formal basic functions are given by $F_\pm$, and $G_\pm$ – certain 'inverse transforms' of $F_\pm$ (see FOC ((i), (ii)), for G_+; and analogues, for G_-).

9.

The full solution contains values of the $g_{a,b}^{(m)}(a,b=\pm)$ and specification of the functions $l_\pm$.

10.

The problem of finding $F_\pm(x,t)$ would be 'circular', were it not possible to introduce an approximation for $\sigma_2(x,t)$ (and hence, also for $F_\pm(x,t)$), off the crack, as found for rapidly-tearing anti-plane shear cracks (see refs [4,5]). This is (partially) justified VIA the formal analogy of the relevant equations for the two cases of crack-propagation (Eqs [(2.7), (2.10)], and, ref [6, Eqs [$(3.2)_2$, (3.5), $(3.10)_2$]]).

The validity of these (iteratively improvable) approximations is assessed by considering the corresponding results for the self-similar, uniform, bilateral propagation of an in-plane shear crack – for which analytical solutions may be found VIA 'functionally invariant procedures' (ref [14]).

11.

Critical-SIF-Criteria for crack-propagation (along $0x^{\pm}$) are obtained VIA expressions for g_+ (reps., g_-) in terms of $\dot{l}_+$ (reps., $\dot{l}_-$); see Eq [(3.9)]. $g_{\pm}$ are called the coefficients of cohesion; in general, $g_{\pm} \equiv g_{\pm}(x,t)$.

12.

These criteria become effective (rather than 'circular') only if $\sigma_2(x,t)$ (and hence, $F_{\pm}(x,t)$) is known, 'off the crack'; and this requires the introduction of an assumed form of $\sigma_2(x,t)$, off the crack – see Eqs [(3.10)].

13.

This yields, also, and 'equation of motion' for $l_+(t)$, in the form:

$$\sqrt{2}\theta(\dot{l}_+/\beta)^2\theta(\dot{l}_+/v_R)^{-1}S(-\dot{l}_+^{-1})^{-1} = -\int_0^{\xi+(\eta_0)} \frac{F+(\xi,\eta_0)}{(\xi_+-\xi)^{\frac{1}{2}}}dy,$$

with F_+ now 'known' (Eq [(4.1)]); and $(\xi, \eta) \equiv$ P-wave characteristic coordinates (Eq [(3.6)]). [An analogous equation for $L_-(t)$ may be derived.] Here, $\theta(\lambda) := (1-\lambda)^{\frac{1}{2}}$, and $S(\gamma)$ is given by Eq [(2.9)]. If $S(\gamma)$ is replaced by a polynomial of degree 5 (Eqs [(4.4/5)]) then $l(t)$ can be calculated, for any fixed x, t (Newton-Raphson); after which, $l(t)$ may be obtained generally (Runge-Kutta).

14.

Refinements of this scheme (e.g., to take account of the influence of the 'back-edge radiation' on the 'motion of the front edge') may be incorporated iteratively (Eqs [(4.6/7)]).

15.

A test of validity of these approximations and assumptions (e.g., of the form of $\sigma_2(x,t)$) is furnished by the exactly soluble problem of bilateral motion of a self-similar (spontaneous) crack (See: Appendix) – which will have constant crack-tip speeds: $|\dot{l}_{\pm}(t)| = V_{\pm}$, provided that the 'cohesions', $g_{\pm}$, and the 'dynamic stress drop', $p(x,t)$, are given by:

$$g_{\pm}(x) = \alpha_{\pm}(\pm x)^{\frac{1}{2}}; p(x,t) \equiv p = \text{ const.}$$

Then $F_{\pm}$ are also constant within the crack domain, $\{x : -V_-t < x < V_+t\}$, and all of the previous calculations may be done in closed-form (Eqs [(5.2)–(5.5)]). Comparisons are made of the graphs for cohesive coefficients versus rupture speeds (for the exact (Eq [(5.14)]) and approximate (Eq [(5.5)]) procedures).

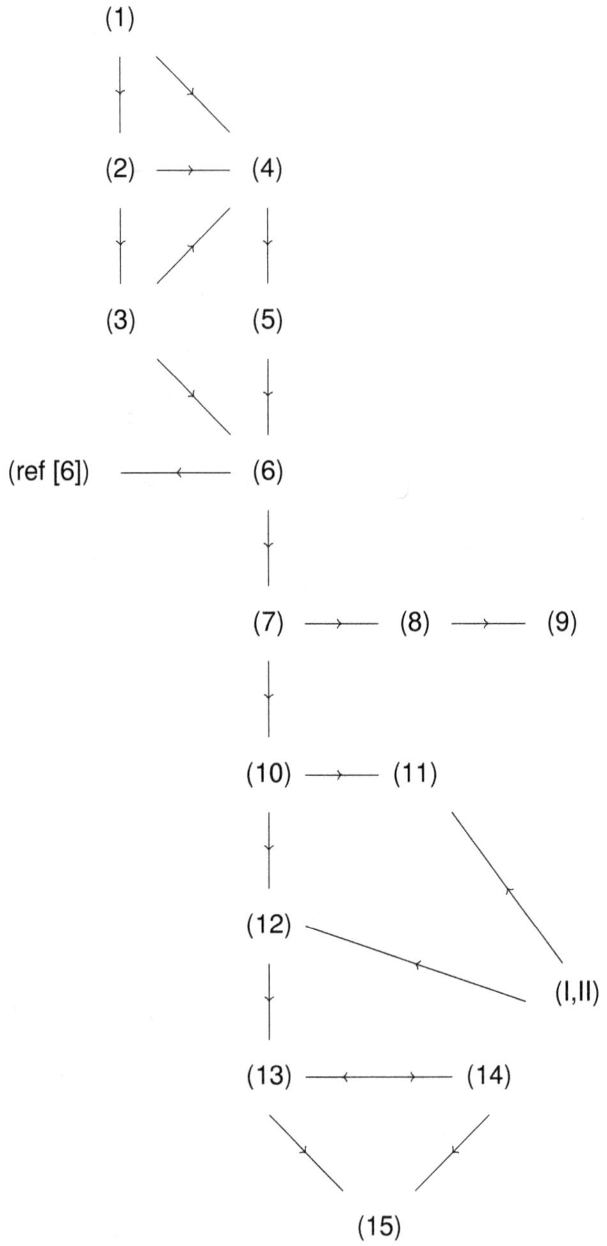

Information flow for CD - 682

Flow-Charts for 'Foundational Papers

In the case of papers whose aims are to establish existence, uniqueness, boundedness, differentiability, etc. (all, in various senses, such as 'strong', 'weak', 'distributional', ...), the flow-charts of Report (2/89) are not wholly appropriate. The most obvious modification involves the replacement of WF, by WP ($\equiv$ wanted properties), and, of FWF, EWF, by FWP, PWP ($\equiv$ formal (reps., proven) wanted properties). All of the other entries in the charts appear to have sensible interpretations for such papers (though many of them may be 'redundant' in specific examples). Consequently, there seems to be no point (at this stage of the project) in developing such charts more thoroughly – especially, since the current emphasis for the ABC is on (directly) applicable results.

Similar comments hold for 'survey papers', which (though of considerable importance) can be described usefully in the ABC only in terms of their scope and general approach.

516 Keogh, PS,
***'High Frequency Scattering by a Penny-Shaped Crack'*,**
Q. J. Mech. Appl. Math. 39 (1986) 535–566

High-frequency, normally incident, plane compressional waves are scattered from a penny-shaped crack. Such 'P-waves' (for the general, axisymmetric problem) admit a representation of the scattered displacement field in terms of inverse Hankel transforms (Eq [(2.6)]) subject to the elastodynamic equations (Eq [(2.1)]), BC (Eqs [(2.2/3)]), and edge conditions (Eq [(2.4)]), with 'radiation conditions' (Eqs [(2.5)]) on the outgoing waves. All of these conditions are found to hold for Eq [(2.6)], provided that (Eqs [(2.7/8)]) interrelating the Hankel transforms of certain displacements, and of associated stress components, hold. The aim is to derive high-frequency, asymptotic approximations to the transform of the normal stress component (VIA expansions of Hankel functions, Watson's Lemma, known behaviour of $\tau_{ZZ}(\rho, 0)$, as $\rho \to \infty$, or $\rho \to 1$: PA Martin, Proc. Roy. Soc. A378 (1981) (263–285), and by analytic continuation arguments (involving the definition of special contours in the complex plane of the transform-variable, α, and a Wiener-Hopf-type factorisation, (see, Eqs [(3.1)–(3.6))])), one may, eventually, derive a contour-integral equation (Eq [(3.15)]) for a function $T_1(\alpha)$, originally defined in Eq [(3.1)]. The arguments (see Eqs [(3.6)–(3.14)]) are intricate (mostly devolving on properties of Hankel functions, of Cauchy-integrals, and on basic function-theoretic results). The final integral equation allows for the existence of Rayleigh surface waves between the crack edges.

On setting $\alpha =: kw, s := kt$ ($k \equiv$ wave-number), one may study solutions 'for large k' (high frequency) of an associated integral equation (Eqs [(4.11)–(4.13)]) whose kernel is shown (Lemma 1) to have finite norm (all k) and, norm $\equiv 0(k^{-1})$, $(k \to \infty)$. Consequently (Lemma 2), the integral equation may be solved iteratively (by representing its solution as an infinite Neumann series, whose terms are defined recursively). On this basis, rigorous approximations are obtained (uniform in $\{-\tau \leqslant w \leqslant \tau\}$) on the transform of the far-field scattered displacements; see PS Keogh, Q. J. Mech. Appl. Math. 38 (1985) 205–232, for relevant asymptotic forms of certain integrals involved here. (Since the results differ for '$|s| \leqslant c$', and for '$|w| \geqslant c$', the notation: '$f = 0_C(k^{-a}; k^{-n})$ as $k \to \infty$' is defined to mean that: $|f(k)| \leqslant Ak^{-a}(|w| \leqslant c)$, and $\leqslant Ak^{-b}(|w| \geqslant c)$, for all $k > 1$. Most of the subsequent estimates use this notation).

Ultimately – through the detailed construction of bounds on several terms (Eqs [(5.1)–(5.19)], Lemma 3, and Appendices) a representation is determined, with error $0(k^{-4})$ as $k \to \infty$, uniformly valid for $-\tau \leqslant w \leqslant \tau$ – provided that v (Poisson's ratio) is 'not small' (Eq [(5–20)]). The limiting case, '$w \to 0$', is handled separately (Eqs [(5.21)–(5.23)]): this corresponds to the 'caustic axis' of ray theory. The other solution is relevant 'away from $w = 0$' – especially, near $w = \pm 1$, or $\pm\tau$.

The asymptotic far-field displacements are found to have the form:

$$u = A_L(\varphi)g(kr)\underline{e}_\tau + A_T(\varphi)g(Kr)\underline{e}_\varphi + 0((kr)^{-2}), \text{ as } kr \to \infty,$$

$$k^2 := \rho w^2/(\lambda + 2\mu), K^2 := \rho w^2/\mu$$

(in spherical polars r, θ, φ) with $g(y) := (y^{-1})\exp\{iy\}$ and A_L, A_T, as Eq [(6.1)]. Graphs of $\mu|A_L(\varphi)|$, and of $\mu|A_T(\varphi)|$, against φ, are given and an approximation to the scattering cross-section is obtained. This paper (together with the principal references cited) makes a significant contribution to the 'asymptotic scattering problem'.

554 Ma, CC and Burgers, P,
'*Mode-III Crack-Kinking With Delay-Time: An Analytical Approximation*',
Int. J. Solids Struct. 22 (1986) 883–899

(1) The principal aim of this paper is to model and explain the propagation of cracks in brittle materials under stress-wave loading – and, at speeds markedly less than the theoretical maximum ('Rayleigh') crack-tip speed. In particular, a delay time for the initiation of kinks, or branches, at existing crack-tips is introduced – and the resulting model presents significant extra mathematical difficulties.

A perturbation scheme is used to calculate the dynamic SIFs of a kinking line-crack in a linearly elastic solid. A discussion of the energy flux into a moving crack-tip is relevant to this approach. The time-delay destroys the self-similarity of the solution – and so excludes methods based on the use of functionally invariant solutions of the wave equation (see, e.g., Cherepanov, GP, et al., Int. J. Engng. Sci. 12 (1974) 665–690 ≡ No 343 of the ABC).

(2) A straight, semi-infinite crack is considered, for a stress-free, homogeneous, linearly elastic medium occupying the cut xz-plane. At time $t = 0$, an incident, horizontally polarised, transverse stress-wave, at angle α, strikes the (stationary) crack-tip. After a short delay, at $t = t_f$, a new crack propagates out of the original semi-infinite crack, at constant speed v_c ($< v_s$, the shear-wave speed), at an angle δ to the main crack. This generates a plane reflected wave and a cylindrical diffracted wave.

(3) Such stress waves propagate according to the PDE

$$\partial_x^2 w + \partial_v^2 w = b^2 \partial_1^2 w \tag{a}$$

where $b := v_s^{-1}$ is the slowness of the transverse wave ($b^2 = \rho/\mu$), and $w(x, z, t)$ is the displacement normal to the xz-plane, ρ is the density and μ is the shear modulus. For Mode-III deformation, the non vanishing stresses are:

$$\tau_{xy} = \mu\partial_x w \qquad\qquad \tau_{xy} = \mu\partial_x w.$$

The incident stress-wave has the form:

$$\tau_{xy}^{\text{ln}} = -\tau_0 H[\mathsf{l}(l, x, z, \alpha)] \quad \mathsf{l} := l + (x/v_s)\sin\alpha - (x/v_s)\cos\alpha. \tag{b}$$

(4) The full solution is obtained by superposition of solutions of two basic problems.

Problem (i): total-field solution for diffraction of a step stress wave (as in (b)) by a stationary, semi-infinite crack.

Problem (ii): dynamic solution for a crack subjected to concentrated forces at the tip, at the initial time of constant-speed tip extension.

Problem (i) involves the use of integral transforms, Wiener-Hopf techniques and Cagnard-de Hoop procedures for the (asymptotic) inversion of Laplace transforms. For Problem (ii), the concentrated forms are of the form $\pm p(l)$, linear in l, moving in the direction of crack propagation at a speed less than the constant speed (and normal to the xz-plane). Methods analogous to those used by Freund (see...) for Mode-I deformations, are required here.

(5) The solution of Problem (i) is known (see, e.g., ...). The diffracted field has the form:

$$\tau_{xy}^D(r, \theta, t) = \int_1^{l/br} \Gamma_l(\theta, \xi)d\xi; \tau_{xy}^D(r, \theta, l) = \int_1^{l/br} \Gamma_2(\theta, \xi)d\xi \tag{c}$$

where the Γ_j are simple combinations of elementary functions of ξ and trigonometrical functions of θ, with r, θ being plane polars about the main crack-tip; see Eqs (2.1–2.3) – and, Eq (2.4), for the asymptotic representation of the total field as (r/bl) tends to 0; from which the 'initial SIF' (for $l < l_f$) is obtained as:

$$K^S(l) = \lim_{x \to 0+} (2\pi x)^{1/2} \tau_{xy}(x, 0, l) =: \{(1 + \sin\alpha) b\pi/2\}^{-1/2} (2\cos\alpha) l^{1/2}. \tag{d}$$

(6) For Problem (ii), Freund's approach is modified. Zero-stress initial conditions are combined with the time-dependent boundary conditions (BC) on the xy-plane:

$$\tau_{sV}^F == (ml + n)\delta(x - ul)H(l) \qquad (-\infty < x < v_c l)$$

$$w^F(x, 0, l) = 0 \qquad (v_c l < x < \infty), \tag{e}$$

where the concentrated forces move (along $0x$) with speed $u(< v_c)$ as the crack tip moves (along $0x$) at speed v_c. The parameters m, n are arbitrary (independent of x, t), to be fixed later, and the basic equation is still (a). By the linearity of (a), it suffices to solve Problem (ii) for the assignments, $\{m = 0, n = 1\}$ and $\{m = 1, n = 0\}$.

(7) For $\{m = 0, n = 1\}$, x is replaced by $\xi := x - v_c l$ (coordinates moving with the crack), and the unilateral and bilateral Laplace transforms (in t, x, respectively) are applied to equation (a), and to the BC, (e). After some manipulations, a Wiener-Hopf (W-H) equation of the form

$$(\beta G)_- = H[\beta_+; \lambda] + (\beta^{-1} F)_+ \tag{f}$$

is obtained (see Eqs (2.8–2.16)), with solution of the form:

$$G_- = J_1[\beta_-, \beta_+, \lambda]; \qquad F_+ = J_2[\beta_+, \lambda]; \tag{g}$$

see Eqs (2.17/8).

(8) Since F is (essentially) the double transform of the stress field, τ_{xy}, formal inversion may be used to obtain the stress field (for $l > b_2\xi$) as:

$$\tau_{xy}^1(\xi, 0, l) = (h/\pi\xi\{h + l/\xi\})\beta(-l/\xi)\{\beta_-(-l/\xi)\beta_+(h)\}^{-1}, \tag{h}$$

where:

$$h := (v_c - u)^{-1}$$

$$\beta(\lambda) := (b^2 - \lambda^2 - b^2\lambda^2/d^2 - 2b^2\lambda/d)^{1/2} =: \beta_+(\lambda)\beta_-(\lambda) \tag{i}$$

$$\beta_\pm(\lambda) := [b \pm \lambda(1 - (/d))]^{1/2}; \qquad d := v_c^{-1}.$$

It follows that:

$$\lim_{\xi \to 0+} \xi^{1/2} \tau_{xy}^2(\xi, 0, l) = h(1 - b/d)^{1/2} \{\pi^2 [b + h(1 - b/d)]\}^{-1/2} l^{-1/2}. \tag{j}$$

Analogous calculations yield, for $\{m = 1, n = 0\}$, the result:

$$\begin{aligned} &\lim_{\xi \to 0+} \xi^{1/2} \tau_{xy}^2(\xi, 0, t) \\ &\quad = \lim_{\xi \to 0+} \xi^{1/2} \int_{b_2}^{l/\xi} \{h/\pi^{1/2}(\lambda + h)\}^2 \beta(-\lambda)\beta_-(-\lambda)^{-1} \partial_h \Omega_+ d\lambda, \\ &\quad = -h^2(1 - b/d)^{3/2} \pi^{-1} [b + h(1 - b/d)]^{-3/2} l^{1/2}, \end{aligned} \tag{k}$$

where a Tauberian theorem has been see to evaluate the last limit, and $\Omega_+(h, \lambda) := (h + \lambda)/\beta_+(h)$.

(9) The fundamental solutions constructed (in (1)–(8) above) may be used to obtain first-order representations for the SIF in Mode-III. This approximation corresponds to the straight, in-plane extension of the crack (except that the associated applied tractions are evaluated at $\theta = \delta$, instead of at $\theta = 0$). In this way, the problems of 'straight extension', and 'kinking at an angle δ', are combined, to first order in δ. Here, (r, θ) are polars about the main crack tip.

The scheme devised by Freund (J. Mech. Phys. Solids 21 (1973) 47–61; ...) is used to construct the approximate solution by 'continuous superposition' (see Eqs (2.20), (2.22), (3.5/6) – and, Eq (2.6), for the basic BC, as given in (e) above). This procedure exploits the linearity by giving special values to the parameters m, n (see (e)), and then integrating the product of τ_{xy}^D, and the associated SIF, K^F (given by Eq (3.5)) over $[0, (1-j/t)v_c]$, with respect to $u := l^{-1}z$. This yields $K^D(l, v_c, \delta)$, the SIF for the diffraction field, to first order. If the approximation for K^F is introduced, and $h := (v_c - u)^{-l}$ is taken as integration variable, then a near-closed form expression (Eqs (3.8/9)) is found for $K^D(l, v_c, \delta)$ – VIA complex integration (see p.891).

The SIF for the incident field alone offers no difficulties; it is given by:

$$\begin{aligned} K^l &= \int_0^{v_c(l-l_f)} -K^F(m=0, n=-1, l-l_f-x_0/v_c)\tau_{0y}^I dx_0 \\ &= 2\tau_0 \cos(\delta-\alpha)[2(1-b/d)(l-l_f)/\pi d]^{1/2}. \end{aligned} \tag{l}$$

The final expression for the SIF is obtained as $k \equiv K^D + K^I$, in the form:

$$\begin{aligned} K(l, v_c, \delta) &= \lambda[A_1 H(l, \Theta_1) + A_2 H(l_2 - \Theta_2) - (l-l_f)d^{-1/2}[A_1 J(\Theta_1) \\ &\qquad + A_2 J(-\Theta_2)] + A_3 \cos\Theta_1; \\ H(l, \phi) &:= [l - d^{-1}b(l-l_f)\sin\phi]^{1/2}; \\ J(\phi) &:= (1-\sin\phi)^{1/2}; \\ \Theta_j &:= \delta + (-1)^j \alpha; \\ \lambda &:= (2^{3/2}\tau_0/\pi)(1-b/d)^{1/2}(1+\sin\alpha)^{-1/2}; \\ A_j &:= \cos[\alpha + (-1)^j \delta/2] - (-1)^j \sin(\delta/2). \end{aligned} \tag{m}$$

NOTE. This expression coincides with the exact result for $\delta = 0$ (i.e., for straight extension).

(10) Numerical calculations are performed for $l_j = 0$ and various values of δ (see Fig. 4). The effect of a delay time ($l_j > 0$) is investigated by taking special values ($\delta = \pm\pi/8, \pm\pi/4$) and then plotting K versus l_j/l (Figs. 5 and 6). The difference between $K(l_j = 0)$ and $K(l \geqslant l_f)$ is found to increase with increases in α, δ, or kinked-crack speed – and to have the sign of δ.

(11) The case of uniform unit step-function loading over the original crack face is also considered. This yields an instantaneous SIF, $K^\bullet$:

$$\begin{aligned} K^\bullet &= \tau_0 \cos\delta[(2/\pi)(1-b/d)]^{1/2}\{M_- + M_+ - 2d^{-1/2}(l-l_f)^{1/2}\}; \\ M_\pm &:= b^{-1/2}(1\pm\sin\delta)^{-1/2}\{l \pm (b/d)(l-l_f)\sin\delta\}^{1/2}. \end{aligned} \tag{n}$$

(12) When the step loading is replaced by a wave pulse with nonzero rise-time, T, the calculations are more complicated – but comparatively simple results may be obtained if $l_f > T$ (see Eqs (3.15/6), where a simple pulse function is chosen). For $l_j < T$, the analysis is more intricate; see '*The extent of the SIF in dynamic crack-growth*', by Ma/Freund (Brown University Report, 1984).

(13) The energy flux into the propagating crack-tip may be related quadratically to the SIF:

$$E = (K^2/2\mu d)(1 - b^2/d^2)^{-1/2} \equiv (\tau_0^2 l/2\mu b^2)E^{\bullet}, \tag{o}$$

where (for step stress loading) $E^{\bullet}$ has a closed form representation (see Eq (4.2)). The kinking angle, δ, and normalized crack speed, $\hat{v}$, at which $E^{\bullet}$ is maximal (for any prescribed α), may be determined numerically (see Fig. 9). It is found that the associated $\hat{v}$ increases with α; and, that 'δ_{max} remains slightly greater than α, for all α'.

(14) Since in-plane problems have combined Mode-I / Mode-II forces on the kink – whereas, for Mode III there is no such 'mixing' – the present scheme cannot be used directly for in-plane deformations; however, the introduction of a delay time remains significant. Moreover, the incorporation of a delay makes possible a more realistic description of the initiation of crack-tip motion. On a 'maximum energy flux' criterion, it is conjectured that, in general, no kinking will occur; instead, a gradual curving appears, until a fixed direction for straight propagation is reached. The next refinement of the model must admit crack propagation before any kinking or bifurcation – but, after a sudden stop, followed by renewed growth (with known initial stresses, etc.) in some direction(s).

(15) Conclusion

This paper sets an important trend, by treating models of crack kinking and bifurcation (branching) which incorporate some of the phenomena observed experimentally. Although the calculations are less tractable than those for the simpler models, it is probable that some species of crack behaviour will be discovered, which would otherwise remain undetected. This situation is reminiscent of that in the theory of oscillations – where the slightest nonlinearity may change the qualitative behaviour drastically.

TAXONOMY (Version 1)

The basic layout involves the 11 categories (cats.) introduced at the outset.

	CATEGORY	ELEMENT
1.	Configuration	c
2.	Model(s) / Materials	m
3.	Required conditions (RC)	r
4.	Wanted functions (WF)	w
5.	Auxiliary functions	a
6.	Solutions	s
7.	Forms (representations) of WF	f
8.	Techniques of principal use	t
9.	Background (references)	b
10.	Outlines of techniques (SDs / charts)	o
11.	Entries (refs-annotations) from the ABC	e

All papers have the basic representation

$$P_o \equiv \langle LMTc; W \rangle$$

All ABC items are equivalent to classes of loaded, modelled, typed configurations – with prescribed kinds of WF.

GENERAL SCHEME FOR TAGGING

The minimal tagging (for a problem / paper, P), say, P_o, is based on the pertinent coding for **L**oared, **M**odelled, **T**yped **C**onfigurations, $c \equiv \{D(0), D(1), \ldots, D(J(c))\}$, with prescribed kinds of wanted functions (denoted collectively by W).

This yields the schematic representation (with **interactions** among components of c covered by L):

$$P_0 \equiv \langle LMTc; W \rangle.$$

In a 'full tagging', the additional categories whose elements are (resp.):

r, a, s, f, t, b, o, e

(see the List of Categories) will be included.

In this way, P_0 may be extended to P. **Note that**, part of r corresponds to L (the underlying paper – or, its annotation, if there is one). The other entries are to be 'coded', as accurately as possible, from the detailed TAX lists.

To form a clear impression of the density / intensity / quasi-resolving-power of Σ , for arbitrary collections of ME, it is important to examine the ABC (Solid Mechanics) scheme in considerable depth – since every established mathematical domain may be comparably treated within the OIB (after the extensions to cover **approach / operationally**). It is the combination of **(i)** taxonomies[499]; **(ii)** topical outlines[500] for fundamental techniques (especially, in fracture problems); **(iii)** corresponding flowcharts[501] for these techniques; and, **(iv)** tagging of annotations[502] of diverse papers, etc., that characterizes the ABS as an IB. This methodology may be transferred, virtually unchanged, to any Σ -subdomain – provided only that the crucial operational capabilities may be compatibly developed[503]. All of these matters are covered in great detail, for the ABC, in several internal reports, so it seems appropriate to include most of this material as **Appendices** to this Essay. More precisely, the **Topical Outlines** and their **Flowcharts** are reproduced in full (with the associated list of **Ba** sic **R**references); while samples of (abbreviated) **taxonomies** and **taggings** are included to illustrate the form of typical Σ -data fields, for which search software must be developed. These **Appendices**[504] are prefaces by explications[505] of the MF-C entries, discussions of '**Rough Models**' (at the ABC / Engineering Interface), and indications of the ABC-coverage of **Numerical / Experimental procedures**[506]. The fact that such an intricate superstructure is required just for Solid Mechanics – even **without operational capabilities** – may suggest that the cognate treatment of essentially arbitrary collections of ME (over suitable combinations of mathematical subdomains within the OIB) must be unfeasible. This, however, is not the case, since the DBMS facilities designed for (say) Solid Mechanics are substantially applicable in **all** mathematical / scientific domains: only the 'data' (of various forms) change with the domains considered. These data fields are, typically, voluminous, but modern IT facilities render issues of size / speed / optimality /... virtually irrelevant. Hence, **the ABC scheme may validly be regarded as a prototype for the (non-operational) Σ -IB.**

Similarly, all of the CD / Annotation material reproduced here (from the Farnborough Fracture Mechanics research project) is only directly relevant to problems in Solid Mechanics, but the **overall scheme**, for classifying / characterising the mathematical techniques and their modes of deployment, is **paradigmatic** for the multifaceted representation of (essentially arbitrary) ME within the Σ -OIB. Moreover, the subsequent imposition of **approach criteria** is central for Σ [507]. Thus, the process of 'locating' ME may be carried out in two stages (say,

[499] Say, $\tau(d)$, for the mathematical subdomain, d (of MR/...)

[500] Say, $T(p)$, for procedure(s), p

[501] Say, $F(p)$, for procedure(s), p

[502] Say, $tg(d)$, for $\tau(d)$

[503] Such a 'SAM/ABC interface' was proposed, as a continuation of the Farnborough project

[504] All taken from Farnborough Reports by JSN Elvey.

[505] Also taken from the above reports

[506] Also taken from the above reports

[507] As already noted, this is just a natural extension of the axiomatic / structural representation of ME – to allow for 'approach phenomena'.

$^1L, {}^2L$), where ^1Le comprises all exact / standard-approximate facets of e, and ^2Le, all facets of e involving abstraction / (inverse) analogy not covered in ^1Le. This view is reflected in the **L**ocational **C**onditions (introduced in E_7, and elaborated in E_6), where the two classes of conditions are distinguished, in the treatment of abstractions / analogies as forms of generalized approximation[508].

It is evident from these 'ABC examples' that the effective location of ME within prescribed environments is typically an intricate and subtle task – even when only standard-approximative criteria are involved[509]. If, instead, the full range of **approach** criteria is introduced, the 'location problem' becomes still more challenging. Moreover, a uniform procedure covers **all** ME – regardless of their ostensible complexity / sophistication. Although such uniformity would be inappropriate for 'hand calculations', it is essential in IT packages, where the search software must be universally applicable. This is in line with the (sometimes formal) representation of **all** ME as (O)T over Σ – the essential basis of Σ -investigations. It follows that (in obvious senses!) Σ , as an 'information medium', is extraordinarily dense / viscous / polarisable / anistropic, etc![510]

One of the basic tenets of Σ is that (at the operational level): **every ME has some LC/OT representation(s)** – ranging from the formal / near-trivial to the arbitrarily complicated. Such representations are (in general) far from unique, but criteria for **quasi-canonical forms** may be specified, to ensure logical correctness. One consequence of this 'democracy' is that the fundamental structure of Σ may be developed through essentially **uniform procedures** – of special importance for the implementation of SAM / IT / AI systems. **The analyses of (sets of) ME as objects, interacting in states of (dis)equilibrium – within prescribed (deterministic / stochastic) environments** – therefore exhibit species of universality[511]. Consequently, the manipulation of arbitrary functions of OT acting on w.f.e may be represented within a uniform framework – to realized an OTC, as considered in other Essays (especially, E_8). The imposition of approach criteria yields an O**A**C, complementing the OTC[512].

NOTE For the arbitrary ME e, the MF-C is denoted by $\mathbb{M}_e(\delta, \xi) \equiv \mathbb{M}(\delta, \xi) \equiv \mathbb{M}(\xi)$ – if e, ξ are 'given'.

It follows from the ABC scheme that **the next stage in analogue-formation**, of e in Σ , may be based on variants of the MF-C – say, $V_\lambda(\mathbb{M}(\xi))$, where each variant of $e \in \mathbb{M}(\xi)$ corresponds to some realization, over an ξ_λ, of the basic MF-C entries[513] (i.e., of S G T : ζ : R B W : $\mathcal{R}$: FR FB FW : M : ER EB EW). It is convenient to put $\mathbb{M}_\lambda(\eta) := V_\lambda(\mathbb{M}(\xi))$ with $e_\lambda \in \mathbb{M}_\lambda(\eta), \lambda \in \Lambda \equiv \Lambda(t)$, though (as usual) the t-dependence is suppressed, unless t-comparisons, or other t-operations, are involved. Here, S denotes **structure** (mathematical / scientific) of the **framework** for ξ; and, $\xi \equiv \{S; G, T\}$, the **environment** for ξ. **NOTE** This is in line with the notation for **diagrams of ME**; see E_5. The set of ME in terms of which e is **represented**, will be called the **domain of** e **in** ξ – say, $\mathcal{D}(e) \equiv \mathcal{D}(e; \xi)$. The elements of $\mathcal{D}(e)$ may be viewed as the **quasi-coordinates**, q_β, for e; so that, e is **located** by the $q_\beta(e)$, within ξ, and $\mathcal{D}(e) =: \{q_\beta(e) : \beta \in B(e; \xi)\}$. Once again, it suffices to write q_β, B, etc.[514]. The representations may be denoted by $\rho_\lambda \equiv \rho_\lambda(e_\lambda, \mathcal{D}_\lambda, \xi_\lambda)$, for $e_\lambda \in \mathbb{M}_\lambda(\eta)$.

Accordingly, for any ME, e, the formal procedure comprises several steps:

(i) specification of ξ**;**

(ii) formation of $\mathbb{M}(\delta, \xi)$**;**

---

508 These LC may be formulated **via** the operations $^1L, {}^2L$ (as well as the 'internal / external conditions' already used).

509 This covers both deterministic and stochastic estimates over all w.d. combinations of the Basic Spaces, $\{_SX\}$.

510 The use of scientific imagery to describe the properties of the Σ -OIB is, of course, both productive and irresistible.

511 Indeed, such 'interaction models' encompass virtually all 'modes of behaviour', yielding w.d. characterizations of ME. The usual circularity thus engendered is harmless, provided that consistency is maintained.

512 See E_7/E_8 for the main ideas; also, earlier remarks in this Essay.

513 Thus V_λ is a type of 'operator', whose formal properties may be studied. **NOTE that**, in full: $\mathbb{M} \equiv \mathbb{M}(\delta, \xi)$ and $\mathbb{M}_\lambda \equiv \mathbb{M}_\lambda(\eta, \xi_\lambda)$

514 The parallel is (mostly) only partial; 'basic components' could also characterize the $q_\beta(e)$. In this sense, $\mathcal{D}(e)$ is akin to a **basis** in a LVS, etc.. Although the accuracy of such descriptions varies wildly with the choice of e, ξ, the terminology is still very useful.

(iii) construction of some $\mathbb{M}_\lambda$ from $\mathbb{M}$;

(iv) identification of the associated e_λ from $\mathbb{M}_\lambda(\eta)$;

(v) determination of $\mathcal{D}_\lambda \equiv \mathcal{D}(e_\lambda, \xi_\lambda)$;

(vi) interpretation of the corresponding representations $\rho_\lambda \equiv \rho(e_\lambda, \mathcal{D}_\lambda, \xi_\lambda)$, to produce $\alpha_\lambda(e)$, the resulting analogue(s) of e.

NOTES

(a) Given e, the specification of ξ is far from unique, but this is unimportant, provided that a definite form is chosen.

(b) As in 'the ABC case', $\mathbb{M}(\delta, \xi)$ contains indications (at least) of all processes required to produce t-max.-operational forms of δ – with e as one possibility.

(c) The relative order of steps (ii), (iv) is not obvious, but it seems that $\mathbb{M}_\lambda$ is obtained from $\mathbb{M}$, initially, by elaborating ξ_λ from ξ; after which possible forms of e_λ become apparent.

(d) The number / variety of representations of e_λ will vary both with the initial choice of e and with the 'structural richness of ξ, but the underlying framework will remain unchanged[515]. Consequently, $\alpha_\lambda(e)$ is formally a **set** – of all w.d. e-analogues, e_λ, **generated** by V_λ from $\mathbb{M}(\xi)$, as **output**, e_λ, of $\mathbb{M}_\lambda(\eta)$.

(e) The OTC procedures may be applied at any stage of analogue-formation – to produce 'extra' variants / extensions. Typically, however, such transformations yield only alternative forms of analogues in sets of **generic types**, rather than intrinsically 'new' types. **This distinction is fundamental for all OTC processes.**

(f) It appears that: **every w.d. e-analogue is of the form specified in (d),** above. That is, **every** analogue procedure for e_λ **could** be formatted in terms of **some** $\mathbb{M}_\lambda(\eta)$ – no matter how e_λ was originally realized. This 'intuitively obvious proposition' is **assumed** here. (The plausible universality of $\mathbb{M}(\xi)$ – broadly interpreted – virtually constitutes a proof but not quite!)[516].

Interim Summary

(a) The emphasis in E_{10} is on standard (deterministic / stochastic) approximation – of all species. Many illustrations are cited in the S/M-List (with diverse sketches in E_{10}/II).

(b) In E_{11}, by contrast, analogues, $\pm\alpha(e)$, are the main concerns – drawing on material from E_5/E_7, especially.

(c) Analogues $\pm\alpha(e)$ are formed **via** the identification / implementation of the **inner / outer conditions for** e over the **environment**, ξ – followed by the (partial) **realization** of (some of) these conditions **adapted to** $\pm\alpha(e)$ over $\pm\alpha(\xi)$.

(d) One possible **route** from e to $\pm\alpha(e)$ involves the **MAP scheme** (see E_9), where every item has (so far) 11 **descriptors**, $(\Delta)_{11}$, to locate and realise it. This is effective for 'sufficiently complicated ME', but less so for 'arbitrary ME'.

(e) Notions of **(internal / external) behaviour** of e in ξ may be invoked to allow the **imitation** of e by $\pm\alpha(e)$, as far as possible.

---

[515]This could allow species of 'type-transitions', where ostensibly 'small changes' in e yield 'large changes' in **card** $\alpha_\lambda(e)$ – for certain forms of V_λ.

[516]A rigorous proof would be tedious and unenlightening, since 'all w.d. pairs' e, e_λ must be considered.

(f) All of the **direct** analogues considered correspond to (partial) realizations of the MAP concepts, c_j (as defined generically in E_9). Since the MAP-List will be regularly extended, this does not limit the potential range of analogues covered. These MAP-based techniques are discussed here at some length.

(g) The role of **(formalized) mathematical / scientific theories** in the study of analogy is also discussed. Such theories are regarded as **structured objects** evolving through **self / mutual interactions**. Relationships with **Metamathematics** are considered, and the notion of **templates** for ME – motivated by earlier research on an IB for Solid Mechanics – is introduced.

(h) It transpires that the ME templates may be modelled closely on the Master Flowcharts, $\mathbb{M}(P)$ for problems P in Solid Mechanics, with structural features essentially transferable from $\mathbb{M}(P)$ – provided that suitable data for e are available – to produce $\mathbb{M}(\xi)$. **This framework covers both direct and inverse analogues** (see (g), above).

(i) To illustrate these complicated processes, **one complete Farnborough report, and excerpts from others, are appended to this Essay** – covering **Topical Outlines / Flowcharts** (for mathematical techniques); and, **Capsule Descriptions / Problem-IB interfaces / Annotations / Taxonomies / Tagging-procedures** (for scanning the literature).

(j) In spite of this formidable complexity, it is possible to represent '**the Σ -superstructure**' in terms of **quasi-products** (over classes, C_k, of ME) **of the template structures** for each of the C_k. Moreover, since $\mathbb{M}(C_k)$ differs from $\mathbb{M}P$ primarily in the data for C_k (rather than in fundamental structure) it follows that (in a sense to be clarified!) **the charts** $\mathbb{M}(P)$ **are typical of all charts** $\mathbb{M}(\xi)$, provided that they are **operational**.

(k) In short: **the underlying scheme for locating ME and developing analogues has a universal quality**. **Overall:** combinations of MAP-based and chart-based forms suffice for arbitrary analogues $\pm\alpha(e)$.

On the Role of Taxonomy / Tagging in Σ .

For the Solid Mechanics ('ABC') IB, the taxonomies and associated tagging procedures are essential components in the identification of configurations / loading / solution-techniq- ues /.... In the overall scheme of Σ , however, the emphasis is somewhat different. Although it is manifestly impossible to produce a complete taxonomy for M_t, it **is** feasible to develop progressively refined classifications (based on, e.g., the MR categories / subcategories /...) as 'natural settings' for essentially arbitrary ME, in Pure / Applies / Applicable Mathematics. These may be termed **pre / quasi-taxonomies**, ${}^q\tau$, with detailed labelling systems. (The term **pre-taxonomy** is used in E_3.) On this basis, charts $\mathbb{M}(\delta, \xi)$ may be formulated succinctly, with variants $\mathbb{M}_\lambda(\eta, \xi_\lambda)$ also economically constructible as 'new elements' of ${}^qT$; and so on. **Note that** both direct and inverse analogues may be covered, provided that $\xi_\lambda, \mathbb{M}_\lambda(\eta, \xi_\lambda)$, are appropriately specified.

The construction of $\mathbb{M}(\delta, \xi)$ – and, thence, of $\mathbb{M}_\lambda(\eta, \xi_\lambda)$ – is based on modifications of the charts $\mathbb{M}(\mathcal{P})$ for problems $\mathcal{P}$ in Solid Mechanics, as already mentioned. Accordingly, the next task is to specify these modifications as precisely as possible. The chart $\mathbb{M}(\mathcal{P})$ has the form: $\sigma s\zeta d; e : f\mathcal{R}g; h : iMj; k : l$.

Here, the symbols have the denotations already assigned – except that σs now replaces abc $\equiv$ EP GP TP, so that σs represents the underlying structure within which $\mathcal{P}$ is defined. Plainly, $\{\sigma, s\}$ is related to the environment, ξ, for $\mathcal{P}$; indeed, in many cases, $\{\sigma, s\}$ coincides with ξ. Both scientific properties, σ, and mathematical characterizations, s, are incorporated in $\mathbb{M}(\mathcal{P})$, and substantially determine the framework for $\mathcal{P}$. In certain configurations there are 'external influences' affecting σ or s, so that $\{\sigma, s\} \underset{\neq}{\subset} \xi$. In Fracture Mechanics, for instance, $\{\sigma, s\}$ corresponds to the 'cracked object', and ξ, to the 'surroundings' **with** the embedded object (each having its own scientific / mathematical facets).

This possible ambiguity may be removed by introducing the **compound structures** $\sigma =: \{\sigma^-, \sigma^+\}, s =: \{s^-.s^+\}$, to cover internal $(-)$ and external $(+)$ structures separately. In these terms, one has: $\xi \equiv \{\sigma, s\}$, where ξ is used in general representations, and $\sigma^{\mp}, s^{\mp}$, in detailed specifications / manipulations. Of course, the scientific / mathematical (in)equations inherent in $\mathcal{P}$ are primarily contained in the entries d e f – whose modifications, in $\mathbb{M}(\delta, \xi)$ and $\mathbb{M}_\lambda(\eta, \xi_\lambda)$, 'produce e, and $e_\lambda \in \pm\alpha(e)$, as (partial) output'. **Notice that** the 'components of ξ' (or, of ξ_λ) need not be mutually independent, since various kinds of interaction may occur – as examples will demonstrate. For instance, σ^-, σ^+ (or s^-, s^+) may influence each other – and it is even possible for $\sigma^{\mp}$ to influence (the forms of) $s^{\mp}$, through consistency (or other) conditions.

For an arbitrary ME, e, the 'generation of e a (partial) output of $\mathbb{M}(\xi, e)$' may be formulated schematically by assigning general interpretations to the symbols $\sigma s \zeta \ldots Mj; k : l$ – for instance, as follows.

Denote by $\tilde{\mathbb{T}}, \mathbb{T}$ (respectively) the families of all w.d. scientific, and mathematical, theories[517] – typically, $\tilde{\tau}$, and τ. Further, if ${}_0\tilde{\mathbb{T}} := \tilde{\mathbb{T}}$, and ${}_0\mathbb{T} := \mathbb{T}$, denote by ${}_k\tilde{\tau}, {}_k\tau$, arbitrary subtheories of ${}_{k-1}\tilde{\tau}, {}_{k-1}\tau$ (respectively), for $k \geqslant 1$, where 'subtheories' are defined through progressively refined classifications in specified Review journals. Then (essentially arbitrary) w.d. combinations of (respectively) scientific, mathematical, and scientific / mathematical subtheories may be represented concisely as: $\tilde{\Gamma}({}_i\tilde{\tau}_j)_P, \Gamma({}_k\tau_l)_Q$, and (say) $\tilde{\Gamma} \perp \Gamma(({}_i\tilde{\tau}_j)_P, ({}_j\tau_k)_Q)$. Here (as in some earlier Σ -notation) the subscripts $i_\alpha, j_\alpha, k_\beta, l_\beta$ are (designated) positive integers, for $\alpha \nearrow \bar{P}, \beta \nearrow \bar{Q}$. The 'admissible forms' of $\tilde{\Gamma}, \Gamma$, and $\tilde{\Gamma} \perp \Gamma$ must have some (Boolean /...) specification(s), for implementations of the operations involved in $\mathbb{M}(\delta, \xi), \mathbb{M}_\lambda(\eta, \xi_\lambda)$; but this need not be pursued yet, since only formal representations are relevant in the present discussion[518] – in which $\mathbb{M}, \mathbb{M}_\lambda$, may, fruitfully, be regarded as mathematical **machines** for **fabricating** ME and their analogues. Thus, even if the aim is to produce analogues, $\pm\alpha(e)$, of a specified ME, e, the machine $\mathbb{M}$ may, in fact, correspond to a **set**, $\{e_\alpha : \alpha \in A\}$, **containing** e – with cognate results for $\pm\alpha(e)$.

In these terms, one prescription for $\mathbb{M}, \mathbb{M}_\lambda$, takes the form:

σ: $\tilde{\Gamma}({}_i\tilde{\tau}_j)_P$

s: $\Gamma({}_k\tau_l)_Q$

ζ: possible frameworks for constituents of δ ('motivated by e') over $\xi' \subseteq \xi$

d: definitions / axioms / (in)equational conditions for δ over ξ

e: auxiliary elements for realizations of δ over ξ

f: maximally constructive form(s) of δ over ξ

$\mathcal{R}$: representations of δ over ξ **via** auxiliary elements

g: formal (in)equational conditions on δ

h: formal auxiliary elements for δ over ξ

i: formal realizations of δ over ξ

M: transformations of the formal realizations

j: effective (in)equational conditions of δ

k: effective auxiliary elements for δ

l: effective realizations of δ (and so, of e).

---

[517] over Σ_t

[518] From now on, $\mathbb{M}$ denotes $\mathbb{M}(\delta, \xi)$, and $\mathbb{M}_\lambda, \mathbb{M}_\lambda(\eta, \xi_\lambda)$ – unless two or more processes are considered; in which case the 'arguments' for each pair e, e_λ are specified.

NOTES

(i) The construction of $\mathbb{M}_\lambda$ parallels that of $\mathbb{M}$ almost completely. The crucial difference lies in **the choice of** $\xi_\lambda \equiv \{\sigma_\lambda, s_\lambda\}$, 'motivated by $\xi \equiv \{\sigma, s\}$', but otherwise virtually unrestricted. **The development of mutually consistent counterparts,** $d_\lambda, e_\lambda, \ldots, k_\lambda, l_\lambda$ – and, of **'filters'**, $\zeta_\lambda, \mathcal{R}_\lambda, M_\lambda$ – offers free play to the 'analogical imagination', to explore novel structural interrelations.

(ii) The potential **complexity** of $\mathbb{M}, \mathbb{M}_\lambda$, is also unlimited, since arbitrarily complicated operations may be introduced at any stage of 'the machine process'.

(iii) Although e is not uniquely determined by $\mathbb{M}, \xi$ is (at least) 'suggested by e'; so the notation $\xi(e)$ may be used, for extra clarity.

(iv) In this set-up, the underlying template – as a collection of (interacting) **mechanisms** – constitutes the 'formal machine', whose **input**, $\mathbb{M}^{\leftarrow}$, comprises ξ and the determining conditions for e – the (partial) **output**, $\mathbb{M}^{\rightarrow}$, of $\mathbb{M}$.

The machines $\mathbb{M}, \mathbb{M}_\lambda$ may be subjected to all w.d. OTC operations. In this context it is convenient to introduce a **uniform representation**:[519]

$$\mathbb{M} =: (m)'_{14}; \mathbb{M}_\lambda =: (m_\lambda)'_{14},$$

where $(x)'_q$ is the **transpose** of $(x)_q$, and the 'components', $m_i, m_{\lambda j}$ correspond to the entries in $\mathbb{M}, \mathbb{M}_\lambda$, respectively – i.e.,

$$m_1 := \sigma, m_2 := s, m_3 := \zeta, \ldots m_{14} := l;$$

$$m_{\lambda 1} := \sigma_\lambda, m_{\lambda 2} := s_\lambda, m_{\lambda 3} := \zeta_\lambda, \ldots m_{\lambda 14} := l_\lambda.$$

The **conversion operators,** c_k, acting separately on the $m_i, m_{\lambda j}$, may then be embodied in **diagonal matrices,** $\mathbb{D}(\hat{c})_{14}, \mathbb{D}_\lambda(\hat{c}_\lambda)_{14}$, with the $\hat{c}_i m_i, \hat{c}_{\lambda j} m_{\lambda j}$ taking all w.d., mutually consistent forms, such that each of the modified machines $\mathbb{DM}, \mathbb{D}_\lambda \mathbb{M}_\lambda$ does produce at least one ME. When only one of the c_i, or $c_{\lambda j}$, changes an element of $\mathbb{M}$, or of $\mathbb{M}_\lambda$, all of the other entries in $\mathbb{D}$, or in $\mathbb{D}_\lambda$ take the value $\hat{1}$ – the identity operator. Other assignments of nontrivial conversion operator correspond to complementary occurrences of $\hat{1}$ in $\mathbb{D}$, or in $\mathbb{D}_\lambda$.

From now on, the general form $\mathbb{M} \equiv (m)'_N$ is used, to allow for possible extensions / refinements of $\mathbb{M}, \mathbb{M}_\lambda, \mathbb{D}, \mathbb{D}_\lambda$, etc.. The formal description of $\mathbb{M}, \mathbb{M}_\lambda$ may be extended by indicating the admissible representations of $\tilde{\Gamma}({}_i\tilde{\tau}_j)_P, \Gamma({}_k\tau_l)_Q$, and $\tilde{\Gamma} \perp \Gamma(({}_i\tilde{\tau}_j)_P, ({}_k\tau_l)_Q)$. For $\tilde{\Gamma}$ and Γ, these may be of **cartesian-product / boolean / composition** types, since some of the ${}_{i_\alpha}\tilde{\tau}_{j_\alpha}$ (or, of the ${}_{k_\beta}\tau_{l_\beta}$) may be deployed within others – consider, for example, **Magnetohydrodynamics** (or, **Probabilistic Number Theory**) as obvious cases. More subtly, **quasi-chemical combinations of (mathematical) subtheories** (with their possible evolutions) offer evocative instances of **associations**[520] involving both the ${}_{i_\alpha}\tilde{\tau}_{j_\alpha}$ and the ${}_{k_\beta}\tau_{l_\beta}$, in nontrivial ways, with diverse sorts of interactions[521] – and this is just one of the multifarious possibilities that arise naturally in Σ -investigations. Many instances of these blends of subtheories, $\tilde{\tau}, \tau$ may be found in the S-M-List[522].

As indicated, the 'coupling of subtheories' ranges from **nil** (cartesian-products) via **weak / medium** (boolean) to **strong** (composition). All levels may occur within the same overall setting. Moreover, the allowable mixtures of scientific, and mathematical, subtheories are only mildly restricted – so the scope for innovation is broad. Even the nil-coupling conveys useful information, by indicating precisely the subtheories involved (at some level(s)). The complete

[519] If the $\sigma^-, \sigma^+, s^-, s^+$ are counted separately, then $\mathbb{M} = (m)'_{16}$, etc.. For expository purposes, $\mathbb{M} = (m)'_N$ allows for 'refinements' of $\mathbb{M}$.

[520] This is a usefully 'neutral' term.

[521] This term is used here in the most general sense(s).

[522] Some of these cases are considered later.

subtheory-specification requires all of these sorts of coupling, since, typically, the constituents, logical combinations of constituents, and compositional couplings of constituents, all reflect different aspects of the environment(s), $\xi = \{\sigma^{\mp}, s^{\mp}\}$, within which e and $\pm\alpha(e)$ may be realized.

The cartesian ('list') aspect of ξ allows the subtheories to be ordered – for parsing / construction routines over appropriate structures. It is assumed that the set, $\mathcal{S}$, of all designated subtheories $\tilde{\tau}, \tau$ is specified[523] in Σ_t (to be maintained / augmented as necessary). The formal framework within which arbitrary ME, e, and analogues in $\pm\alpha(e)$, may be generated, is, therefore, **the family** (say, $\mathcal{F}$) **of all w.d. unions of** (finite) **compositions of elements chosen from** $\mathcal{S}$. Here, definitions of (formalized) Mathematical / scientific theories[524] must be introduced, so that **realizations** of mcF may be obtained.

If $\mathcal{L}$ is a **formalized language**, with **alphabet** A, **terms** T, and **formulae** F, then a **theory** $\tau = \{\mathcal{L}, C, \mathcal{A}\}$ comprises $\mathcal{L}$, a **consequence operation**, C, on $\mathcal{L}$, and a collection, $\mathcal{A}$, of **axioms**. This deceptively simple specification covers virtually all of the subtheories encountered in 'general mathematical activity'. The possible exceptions (necessarily somewhat 'pathological') are excluded from $\mathcal{S}$. The modifications / extensions required to handle such exceptions, as well as certain **higher-order theories**[525], may be studied in works on Mathematical Logic. Moreover, detailed discussion of the structures of $\mathcal{L}, C, \mathcal{A}$, and of their interrelations are given by Rasiowa / Sikorski, op. cit.. All that matters here is that $\mathcal{S}$ – and so, $\mathcal{F}$ – may be represented quasi-effectively over Σ_t. To achieve this, the notion of **composition of subtheories** must be clarified.

For Σ , it appears that compositions of the forms $\tau_1 \circ \tau_2, \tilde{\tau}_3 \circ \tau_4, \tilde{\tau}_5 \circ \tilde{\tau}_6$, for suitable $\tau_i \in \mathbb{T}, \tilde{\tau}_j \in \tilde{\mathbb{T}}$ may be defined only 'locally', since (e.g.) procedures in τ_2 could be introduced into τ_1 at various 'levels' – to model / realize / implement procedures in τ_1. Denote by $\bigwedge_{\tau_1(\tau_2)} \subset \mathcal{L}_1 \cup \mathcal{A}_1$ the set of τ_1-elements at which (facets of) τ_2 may be 'injected' – types of quasi-valency criteria (depending on the detailed structures of τ_1, τ_2) govern this process. Further, if $\mathcal{D} := \mathcal{L} \cup \mathcal{A}$ and $W' := \mathcal{D} \backslash W$, for $W \subset \Lambda$, then (formally, at least):

$$\tau_1 = \tau_1|_{W'} \cup \tau_1|_W(\tau_2) =: \tau_1|_{W'} \cup \tau_1 \circ_W \tau_2$$

where $\circ_W$ indicates 'composition over $W \equiv W_{12}$'. Since τ_1, τ_2 are essentially arbitrary, this formalism is general. Indeed, it may be extended to cover $\tilde{\tau}_3 \circ \tau_4$ and $\tilde{\tau}_5 \circ \tilde{\tau}_6$ – and even $\tau_4 \circ \tilde{\tau}_3$, when this is w.d.!

Notice that $\tau_1 \circ_W \tau_2 \equiv \tau_1|_W(\tau_2)$ is not uniquely determined, since no **prescription** is given for the introduction of τ_2-facets at the elements w of W, over τ_1. However, an adequate scheme may be based on 'pointwise assignment', through the functions ψ (say), where $\psi : W \to \Omega, \psi(w) := (w, \tau_2^w) =: w$, for $W \subset \Lambda_{\tau_1(\tau_2)} \subset \mathcal{D}_1$. Of course, there may be families, $\Psi \equiv \{\psi : W \to \Omega\}$, for each choice of τ_1, τ_2, but individual functions, $\psi \in \Psi_{\tau_1\tau_2}$ are still w.d.. Thus, no constructive procedure (over $\mathbb{T} \times \mathbb{T}$, etc) may be found for elements of $\Psi_{\tau_1\tau_2}$ – nevertheless, every assignment function has a systematic representation, in a quasi-universal format. Parallel definitions of (say) $\tilde{\tau}_3 \ ' \circ \ \tau_4, \tilde{\tau}_5 \ '' \circ \ \tilde{\tau}_6$, with assignment functions $'\psi,'' \psi$, present no problems of principle. Although the operations[526] $\circ, '\circ, ''\circ$, are obviously noncommutative; they are all governed by a set of **general** (consistency) criteria. For the reverse composition (say) $\tau_4 \ ''' \circ \ \tilde{\tau}_3$, the conditions are less precise, since possible inverse analogues may arise – or particular importance in Σ . Indeed, the implementation(s) of $'''\circ$ constitute one approach to the creation of nontrivial (inverse) analogues. Some items in the S/M-List illustrate this strategy[527]

---

523 Thus $\mathcal{S}$ is a refinement of $\tilde{\mathbb{T}} \cup \mathbb{T}$

524 See Rasiowa / Sikorski, op. cit. for detailed treatment of Mathematical theories; and Härtkamper / Schmidt (eds.), '*Structure and Approximation in Physical Theories*' (Plenum, 1981), for Physical (and other scientific) theories.

525 See, e.g., Enderton, HB, '*A Mathematical Introduction to Logic*' (AP, 1972) for basic ideas; also, Boolos / Jeffrey, op. cit. and Mendelson, op. cit. (2nd ed., footnote, p59, for references etc.)

526 It is understood (from now on) that $\circ \equiv \circ_\Omega, '\circ \equiv {}'\circ_{'\Omega}, ''\circ \equiv {}''\circ_{''\Omega}$, etc., for sets $\Omega,' \Omega,'' \Omega$. For intricate compositional manipulations, a more transparent notation will be introduced.

527 When there is no danger of confusion, composition may be denoted simply by $\circ$ (it being understood that the processes involved are w.d.). If at least two types of composition occur in an expression, then they will be distinguished – e.g., as $\overset{m}{\circ}$, for $m = 1, 2, \ldots$.

The representation of $\mathbb{M}, \mathbb{M}_\lambda$ allow various forms of **machine-algebra**, and of **machine-calculus** to be investigated, where (generalized) algebraic, or analytical, operations are applied to $\mathbb{M}, \mathbb{M}_\lambda$, through suitable combinations of OT in the elements of the diagonal matrices $\mathbb{D}, \mathbb{D}_\lambda$. **NOTE that** the apparent increase in generality if $\mathbb{D}, \mathbb{D}_\lambda$, were replaced by **non**diagonal matrices, may be shown to be illusory. In fact, the 'constituents' of $\mathbb{M}, \mathbb{M}_\lambda$ may be transformed in all admissible ways by cognate specification of the operators $\hat{c}_i, \hat{c}_{\lambda j}$.[528]

It is convenient to denote by $\mathbb{D}^{\pm\alpha}, \mathbb{D}_\lambda^{\pm\alpha}$, the transformations producing the analogues $\pm\alpha(e)$, $\pm\alpha(e_\lambda)$, from the machines $\mathbb{M}^{\pm\alpha} \equiv \mathbb{D}^{\pm\alpha}\mathbb{M}, \mathbb{M}_\lambda^{\pm\alpha} \equiv \mathbb{D}^{\pm\alpha}\mathbb{M}_\lambda$, respectively – where $\mathbb{M} \equiv \mathbb{M}(\xi, \xi)$, and $\mathbb{M}_\lambda \equiv \mathbb{M}(\eta, \xi_\lambda) =: V_\lambda\mathbb{M}$. Here, as always in these Essays, $\pm\alpha(e) \in \pm\text{An}(e)$, the class of all direct / inverse analogues of e (over Σ_t), and $e_\lambda \in \mathbb{V}(e)$, the class of all Σ_t-variants of e[529]. Of course, since the machines $\mathbb{M}, \mathbb{M}_\lambda$, are themselves ME, the usual 'Σ -circularity' is present in these calculational schemes – but this does not significantly compromise the viability of the OIB, provided that processes of local-consistency continuation (l-c-c) are followed at all stages of investigations[530], to ensure that an adequate type of **global** consistency obtains[531]. This formal framework for analogue-development underlies all of the items in the S/M-List – indeed, it furnishes an **anatomy of analogy** within Mathematics and its applications. This, in turn, explicates the **modus operandi** of every analogue considered.

The LC-representations of ME are hinted at in E_1, and developed in E_5/E_7. In principle, 'LC information for e' should be equivalent to 'machine-specification of e' – since each of these sets of conditions determines e; but this is not quite so, if issues of (non)uniqueness, or precise form, are taken into account. Nevertheless, these 'versions of e' **are** closely interrelated, and may be used alternatively, according to contexts[532]. In E_5, the emphasis is on 'locational manifolds' as frameworks for general discussions of approach; whereas, in E_7, **distinctions** – among variants, generalizations, abstractions, analogues, and inverse-analogues, of e – are the main concerns. Perhaps the clearest characterization of $\mathbb{M}, \mathbb{M}_\lambda$ (with $\mathbb{D}, \mathbb{D}_\lambda$) involves the **effective determination** of (approaches to) e. Condensed versions of LC, or $\mathbb{M}$, types may be introduced, to discover / explore 'new' (inverse) analogues. Frequently, **blends** of these types are required, to produce adequate descriptions / analyses of analogues 'hidden within sophisticated mathematical / scientific structures'.

On this basis, possible (inverse) analogues may be identified – and then investigated, if their 'credentials' are sound. Such processes are very subtle, so it is probably inevitable that the underlying apparatus may (sometimes) be of byzantine complexity! As usual, however, subsequent formulations of inelegant, new results may be simplified / extended, before being reintroduced into the OIB. This, of course, is in the nature of mathematical research, but the enhanced SAM procedures extend greatly the scope of these procedures.

In E_{10} Part I, a broad selection of quasi-operational schemes in Pure / Applied / Applicable Mathematics is outlined – to illustrate the (essentially, unlimited) purview of enhanced SAM packages. Diverse examples of (generalized) 'standard approximation' – deterministic or stochastic – are sketched in E_{10} Part II (mostly, chosen from the S/M-List). Here, the 'approximation mechanisms' are exhibited, in relation to the Basic Spaces of E_1 and the techniques of E_6. Some analogues are noted, 'in passing' – but **the intensive study of mathematical analogy** is reserved for E_{11} Part II.

---

[528] In this sense, the constituents of $\mathbb{M}$ (or, of $\mathbb{M}_\lambda$) are effectively mutually independent (primarily, through their lying in 'disjoint categories' of ME).

[529] This compact notation is well adapted to analogical exploration.

[530] This amounts to the local matching of all procedures involving overlapping / 'contiguous' sets of ME.

[531] In Mathematical Logic, the properties of consistency / confirmation / satisfiability / deducibility /... are analysed in detail; but such distinctions (see, e.g., Mendelson, op. cit.), are rarely relevant for Σ.

[532] More precisely, $LC(e) =: L(e)$ determines e **in principle**, but contains no information on **effective calculation** of (or, **approach** to) e. The same comment applies to $LC(s) \equiv L(s)$, for some SE s.

Calculational Flow-Charts

In spite of the diversity of types of calculations in FM (both LEFM and nonlinear FM), a **single, 'master flow-chart' may be given to exhibit all of the fundamental features of even the most general calculations**.

The **flow-charts for particular problems** may be constructed by expanding selected items in the master chart. This approach to the analysis of calculations will help to organise the material in the Information Base ≡ IB (of which the Annotated Bibliography / Commentary (ABC) is the principle element). The 'filters', schematically denoted as C (specification of coordinate system(s)), R (analytical representation(s)) and M (mathematical modification(s)) cover all of the operations encountered in general FM problems – whatever their apparent intricacy. In any particular calculational procedures, only a small proportion of the operations in the filters will be 'activated'. Thus, in a sense, **each specific problem corresponds to some 'section' through the master flow-chart**.

Although this approach to the description of calculations cannot be made absolutely precise or unambiguous, it is, nevertheless, most helpful in the context of a large-scale information system. Ultimately it may be possible to classify 'arbitrary calculations' by matching them with a relatively small number of '**generic flow charts**'. Consequently, there are, essentially, **three classes of flow-charts**: 'general', 'generic' and 'specific'. The generic charts correspond (roughly) to 'tool-kits', Kp: while, the specific flow-charts represent comparatively detailed calculational prescriptions, $K{*_P}$ (where $*_P(M)$ is as defined in $\mathcal{R}_2$)[533].

In any detailed calculation, the 'layout and organization' may well be ostensibly different from those of the 'master flow-chart'; however, my claim is that: **it is always possible to reformulate a calculation as a collection of subcalculations** (some of which may be interdependent) of **the basic 'C-R-M type'**. Often, the comprehensibility of arguments may be improved through such reformulations, since they correspond 'intuitively' to a natural evolution from problem specification to maximally explicit solution[534].

A substantial task in the present project is that of constructing 'typical flow-charts' (generic and specific) covering classes of calculations of (potential) importance in engineering practice. In the remainder of this report, some preliminary indications are offered for the **general taxonomy of calculations** (along these lines).

---

[533] $\mathcal{R}_2$ ≡ Farnborough research report (2/89) by JSN Elvey. Here, $*_P$ ≡ min. no. of 'basic operations' to solve P, and $K{*_P} \in K_P \times \cdots \times K_P(*P$ 'factors')

[534] Indeed, this explication of solution procedures is directly applicable to very broad classes of scientific and engineering problems – and to problems in 'pure mathematics' – provided that the entries in the chart are given cognate interpretations. (I hope to make such applications to other IBs; e.g., for **nonlinear oscillations**)

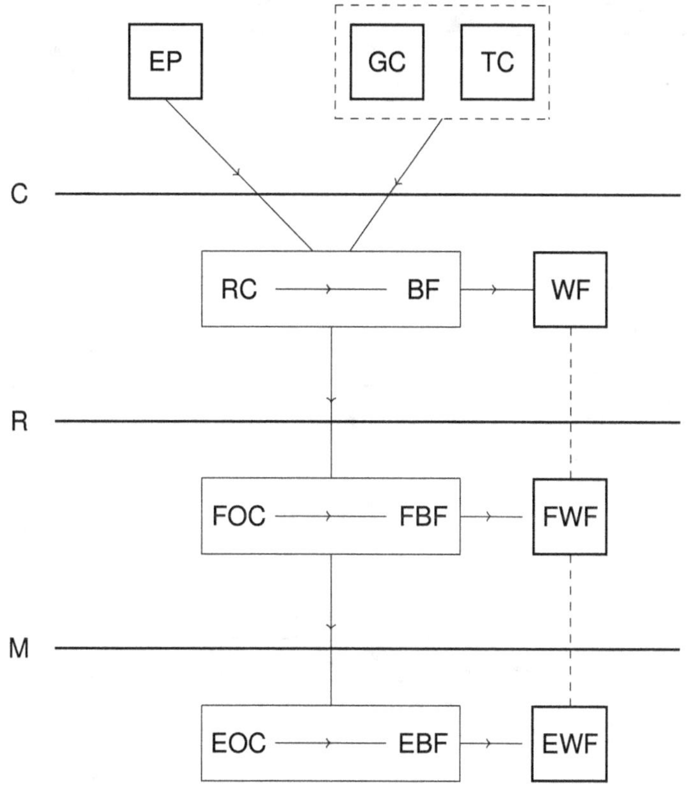

Figure 3. Schematic 'master flow-chart' for the solution of general FM problems. (Also more widely valid).

C: coordinate specification

R: representation (of BFs and auxiliary functions)

M: mathematical modification (to 'effective forms')

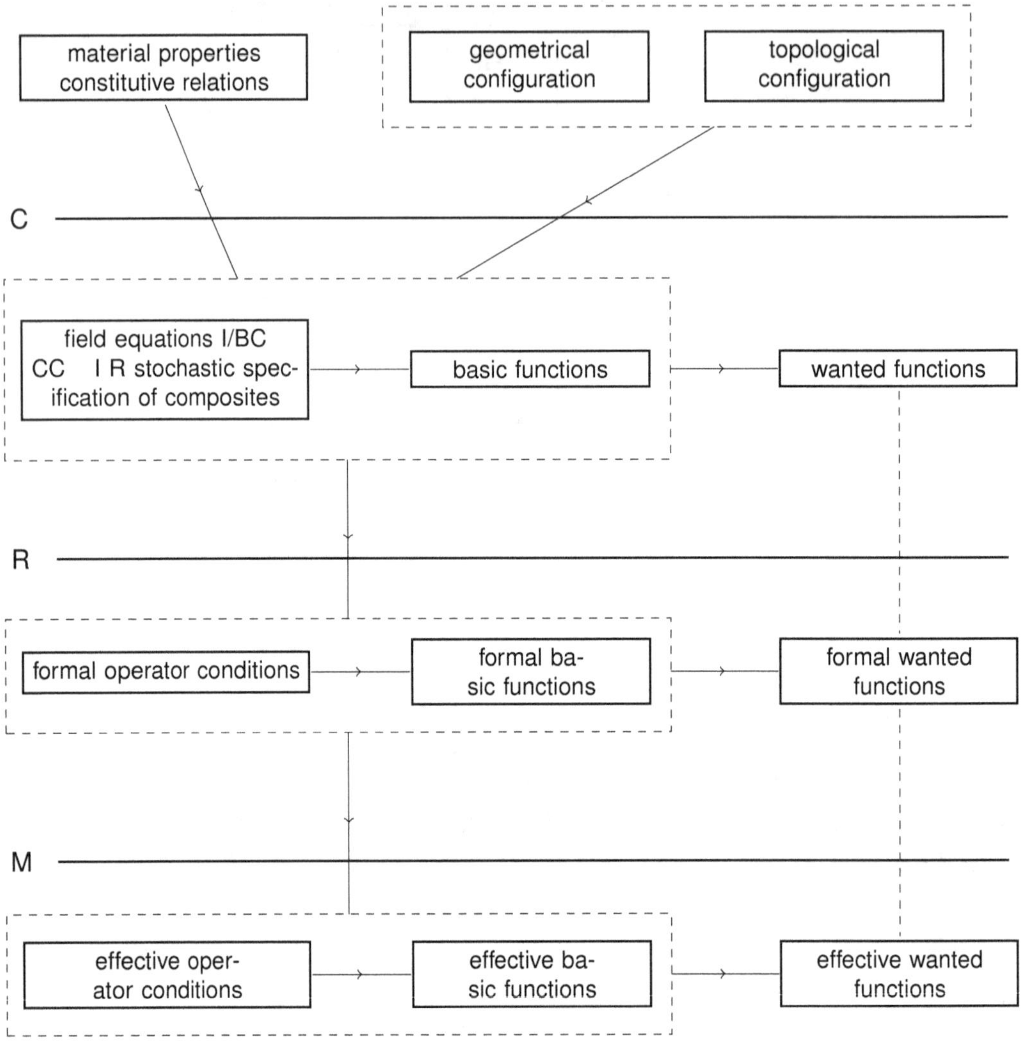

Figure 4. Explicit Version of the Schematic 'Master Flow-Chart' For Solution Procedures.

4. Indication of Preliminary Taxonomic Approach

4.1 The Basic Idea

It is convenient to identify (and list) several very broad domains of mathematics, so that techniques based mainly on results from these domains may be listed, and then 'combined' to produce various 'hybrid procedures'.

For example, introduce the 'calculational domains' denoted generally as:

- Real Analysis
- Complex Analysis
- Operational Calculus
- Asymptotic Analysis
- Tensor Analysis
- Numerical Analysis

- Functional Analysis
- Linear Algebra
- General Algebra
- Approximation Theory

A framework for the description of diverse solution procedures will be developed, gradually, by expanding and analysing the items for the following bare lists of topics.

4.2 Fundamental Mathematical Processes for FM Calculations

4.2.1 Series expansions (real)

Taylor; Fourier; orthogonal functions, general eigenfunctions,...

4.2.2 Series expansions (complex)

Taylor; Laurent; Complex-Fourier; Lagrange; Teixeira ($\equiv$ generalized Laurent),...

4.2.3 Integral representations

Cauchy (planar); potential-theoretic (planar / spatial); (in)finite integral transform (planar / spatial),...

4.2.4 Boundary problems for analytic function

Riemann-Hilbert; Keldysh-Sedov; Wiener-Hopf; generalized Wiener-Hopf,...

4.2.5 Inversion theorems for integral transforms

Fourier; Laplace; Mellin; Hankel; Kontorovich-Lebedev; Mehler-Fock; Radon; Weber,...

4.2.6 Existence and uniqueness theorems

Infinite systems of linear equations; Fredholm / Volterra integral equations; systems of (simultaneous) n-integral equations ($n = 1, 2, 3, \ldots$); potential-theoretic systems of integral equations; (systems of) elliptic, or parabolic, or hyperbolic, PDEs,.... Here, '$n = 1$' corresponds to 'standard', '$n = 2$', to 'dual', '$n = 3$', to 'triple', integral equations; and so on.

4.2.7 Regularization theorems

Infinite systems of linear algebraic equations; singular integral equations (planar / spatial),...

4.2.8 Assignment of coordinate system(s)

Rectangular cartesian (the BASIC SET); plane, spherical, cylindrical polars; bi-polars; spheroidal, ellipsoidal, paraboidal, coordinates;...

General separable systems of coordinates; general curvilinear coordinates with level curves matching prescribed boundaries; coordinates obtained VIA conformal mapping (planar), and VIA analogues of conformal mapping (spatial),...

4.2.9 Application of all inverse mappings from the chosen coordinates to the BASIC SET

Specification of all relevant transformations.

4.2.10 Completeness theorems for basic functions

Complex 'potentials' (Kolosov / Muskhelishvili, Westergaard,...)

Stress functions (Airy, Stokes,...)

Strain energy (density) functions

Displacement potentials

Papkovich-Neuber functions

(Williams) eigen-functions,...

4.2.11 Representation theorems for BFs

real / complex, closed-form functions (planar / spatial)

real / complex series (of elementary, special, orthogonal,... functions)

real / complex (non)singular integrals involving

(Elementary / Special / orthogonal,... functions)

integral transforms (Fourier, Laplace, Mellin, Hankel, Kontorovich-Lebedev, Mehler-Fock, Radon, Weber,...)

'mixed constructions',....

4.2.12 Admissibility theorems for sets of RC

For problems $P \in$ where all of E, G, T, W are prescribed, all classes of required conditions, $R \in$ are specified, such that the resulting problems P have unique solutions (possibly with extra smoothness requirements...). The RC must be formulated i terms of each complete set of BFs. Such sets of RC are called admissible (for P).

4.2.13 Derivation of formal operator conditions

If the required conditions are admissible for some problem then they may be converted into some system of formal operator conditions for the chosen basic functions (or, possibly, for certain transformed basic functions). These conditions, however, are not, in general amenable to direct solution....

4.2.14 Expression of all RC in the chosen coordinates

This involves application of the invertible, coordinate mappings to the RC expressed in terms of complete sets of BFs in admissible representations. This yields a set of formal relations interconnecting functionals of various (transformed) basic functions.

4.2.15 Derivation of effective operator conditions

Among the more familiar species of effective operator conditions are those corresponding to:

(i) (in)finite systems of real / complex linear algebraic equations;

(ii) systems of (real / complex) recurrence relations among terms of general series;

(iii) other types of relations among series 'coefficients';

(iv) various classes of boundary problems for analytic functions;

(v) sets of real / complex, (non)singular, (non)linear, simultaneous, n-integral equations (for planar / spatial domains);
NOTATION: $1 - \int$ equations (standard)
$2 - \int$ equations (dual)
$3 - \int$ equations (triple), etc.

(vi) generalized Wiener-Hopf procedures;

(vii) (elliptic, hyperbolic or parabolic) systems of PDEs;

Each of these sets of EOCs has its associated general solution procedures – with specializations for particular problem classes.

4.2.16 Effective Determination of Transformed BFs

The EOCs correspond to various types of operator equations (or, operator inequalities). The theorems, and associated procedures, developed to handle EOCs are of many types; but the following list contains the most fundamental approaches:

- general manipulation of closed form (real / complex) functions;
- (approximate) solution of (in)finite system of linear algebraic equations;
- (approximate) solution of (non)linear recurrence relations;
- application of (in)finite integral transforms;
- regularization of (systems of) singular integral equations;
- reduction to boundary problems for (sets of) analytic functions;
- introduction of auxiliary mappings and functions (e.g., to represent formal solutions);
- use of (known) singular solutions [e.g., 'Green's', 'weight', or 'influence', functions] to obtain general solutions;
- general schemes for solution of systems of singular, integral equations;
- general procedures for (systems of) Fredholm, Volterra or Abel integral equations
- standard quadrature schemes;
- specialized quadrature schemes;
- solution of variational inequalities for crack problems equivalent to 'moving-boundary problems';
- identification of generic singularities in solutions of (classes of) PDEs (especially, for elliptic systems), where (analogues of) SIFs appear 'naturally';
- qualitative schemes for assessing degrees of regularity or singularity of solutions (depending on the configurations, and the RC); see, e.g., 'Qualitative Methods in Elasticity', by P Villaggio (Noordhoff, 1977);
- Various combinations (to be specified) of procedures of the above types.

4.2.17 Derivation of Analytical / Numerical EFs

The following classes of methods are worth distinguishing.

- Analytical schemes
- Analytical / numerical schemes
- Numerical schemes (e.g. FEM, BEM)
- Numerical / graphical schemes
- Analogues schemes (e.g. via caustics), or via photo-elastic methods;
- other 'hybrid schemes' (see the recent book by PS Theocaris (Springer));
- the derivation, and use, of path independent integrals.

In particular, the analysis of general classes of (quasi-)path-independent integrals is central to the study of many nonlinear fracture phenomena. The various types of PIIs must be classified – so that appropriate calculational techniques may be discussed, and formulated for the IB.

Problem Interface

Practical taxonomy: Paths through the ABC

1. This material is developed on the following assumptions. **The annotations are stored within a computer-based system,** in which **the TITLES may be extracted** (at time t) **as (i) an alphabetical list** (non-chronological), or, **(ii) sequences of A/B chronological lists,. . . . More refined classifications are suggested – as aids for locating the most relevant information in the ABC** for the investigation of particular, 'practical problems'. (See: the following subsections).

2. The annotations may be grouped in various other ways – e.g.:

(a) n-dimensional problems $(1 < n < 3)$

(b) finite / infinite bodies

(c) linear / nonlinear elasticity

(d) viscoelasticity

(e) elastoplasticity / plasticity

(f) thermoelasticity

(g) special (approx.) theories, such as: 'thin / thick plate / shell', 'membrane',. . .

(h) types of loading

- **(i)** local / remote: Mode-I, II, III;
- **(ii)** more generally: tension / in-plane shear / longitudinal, shear;
- **(iii)** thermal;
- **(iv)** other (E/M,. . .);
- **(v)** 'irregularities' [e.g., dislocations, other cracks, inclusions, pre-stress / strain,. . .];
- **(vi)** time-dependent loading [e.g., impinging waves,. . .].

NOTE that all of these classifications (and many more) are inherent in the flow-charts ('level 1); see 4^0. At this stage, various 'topical clusters of annotations' are offered to the user; and, on this basis, appropriate mathematical models may be selected (or, further developed).

3^0. The fundamental techniques for handling mathematical problems involving 'loaded flawed bodies' may be found in a collection of basic references [available[535] to 'users' (in some sense)] listed in the ABC Introduction (occasionally augmented, or modified as new works appear. . .). Call this list $R \equiv (r_k : 1 \leqslant k \leqslant N(t))$. Citations from R have the forms $[r_k : p_1 - p_2]$ or $[r_k :$ Sec $s]$, etc.. These citations must be 'built into the ABC', in the sense that the fundamental techniques, frameworks, and representations, for a range of 'basic generic problems' have been identified and adequately covered by selected items from R.

The choice of r_k must cover both the 'rough models' (see Section 4^0) and the most refined mathematical models: and the reference lists within the r_k may also be cited. The categories: 'practical', 'computational', 'analytical', and 'foundational', aptly characterise the (largely, complementary) approaches to any nontrivial applied – mathematical problem (and for 'pure mathematics, also – with the possible exception of 'practical'!).

[535] It is anticipated that any institution subscribing to the ABC would have access to (or, obtain) all of the r_k as reference sources.

4^0. Rough Models

Prototypical problem: **given a flawed body, B, under loading, $\mathcal{L}$ (internal; or, external – via interaction with other parts of some 'machine', and with the larger 'environment').**

Question: **Is B 'safe', now, for a least, T? If so, for how long will B remain safe for T? If not, how can B be made safe for T?** NOTE. It is assumed that:

(i) The 'criteria for safety' are **known** (e.g., in terms of SIFs, ERRs, fracture toughness, bounds on various stresses, etc.).

(ii) 'Practical definitions of '(in)finite', 'thick', 'thin', ..., are available; and, **all relevant basic material properties of B are known.**

(iii) The relative importance of **interaction effects** (within B, and between B and its environment) has been assessed (to 'orders of magnitude').

(iv) The overall shape of B is known (e.g., cylinder, ellipsoid, half-plane, polygonal prism, ..., 'general', convex, etc.). NOTE. A **catalogue of basic shapes** (cf. 'Murakami') is supposed 'given' [available to users for the specifications of rough models]. More intricate shapes may be described approximately as suitable combinations of basic shapes – and the flaws may be specified analogously. This process is comparable with those required in applications of the BEM, or FEM – approximations are involved, which may introduce (or, eliminate) singularities; great care must be taken to monitor such effects!

On this basis, the rough model(s) may be formulated – and, clusters of pertinent annotations may be identified for the practical problem, P_0.

5^0. Summary, so far.

A practical problem, P_0, is discerned in a 'structure', S [e.g., an aircraft, a bridge, a nuclear reactor, a crane,...]. A list of (typical) engineering properties (EP), safety criteria, and a catalogue, $\mathcal{C}$, of basic shapes (and flaws) (cf., Murakami) are available to the 'user' – who must formulate a rough model, M_0, of P_0. From M_0, clusters, $\mathcal{A}_0^-$, of (titles of) pertinent annotations may be obtained, and used to formulate a (more precise) mathematical model, M, of the idealised problem, P (i.e., a set of equations, inequalities, I/BC,...).

NOTE that, M_0 must contain enough information for suitable (classes of) theories to be selected (e.g., (in)finite, think / thick, plate / shell; (non)linear elasticity; small / large deflections; specified 'wanted functions' (WF), etc.; 'membrane theory', elastics theory, elastoplasticity, viscoelasticity,...).

Schematically, this may be depicted as follows:

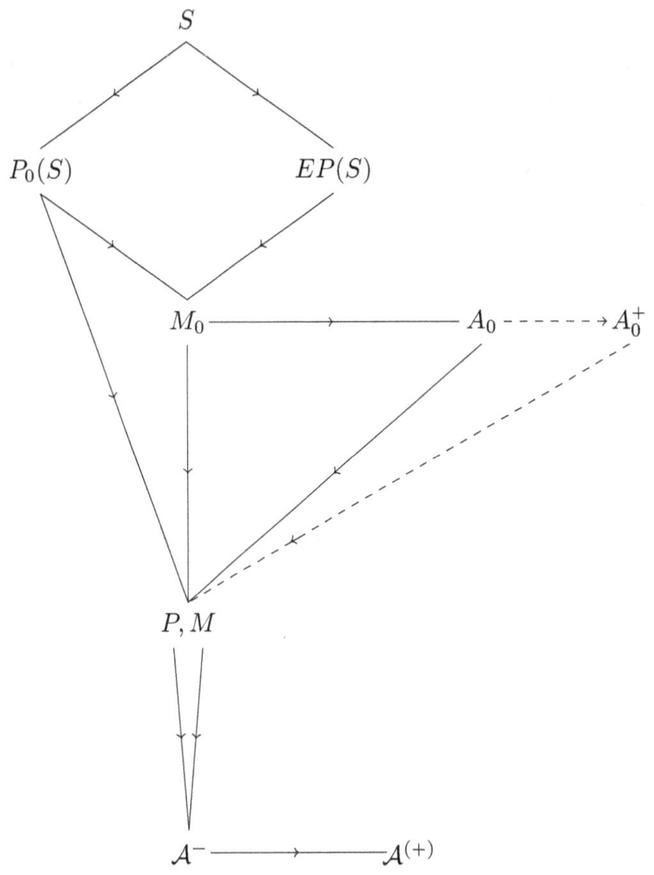

$\mathcal{A}^- \equiv$ references (titles only), from ABC
$\mathcal{A} \equiv \mathcal{A}^-$ plus annotations
$\mathcal{A}^+ \equiv \mathcal{A}$ plus all flow-charts, and extra references to MR, AMR, etc..

This brings the user to the interface with the ABC in such a way that a ('small' – possible, empty!) list, $\mathcal{A}^-$, or highly pertinent (titles of) annotations may be extracted (VIA a 'pattern-matching procedure over P, and the model specification for M').

NOTE that (i) A_0 and A may have little in common (indeed $\mathcal{A}_0 \cap \mathcal{A}$ may be empty!). (ii) $\mathcal{A}_0^-$ could (usefully) be derived in part from AMR (Applied Mech. Revs.).

6^0. Matching procedures.

The ABC offers the following 'facilities' (for investigating P and M).

(i) Fundamental mathematical information (VIA the list, R, of – accessible – sources).

(ii) Property-orientated title lists, $\mathcal{A}^-$.

(iii) Corresponding annotations, $\mathcal{A}$.

(iv)[536] Associated flow-charts / strategies, $\mathcal{A}^+ \backslash \mathcal{A}$.

The information in (i) may be consulted, either, for general orientation; or, to gain familiarity with the standard properties of inequalities / equations occurring centrally in M – and, with the species of required conditions (RC), WF, etc., involved (along with related material on possible representations: BF, etc.).

Ultimately: each central mathematical procedure will be encapsulated within the ABC, so that its principal characteristics, and mode(s) of use may be readily ascertained. The initial matching (from (ii)) involves RC (and WF), mainly – together with configurations.

The aim is either (a) to achieve a sufficiently close match for a solution to P (and hence, to P_0) to be obtainable (formally, or via 'approximation') – essentially by transcription; or else, (b) to match a subproblem of P, so that 'the rest of P' may be tackled as a (reduced) 'new problem' – either 'from scratch', or by further attempts at (partial) matching; or else (c) to identify P (relative to the ABC(t)) as an 'essentially unsolved problem', for which new techniques (or, new combinations of existing techniques) appear to be required (since no 'sufficiently close variant of P' is covered by ABC(t)). Eventually, after this problem, P, has been solved, the solution of P – and all of the related mathematical procedures, etc. – will be incorporated (say, in ABC($t+\tau$)).

7. Preliminary data processing.

(i) In its present form, the ABC comprises a collection of annotations, with some unifying information, and criteria for interpretation. The annotations are identified by numbers, titles, and author(s); but, so far, they are not retrievable (except by numbers), since they are not stored within a computer system. Once this is done, 'search routines' may be used to generate property-orientated sub-lists, $L_k \equiv L(P_k)$; $\bigcup_k L_k = L$ – where P_k are the possible properties characterising P (e.g., thick plate; anistropic material; inclusions, edge-crack(s),...). If P is characterised by $(P_j : j \in J_P)$, then the relevant list, L_P, for P:

$$\mathcal{A}^- \equiv L_P - \bigcap_{j \in J_P} L_j$$

NOTE As for $\mathcal{A}_0^-$, parts of $\mathcal{A}^-$ could be derived from MR (Math. Revs.); some of these papers may not be in the ABC ... but $\in \mathcal{A}^+$.

(ii) In order for (i) to be implemented, the 'master list', L, of (numbered titles of) annotations must itself be 'marked up' to correspond to an overall list of possible problem characteristics e.g., for (current) item n:

n AUTHOR(S), TITLE; REFERENCE $(P_j : j \in J_n)$.

(J_n along may suffice, since if $J_n \subset L$ then all relevant P_j are reclaimable from J_n – provided that all properties in L are numbered and listed.)

THUS:

n; AUTHOR(S); TITLE; REFERENCE; J_n

is enough – with J_n specified:

$$J_n := \{j_{n\sigma_1}, \ldots, j_{n\sigma_{l_n}}\} \qquad \{\sigma_1, \ldots, \sigma_{l_n}\} \subset |L|$$

$$|L| := \{1, \ldots, \#L\}$$

8. An example.

In the present state of the ABC (and for the remainder of the current project), no realistic attempt may be made to use the system fully – since the preliminary data-processing will still have to be done. This will be the first priority in the next phase of development of the ABC. However, the initial steps (formulation of P_0, EP, and $M_0\ldots$) may be attempted for a problem of known engineering interest as follows.

$S \equiv$ aircraft wing with rows of rivets, and regularly spaced (mutually parallel) stiffeners – under remote, longitudinal tension. Observed problem: cracks tend to emanate from the rivet-hole boundaries; the rivets act essentially as inclusions (with negligible load transfer).

P_0**:** **(i)** monitor the crack-growth;

(ii) determine the crack-tip SIFs;

(iii) estimate the interaction of the cracks with the stiffeners.

$EP(S)$**:** homogenous, isotropic, linearly elastic materials (for main body and rivets)

M_0: thick(ish), wide strip, of infinite length, with regular longitudinal arrays of circular-cylindrical inclusions interspersed with parallel stiffeners (each of about 10 (?) cylinder diameters in width), under constant longitudinal tension.

obtained (e.g.) VIA key-word search in the list of titles of the ABC.

$$A_0^- \equiv L_{P_0} - \bigcap_{j \in J_{P_0}} L_j$$

Here, this could involve: cracked thick plates; inclusions in cracked plates; stiffened (thick) plates with cracks; interaction of cracks (with (i) boundaries, (ii) flaws, (iii) interfaces, (iv) inhomogeneities, (v) stiffeners,...). If the inclusions do not interact with each other, or with the (non-neighbouring) stiffeners; and, if the stiffeners are (effectively) mutually independent, then the basic problem becomes one of a single inclusion in a semi-infinite, moderately thick (wide) strip (under longitudinal tension), with a longitudinal stiffener. A crack emanates from the inclusion boundary, and one wants to estimate the associated SIF, the effects of the nearest stiffeners, the crack path, etc.. As usual, this entails finding an asymptotic representation for the appropriate stress component, in a neighbourhood of the crack tip.

The aims include: prediction of rate, and direction, of crack growth; effect of plate thickness, and of (location of) stiffener. Until the Data-Processing facilities are established (in the next stage of the ABC development) this selection scheme cannot be tried in practice; however, purely for illustration, a small selection, $\mathcal{A}_0^-$, is indicated, for this problem, P_0. From A_0^- (and, in general, the actual papers in $\mathcal{A}_0^-$), information may be accumulated for the precise formulation of a mathematical model, M, and associated problem, P (in terms of (e.g.) PDEs, I/BC, etc.). At this point, a second selection, A^- (involving key-words referring to various mathematical techniques pertinent to the solution of the PDEs, etc. (e.g., representations in terms of LT, FT,...,Cauchy integrals, single / double-layer potentials, ..., complex 'potentials', P-N potentials, ...)) may be obtained. On this basis, a solution strategy, say, S_P, may be formulated – yielding, in favourable cases, an associated solution-process, Π_P; which may be effectively implemented as a procedure, say, $I(\Pi_P)$ (allowing maximum analytical / numerical results to be derived).

NOTE: At any stage of the above process it may be impossible (VIA the ABC alone) to proceed further. For instance, although P_0 is always capable of some formulation, it may be very hard to produce a suitable M_0 – especially for 'unusual problems'. Again, even if M_0 is available, it may still be difficult to extract $\mathcal{A}_0$; and thence, to formulate P, and M; and, to obtain $\mathcal{A}$. This is less a reflection on the likely adequacy of the ABC than on the well-known difficulty of combining realistic modelling with mathematical tractability – even at the highest level of technical sophistication. Thus: from the foregoing discussion, it appears that 'effective entry to the ABC' can be gained only VIA 'the $P_0/M_0 \to \mathcal{A}_0 \to M/P \to \mathcal{A}^- \to \mathcal{A} \to \mathcal{A}^+$ scheme'. If a 'bridge' is needed between the 'concrete', practical engineering problems and the 'abstract' ABC formulations, then it must simulate activities of this type: 'practical problem description'; 'rough modelling'. Within the ABC, some help (for formulation of the M_0) could be offered through a 'glossary' of practical terms and their 'model equivalents' (including quantitative guidelines for the designations (non)linearly elastic, visco-elastic, elasto-plastic... 'thick', 'thin', '(an)isotropic', 'orthotropic', '(semi)(in)finite', solid / plate / shell / cylinder..., '(negligible) interactions' (among various 'irregularities'), etc., etc.).

This would help 'users' to formulate rough models, M_0, in ways consistent with the ABC conventions.

9. The M, P/ABC Interface.

From P and M, the list $\mathcal{A}^-$, and full set of annotations $\mathcal{A}$; or, augmented (by flow-charts, MR, etc.) set, $\mathcal{A}^+$, may be obtained – with maximal ABC-characterisation of P corresponding to $\mathcal{A}$, and 'extra information', to $\mathcal{A}^+ \backslash \mathcal{A}$ – including all papers treating (highly) pertinent mathematical techniques, for P, in isolation; and skeleton solution-strategies which may be 'tried out' on the problem at hand, P, and (again, in favourable cases) extended or adapted to complete the solution, exactly, or approximately.

If none of the solution strategies generated by $\mathcal{A}^+$ proves adequate, it will be clear that no direct extension of the schemes within ABC(t) will suffice; and that, consequently, some fundamentally different approach is required – possibly VIA modifications in M, P – to obtain a more tractable mathematical problem; otherwise, through the (development and) introduction of fundamentally different (existing, or new) mathematical techniques.

At this stage, the 'user' would reconsider all aspects of the problem (i.e., forms of P_0, M_0 – yielding a modified list $\mathcal{A}_0^-$, etc.).

10. It follows from the previous discussion that: all items in the ABC must be 'tagged', as comprehensively as possible, to permit their optimal matching with the characteristics of P. This can be done only by studying the structure of each paper (e.g., VIA its annotation), and thus identifying the properties in question (e.g., by their numerical labels in a (regularly maintained) 'invariant list', $\mathcal{L}$, covering the entire ABC). This would produce entries of the form (cf. (7), above):

m	AUTHOR(S)	TITLE	$J_m \subset \mathcal{L}$

Here, $\mathcal{L}$ contains (i) general material properties; (ii) type (and shape) properties of flawed bodies; (iii) types (and shapes) of irregularities; (iv) types of loading;... (see Section 2^0 for some examples).

NOTE. $\mathcal{L}$ should be developed from a detailed form of the solution flow-charts (on which the CDs are based). The possible coordinate systems, modes of representation, associated operator conditions, basic / mathematical procedures, etc., must be specified and labelled – in sufficient detail for all of the main entries in the ABC to be suitable tagged.

NOTE. This process of classification / labelling will require several 'iterations' before anything resembling either completeness or optimality is even approached!

11. From the list of titles recorded so far (currently, just over 2000) a preliminary selection was made 'by hand' (a laborious task!) for the rough, M_0, and problem, P_0. (i) Some 10 items were judged '(highly) pertinent', and (ii) between 20 and 30 items, '(fairly) relevant', with (iii) around 30 other items, 'marginal'. Examples of (iii) are ## 1145, 1724, 2117; of (ii), ## 454, 324, 55. The full list (i) is: ## 171, 214, 221, 245, 341, 360, 470, 1310, 1322, 1628, 2020, covering

- **(a)** analytical procedures for thick plates;
- **(b)** theoretical results for (elliptical) inclusions;
- **(c)** results on the transfer of load from (riveted) stringers to (thin) plates;
- **(d)** results on the effect of stringers on cracks;
- **(e)** results on the interaction of inclusions with crack(s) emanating from the boundary;
- **(f)** results on crack-paths in thick plates.

In view of the linearity of underlying equations, all of the above results may be combined to obtain information about M_0/P_0..

NOTE. References 1–1123 are as listed in the Report on Research Agreement 2004/12. The other items in list (i) are as follows.

 # 1310: MAITI, M., '*On the equilibrium of a Griffith crack opened by a rigid inclusion*'. SIAM J. Applied MAths. 38 (1980) 209–214.

 # 1322: BUDIANSKY, B., '*Transfer of load to a sheet from a rivet-attached stiffener*', J. Math. and Phys. 40 (1961) 142–162.

 # 2020: ISIDA, M., '*Method of Laurent series expansion for internal crack problems*', Ch. 2 in 'Mechanics of Fracture' (ed. SIH, G.C., Noordhoff, 1973).

The full list is as follows.

$55, 90^{-}, 101^{(-)}, 106^{(+)}, 114^{-}, 115, 119^{-}, 171^{-}, 173, 214^{(+)}, 221^{-},$
$245^{-}, 248, 264^{-}, 269^{-}, 281^{-}, 318^{-}, 324, 326, 319^{--}, 333^{-}, 341^{-}, 360,$
$393^{-}, 413^{-}, 454, 470, 534, 575, 611, 766, 812, 883, 954, 955, 1004, 1008,$
$1031^{-}, 1061^{-}, 1070^{-}, 1105^{-}, 1145^{-}, 1149, 1198, 1310, 1322^{(+)}, 1392^{--},$
$1485^{(-)}, 1566^{--}, 1614^{(-)}, 1628^{-}, 1654^{--}, 1724^{--}, 1731^{(-)}, 1784^{-}, 1790,$
$1969^{-}, 2117^{-}, 2020, 2131^{(-)}, 2149^{-}, 2153^{(-)}, 2196^{-}, 2202^{-}.$

Underlined items are of type (i), items with no marking, of type (ii), and items with (−), of type (iii), if not underlined.

NOTES

(a) The initial search depended upon the formulation of M_0/P_0 – and on the consequent choice of key-words. The classification into types ((i), (ii), (iii)) corresponds to my subjective reaction to the titles of papers (without seeing the annotations). In the fuller version, this initial classification may be based on a perusal of the annotations whenever there is any doubt.

(b) On the basis of this preliminary information, the mathematical model, M, and detailed problem, may be formulated.

(c) This permits a second interaction with the ABC, where the inequalities / equations of P are matched with solution procedures (as classified along the lines indicated in recent Reports).

(d) At this point, the 'user' may consider the whole problem in detail, and try to develop a complete solution in maximally effective form. If any radically new approach(es) or mathematical technique(s) be required, then these should be incorporated (in available) or, at least, noted, in supplements for the ABC.

The exact form of classification scheme for mathematical procedures will be decided during the next phase of development of the ABC, when the primary data may be processed within a computer system. However, the 'tagging' scheme will be used to identify the main mathematical characteristics (frameworks, representations, and modifications) associated with the 'filters' $\mathcal{C}, \mathcal{R}, \mathcal{M}$, in the 'master flow charts'.

APPENDIX 4
Analogue Synthesis and Analysis

Analogue-Mechanisms: Models / Examples

The S/M-List contains items from all of the main subdomains of Mathematics and its applications. In $\mathrm{E}_{10/II}$, outlines are given of diverse instances of generalized (deterministic / stochastic) '**standard approximation**'. The **modes** of approximation are exhibited, along with 'natural settings' – e.g., within variants of the spaces S^X introduced in E_1 and discussed in E_6.

NOTE. The **Bibliography on non-numerical distance-functions** (U. of Vienna, 1984, op. cit.) is also relevant here. The detailed **realizations** of approximation procedures form the bulk of this material. Such procedures are often obscured within complicated environments – so that even their **identification** is problematical. The Σ -database of **implemented approximation schemes** (founded on E_5/$\mathrm{E}_{10/II}, \ldots$) constitutes a fundamental research resource – especially, in **operational form**. Many of these implementations are surprisingly intricate, requiring subtle interpretations and ingenious constructions – to be specified (as far as possible) in terms of 'unit-Σ -operations'[537].

By contrast, the recognition, validation, and exploitation of **analogues** are almost **poetic** activities! – constrained by the demands of rigour – for which the elaborate frameworks of $\mathrm{E}_{11/I}$ are designed. It is convenient to introduce the sets: $G_M := \{\text{generic ME in } M\}$; $G_S := \{\text{generic SE in } \mathcal{S}\}$; $Y_{G_M} := \{\mathrm{LC}(e), \mathbb{M}(e) : e \in G_M\}$, and to denote by $?\alpha(d), \alpha(d)$, respectively, an arbitrary **conjectured**, or **validated** analogue of d – and, by $\eta^{\leftarrow}, \eta^{\rightarrow}$, the input, or output of the assignment η. Then the relation of $?\alpha$ to α takes the form (with $G := G_M \cup G_S$):

$$\models \alpha(d) \Leftrightarrow \exists \eta \in Y_G : ?\alpha(d) \in \eta^{\rightarrow}.$$

Here, $d \equiv e$, for **direct** analogues, and $d \equiv s$, for **inverse** analogues.

In these terms, there are two (mutually inverse) **Basic Problems for Analogues.**

Π_1: **Given** $\eta(d)$ and a set $\mathbb{B}$ of w.d. perturbations of $\eta(d)$; **to determine** the set $\eta_{\mathbb{B}} := Y_G \cap \{b\eta(d) : b \in \mathbb{B}\}$.

Π_2: **Given** $?\alpha(d)$; **to determine** some $\eta \in Y_G$ such that $?\alpha(d) \in \eta^{\rightarrow}$.

NOTES.

(i) Here, d and b cover **both** direct **and** inverse analogues.

(ii) The set Y_G covers all Σ_t**-solvable assignments over** G (or, over countable subsets of G, etc., if logical problems arise!).

(iii) $\eta^{\leftarrow}, \eta^{\rightarrow}$ correspond, mainly, to $\mathbb{M}^{\leftarrow}, \mathbb{M}^{\rightarrow}$, since the locational conditions, LC(d), alone, give (at most) formal determinations of d, with no information as to **how** d is to be approached / calculated.

For **direct** analogues, the calculational prescriptions in $p\eta(e)$ are essentially unambiguous (even if technically sophisticated / complex); but the **inverse** analogues raise more recondite problems of selection and interpretation. Of course, every SE, s, has some LV/$\mathbb{M}$-characterization(s), where concepts from the relevant sciences are blended with formal conditions / (in)equations – but the key (highly **non**algorithmic) task is **to 'translate' these concepts**

[537] Here, and in many other contexts, the 'unit-operator' terminology of Chemical Engineering is suggestive for indicating basic processes and their interconnections. See, e.g., Foust, AS, et al., '*Principles of Unit Operations*' (Wiley, 1960; 2nd ed., 1980)

into new / stimulating mathematical structures. It follows that the problems Π_1, Π_2 must be formulated extra carefully in these cases – say, as $^-\Pi_1$, $^-\Pi_2$, to distinguish them from the 'direct forms', with $^-\eta$, $^-\mathbb{M}$, etc., as 'constituents'. This should clarify the processes involved in producing $\pm\alpha(d)$ from d – as well as highlighting the quintessential distinctions between the two types of analogues.

The radical distinction between direct, and inverse, analogue-formation is reflected in the typical **mechanisms** for these processes. For $\eta_\alpha(\alpha(e)) =: p_\alpha\eta(e)$, the mechanism – say, $\mu(\eta_\alpha)$ – is quasi-algorithmic, since p_α acts on the constituents of $\eta(e)$ in w.d. (deterministic / stochastic) ways – specifiable in terms of suitable combinations of OT. The assignment $\eta_{-\alpha}(-\alpha(s)) =: q_{-\alpha}\eta(s)$, however, contains some elements that may be merely '**suggested**' by constituents of the SE s. Hence, the crucial task is **to characterize such modes of suggestion**, in nontrivial ways, by investigating possible mechanisms, $^-\mu(\eta_{-\alpha})$. Suppose now that s is defined over a scientific theory, τ_s, through axioms / models – based on observation / experiment and comparisons with other (empirical) models – incorporated, as constituents / environment (ξ_s), in $\eta(s)$. Obvious examples are furnished by the templates for calculations in Solid Mechanics, which 'produce' fundamental parameters / functions /... through mathematical / calculational flowcharts, etc., as discussed in $\mathsf{E}_{11/I}$.

The interrelationships among models / techniques / parameters /... are represented mathematically in the flowcharts / annotations of the Fracture-Mechanics 'ABC'[538]. More generally, for arbitrary SE, there are m**-ary relations**, say, $r_i(c_{\sigma^i})_m$ among the constituents, c_{σ_k}, of $\eta(s)$. The possible inverse analogues are conjectured (conjured?!) by imitating aspects of the $r_i(c_{\sigma^i})_m$, for $m \geqslant 1$, over purely **mathematical** environments[539]. It is in these processes that novel operations / structures may emerge – though the initial efforts will be, mostly, highly experimental.

The first task is to identify the constituents of s in 'suggestive ways'. There is, of course, no unique decomposition of s (unless s is 'extremely simple'), so it is necessary to formulate criteria for 'favourable decomposition' – with a view to inverse analogy. This may be achieved by **extrapolating from diverse examples**, taken from ostensibly disparate scientific domains[540].

These extrapolations furnish the basis of general schemes of analogue-formation – direct or inverse. Some of the fundamental ideas are outlined in $\mathsf{E}_{11/I}$; other issues will emerge as examples are treated, and the overall 'theory of analogues' evolves – to a stage where it becomes part of the Σ -superstructure[541] for exploring the frontiers of Mathematics, by 'surveying neighbourhoods in conceptual space'[542]. **Notice that**, in Σ , the emphasis is always on **MA**; but the principles / techniques apply, largely unchanged, to arbitrary **SA** – with mathematical frameworks, in these cases, forming the **background** within which 'scientific transactions' are effected. Sometimes, the SA-processes derive from unobvious interrelations among **mathematical** formalisms – but such instances do lie properly within the purview of Σ [543].

In the examples, only key elements are identified. The scientific / mathematical background is assumed – but references are given to accounts especially pertinent to the aims pursued here. **First, the examples are listed / sketched** (with the types of analogue(s) involved, and the citations)[544]. The intention is to present a diverse / nontrivial (but, necessarily, very small) collection – drawn, mainly, from the S/M-List. Although this is only a minute sample of the – potentially, unlimited – possibilities, it may be augmented, at any stage (if necessary), until it produces adequate characterizations of 'the Σ -approach to analogues' – as a 'standard OIB facility'[545]. This is in line with the entire Σ -scheme, where the overriding priority is: **to establish a universal environment for the (quasi-optimal) organization / development of**

---

[538] See $\mathsf{E}_{11/I}$ for some examples.

[539] By convention, **unary** relations, $r_j(c_\sigma)_1$, exhibit properties of c_{σ_1}. Here, the $(c_\sigma)_m$ are perturbations of (subsets of) the constituents (say, $(c)_L$) of $\eta(s)$.

[540] The illustrations (S/M) given later indicate what is involved.

[541] recall that, in a sense, Σ 'is' a universal / operational environment for 'approach'

[542] the term 'surveying', with its topological connotations, seems apt, here.

[543] A good example is given by dislocation dynamics, and, relativistic electrodynamics: see, e.g., MR 22 # 5204–6 (papers by EF Holländer).

[544] There is some parallel with the style adopted in $\mathsf{E}_{10/I}$, but the treatment is even briefer.

[545] Of course, no definitive characterization of 'global validity' is attainable.

mathematical knowledge.

In **the second phase**, the chosen examples are exhibited (as far as possible) in the '**constituent-assignment format**' discussed in $E_{11/I}$. It is evident that the analogue-construction procedures (based on perturbation of the corresponding assignment flowch- arts) are **theoretically** exhaustive; but, in practice, only samples are produced. This underlines another seminal objective of Σ : **to promote 'fruitful prospecting' for stimulating analogues – in all kinds of mathematical / scientific terrain** – a fundamental facet of all rigorous research activity (but often obscured by formalism). So far, the classes of **admissible perturbations** of members of Y_G have not been examined – for the excellent reason that there are virtually no **global** limitations! Even more: the **local** conditions (on some item(s) in a fixed assignment, $\eta(d)$) depend sensitively on the nature of the constituents involved. The '**perturbation operators**' $p_\alpha, q_{-\alpha}$, acting (respectively) on $\eta_\alpha(\alpha(e))$, and, on $\eta_{-\alpha}(-\alpha(s))$, are called Σ_t**-solvable** whenever they produce Σ_t-solvable assignments[546].

Hence, the only **general** conditions on perturbation operators $b \in \mathbb{B}$ are:[547]

(i) that $b\eta(d)$ **is w.d.;**

(ii) that $b\eta(d)$ **is logically consistent.**

The **interpretation** of 'solvability of $b\eta(d)$ over M_t (via G)' may then take two (distinct) forms:

(a) $\exists\alpha(d) \in M_t : \eta_\alpha(\alpha(d)) \in [b\eta(d)]^{\rightarrow}$;

(b) an $\alpha(d)$ **may be w.d. over** M_t **such that** $\eta_\alpha(\alpha(d)) \in [b\eta(d)]^{\rightarrow}$.

Otherwise, $b\eta(d)$ is **nonsolvable** over M_t.

A further distinction must be drawn between **constructive**, and **nonconstructive existence / definition** – denoted (respectively) by $^c\exists$,^{c}w.d., and $\exists$, e.d. – as well as between constructive, and λ-**operational** existence / definition, where λ indicates the level of operationality, for the procedures concerned[548]. Plainly, the whole scheme for (non)solvable assignments cannot be rendered **totally** rigorous – these must always be heuristic elements;[549] but, within the overall Σ -OIB context, this formalism is extremely useful in the systematic investigation of analogues. Indeed, no OIB of the complexity / sophistication of Σ could be free from **minor** global inconsistencies and heuristic elements. What matters is, rather, **quasi-local soundness**[550]. The exceptions, where local informalities / circularities accumulate[551] – or, are globally propagated[552] – may be controlled at the user-interface, through critical scrutiny, together with intuitive response. In this sense, Σ should be regarded as a vast collection of formalized theories, whose combined structure is completed by labyrinthine (heuristic) interconnections[553]. Although it is probably that Σ , as a whole, could be formalized over some hybrid (standard / fuzzy) logic, neither power nor clarity would be gained, and the 'interactive perspective' might be obscured. Consequently, the development of Σ is best pursued by exploiting the mixture of formal and informal substructures to derive 'new results of quantifiable validity' – a strategy that mirrors current research procedures!

So far, there is some imprecision in the action of perturbation operators on assignments, $\eta(d)$, of the ME/SE, d. This arises from the qualitative difference between locational conditions, $L(d)$, and machines, $\mathbb{M}(d)$, for d. Here, the cardinality of the **set** $L(d)$ is not fixed; whereas, for the templates embodying $\mathbb{M}(d)$, the number of components is specified, independently of d, as N[554]. The simplest way of removing this 'imbalance' is to consider the conditions in $L(d)$

[546] that is, their output sets contain some w.d. ME

[547] of course, diverse special conditions must be imposed 'locally' to ensure that (i)/(ii) are valid.

[548] the notation: $^\lambda\exists$,$^\lambda$w.d. seems apt here.

[549] Individual assignments **may** be treated rigorously, but some residual imprecision' appears to be unavoidable.

[550] See, e.g., Boolos, GS / Jeffrey, RC, '*Computability and Logic*' (CUP, 2nd ed., 1980)

[551] e.g., through iterative / recursive procedures

[552] e.g., in quasi-continuation schemes, of diverse types

[553] See, e.g., the various sorts of 'diagrams' in E_7.

[554] with $N = 14$ (possibly, 16) in most instances

individually: $L(d) =: \bigwedge_{1\leqslant i\leqslant N_d} c_i(d)$, with associated perturbations $bL(d) := \bigwedge b_i c_i(d)$, etc.[555]. Then[556] $bL(e/s)$ may be called **significant** whenever it characterizes some w.d. ME/SE – say, bd. The perturbation b therefore has N_d components, b_i, and $b_i c_i(d)$ may have any w.d. form 'derivable from $c_i(d)$'. This covers all w.d. **direct** analogues of d (i.e. MA of ME and SA of SE). The treatment of **inverse** analogues in terms of LC (rather than machines) requires further consideration.

Comparison of the 'LC' and '$\mathbb{M}$' frameworks shows that (for the inverse ME, $-\alpha(s)$, of the SE, s), generation of the **solvable** LC, $L^{-1}(-\alpha(s))$, from the **significant** LC, $L(s)$, requires (**non**algorithmic!) substitution operations on some of the $c_i(s)$. These substitutions may be 'suggested by $L(s)$' but will not, in general, be derivable from $L(s)$. Comparable transformations of $\mathbb{M}(s)$ are required to produce inverse MA. The distinction between these representations remains as '**existential**' versus '**quasi-constructive**'. They are **combined** in the **assignments** $\eta(e), \eta(s), \eta(d)$ [solvable], and $b\eta(d)$ [solvability open].

It follows that the mechanisms, $\mu(\alpha(d))$, of direct analogues of d may be determined from $\mathbb{M}(d)$ – and 'inferred from $L(d)$'; but the mechanisms $\mu^{-1}(-\alpha(d))$ may be explicated only after the 'suggestive' elements have been specified. The 'new ME', $-\alpha(s)$, may then itself be represented in solvable forms $L(-\alpha(s)), \mathbb{M}(-\alpha(s))$, incorporating the 'new structure(s) /...' introduced to define 'the ME $-\alpha(s)$'; and so on! This is a quasi-regularization process, of a recursive nature.

Summary

Distinctions are drawn between the 'L' (**locational / existential**), and $\mathbb{M}$ (**machine / quasi-constructive**) representations of analogues[557]. For **direct** MA/SA, $\alpha(d)$, the underlying **mechanism(s)**, $\mu(\alpha(d))$, may be '**understood**' from $L(\alpha(d))$, and (partially) **implemented** from $\mathbb{M}(\alpha(d))$[558]. For **inverse** analogues, $-\alpha(s)$, however, the mechanisms may be explicated **only if** various **nonalgorithmic substitutions** ('suggested by s') are introduced[559] – to convert **significant** assignments into **solvable** assignments – producing **new ME**, to be incorporated within the Σ -OIB. All analogues, $\pm\alpha(d)$, are 'generated' by associated **perturbation operators** (p_α, or $q_{-\alpha}$, acting on $\eta(d)$) whose formal properties are worthy of investigation, in the development of a rigorous 'Theory of Mathematical Analogy'[560].

Properties of Assignment-Perturbation Operators

Operations on $\eta(d) := (L(d), \mathbb{M}(d))$ are necessarily of very general scope, since the constituents – $c_w(d)$, of $L(d)$, and $m_j(d)$, of $\mathbb{M}(d)$ – are restricted only by **realizability / consistency criteria**[561]. Moreover, **card**$\mathbb{M}(d) =: N$ is independent of d. A **unified scheme** may be based on the consideration of individual 'components' of $L(d), \mathbb{M}(d)$, through the representation: $b =: (b_L, b_\mathbb{M}), b\eta(d) := (b_L L(d), b_\mathbb{M}\mathbb{M}(d))$, where each of $b_L, b_\mathbb{M}$ acts 'diagonally' on the components of $L(d), \mathbb{M}(d)$, respectively: $b_L L(d) = \bigwedge_w b_{Lw} c_w(d); b_\mathbb{M}\mathbb{M}(d) = \{b_{\mathbb{M}j} n_j(d) : j \in \bar{N}\}$[562]. It is understood that $b_\mathbb{M}\mathbb{M}$ is 'constructed' in the same way as $\mathbb{M}$, but with perturbed

555 The possibility '$N_d = \infty$', for some sequential conditions (or even; of **un**countable families of conditions) is allows, for 'completeness'!

556 Here, $L(e), L(s)$ are the two possible forms of $L(d)$, etc. The symbol b covers both p_α and $q_{-\alpha}$ (above), as appropriate.

557 See $\mathrm{E}_{11/I}$ for motivation, etc..

558 Again, see $\mathrm{E}_{11/I}$ for a discussion of mechanisms / machines for analogues.

559 i.e., the substitutions involve intuition / imagination / creativity – rather than pure implementation, however complex.

560 The notation $b\eta(d)$ covers both cases.

561 The $c_w(d)$ are $v_w(d)$-place predicates (for $w \in W \equiv \bar{N}_d, \mathbb{P}$, or a family, $\{w\}$).

562 This is an obvious extension of 'scalar products' of vectors; or of the action of diagonal matrices.

elements, $b_{\mathbb{M}j}m_j(d)$ – an entirely natural process in engineering contexts![563] The main difficulties stem from the requirement that **perturbed machines should have w.d. outputs**, since, plainly, this will not be the case for 'arbitrary perturbations of $\mathbb{M}(d)$'[564].

One way to obviate this problem is **to define** $(b_{\mathbb{M}}\mathbb{M}(d))^{\rightarrow}$ **to be** ϕ (the **empty set**) **whenever there is no w.d. ME 'rigorously obtainable from** $b_{\mathbb{M}}\mathbb{M}(d)$**'**[565]. The **set** of all such **'inadmissible perturbations'** thus constitutes **the kernel of the mapping**[566] (with $\mathbb{M} \equiv \mathbb{M}(d)$, etc.):

$$\varphi : \mathcal{B} \equiv \underset{j \in \bar{N}}{X} \mathcal{B}_j \rightarrow (\mathcal{B}\mathbb{M})^{\rightarrow} \equiv \{(b\mathbb{M})^{\rightarrow} : b \in \mathcal{B}\},$$

where the $\mathcal{B}_j$ are defined as follows.

Let $M_j (j \in \bar{N})$ comprise **'all machine constituents** m_j **of the same type'** – as designated in the specification of ME templates (see $E_{11/I}$, and the research reports cited there). **Then** $\mathcal{B}_j \equiv \{b_j\}$ **is the set of all single-valued self-maps of** M_j. Although this definition is ineffective (and partially retrospective) it **is** tenable, and allows the exploration of possible (inverse) analogues to be pursued within a formal framework – through processes of **machine-analysis**. There is, of course, an unavoidable element of 'trial and error' in this scheme – as a means for producing 'new' analogues. Rather, it is a powerful prescription for the (Σ_t-maximal) **solution of 'well-set problems'**, in all mathematical / scientific domains – and, for the systematic investigation of **'near-variants'**, $v(d)$, of ME/SE d already represented in 'template forms'. In spite of these limitations, it should be emphasized that – in the overall task of **conjecturing / identifying / justifying / proving** properties of (inverse) analogues – these techniques are of fundamental importance. Moreover, it is argued in $E_{11/I}$ that **every** (inverse) **analogue** (no matter how intricate / sophisticated / arcane) **has some w.d. template representation(s)**.

The main alternative to machine / template-**analysis** is machine **synthesis**. In this process, properties of any 'conjectured (inverse) analogue of d' are **prescribed** at the outset, and the basic problem is to design / construct the machine(s) for which these properties (partially) characterize the output. Obvious examples occur in **Circuit Theory, Dynamical Systems Theory, Chemical Engineering** and, especially, **Control Theory**. Naturally, precise synthesis criteria in Σ , for $\mathbb{M}(\eta, \xi_\lambda)$ to 'produce' e_λ, are derivable only for environments ξ_λ 'of sufficiently rich structure'. Nevertheless, the ideas may be extended to furnish weaker – but still useful – results for more general classes of systems[567]. Therefore, the primary task is to develop quasi-synthesis procedures for machines generating approaches to prescribed species of ME – with reductions (wherever possible) to standard procedures, under additional conditions[568]. There are several broad schemes for synthesis, of various degrees of completeness, according to the context(s). In all cases, however, the 'analysis', and 'synthesis', problems are mutually inverse – within the framework of General Systems Theory. It is therefore necessary to represent the templates / machines for ME as (species of) general systems[569].

Although the **analysis** and **synthesis** of e are, in a sense, 'opposite processes', they are subtly interrelated in the in the investigation of analogues, based on templates / machines, $\mathbb{M}(\delta, \xi), \mathbb{M}_\lambda(\eta, \xi_\lambda)$[570]. To understand this, suppose that e is w.d.[571] within a theory, τ. Then, procedures for **calculating / approaching** e, over τ[572], amount to forms of **synthesis** of $\mathbb{M}(\delta, \xi)$; whereas, **determination of the properties of** e, from a **specified** $\mathbb{M}(\delta, \xi)$, involves problems of **analysis**[573].

[563] E.g., components may be reduced / enlarged, shapes may be modified.
[564] The associated problems for $b_L L(d)$ are discussed later.
[565] Through mathematical derivation / approach procedures.
[566] In general, **ker**φ may be of any 'size', from 'negligible' to 'almost all of $\mathcal{B}$'.
[567] The 'strength' of such results depends critically on the forms of ξ_λ – but **some** machine representation may always be obtained.
[568] A selection of standard procedures is discussed later.
[569] References are given later.
[570] See $E_{11/I}$ for details / references.
[571] i.e., some LC for e over τ are specified.
[572] e.g., within the frameworks discussed in $E_5/E_6/E_7$.
[573] i.e., the structure of $\mathbb{M}(\delta, \xi)$ is prescribed, and properties / behaviour of e must be derived from $\mathbb{M}(\delta, \xi)$.

Notice that the techniques of **analysis** for $\mathbb{M}(\delta, \xi)$, to calculate e, are essentially **prescribed** in the template – but refinements / enhancements / variants may still be developed, to increase effectiveness of to enlarge scope[574]. By contrast, the **synthesis** of some $\mathbb{M}_\lambda(\eta, \xi_\lambda)$, to produce ME, e_λ, having specified characteristics (even if it is known to be attainable), typically offers only **hints** as to the procedures to be followed[575].

This is, of course, a situation common to most nontrivial mathematical '**inverse problems**' – of 'pure', or 'applied', types[576]. The **levels of synthesis** in particular procedures may be gauged (roughly) by the **extent of replication** of the prescribed characteristics; or (in less favourable cases), by the **approaches to replication**[577]. Plainly, there is a strong connection between 'inverse problems' and 'converse theorems'. Indeed, in certain formulations, converse theorems furnish **solutions** to inverse problems. This (obvious) interrelation is of special importance for the OTC aspects of Σ .[578]

Accordingly, it is worthwhile to outline some fundamental synthesis schemes – initially, within the following domains[579].

(1) Electrical circuits (and their equivalents).

(2) Sampled-data control systems.

(3) Optimal feedback / adaptive control systems.

(4) Dynamical systems.

(5) Automata.

(6) Threshold logic gates.

(7) Stochastic control systems.

(8) Antenna design.

(9) Spectral function theory.

(10) Integral Geometry.

(11) Chemical Engineering (Plant Design).

(12) Multivariable Control Systems.

(13) Chemical Structure-Synthesis / Analysis.

(14) Generalized / Abstract Harmonic Analysis.

(15) Machine Design.

(16) Miscellaneous Inverse Problems.

NOTE. For all forms of systems / networks, the 'structured acronym'

(N)LT(IN)V(N)D

is introduced – for (non)linear, time-(in)variant, (non)deterministic – from which all relevant combinations of system-properties may be succinctly denoted.

The range / scope of synthesis schemes is diverse – even within each domain cited so far – but this is mainly a matter of implementation, rather than of fundamental principles. Further, some of the listed domains overlap considerably; and, many 'new topics' will be added as Σ

[574] See the 'ABC' (Solid Mechanics) examples in $\mathrm{E}_{11/I}$ for basic ideas.
[575] The paradigms involved are illustrated later.
[576] This class of problems is considered later.
[577] This measure of 'level' is clarified in examples.
[578] Recall that, for OT, T^{-1} may be a converse of T.
[579] All references for this list are given in the outlines.

evolves[580]. The aim here is to support the claim that: an (essentially) arbitrary ME may be **analysed** / (conceptually) **perturbed**; or, **synthesized** (e.g., via forms of **feedback**) in terms of appropriately structured **templates / machines**. This claim is, in turn, based on the (nonunique) representability of ME as (partial) machine output[581].

The dominant **models** for the synthesis of (analogues of) ME/SE are based (broadly) on techniques for the various types of synthesis listed above (as well as some other types still to be identified). The **machines** corresponding to **general** classes of ME/SE are, of course, only **motivated** by the precise forms for mathematical / scientific paradigms (such as **electrical circuits**). Nevertheless, the underlying structure is pervasive – even in the most generalized models, whose very **formulations** involve analogical ideas. Another fundamental issue here is: **the relation between (quasi-)constructive procedures are synthesis**. In an obvious sense, these processes are virtually synonymous but, for Σ , the interpretations exhibit important distinctions – mainly concerning '**design**' versus '**implementation**'. In any case, the purview of Σ extends diversely beyond the confines of strictly mathematical construction; so, considerable explication of the fundamental processes is necessary, to fashion a coherent scheme. Again: syntheses amy be viewed as hierarchical constructions of entities from classes of 'similar, but less complex, entities' – rather than as **as hoc** procedures. This is a basic distinction.

For instance, in Chemistry, **synthesis** (of compounds /...) may be governed by **(non)autonomous reactions** – of both **organic** (carbon-based), or **inorganic**, types[582]. Reactions may be n**-molecular**[583] ($n \in \mathbb{P}$), with various species of **driving-forces**[584], promoting **stable chemical (re-)combination**[585] – subject to **quantum-mechanical / thermodynamic / kinetic / quasi-Hamiltonian /...** criteria, over associated (in)formal frameworks[586]. Accordingly, the overall mechanisms of such reactions may be studied within Statistical Mechanics, complemented by relevant techniques from other fields (e.g., Solid-State Physics)[587]. There is a further distinction between small-scale and large-scale synthesis – which may suggest interesting inverse analogues[588].

Another aspect of synthesis in Chemistry involves **determination of structure from spectroscopic data** – a class of inverse problems with (in general) nonunique solutions[589]. Insofar as every chemical synthesis is the product of some **reaction(s)**, the systematic investigation of **reaction paths** is of fundamental importance in Chemistry[590] – and (suitably generalized) in virtually all 'scientific / mathematical processes', where they could be regarded as quasi-geodesics![591] Further, every inverse problem (in whatever field) has a natural interpretation in terms of cognate synthesis procedures. It may even be the case that the two classes of problems (broadly interpreted) coincide[592]. For the corresponding **analysis** problems (again, in **any** scientific / mathematical field) the solution techniques stem directly from the underlying

[580] Many of these examples stem from 'inverse problems' in Pure / Applied Mathematics; such problems formally cover all types of synthesis.

[581] This is postulated here; a justification is outlined in $E_{11/I}$.

[582] See, e.g., '*Mechanism and Structure in Organic Chemistry*' by ES Gould (HR&W, 1959); '*Principles of Organic Synthesis*', by ROC Norman (Methuen, 1968); '*Inorganic Chemistry*' by KF Purcell / JC Kotz (Saunders, 1977).

[583] For $n \geqslant 3$, n-molecular reactions generally involve n**-fold collisions**; the other route has **successive binary collisions** of different species of molecules.

[584] E.g., thermal / E-M /....

[585] Both thermodynamic and kinetic stability may be considered – along with other forms (e.g. stability against specified types of perturbations). **Also:** a stable compound may still react with **other** elements; etc..

[586] See, e.g., Levine, RD, '*QM of Molecular Rate Processes*' (OUP, 1969); Levine, RD / Bernstein, RB, '*Molecular Reaction Dynamics*' (OUP, 1974); Nikitin, EE, '*Theory of Elementary Atomic and Molecular Processes in Gases*' (OUP, 1974).

[587] See, e.g., Simpson, WT, '*Theories of Electrons in Molecules*' (P-H, 1962); Hannay, NB, '*Solid State Chemistry*' (P-H, 1967); Tiller, WA, '*The Science of Crystallization*' (CUP, 2 Vols, 1991).

[588] Issues of realizability / efficiency arise here.

[589] See, e.g., Levine, IN, '*Molecular Spectroscopy*' (Wiley, 1975)

[590] See, e.g., Purcell / Kotz (op. cit., esp. Ch 7) for inorganic reactions; and, Gould (op. cit.) for organic reactions. **Mechanisms** and **pathways** are closely related.

[591] Many instances are identified in outlines / examples from the S/M-List.

[592] This connection is clarified in later discussions / examples.

models; so the scope for interesting **inverse** analogues is comparatively limited[593].

Key References for Synthesis-Paradigms.

The formulation of synthesis-paradigms is best achieved by citing (collectively) key references for all of the main topics – and then outlining the associated procedures, to furnish a flexible framework for the discussion of diverse examples, allowing other topics to be added at any time. The **bibliographies** of the cited books / papers cover all of the fields considered in detail. The aim here is merely to indicate basic problems / techniques – from which possible (inverse) analogues may be investigated.

1. Electrical Circuits and Their Simulations.

There are several procedures for the synthesis of (a) passive, and (b) active networks – via digital / analogue schemes. The simulations may be of electrical / mechanical / thermal / acoustic /. . . types. Some of the techniques extend to classes of more general (linear) systems. **NOTE that** the concepts 'realization', and 'synthesis', are closely related (even coincident, in certain contexts).

1.1 Bode, HW, '*Network Analysis and Feedback Amplifier Design*' (Van Nostrand, 1945)

1.2 Weinberg, L, '*Network Analysis and Synthesis*' (McGr-H, 1962)

1.3 Ferris, CD, '*Linear Network Theory*' (Merrill, 1962)

1.4 Haring, DR, '*Sequential Circuit Synthesis*' (MIT, 1966)

1.5 Kuo, FF, '*Network Analysis and Synthesis*' (Wiler, 1966)

1.6 Antoniou, A, '*Digital Filters. . .*' (McGr-H, 1979)

1.7 Papoulis, A, '*Circuits and Systems*' (Holt-Saunders, 1980)

1.8 Chen, W-K, '*Broadband Matching. . .*' (World Scientific, 1988)

1.9 Cheng, DK, '*Analysis of Linear Systems*' (A-W, 1959)

1.10 Kin, WH / Chien, RT-W, '*Topological Analysis and Synthesis of Communication Networks*' (Columbia UP, 1962)

1.11 Truxal, JG, '*Intro. Systems Engineering*' (McGr-H, 1972)

1.12 Paul, CR, '*Analysis of Multiconductor Transmission Lines*' (Wiley, 1994)

1.13 Olson, HF, '*Dynamical Analogies*' (V. Nostrand, 1958) [See also 1-3, Ch X, 1-9, Ch 4]

1.14 Kalman, Re / Falb, PL / Arbib, MA, '*Topics in Mathematical Systems Theory*' (McGr-H, 1969) (especially, for **realization** criteria).

1.15 Oster, G / Perelson, A, '*Chemical Reaction Networks*' (IEEE Trans. Circ. Syst. Vol. CAS–21 (11/74) 709–721)

1.16 Shigley, JE, '*Simulation of Mechanical Systems*' (McGr-H, 1967)

1.17 Boite, R / Dewilde, P (eds.), '*Circuit Theory and Design*' (Proc.), Delft UP / N-H (1981)

1.18 Fuhrmann, PE (ed.) '*Math-Theory of N-W and systems*' (LNCIS #58, Springer, 1984)

[593]**Direct** analogues of the analytical procedures are, however, readily investigated. The templates for **scientific** problems do offer possibilities for inverse analogues.

1.19 Budak, A, '*Passive and Active N-W Analysis and Synthesis*' (Houghton-Miffling, 1974). See also: Several Proc. of MRI Symposia (Poly. Inst. Brooklyn (now 'NY')) 1952 –, and the RAAG Memoirs (ref 5.3 here).

1.20 MacFarlane, AGJ, '*Dynamical System Models*' (Harrap, 1970)

1.21 Bell, DJ, ed., '*Recent Mathematical Developments in Control*' (AP, 1973)

2. Sampled-Data Control Systems.

NOTE. Since all types of control systems are variously interrelated, the citations here are limited to a few mathematically oriented treatments for each type of system. Discussions of these interrelations (with cognate references) are given later.

2.1 Lindorff, DP, '*Theory of Sampled-Data Control Systems*' (Wiley, 1965)

2.2 Ragazzini, JR / Franklin, GF, '*Sampled-Bata Control Systems*' (McGr-H, 1958)

2.3 Maybeck, PS, '*Stochastic Models...*' Vols 1–3 (esp., Vol 3) AP, 1982

2.4 Bertsekas, DP / Shreve, SE, '*Stochastic Optimal Control – The Discrete Time Case*' (AP, 1978)

2.5 Williamson, D, '*Digital Control...*' (P-H, 1991)

2.6 Gössel, M, '*Nonlinear Time-Discrete Systems*' (LNCIS #41, Springer, 1982)

2.7 Halanay, A / Ioneseu, V, '*Time-Varying Discrete Linear Systems*' (Operator Theory #68; Birkhäuser, 1994)

2.8 Netushil, A (ed.), '*Theory of Automatic Control*' (MIR, 1978)

2.9 Maciejowski, JM, '*Multivariable Feedback Design*' (CUP, 1989)

2.10 Newton, GC, et al., '*Analytical Design of Linear Feedback Controls*' (Wiley, 1957)

3. Optimal / Adaptive Control Systems.

NOTE. The distinction between **optimal** and **adaptive** control systems is rather subtle. One view is that optimal systems are adaptive – with directly measurable goals. Further: **feedback** is a form of **adaptation** (in **closed-loop systems**).

3.1 Pallu de la Barrieère, R, '*Optimal Control Theory*' (Saunders, 1967; Dover, 1980)

3.2 Lee, EB / Markus, L, '*Foundations of Optimal Control Theory*' (Wiley, 1967)

3.3 Young, LC, '*... Calculus of Variations and Optimal Control Theory*' (Saunders, 1969)

3.4 Sworder, D, '*Optimal Adaptive Control Systems*' (AP, 1966)

3.5 Junt, KJ (ed.) '*Polynomial Methods in Optimal Control and Filtering*' (Peregrinus, 1993)

3.6 Newton, GC / Gould, LA / Kaiser, JF, '*Analytical Design of Linear Feedback Controls*' (Wiley, 1957)

3.7 Doeblin, EO, '*Control System Principles and Design*' (Wiley, 1985) [for aspects of **nonlinear** feedback].

3.8 Benveniste, A, et al., '*Adaptive Algorithms and Stochastic Approximations*' (Springer, 1990)

3.9 Gabasov, R / Kirillova, F, '*The Qualitative Theory of Optimal Processes*' (Dekker, 1976)

3.10 Landau, ID, '*Adaptive Control...*' (Dekker, 1979)

3.11 Netushil A (ed.), op. cit.

4. Dynamical Systems.

The term 'dynamical system' covers diverse processes evolving in time. The formal (axiomatic) definition(s) may seem pedantic, but subtle problems make it necessary to exercise extreme caution. Time-invariant systems are static; all others are dynamical. There are, of course, quasi-transcriptions converting 'static' systems into dynamic systems, by 're-naming' some variable as 'time'. Such cases may be identified by checking in detail the basic axiomatic formulation(s).

4.1 Kalman, RE, et al., (op cit.), '*Topics in Mathematical System Theory*' (McGr-H, 1969) [see esp., pp2–22 for the basic definitions]

4.2 Porter, B, '*Synthesis of Dynamical Systems*' (Nelson, 1969)

4.3 Šiljak, DD, '*Large-Scale Dynamic Systems*' (N-H, 1978)

4.4 Zadeh, LA, et al. (op cit.) '*System Theory*' (McGr-H, 1969)

4.5 Porter, WA, '*Modern Foundations of Systems Engineering*' (Macmillan, 1966)

4.6 MacFarlane, AGJ (op cit.) '*Dynamical System Models*' (Harrap, 1970)

5. Automata.

The general definition of automata – as machines converting input into output – formally covers the Σ -templates / machines for ME/SE, from which (inverse) analogues are to be 'fabricated' through the application of suitable 'perturbations of the template constituents'. Further, most classes of networks / systems may be formulated in terms of automata. Although there is still intensive research in this area, the most relevant results for Σ (on algebraic / analytical / structural aspects) may be found in a few key references.

5.1 MRI Sympos. Proc. #XII (1962), '*Mathematical Theory of Automata*' (Interscience, 1963) (This contains several basic papers)

5.2 Booth, TL, '*Sequential Machines and Automata Theory*' (Wiley, 1967)

5.3 Kondo, K (ed.) RAAG Memoirs, Vols 1–3 (U. of Tokyo, 1962–1968)

5.4 Hartmanis, J / Stearns, RE, '*Algebraic Structure Theory of Sequential Machines*' (P-H, 1966)

5.5 Kobrinskii, NE / Trakhtenbrot, BA, '*Intro. to the Theory of Finite Automata*' (N-H, 1965)

5.6 Schuh, JF, '*Algemene Theorie der Automaten*' (Philips, Eindhoven, 1963 [Dutch])

5.7 Haring, DR, '*Sequential Circuit Synthesis*' (MIT, 1966)

5.8 Lee, SC, '*Modern Switching Theory...*' (P-H, 1978)

6. Threshold Logic Gates.

Although this is a somewhat specialized form of synthesis, the domain of applicability is broad, covering many sorts of 'information-processing without memory' (e.g., adaptive control, digital computer circuits, neuro-biological models).

6.1 Destouzos, ML, '*Threshold Logic: A Synthesis Approach*' (MIT, 1965)

6.2 Lewin, D, '*Computer-Aided Design of Digital Systems*' (Crane Russak, NY, 1977)

6.3 Ádám, A, '*Truth Functions and ... Their Realization ...*' (Akad. Kiadó, Budapest, 1968)

6.4 Lewis, PM / Coates, CL, '*Threshold Logic*' (Wiley, 1967)

6.5 Hu, S-T, '*Threshold Logic*' (U. Calif. Pr., 1965)

6.6 Sheng, CL, '*Threshold Logic*' (AP, 1969)

6.7 Lee, SC, '*Modern Switching Theory*' (P-H, 1978)

6.8 Schuh, JF, op. cit.

6.9 Murgai, R, et al., '*Logic Synthesis...*' (Kluwer, 1995)

7. Stochastic Control Systems.

All types of systems may be considered here, since it is a matter of whether the input and the parameters are specified ('*deterministic models*'); or, some are given stochastically ('*stochastic models*'). Thus it is appropriate to list some sources where all types of systems are treated within unifying frameworks – together with accounts of the extra techniques developed to handle non-deterministic models.

7.1 Jacobs, OLR, '*... Control Theory*' (OUP, 1974)

7.2 Ogata, K, '*State-space Analysis of Control Systems*' (P-H, 1967)

7.3 Kuo, BC, '*Automatic Control Systems*' (P-H, 3rd ed., 1975)

7.4 Netushil, A (ed.), '*Theory of Automatic Control*' (MIR, 2nd ed., 1978)

7.5 Aström, KJ, '*... Stochastic Control Theory*' (AP, 1970)

7.6 Bertsekas, DP, et al., (op cit.), '*Stochastic Optimal Control...*' (AP, 1978)

7.7 Laning, JH / Battin, RH, '*Random Processes in Automatic Control*' (McGr-H, 1956)

7.8 Nagrath, IJ / Gopal, M, '*Control Systems Engineering*' (Wiley, 2nd ed., 1982)

7.9 Merriam, CW, III, '*Automated Design of Control Systems*' (G&B, 1974)

7.10 Maybeck, PS, '*Stochastic Models...*' Vols 1,2,3 (AP, 1982)

8. Antenna Design.

The most complete results cover **planar** source synthesis, where exact representations may be derived. For 'non-planar cases', various approximation schemes have been developed.

8.1 Rhodes, DR, '*Synthesis of Planar Antenna Sources*' (OUP, 1974)

8.2 Jones, DS, '*Methods in E-M Wave-Propagation*' (OUP, 1979)

8.3 Jordan, EC / Balmain, KG, '*E-M Waves and Radiating Systems*' (P-H, 2nd ed., 1968)

8.4 Comton, RT, '*Adaptive Antennas...*' (P-H, 1988)

8.5 King, RWP, et al., '*Arrays of Cylindrical Dipoles*' (CUP, 1968)

8.6 Jones, DS, '*The Theory of E-M*' (Pergamon, 1964)

8.7 Hansen, RC (ed.), '*Microwave Scanning Antennas*' (AP, repr. Peninsula, publ. 1985)

8.8 Moulin, EB, '*Radio Aerials*' (OUP, 1949)

9. Spectral Function Theory.

The underlying theory involves (families of) linear transformations over vector spaces, and some aspects of abstract harmonic analysis. In each case, an operator is to be synthesized from (simpler) 'canonical operators'. The basic problem stems from standard Fourier Analysis, but there are many generalizations – for instance to HS, BS (for algebraic settings).

9.1 Kharin, VP, et al. (eds.), '*Commutative Harmonic Analysis*' (EMS Vol. 15, Springer, 1991)

9.2 Nikol'skii, NK, '*Treatise on the Shift Operator*' (Springer, 1986)

9.3 Meyer, Y, '*Recent Advances in Spectral Synthesis*' (in LNM #266, Gulick, D, et al., eds., Springer, 1972)

9.4 Edwards, RE, '*Fourier Series, Vol 1*' (H,R&W, 1967)

9.5 Lee, YW, '*Statistical Theory of Communication*' (Wiley, 1960)

9.6 Dunford, N / Schwartz, JT, '*Linear Operators, Part II*' (Wiley, 1963)

10. Integral Geometry.

The synthesis aspects of integral geometry correspond to inverse problems.

10.1 Santaló, LA, '*Integral Geometry and Geometric Probability*' (A-W, 1976) (with copious references)

10.2 Ambartzumian, RV, '*Factorization Calculus and Geometric Probability*' (CUP, 1990)

10.3 Sharafutdinov, VA, '*Integral Geometry of Tensor Fields*' (VSP, Utrecht, 1994)

10.4 Sabatier, PC, et al., eds. (op cit.), '*Tomography and Inverse PRoblems*' (Hilger, 1987)

10.5 Deans, SR, '*The Radon Transform...*' (Wiley, 1983)

11. Chemical Engineering Plant Design.

The role of synthesis here is primarily in the realization and identification of linear and nonlinear system – that is, in the construction of models of systems 'directly' from specified I/O data.

11.1 Seinfeld, JH / Lapidus, L, '*Mathematical Methods in Chemical Engineering*', (Vol. 3 (P-H, 1974)) [Ch. 9 covers **linear** systems; **nonlinear** systems are treated in Ch. 10]

11.2 Wiener, N, '*Nonlinear Problems in Random Theory*' (MIT, 1942)

11.3 Schetzen, M, '*Volterra and Wiener Theories of Nonlinear Systems*' (Wiley, 1980)

11.4 Billings, SA, '*Identification of Nonlinear Systems: a Survey*', IEE Proc. 127D (11/80), 272–285

12. Multivariable Control Systems.

The use of transfer functions (characterized by pole / zero distributions) is the dominant procedure for **feedback** control synthesis, but other forms of control require different techniques.

12.1 Maciejowski, JM, '*Multivariable Feedback Design*' (CUP, 1989) [with substantial reference-lists for each chapter].

12.2 Atherton, DP, '*Nonlinear Control Engineering*' (Van Nostrand, 1975)

12.3 Siljak, DD, '*Large-Scale Dynamic Systems...*' (N-H, 1978)

12.4 Bose, NK, ed., '*Multidimensional Systems...*' (IEEE Press, 1979)

12.5 Isidori, A, '*Nonlinear Control Systems...*' (LNCIS #72, Springer, 1985)

13. Chemical Structure-Synthesis / Analysis

This is a basic form of structural synthesis. Another aspect of chemical synthesis involves n-molecular reactions – including various types of **reaction paths** for organic / inorganic substances.

13.1 Creswell, CJ, et al., '*Spectral Analysis of Organic Compounds*' (Longman, 2nd ed., 1972)

13.2 Norman, ROC, '*Principles of Organic Synthesis*' (Methuen, 1968)

13.3 Levine, IN, '*Molecular Spectroscopy*' (Wiley, 1975)

13.4 Purcell, KF / Kotz, JC, '*Inorganic Chemistry*' (Saunders, 1977)

13.5 Lee, HH, '*Heterogeneous Reactor Design*' (Butterworth, 1985)

13.6 Gould, ES, '*Mechanism and Structure in Organic Chemistry*' (H,R&W, 1959)

13.7 Levine, RD, '*QM of Molecular Rate Processes*' (OUP, 1969), Levine, RD, et al., '*Molecular Reaction Dynamics*' (OUP, 1974)

13.8 Nikitin, EE, '*Theory of... Molecular Processes*' (OUP, 1974)

14. Generalized Harmonic Analysis.

Under this heading, all of the extensions of classical harmonic analysis are covered – including all types of 'eigenfunction expansions' and all versions of 'the spectral theorem', for various classes of algebraic / analytical structures. All of these representations are species of synthesis.

14.1 Wiener, N, '*Generalized Harmonic Analysis*' / '*Tauberian Theorems*' (MIT, 1964) [original papers: 1930 / 1932]

14.2 Lee, YW, '*Statistical Theory of Communication*' (Wiley, 1960)

14.3 Kato, T, '*Perturbation Theory of Linear Operators*' (Springer, 1966)

14.4 Titchmarsh, EC, '*Eigenfunction Expansions...*' Vol 1 (OUP, 1946); Vol 2 (OUP, 1958)

14.5 Levitan, BM, et al., '*... Spectral Theory*' (AMS, 1975)

14.6 Edmunds, DE, et al., '*Spectral Theory and DO*' (OUP, 1987)

14.7 Berezansky, YM, et al., '*Spectral Methods in Infinite-Dimensional Analysis*' (Kluwer, Vols 1,2, 1995)

14.8 Pastur, L, et al., '*Spectra of Random / AP Operators*' (Springer, 1992)

14.9 Cotlar, M, et al., '*... Functional Analysis*' (N-H, 1974)

14.10 Zygmund, A, '*Trigonomietric Series*' (CUP, repr. 1968)

14.11 Khavin, VP, et al. (eds.), '*Commutative Harmonic Analysis*' (EMS Vol. 15, Springer, 1991)

15. Machine Design.

This is, perhaps, the archetypal synthesis paradigm, since the aims and procedures are transparent. There are many variants, from space linkages to complicated cams / gear-trains / screw-motions. Among the techniques there are 'graphical', 'analytical' and (increasingly) 'CAD'. The most intricate schemes may involve combinations of several 'basic mechanisms'. As always, for Σ , analogues may be sought by 'perturbing the constituents of global devices'.

15.1 Kimbrell, JT, '*Kinematics, Analysis and Synthesis*' (Mc-Gr-H, 1991)

15.2 Hunt, KH, '*Kinematic Geometry of Mechanisms*' (OUP, 1978)

15.3 Norton, RL, '*Design of Machinery*' (McGr-H, 2nd ed., 2001)

15.4 Gonzalez-Palacios, MA / Angeles, J, '*Cam Synthesis*' (Kluwer, 1993)

15.5 Koenigsberger, F / Trusty, J, (eds.), '*Machine Tool Structures, Vol 1*' (Pergamon, 1970)

16. Miscellaneous Inverse Problems.

Since every form of synthesis involves the solution of some inverse problem(s) (suitably formulated), all of the examples listed so far could have been (formally) covered here. Nevertheless, it is more instructive to consider each species of synthesis within its 'natural context(s)' – rather than as an instance of very general classes of problems.

16.1 Sabatier, PC, ed., '*Inverse Methods in Action*' (Springer, 1990)

16.2 Sabatier, PC, ed., '*Tomography and Inverse Problems*' (Hilger, 1987)

16.3 Petkov, V, et al., eds., '*Integral Equations and Inverse Problems*' (Pitman, 1991)

16.4 Newton, RG, '*Inverse Schrödinger Scattering...*' (Springer, 1989)

16.5 Venikov, VA, '*Theory of Similarity and Simulation*' (Macdonald, 1969)

16.6 Deans, SR, op. cit., '*The Radon Transform...*' (Wiley, 1983)

16.7 Kaplan, A, et al., '*Stable Methods for Ill-Posed Variational Problems*' (Akad. Verlag, 1994)

16.8 Bakushinsky, A, et al., '*Ill-Posed Problems: Theory and Application*' (Kluwer, 1994)

16.9 Romanov, VG, '*Integral Geometry and Inverse Problems...*' (Springer, 1974)

Interim Summary

Most of the 'apparatus developed in E_1–E_9 is designed to exhibit arbitrary ME in 'conceptually versatile forms', within approach environments. The (inconsequential) formal inconsistencies in this scheme are compensated by the gain in unification: **all generic ME are represented as (partial) output of templates / machines**. This correspondence: $e \leftrightarrow \mathbb{M}(e, \xi)$, is built up gradually in $E_{11/I}$ – as an extension of the representations implemented in the IB for Solid Mechanics[594]. This formulation is further extended, in $E_{11/II}$, to indicate calculational procedures for ME/SE (from their associated 'machines'); and, to discuss 'admissible assignment-perturbations' – for the exploration of possible (inverse) analogues. The determination of w.d. machines $\mathbb{M}(e, \xi)$, from e, involves diverse species of **quasi-synthesis** – based on mathematical / scientific synthesis-paradigms (some of which are listed in $E_{11/II}$, with basic references). Other paradigms will be included as Σ evolves.

[594] See $E_{11/I}$ for references

Capsule Descriptions of Synthesis Schemes.

In accordance with the overall plan, the list of key references for the paradigms #1 to #16 (above) is complemented here by CD of some typical procedure(s) for each paradigm[595]. The background scientific / mathematical frameworks are assumed, so that the synthesis schemes may be specified as succinctly as possible. Mostly, only a single technique is outline – variants / alternatives may be found in the cited books / papers. Types of synthesis not so far covered will be incorporated as the OIB evolves. **Compound syntheses** (involving more elementary syntheses as 'constituents') will be discussed later. **Combinations of approach and synthesis, for arbitrary ME**, essentially **characterize** Σ as a research environment. The detailed structures / interrelations / constructions /... constituting Mathematics and its applications are largely products of **analytical** techniques, but the fruitful development of nontrivial, general theories requires imaginative use of **synthesis**, in many formulations, at fundamental levels.

The synthesis-paradigm CD are labelled S-P:$m(1 \leqslant m \leqslant 16$, so far), or S-PL$m \cdot n$ (if two, or more forms of S-P:m are considered). The basic references are cited as already listed – e.g., 1.8 ≡ Chen, W-K, '*Broadband Matching...*'; 5.3 ≡ Kondo, K (ed.), RAAG Memoirs, Vols 1–3.... Detailed references to these sources – e.g., 11.1, Sec. 10.1 – may also be given, to explicate synthesis-routines (S-R); and so on. Insofar as **each S-R is itself a compound ME**, it must have some (modified) 'template / machine representation', corresponding to a synthesis of the S-R! In principle, such representations may be obtained in the standard way – by assigning appropriate interpretations to the symbols in the 'abstract template' – to accumulate a fundamental (regularly augmented) collection of S-R templates, each characterizing some approach(es) to the associated class(es) of ME. To some extent the MAP #42: (RE-)CONSTRUCTABILITY / RETRIEVABILITY, embodies 'the synthesis view of Mathematics'. Within Σ , however, **all ME are synthetically approachable** – over suitably specified structural frameworks.

Since Σ is concerned, substantially, with **global synthesis and approach**, the approach-paradigms (A-P) are treated multifariously in these Essays – in E_1–E_9, with diverse examples of 'standard approximation' (deterministic / stochastic); in $E_{10/I}$ (for various mathematical / scientific domains) and in $E_{10/II}$ (for individual processes). Primary notions of analogy are discussed in E_6 – mainly, through 'deformations of LC'. Here, the systematic development of analogy (in terms of 'perturbed synthesis-representations') is based on elaborate schemes originally designed for problems in Solid Mechanics – but later, recognized as 'universal'. Accordingly, the longterm exploration of S-P – exemplified by the forms already indicated – is of fundamental importance in the construction of the OIB.

> 'Sir – The manifold logical circularity inherent in the use of syntheses for the determination of (arbitrary) ME / analogues does not detract significantly from the essential soundness of Σ as a research environment': **Discuss**!

The concepts of **identification, realization, synthesis** (and variants of these processes) were introduced – along with **stabilization, controllability, observability** – to construct, monitor, adapt, utilize classes of linear and nonlinear, deterministic and nondeterministic, automonous and nonautonomous systems, of diverse sorts. The inverse processes (to synthesis) of **analysis / prediction** (for quasi-stochastic systems) furnish paradigms for direct calculation. Within Σ , the term 'calculation' covers combinations of **axiomatic deduction, algorithmic construction, generalized (non)deterministic approximation** (of all w.d. types) and **(inverse) abstraction / analogy** (over arbitrary theories). In this framework, **every ME is associated (interchangeably) with some (class of) OT; or, of machines / systems**. These extremely (perhaps, 'ultimately') general interpretations are designed **to maximize the purview of nontrivial calculation**. Accordingly, it is important to develop adequately broad formulations of the underlying concepts – say, $\{\tilde{C}\}_Q$.

It transpires that all of these basic system-concepts do have natural extensions from linear / deterministic models to broad classes of nonlinear / nondeterministic models[596]. The further

[595]These CD are highly abbreviated; only essential steps are mentioned.

[596]See, e.g., refs: 2.3, 3.2, 3.9; Stern, TE, IEEE Trans. Circ. Th. CT-13, 74–81, 1966

extensions, to arbitrary ME, e, may be explored by **associating a system** (say, $S_e(\xi, \xi)$, etc.) **with each machine** $\mathbb{M}_e(\xi, \xi)$ – and then studying the behaviour of S_e from a control perspective. This seems to be viable, since the production of e, as (partial) output of a process over $\mathbb{M}_e(\xi, \xi)$, constitutes a quasi-control problem[597] – for which possible variants of fundamental 'regulation criteria' may be investigated. This scheme – of modelling e by a process, P_e, over a system, S_e, characterizing / governing all of the mathematical properties of e – **exhibits 'conceptual perturbations of e' as (partial) output of w.d. modifications of S_e.**

For Σ , a natural starting point is furnished by mathematical frameworks for General Systems Theory – for instance, as formulated in the book: '*General Systems Theory: Mathematical Foundations*', by MD Masarovic and Y Takahara (AP, 1975), op. cit.[598]. In the following discussions, this formalism is assumed, along with all of the fundamental results obtained. Since projects of the immense size / complexity of Σ inevitably involve diverse references to external sources (e.g., books / papers), the specification of 'key references', for the development of fundamental arguments, is essential. The formal logical circularity thereby introduced has been noted many times in these Essays – with the contention that it is 'harmless in mathematical practice', provided only that all of the cited sources are themselves 'mathematically sound'.

In this formulation, a **system** is defined by $S \subset \underset{i \in I}{X} V_i$, where $\bar{V} := \{V_i : i \in I\}$ is any **prescribed family of sets,** V_i (the **components / system-objects** of S). In particular, if $|I| = 2, V_1 \equiv X, V_2 \equiv Y$, then $S \subset X \times Y$ is a generic system with **input** X and **output** Y. If the elements $v_i \in V_i$ are sets, S is a function, and I is linearly ordered, then an **abstract time-system** is obtained. Alternatively, if each V_i has its own '**global structure**', then an **algebraic system** is defined[599]. Plainly, time-systems include **dynamical systems**. The general form may be reduced to 'the case $|I| = 2$' by taking $X := \underset{i \in I_x}{X} V_i, Y := \underset{j \in I_y}{X} V_j$, for any **partition,** $\{I_x, I_y\}$, of I, with 'input' X and 'output' Y – an **input-output** (I/O) **system**. This is taken as the generic / canonical form. S is of **function type** IFF S (as a relation) is a function (i.e., single-valued). **The overall aim** is to show that **the General Systems Theory formalism is adequate for the representation of arbitrary ME by associated systems (machines), whose w.d. perturbations yield all possible (inverse) analogues.**

This entails, in particular, that all of the standard types of systems considered in Control Theory, Mathematical System Theory, Network Theory, etc., may be obtained as 'special cases of the abstract forms'[600]. Although this aim is largely achievable for **deterministic** models[601], **extra structures** must be introduced to cover **stochastic** models[602]. The resulting theories – even if they do not satisfy all criteria of axiomatic-rigour / comprehensiveness / elegance /... – are highly pertinent to the formalized heuristics schemes characterizing Σ as an OIB. Possibilities for introducing basic concepts from Set Theory, Algebraic Logic, Recursive Function Theory, etc., into the foundations of GST, have been noted[603], and, no doubt, many other modifications have been made to the theory in M/T (GST). This evolution of GST will, of course, be incorporated within Σ_t, but (as for all of the diverse theories discussed in these Essays) the original version suffices here, to illustrate the structures underlying the ME and their (inverse) analogues.

The machines / systems associated with 'nontrivially **stochastic** ME' must be treated by techniques of **Stochastic Control Theory** – a highly developed field based (mainly) on the (qualitative) analysis of **stochastic (in)equations** (difference / differential / integral / functional, and all w.d. combinations of these types)[604]. The crucial tasks involve the definition

[597] See, e.g., Mesarović, MD / Takahara, Y, '*General Systems Theory: Mathematical Foundations*' (AP, 1975)

[598] This is abbreviated from now on as M/T (GST). The earlier book, '*Theory of Hierarchical Multilevel Systems*', by MD Mesarovic / D Macko / Y Takahara is denoted as M/M/T (HMS) (AP, 1970). Mostly, the designations (GST), (HMS), alone, will suffice.

[599] The motivation for this terminology will become clear later.

[600] These 'cases' include types #1–#16, already listed.

[601] See, e.g., M/T (GST) and M/M/T(MHS), for basic examples.

[602] Some possibilities are considered later.

[603] See, e.g., the review by AA Mulhin in BAMS 81 (1975) 1042–4.

[604] See, e.g., Aström, KJ, '*Intro. to Stochastic Control Theory*' (AP, 1970); Johnson, A, '*Process Dynamics...*' (Pere-

of 'comparable systems' for **arbitrary ME** – and the development of cognate criteria for (quasi-)controllability / observability / stability, etc..

For linear systems, or nonlinear systems, of deterministic type, all of the necessary apparatus may be found in (e.g.) the books of Ogata, op. cit., and Lee / Markus, op. cit., Gabasov / Kirillova, op. cit., with related formalism for classes of network problems (see, e.g., the references listed for Topic 1, above).

NOTE. The interrelations among general Control problems, and network models (possibly, with time-dependent / nonlinear / stochastic components), are discussed in (e.g.) the RAAG Memoirs (ref 5.3, above), in various MRI Symposia Volumes (distrib. by Wiley) and in the journals: Mathematical Systems Theory (Springer), and IEEE Trans. on Circuits and Systems.

The formal connections between **Circuit Theory** and **Control Theory** are furnished by **the state-space representation of (electrical) networks**[605]. The associated (differential) equations may be established directly for **linear network systems, with constant components**[606]. Extensions to cover **time-dependent linear systems**[607], and **time-(in)dependent nonlinear systems**[608] – all, of various types – may be found in the literature. **NOTE that**, although the underlying DE involve (typically, unobservable) **state variables**, effective control criteria **are** derivable within this framework. Systematic procedures for defining state variables – with associated '**normal-form DE**' – have been developed for general classes of (non)linear / time-(in)dependent networks[609].

For **stochastic control systems**, the situation is less clear, since there are several possible calculational frameworks[610]. Moreover, the links between **stochastic networks** and **control systems** require careful consideration[611]. The most direct approach seems to be in terms of **stochastic state models** of (dynamical) systems[612] – since **network equations** may be formulated over **state space**, for deterministic systems[613]. On this basis, definitions of stability / controllability / observability /. . . , may be formulated for 'stochastic circuits / networks', defined to cover such systems with some stochastic component(s) / input(s). The standard classes of components comprise[614] **resistors, inductors,** and **condensers** (in all w.d. combinations) (respectively), though various types of transistors, or filters, etc., may also be regarded as components – in certain contexts[615].

The **driving forces** for these circuits / networks may be **direct / alternating**[616] **voltages / currents**[617]. Several **special circuit components**[618] (e.g., **varactors**) may also be incorporated in circuits, but the underlying principles / techniques are unaffected. Examples of control analyses for classes of linear deterministic circuits, in state-space representations, indicate that the **state descriptions** must satisfy various **controllability / observability conditions**[619]

grinns / IEE, 1985), and Maybeck, ref. 2.3.

[605] For linear systems, see, e.g., Ogata, op. cit; Zadeh / Desoc, op. cit; and, for nonlinear systems, Isidon, A, '*Nonlinear Control Systems. . .*' (Springer, 1985). The **network forms** are given in (e.g.) Stern, TE, IEEE Trans. on Circuit Theory, CT-13, 74–81, 1966 – for general nonlinear networks; and, for linear networks, in ibid. CT–9, 303–6 (PR Bryant).

[606] See, e.g., Schwarz, RJ / Friedland, B, '*Linear Systems*' (McGr-H, 1965), esp., Ch 2.

[607] See, e.g., Schwarz / Friedland, op. cit.

[608] See, e.g., Stern, op. cit., and Lee / Markus, '*Foundations of Optimal Control Theory*' (Wiley, 1967), op. cit..

[609] The basic scheme is given in Stern, op. cit.; see also Ogeta, op cit. (for linear models).

[610] For instance, the underlying dynamical systems may be specified in I/O form; or else, in **state-variable form**. The stochastic processes may be represented **via covariance functions**, or, **spectral densities**; or, **state models** – and so on!

[611] Not only the stochastic input / components, but also the **network topology** must be taken into account. . . .

[612] See, e.g., AströM, KJ, '*Intro. to Stochastic Control Theory*' (AP, 1970), Ch. 3.

[613] See, e.g., Stern, op. cit., for a general formulation.

[614] Condensers are also called **capacitors** (for storing **electrical energy**). The inductors store **magnetic energy**.

[615] Systems with 'random interconnections' may also be treated.

[616] The basic distinctions are discussed in (e.g.) '*Techniques of Circuit Analysis*', by GW Carter / A Richardson (CUP, 1972).

[617] Again, see Carter / Richardson, op. cit.. The standard terminology requires a **circuit** to contain at least one **loop** – whereas, an **arbitrary network** need not do so.

[618] Most of these have nonlinear characteristics. See, e.g., Löcherer, KH / Brandt, CD, '*Parametric Electronics. . .*' (Springer, 1982); Decroly, JG, et al., '*Parametric Amplifiers*' (Macmillan, 1973)

[619] See, e.g., Gökner, IC, in Boite / Dewilde (ref. 1.17 in Key Refs, above)

– with cognate criteria for **non**linear **networks**[620]. Possible extensions of these criteria to cover **stochastic** networks may be envisaged in terms of appropriate 'mean values' over the stochastic elements (including driving forces) in the networks.

NOTE. this entire discussion is intended to show that **'control paradigms' may be generalized to cover essentially arbitrary networks, as well as more familiar types of 'dynamical systems'**.

For **linear** systems, diverse conditions for controllability have been derived – including **estimates of the minimal distance between controllable systems**. These results concern, mainly, time-invariant, deterministic systems – but the validity of such results makes it plausible (at least) that comparable results may be attainable for time-varying / nonlinear / nondeterministic systems. (Initially, such extensions, might be obtained '**locally**', via various forms of **perturbation**. The '**global versions**' are likely to depend on **operator-theoretic techniques** and **stochastic calculus**.)

An extensive collection of papers in this area (many with associated numerical implementations) is given in: '*Numerical Linear Algebra Techniques for Systems and Control*' (RV Patel, et al., eds.) (IEEE Press, 1994).

There is, of course, no **unique** framework for the production / synthesis of ME – and the subsequent exploration of their admissible (inverse) analogues. Nevertheless, the (perturbed) machine-representation, with the associated 'control criteria' (of all deterministic / stochastic types) seems to provide near-maximal scope, subject to versatile monitoring – from ultra-delicate to strong, and from local to global in range. **In short**, 'the control paradigm' (very broadly interpreted) is potentially adequate for **all** Σ -transactions. In this context, interpretations over GST appear to be sufficient, for approaches to arbitrary ME – though the detailed correspondences will be tenuous, in many instances. Moreover, such representations may be modified, at any time (in the light of improved techniques, etc.) without affecting the overall structure of the OIB.

[620] See, e.g., refs in Gökner, op. cit.

APPENDIX 5
Analogue Exploration

$E_{11/I}$ and $E_{11/II}$ are concerned with a general formalism for the representation, analysis and synthesis of analogues of arbitrary ME/SE, β. This elaborate scheme has evolved from the 'template forms' designed to facilitate the systematic **tagging** of research procedures in Solid Mechanics – but later found to be **universally appropriate** within Pure / Applied / Applicable Mathematics.

Here (in $E_{11/III}$) the scheme is further extended to cover **approach criteria** – so that the Σ -Lists may be '**approach-tagged**', and the templates may be rendered **operational**, as **I/O machines** of the types considered in $E_{11/I}$. Another class of representations involves (generalized) **computer programs** (over processors with any w.d. architecture(s)) embodying detailed **implementations** of the enhanced SAM / graphics /... algorithms involved. The study of '**perturbations**', of any of the (standard) representations of β, allows the development of criteria for the '**generation**' of **direct** (algorithmic) or (quasi-)**inverse** (**non**algorithmic) **analogues** of β. On this basis, a broadly representative set of examples may be treated – as an indication of future 'canonical Σ -practice'.

The framework discussed in $E_{10/II}$, for overall tagging of the **items** on the Σ -Lists, is complemented by the templates / machines, for all of the Σ -List items **and all ME / SE** – as introduced in $E_{11/I}$. Since these diverse items involve intricate mathematical / scientific processes (as well as collections of ME), their effective tagging must reflect this intricacy – as the descriptors, $(\Delta^n)_{11}$ indicate. Indeed, the varied selection of CD in $E_{10/II}$ has (even in its present limited form) an extremely broad compass. Ultimately, the aim is to make this compass 'effectively dense in M_t, for large t'! Consequently, it is important to devise a highly condensed **partial-tagging** format for the **essential characteristics** of each item – for instance, as follows[621].

Denote by $\delta\langle e\rangle$ **the (exact) determination(s) / representation(s)** of the (possibly, multi-valued) ME e; by $@\langle e\rangle$, **abstractions of** e; and, as before, by $\mathcal{A}\langle e\rangle, A\langle e\rangle$, respectively, the **(non)deterministic standard-approximation(s)**, and **(other) approach(es)**, to e – all over Σ_t – with[622] $\mathbb{A}\langle e\rangle := (\delta \vee \mathcal{A} \vee A \vee -A \vee @)\langle e\rangle$. Then one may also use the notation: $\mathcal{A}\langle e_2|e_1\rangle; A\langle e_2|e_1\rangle; -A\langle e|s\rangle; @\langle e|b\rangle$, for the (respective) statements: 'e_2 **is an approximation to** e_1'; 'e_2 **is a direct analogue of** e_1'; 'e **is an inverse analogue of (these)** s'; 'e **is an abstraction of (the ME / SE)** b'. The interpretations of 'δ' and '$\delta\langle e\rangle$' require additional explanation. In the simplest cases they mean: **exact calculation(s)** $\langle$**of** $e\rangle$; but, in many S/M-items, they mean, further: **extension(s) of domain(s) of exact calculability** $\langle$**of extensions of** $e\rangle$. Obvious examples are: **integration over classes of groups[623]; distance-functions on probability spaces[624]**; and, **generalized inverses of (linear) operators[625].**

Slightly less obvious examples are: **the Hahn-Banach extension theorem** (in several forms)[626]; **the Central Limit Theorem over BS[627]**; **multivariate Chebyshev Inequalities[628]**; and, **differentiability / analyticity over T. Groups[629]**. Moreover, all **constructive procedures** for (extensions of) e are also covered by $\delta\langle e\rangle$, while the approximative / (inverse-)analogical representations of (extensions of) e are classified through $\mathcal{A}\langle e\rangle/A\langle e\rangle$. Finally, general approaches to (extensions of) e involve $\mathbb{A}\langle e\rangle$. This notation has, of course, been implicit (at least) in all of the Essays. It is reiterated (and expanded) here because it provides an adequate basis for

---

[621] Implications of this classification for the treatment of **arbitrary ME / SE**, β, are considered later.
[622] **NOTE that** δ is formally covered by $\mathcal{A}$, in some contexts, but in general these classes are treated as disjoint.
[623] See, e.g., Hewitt, E / Ross, KA, '*Abstract Harmonic Analysis, Vol. 1*' (Springer, 1963)
[624] E.g., Lukacs, E, '*Stochastic Convergence*' (Heath, 1959)
[625] See, e.g., Ben-Israel, A / Greville, TNE, '*Generalized Inverses...*' (Wiley, 1974)
[626] See, e.g., Edwards, RE, '*Functional Analysis...*' (H, R&W, 1965) for several forms.
[627] See, e.g., Kuelbs, J (ed.) '*Probability on BS*' (Dekker, 1978)
[628] See, e.g., Marshall, AW / Olkin, I, Ann. Math Stat 31(1960) 1001–1014
[629] See, e.g., PCPS 61 (1965) 347–379 (GA Reid)

the **abbreviated tagging** of the S/M, and associated, lists. Such tagging should be founded on the **Basic Spaces**, ${}_SX$(see E$_1$ / E$_5$), the **MAP Concepts**, c_j (see E$_9$), and **all cognate (w.d.) types of approach** – together with indications of the **key objects** from the (permanently ordered / augmented) lists U_t, S_t, involved in 'item n'[630].

Recall (from E$_1$, and especially E$_3$) that Σ_t is partially characterized by a – progressively constructed – list, U_t, of **underlying ME** (say, u_k), added as necessary to specify 'new' ME / concepts / procedures / theorems /... introduced as the OIB evolves. Plainly, this is a 'practical process', in the sense that **descriptions of** (operational) ME determined over (say) $U_{t_1} \equiv (u)_{K_1}$ may become **elements of** $U_{t_2} \equiv (u)_{K_2}$, for some $K_2 > K_1$. All that is required, in this scheme, is the adequacy of U_{t_2} for the formulation of ME in M_{t_1} – with t_2 ($> t_1$) suitably chosen. In general, the elements u_k are '**of noun type**' (even though they are formally regarded, in Σ , as OT!); whereas, the genuinely operational ME ('**of verb type**') include **implementations** of some of the u_m (e.g., for[631] u_m :=**linear integral operator**). Although this hierarchical structure over the set of all ME in Σ is heuristically specified here, a more rigorous (but less useful!) specification could, in principle, be given. Since t increases indefinitely, the fact that M_{t_1} is effectively specifiable only over all U_t with $t \geqslant t_2 > t_1$ is not problematical – at least, on the level of 'research activity'.

On this basis, every ME in Σ_{t_1} is **precisely formulatable**[632] over some underlying set U_{t_2} – and so, over all U_t with $t \geqslant t_2$. Schematically, one has: $e =' \varphi^e(u^e_{\sigma^e})_{L^e}$, where σ^e selects appropriate elements, $u^e_{\sigma^e_k}, 1 \leqslant k \leqslant L^e$, in the descriptively-precise formulation of e[633]. It follows that: **every S/M-item has some representation(s) over the underlying sets U (of ME) and $\mathcal{S}$ (of SE)**. Consequently, all levels of tagging may be specified through the symbols $\mathcal{A}\langle e|V\rangle, \mathcal{A}\langle e|W\rangle$, and, $-\mathcal{A}\langle e|Y\rangle$, denoting (respectively) deterministic and nondeterministic, direct-analogical, and, inverse-analogical approaches to e over $V \subset U, W \subset U$, and $Y \subset \mathcal{S}$ – along with $\delta\langle e|V\rangle$ and (exceptionally) $\delta\langle e|Y\rangle$[634].

A typical reduced tagging (for 'item n' on one of the Σ -lists) is thus of the form:

$$t_n \approx [n; (c_{\psi^n})_{p_n}; ({}_SX_{x^n})_{q_n}; V_n; W_n; Y_n; V'_n; Y'_n.$$

Here, the n-th listed item exemplifies the concept(s) $\{c_{\psi^n}\}_{p_n}$ over (some of) the spaces $\{{}_SX_{x^n}\}_{q_n}$ through 'procedures' involving $\mathcal{A}\langle \alpha_n|V_n\rangle, \mathcal{A}\langle \beta_n|W_n\rangle, -\mathcal{A}\langle \gamma_n|Y_n\rangle, \delta\langle \xi_n|V'_n\rangle$, and (exceptionally) $\delta\langle \eta_n|Y'_n\rangle$, with $V'_n \subset U$ and $Y'_n \subset \mathcal{S}$[635]. Although this representation is superficially complicated, the associated **interpretation** is remarkably simple. The 'selectors' ψ^n, χ^n pick out the relevant concept(s), and space(s), while the subsets $V_n, \ldots Y'_n$, indicate the (generic) ME in terms of which the approximation(s) to α_n, direct analogue(s) of β_n, inverse analogue(s) of γ_n, and absolute determination(s) of ξ_n, or η_n, may be realized[636]. By distinguishing among these 'fundamental routines' (of (non)deterministic approximation, (inverse) analogy, and construction) one may characterize both the modes, and the types, of conceptual approaches pellucidly – and with 'high resolution'. This constitutes an 'intermediate level of tagging', between the most basic, $[n; (c_{\psi^n})_{p_n}]$ and **quasi-maximal** versions reflecting the descriptors, $(\Delta^n)_{11}$ (introduced in E$_{10/II}$). In particular, the CD (=: Δ^n_4) are exemplified in E$_{10/II}$, for a (small but diverse) set of topics, and the (broad) techniques (of approximation / analogy) are covered as Δ^n_{11}: Model(s) / FORMULATION(S). It may be necessary to augment $t_n \approx [\ldots]$, above, by including 'indications of Δ^n_{11} – 'content' – to produce (say) $\tilde{t}_n := (t_n, \tilde{\Delta}^n_{11})$, where $\tilde{\Delta}^n_{11}$ denotes an extreme summary of Δ^n_{11} – or, in some cases, merely a list of the procedures used. The resulting format(s) will be called **standard taggings**.

As a guide to the 'approach content' of the items on the various Σ -lists, the following indicators were introduced.

---

[630]A further extension, to cover very short outlines of the procedure(s) used – or in some cases, just a list of these procedures – produces a '**reduced** tagging' (as explained later)

[631]In fact, **every** names ME with nontrivial operational content will, eventually, be so implemented.

[632]i.e., all ME in Σ_{t_1} may be rigorously specified.

[633]This contains purely linguistic (e.g., English) parts, and some combinations of symbolic and linguistic parts.

[634]This covers all possibilities.

[635]As usual, $(\lambda)_{p_n} := (\lambda_1, \ldots, \lambda_{p_n})$ is an ordered **list**; and $\{\mu\}_{q_n} := \{\mu_1, \ldots, \mu_{q_n}\}$, is an **un**ordered **set**.

[636]Later in E$_{11/III}$, β denotes some ME/SE, and **quasi-inverse** analogues of ME are introduced.

$\delta :=$ determination / deduction / calculation / exposition / representation.

$\mathcal{A} :=$ standard approximation (of any (non)-deterministic type)

$@ :=$ abstraction / generalization

$(-)A :=$ (inverse) analogy (of SE) / quasi-inverse analogy (of ME): **see later remarks**.

$\mathbb{A} :=$ approach (covering combinations of $\delta, \mathcal{A}, @, \pm A$)

NOTE. Recall that $-A(s)$ refers to inverse analogues of the SE s, while $+A(e) \equiv A(e)$ denotes direct analogues of the ME e. Mostly, the context distinguishes between A and $-A$, but, in a more complete scheme, both the sign(s) and the arguments will be specified.

For all items, the approach **types** are written (if not omitted) in the fixed order $\delta; \mathcal{A}; @; A; -A; \mathbb{A}$ – simply to indicate the dominant types exemplified. Thereafter, detailed realizations may be investigated, within the frameworks developed in $\mathsf{E}_{10/I}$, $\mathsf{E}_{11/I}$ and $\mathsf{E}_{11/II}$. If this 'juggernaut' seems over-elaborate, it should be remembered that arbitrarily intricate / sophisticated ME/SE may be involved.

Where two or more indicators label the same item, the 'standard order' may be varied. For instance, a constructive procedure with some approximate implementation(s) would be labelled as $\delta/\mathcal{A}$; whereas, an approximation scheme which (sometimes) allows constructive determinsation(s) would correspond to $\mathcal{A}/\delta$ – and so on, for other pairs of indicators.

In this way, each item on a Σ -list is labelled by its (main) approach form(s) – say, as ${}_k f_n$ for item n on list k, with k suppressed unless there is possible ambiguity. The most basic tagging representation then becomes: $[n; (c_{\psi^n})_{p_n}, f_n]$, with cognate representations for all higher-level taggings. For initial scanning, each of the items ${}_k i_n$ on the Σ -lists contains ${}_k f_n$ within the standard citation. The assignment of ${}_k f_n$ to item ${}_k i_n$ requires fundamental understanding of the process(es) underlying ${}_k i_n$. For the lists S/M, G, and CM (each, so far, containing between 2000 and 3000 items) this task alone is daunting – but crucial. Although it is **non**algorithmic 'essential consistency' **is** attainable. Such compromises typify Σ -developments!

Here, f_n corresponds to Δ^n_q in $(\Delta^n)_{11}$. At this stage it is convenient (and more elegant!) **to reorder the original entries Δ^n_i so that $(\Delta^n)_q$ is the level-q tagging of i_n.** Although there is, inescapably, an element of arbitrariness in this process, it is important for scanning / taxonomic purposes. Moreover, **all** w.d. tagging assignments may be covered by the general notation[637]

$$\{\Delta^n\}_S := \{\Delta^n_j : j \in S \subset \#\bar{I}\}$$

where the set I has $\#I$ elements ($\#I = 11$, in the current formulation), and, if S is ordered, then so is $\{\Delta^n\}_S$. This ordering, like that within f_n, may be used to designate the relative importance ascribed to various characteristics of i_n – and also to extend (globally over all lists of examples) the underlying sets of criteria Δ^n, should this become necessary. **This tagging format may be combined with the template / machine representation.**

The **operational modes** of $\mathbb{M}, \mathbb{M}_\lambda$ are based on the **solution** of **quasi-LC/D** for e, e_λ – via (generic) modifications of the elements m_k. Here (as in $\mathsf{E}_{11/I}$) $M_1 \equiv \sigma := \tilde{\Gamma}({}_i\tilde{\tau}_j)_P$ – combinations of **scientific** theories; $m_2 \equiv s := \Gamma({}_k\tau_l)_Q$ – combinations of **mathematical** theories; $m_3 := \zeta; m_4 := d; m_5 := e; m_6 := f; m_7 := \mathcal{R}; m_8 := g; m_9 := h; m_{10} := i; m_{11} := M, m_{12} := j; m_{13} := k; m_{14} := l$.

The admissible forms of classes of **modifications** of the m_k are discussed in $\mathsf{E}_{11/I}$, $\mathsf{E}_{11/II}$ (and also in E_6, E_7). The 'generalized machines', $\mathbb{M}, \mathbb{M}_\lambda$ –for the fabrication of ME e, e_λ – may usefully be modelled by **'abstract engineering systems'**. Very general views of such systems (with 'practical analogues' in mind) are formulated in the books: '*Analysis and Design of Engineering Systems*' (Class Notes, MIT, 1961) by HM Paynter; and, '*Dynamical System Models*' (Harrap, 1970) by AGJ MacFarlane: **Electric-Circuit models** of diverse scientific phenomena were introduced long ago by G Kron – e.g.: for **Maxwell's equations** (Proc. IRE 32 (1944) 289–299); **Schrödinger's equation** (Phys. Rev. 67 (1945) 39–43); **(non)linear elasticity** (J.

[637] Recall that $\forall m \in \mathbb{P}, \bar{m} := (1, \ldots, m)$; see E_{-1}.

Appl. Mech 11 (1944) 149–161); and for **(in)compressible flow** (J. Aeronaut. Sci. 12 (1945) 221–231). See also: '*Electromechanical System Theory*', HE Koenig / Blackwell (McGr-H, 1961).

Other species of models of high pertinence for synthesis / fabrication procedures in Σ include diverse processes of **chemical** / **biological** types – for instance, in **chemical bonding, chemical reaction analysis, isomerism, chemical plant operation, molecular biology, cellular biology, chemical morphogenesis, population dynamics**. A variety of phenomena in **chemical physics** may be found in the S/M-List; and so on! Again, all branches of **Engineering** are replete with analysis / synthesis schemes applicable to the generation of (inverse) analogues, through associated machines $\mathbb{M}, \mathbb{M}_\lambda$. A further (almost unlimited) source of 'interesting models' is furnished by **(Theoretical) Physics**, in its may subdivisions. These observations are, of course, obvious – but still worth making, to emphasize the rich analogical **milieux** barely even recognized for their immense **mathematical** potential. The gradual development of the Σ -OIB will forge fundamental **symbiotic relationships** between **Theoretical Science** and **Mathematics**. Indeed, the entire template / machine / LC-deformation /... apparatus / formalism has been designed for the effective exploration of such **analogical interfaces**.

Some basic references for the chemical / biological examples may be found in (e.g.): Segel, LA (ed.) '*Mathematical Models in Molecular and Cellular Biology*' (CUP, 1980); L Edelstein-Keshet, '*Mathematical Models in Biology*' (Birkhauser / McGr-H, 1988); King, RB (ed.) '*Chemical Applications of Topology and Graph Theory*' (Elsevier, 1983) – especially, for **synthesis design** (pp 75–98: PG Mezey) and **chemical manifolds** (pp 379–391: L Peusner).

For **information flow**, in networks, see, e.g., papers in the RAAG Memoirs (K Kondo, ed.) op. cit.. The analysis of **chemical reaction potential-energy surfaces** (using reaction-coordinates / quasi-Hamiltonian-techniques) is pursued in great detail in papers by RA Marcus (et al.), mostly in J. Chem. Phys., from about 1960 onwards. Applications of **Graph Theory** to **chemical reaction networks** are given in J. Chem. Phys. 60 (1974) 1481–1492; 1493–1501 (BL Clarke). A **topological characterization of chemical networks** may be found in SIAM J. Appl. Math. 15 (1967) 13–68 (P Sellers). **NOTE that** comparable procedures for the synthesis of (minimal) **switching systems** are given in TAMS 88 (1958) 301–326 (JP Roth).

The exploration of (inverse) analogues, for 'arbitrarily prescribed ME, e (or SE, s)', therefore entails the application of all admissible **perturbations** to the components of $\mathbb{M}(\xi)$ (or[638] $\mathbb{M}'(\xi')$), together with associated **interpretations** of the '**output**' of $\mathbb{M}_\lambda(\eta)$ and $\mathbb{M}'_\lambda(\eta')$. The procedures for $\mathbb{M}_\lambda(\eta)$ are essentially **algorithmic** – since, even the various possible interpretations may be fully correlated with relevant LC. For $\mathbb{M}'_\lambda(\eta')$, by contrast, the analogues $-\alpha(s)$ are merely '**suggested**' by (aspects of) $\mathbb{M}'_\lambda(\eta') \equiv \mathbb{M}'_\lambda(\eta', \xi'_\lambda)$, and so must be individually '**validated**' in each case, by specification of the associated LC / environment(s). This process may, in turn, yield purely **scientific** analogues – say, $\alpha'(s)$; either, directly defined, or else (partially) suggested by $-\alpha(s)$. Moreover, such 'cycles' will, in similar forms, be **iterated**, as Σ evolves!

The full power of approach furnishes a formalism for **generalized continuation / convergence criteria**, over substantially arbitrary domains of (mutually compatible) ME/SE – rendering the **Mathematical Universe** both **globally navigable** and **locally investigable**, at any depth / level.

At this stage it is important to emphasize (yet again) that the **primary aim** of these Essays is to **unify** the multifarious aspects of 'mathematical activity', by **locating** them within a **universal approach environment** – potentially encompassing all facets of Pure / Applies / Applicable Mathematics[639], with its immense complexity and boundless possibilities for growth. It is not surprising, therefore, that a proper understanding of the vast Σ -superstructure (as adumbrated in terms of various novel substructures) is attainable only through considerable effort. Again, only a maximally versatile (computer-based) IB with highly sophisticated SAM facilities (corresponding to approach-formulations) can remotely aspire to achieving such comprehensive

[638] $\mathbb{M}'(\xi', \xi')$ corresponds to SE, s, for inverse analogues, $-\alpha(s)$, and $\mathbb{M}'_\lambda(\eta')$, to perturbations of $\mathbb{M}'(\xi')$.

[639] The Σ -view (that the conventional boundaries within / among these subdomains of Mathematics are illusory) has been propounded (especially for nonlinear Functional Analysis) by E Ziedler in expository books (op. cit.). The Σ -scheme substantially realizes this aim.

unification. Imperfections will, inevitably, remain, but the increase in **connectedness** within the **conceptual realm** will be very substantial – and **routinely susceptible of improvement**. As a collection of **research milieux**, Σ appears to provide optimal (and unprecedented) facilities for innovative calculation.

The relation of Σ **-operators** to **theoretical CS** stems from the representation of every ME, e (or SE, s) as a (collection of) **subroutine(s)** for the construction / evaluation / deduction /... of e (or s). The associated **LC-forms** are of an **existential** nature, since they do not (in general) contain **mechanisms** to **produce** the entities concerned. The subroutines correspond to **machines** built over **templates** – as outlined in $E_{11/I}$. These $\mathbb{M}$**-type realizations** of ME/SE would, of course, be utterly intractable for 'hand calculation', but they are ideally fitted for the Σ **-OIB**, since they embody all mathematical properties of e (or s) effectively and succinctly – culminating in the OTC/OAC as a model of **General Mathematical Activity**. The crucial characteristic distinguishing Σ from any theorem-proving package (however sophisticated) is the **interactive mode**, which allows users to monitor and guide the OTC procedures, in **heuristic exploration** and **proof-generation**, using conventional notation.

Perhaps the most distinctive quality of Σ as a calculational system is that it may be viewed as a quasi-**medium** of (mutually) **interacting** atomic / molecular /... **ME**, capable of diverse modes of **combination** – of various degrees of stability / complexity / sophistication / power /.... In this model of **G**eneral **M**athematical **A**ctivity, no ME can exist in isolation; rather, it is defined **in terms of** other ME (or else, **axiomatically**, in relation to basic properties), and consequently, **every ME (in)directly influences all other ME**. The possible **modes** of interaction / combination **cover all aspects of mathematics**; many examples are discussed in these Essays. The essential point is that **the** Σ **-superstructure is universal**, admitting all forms of association / combination within a general, unifying representation. The fundamental challenge – **substantially met** in the Σ -scheme – is to **realize** such unification **nontrivially**.

In $E_{11/II}$ it is emphasized that the most direct route to a formalism for the analysis / synthesis / exploration of (inverse) analogues for arbitrary ME/SE may be based on procedures of **General Systems Theory** (GST). The engineering systems treated in the books by Paynter, op. cit. and by MacFarlane, op. cit. furnish examples of general systems for which correspondences with GST are readily established. The basic argument is very simple:

(a) Every ME/SE is 'obtainable' as (partial) output of some GS;

(b) broad classes of GS are **realizable** (hence, synthesizable); so

(c) 'most' species of synthesis may be subsumed by some quasi-universal GST-realization procedure(s); hence,

(d) analogues of ME/SE may be investigated within the GST scheme – typically, for I/O models over state spaces, for all w.d. classes of 'perturbations'.

In this framework, a fundamental role is played by the links between **models** and **machines**, for ME/SE.

Several questions arise from the outline prescription in (a)–(d), namely:

(a)′ **how** are the machines, $\mathbb{M}_e$, etc., related to GS?

(b)′ does realizability always guarantee synthesizability?

(c)′ **how** are species of synthesis subsumed by a GST process? and,

(d)′ **how** are the machines representable within GST?

A (general) machine may be regarded as an I/O system. For instance, the input may be prescribed I/BC, and the output, some (class of) function(s). This simple example is, nevertheless, somewhat typical, since 'arbitrary ME' may be 'produced' as solutions of (sets of) operator (in)equations, over diverse (axiomatic) domains – for which the input consists of I/B data, various constraints, domain-specification, etc.. The claim, for Σ , is that: **given I/O assignments**

for any ME/SE, there is a procedure for realizing a class of corresponding machines (in 'template representations'), with appropriate operational components.

For **inverse** analogues of **SE**, the perturbations of machine components must be complemented (at some stage) by 'suggestive transformations' – to motivate the definition of novel mathematical structures, by fully rigorous procedures. Of course, such transformations – say, v : {SE} → {novel ME} – cannot be specified generally, but significant qualitative results may be established, and gradually developed within a unified setting. This constitutes an important longterm aim for Σ .

In GST there are realization schemes for generalized time-(in)variant dynamical systems (of which **static** systems are special cases). The relevant procedures are given in the monograph: '*GST: Mathematical Foundations*' by MD Mesarovic / Y Takahara (AP, 1975), op. cit.. Since any machine is essentially an I/O system, the main issue here devolves on the machine-representability of arbitrary ME/SE. This, in turn, however, seems evident – at least in the sense that every ME/SE is specified / determined by sets of operational conditions over suitable mathematical / scientific structures. The template / machine formulations simply reflect this situation quasi-systematically. For the present discussion it is assumed that realization procedures **are** w.d., and yield syntheses, for all ME/SE, in principle. On this basis, **direct** analogues may be identified by applying perturbation / modification (P/M) operations to the components of ME/SE-machines (derived from models of the relevant structures / phenomena) as considered in $E_{11/I}$, $E_{11/II}$, while preserving (weakly) the essential characteristics of the ME/SE originally synthesized. On the other hand, the development of **inverse** analogues requires (felicitous) associations of novel mathematical properties with scientific attributes of the underlying SE – often violating fundamental scientific constraints – in a brand of **(quasi-)formalized heuristics**.

The concept of **model** applies equally to ME and SE, so one effective scheme for studying analogues of ME/SE may be characterized by **progressive perturbation / modification** (P/M) of 'initial **models**' – of which the corresponding machines are, essentially, constructive **realizations**: say, $\mathbb{M}_{e/s} \approx \mathbb{R}_C[m_{e/s}]$. The realization processes, $\mathbb{R}_C$, in particular examples, naturally depend on the detailed nature of e, s, and their 'host structures / theories'. Some of these processes are indicated (with key references) in $E_{11/II}$, for diverse scientific contexts, and for a few purely mathematical cases. For direct analogues, this 'P/M-process' is basically the same for **ME** and **SE**, since **some** crucial behavioural characteristics are quasi-conserved, at each stage (to justify the term 'analogue', at all). The **inverse**-analogue procedures eventually involve **deliberate divergence** from **SE**-behaviour – and, even from 't-standard **ME**-behaviour', when existing mathematical structures are '**almost reproduced**' in some unfamiliar context(s). Many examples from the S/M-List illustrate this type of 'analogical evolution'.

From these remarks one may conclude that, although '**strict** inverse analogues' must derive (ultimately) from **SE**, species of '**quasi**-inverse analogues' – produced, for instance, by extending the purview of 't-standard' formulae / representations / structures /... suggestively, and then supplying the necessary rigour – may originate in various **ME**.

NOTE that, since **some** aspect(s) of (almost) arbitrary SE may be (partially) modelled, the 'P/M-process' typically comprises several 'stages', each of direct type – followed by 'imaginative substitutions / transformations', for (quasi-)inverse analogues. At higher levels of 'detachment from the initial data' (i.e., from e/s, for (quasi-)inverse analogues) one encounters abstract / analogical **interpretations** of SP associated with s – for instance: viscosity; torsion; isomerism; phase transition; subsidence! The mathematical-structural versions of these SP offer ample scope to the poetic imagination – tempered by logic, consistency, and (ultimately!) rigour. In all aspects of mathematical analogy, issues of (non)conservation of **centrality** (under inter-structure / theory transformations) are of fundamental importance.

For some SE/SP, virtually all aspects defy nontrivial, rigorous modelling – typically in the Biological, Social and Economic Sciences, but also in 'exotic areas' of Engineering and the Physical Sciences; examples are given later. In such extreme cases, essentially **intuitive responses** to the perceived nature of the SE/SP could well 'generate' fascinating results – though schools of red herring may be found in these waters! The resulting innovations must, of course, satisfy al of the local / global consistency conditions arising from 'interactions' with existing ME/SE – as far as these conditions are ascertainable.

In general, combinations of t-maximal modelling (augmented by 'nonscientific mathematical behaviour') and (purely) 'intuitive response' may be deployed, to explore the full spectrum of possibilities. The Σ **-superstructure** is designed, partly, to facilitate these exploratory routines – with **feedback** of the resulting (quasi-)inverse-analogue ME completing 'the creative cycle', in which 'new' (quasi-)inverse analogues are 'produced' from existing ME/SE.

On general grounds it is to be expected that (quasi-)inverse analogues obtained from 'marginal deformations of current mathematical / scientific conventions / principles' will tend to be less stimulating than analogues motivated primarily by intuitive responses. Indeed, it is plausible (at least) that such 'marginal analogues', say, $-\alpha(e), -\alpha'(s)$, are – in various senses – weakly equivalent to e, s; whereas, the 'conceptual separation' of the corresponding 'intuitive analogues' (from e, s) is of a qualitatively different order – perhaps, reminiscent of 'orders of infinity' in Analysis[640], or, of 'measures of complexity' in the Theory of Algorithms[641]! At all events, it is clear that the modes of imitation inherent in direct / (quasi-)inverse analogues may range from faithful (transcription / isomorphism /. . .) to 'arbitrarily faint resemblance' – where the poetry really begins!

The models, m_e, m_s and machines, $\mathbb{M}_e, \mathbb{M}'_s$ (for ME, SE respectively) have essentially similar basic forms. The crucial difference is that all definitions / specifications / procedures /. . . in $\mathbb{M}_e$ stem from formal mathematical structures; whereas, the 'components' of $\mathbb{M}'_s$ reflect only mathematical representations of conjectural / empirical / observational /. . . scientific relationships – within some theories. In spite of differences in detailed structure, both $\mathbb{M}_e$ and $\mathbb{M}'_s$ may be represented as **templates of operational components** – as discussed fully in $E_{11/I}$. At the next stage, various classes of 'canonical P/M-processes' may be applied (in all w.d. combinations) to specified collections of machine-components. This stage potentially covers all types of **direct** analogues. Full characterization of the class of 'admissible P/M-processes' is problematical; but broad specification (e.g., in terms of 'implicatory deformations', etc. – as in LC/LD: see, especially, E_6) may be attempted.

NOTE. From now on, the notation $\mathbb{M}_\beta$ will be used to cover both of $\mathbb{M}_e, \mathbb{M}'_s$ – unless distinctions must be made.

The transformations / substitutions invoked in the development of **inverse** analogues from **SE** – or, of **quasi-inverse** analogues from **ME** – are, by contrast, certainly **not** amenable to systematic classification. The most promising scheme involves the gradual search for significant **patterns** in collections of examples – a process that will, eventually, yield new inverse analogues (and subtly modified patterns). In spite of these obstacles, it is possible (and worthwhile!) to formulate **all species of analogues** within a unified framework founded on template representations of ME/SE – that is, on **machines,** $\mathbb{M}_e, \mathbb{M}'_s$. The changes in components may be monitored at all 'direct stages' (common to direct and inverse analogues). The 'intuitive transformations' inherent in fashioning **inverse** analogues may then be 'recorded' – together with their associated constructions in the novel structures developed. When these 'new ME' and **their** main properties become familiar, they may be used, in turn, to create yet more (inverse / quasi-inverse) analogues; and so on!

Recall that variants of the machines $\mathbb{M}_e(\xi, \xi)$ are denoted by $V_\lambda(\mathbb{M}_e(\xi, \xi)) =: \mathbb{M}_{e\lambda}(\eta, \xi_\lambda)$ – usually abbreviated to $\mathbb{M}(\xi, \xi)$ and $\mathbb{M}_\lambda(\eta, \xi_\lambda)$; or, even to $\mathbb{M}(\xi), \mathbb{M}_\lambda(\eta)$, when the 'initial ME', e, and the environments ξ, ξ_λ have been specified. In the general context, $\xi \equiv \{S; G, T\}$, where S denotes the mathematical / scientific **structure(s)** involved, and G/T, the **geometrical / topological** data (broadly interpreted). The basic 'entries' in the MF-C will now be considered, in turn, to assess their possible ranges of P/M-operations.

As a framework for the application of P/M-operations, the Σ -setting for (arbitrary) **M**athematical **I**nformation may be used. The relevant formulations are discussed in E_3 – in terms of: **Nomenclature / (basic) Definitions / Axioms / formal Theorems / t-Operational Theorems / Formalism(s) / Structures** – all for each **subtheory** considered. In the following scheme, the material of E_3 is used, with minimal further explanation.

---

[640] See, e.g., Hardy, GH, '*Orders of Infinity*' (CUP, Cambridge Tracts. . . Np. 12)

[641] See, e.g., Papadimitrion, CH / Steiglitz, K, '*Combinatorial Optimization: Algorithms and Complexity*' (P-H, 1982)

NOTE that the **general** ME/SE **notation** is now: $\mathbb{M}_\beta(\mathcal{J},\xi_\mathcal{J})$, with **variants** $\mathbb{M}_{\beta\lambda}(\mathcal{J},\xi_\mathcal{J})$.

A solid foundation for the (O)IB aspects of Σ may be established as follows. Consider M_t as a quasi-maximal collection of (interrelated / exhaustive) **subtheories** for Pure / Applied / Applicable Mathematics – say, as $M_t \approx \{\Theta_h : h \in \bar{H}\}$. Suppose that Θ is **characterized** by the sets $\mathcal{N}_h$ (**Nomenclature**), $\mathcal{D}_h$ (**Definitions**), $\mathcal{A}_h$ (**axioms**), ${}_F\mathcal{J}_h$ (**Formal Theorems**), ${}_O\mathcal{J}_h$ (**Operational Theorems**), $\mathcal{F}_h$ (**Foundations**), $\mathcal{S}_h$ (**Structures**), $\mathcal{R}_h$ (**Rudiments**). Then $\Theta_h = \{\mathcal{N}_h, \mathcal{D}_h, \mathcal{A}_h, {}_F\mathcal{J}_h, {}_O\mathcal{J}_h,$ $\mathcal{R}_h, \mathcal{S}_h, \mathcal{F}_h\}$. The classification of subtheories may be based on the scheme in **MR**, and (for scientific subtheories) systems in relevant Review Journals. Here, details are not important – provided that definite, (mutually) consistent frameworks are adopted (once and for all time) and maintained / augmented as necessary. Again, the specification of Rudiments and Foundations, for each Θ_h, must be partially subjective, but this raises no problems of principle, so long as the procedures used are found and compatible.

Here, the **Foundations** of Θ_h may be taken as $\mathcal{F}_h := \mathcal{R}_h \cap (\mathcal{N} \cup \mathcal{D}_h \cup \mathcal{A}_h)$; see E_3 for the basic ideas. Then the Foundations of Σ_t will be given by $\mathcal{F}_{\Sigma_t} = \{\mathcal{F}_h : h \in \bar{H}\}$, where all of the sets Θ_h are formally t-dependent – through the subsets $\mathcal{N}_h, \mathcal{D}_h, \mathcal{A}_h, {}_F\mathcal{J}_h, {}_O\mathcal{J}_h, \mathcal{F}_h, \mathcal{S}_h, \mathcal{R}_h$. In principle, H could increase with t – if the quasi-partition of M_t into subtheories were refined; or, if qualitatively 'new' subtheories were created (for instance, through patterns motivated by inverse analogues). This, however, does not affect the formulation of general schemes to implement / classify P/M-operations on Σ -machines for ME.

Recall that the fundamental issue in ME-Representation concerns the (mathematical / scientific) environment(s) over which the ME must be realized. Such environments may be viewed as w.d. combinations of 'component subtheories' (as discussed in $E_{11/I}$) with each subtheory formulated in the 'Θ_h-representation'. The admissible **modes of combination** of (different) subtheories are diverse – but they are essentially reducible (at each stage) to one of three basic types: **Cartesian-product, boolean**, and **composition**, whenever these types are w.d. – to produce the mathematical / scientific **environments** within which the (inverse) analogues are to be 'fabricated' / explored. Such **theory-combination procedures** are discussed in $E_{11/I}$ – in the broad context of analogue-representation, DBMS design, and the MAP scheme (as outlined in E_9). The underlying **models** for 'analogue-machines' are exhibited as **operational extensions** of the **Master Flow-Charts** developed for problems in Solid Mechanics – with associated **tagging**-codes, to monitor the techniques used and the solutions obtained. Within this framework, **direct** analogues are identified as (partial) output of 'perturbed machines' – whereas, **inverse** analogues require the 'injection' of further (generally, non-algorithmic) transformations / interpretations, 'suggested' / motivated by the original scientific phenomena. In addition, a class of **quasi-inverse** analogues may by obtained (entirely within M_t), as already mentioned in this Essay.

The Overall Structure

In $E_{11/I}$, $E_{11/II}$, an elaborate scheme for the representation, analysis and (partial) synthesis of (inverse) analogues of ME/SE is developed – based on **operational extensions** of the '**template representations**' introduced for problems in **Solid Mechanics**. The operational versions of the templates are viewed as (quasi-)**machines** to 'fabricate the ME/SE' as (partial) **output**. The **input** (for Solid Mechanics) includes: material properties / configurations / I/BC for underlying (in)equations. The 'core of the machine' includes: Basic Functions / Representations / Operator-Conditions / Transformations / Mathematical Procedures – to produce (ultimately) Effective Wanted Functions.

These templates are closely linked with detailed **taxonomies** / **taggings** / **annotations** of the relevant literature – but: the procedures involved are not (in the Solid Mechanics package) realized over any **O**IB.

In spite of this limitation, it is evident that: the template framework may be **extended** to cover **arbitrary** ME/SE over Σ , with associated operational content. Indeed, such **operational im-**

plementation is fundamental to almost all aspects of Σ , as discussed from many points of view throughout these Essays. Again, the processes of **analysis / prediction** are quasi-**inverse** to those of **synthesis**. Hence, the exploration of analogues involves subtle combinations of mutually inverse procedures – syntheses, to 'construct' machines for 'generic forms of ME/SE'; and, analysis / prediction, to study the effects of admissibly perturbed machines on the forms of the ME/SE 'produced', as (inverse) analogues. The overall structure of the ME/SE machines has been specified in some detail in $\mathrm{E}_{11/I}$ and $\mathrm{E}_{11/II}$ – in terms of the allowable forms of the 'machine components', and the associated classes of perturbations that (eventually) yield the **direct** analogues. The (quasi-)inverse (ME)/SE entail (non-algorithmic) mathematical interpretations at some stage(s) of the direct analogue procedure(s).

There are (broadly) two classes of syntheses: $\mathcal{S}_{\leftrightarrows}$, with prescribed input **and** output; and $\mathcal{S}_{\rightarrow}$, where only the input is prescribed. For $\mathcal{S}_{\leftrightarrows}$, the task it so construct the corresponding machine(s); whereas, for $\mathcal{S}_{\leftarrow}$, the whole range of {machine, output} paris must be investigated, if some (class of) input(s) is 'given'. These types of syntheses may be designated as **closed** ($\mathcal{S}_{\leftrightarrows}$), or **open** ($\mathcal{S}_{\leftarrow}$). The classical synthesis problem (in all domains) is of closed type, since the I/O is specified – though there are variants where the I/O may be 'incompletely specified', in certain ways. The examples (##1–16) of synthesis domains listed (with basic references) in $\mathrm{E}_{11/II}$ cover a wide range, and #16: '**Miscellaneous Inverse Problems**', leaves open the possibility of including arbitrary synthesis schemes, as required.

Accordingly, the principal tasks to be addressed here are as follows.

1. Standard representation of **generic** ME/SE **machines**, in terms of:

- **(a)** basic **structure(s)**;
- **(b)** allowable classes of **components**;
- **(c)** allowable classes of **perturbations**;
- **(d)** generic **mechanisms**;
- **(e)** typical **representations** in domains $\mathcal{D}_k$;
- **(f)** GST-representations for (a)–(e).

2. **Demonstration** that this framework really **can** produce (potentially arbitrary) ME/SE, **via** w.d. mechanisms.

3. '**Generation**' (via perturbation) of some direct analogues of ME/SE.

4. **Investigation** of some (quasi-)inverse analogues derived from direct analogues.

NOTE.

Variants of β 'become' (direct) **analogues** of β whenever their domains of definition '**differ significantly** from that of β'. For certain classes of ME/SE this distinction could be made rigorous. Mostly, however, it is treated as intuitively / heuristically clear!

NOTES.

1. **(a′)** The basic structure of ME/SE-machines is explicated in $\mathrm{E}_{11/I}$ (mainly) and $\mathrm{E}_{11/II}$. This will be summarised, and then extended, here.

- **(b′)** The allowable classes of components are extremely diverse – as may be shown by considering each form of component in turn, in all w.d. mathematical / scientific contexts.
- **(c′)** The only significant limitations on forms of perturbation correspond to consistency / realizability, for the 'new' machines obtained, and for **their** I/O.
- **(d′)** The range of w.d. **mechanisms** corresponds to the effective **span** of the Σ_t-CB.

(e′) The scientific / mathematical domains, listed as ##1–16, admit **characteristic synthesis schemes**, say, $\mathcal{S}^k_{\leftrightarrows}$ – for which CD must be given.

(f′) The conjectured **adequacy of GST** as a formalism subsuming all of the specialized models may eventually be demonstrable (as GST evolves). Meanwhile, it provides a suggestive **paradigm**, provable for certain

classes of systems / machines.

2′. This, also, cannot be proven in complete generality, but the treatment of diverse examples provides strong plausibility arguments. In this sense, 'intuition' is not out of place within Σ .

3′. Patterns will emerge, as Σ evolves, for classes of perturbations producing certain types of direct analogues. Eventually, rigorous results may be established – as part of the **O**perational **A**nalogue **C**alculus (see also, E_6).

4′. At any stage of the direct-analogue process, transformations may be introduced to fashion novel mathematical structures – motivated by the underlying SE, for **inverse**-analogues. If, instead, the initial machine defines an **ME**, then the use of suggestive transformations may yield some **quasi-**inverse analogue(s).

All of these matters are investigated in the remainder of this Essay.

Recall that the **machines** for ME/SE are derived from those in the Solid Mechanics IB, through a combination of **generalization** and **operational extension** – within a framework including **taxonomies / tagging / flowcharts** for basic techniques (with associated 'topical outlines'). All of these aspects are covered fully in various (internal) reports for RAE/DRA (Farnborough)[642]. For 'generic ME/SE machines', an **abbreviated format** may be found in $E_{11/I}$, namely (for the ME/SE β):

$$\mathbb{M}_\beta(\mathcal{J}) \sim \mathbb{CA}M(\mathcal{P}_\beta),$$

where $\mathcal{P}_\beta$ denotes the formal specification problem for β, $M(\mathcal{P}_\beta)$ is the associated 'Master Flow-Chart', the (partially schematic) operators $\mathbb{C}, \mathbb{A}$ produce some constructive / approach formulation(s) of $M_\beta \equiv M(\mathcal{P}_\beta)$, and

$$M_\beta \approx \sigma s\zeta d; e : f\mathcal{R}g; h : iMj; k : l.$$

Here, $\zeta, \mathcal{R}, M$ stand for **coordinates, representations** and **mathematical procedures** (respectively). It is argued, in $E_{11/I}$, that **this format is applicable to completely arbitrary mathematical / scientific problems** – in particular, to the investigation of arbitrary ME/SE.

The **interpretation** of the 'components' of M_β involves combinations of **formalized** scientific / mathematical **theories** – using **cartesian-products, boolean-connectives and functional-composition**. Typical combinations of such theories are denoted as $\tilde{\Gamma}({}_i\tilde{\tau}_j)_P$, **scientific**; $\Gamma({}_k\tau_l)_Q$ **mathematical**; and, $\tilde{\Gamma} \perp \Gamma(({}_i\tilde{\tau}_j)_P, ({}_k\tau_l)_Q)$, **scientific / mathematical**. The detailed **forms** are elucidated in diverse examples. **NOTE that**, if $\sigma =: \{\sigma^-, \sigma^+\}$, $s =: \{s^-, s^+\}$ account for 'internal' ($-$) and 'external' ($+$) structures, separately, then the scientific / mathematical **environment** for M_β – and thus also for $\mathbb{M}_\beta(\zeta)$ – is given by $\xi \equiv \{\sigma, s\}$; or, for 'hybrid forms', by $\xi \equiv \sigma \perp s$, where $\sigma \equiv \tilde{\Gamma}$ and $s \equiv \Gamma$. With these preliminaries, generic machines, $\mathbb{M}_\beta(\zeta)$, may be specified through the 'components', $\sigma, \ldots, l$ – as in $E_{11/I}$. This appears to be a very general representation, covering (at least) all currently w.d. ME/SE. If (somehow) entities outside the purview of these machines were to be introduced, then routine modifications could incorporate the new entities consistently.

The processes embodied in the schematic operators $\mathbb{A}$ and $\mathbb{C}$ are discussed in detail in several of the other Essays, with copious references. It has been remarked (even in E_1) that Σ has the principal aim of **the systematic reformulation of Mathematical Knowledge from**

[642] See $E_{11/I}$ for references

'the $\mathbb{A}/\mathbb{C}$-perspective' – indeed, most of the developments in E_1–E_9 are motivated by this aim. For the representation of $\mathbb{M}_\beta(\zeta) \equiv \mathbb{M}_\beta(\zeta, \xi_{\cdot\zeta})$, to fabricate the ME/SE β **via** ζ over the environments ξ_ζ, the following **identifications** are made (in $E_{11/I}$).

σ: $\tilde{\Gamma} \equiv$ formalized **scientific theories**

s: $\Gamma \equiv$ formalized **mathematical theories**

ζ: **frameworks** for ζ ('motivated by β')

d: **LC** for ζ over ξ_ζ

e: **auxiliary elements** for realization of ζ in ξ_ζ

f: maximally **constructive forms** of ζ in ξ_ζ

$\mathcal{R}$: **representations** of ζ **via** auxiliary elements

g: **formal (in)equational conditions** on ζ

h: **formal auxiliary elements** for ζ in ξ_ζ

i: **formal realizations** of ζ in ξ_ζ

M: **mathematical transformations**

j: **effective (in)equational conditions** on ζ in ξ_ζ

k: **effective auxiliary elements** for ζ in ξ_ζ

l: **effective realizations** of ζ (and so, of β)

In this context, the perturbed machines are denoted[643] as $\mathbb{M}_{\beta\lambda}(\zeta, \xi_{\zeta\lambda})$, where λ indexes the family $\{\mathbb{M}_{\beta\lambda}\}$. The relationship of $\mathbb{M}_{\beta\lambda}$ to $\mathbb{M}_\beta$ is also discussed in $E_{11/I}$. The crucial ingredient is the (almost unrestricted) choice of $\xi_{\zeta\lambda}$, 'motivated by ξ_ζ'. The other components of $\mathbb{M}_{\beta\lambda}$ closely parallel those of $\mathbb{M}_\beta$ – but their realizations may involve (and 'promote') highly irregular **quasi-**mathematical structures, whose validation (where possible) may yield novel **inverse** analogues. The 'admissible forms of perturbation' of the components of $\mathbb{M}_\beta(\zeta, \xi_\zeta)$ are determined / governed primarily by the underlying constraint that all associated operations should be w.d.[644]. Plainly, the detailed restrictions vary with β. For **direct** analogues, **all** operations / operands must be (quasi-) w.d., at each stage of analogue-formation. Divergence from this fundamental requirement entails the introduction of 'new' SE/ME whose mathematical properties may be arbitrarily specified – subject only to consistency with the (then) current OIB. It is this – highly non-algorithmic – class of processes that embodies the essence of **inverse** analogy.

At this point it is useful to recall the notation, for **perturbed machines**, from $E_{11/I}$. There, $\mathbb{D}^{\pm\alpha}, \mathbb{D}_\lambda^{\pm\alpha}$ denote the transformations producing analogues (of e) $\pm\alpha(e), \pm\alpha(e_\lambda)$, respectively, from the machines $\mathbb{M}^{\pm\alpha} \equiv \mathbb{D}^{\pm\alpha}\mathbb{M}, \mathbb{M}_\lambda^{\pm\alpha} \equiv \mathbb{D}_\lambda^{\pm\alpha}\mathbb{M}_\lambda$, where $\mathbb{M} \equiv \mathbb{M}_\beta(\zeta, \xi_\zeta)$, and $\mathbb{M}_\lambda \equiv \mathbb{M}_{\beta\lambda}(\eta, \xi_{\eta\lambda})$ $=: V_\lambda\mathbb{M}$. The matrices $\mathbb{D}, \mathbb{D}_\lambda$ may, WLOG, be assumed **diagonal**, of order N (=14, or 16, in the present formulation). The **variants**, β_λ, of β are obtained as (partial) output of $\mathbb{M}_\lambda$, and $e_\lambda \in \mathbb{V}(e)$, the set of all w.d. variants of e over Σ_t[645]. The (diagonal) elements of $\mathbb{D}^{\pm\alpha}, \mathbb{D}_\lambda^{\pm\alpha}$ covers all relevant combinations of OT deployed in fabricating machines for $\pm\alpha(\beta), \pm\alpha(\beta_\lambda)$. The corresponding (compound) operations may be deterministic or stochastic, of arbitrary complexity / sophistication. For **direct** analogues, the operations are **(quasi-)algorithmic**; but, in the

[643] This notation also covers **variants** of $\mathbb{M}_\beta$ (as already defined).

[644] The general requirement in Σ , that all procedures should be t**-max. operational**, imposes extra conditions; see **f:** later in this essay, for instance.

[645] Recall that β covers both ME and SE; that **direct** analogues 'stem' from e, and **inverse** analogues (ultimately) from s – while some '**quasi-inverse**' analogues may also be derived from e with 'proper interpretation'.

formation of **inverse** analogues, 'imaginative interpretations' may be 'conjured up' – to be retrospectively justified – often through the development of nontrivial 'new structures'. In spire of the (mostly) non-algorithmic nature of the key inverse-analogue transformations, they may be formally included in the matrix representations of perturbed machines.

It follows that: all of the 'principal tasks' – say, $t_1(a), \ldots, t_1(f), t_2, t_3, t_4$ – identified at the beginning of this Essay, may be investigated in terms of the classes – say, $C_i^M, C_{i\lambda}^M$ (of admissible components of $\mathbb{M}, \mathbb{M}_\lambda$) and $C_j^D, C_{j\lambda}^D$ (of admissible operators for the elements of $\mathbb{D}^{\pm\alpha}, \mathbb{D}_\lambda^{\pm\alpha}$), where $1 \leqslant i, j \leqslant N$. Plainly, the scope of these classes and operators will be limited only by mutual compatibility constraints among the (w.d.) combinations of ME/SE involved – and these ME/SE may be selected from virtually arbitrary subdomains of Mathematics / Science. Nevertheless it transpires that significant progress may be made, even within this very general framework.

In the following sections, the key properties of components of $\mathbb{M}, \mathbb{M}_\lambda$ are outlined. **NOTE that block-diagrams**, and **signal-flow graphs** (as descriptions of control systems) are both applicable to the schematic representation of Σ -machines.

ξ**:** The **environment** of $\mathbb{M}_\beta$, or, of $\mathbb{M}_{\beta_\lambda}$, may be **mathematical** (Γ), **scientific** ($\tilde{\Gamma}$), or **mathematical / scientific** ($\Gamma \perp \tilde{\Gamma}$) – comprising (at various stages) combinations of **formalized theories** of appropriate types. The allowable modes of combination have been characterized (in $\mathrm{E}_{11/I}$) as **Cartesian-product, boolean**, and **composition**. The resulting compound formalized theories (with associated **tree structures**) blend scientific and mathematical formalisms in diverse ways. Of course, the **scientific** notions must be expressed through sets of **(in)equations** – but the **forms** of these (in)equations are motivated by experiment / observation / hypothesis, rather than by axiomatics / aesthetics / formal-relations. The compound theories may be specified both **locally** (at nodes of the underlying tree / graph-'skeletons') and **globally** (in terms of the primary mathematical / scientific subdomains encompassed). The detailed 'interactions' within collections of subtheories may be very intricate, requiring monitoring apparatus of cognate complexity, so that 'emerging (inverse) analogues' may be analysed / synthesized systematically. These observations summarize the **general properties** of the 'environment' of $\mathbb{M}_\beta$, or of $\mathbb{M}_{\beta\lambda}$ – in terms of admissible combinations of compound theories. Discussions of **specific cases** may be found in the treatment of examples from the Σ -Lists, which range widely over Pure / Applied / Applicable Mathematics. The ultimate purpose of such 'blending' of disparate (non)deterministic theories is to exploit the fecundity of scientific / Mathematical formalism to develop universal **O**perational-**S**ystem **C**alculus.

ζ**:** The next class, ζ of $\mathbb{M}$-components covers generalized coordinate systems and their 'host spaces', together with certain aspects of '**approach milieux**'. Much of this material (e.g., standard coordinate systems, tensor algebra / calculus, MS) is treated exhaustively in surveys / books. Many other topics, of particular importance for Σ , are discussed (with copious references) in other Essays. A 'running compendium' of coordinate schemes, in all mathematical domains, will play a significant role in the longterm development of Σ . This involves the classification of all (quasi-)coordinate systems – of whatever origin - -to account for areas of application, interrelations, generalizations, etc.. As usual, the aim is to cover arbitrary subdomains of Pure / Applied / Applicable Mathematics, within approach frameworks. The constructions involved in the definition / use of these systems are often 'surprising' / intricate / subtle – underlining the fact that nontrivial examples of general theories in Applied / Applicable Mathematics may be technically far more demanding than the underlying theories themselves. This is illustrated by many items in the Σ -Lists.

d**:** The LC, $l_\zeta \equiv l_\zeta(\xi_\zeta)$, for a specified ME/SE, β, will be 'motivated by β' but, in general, not fully determined by β. Indeed, typically, a **family**, $\{\mathbb{M}(\chi)\}$, of machines corresponds to β – the principal criterion being that β must be **algorithmically** obtainable[646] from every

[646]This may involve 'controlled approximation' – but **not** analogy (which is treated separately).

machine $\mathbb{M}(\chi)$. The range of admissible LC in the $\mathbb{M}(\chi)$ is, therefore, very wide, and so is the variety of algorithms. Nevertheless every LC for e is representable by a collection of (in)equations within some combination of (formalized) theories. In the most extreme cases, the LC-set need not be finite, or even countable. The types of (in)equations involved depend on the theories over which LC are formulated – for instance, algebraic / differential / integral / boolean / 'operator' /..., all in (non)deterministic settings 'suggested by characteristics of e'. It is thus $\xi_\zeta, l_\zeta(\xi_\zeta)$, and their variants, $\xi_{\zeta\lambda}, l_{\zeta\lambda}(\xi_{\zeta\lambda})$, that largely govern the possible forms of $\mathbb{M}_\beta, \mathbb{M}_{\beta\lambda}$.

e: Whenever ξ_ζ, l_ζ are specified, the w.d. **representation(s)** of ζ over the resulting framework will involve associated **auxiliary elements**. Familiar examples include: **E-M potentials**, 2-D-elasticity **complex potentials**, species of **Green's functions, group characters, (co)homology groups** (of TS, etc.), **generating functions** in Analysis / Combinatorics. Although these examples may appear disparate, brief consideration indicates that they illustrate a common theme. The Σ -Lists contain numerous (mainly, less familiar) examples of these processes – many embodying highly ingenious constructions. One of the Σ -tenets asserts that: **every ME has some** (possibly trivial) '**auxiliary representation(s)**'. This tenet, and the (formal) identification of each ME with some OT – both within approach milieux – are among the most distinctive features of Σ as an OIB, essentially guaranteeing that **an arbitrary ME has some associated** (w.d.) **machine(s)**, as (tacitly) assumed, so far.

f: A fundamental facet of 'approach' in Σ is that every procedure used is maximally **constructive** (over the relevant framework(s)). For the machines $\mathbb{M}_\beta, \mathbb{M}_{\beta\lambda}$, this entails that each representation of ζ, in ξ_ζ of $\xi_{\zeta\lambda}$, must be t-max.-operatio- nal; so the same criteria hold also for all auxiliary elements introduced in associated 'machine processes'. Consequently, the admissible perturbations of any machine must conserve constructiveness (as well as producing w.d. output); this greatly restricts the choice of perturbations – although weaker restrictions may be acceptable when 'vaguer' (inverse) analogues are investigated. See E_6 for Comments on the relative strength of analogues. The criteria for various species of constructivity are discussed, especially, in E_5, where several types of approximation frameworks are covered, with diverse references. In particular, a broad selection of **non-numerical distance functions** (with brief comments) is given in the (U. of Vienna) Bibliography compiled by R Schmid-Zartner / H-C Reichel / B Pötscher (op. cit.). Many arcane approach schemes are exemplified in the Σ -Lists.

$\mathcal{R}$: Whereas e: is concerned with the auxiliary elements required to represent ζ over ξ_ζ, etc., the 'filter' $\mathcal{R}$, in the MF-C for the ME/SE β, covers the corresponding representation **procedures** – which are extremely diverse, even in their most general forms. For instance, the topic of **group representations**, alone, spans vast mathematical territories. Moreover, the claimed ubiquity of auxiliary elements implies that no systematic classification of procedures will be tractable. Nevertheless, the overall OIB structure can function effectively on a coarser-grained basis; the finer details may be filled in as they become relevant, rather than routinely. In this way, the labyrinthine interconnections may be simplified to practically implementable forms, allowing the use of DBMS/GST techniques.

NOTE. For the (non-operational) Fracture-Mechanics IB, the collection of **T**opical **O**utlines exemplified representation procedures, covering only the most basic ME involved. This focuses attention on the vastness of the corresponding task for the Σ -OIB. (These **TO**, with associated **C**alculational **F**low-**C**harts may be found as Appendices to E_{11}).

g: The combinations of d:, e:, f:, and $\mathcal{R}$:, corresponding to various sets of LC, etc., yield cognate **F**ormal **C**onditions of ζ (and, possibly, on associated parameters / functions). In the Solid Mechanics problems, these conditions are realized as coupled / (non)linear / difference / differential / integral / transcendental /... (m)equations – of (non)deterministic ODE/PDE/...types. Although this covers a wide range of 'operator equations', it also underlines the challenge of forging global frameworks for the MF-C of **arbitrary** ME/SE, β; with the additional requirement of nontrivial operational content. It is this development of t-max.-operational milieux in the most general settings that (above all) distinguishes Σ from all existing SAM-based packages. The extreme intricacy of the resulting information structures, with diverse / subtle interrelationships, raises multifarious issues which are discussed in the other Essays.

Theoretically, every combination of (in)equational conditions could arise as the F(O)C of some β – whose potential 'generability as (partial) machine output' depends on consistency / solvability /... criteria.

h: The F(O)C for ζ, over ξ_ζ, etc., may be t-max. 'solved' in terms of formal auxiliary elements (at least, 'in most cases'), which are assumed to satisfy all of the relevant formal (/defining) relations. These elements may be combined in all Σ -admissible ways, to obtain 'new' elements 'specially tuned to the problem(s) at hand'. At this stage, no detailed calculation is involved, since all of the expressions are formal; but the 'calculational challenges' are exhibited (at the next stage) when the forms in h: are used in g: – to 'generate' the formal realizations, in i:.

i: When the conditions g: are expressed in terms of the (formal) elements h:, it is frequently possible to obtain (formal) realizations of ζ (and, if necessary, associated parameters / functions), so that the obstacles to 'complete calculation' may be assessed and (at later stages) overcome, to some degree. In the FM 'ABC', common tasks include: the (approximate / asymptotic /...) inversion of integral transforms (of several types), the solution o coupled integral equations, and the use of diverse techniques for 'asymptotic analysis'. See $\mathrm{E}_{11/I}$ (and its Appendixes) for examples.

M: The word 'transformation' is used very broadly here, to cover all procedures deployed to alter the internal / external structure / interpretation of any (set of) wfe encountered in a Σ -investigation. Thus, the integral transforms, etc., common in Applied / Applicable Mathematics, are just among the most obvious illustrations of the fundamental strategy of viewing an arbitrary ME/SE β in several (partially) equivalent frameworks. Plainly, the scope of such 'activities', over Σ_t, is essentially M_t itself, since β is unrestricted (provided that all 'constituents' involved are w.d. / consistent). This apparent vagueness, however, raises no problems of principle: rather, it highlights the intrinsic unity of Mathematics as a calculational system, where every ME (in)directly influences the properties of other ME – as may be seen in many examples in the Σ -Lists.

Consequently, the (families of) admissible transformations may best be (partially) classified over the mathematical subdomains where they are 'most naturally located'.

j: The next stage in the process of producing β, as the (partial) output of a 'machine' $\mathbb{M}_\beta(\zeta, \xi_\zeta)$, converts the F(O)C into (quasi-)**effective conditions** which (when expressed in terms of effective auxiliary elements) admit (quasi-)**algorithmic / approach 'solutions' / realizations** of ζ over ξ_ζ – and thence (properly interpreted) of β.

The entire scheme embodied in the (operationally extended) MF-C, 'for β via ζ over ξ_ζ', is pellucid, if it is viewed as an elaborate development of the (non-operational) Flow-Charts for Solid Mechanics problems (where the motivation, and interrelations among the successive stages, are comparatively obvious). The universality of this scheme (implemented with 'approach criteria') is fundamental for the entire Σ -Project – even though it is **explicitly** discussed only in E_{11}! Indeed, the structure of Σ -investigations, as diversely studied in the other Essays, is **implicitly** governed by 'variants of machine representations'.

NOTE. See $\mathrm{E}_{11/I}$ for extracts from research reports on the MF-C; and **Appendices** to E_{11}, for reproductions of other reports (with brief notes on their relevance for Σ).

k: The auxiliary elements introduced in e:, and formalized in h:, must still be 'converted into pre-operational forms' before they are **effectively usable** to realize ζ (and thence, β)

over ξ_ζ, through the conditions j:. Here, again, this strategy is transparent in (e.g.) Solid Mechanics contexts, but far from obvious for arbitrary ME/SE, β. One of the major tasks of these Essays is to develop, in great detail, a (quasi-)**universal set of procedures for the approach-realization of** β over some prescribed environment(s), where β may range over $M_t/\mathcal{S}_t$. In view of the extreme generality of this task, it is not surprising that the fundamental issues are very subtle; and, that the 'solutions' are both highly sophisticated and technically intricate, as well as (in many instances) decidedly arcane – a fault that, it is hoped, will be moderated in subsequent formulations of the '**Σ -Theory of Mathematical Synthesis**'.

l: The **culminating stage** of the MF-C process to produce (approaches to) the ME/SE β – via ζ, over ξ_ζ – will, in general, be multi-faceted and technically challenging, no matter which (sub)domains of Mathematics / Science are involved. The operational paradigm (central to all Σ -activity) entails that every 'component' of the realization process(eS) must be in maximally constructive form(s). Even on its own, this requirement makes heavy demands, since many mathematical results currently lack constructive formulations / proofs[647].

NOTE. It is important to stress, again, that, while $\mathbb{M}_\beta, \mathbb{M}_{\beta\lambda}$, produce (**nonanalogical** approaches to) the ME/SE β, and its variants, β_λ, the (inverse) **analogues** of β (or of β_λ) are obtained through the actions of **more general** (classes of) **perturbations**. The essential distinction is that **variants** are (broadly) **type-preserving**; whereas, **direct** analogues **reproduce** some behavioural **facet(s)** of β, in entities of demonstrably **different** types from β. Lastly, for the (quasi-)**inverse** analogues of β, only (often, exotic) '**flavours**' of the type / behaviour of β are preserved – in (typically) **non**algorithmic transformations.

Synthesis CD / Flow-Charts

The synthesis paradigms (#1–#16), identified (with basic references) in $\mathrm{E}_{11/II}$ are far from exhaustive – though #16 ('Miscellaneous Inverse Problems') formally covers all 'new' synthesis domains, with 'proper formulation'. A cumulative set of {**synthesis, domain**} **examples** will form an important part of the OIB – ranging over all w.d. subdomains of Science / Mathematics, at 'coarse-grained levels', as indicated in (e.g.) MR, and various scientific review journals. The diverse procedures for 'fine-grained fabrication' or arbitrary elements from these subdomains may be developed by specializing the general schemes, as appropriate, in adequate detail.

The fundamental question, here, is whether there are **quasi-universal synthesis processes** subsuming those in all particular subdomains – potentially, at all levels of detail. No matter what (scientific / mathematical) field(s) may be involved, the associated synthesis problem(s) have the overall character of **multivariable control problems** for (arbitrarily) **complex systems**, of (non)discrete / (non)linear / (non)deterministic form(s). In spite of this extreme generality, various algorithms for types of (optimal) synthesis may be developed for such systems – without unduly restrictive specification of the I/O, or of the various components. Indeed, it appears that all of the classes C_k, for domains D_k, as listed in $\mathrm{E}_{11/II}$, may be (at least, formally) related within this (abstract) framework. The highest level of abstraction at which nontrivial calculations are w.d. corresponds to **GST** (initially, for quasi-linear systems, though there is scope for extension to handle nonlinear behaviour). The actual **algorithms**, however, can produce testable results only when the system components and input have been **instantiated**. This is, of course, a standard situation in Mathematics – but it is especially relevant here, since the aim is to **demonstrate** that (variants of) **all species of synthesis** in Pure / Applied / Applicable Mathematics **are** coverable by a single 'global' scheme.

It is therefore useful to proceed on the **assumption** that a **universal-synthesis clock-diagram** (way, $\mathbb{D} \equiv \mathbb{D}(t)$, for $\mathbb{M}(t)$ in Σ_t) may be developed, with the key property that, for $1 \leqslant k \leqslant 16$ (so far), there are '**specialization-transformations**', $\mathcal{S}_k$, such that $\mathcal{S}_k\mathbb{D}(t) =: \mathbb{D}_k(t)$

[647] However, the gap is closing! See Bishop / Bridges, '*Constructive Analysis*' (Springer, 1985), op. cit., and its Bibliography. The 'constructive activity' has certainly continued since 1985.

covers the class C_k of synthesis procedures. For this to be possible, it is necessary for $\mathbb{D}(t)$ to have a highly complex structure, where (typically, for some C_k) 'most of the components' act as identity operators[648], while other components perform the (minimal) general operations required to realize C_k.

Since the aim is to 'fabricate' (at least, approaches to) arbitrary (generic) ME/SE, β (over Σ_t), the fundamental question arises: whether $\mathbb{D}(t)$ is **constructible – even in principle**. From one point of view it is a matter of **program-synthesis** (for some 'generalized computer')[649] with $\mathbb{D}(t)$ specifiable (somehow) from an 'extended program flow-chart'. This, in turn, is formally / mathematically equivalent to **program-construction / verification**. The 'optimal choice of framework', in any particular case, obviously depends on the mathematical / scientific fields involved, but the issue here is one of universal **representation**, rather than efficient calculation. One possible 'design' for $\mathbb{D}(t)$ could involve countable arrays of components (in subsets of specified types of elements) with arbitrarily specifiable combinations of (series / parallel /...) **connections / iterative**-operations / (prescribable) **algebraic-analytical-topological-...** operations /... – all, over variously defined domains (e.g., **manifolds, semigroups, TS**). **NOTE that** the models of scientific entities (from which inverse analogues may emanate) are always formulated mathematically (even if they have empirical precursors). Consequently, the 'exotic design' just outlined for $\mathbb{D}(t)$ could, in principle, suffice to synthesize **all of** $M_t/\mathcal{S}_t$ (rather than 'merely' covering the classes C_k). In other words, **the entire edifice of Mathematics / Theoretical-Science** (as formulated over Σ_t) **is (quasi-)synthesizable** within some class of universal machines containing countably many 'components', interconnected in various ways.

Remark. Necessary and sufficient **realizability conditions** for network / system functions (to guarantee the **possibility of synthesis**) are discussed extensively in Weinberg, op. cit., Chs 7–10. See also: Zadeh / Polak, op. cit.; Kalman / Falbi / Arbib, op. cit..

In practice, for 'most ME/Se β, $\mathbb{D}_\beta(t)$, the 'β**-active part** of $\mathbb{D}(t)$', will be finite, and of 'comparatively simple structure' – where $\mathbb{D}_\beta(t)$ is viewed as a finite section of (some realization of) $\mathbb{D}(t)$, and $\mathbb{M}_\beta(t;\zeta,\xi_\zeta)$ is the associated machine producing (approaches to) β as (partial) output. Moreover, by introducing appropriate **recursion / iteration** procedures, it may even be possible to obviate the need for countable (rather than finite) structures altogether, since **proofs** over countable frameworks may be handled by recursion; **sequences**, by iterative techniques; and, uncountable **families** of ME/SE, by 'calculus-based schemes', over suitable (**analytical / algebraic / topological / combinatorial /...**) domains. Although this is only a plausibility argument, it seems to be well-founded: the **non**recursive / **non**iterative aspects of 'countable ME' necessarily involve limiting operations (of various kinds) and so may be treated by methods similar to those for families of ME/SE – again, within corresponding domains. All of these matters will be clarified – if not completely resolved – as Σ develops.

It is important to recall, in characterizing the OIB, that Σ is primarily concerned with (increasingly refined versions of) **formalized heuristics** – rather than with exact computation of 'arbitrary ME/SE', as in current SAM packages (however 'advanced'). Indeed, the aim is to **embed** SAM operations, cumulatively, within the Σ -OIB – subject to 'approach paradigms' in all Σ -investigations. As the purview of (constructive) SAM procedures grows, there will be progressive overlaps in these styles of 'calculation' – all of them being, in essence, **facets** of overall 'mathematical activity' – to be unified, as far as possible.

The foundations of 'Theoretical CS' (subsuming **IT / AI / SAM / automatic-theorem-proving /...**) combine diverse results from **Mathematical Logic, Lattice Theory, Boole- an Algebra, General Topology, Category Theory** (among other fields). The efficient implementation of algorithms (in the development of complex software systems) draws on several further topics (e.g., **Graph Theory, Optimization Theory**, species of **Algebras**). Moreover, aspects of **Stochastic-Process Theory** (for statistical evaluation of routine / queues / performance) in (random) computer networks, with serial / pipeline / parallel /... links, are dominant in many industrial contexts, where species of syntheses are key objectives.

---

[648] or else, are **circumvented**

[649] These ideas, for Σ , are broadly discussed in E_8.

All of this emphasizes, yet again, that Σ **functions of several** (largely complementary) **levels**, where procedures are implemented in varying degrees of detail – from schematic outlines of 'proof-strategies' to maximal realizations of diverse OT (where the continuing development of SAM algorithms is essential). The formalism created in these Essays allows OT to be combined / manipulated / perturbed /... in intricate / subtle ways. It also furnishes a consistent methodology for the construction of t-max.-operational versions of '**known** theorems', for use in the OIB. In the case of '**new** theorems', constructivity will remain the dominant aim – but, 'in the nature of things', conjectures / heuristics will motivate the initial formulations / proofs. This balance may change, very gradually, as the 'Σ -paradigms' become established, but intuition / imagination will (and should) continue to provide 'inspiration' that purely technical schemes generically lack. The subsequent **re**formulation of such inspirational results, for routine use in the Σ -OIB, is, however, essential – and produces fertile environments for further speculative activity! In this way, Mathematics (and its applications) will flourish, and the (unavoidable) effects of **hysteresis** (in levels of operationality) will be minimized. In this multi-level framework, the study of machine-synthesis / analysis / perturbation is of fundamental importance.

In the area of **proof-development** (over specified logical domains), as opposed to β-synthesis, etc., several computer-based systems may be used – some of which are mentioned in E_8, where their current limitations are discussed. Ultimately, though, variants of such systems will become important – for instance, in **testing proposed sets of axioms** for consistency / novelty / scope /..., and in **checking the validity** of (quasi-)formal 'segments' of proposed proofs.

Among the references (potentially) relevant for Σ are: Constable, RL, et al., '*Implementing Mathematics with the Nuprl Proof Development System*' (P-H, 1986); Nordström, B, et al., '*Programming in Martin-Löf's Type Theory*' (OUP, 1990); Huet, G, et al., eds., '*Logical Frameworks*' (CUP, 1991); Nivat, M, et al., eds., '*Algebraic Methods in Semantics*' (CUP, 1985); Bloom, SL, et al., '*Iteration Theories*' (Springer, 1993). A more directly mathematical package (still under development) is outlined in: McAllester, DA, '*Ontic – A Knowledge Representation System for Mathematics*; (MIT, 1989). For **software design** / **synthesis** (to specifications), which underlies theories of ME/SE-Machine-Synthesis, see, for instance: Stone RG, et al., '*Program Construction and Verification*' (P-H, 1986); Hehner, ECR, '*The Logic of Programming*' (P-H, 1984); Gries, D, '*The Science of Programming*' (Springer, 1981); Loeckx, J, et al., '*The Foundations of Program Verification*' (Wiley, 1984); Kubiak, R, er al., '*An Introduction to Programming with Specifications: A Mathematical Approach*' (AP, 1991); Tanik, MM, et al., '*Fundamentals of Computing for Software Engineers*' (van Nostrand, 1991); Gurari, EM, '*An Introduction to the Theory of Computation*' (CS Press, 1989).

Although all of these (CS-based) schemes are formal / theoretical, in principle, the cover (efficient) **procedures** for performing specified **calculations** – from the most abstract to the extremely practical. As such, they embody **mechanisms** for these calculations; hence, also, formal **machines**. Indeed, every Σ -machine, $\mathbb{M}_\beta(\zeta, \xi_\zeta)$, with all 'components' specified, may be represented as a **program** for some computation – with (partial) output β, 'via ζ, over ξ_ζ'. More precisely, $\mathbb{M}_\beta$ **implements** the program for β, etc., where the **structure** of $\mathbb{M}_\beta$ corresponds to the **algorithm(s)**, Π_β, involved.

It follows that there is a (detailed) correspondence between the set(s) of CS-based terms / constructions and the set(s) of mathematical terms / operations 'defining Π_β in $\mathbb{M}_\beta$'. Since Σ is concerned, ultimately, with **β-generation** over the OIB (where β ranges over $M_t/\mathcal{S}_t$), and the OIB functions within a (very sophisticated) computer environment, the importance of the **CS formalisms** is evident. Moreover, in the **implementation** of diverse algorithms (over associated frameworks) issues of **realizability** / **modes** / **optimization** of computations loom large – raising a wide variety of recondite problems of hardware / software design. In spite of the great emphasis, in Σ , on the enhancement of **mathematical** research, all of these technical matters are of the utmost significance in increasing speed / efficiency / scope of Σ -investigations. Consequently it is worthwhile to include here a short selection of **source-books** – with occasional comments.

Hockney, RW, et al., '*Parallel Computers 2*' (Hilger, 1988): Architecture, Programming, and

Algorithms. Fox, G, et al., '*Solving Problems on Concurrent Processors, Vol. 1 – General Techniques and Regular Problems*' (P-H, 1988) [corresponding **programs** are given in '*Vol. 2*']. Evans, DJ, ed., '*Parallel Processing Systems: An Advanced Course*' (CUP, 1982) [covers many aspects of parallel / distributed systems, and conversion of sequential programs to parallel forms]. Chandy, KM, et al., '*Parallel Program Design: A foundation*' (A-W, 1988) [extends to distributed / concurrent algorithms the principles introduced by EW Dijkstra for sequential algorithms, in the book, '*A Discipline of Programming*' (P-H, 1976)]. Muchnick, SS, et al., '*Program Flow Analysis – Theory and Applications*' (P-H, 1981). Ullman, JD, '*Computational Aspects of VLSI*' (CS Press, 1984). Sherwani, N, '*Algorithms for VLSI Physical Design Automation*' (Kluwer, 1993).

Raynal, M, et al., '*Synchronization and Control of Distributed Systems and Programs*' (Wiley, 1990). Gelenbe, E, '*Multiprocessor Performance*' (Wiley, 1989). Garey, MR, et al., '*Computers and Intractability*' (Freeman, 1979). '*Complexity Theory – Current Research*', Ambos-Spies, K, et al., eds. (CUP, 1993). Traub, JF, ed., '*Analytic Computational Complexity*' (AP, 1976). [Many other books / papers 'at the interface of SAM / constructive-procedures / software-design' are covered in J. Symbolic Computation (JSC). A wise selection of earlier papers / books is given in the Bibliography of: Elvey, JSN, '*Symbolic Computation and Constructive Mathematics*' (U. of Waterloo Research Report CS-80-49 (pp226–249), 1980, op. cit.)]

NOTE that the implementation of (quasi-)**numerical** procedures – either independently, or else, as parts of hybrid SAM routines – is a vast field, with many current journals / books. For instance: ACM Trans. Math Software; Math. of Computation; Numer. Math.; J. Algorithms; SIAM J. on Num. Analysis; and the books: Press, WH, et al., '*Numerical Recipes – The Art of Scientific Computing*' (CUP, 1989); Golub, GH, et al., '*Matrix Computations*' (2nd ed., The Johns Hopkins UP, 1989); Marchuk, GI, '*Methods of Numerical Mathematics*' (2nd ed., Springer, 1982); Acta Numerica: 1992– (annual compilations of survey articles) (CUP). Baker, CTH, '*The Numerical Treatment of Integral Equations*' (OUP, 1978); Blum, EK, '*Numerical Analysis and Computation: Theory and Practice*' (A-W, 1972) [with extensive discussions of approximation frameworks / criteria]. Meis, T, et al., '*Numerical Solution of PDE*' (Springer, 1981). Reinhardt, H-J, '*Analysis of Approximation Methods for Differential and Integral Equations*' (Springer, 1985).

These brief lists give only the barest indication of the voluminous literature on 'generalized computation' – but they do show a clear tendency towards **unification**, especially in combination with the elaborate **graphics packages** now used in SAM, the BE/FE methods for classes of operator equations, and the widely-applicable techniques of 'stochastic calculus'. Further, substantial parts of these Essays are concerned with the systematic use of **in**equations (over arbitrary 'ordered domains') to extend the purview of 'standard approximation'; and, with the introduction of other 'approach criteria', based on abstraction / (inverse) analogy, to produce (seemingly) ultimately general conditions for 'closeness' in Mathematics – with specializations to cover all w.d. '(non-)numerical distance-functions' (as exemplified in the briefly-annotated **Bibliography**... by Schmid-Zartner, et al., op. cit.).

Interim Summary

Several realizations, $\rho_k(\beta)$, of the (arbitrary) ME/SE β have been defined within Σ , so far. They include:

$\rho_1(\beta) \equiv$ LC/D(β) **[Locational Conditions / Descriptions]**

$\rho_2(\beta) \equiv$ MF-C(β) **[Master Flow-Chart form]**

$\rho_3(\beta) \equiv \mathbb{M}_\beta(\zeta, \xi_\zeta)$ **[machine-synthesis form]**

$\rho_4(\beta) \equiv \Pi_\beta$ **[computer program with specification]**, together with w.d. **hybrid forms** – say, ${}_H\rho(\beta)$. Although all of these realizations are quasi-equivalent – in the sense that, suitably interpreted / operationally-extended, they all embody essentially the same mathematical information – they nevertheless 'inhabit **distinct formal frameworks**' – each motivating associated **generalizations / perturbations**, and hence, also, characteristic ((quasi-)inverse) **analogues**.

It is useful to recall the following features of the $\rho_k(\beta)$.

(i) $\rho_1(\beta)$ is typically purely **descriptive** (non-operational) and formal – yet admitting (non) implicatory **logical deformations**.

(ii) $\rho_2(\beta)$ is presented as **descriptive**, but may be enhanced to t**-max.-operational** form.

(iii) $\rho_3(\beta)$ is viewed as an **I/O system** with β as (partial) output (of an **active machine**).

(iv) $\rho_4(\beta)$ is a set of **logical procedures**, whose **specification** (as a program) includes β, as (partial) output.

Notice that, here, a 'computer program' is understood as a (collection of) subroutine(s) over some SAM system(s) – suitable enhanced, 'in the Σ -style', to cover 'approach phenomena'. However, ultimately, the underlying processes are governed by the theories of (digital) computation, for serial / pipeline / parallel /... computers, realized over appropriate combinations of '**processors**', with VLSI design, etc.. The 'SAM level' applies to $\rho_3(\beta)$, but the **mechanisms** in the $\rho_3(\beta)$-machines have the detailed structures of fundamental logical functions, etc., as understood in the $\rho_4(\beta)$-representations. Although these remarks may be 'obvious' in a general setting, many subtle distinctions arise 'in mathematical practice', where two, or more, of these representations may occur naturally in some Σ -investigation(s). Plainly, $\rho_3(\beta)$, with all components t-max.-implemented, is the 'preferred representation' for ((quasi-)inverse) analogue-synthesis / analysis; but, in principle, any of the $\rho_k(\beta)$ – or, even ${}_H\rho(\beta)$ – could be adopted as 'basic'.

As a fundamental step towards the synthesis of arbitrary ME/SE, β, the classes C_k $(1 \leqslant k'leq16)$ with diagrams / flow-charts $\mathbb{D}_k$, as in $\mathsf{E}_{11/II}$, will now be considered, in turn, to pursue the primary aim of exhibiting a broadly common pattern among the $\mathbb{D}_k$, reflected in fundamental structural similarities. The notation / frameworks / techniques introduced must be flexible enough to accommodate all of the C_k ('suitable formulated'); and, to suggest possible extensions covering essentially arbitrary β – in potentially effective ways. Of course, this gargantuan challenge can be met (even partially) only within 'approach milieux', since some β are (provably) outside the purview of exact calculation – over existing (G)SF (see E_2 for remarks on **G**eneralized **S**tructured **F**rameworks). Indeed, the $\mathbb{D}_k$ themselves may embody approaches to β, rather than terminating algorithms for constructive processes. This, however, does not detract significantly from the intrinsic soundness of the overall synthesis scheme.

In the following sections, CD of synthesis paradigms (SP$_k$, $1 \leqslant k \leqslant 16$) are given. Only the most central methods are outlined; variants / extensions may be found in the extra references provided. Associated diagrams / flow-charts are developed later.

Recall that the main goal of this discussion is to demonstrate, as far as possible, that all of the SP$_k \equiv s_k$ are representably by machines, $\mathbb{M}_{s_k}$, as specified in $\mathsf{E}_{11/I}$, and elaborated in $\mathsf{E}_{11/III}$. This may be accomplished by showing that: **the operational scheme** in s_k is **fully reproducible by** $\mathbb{M}_{s_k}$, 'with appropriately defined components', $m_1, \ldots, m_N$ $(N = 14$ or 16, so far). To systematize this process, **the** s_k **must be formulated quasi-uniformly** (in spite of their disparate environments). This, in turn, is achievable only after **preliminary versions** of CD for the s_k have been developed. Although there are, in general, several types of synthesis in each of the classes C_k, it suffices, for now, to consider only typical cases of each class. In the eventual implementation of the Σ -OIB, all essentially distinct cases must be treated, but such an exhaustive procedure is unnecessary here.

Rather, the **structural** aspects of the s_k are to be compared with various **mechanisms** in 'associated machines', whose components are 'suggested' by the s_k, until adequate 'matches' are attained. Denote by s_k^* the quasi-uniform realizations of the typical elements of C_k, and, by s_k (as before) any precursor of s_k^*.

General Comments of STABILITY

In all of the S-P, conditions for **stability** are fundamental – and must be incorporated within the synthesis procedures. The stability criteria are diverse – depending on the underlying frameworks used to represent the control models. As a partial background for the s_k, references for the main species of stability conditions are collected here. Many of these references are cited again in the s_k, but it is convenient to list them separately, for easy comparison. See also $\mathsf{E}_{11/II}$ for citations.

For linear models, an extensive discussion of types / conditions is given in Schwarz / Friedland, '*Linear Systems*' (McGr-H, 1965); see also, Maciejowski, '*Multivariable Feedback Design*' (CUP, 1989). Several stability criteria for (**non**)-linear systems are discussed in: Ogata, '*State Space Analysis of Control Systems*' – for instance, the methods of Liapunov, and of Popov; see also: Letov, '*Stability in Nonlinear Control Systems*' (Princeton (transl.), 1961); Zubov: '*Mathematical Methods for the Study of Automatic Control Systems*' (Pergamon, 1962). For (electrical) **networks**, the stability problems are somewhat different (though ultimately equivalent): see, e.g., Kuo, FF, '*Network Analysis and Synthesis*' (Wiley, 1966), Ch. 10; Chen, W-K, '*Broadband Matching*' (World Scientific, 1988); Bode, HW, '*Network Analysis and Feedback Amplifier Design*' (van Nostrand, 1945); Janusz, A, et al., '*Computer-Aided Analysis, Modelling and Design of Microwave Networks – The Wave Approach*' (Artech House, 1996); Compton, Jr., RT, '*Adaptive Antennas*' (P-H, 1988); Paul, CR, '*Analysis of Multiconductor Transmission Lines*' (Wiley, 1994); Messerle, HK, '*Dynamic Circuit Theory*' (Pergamon, 1965); Grünbaum, FA, et al., eds., '*Signal Processing* IMA Volumes ##22, 23 (Springer, 1990), especially, #23; Maybeck, PS, '*Stochastic Models, Estimation and Control*', Vols. 1–3, (AP, 1982); Takayama, A, '*Mathematical Economics*' (CUP, 2nd ed., 1985).

Many other references could be given, but this is better left to individual sections (if necessary). Notice that both 'lumped' and 'distributed' systems are covered – and that, in accordance with circuit representations of scientific phenomena (of which G Kron's results, cited earlier in this Essay, are prime examples) the potential scope of the present discussion is extremely wide.

SP$_1$: Electrical Circuits and Their Simulations

These topics are subsumed by Modern Network Theory, which is usually formulated for classes $\mathcal{N}$ of (generic) networks N possessing (some of) the following properties.

π_1 : **real-valued** excitations and responses.

π_2 : **stationary** excitations and responses.

π_3 : **linear** components.

π_4 : **passive** components.

π_5 : **casual** responses.

π_6 : **reciprocal** responses.

For any set S, write $v_S := \{v_\lambda : \lambda \in S\}$. Recall the Σ -convention, $\bar{n} := (1, \ldots, n)$, for every positive integer n. Then: a standard network may be denoted as $N \equiv {}^{\phi}N \equiv N\{\pi_{\bar{6}}\}$; and its S-variants, by ${}^{S}N \equiv N\{\sim \pi_S, \pi_{\bar{6}\setminus S}\}$, where $\sim \pi_k$ stands for the negation of π_k. The **interpretations** of these properties are discussed in (e.g.) Chen, W-K, '*Broad-Band Matching...*' (World Scientific, 1988), with (necessary and) sufficient conditions – mostly, in terms of criteria for 'admissible signal pairs', as functions of t (time). See also: Weinberg, L, '*Network Analysis and Synthesis*' (McGr-H, 1962), op. cit., for wide-ranging treatments of most standard NW problems – within function-theoretic / graph-theoretic / (linear) algebraic / computational frameworks.

A succinct, yet discursive, presentation covering several types of passive / active, analog / digital (system) synthesis is given in: Papoulis, A, '*Circuits and Systems – A Modern Approach*'

(H,R&W, 1980). An important **extensions** of basic (E-M) NW theory, to cover **nonlinear** behaviour, is developed in the paper: '*The Generalized Lagrange Formulation for Nonlinear FLC Networks*', by HG Kwatny / FM Massimo / LY Bahar (IEEE Trans. Circ. Syst. CAS-29 (1982) 220–233), where a state-space representation is used. This formalism produces sets of coupled (non)linear (P)DE whose solutions furnish the complete solutions of the underlying network, $\mathcal{N}$, when suitable I(/B)C are specified. It follows, therefore, that such classes of network problems may be represented in terms of I/O assignments for (generalized) machines, $\mathbb{M}_{\mathcal{N}}$, corresponding to the set of (P)DE constructed. In other words: **this whole scheme yields full analysis / canonical-representation** for such classes of systems. The challenge of developing general **synthesis routines** on the basis of these **state-space formulations** raises many recondite problems, some of which will be considered later. See also, MacFarlane, AGJ, '*Dynamical System Models*' (Harrap, 1970), op. cit., for an extensive discussion of state-space models.

Thus: diverse types of (non)passive / (non)linear circuits / networks are rigorously representable as Σ -machines whose mechanisms are characterized by sets of coupled (non)linear (O)DE – or, in some cases, integro-differential equations. Indeed, this might even be extendable to cover sets of **stochastic** DE, etc., for **non**deterministic systems. Yet more generally, some networks / systems with **HS-operator components** may be treated. See, e.g., Zemanian, AH, '*Passive Operator Networks*', IEEE Trans. Circ. Systm. CAS-21 (1974) 184–193, and references cited there.

One possible path towards quasi-algorithmic **synthesis** of general systems, in terms of machines modelled by sets of coupled operator equations, may be sketched as follows. Since the extended Lagrangian / Hamiltonian representations (amongst others) have been shown to cover almost all NW/systems, $\mathcal{N}$, of scientific interest (including those with nonlinear / stochastic / time-varying components) it appears that: all systems constructed from **other** synthesis procedures (of whatever type(s)) may be **recast in 'coupled-operator' form(s)**, $\mathcal{N}_{\text{op}}$, with associated I/B data. It then follows (in principle) that **each step** of any 'original' synthesis algorithm may also be **recast** in operator format – to produce (retrospectively) some **direct algorithm** for $\mathcal{N}_{\text{op}}$.

One scheme of this kind is outlined in: '*Preliminary Simplifications in State-Space Impedance Synthesis*', by PJ Moylan (IEEE Trans. Circ. Syst. CAS–21 (1974) 203–205), where the principal reference is to: Anderson, BDO / Vongpanitlerd, S, '*Network Analysis and Synthesis*' (Prentice-Hall, 1973), for a variety of 'state-space routines'. Although this paper is concerned mainly with linear passive networks – through matrix transformations – the underlying procedures suggest possibilities (for suitable **non**linear operators) in more general NW/System settings. In the matrix formulation, the interrelations of the frequency-domain, and state-space, representations is transparent – since each is algorithmically obtainable from the other through relatively straightforward manipulations. A detailed example of state-space synthesis of linear, time-varying, passive networks with prescribed impedance matrices $Z \equiv Z(t,\tau)$ is given in the paper by Anderson, BDO and Moylan, PJ, IEEE Trans. Circ. Syst. CAS–21 (1974) 679–687, where the generic form of the state-space equations for Z is taken to be: $\#\{\dot{x} = F(t)x + G(t)u; v = H'(t)x + J(t)u\}$ – so that Z is expressible as: $Z(t,\tau) = J(t)\delta(t-\tau) + H'(t)\Phi(t,\tau)G(\tau)1(t-\tau)$.

Here, $u(.), v(.)$ denote the port voltage, and current, vectors of the network, N, and $\delta, 1$, the unit impulse, unit step function, respectively; while Φ is the **transition matrix** of F. The synthesis problem is to construct some network(s) N with state-space equations # reproducing the assumed form of Z.

NOTE that Φ is also called the **fundamental matrix**, satisfying: $\dot{\Phi}(t,\tau) = F(t)\Phi(t,\tau)$, $\Phi(\tau,\tau) = I$ (the Unit matrix). This formalism is treated extensively in (e.g.) '*State Space Analysis of Control Systems*', by K Ogata P-H, (1967). It is not necessary for **both** Z **and** $\{F, G, H, J\}$ to be specified, since each may be obtained from the other – by standard techniques of Linear Systems Theory. The construction of N from Z, though nontrivial, is still essentially routine (for appropriate selections of components). See, for instance, Anderson / Vongpanitlerd, op. cit., for some versions of these procedures – most of which are state-space versions of time /

frequency-domain schemes (rather than 'new' procedures). More general (state-space) realizations of various system (matrix) functions for classes of (non)linear networks are still being developed, and may be found in (e.g.) the **IEEE Trans. Circ. Syst.**, with further references.

A different approach to the representation of **nonlinear** systems is developed by Y Kamamura, '*A Formulation of Nonlinear Dynamical Networks*', IEEE Trans. Circ. Syst. CAS–25 (1978) 88–98, where each class of network variables is transformed so that it ranges over some 'standard **function space**'. The associated network (operator-) equations may then be studied / solved, through 'fixed-point solvability criteria'. The **scope** of this method includes **RLC-networks** with couplings among branches of different kinds (e.g., currents, voltages, charges and fluxes); networks with **active elements; distributed networks** (e.g., transmission lines); and, (dis)continuously **time-varying** systems. Solutions (if they exist) may be approximated **algorithmically**.

Although the issues of **synthesis**, for such networks, cannot be fully resolved by this formulation, it appears that (as for the quasi-Lagrangian scheme) syntheses obtained by other techniques may be **recast** in terms of the underlying (differential / integral) transformations, to furnish a basis for systematic methods.

Remark.

As always, the emphasis in Σ is on **general** schemes, subsuming (as far as possible) the multifarious syntheses routines in the (past and current) 'circuit / system literature' to investigate diverse theoretical / implementational problems. Further, since the principal aim is to **unify** all of the (paradigmatic) syntheses discussed in $\mathrm{E}_{11/II}$ – with a view to the ultimate formulation of **mechanisms** for β-**machines** – maximal **abstraction** is essential. In the following sections, general versions of synthesis procedures (for the other SP_m, $2 \leqslant m \leqslant 16$, as identified in $\mathrm{E}_{11/II}$) are outlined. After this, attempts are made to develop representations formally) covering **all** of the SP_k – so that tentative mechanisms, say, μ_β for $\mathbb{M}_\beta$, may be envisaged. For this task of unification, the most appropriate framework appears to be **the state-space representation of (non) linear / time-(in)variant / (non)deterministic dynamical systems**, of **(non)discrete** types. Plainly, such generally applicable representations will not always by the most elegant / effective – but they should, in principle, furnish a quasi-universal scheme, for control problems, potentially adaptable / extendable to arbitrary β-machines.

Notice that both digital and analog **feedback** (F-B), of sampled data from various 'signals', are used routinely to produce **compensators** for the classes of (technological) systems where many of the basic control concepts originated. These practical aspects of design actually **enhance** the Σ -potential of control procedures – by allowing for **imperfections** of components, **noise, perturbations**, etc., as essential elements of the design process. Again, there are significant overlaps in the aims / techniques / results of optimal / adaptive / stochastic control theory, for (non)discrete / (non)linear systems. On the other hand, the distinction between **open-loop** and **closed-loop** subsystems must always be recognized. These similarities / differences – along with the interrelations among **networks / dynamical-systems / (finite) automata** – must be rigorously observed in all calculations, within suitable defined **time / frequency / state** domains.

All of the main types of control problems (**continuous / digital / optimal / adaptive / stochastic**) for **linear** / time-dependent systems, are discussed in: Johnson, A, '*Process Dynamics, Estimation and Control*' (Peregrinus / IEE, 1985). This (partially simplified) scheme also covers **time-delay** effects and **linearization**. Stochastic components are incorporated through 'additive **white noise**'. The underlying frameworks are finite-dimensional **state-spaces**. The corresponding **rigorous** schemes for **nonlinear / distributed / delay / stochastic /...** models, require **infinite-dimensional** state-spaces, and involve **functional / (P)DE, Viewner processes** and **stochastic (P)DE** – but the **design aims**, and **behavioural scope**, differ significantly, from those of the simplified models, relatively rarely – for 'practically important systems' in diverse scientific / mathematical domains. Moreover, **differential geometric** techniques allow the **exact linearization** of large classes of nonlinear models (through **synthesizable feed-**

back procedures); see Isidori, A, '*Nonlinear Control Systems: An Introduction*' (LNCIS #72, Springer, 1985).

In cases where nonlinearities / argument-deviations / stochasticity /... cannot be simplified, without radical change of models, more elaborate mathematical apparatus must be used, within the overall formalism of **Dynamical Systems Theory** – where state-space representations are standard, and may be adapted to model diverse networks, (sequential) machines / automata, and other sorts of I/O processors – of which the Σ -machines, $\mathbb{M}_{\beta}$, are (very general) examples, many combining a selection of basic submachines, of various types, to produce the ME/SE β as (partial) output. It is important to realize that, no matter what type of control model is considered, issues of (quasi-)**stability** of synthesized systems must be resolved – within the time / frequency / state /... domains, for which a variety of criteria have been developed. Although this aspect of Control Theory may appear to be (at best) tenuously related to the analysis / synthesis of (inverse) analogues, appearances can be highly deceptive – as examples will show!

The basic representations may be briefly characterized as follows. **Time-domain** procedures incorporate the 'history' / IC and seem natural for many (mechanical) engineering applications. **Frequency-domain** (or, '**transform**') schemes are especially appropriate for electrical / communications networks (e.g., via manipulations involving pole / zero distributions of **transfer functions**, and inverse transforms).

NOTE. A very general scheme, subsuming many classes of synthesis problems for (some) **linear, multivariable systems** is outlined in LNCIS #58, pp1–15 (AC Antoulas).

Whereas the interrelations of the time-domain and frequency domain schemes are transparent for **linear** (multivariable) systems, the general treatment of **nonlinear** systems, in either of these domains, is problematical and typically intractable, without complicated approximation procedures. The **state-space equations** are essentially 'equations of motion' (sets of coupled, first-order (P)DE) – counterparts of the Lagrange / Hamilton equations for dynamical systems, and so applicable to arbitrary **nonlinear** models. The primary distinction between 'state' and 'time' analyses lies in methodology: **I/O relations**, for time-domain, and **DE** for state-vectors (or state-functions). For linear models, time-response is studied partly through (Laplace) transform techniques – rather than by the (direct) solution of sets of coupled DE (as in the state-space approach).

In all **closed-loop** models there is some form of **data-sampling**. The prime distinction between general **feedback-control** and **adaptive control** is that adaption must produce high-performance results even under 'extreme' changes in system-parameters and 'large' (external) perturbations. Further classifications, to incorporate various **optimization criteria** – and extensions to cover diverse **stochastic behaviour** – may be introduce. The resulting, comprehensive, Control Theory accommodates all of these model-types as essentially separate, yet multiply-interrelated, (partial) realizations. The ultimate step, for Σ , embodies all of the control paradigms to 'suggest' possible mechanisms for the machines $\mathbb{M}_{\beta}$. On this account, the principal defining features of each control type are outlined separately – with the eventual aim of formulating a quasi-universal scheme, from which each of the more specialized models may be derived, on the basis of precise conditions.

The inclusion of stochastic effects requires the use of **S**tochastic **C**alculus – especially, the solution of (sets of) (non)linear **SDE**, of various types. The relevant techniques depend on several preliminary results – to establish the validity of counterparts to the standard processes of 'classical calculus': for instance, '**differentiation**' of (composite) functions; '**integration**' (in specified senses); properties of **sequences / series / substitutions**; interchange of limit operations.

As it turns out, most of the 'differential properties' are specified in terms of corresponding 'integral properties' (as in the theory of Generalized Functions). Moreover, there are three basic types of SDE, governed by (i) random I/BC; (ii) random 'forcing functions'; (iii) 'random elements' in the (non)linear SDO. Typical problems are of mixed character. The case of linear (or, linearized) O(S)DO has received most attention, but general theories of random operator equations (over specified classes of spaces) have also been investigated – allowing for 'arbitrary nonlinearities', as well as a variety of exotic algebraic / analytical / topological /... behaviour in the (S)DO (though the main results are, obviously, less complete than those for the 'simplified models'). For the purposes of Control Theory, it is assumed that all of the SDE encountered are (at least, formally) solvable – in appropriate senses, such as 'mean-square' / 'sample-function' /.... The 'solutions' are stochastic processes whose characteristics are representable in terms of those in the I/BC, 'forcing', or SDO. The following references cover the main techniques / results.

McShane, EJ, '*Stochastic Calculus and Stochastic Models*' (AP, 1974); Miller, KS, '*Complex Stochastic Processes*' (A-W, 1974); Balakrishnan, AV, '*Applied Functional Analysis*' (Springer, 1976); Maybeck, PS, '*Stochastic Models, Estimation and Control*', Vols. 1–3 (AP, 1982).

SP$_2$: Sampled-Data Control Systems.

These models are especially pertinent when (digital) computers are among the components (e.g., for 'real-time processing'). Although there is some overlap with feedback / optimal control, a distinctive theory of sampled-data (S-D) systems may be developed – over time / frequency / state-variable frameworks. A typical S-D system has both **discrete** (computer) and **continuous** ('generated') variables – and thus requires a combination of finite-difference and differential (in)equations to produce adequate models. In practice, the so-called **z-**, and **w-, transforms** are extensively used – in conjunction with other analytical / algebraic procedures – to derive (e.g.) the associated **pulse-models**, and to employ **compensation** (to achieve prescribed response criteria). As in SP$_1$, only the fundamental schemes are of interest for Σ – as possible paradigms of the mechanisms of machines $\mathbb{M}_\beta$. This view is taken for all of the SP$_k$ treated here, since detailed implementation is of relatively minor importance in the study of general analogues.

Techniques for **the design of sampled-data systems**, within a state-space framework (using z-transforms, confound mapping, contour-integral representations, etc. to complement the 'state scheme') yield various classes of **transfer functions**, for which **pole-zero distributions** (determining response characteristics) may be analyzed / synthesized. See, e.g., the references cited in E$_{11/II}$; and, for a unified treatment of **invariant, deterministic** models, Nagrath, IJ / Gopal, M, '*Control Systems Engineering*' (2nd ed., Wiler, 1982), op. cit.[650]. For **adaptive** / **stochastic** / **optimal** control, see the comments / citations in SP$_3$, SP$_7$. The (general) case of **time-varying linear models** may be treated systematically[651]: see, e.g., Stubbard, AR, '*Analysis and Synthesis of Linear Time-Variable Systems...*' (U. Calif. P., 1964). See also: Maciejowski, JM, '*Multivariable Feedback Design*' (A-W, 1989), for linear, invariant, **continuous** systems)[652]. Partial extensions of (some of) these schemes to cover time-variable / 'strongly-nonlinear' models have been explored, but (mainly) it is necessary to introduce different techniques, for which references will be given later. The most pragmatic approach involves **combinations** of **stubberud's** procedures (and their more recent developments) with the **exact linearization** techniques due, for instance, to A Isidori: '*Nonlinear Control Systems...*' (LNCIS #72, Springer, 1985), op. cit.. Thereafter, blends of such **exact** schemes with **rigorous approximations**, of various forms, will extend the purviews of synthesis routines to 'arbitrary' systems. The attainable levels of precision (of analysis / synthesis) obviously depend on the detailed structures of the sets of (in)equations specifying the systems / networks considered – on which the general Σ -machines, $\mathbb{M}_\beta$, are modelled. **These remarks apply (m.m.) to all species of control network problems**. All of the resulting (regulated!) imprecisions entailed by this strategy are reflected in the enlarged scope of w.d. (inverse) analogues 'derivable' from the associated machines. On this basis, therefore, 'arbitrary deterministic models' may be regarded as 'effectively analysable / synthesizable'.

[650] See also, MacFarlane, AGI, '*Dynamical System Models*' (Harrap, 1970), op. cit.

[651] Another approach is given in; Liang, TJ, IEEE Trans. Circs. Syst. CAS–25 (1978) 31–40; see also Balakrishnan, in Zadeh / Polak, op. cit.

[652] Most of these procedures are readily modified for (partially) discrete (e.g., S-D) models

The essence of S-D systems lies in the blend of discrete and continuous ('analog') signals – which must be reflected in the mathematical formulation(s). The transfer-function (frequency) method is (strictly) limited to linear, invariant systems; whereas, the time-domain / state-space formalism(s) may be deployed for all types of models. The crucial consideration, however, is that transfer functions yield (almost) immediate syntheses, while, state-space design procedures are typically more intricate, but potentially applicable to nonlinear / time-variable / multivariable models. Examples of time / state routines for S-D design are given in (e.g.): Franklin, GF / Powell, JD, '*Digital Control of Dynamic Systems*' (A-W,1980), where many other aspects of digital control are also treated. In particular, it is shown that: in general, an S-D system is **inherently time-varying** – and that (hence), not every S-D system admits a transfer function. For an extensive treatment of the transfer-function formalism, covering a wide range of linear models, see: Schwarz, RJ / Friedland, B, '*Linear Systems*' (McGr-H, 1965). **NOTE that** most of the synthesis algorithms for **continuous** models are readily adapted to cover **discrete**, or **discrete / continuous** counterparts. This appears to extend even to the (exact) linearization procedures (e.g., via 'suitably designed feedback'), and so to include nonlinear S-D systems – which are thus effectively synthesizable.

The calculation of linearizing feedback for specified classes of nonlinear systems is based on certain decomposition / invariance properties of linear systems, and the (partial) generalization of these properties to the 'flows' generated by nonlinear systems. This is accomplished through a blend of nonlinear system theory and differential geometry over manifolds, as expounded in (e.g.): Warner, FW, '*Foundations of Differential Geometry and Lie Groups*' (Scott Forsman, 1971 / Springer, 1983), where many (complementary) references are given, for cognate aspects of geometry / control-theory; and, Schetzen, M, '*Volterra and Wiener Theories of Nonlinear Systems*' (1st ed., Wiley, 1980). Notice that all of the procedures cited in this section may also be applied in cases where optimal / adaptive / stochastic criteria are involved – as will be assumed in the subsequent discussions. The main results, for (linear) **S-D models**, are, typically algorithms for some transfer function(s) – from which syntheses are readily implementable (see, e.g., Budak, A, '*Passive and Active Network Analysis and Synthesis*' (Houghton Mifflin, 1974)).

SP$_3$: Optimal / Feedback / Adaptive Control Systems.

Optimality criteria may be formulated for any type of control model. The satisfaction of such criteria depends on the detailed structure of the model – and leads to corresponding **synthesis** routines. The (continuous) monitoring of various intermediate / final outputs, to ensure that prescribed performance conditions hold, may be viewed as optimal **adaptive** control. This is an extreme form of **feedback** control, covering large / unpredictable changes in system / environmental parameters. In this section, several variants / combinations of these models are considered. It is therefore convenient to begin with 'basic feedback models'.

Remarks.

(a) One rigorous synthesis scheme devolves on the **realization** of systems with **prescribed state-space equations**, corresponding to some **rational transfer functions** – where such realizations incorporate specifications of system components and their interconnections. These procedures hold formally only for linear time-invariant models, but some extensions to classes of time-varying models may be developed. See: T. Takahashi, et al., IEEE Trans. Circs. Syst. CAS–25 (1978) 79–88, for the basic formulation of this technique.

(b) All of the schemes discussed so far – for **(non)linearizable / time-(in)variant models** may involve **feedback**, in various forms, and hence may be considered within **unified frameworks** for **the synthesis of feedback controls**. Recall, also, that **effective equiva-**

lences between '**circuit / network**' and '(dynamical) **system**' models may be established through appropriate state-space representations. See: Stern, TE, '*On the Equations of Nonlinear Networks*', IEEE Trans. Circ. Th. CT–13 (1966) 74–81 (op. cit.) for the original procedure.

(c) Extensive discussions of **pole / zero distributions** of transfer functions (of several kinds) may be found in books on linear, time-invariant networks / systems. In particular, procedures for 'shaping the response' to various types of signals are formulated in terms of 'pole / zero modifications'. See, e.g., Budak, A, '*Passive and Active Network Analysis and Synthesis*' (Houghton Mifflin, 1974); Weinberg, L, '*Network Analysis and Synthesis*' (McGr-H, 1962); Schwarz, RJ / Friedland, B, '*Linear Systems*' (McGr-H, 1965); Maciejowski, JM, '*Multivariable Feedback Design*' (A-W, 1989). Related results for **digital filters** are discussed in (e.g.): Antoniou, op. cit., and Williamson, op. cit..

As already mentioned, various types of **feedback** are ubiquitous in synthesis procedures – since all feedback embodies 'corrections to undesirable system (-output) behaviour'. It is, therefore, neither viable nor desirable to treat feedback techniques in isolation. Rather, the 'feedback aspects' of optimal / adaptive / stochastic /... control schemes are briefly identified here, and reintroduced in the outlines of compound control schemes. Moreover, broad equivalence between specified (classes of) closed-loop systems, and corresponding open-loop systems, somewhat complicate the general classifications. The issues of **realizability, controllability, observability** and **stability**, are diversely interconnected. The original definitions were for deterministic, linear, invariant models; but subsequent extensions cover nonlinear, time-variant, stochastic models. Further, 'transfer-function models' assume null initial conditions, and refer only to I/O, ignoring 'internal states' – which may develop 'local instabilities', requiring modification through corresponding localized feedback. These internal processes may also enhance overall system performance. The most serious limitations of 'standard transfer-function procedures', however, lie in their **inconclusiveness**, and their inability to identify unrealizable specifications. Thus, there is in general no rigorous prescription for approaching an arbitrarily prescribed system – even if, by chance, it **is** realizable. Again, for complicate (e.g., multiple I/O) systems, the block-diagram reductions become progressively intractable; but, in such cases, the use of signal-flow graphs obviates the need for reduction.

On the other hand, if some appropriate system function is (somehow) completely specified, then realizability criteria may be applied, and (in favourable instances) the corresponding system(s) may be exactly synthesized – possibly, in several distinct ways. If only the I/O relation, say, $O = f(I)$, is prescribed, then the system-function method may be hard to implement, even approximately, through iteration; then, time-domain / state-space procedures must be used. For Σ , where the underlying aim is to investigate mathematical analogy on the basis of arbitrary synthesis-paradigms for (quasi-)machines, $\mathbb{M}_\beta$, the relative (dis)advantages of various techniques are not of paramount importance – compared with the associated possibilities for exploring stimulating (inverse) analogues of ME/SE, β.

Another fundamental issue arises from the **physical limitations of implementation** (through circuits / systems) of prescribed / calculated system / model functions – especially where **non**-rational functions are involved. The resulting problems require that the prescribed behaviour be **approximated** by combinations of rational elements: the so-called **Approximation Problem**. Thereafter, the main task becomes that of **synthesizing** some network(s) / model(s) corresponding to these rational approximations – with the extra constraint(s) of using (only) **specified types** of model-components. The general cases of **active**, and **passive**, synthesis are treated separately. Typically, non-rational system characteristics are obtained graphically, but other possible sources include formulations in terms of (e.g.) algebraic / transcendental functions (including various Special Functions).

The reduction / elimination of (quasi-)experimental content in feedback (and other) design schemes may be achieved by various optimal control procedures, characterized by (e.g.) the determination of control functions (say, $u \equiv u(t)$) to minimize some performance index, $J \equiv \int_{t_0}^{t} F[\underline{x}(t), \underline{u}(t), t]dt$, where $\dot{\underline{x}}(t) = f[\underline{x}(t), \underline{u}(t), t]$, with $\underline{x}(t_0) = \underline{x}_0$. This formulation covers most of the standard engineering criteria, and may even be extended (through the use of 'mean-square integrals') to problems of stochastic control.

For deterministic, linear, invariant systems with single I/O, the **A**nalytical **D**esign scheme (within the frequency domain) has $J = J_e + J_u$ – the sum of integral squares of the error signal, e, and the control signal, u. In variational problems, where one of the J_e, J_u is to be held fixed, the index becomes $J_e + k^2 J_u$, or $J_u + k^2 J_e$, where k^2 is a 'Lagrange multiplier'. The method is based on using Parseval's theorem, to express J_e and J_u in terms of the LT, E, U, of e, u : $J_e = \frac{1}{2\pi i}\int_{i\mathbb{R}} E(s)E(-s)ds; J_u = \frac{1}{2\pi i}\int_{i\mathbb{R}} U(s)U(-s)ds$. After this, the minimization of J involves function-theoretic techniques, including **spectral factorization**[653], to obtain system functions with constrained pole/zero patterns. For **multi**-I/O models, the spectral factorization of **matrix-valued** functions[654] must be obtained. In both cases, types of Wiener-Hopf integral equations become solvable over the frequency domain, and the syntheses may be implemented in terms of optimal closed-loop transfer functions. For single I/O models, a full treatment is given in Newton / Gould / Kaiser, op. cit.. The **limitations**[655] of this 'analytical method' – to linear, invariant systems, with quadratic performance index, null IC / no time-constraints – made it essential to develop more widely applicable techniques: the **state-variable procedures**.

It transpires that state-variable formulations are valid for nonlinear, time-varying multi-I/O models, with a broader class of performance indices – and even some extensions to stochastic models.

The relation of general (rational) **transfer functions** to **state-space representations** is exhibited in (e.g.) Nagrath / Gopal, '*Control Systems Engineering*' (2nd ed., Wiley, 1981); see also, Ogata, '*State-Space Analysis of Control Systems*' (P-H, 1967) for a detailed treatment. Although this covers only linear, invariant models, the state-space formalism is (as already mentioned) valid also for nonlinear / time-variable models. For optimal control problems, there are, essentially, two types of procedures – apart from the direct, or 'calculus', methods, which are rarely applicable: (a) by extension of the **Calculus of Variations**; (b) through the use of **Dynamic Programming**. Additional issues, such as **stability** may be handled by other methods (e.g., **Liapunov** criteria). For a **direct** method, see, e.g., Ogata, op. cit., Sec. 9.2. An example of '**calculus-based minimization**' may be found in Nagrath / Gopal, op. cit., Sec. 13.2; see also, ibid., Sec 13.3, for a **transfer-function procedure** involving **spectral factorization**. For **stochastic** systems, the variational (Pontryagin) scheme cannot be used, but variants of dynamic programming remain valid. Both of these approaches are treated (briefly but soundly) in: Intriligator, MD, '*Mathematical Optimization and Economic Theory*' (P-H, 1971).

More detailed discussions may be found in (e.g.): Pontryagin, LS, et al., '*The Mathematical Theory of Optimal Processes*' (Wiley, 1962); Larson, RE / Casti, JL, '*Principles of Dynamic Programming*', Part II (Dekker, 1982); Tu, PNV, '*Introductory Optimization Dynamics*' (Springer, 1984). Models involving random processes are considered later – though certain types of stochastic input may be covered by the methods given here; notably, even in the original 'AD scheme' in Newton / Gould / Kaiser, op. cit.. The main OC schemes correspond, respectively, to 'classical **C**alculus-**o**f-**V**ariations, Hamilton-Jacobi (in)equations (for **D**ynamic **P**rogramming) and **E**xtended **C**alculus-**o**f-**V**ariations (Maximum Principle). There are many cases where more than one scheme is applicable; but only DP is valid for general stochastic models. The OC of deterministic systems represented by ODE is treated rigorously, and fundamentally, in: Lee, EB / Markus, L, '*Foundations of Optimal Control Theory*' (Wiley, 1967), where the interrelations among the main approaches are discussed systematically, from the viewpoint of the **Qualitative Theory of (O)DE**. Various techniques for adaptive control may be linked with **feedback, learning, optimization**, or **model-following**. (MRAS: see Landau, YD, '*Adaptive Control...*' (Dekker, 1979)).

---

653 See, e.g., Carrier, GF, et al., '*Functions of a Complex Variable...*' (McGr-H, 1966), Ch. 8; Youla, DC, et al., IEEE Trans. Aut. Control AC–21 (1976) 3–13

654 Youla, DC, et al., IEEE Trans. Aut. Control AC–21 (1976) 318–338; Jones, DS, PRSL A434, 419–433 (1991)

655 Even servo-mechanisms are excluded

SP_4: Dynamical Systems.

This class of systems essentially coincides with the class of state-space formulations. A DS is represented by a set of constrained (O)DE, possibly depending on various parameters. In this sense, every DS could be associated with some control problem(s) – for instance, in investigations of bifurcation-phenomena / chaos. The relative scope / utility of state-space, and frequency-domain formulations of Control Theory must be assessed for each broad species of models considered. The scheme developed for (non-)linear multivariable feedback design in the book by Maciejowski, op. cit., demonstrates that the advantages of time-domain / state-space procedures is not universal. Nevertheless, for versatility / range of applicability, the '**DS-view**' of analysis / synthesis is unsurpassable – even if other frameworks may seem 'more natural', in certain contexts. The case of deterministic, linear, time-(in)variant models is treated comprehensively in: Ogata, K, '*State Space Analysis of Control Systems*' (P-H, 1967), while detailed results for some aspects of **nonlinear** models are given in Lee/Markus.

Recall that broad classes of scientific phenomena may be modelled by (electric, or equivalent) circuits; and, further, that these circuits may be represented over state space(s) by (sets of coupled) **differential equations** (see, e.g., the results of TE Stern, '*On the Equations of Nonlinear Networks*', IEEE Trans. Circ. Th. CT–13 (1966) 74–81, op. cit.).

It follows that the state-space approach essentially covers all dynamical ME/SE – with appropriate modelling / interpretation. Consequently, all of these ME/SE may be associated with Σ -machines, and so they may, in principle, be (quasi-)synthesized through generalized state-space schemes – and then investigated for (inverse) analogue / (quasi-inverse) analogue, structure.

SP_5: Automata.

Although the formal definitions are typically framed in terms of countable/finite I/O, and much of the basic theory is algebraic in character (involving finite fields, semi-group realizations, etc.), it still transpires that many important aspects of circuit / systems may be investigated within **G**eneral **A**utomata frameworks. The references given in $E_{11/II}$ [briefly: MRI Symp. Proc. XII (Wiley, 1963); Booth, TL, (Wiley, 1967); Kondo, K (ed.), RAAG Memoirs (U. Tokyo 1962–1968); Hartmanis / Stearns (P-H, 1966); Kobrinskii / Rakhtenbrot (N-H, 1965)] are elaborate / supplemented here, and some remarks on realization / synthesis are made. The elaboration consists, mainly, of the identification of specially pertinent papers (in Proc.), or chapters (in books); the supplementary references broaden the discussion.

The MRI Symp. Proc. #XII contains several seminal papers on the scope / structure of automata. In particular, the work of T Sunaga views automata as dynamical systems (DS), for which state-space techniques are applicable – as well as the representation theory of semi-groups / hypercomplex-numbers. Further, the paper by M Iri (pp. 415–435) treats 'general information networks within topological frameworks' – including **linear electrical circuits** (of both **nodal**, and the dual **mesh** types). The techniques for electrical networks are expounded more formally in: Slepian, P, '*Mathematical Foundations of Network Analysis*' (Springer, 1968). Major developments of electromagnetic network theory may be found in the RAAG Memoirs, Vols. I–IV – together with topological / algebraic procedures for the analysis / synthesis of oriented switching circuits, matroid-theoretical schemes, and many closely related topics. Another approach to the 'structural solvability, by graph / matroid methods, of arbitrarily complex (non)linear systems, and of electrical networks, may be found in the books: Murota, K, '*Systems Analysis by Graphs and Matroids*' (Springer, 1987), and Recski, A, '*Matroid Theory and its Applications*' (Springer, 1986).

Another link between **A**utomata **T**heory and **General System Theory** is discussed in: MRI Symp. Proc. XXI, pp561–589, by BP Zeigler: '*Modelling and Simulation: Structure-Preserving Relations for Continuous and Discrete Time Systems*'. In particular, realizations of discrete-time systems by continuous-time systems are considered. Questions of information-flow, decomposition, and synthesis, for sequential machines are treated in Hartmanis / Stearns, op. cit.. A modified theory of transfer functions, and an account of random processes in sequential machines are among the diverse topics considered in Booth, op. cit..

SP$_6$: Threshold Logic Gates.

In its broad interpretation this topic encompasses much of the logical design of (general) **switching** circuits, where synthesis procedures are a principal concern. Accordingly, the reference list in E$_{11/II}$ is augmented here, and comments are made on the nature (and interrelation) of fundamental issues in this area. As usual, the extra bibliography could be far more extensive, but it suffices to indicate the primary research challenges – and their connections with Σ -machines. Various links with adaptive control and decision theory are well known. The emphasis here is on **syntheses** of (digital) **logic systems** – for which a wide range of schemes may be found in the literature.

The most basic scheme involves **S**ingle-**T**hreshold-**E**lement **realization** of n-variable **B**oolean functions, where (apart from existence theorems) there may be constraints on **reliability**, the maximum value of n, etc.. The central problem is that of **synthesis of TE-networks** (for which the STE realization is prerequisite). Necessary and sufficient conditions for STE-realizability of (arbitrary) B-functions involve intractable nonlinearities, so **approximation** procedures, allowing **exact iterative realization**, are developed. On this basis, network-synthesis routines may be formulated, with adequate characteristics (for the number of elements, mode(s) of interconnection, etc.). Several methods may be used, but there are, broadly, two frameworks: **B-Algebra**, and **Rademacher-Walsh transforms**. These aspects are covered by Dertouzos, op. cit..

Somewhat more general schemes – for sequential (switching) circuits (with 'memory') depend on properties of **state assignments / transitions**; 'abstractions' of this specification are classified as (finite) **automata**. A detailed treatment of state-assignment routines is given by Haring, op. cit., and by Lewin, D, '*Design of Logic Systems*' (Van Nostrand, 1985). The treatment by Lee, op. cit., utilizes **B-calculus, multi-valued logics**, and several other mathematical subtheories. A near-exhaustive discussion, using **function theoretic results over prime-power fields**, is given in: Moisil, Gr. C, '*The Algebraic Theory of Switching Circuits*' (Pergamon, 1969), where diverse types of elements are considered. Some links with **adaptive control** may be found in: Tsypkin, Ya X, '*Adaptation and Learning in Automatic Systems*' (AP, transl., 1971). Since automata may be viewed as dynamical systems, the potential links with Σ -machines are evident.

SP$_7$: Stochastic Control Systems.

This class of systems allows for random elements in the model (parameters), the model environment, or the presence of 'noise' and other uncertainties. This is not the same as the cases of random signals in otherwise-deterministic systems – for which the standard (time / frequency / state-space) frameworks may still be used. Nevertheless, it is convenient to include random inputs here, even though they do not necessarily require stochastic-calculus procedures. Typically, the output of a SCS is a stochastic process with desirable characteristics – produced by feedback, or by other devices, all to be calculated from the model / environment data. The generality of this specification entails that there are several (partially) distinct types of SCS problems, with a cognate collection of solution techniques.

The associated design problems, covering all forms of (quasi-)synthesis, include the following tasks.

(a) Formulation of models via Stochastic (O)DE

(b) Parametric optimization of specified $\{$system, regulator$\}$ pairs, with unknown parameters

(c) Construction of some control law(s), for given {system, criterion} pairs.

Among the prescription admitting rigorous solutions are: the **Linear / Quadratic / Gaussian** problem (treated in detail by Maybeck, op. cit., Vol. 3); the '**controlled diffusion** model' (see, e.g., Krylov, NV, '*Controlled Diffusion Processes*' (Springer, 1980)) and **controlled Markov processes** (see, e.g., Fleming, WH / Soner, HM, '*Controlled Markov Processes and Viscosity Solutions*' (Springer, 1993)). For the class of Sequential Decision Models, rather complete treatments of several systems (via dynamic programming schemes) may be found in: Bertsekas / Shreve, '*Stochastic Optimal Control: The Discrete-Time Case*' (AP, 1978). In all of the above models, issues of **calculability / approximation / convergence / stability** are resolved.

The connections between **filtering / prediction** and **stochastic control** are discussed in: Åström, KJ, '*Introduction to Stochastic Control Theory*' (AP, 1970). The **Wiener-Kolmogorov scheme** requires the solution of (n-dimensional) **Wiener-Hopf integral equations** (mostly, an intractable task, even for analytical approximation). A generalization of Wiener-Kolmogorov theory, embodying a **recursive** procedure, and valid even for **non-stationary** stochastic processes, is known as the **Kalman filter technique** – as presented in Kalman, RE, ASME J. Basic Engrg. 82 (1960) 34–45; Kalman / RS Bucy, ibid. 83 (1961) 95–107 – which requires the solution of (matrix-)**Riccati equations**. Questions of stochastic controllability / observability for (non)linear models are treated in Maybeck, op. cit. (esp. Vol 3).

SP$_8$: Antenna Design.

There are two broad classes of design problems here:

(a) design to produce sufficiently sharp **reception** of prescribed types of 'signals';

(b) design to (re)produce specified (transmitted) **radiation patterns** to prescribed tolerances.

Both of these 'prototypes' are rich in analogical potential. The **planar** problem may be treated rigorously in terms of 'prolate spheroidal' (and related) functions; see Rhodes, DR, '*Synthesis of Planar Antenna Sources*' (OUP, 1974). Other types of problems may be classified according to the **form(s) of radiation** involved, and the **dimension**: linear / planar / spatial. In addition, issues of **control** may arise – to maintain / modify antenna behaviour under (non)deterministic variations of the environment(s) / component(s). A variety of types / techniques may be found in: Jones, DS (1964, 1979) op. cit., Jordan, EC / Balmain (1968) op. cit., Hansen, RC (ed.) (1985), op. cit., Moulin, EB (1949), op. cit..

SP$_9$: Spectral Function Theory.

This topic, unlike most of the other SP, is primarily mathematical – though there are important implications for Applied / Applicable Mathematics. The basic ideas involved linear transformations of VS, and aspects of Generalized Harmonic Analysis (discussed in SP$_14$). The underlying syntheses stem from Fourier Analysis / Synthesis, but the purview is far broader – as may be appreciated from the fundamental reference: Nikol'skii, NK, '*Treatise on the Shift Operator...*' (Springer, 1986). For an operator T acting on BS X, elements $x \in X$ of the set $\bigcup_{\lambda,n}\{\text{Ker}(T-\lambda I)^n : n \geqslant 1, \lambda \in \mathbb{C}\}$ are called **root-vectors** of T; and T is called **complete** IFF:

$$V(\text{Ker}(T-\lambda I)^n : \lambda \in \sigma_p(T), n \geqslant 1) = X,$$

where $V(\ldots) \equiv$ **closed linear hull**, and $\sigma_p(T) \equiv$ **point-spectrum** of T. Then: T **admits spectral synthesis** IFF every T-invariant subspace $Y(\subset X)$ is generated by root-vectors of Y.

Another form of spectral synthesis may be defined within the frameworks of diophantine approximation and almost-periodic (AP) functions. The so-called **Pisot numbers** (real algebraic numbers $\theta > 1$, all of whose **associates** have norms > 1) play a basic role in this theory. In all variants of spectral synthesis, the **spectra** of certain functions φ_λ are defined to be the **supports** of the Fourier transforms (FT) of the φ_λ. In general, the problem of **harmonic / spectral synthesis** may be formulated as:

> Can an ME be reconstructed from the set of 'its harmonics' – and, if so, how may this be done?

Although Fourier series / integrals furnish the underlying model (for real-valued functions), notions of harmonics / reconstruction may be defined very broadly, in relation to transformation groups, **B**anach algebras, etc. – with corresponding types of synthesis. The relevance of such procedures for Σ list in the overall methods, which may be interpreted multifariously in the contexts of β-machines. See, e.g.: Meyer, Y, '*Algebraic Numbers and Harmonic Analysis*' (N-H, 1972); Khavin, VP / Nikol'skii, NK (eds.) EMS Vol. 15, '*Commutative Harmonic Analysis: I*' (Springer, 1991).

SP$_{10}$: Integral Geometry.

In the synthesis context, all variants of Integral Geometry may be viewed as **inverse problems**. It is treated here as a separate topic because of the diverse literature on this subject. Indeed, every procedure for solving one of the associated inverse problems constitutes a w.d. synthesis technique. A good reference for this point of view is: Romanov, VG, '*Integral Geometry and Inverse Problems for Hyperbolic Equations*' (Springer, 1974). The syntheses are, typically, in the form of integral representations of projections onto families of subspaces (hyperplanes, in the simplest cases). Although this is a rudimentary sort of synthesis, there is scope for the reproduction of extra structural features.

Obvious applications include: **Tomography Geophysics, Image Reconstruction, Stereology**. See: Serra, J, '*Image Analysis and Mathematical Morphology*' (AP, 1982); Wechsler, H, '*Computational Vision*' (AP, 1990). **See also** the **BR** cited in E$_{11/II}$ (esp., Santaló, and Ambartzumian).

S$_{11}$: Chemical Engineering Plant Design.

The synthesis schemes here are primarily concerned with the realization and identification of diverse (non)linear processes of industrial importance. It transpire that comparable procedures govern '**Scientific System Design**' in general – and the BR in E$_{11/II}$ cover (non)linear / (non)deterministic models in many fields. Here, indications are given of some fundamental techniques.

The **Volterra Theory** is based on extensions (to functionals) of Taylor series (for functions). The Fréchet version (1910) shows that: any continuous functional is representable by a series of (canonical) functionals, of integer orders – and, that the convergence is uniform over each compact set of continuous (argument-)functions. The Volterra canonical functionals have the form of multiple integrals:

$$H_n[x(t)] := \int_{\mathbb{R}} \cdots \int_{\mathbb{R}} h_n(\tau)_n \prod_{1 \leqslant k \leqslant n} (x - \tau_k)(d\tau)_n,$$

and a natural generalization of the **Weierstrass Approximation Theorem** demonstrates that the set $\{H_n[x(t)]\}$ is **complete**.

This completeness makes it possible to model broad classes of systems by functional expansions involving extra **parameters** – to be 'evaluated' by various procedures: **model identification**. However, both convergence and identification have significant limitations in Volterra's theory, though it is still widely useful. Some of these limitations are overcome in the theory initiated by Wiener, in the 1930s, which is based on **Wiener-integral representations** and **orthogonal functionals** (constructed from the $H_n[x(t)]$) for White-Gaussian input. This allows a much broader class of nonlinear systems to be treated, within the extended framework.

Applications to chemical plant design are discussed in Seinfeld / Lapidus (1974), which contains copious references. The theory / applications of both schemes may be found in Schetzen (1980), in considerable detail, for a variety of models in The Sciences. The emphasis on chemical processes in this SP stems mainly from their prototypicality for plant design. Chemical reaction theory is covered by SP_{13}.

Another form of chemical-plant design – logically preceding the Volterra / Wiener schemes – involves the selection / combination of basic processes to perform specified tasks. This, too, may be extended to tasks within any scientific / mathematical field, but the terminology and development of this point of view originated in **Chemical Engineering**, where the basic processes are (usually) called **unit operations** (UO). In this way, and ME/SE may be broadly represented as some (weak) combination of basic procedures, selected from lists of admissible operations.

The main distinguishing feature of the UO (compared with the elements of **C**alculational **B**ases: see E_4) is that each of the UO corresponds to the physico-chemical process(es) characteristic of a single subdomain of Chemical Engineering. An excellent explication of this scheme is given in: Foust, AS, et al., '*Principles of Unit Operations*' (Wiley, 1960; 2nd ed., 1980). It is notable that the number of distinct UO is comparatively small – though extra UO may be designated, sporadically, as new methods / materials /... are discovered. The UO are, however, typically nontrivial, requiring sophisticated (mathematical) models for rigorous treatment. Moreover, they are (subtly) interrelated, and so the various interactions must be investigated. Many of the theories involve non-uniformity / instability / turbulence – as well as diverse behaviour of phases of matter under extreme conditions (e.g., of temperature / pressure). Complete analyses of such phenomena cannot always be produced, but restricted (e.g., asymptotic) results usually suffice, for design purposes. Many of the UO – along with topics related to chemical reaction / synthesis: see SP_{13} – are developed, in great detail, in J. Chem. Phys. (a rich source of inverse analogues). The challenge, for Σ , is to develop / classify **quasi-UO representations of arbitrary mathematical / scientific subtheories** – and thus to exhibit **quasi-syntheses** of these subtheories. The enormous potential scope of this approach to Σ -synthesis may be better appreciated by perusing the (voluminous) Chemical Engineers' Handbook (e.g., 5th ed., McGr-H, 1973, pp 1958), where copious references are listed. See, in particular, the entries on **computer control of unit processes**.

SP_{12}: Multivariable Control Systems.

Multivariable (MV), or equivalently, **Multidimensional (MD), systems**, encompass virtually all areas of Systems / Control Theory – in particular, (non)linear **network synthesis, signal processing** (especially, **filtering**), and associated **approximation problems** arising in the efficient implementation of the resulting algorithms. For the linear theories, **algebraic procedures** for multivariate polynomial / rational-functions, and matrix reductions / representations, must be developed. Most of these procedures are already available as **SAM routines: see E_2**, where many references are cited.

The notion of **multivariable positive-real functions** (introduced by H Ozaki / T Kasami, IRE Trans. Circ. Th. CT–7 (1960) 251–260) extends the original concept (see, e.g., L Weinberg, '*Network Analysis and Synthesis*' (McGr-H, 1962), op. cit., Ch. 6) to broader classes of networks with **variable components**, or **lumped / distributive character**. In most cases, **the synthesis procedure(s) must produce matrices** (with functional elements) satisfying a variety of conditions – from which corresponding **circuits** (with prescribed types of components) **may be realized. Nonlinear MV-systems** are treated, mainly, by variants of **Volterra functional expansions**, as in Schetzen, op. cit., for instance.

In **MD-signal processing**, issues of **stability** are fundamental. Some of the **(in)stability**

criteria are reminiscent of the **Yang / Lee Circle Theorem** in **Statistical Mechanics**. The principal aims centre around **the effective design / realization of stable filters**, for specified classes of signals. **Other approaches** include: **generalizations of the SISO (Nyquist-type) techniques** to MV systems (see Maciejowski, JM, '*Multivariable Feedback Design*; (A-W, 1989), op. cit.); **extensions of LQG method**; and **Optimal design**, via **Youla parametrization** (both also covered by Maciejowski).

Excellent sources for all aspects of MV-System design may be found in the **IEEE reprint volumes**: Bose, NK (ed.), '*Multidimensional Systems: Theory and Applications*' (IEEE Press, 1979); Palmer, JD, Saeks, R, (eds.) '*The World of Large Scale Systems*' (IEEE Press, 1982). Diverse topics in signal processing are discussed in the volumes: '*Signal Processing: Vol. 1: S-P Theory* (L Auslander, et al., eds.); *Vol 2: Control Theory and Applications*' (FA Grünbaum et al., eds.), IMA Vols. nos 22, 23 (Springer, 1990, 1990). **See also** the proceedings: Boite, R / Dewilde (eds.), '*Circuit Theory and Design*' (Delft UP / N-H, 1981; pp 1090), op. cit.. The **linear**-algebraic techniques for System / Control Theory are covered very thoroughly in: Patel, RV, et al., eds., '*Numerical Linear Algebra Techniques for Systems and Control*' (IEEE Press, 1994). Other aspects of MV-Systems are covered in the extra references given in $E_{11/II}$.

SP_{13}: Chemical Structure Synthesis / Analysis.

There are two (related) forms here: (a) **structural synthesis** of chemical compounds, and (b) **chemical reaction dynamics** (including species of **reaction-paths**) for classes of (in)organic substances. Although the 'synthesis aspects' are of primary interest for Σ , the (quasi-inverse) **analysis** procedures are obviously of importance in explications of the realization schemes. **Case (a)** corresponds to '**closed** synthesis' of machines; and **(b)**, to '**open** synthesis' – in the terminology introduced earlier in this Essay.

The distinctions between **organic** and **inorganic** syntheses are mainly 'technical': comparable principles govern **all** types of realization. Moreover, n**-molecular** reactions may (typically) be replaced by rapid successions of **bi-molecular** reactions. From '**the** Σ **-perspective**', the principal modes of **organic synthesis** are excellently characterized in: Norman, ROC, '*Principles of Organic Synthesis*' (Methuen, 1968), op. cit.; while procedures for **inorganic synthesis** are covered fully in: Purcell, KF / Kotz, JC, '*Inorganic Chemistry*' (Saunders, 1977), op. cit.. Structural properties of **inorganic** compounds are treated in Purcell / Kotz, and in Wells, AF, '*Structural Inorganic Chemistry*' (4th ed., OUP, 1975). Corresponding issued for **organic** compounds are considered in (e.g.): Eliel, EL, a*Stereochemistry of Carbon Compounds*' (McGr-H, 1962).

The study of **reaction pathways** is discussed in some detail in Purcell / Kotz – partly, as reaction **mechanisms** (see, Gould, ES, '*Mechanism and Structure in Organic Chemistry*' (H,R&W, 1969), for a detailed account). The (quantum-mechanical) investigation of chemical reactions is considered in the books by Levine, RD (et al.), and by Nikitin, EE, as cited in $E_{11/II}$. The **analysis** of organic structures by **spectroscopic** techniques is covered in the references by Creswell, CJ, et al., and by Levine, IN, in $E_{11/II}$. The **quasi-Hamiltonian theory of reaction dynamics** has been developed mainly by RA Marcus (and coworkers) in papers in J. Chem. Phys. (see Levine, RD, et al. for some references). The (industrial) implementation of diverse reaction processes involves systematic reactor design: see Lee, HH, '*Heterogeneous Reactor Design*' (Butterworth, 1985) for a wide-ranging treatment.

NOTE. The correlations between Σ -machines and the multifarious schemes mentioned here are often very tenuous; but, in many instances, highly suggestive correspondences may be established – yielding far-reaching (inverse) analogues.

SP_{14}: Generalized Harmonic Analysis.

It is evident that Classical Harmonic Analysis – Fourier series / integrals – embodies (linear) synthesis, in very direct forms, for classes of real-valued / (non)periodic functions. These topics are covered in (e.g.) Zygmund, A, '*Trigonometric Series*' (CUP, 2nd ed., 1959; repr., 1968) op. cit., and Khavin, VP, et al., (eds.) '*Commutative Harmonic Analysis: I*' (EMS Vol. 15, Springer, 1991), op. cit., Chapter II. For such cases, the problems of analysis / synthesis have been studied intensively, and diverse precise / efficient procedures are now 'standard tools'.

The 'generalizations' of basic 'Fourier theory' are of several types. For instance: (i) involving **eigenfunctions of (linear) / (O)DO**, instead of trigonometric functions; (ii) involving **AP functions**; (iii) involving '**statistical analysis' of (classes of) functions**; (iv) involving '**infinite-dimensional spectral analysis**' (v) involving **spectra of 'random operators'**; (vi) involving **non-commutative (harmonic) analysis**; (viii) involving **'other' operators**. The term 'Generalized HA' was introduced by N Wiener (Acta Math. 55 (1930) 117–258), for variants of type (iii); **here**, it covers **all** of types (i)–(vii), through (e.g.) the references cited in $E_{11/II}$. For (iii), **see also** Wiener, N, '*Extrapolation, Interpolation and Smoothing of Stationary Time Series*' (MIT / Wiley, 1949); and for (vi), e.g.: MPCPS 77 (1975) 91–102 (FJ Yeadon); ibid. 110 (1991) 365–383 (S Goldstein); LNM##587, 880 (Springer, 1977, 1981). These conference proceedings contain diverse reference lists. The basic topics involve non-commutative analogues of (abstract) integration over various (non-commutative) structures of (non)deterministic types. **NOTE that** the treatise by G Warner, '*Harmonic Analysis on Semisimple Lie Groups*', Vols. I, II (Springer, 1972) is also highly pertinent, here. In particular, it contains a detailed treatment of **invariant integrals on** (classes of) **Lie algebras / groups**. The types (vii) allow for all forms of (quasi-) harmonic synthesis not covered by types (i)–(vi). This (obvious) device is used frequently in these Essay; indeed, the extreme heterogeneity of the subject-matter makes it (or some equivalent) essential. Recall, for instance, the classification of 'diagrams' in E_7. In this way the overall 'content of Σ ' may be enlarged indefinitely, without fundamental alteration of the underlying 'information structure' – a crucial characteristic in a largescale OIB.

SP_{15}: Machine Design.

It is noted in $E_{11/II}$ that machine design is the archetypal form of synthesis – whatever the types / functions of the machines may be. The corresponding interrelations with Σ -(inverse) analogues are transparent, in most examples, since the underlying sub-mechanisms have w.d. actions. Consequently, the scope for creation / investigation of (inverse) analogues, within this framework, is considerable.

The references listed in $E_{11/II}$ cover general **kinematic geometry** (in particular, for **cams**), the **structure** of machine tools, and the (transmission of) **forces on components** (in diverse applications). The structural aspects concern, primarily, **static / dynamic stiffness** properties of (arbitrarily complex) machines (e.g., for metal-cutting processes – though comparable principles govern multifarious fabrications in Mechanical / Civil / Chemical / Electrical... Engineering, and elsewhere). See, e.g., Major, A, '*Dynamics in Civil Engineering* (Vols I–IV, Akadémiai Kiadoó, Budapest, 1980) as well as the two volumes (edited) by Koenigsberger / Tlusty, op. cit.. Basic schemes for (mainly) Mechanical Engineering are clearly treated in (e.g.), Norton, RL, '*Design of Machinery...*' (2nd ed., McGr-H, 2001).

The variety of the possible 'component (sub)mechanisms' involved in designs is also extremely diverse – as exemplified by the (systematically classified) collection: Artobolevsky, II, '*Mechanisms in Modern Engineering Design*' (Vols. I–V; transl., MIR, Moscow, 1976), op. cit.. Volumes I, II cover **lever**-based devices; Volume IV covers **cam / friction / flexible-link** mechanisms; and, Volume V broadens the purview to include **hydraulic / pneumatic / electrical** devices. The applications range from purely geometrical transformations to highly sophisticated control equipment, for machines in all engineering / scientific domains. Even the **simulation** of classes of n-**variable** mathematical **functions** may be achieved. The individual mechanisms are numbered (from 1 to 4745!), and classified according to action / construction. **NOTE that** a **second level of synthesis** is involved here, since all of the mechanisms $\mu_k, 1 \leqslant k \leqslant 4745$, may be regarded as **realizations** over some 'set(s) of **primitive ingredients**' (e.g., **levers, ratchets, pistons, gear-wheels, resistors, capacitors, inductors, transistors**). The overall classification is summarised in several **descriptive tables** comparable with the preliminary phases of the development of **taxonomies**, where many 'hybrid mechanisms' may be classified by quasi-interpolation.

Two less formal, but very suggestive, collections of mechanisms / processes are given in: Chironis, NP, '*Mechanisms, Linkages, and Mechanical Control*' (McGr-H, 1965); and, '*The Way Things Work*' (Vols. 1,2; Allen & Unwin, 1967, 1971 – translated / adapted from German originals of 1963, 1967). Both of these compilations have been edited, with contributions from many sources – Chironis', from the journal **Product Engineering** (mainly); the other, from nontechnical (but nontrivial) vignettes of artefacts / mechanisms / processes /... 'spanning the modern technological world'. Of course, extra volumes will appear regularly to keep pace with new developments; but the **modes of representation** are of primary interest for Σ – as 'pre-templates' for wide-ranging syntheses, from which possible inverse analogues may be investigated. This scheme includes some phenomena mentioned in other SP (e.g., Chemical reaction processes) but the emphasis is on industrially produced 'apparatus', viewed somewhat informally. Although such specifications are far from rigorous / complete, they are still valuable to offer initial indications of unfamiliar phenomena.

NOTE. A detailed treatment of closed / open-loop mechanisms / manipulators – of planar / spatial types – is given in: Duffy, J, '*Analysis of Mechanisms and Robot Manipulators*' (Arnold, 1980), op. cit.. The systematic notations / representations suggest corresponding **synthesis** / design procedures.

SP$_{16}$: Miscellaneous Inverse Problems.

The notion of 'inverse problem' originated in Potential Theory, E-M Theory and Scattering Theory, where the broad aim was to determine the shapes of gravitating / radiating bodes, or of 'obstacles'. The most obvious applications were in Geophysics, Tomography and 'non-destructive investigation'. Further consideration indicates that: **any inference of causes from effects** constitutes a species of **(generalized) inverse problem** – and, that: **such problems occur in all domains of Pure / Applied / Applicable Mathematics**, as well as in many parts of Engineering and The Sciences. Since 'output' is produced by some machine(s) with prescribed 'input(s)', it follows that: **the synthesis of the** $\{$**machine, input**$\}$ **pair(s)**, corresponding to the observed / measures / postulated /... output, **furnishes the solution(s) of some** (class of) **inverse problem(s)**. Conversely: **every 'given' inverse problem** (regardless of its origin) **may be (re)formulated as a (quasi-)synthesis problem**.

This (essential) equivalence between $\{$inverse problems$\}$ and $\{$syntheses$\}$
shows that there can be no uniform (set of) procedure(s) for the solution of inverse problems (IP). Nevertheless, if the IP are classified according to domains of origin, and structural characteristics, then various 'quasi-locally valid' schemes may be developed. One obvious example of this is given by: Lax, PD / Phillips, RS, '*Scattering Theory*' (AP, 1967), where functional-analytic formulations are shown to cover several, physically distinct, forms of scattering. See also: Agranovich, ZS / Marchenko, VA, '*The Inverse Problem of Scattering Theory*' (G&B, transl., 1963). The **S/M-List** contains diverse IP from virtually all areas of Mathematics. **The identification of these IP**, and their **(continuing) classification**, will add greatly to the power of Σ . For Applied / Applicable Mathematics, the journal **Inverse Problems** (IOP Publishing, Ltd. 1985–) may be consulted – though there are, of course, copious examples of IP treated in other journals. The situation for Pure Mathematics is less clear, and a **systematic compilation of IP** – often 'hidden' in ostensibly unrelated frameworks – would be extremely valuable (and instructive for the compilers).

In **SP**$_{10}$, some connections between **Integral Geometry** and **Inverse Problems** are mentioned. There, the emphasis is on probabilistic applications, but the references are also pertinent to SP_{16} – and cover (mainly) IP for hyperbolic PDE. Comparable methods are also known for some (non)linear parabolic / elliptic (systems of) PDE. Of course, the techniques of Integral Geometry constitute one branch of the **theory of IP** – even though there appears to be no coherent 'general theory' covering all aspects.

What does emerge, however, is that 'typically', DP(π) is more completely solvable than IP(π), where π denote a '**problem environment**', and DP(π), {**direct problems specified over** π}; while, IP(π) denotes {**inverse problems specified over** π}. The obvious vagueness of this description may be (substantially) removed – to establish a **quasi-taxonomic framework** for (say) DP(Σ), IP(Σ), the 'collections of w.d. DP, IP, over Σ_t'. Detailed classifications of the w.d. types of DP, IP (over prescribed (sub)theories) would thus form part of such taxonomic investigations. There is also some overlap with topics in the Philosophy of Science: see, e.g., Härtkamper, A / Schmidt, HJ (eds.), '*Structure and Approximation in Physical Theories*' (Plenum Press, 1981), where many references are given.

Further consideration of the 'causes-from-effects' definition of 'general IP' shows that: **every solution of a general IP is equivalent to the proof of some corresponding (quasi-)converse theorem**. This statement extends even to ostensibly **non**mathematical solution-schemes (e.g., species of simulation) for which heuristic, or systematic, models may be formulated. The representation of (quasi-)**converses** as types of **operational inverses** enhances the unification embodied in Σ .

The most undesirable characteristics of IP include (typically): ill-posedness; non-uniqueness of solution(s); and comparative reliance on **ad hoc** procedures. As a result, the **classification** of IP – in terms of analytical / topological / algebraic / stability /... properties of their 'canonical solutions' – is, so far, very rudimentary. Nevertheless, the classes IP(Σ), DP(Σ) are of such fundamental importance that their rigorous investigation (as 'structured quasi-spaces') is crucial for the effective development of the OIB. Some progress in this direction for: Mechanics / Geophysics / Scattering / transmission / sources /... is discussed, within a broad framework, by PC Sabatier, pp469–642 in: PC Sabatier, Ed., '*... Tomography and Inverse Problems*' (Hilger, 1987); see also: other chapters, for diverse applications / techniques, and large bibliographies.

In spite of the limitation of Sabatier's scheme to various (quasi-)field problems, the treatment of the underlying procedures does produce preliminary characterizations of solution types / properties / techniques – with the potential for extension / modification to cover arbitrary mathematical domains. The initial phase of this program involves the identification of IP, wherever they may occur – so that a skeletal (sub)taxonomy may be constructed, exhibiting classes of IP procedures (from distinct fields) possessing certain common structural features, when appropriate analogues are invoked.

The preliminary **tagging** of the Σ -Lists (in terms of the schematic 'operators' $\delta, \mathcal{A}, A, -A$, $@, \mathbb{A}$ – as explained at the start of this Essay) furnishes a basis for specifying IP – for instance, by adding some extra symbol(s) to the tagging ciphers. In this way, subcollections of 'similar IP' (over disparate domains) may be compiled, as a step towards the (asymptotic!) development of IP(Σ). The essential equivalence of {IP}, {converses} and {syntheses}, shows that $\mathcal{J}(\Sigma)$ is partitioned as DP(Σ)$\cup$IP(Σ), etc.. Consequently, most investigations will involve only the quasi-local properties of these sets.

Since every (O)T possesses many types (even **families**) of converses – of varying degrees of completeness / strength – cognate relationships hold among subsets of DP(Σ), and subsets of IP(Σ). The connections among DP and (arbitrary) ME/SE stem from the associated LC/LD. For instance, the determination of e from some LC(e) may be formulated as a (set of) OT, whose (partial / weak /...) converse(s) will be interpretable as the solution(s) of IP for (aspects of) e. Some {DP,IP} pairs, of basic importance, may be found in **Constructive Approximation Theory** (CAT): in particular, the direct theorem of D Jackson (a constructive improvement of the original **Weierstrass Approximation Theorem**) and the '**converse** / **inverse theorem**' of S Bernstein (on a differential properties of functions which have prescribed (monotonic) sequences of 'best approximations' within associated function spaces). See, e.g., Timan, AF,

'*Theory of Approximation of Functions of a Real Variable*' (Pergamon, 1963; repr., Dover), where these classes of theorems are treated in detail. More recent results are covered in (e.g.) De Vore, RA / Lorentz, GG, '*Constructive Approximation*' (Springer, 1993). Although these, and cognate, examples have comparatively narrow scope, they do allow **precise characterization(s)** of the DP and IP involved; and, of the possible effects of perturbing the conditions for such problems, to obtain variants / generalizations.

NOTE. The word 'inverse' has distinct interpretations in relation to 'IP' and 'inverse analogue(s)'. Moreover, the IP of Approximation Theory and (sometimes!) associated with **converse** theorems (cf., e.g., Timan, op. cit.), as well as with **inverse** theorems (cf., De Vore / Lorentz, op. cit.). This 'variable notation' is, of course, a common hazard in mathematical investigations. The differences in meaning for 'inverse' are, however, especially subtle – as examples will show. For general ME, the sharp results of CAT are lacking, but the overall distinctions remain (qualitatively) valid.

To obviate these potential ambiguities it is important to explain (or, in some cases, recall) the relevant definitions of 'inverse' in various Σ -contexts.

Direct OT: and OT, $\tau : p \Rightarrow q$, where p, q are operationally represented, and the proof of '$p \Rightarrow q$' is t-max.-operational (over Σ_t).

Inverse OT: any operational converse, τ^{-1}, of τ, such that (whenever the terms are w.d.) $\tau^{-1}(\tau e) = \tau(\tau^{-1} e) = e$.

Direct OA: any operational interpretation $\alpha(e)$, of $\hat{\eta} L(e)$ – some perturbation of LC(e).

Inverse OA: any operational **mathematical** interpretation, $-\alpha(e)$, of $-\hat{\eta} l(s)$ some perturbation of LD(s), for the **SE**, s, and the associated **ME**, e.

Direct Problem: investigation / determination of effect(s), $\hat{E}$, of specified cause(s), $\hat{C}$.

Inverse Problem: characterization / construction of the cause(s), $\hat{C}$, producing specified effect(s), $\hat{E}$.

In CAT, the classes $\mathbb{D}T, \mathbb{C}T$, of (respectively) **Direct Theorems**, and **Converse Theorems** are fundamental – each formulated in terms of properties of the **moduli of continuity** of the (sequences of) function involved. For instance, **smoothness** properties of f may be derived from rates of decrease (to O) of the sequence(s) δ, **best approximation(s)**, $\beta_k(f)$, to f – and **conversely**, given $\{\beta_k(f)\}$, some corresponding (class of) function(s) may be obtained.

In more general frameworks, the precision typical of CAT is lacking, but weaker versions of the CAT results may be explored. Another field where strong realization procedures have been developed is that of **logic synthesis for field-programmable gate arrays** (see also SP$_6$), as treated comprehensively in the book by Murgai / Brayton / Sangiovanni-Vincentelli, op. cit. (Kluwer, 1995). The (partially complementary) problems of '**physical design**' (e.g., relative locations / routing for components on a microchip) are covered in detail by N Sherwani, '*Algorithms for VLSI Physical Design Automation*' (Kluwer, 1993) and, in the Proceedings: Korte, B, et al. (eds.), '*Paths, Flows and VLSI Layout*' (Springer, 1990). All of these topics involve **the solution of various IP** – albeit, somewhat narrow in scope.

Further, diverse IP stem from associated **ill-posed problems** (I-PP). See, e.g., Bakushinsky, A / Goncharsky, A, '*All-Posed Problems: Theory and Applications*' (Kluwer, 1994), op. cit., where many references are listed. Plainly, the techniques used for IP, and for I-PP, are often interrelated (mostly involving Applied / Applicable Mathematics). The IP pertaining to Pure Mathematics are somewhat different in nature, so it is worthwhile to characterize such IP, as far as possible. It has been noted that many IP are also I-PP – especially where numerical computation is involved, with the associated issues of **stability** / **consistency** / **convergence** / **optimality** /.... This is one of the key distinctions between 'non-Pure IP' and 'Pure IP' – say, $\mathcal{P} \equiv \{\pi\}$, and, $\mathcal{P}' \equiv \{\pi'\}$, respectively, as sets of generic problems π, and π'. Typically, then, all operations in the formulation(s) / solution(s) of π' are **symbolic** / **exact**. Even the approach facets of π' must be precisely representable – for instance, **via** the Landau O/o/$\sim$ symbols; or, within the approach-structures discussed in other Essays. Alternative broad representations, in terms of **converses**; or, of **syntheses**, have been partially explored in E_8, and earlier in this Essay. There frameworks cover **pure deduction**, all types of **constructions** and all (convergent) **algorithms**. Finer classifications rank **direct** deductions above '**proofs-by-contradiction**'; and, **finitary processes** / **algorithms**, above **infinitary** ones. The detailed development of such ranking procedures constitutes another important aspect of the overall **OIB design**.

Evidently, **every (O)T, τ, potentially 'generates' IP of the form:**

$$\tau_\lambda : {}^-\varphi_\lambda(q) \Rightarrow {}^-\psi_\lambda(p), \text{ **where**}$$

$$\tau : p \Rightarrow q.$$

Here, $p, q, {}^-\psi_\lambda(p), {}^-\varphi_\lambda(q)$, are, in general, **multivariable predicate functions** of various types, with ${}^-\psi_\lambda(p), {}^-\varphi_\lambda(q)$ of forms 'suggested' / motivated by p, q. The (partially) operational **proofs** of the τ_λ then constitute **solutions** to corresponding **IP**. Indeed, this formulation is broad enough to cover **all w.d. IP** over Σ_t, provided that '$\Rightarrow$' is interpreted to include **constructive approach**, as well as **logical derivation** / **deduction** over (at least) the first-order Predicate Calculus, ${}_1$PC. If ${}^-\varphi_\lambda(q) \Leftrightarrow q, {}^-\psi_\lambda(p) \Leftrightarrow p$, then τ_λ is a full converse of τ; otherwise the most byzantine modifications / deformations of τ may produce essentially arbitrary partial / weak /... **quasi-converses** of τ – whose relative strength / power /... could be studied systematically. Such procedures for the comparison of these **quasi-IP** are unnecessary at this stage, but it **is** worthwhile to indicate some of the principles upon which they may be based.

First, there is the (obvious) **implicatory ordering** for each of $\{{}^-\varphi_\lambda\}, \{{}^-\psi_\lambda\}$. Let, p, q be specified (compound) predicates, and denote by $\mathcal{J}_\lambda \equiv \mathcal{J}_\lambda(q, p)$ the IP: ${}^-\varphi_\lambda(q) \Rightarrow {}^-\psi_\lambda(p)$. Then the following order relations may be introduced:

$$\mathcal{J}_{\lambda'} \;{}_1\!> \mathcal{J}_{\lambda''} \Leftrightarrow {}^-\psi_{\lambda'} \Rightarrow {}^-\psi_{\lambda''}; \text{ and,}$$

$$\mathcal{J}_{\lambda'} \;{}_2\!> \mathcal{J}_{\lambda''} \Leftrightarrow {}^-\varphi_{\lambda''} \Rightarrow {}^-\varphi_{\lambda'}.$$

Typically, such orderings will obtain only among certain **sub**predicates of the ${}^-\varphi_\lambda(q)$; or, of the ${}^-\psi_\lambda(p)$ – which yields more subtle orderings over $\{\mathcal{J}_\lambda\}$; and so on! Again, the **cast(s)** / **task(s)** associated with τ (see E_8 for this terminology) may themselves be quasi-ordered through **weighting** of the ME involved. Some attempts in this direction, for 'scientific predicates', were made long ago, by H Jeffreys: '*Scientific Inference*' (CUP, 2nd ed., 1957), but no comprehensive scheme remotely adequate for Σ seems to have been even considered. In any case, only schemes evolving with Σ could be viable – though, intuitive judgements of the relative significance of carious ME in calculations are routinely made in all nontrivial mathematical investigations.

It is clear, from the discussions in E_{11} for far, that: **the families $\mathcal{R}_k, 1 \leqslant k \leqslant 4$, of representations**, $\rho_k(\beta)$, by **machines, syntheses, converses, inverse-problems** (respectively) for arbitrary ME/SE, β, **are essentially equivalent**. Here (as before) $\mathcal{B} \equiv \{\beta\}$ covers all domains of Pure / Applied / Applicable Mathematics, and The Sciences, within a generalized Approach environment – where the **SE** are defined / specified through (mathematical interpretations of) **L**ocational **D**escriptions (the scientific counterparts of **L**ocational **C**onditions for **ME**). In this scheme, all 'approach transactions' for ME/SE are reflected in (equivalent) transactions among their corresponding representations. The interchangeability of $\rho_j(\beta)$,
$\rho_k(\beta)$, 'dynamically' – in any calculation – adds considerably to the power of Σ as an OIB, since the identification / exploitation of optimal operational frameworks is often crucial. Further, the possible discovery of 'new' modes of representation is readily accommodated within this formalism, without introducing inconsistencies.

The multi-faceted view of β-representation has now been discussed in some depth. On this basis, potential (inverse) analogues of β may be investigated through the application of essentially arbitrary OTC transformations to (collections of) w.d. combinations of the $\rho_k(\beta)$ –

with sets, say, $S_\beta(\theta_\beta)_L$, of suitable chosen auxiliary ME/SE, $\theta_{\beta i}, 1 \leqslant i \leqslant L$. The 'approach aspects' of such transformations are paramount in the formation of analogues. Although this scheme reflects the evolution of analogues in '**S**tandard **M**athematical **A**ctivity', it often entails lengthy / intricate manipulations – of combinatorial / algebraic / analytical /. . . types – many of which would be intractable 'by hand'. Within the Σ -OIB, however, the facilities for formalized heuristics are (potentially) almost unlimited, and so the 'β-formalism' becomes both natural and productive.

The Σ -Lists contain diverse examples of IP over the spaces $_SX$– from which (inverse-)analogues may be derived. The elaborate 'apparatus' developed in E_{11} facilitates the systematic pursuit of such procedures, within the approach environments discussed in the other Essays. It should be evident, by now, that the overall 'Σ -environment' is flexible enough to accommodate proofs in all domains of Pure / Applied / Applicable Mathematics – in t-maximally operational forms. Nevertheless, the approach / operational paradigms make extreme demands on the representational / manipulative / deductive / constructive /. . . capacities of the OIB. The detailed treatments in E_1–E_9 (with cognate references) cover a vast terrain of (non)deterministic MP/SP, broadly exemplified by the items in the Σ -Lists. Indeed, there are well over 5,000 items, so far – each minimally tagged for 'approach content', with the potential for **full tagging** by the descriptors ${}^k\Delta_i^n, 1 \leqslant i \leqslant 11$, for item n on list k. See $E_{10/II}$ for specifications of the Δ_i, and for basic motivation.

The projected evolution of Σ_t, from some initial form, $\Sigma_{\,t_0}(\equiv \Sigma_0)$ is considered, broadly, in E_1, where several complications / (encyclopaedic) dictionaries are mentioned as partial sources for the underlying IB, at 'research level'. At a more expository level, there are other useful sources. For instance: Borowski / Borwein, '*Dictionary of Mathematics*' (pp659, 1st ed., Collins, 1989); Behnke, H, et al., eds., '*Fundamentals of Mathematics*', Vols. I–III (Vandenhoek & Ruprecht, 2nd ed., 1962; English transl., MIT, 1974); various eds., '*Handbook of Applicable Mathematics*', Vols. I–VI (Wiley, 1980–1984).

The more formal / logical aspects of **M**athematical **I**nformation are covered in E_3 – in terms of **N**omenclature / **D**efinitions / **A**xioms / **R**udiments /. . . , with cognate references. There, too, the notion of **C**alculational **B**ase is introduced. Then every theory, τ, may be characterized over $\mathcal{N}, \mathcal{D}, \mathcal{A}, \mathcal{R}$ – within the terminological setting, $\mathcal{B}$, of 'basic elements'. The **representation** of (O)T takes account of the notation / (decision-) procedures of **A**utomatic **T**heorem-**P**roving: see, especially, Siekmann / Wrightson, eds., '*The Automation of Reasoning*' (Springer, 1983) op. cit., and Robinson, JA, '*Logic, Form and Function*' (Edinburgh UP, 1979), op. cit..

The Σ -formulations of (O)T, as $T \equiv T(\theta_T) =: T({}_kx_\sigma)_n$, with the ${}_{k_i}x_{\sigma_i}$ suitably selected from the (permanently ordered) set, $\mathcal{U}_t$, of 'underlying ME', is combined with the AT-P versions. Transformations of (O)T are defined by: $f[T(\theta_T)] := T(f(\theta_T))$ – subject to obvious consistency conditions – whenever $f(\theta_T)$ is w.d.. See E_1 for details. As before, the **approach environment** encompasses all types of (non)deterministic approximation / abstraction / (inverse-)analogy, over the Basic Spaces, $_SX$. This appears to encompass all known forms of '(quasi-)**closeness**' encountered in **G**eneral **M**athematical **A**ctivity; but there is, in principle, no difficulty in incorporating other (distinct) closeness criteria – should this prove necessary.

The extremely (ultimately?) general frameworks / tools forged in these Essays have been designed to promote both unification and extension of arbitrary mathematical fields – and so to initiate comprehensive syntheses of Operational Mathematical Knowledge. This aim has been motivated – over some 30 years – by the growing disconnection of Mathematics (as a vast 'calculational system'), intensified by increasing productivity in multifarious specialized fields. What was, originally, mainly a largescale project to broaden / enrich the realms of **calculability** – especially in 'unobvious contexts' – has now encompassed **all areas of Pure / Applied / Applicable Mathematics, and 'The Sciences'**. This intentionally grandiose scheme has become progressively more feasible since its inception – primarily through two (linked) trends: the continuing development of highly sophisticated **SAM packages**, and the (relatively recent) concentration on research into **effective / constructive procedures**, in virtually all mathematical / scientific subdomains.

The interdependencies among Σ -facilities are best appreciated simply by recalling the titles of the other Essays.

E_1: The Σ -Mathematical Processor

E_2: Fundamental SAM Algorithms

E_3: Mathematical Information

E_4: Operational Theorems

E_5: Approximation Frameworks

E_6: Abstraction / Analogy as Approximation

E_7: Approaches to Mathematical Information

E_8: Proofs as Combinations of Operational Theorems

E_9: 'An Approach View' of Mathematics

$E_{10/I}$: Some Quasi-Operational Schemes

$E_{10/II}$: Diverse Examples of Approach

The **Representation / Synthesis / Analysis / Exploration of (inverse) analogues** – as considered in $E_{11/I,II,III}$ – will require all of the techniques explicated in E_1–E_{10}, together with all w.d. synthesis procedures – based on both scientific and mathematical paradigms (see $E_{11/II}$, for fundamental examples). Viewed in this light, Σ emerges as a multi-faceted system within which Mathematics may be 'optimally unified / developed'. The staggering volume / scope of the material in research journals precludes its effective used in the derivation of 'new results' – as well as tending to produce duplications / trivialities. These limitations, however, are substantially transcended in the Σ -environment, where **OT** may be **applied** systematically to **expressions** generated in interactive calculation. Moreover, all w.d. forms of **approach** to ME, e allow 'neighbourhoods in conceptual space' to be explored (through the formalism of E_1,...,E_{10} – with extensions to analogues, as discussed here, in E_{11}). Further, calculations intractable 'by hand' may be performed as **SAM routines** – of regularly increasing power / sophistication. The representations of (operational) proofs as combinations of suitable transformed (O)T produce '**environments for investigation**' that 'fit' specified classes of problems. This, in turn, motivates the study of '**diagrammatic** aspects of operational proofs'; and, of approaches to (diagrams of) OP. The various resulting 'calculational **milieux**' offer unprecedented facilities for the 'free-wheeling', **formalized heuristics** characteristic of Σ .

APPENDIX 6
Analogue Examples

The elaborate system for representing / analysing / synthesizing / exploring ((quasi-)inverse) analogues, $\pm\alpha(x)$, of ME/SE (say, x) has been developed in some detail in $\mathsf{E}_{11:I,II,III}$. The matters to be addressed here, in $\mathsf{E}_{11/IV}$, involve the interplay of this formalism with a variety of specific prescriptions – or, **locators** – $\Lambda_x(x)$, where

$$\Lambda_x(x) := \begin{cases} LC(x) \equiv C(x), & \text{for } x \in \tilde{\mathcal{M}} \\ LD(x) \equiv D(x), & \text{for } x \in \tilde{\mathcal{S}} \end{cases}$$

Here, $\tilde{\mathcal{M}} \equiv \{\text{ME in } \Sigma\ \}$; $\tilde{\mathcal{S}} \equiv \{\text{SE in } \Sigma\ \}$; $C(x) \equiv (C_x(x))_{n_x}$; $D(x) \equiv (d_x(x))_{q_x}$, and $\Lambda_x(x) \equiv (\lambda_x(x))_{r_x}$ – for **countable locators**. For **un**countable locators, the notation is modified to: $\Lambda_x(s) \equiv \{\lambda_{x\gamma} : \gamma \in G_x\}$, etc.. Possible **implementations** of any $\Lambda(x)$ – after suitable identifications have been made – correspond to (quasi-)**machines**, $\mathbb{M}_x$, as considered in $\mathsf{E}_{11/I}$ – for which x is 'obtainable as (partial) output'. Intermediate forms 'between Λ_x and $\mathbb{M}_x$' are furnished by (extensions of) the **M**aster **F**low-**C**harts (originally developed for research in Solid Mechanics (see $\mathsf{E}_{11/I}$) but later recognized as being 'quasi-universal').

The forms of $\Lambda_x(x)$ depend on the generality / complexity / type /... of x; but, in all cases, $\Lambda_x(x)$ comprises some (un)countable collection of conditions on (facets of) x – which, together, 'locate x (non)uniquely within the relevant conceptual space(s)'. Thus, locators for x may (but need not) constitute defining conditions for x – within classes of frameworks – though minimality criteria will not be imposed routinely The **M**aster **F**low-**C**hart scheme (originally developed for problems in Solid Mechanics) is discussed extensively in $\mathsf{E}_{11/I}$ – with various extensions, to cover 'arbitrary ME/SE', and to incorporate t-maximal operationality. These schemes involve (nested) families of formalized mathematical / scientific theories, as well as generalized operational / approach procedures, in all w.d. areas of Mathematics / Science. As expounded in $\mathsf{E}_{11/I}$, these aims are achieved in several stages. In the initial phase, the entries in the 'Solid-Mechanics MF-C' are (individually) generalized to encompass arbitrary ME/SE, x. In the second phase, the resulting **G**eneralized MF-C is rendered **potentially operational** – by representing its entries in terms of combinations of OT. In the third phase, the enhanced MF-C is 'converted' into a (quasi-)**machine**, $\mathbb{M}_x$, embodying algorithms for the 'production' of x as (partial) output. This conversion process is a sort of implementation.

The three broad phases (P_1, P_2, P_3), characterized as: $\Lambda_x \overset{\mathsf{P}_1}{\to}$ MF-C$(x) \overset{\mathsf{P}_2}{\to}$ GMFC$(x) \overset{\mathsf{P}_3}{\to} \mathbb{M}_x$ are treated in $\mathsf{E}_{11/I}$. On this basis, there are five types of representations of (the ME/SE) x; namely, in terms of

(i) Λ_x: mostly ineffective, but formally complete;

(ii) MF-C(x): mainly indicative (of limited scope);

(iii) GMF-C(x): an enhanced / graphic version of Λ_x;

(iv) $\mathbb{M}_x(\zeta, \xi_\zeta)$: a constructive / approach implementation of (iii);

(v) various classes of MODELS, μ_x.

The locators, Λ_x, are, of course, far from unique. They are sets of conditions sufficient to 'determine' x – within the prescribed framework(s), which may have stochastic facets. The symbol Λ_x is, therefore, to be interpreted as '**any** w.d. locator of x'. In applications, the sub-conditions $\lambda_{xi}(x), 1 \leqslant i \leqslant r_x$, (or $\{\lambda_{x\gamma} : \gamma \in G_x\}$), must be fully specified. The **models** are of two principal kinds: **logical** (in formalized theories); and, **reproductive** (in Applied / Applicable Mathematics – and, less commonly, in Pure Mathematics). Here, a **logical MODEL of a theory** T is any structure within which every T-statement is true.

It is useful to recall the aims of P_2 – to produce $\mathcal{F}_x(\mathcal{P}_x) := $ GMF-C(x) – from which $\mathbb{M}_x$ is obtained as: $\mathbb{M}_x(\zeta, \xi_\zeta) \sim \mathbb{CM}\mathcal{F}_x(\mathcal{P}_x)$; that is, $\mathbb{M}_x$ embodies the constructive approach(es) to $\mathcal{F}_x(\mathcal{P}_x)$, with (partial) output constructively approaching x. Here (schematically), one has:

$$\mathcal{F}_x \sim \sigma s \zeta d; e : f\mathcal{R}g; h : i\mathcal{M}j; k : l$$

These 'constituents' of M_x are discussed in (individually) in $E_{11/III}$, under the following designations.

σ: $\tilde{\Gamma} \equiv$ formalized scientific theories

s: $\Gamma \equiv$ formalized mathematical theories

ζ: frameworks for ζ ('motivated by x')

d: LC/LD for ζ (over environment(s) ξ_ζ)

e: auxiliary elements (for realizations of ζ in ξ_ζ)

f: t-max.-constructive forms of ζ (in ξ_ζ)

$\mathcal{R}$: representations of ζ '**via**' auxiliary elements

g: formal (in)equational conditions on ζ

h: formal auxiliary elements for ζ (in ξ_ζ)

i: formal realizations of ζ (in ξ_ζ)

$\mathcal{M}$: 'mathematical transformations'

j: effective (in)equational conditions on ζ (in ξ_ζ)

k: effective auxiliary elements for ζ (in ξ_ζ)

l: effective realizations of ζ (and so, of x)

Denote this collection of (generic) 'constituents' by $\mathbb{K}_x$. Then **the basic claim** is that the compound scheme $x \to \mathbb{K}_x \to \mathcal{F}_x(\mathcal{P}_x) \to \mathbb{M}_x \equiv \mathbb{M}_x(\zeta, \xi_\zeta)$ is w.d., for all ME/SE within Σ – and, that **approach / OTC operations** may be applied to all 'components' of $\mathbb{M}_x$.

NOTE that Λ_x is primarily a **locator**, and so furnishes an existential / non-effective quasi-representation of x; but, the development of Σ makes use of both Λ_x and $\mathbb{M}_x$, in 'complementary modes' – especially, in the study of analogy. (See **E**$_6$**: 'Abstraction / Analogy as Approximation'**) The roles of the (schematic) operators $\mathbb{C}, \mathbb{A}$ obviously vary with x – but the aim (as always in Σ) is to maximize the $\mathbb{C}/\mathbb{A}$ content of x, $\pm\alpha(x)$.

If x **is** w.d., then **any** Λ_x formally determines x. In such cases, (approaches to) x may be obtained as (partial) output of associated machines $\mathbb{M}_x$. Another, closely related, 'manifestation of x' is provided by admissible **models**, μ (defined over diverse frameworks). This is especially pertinent to simulations / synthesis procedures – where the logical models admit ALGORITHMS.

All of these species of representations of x may be used to explore possible ((quasi-)inverse-)analogues of x – by monitoring the (partial) 'output' associated with $\Lambda_x, \mathcal{F}_x, \mathbb{M}_x, \mu_x$ – under various types of 'perturbations' of the constituents.

NOTE. The (basic) ME-C(x), say, F_x – as originally envisaged for Solid Mechanics problems – still 'makes sense' for arbitrary ME/SE x, provided that: (i) the initial entries (**E**ngineering **P**roperties / **G**eometrical **P**roperties / **T**opological **P**roperties) are replaces by σ and s (spanning (formalized) scientific / mathematical theories); and (ii) the relevant data for 'problems' are extended, by viewing 'the generation / production of any x' as a **problem**. The transition from F_x to $\mathcal{F}_x$ is, therefore, mainly concerned with elaboration, to cover the detailed properties / behaviour of x as an element of Σ_t. **Notice, also**, that the notation used here (and from now on) differs from that in $E_{11/I}$ (where $\mathbb{M}$ has several interpretations – albeit, essentially equivalent). The symbols $F_x, \mathcal{F}_x$ (for the 'basic' and 'elaborated', x-flowcharts), with $\mathbb{M}_x$ for the associated machine(s), furnish an adequate scheme for the systematic investigation of analogy – in all of its mathematical forms.

The 'entries', $\sigma, \ldots, l$, are (initially) **non**-operational, but of broad enough scope to cover 'outlines of any x'. The processes of developing (quasi-)operational forms may involve any w.d. combination of the diverse / general /... procedures discussed (from many points of view) in E_{10}. The fact that this often yields cumbersome / intricate / arcane /... formulations is not important here. What matters is **the** (potential) **adequacy** of this formalism for producing x-machines $\mathbb{M}_x(\zeta, \xi_\zeta)$.

NOTE. Each collection of conditions / procedures sufficient for the (in)effective specification of x is called a **SPECIFIER of** x (e.g., $\Lambda_x, \mathcal{F}_x, \mathbb{M}_x, \mathcal{M}_x$).

The classes of **admissible perturbations** are examined in $E_{11/II}$, where it is found that compatibility and consistency are the basic restrictions – after which the 'products of perturbed x-machines' (if these produces are w.d.) may be considered in current, or possibly 'new', frameworks. This procedure constitutes a type of machine **analysis**, for the exploration of **direct** analogues, 'generated from x-machines'. A largely complementary set of techniques, applicable to both direct, and (quasi-)inverse, analogues, deals with diverse **synthesis** processes. For **direct** analogues, synthesis is used to (re)construct machines to produce conjectured, of 'somehow discerned', results – through a variety of techniques, including the imposition of approach criteria. By contrast, the **(quasi-)inverse** analogues are compounded of rigorous analytical / synthetic 'process-segments', bridged by various mathematical **(re-)interpretations** of (almost) arbitrary scientific (or, mathematical) 'facts'. Although these interpretations cannot be established rigorously, the resulting **structures** must be treated purely on their logical / mathematical merits – as examples will show.

A wide range of synthesis schemes is identified (with substantial references) in $E_{11/II}$ – and discussed, in variable detail, with further references, in $E_{11/III}$. The overall aim is to develop synthesis procedures within **G**eneral **S**ystems **T**heory, subsuming all of the variants for particular applications. Some progress in this direction has been made (see the remarks in $E_{11/III}$, for instance); but a comprehensive representation on these lines can be attained only by combining effective solutions, for all of the variants, within a uniform framework. So far, this aim has not been fully achieved, so the investigation of diverse models, and the search for abstract formulations, remain important tasks for Σ .

The underlying formalism must fulfil (at least) the following conditions.

(a) Every item in the Σ -Lists (collectively $\mathbb{L}_t$, say) has some w.d. machine-representation(s) – at various levels.

(b) To every w.d. locator, Λ_x, there corresponds a set of n ($\geqslant 1$) machines, $\mathbb{M}_{xi}$ – from which some analogue(s) may be 'generated'.

Evidently, (b)$\Rightarrow$(a); but the separate designation is convenient.

(c) If $\alpha(x)$ is a w.d. analogue of x, then perturbations of $\mathbb{M}_{\alpha(x)}$ may be used to seek analogues of $\alpha(x)$; and so on!

NOTE. Here, as before, 'nonstandard associations' (bridging calculational segments) may be introduced in the formation of **(quasi-)inverse** analogues. Possible violations of existing scientific / mathematical theories may, nevertheless, 'with proper interpretation', yield rigorously valid 'new' mathematical structures – which may then be used to explore further analogical terrain. This is a characteristic Σ -strategy, though it is, typically, masked in the (retrospectively!) axiomatic presentations of such results.

If $x \in \mathbb{L}_t$ is any item in a Σ -list, then $\mathbb{M}_x$ is, generally, an intricately inter-connected **collection** of **sub**machines, together embodying the structure(s) / behaviour /... inherent in x. Machines of this type are, therefore (highly) **decomposable** – though the degree / form of decomposability is probably best defined 'locally', in prescribed contexts – for instance, as follows.

Every $\mathbb{L}$-item, say $x \equiv \langle i \rangle$, has w.d. 'approach context' (general-approximative / analogical), expressible in terms of active / formal OT-families – say, T^a_B, T^f_Γ – for suitable finite / (un)countable sets B_x, Γ_x. Mostly, B_x, Γ_x are 'small'; the active OT characterize the primary operations involved, and the formal OT cover all other ME/SE in the enunciation of x. (Recall the Σ -convention that **every** ME/SE is treated as an OT: see, e.g., E$_3$). It follows that the obvious 'initial decomposition' of x has the form:

$$x \approx \{T^a_{x\beta}; T^f_{x\gamma} : \beta \in B_x; \gamma \in \Gamma_x\},$$

where each of the OT $T^a_{x\beta}, T^f_{x\gamma}$, has its own cast / test / explicit version(s) – for which appropriate active / formal OT must be identified. (See, e.g., E$_8$.)

This identification process, for active / formal OT, is empirical in nature, but adequate for locator / machine-perturbation. The basic notions are covered in E$_3$, where 'constituents of (formal) theories' are discussed. The aim is to present adequate information about $x \equiv \langle i \rangle \in \mathbb{L}_t$ in terms of 'quasi-minimal sets' of constituents of x. The scheme in E$_3$, covers the **Nomenclature** ($\mathcal{N}$), **Definitions** ($\mathcal{D}$), **Axioms** ($\mathcal{A}$), and **Rudiments** ($\mathcal{R}$) – together constituting **Foundations** (Φ), for the underlying (formal) **Theory** ($\mathcal{T}$) – within which general (O)T may be formulated. The same criteria apply to the treatment in Σ of **any** ME/SE, x – whatever its degree of complexity may be. Thus, in this framework, maximal / minimal levels of decomposition may be specified, for arbitrary x in Σ , with 'empirical consistency' (to be monitored in interactive calculation). Although full consistency is (in principle) rigorously attainable, the resulting intractability rules out this option. Indeed, such excessive formalization impedes mathematical progress / discovery. Of course, when ostensibly important results have been (somehow) produced, their 'ultimate validity' must be established – including all species of 'book-keeping'. For Σ -investigations, monitoring / feedback is fundamental – especially in the exploration of possible ((quasi-)inverse) analogues through representation / decomposition and perturbation of x-locators / machines.

For the purposes of analogue-investigation, '**Mathematics at time** t', M_t, is simply '**the content of** Σ_t (as an OIB)'. If (on the Platonist view, at least) M^*_t encompasses M_t and all results 'inherent in M_t' (some of which might 'never' be consciously proven!) then the underlying conjecture in Σ is that (up to ultimately inherent results!) $\overline{\mathbb{A}M_t} \sim M^*_t$ as $t \to +\infty$ – where $\overline{\mathbb{A}M_t}$ denotes **the approach-closure** of M_t. For an early discussion of this idea (since, greatly amplified) see: '*Varieties of Approximation*', by JSN Elvey (pp77–94 in: J Johnson / S McKee / A Vella, eds., '*Artificial Intelligence in Mathematics*', OUP, 1994).

The (running) Σ -lists, $\mathbb{L}_t$, of mathematical topics / phenomena (MT/P), have been compiled mainly for 'strong approach-significance'. Although $\mathbb{L}_t$ is '**small**' in M^*_t, it is **not negligible**, since $\mathbb{A}\mathbb{L}_t$ – with the elements of $\mathbb{L}_t$ 'suitable selected' – **could** (as $t \to \infty$) be **quasi-dense** in the set {w.d. MT/P in Σ_t}! In this form, such assertions may seem purely speculative; but, within the formalism of Proof-Theory / Model-Theory, combined with approach criteria, the arguments become far more compelling. Of course, there is no intention to 'reproduce' M^*_t (except, **asymptotically**). Rather, it is a matter of rendering every conceivable w.d. MP 'potentially amenable to Σ -investigation'. It should be clear, from the multifarious arguments in $\{\text{E}\}_{10}$, that the Σ -superstructure / formalism has essentially unlimited scope – apparently covering all known mathematical procedures, and even allowing the incorporation of possible 'new procedures', without invalidating the existing scheme. Moreover, these capabilities are (by design) implemented at t-max.-operational calculational levels. In spite of its embodying elements of logical informality, the Σ -framework is almost self-evidently adequate for the systematic prosecution of mathematical investigation(s), MI.

The considerable generality of the prescriptions embodied in $\mathcal{F}_x(\mathcal{P}_x)$ – and so, also, in $\mathbb{M}_x(\zeta, \xi_\zeta)$ – almost furnishes a proof of machine-representability, for arbitrary x. In the realm of quasi-formalized heuristics (typical for Σ) this level of justification **is adequate**. Fully rigorous proofs (probably attainable) would not enhance the efficacy of Σ -investigations, where interactive monitoring (mostly) ensures consistency, etc.. This attitude to 'unnecessary formal proof' is also adopted in the other Essays. As the Σ -OIB evolves, consistency / rigour will, of course, be imposed at every level – but hysteresis will be inevitable.

The overall format for this Essay, $E_{11/IV}$, may now be indicated – for 'arbitrary ME/SE', x, including any set of items from $\mathbb{L}_t$ (interpreted as propositions, etc.). On this basis:

(i) every ME/SE x has some locator / chart / machine / model representation(s), each of which must be 'appropriately specified', at various levels of constructivity / operationality – say, as $\Lambda_x, \mathcal{F}_x, \mathbb{M}_x, \mu_x$;

(ii) for each of $\Lambda_x,' mcF_x, \mathbb{M}_x, \mu_x$, classes of admissible perturbations (of constituents / components /. . .) may be examined;

(iii) initially, this may yield **Gen**eralizations / **Ab**stractions of x – and thence, also

(iv) 'progressively distant' **direct** analogues of x, with diverse dominant behaviour.

NOTE that: all of the stages (i)–(iv) may be affected (exactly, or approximately) **rigorously**.

(v) The **(quasi-)inverse** analogues of x (by contrast) involve **non**algorithmic / irregular '**associations**' connecting pairs of 'algorithmic segments' – the resulting chain 'emanating from x' and 'terminating at $-\alpha(x)$'.

These 'associations' are (somehow) motivated by properties of x (as an **SE**, for **inverse** analogues; or else, as an **ME**, for **quasi-**inverse analogues). Although such 'associations' cannot (by their nature) be systematized / generated /. . . in any prescribed way, cumulative collections will constitute a valuable research resource in Σ – from which **qualitative theories of analogy** will emerge. Since even the Σ -Lists, $\mathbb{L}_t$, contain(so far) over 1000 items 'of analogical type' (as the abbreviated taggings indicate), the only viable procedure is to treat a small selection of highly nontrivial examples, as thoroughly as possible – on the understanding that, as Σ evolves, all approaches to every x will (eventually) be incorporated in the IOB, providing the unrivalled **Environments for Investigations** (see E_8) that distinguish Σ fundamentally from all other SAM-based systems – by identifying (for each 'properly formulated problem $\mathcal{P}$) a diversely-ranging set of 'OT pertinent to $\mathcal{P}$'.

The obvious (and probably best) plan is to use the primary classifications of ME and other (science) review journals to make a small, but arguably diverse, selection of topics – from which specific examples of analogue-formation are to be developed. Although the ME (and science) lists are revised, sporadically, the extensions (and deletions) have been relatively modest – since the overall forms of Mathematical / Scientific Knowledge have reached quasi-stable states (say, Level-0). The Level-1 items are typified by the ME entries coded as 00, 01,. . . 94 (in the 1985 version). Comparable items in The Sciences, etc., may be found in the relevant review journals. For Σ -purposes, the Level-0 headings may be taken to be: **Mathematical Sciences [Logic; Algebra; Geometry; Analysis; Topology; Probability Theory; Classical Applied Mathematics; Applicable Mathematics]; Physical Sciences; Chemical Sciences; Engineering Sciences; Biological Sciences; Social Sciences.**

The (minimally) expanded Level-0 heading for Mathematical Sciences reflects the aims of Σ . Of course, the classifications at Level-0/-1/-2/. . . are far from unique – and even within the items at Level-k, there will always be overlaps, ambiguities, etc.. Nevertheless, sound DBMS schemes **may** be built on such bases! Indeed, the DBMS formalism covers knowledge-based systems at all levels of complexity / sophistication; and formally (at least) M_t is a (dynamical!) system – of immense complexity. It is also representable as a **general** system. Consequently (as remarked in earlier Essays) both DBMS Theory and **G**eneral **S**ystems **T**heory are highly pertinent to the 'design aspects' of Σ . The relevant references may also be repeated here: Kanellakis, PC, ed., '*Advances in Computing Research, Vol. 3, The Theory of Databases*' (JAI Press, 1986); Masarovic, MD / Takahara, Y, '*General Systems Theory: Mathematical Foundations*' (AP, 1975); Mesarovic, MD / Macko, D / Takahara, Y, '*Theory of Hierarchical Multilevel Systems*' (AP, 1970).

Although efficient DBMS routines are of crucial importance for Σ , they are not treated in these Essays. Developments in the mathematical theory of DBMS algorithms – and current implementations – may be found in the CS literature, and incorporated (as appropriate) into the corresponding Σ -facilities. As usual, the underlying mathematical results will (eventually) appear in the OIB! The DBMS routines essentially define the IT capabilities of the processor, P_Σ (see E_1), while admitting all possible AI tools – subject only to overall compatibility / consistency. As the depth, capacity, and scope of the AI schemes are increased, their roles in (interactive) Σ -investigations will grow in (interactive) Σ -investigations will grow correspondingly. The remarks made (in earlier Essays) on the strong limitations of existing theorem-proving packages, almost certainly remain well-founded: very few nontrivial results have been obtained. With certain modifications (as yet unknown), however, it is very plausible that such packages could furnish extremely powerful additions to the Σ -arsenal. The overall form of Σ is so organized that AI operations may be consistently incorporated, at arbitrary times.

Recall that every ME/SE x has (at least) the interrelated representations denoted by $\Lambda_x, \mathcal{F}_x, \mathbb{M}_x, \mu_x$; that is, as: locational relations, generalized master flow-charts, machines, models (respectively). These classes of representations cover **all** ME/SE, from the most primitive ('quasi-irreducible') entities to arbitrarily complicated / sophisticated (O)T – and even 'randomly formed (w.d.) theories'! For the more complex entities, most representations are, of course, highly reducible / decomposable (as already remarked).

It is useful to denote the collection of **all** (nontrivially distinct) **locators of** x by ${}^\Sigma\Lambda_x =: \{\Lambda_x^\delta : \delta \in \Delta_x\}$, where the finite / (un)countable set $\Delta_x(t)$ is 'w.d. in principle' – provided that criteria for distinctness (of $\Lambda_x^{\delta_1}$ from $\Lambda_x^{\delta_2}$, $\delta_1, \delta_2 \in \Delta_x$) have been specified: a logical / metamathematical problem.

NOTE that this is not a calculational scheme (there is no question of defining / exhibiting the elements of ${}^\Sigma\Lambda_x$); rather, a matter of descriptive organization. Nevertheless, it is assumed that pertinent locators of any (fixed) x can be produced – for calculations / investigatory purposes. In other words, in the course of interactive computation, several Λ-representations of 'the same x' may be deployed – to examine various perspectives / aspects / facets /... of x; but the totality of locators of x, i.e., ${}^\Sigma\Lambda_x$ (compounded of **generic constituents**) is never used explicitly. This is in line with the other 'global' Σ -sets – of OT, OP, etc.. The same considerations hold also for variants of $\mathcal{F}_x, \mathbb{M}_x, \mu_x$ – in Σ -investigations – where the sets / families ${}^\Sigma\mathcal{F}_x$, ${}^\Sigma\mathbb{M}_x$, ${}^\Sigma\mu_x$ correspond to ${}^\Sigma\Lambda_x$.

In principle, all elements of ${}^\Sigma S_x$ are (logically) **equivalent** – in each of the (separate) 'cases' $S = \Lambda, \mathcal{F}, \mathbb{M}, \mu$ – provided only that every element is 'mathematically adequate for its specified task(s)'. This (quasi-tautological) statement is still significant for Σ , since the **procedures** involved in achieving these tasks are **multifarious** – as are the associated **purviews** of the underlying calculational schemes (the **specifies**, $\Lambda_x^\delta, \ldots, \mu_x^\eta$).

It follows that the detailed properties of any specifier, S_x^θ, strongly influence both the degree of effective determination of x, and the facility for 'analogical generation' from x. Although the perturbation operations – for any specifier – are w.d., and may be handled rigorously, the crucial issue is to estimate (if possible) the 'inherent **analogical fecundity**' of arbitrary ME/SE x. Some remarks on this matter may be found in E_6 – mainly for direct analogues. Since (quasi-)inverse analogues are comprised of algorithmic segments linked (at 'nodes') by **non**algorithmic 'associations', the processes for direct analogues are incorporated into those for inverse analogues. Nevertheless, within Σ_t, it is possible for both of the conditions: $v_t^+(x) > v_t^-(x); v_t^+(y) < v_t^-(y)$ to hold (for suitable ME/SE x, y), where $v_t^+(w), v_t^-(w)$ denote (respectively) the number of direct, and (quasi-)inverse, analogues of w 'validated in Σ_t'. Thus, subtle effects govern 'analogical productivity', so no facile solutions should be expected. Moreover, the **quality** of analogues is more significant than the **number** of analogues 'generated' from x – though it is difficult to quantify such judgements. Again, comments on this problem are made in E_6, where a tentative formalism for the representation of abstraction / analogy as approximation is developed. The primary obstacles lie in the elusiveness of the underlying concepts.

In spite of these difficulties, it is possible to formulate **two basic problems**.

P_1: The analogue, $\alpha(x)$, is '**suggested**', and some modified machine components $\alpha(m_i)$ are (partially) conjectured. The problem is **to determine corresponding specifiers** $\Lambda_{\alpha(x)}\mathcal{F}_{\alpha(x)}$, and the rigorous form(s) of $\mathbb{M}_{\alpha(s)}$ – and hence, of $\alpha(x)$.

P$_2$**:** For given x, the $m_i(x)$ are (somehow) 'known' – but analogues of x are sought, quasi-**heuristically**, by **perturbing** the $m_i(x)$ until 'new', 'interesting' algorithms, $C_{\alpha(x)}$, are realized by the modified machines; so producing, as (partial) output, ME/SE denoted by $\alpha(x)$, to be viewed as some w.d. analogues of x. The problem is **to identify the pairs,** $(\alpha(x), C_{\alpha(x)})$.

Thus, **Problem 1**, for some 'proposed analogue(s)', $\alpha(x)$, corresponds to the **synthesis** of $\mathbb{M}_{\alpha(x)}$ – with the initially guessed forms of certain components – from the other specifiers, $\Lambda_{\alpha(x)}, \mathcal{F}_{\alpha(x)}$, in t-max.-determined forms. In **Problem 2**, $\mathbb{M}_x$ is specified, and possible (direct) analogues, $\alpha(x)$, are sought by heuristic perturbation of the components of $\mathbb{M}_x$, and consideration of the resulting (partial) output. A third class of problems (**Problem 3**) is concerned with **the combination** of n ($\geqslant 2$) **algorithmic segments (a.s.)** $\sigma_j(C_j), \sigma_k(C_k)$, at 'nodes', v_{jk}, where σ_k is regarded as **generated** from σ_k **via** inter-theoretic transformations 'applied at v_{jk}'. This prescription covers both inverse, and quasi-inverse analogues. In these terms, **Problem 3** may be stated as:

P$_3$**: Given** $\alpha_j(C_j) : x \mapsto \alpha_j(x)$, **over** ξ_j.

> **To produce a transformation** $\tau_{jk} : \xi_j \mapsto \xi_k$ **and a mapping** $\sigma_k(C_k) : \alpha_j(x) \mapsto -\alpha_k(x)$, **over** ξ_k. Here, ξ_j, ξ_k, are (calculational) **environments**, and $-\alpha_k(x)$ denotes some **(quasi-)inverse analogue** of x.
>
> **NOTE that** problems involving $p(\geqslant 2)$ initial a.s., 'generating' $q(\geqslant 2)$ (quasi-)inverse analogues, are essentially reducible to 'the case $p = 1 = q$'.

Taken together, Problems (1)/(2)/(3) cover all of the processes of (quasi-)inverse-analogue formation. The **analytical** tools correspond to boolean combinations of compositions of OT (spanning all areas of Mathematics). The varieties of **synthesis** include all of the prototypes outlined in E$_{11/II}$, as well as other procedures not, so far, identified. All of the routines deployed are to be realized over **approach environments**. A crucial 'procedural question' in analogical investigations is 'the proper selection of **levels of detail** (in appropriate senses) of the **specifiers** introduced in the formulations of problems. Here, it must be appreciated that Σ -milieux support **interactive** computation; primarily, within **environments-for-investigation** (EfI). The finer details are, mostly, embedded in the OT-implementati- ons – making the EfI comparable with elaborate, specialized tool-kits. The structure of formalized mathematical (/scientific) theories is of fundamental importance in Σ – as expounded in several other Essays. In particular, the roles of **foundations, extensions, calculational procedures** and (partial) **compositions** with other theories, are all crucial in the operational development of **specifiers** – upon which analogical developments depend. For the Σ -exploration of analogues (of all species) the level of initial specifiers will (typically) be low, since the **key aim** is to establish **plausibility** of the possible analogues – conjectured, or 'fortuitously glimpsed'! Thereafter, attempts to solidify the arguments (via Problems (1)/(2)/(3)) will require progressively refined / powerful operations – until (at least) 'formal proofs' are obtained. In the final stages, total consistency / rigour must be imposed. At each stage, from preliminary to final, the full OTC-resources may be deployed. Hence, the potential scope is unlimited – a crucial design feature of Σ , motivated by the goal of (asymptotically) encompassing M_t^*.

On this basis, typical specifiers for (nontrivial) x will be interpretable as operational or nonoperational **block-diagrams** for the proof / construction / calculation / estimation /. . . of (some facet(s) of) x using (non)deterministic procedures (at high levels). The complexity of x-specifiers varies widely with the type / nature / structure of x, but the guiding principle is to maximize the levels of constituents, as far as overall constraints allow. Such (non-unique) prescriptions suffice for Σ -analogue investigations. The basic idea is as follows. Denote by $S_x \equiv (s_x)_{P_x}$ any x-specifier, with (finite / (un)countable) index-set P_x. **The initial aim** is to convert S_x into some **quasi-finite form(s)** – by introducing suitable spaces / limit-operations /. . . : say, as $S_x \twoheadrightarrow \bar{S}_x := (\bar{s}_x)_{N_x}$, where N_x is a (typically, 'fairly small') positive integer. Thereafter, the $\bar{s}_{xn}$ are to be examined (individually), or in 'clusters') to construct (subject to consistency

/ realizablity /...) higher-level representations with fewer constituents. In principle, these processes may be repeated until (non)unique 'near-optimal representations' are obtained. Frequently, the required transformations may be effected 'intuitively', with minimal calculation; otherwise, Σ -OTC operations may be used – over relevant EfI. The resulting 'optimized specifiers', say, S_x^*, would be derived in the order: $\Lambda_x^*; \mathcal{F}_x^*; \mathbb{M}_x^*$ – with the model(s), μ_x^*, either derived independently (from various mathematical / scientific theories); or else, from combinations of the other specifiers.

NOTE. In practice, the x-dependence will be shown only if ambiguities arise.

It is now convenient to **systematize** the notation for specifiers and their transformations. Let $\Lambda \equiv S_1, \mathcal{F} \equiv S_2, \mathbb{M} \equiv S_3, \mu \equiv S_4$, and define $\bar{\rho}_{xk}, \rho_{xk}^*$ by: $\mathrm{S}_{xk} \twoheadrightarrow \bar{S}_{xk}, \bar{S}_{xk} \twoheadrightarrow S_{xk}^*$ (respectively), where S_{xk}, ρ_{xk}, etc., refer to the ME/SE x, when such dependence must be shown. It is understood, of course, that these transformations are broadly schematic – but (in principle) realizable in arbitrary detail, provided that at least one of the S_{xk} (typically, S_{x1}) is w.d. over some environment(s). Indeed, the entire Σ -scheme is founded on the premise that all ME/SE may be represented by **locators** (LC/LD, in the other Essays). **Notice that** the **models**, S_{x4} (giving 'settings for x') 'usually' embody philosophical / logical / scientific / mathematical structures – rather than just minimal information for the definition / 'production' /...of x. Consequently, the interrelations of $S_4, \bar{\rho}_4, \rho_4^*$, with the rest of the S_k, etc., are mostly not susceptible of systematic characterization. Their importance for Σ -analogical investigation, however, is central, since they determine the context(s) for valid development / extension of any ME/SE x, over Σ_t.

At this stage, the (again, initially schematic) operators, $\hat{\rho}_{xpq}, S_{xp}^* \twoheadrightarrow S_{xq}^*$, for $1 \leqslant p \neq q \leqslant 3$, may be introduced. This is especially pertinent, here, since the formation / exploration of direct analogues is (principally) founded on the (admissible) perturbation of x-machine components. Since each type of specifier incorporates all of 'the essential properties of x', it is intuitively clear that all of the ${}_x\hat{\rho}_{pq}$ are w.d. – and, that the discussions of $\Lambda_x, \mathcal{F}_x$ and $\mathbb{M}_x$ (especially, in $\mathrm{E}_{11:I/II/III}$, for $\mathcal{F}_x, \mathbb{M}_x$; and, diversely in E_{10}, for Λ_x) offer strong indications for the **construction** of all of the operators $\bar{\rho}_{xk}, \rho_{xk}^*, \hat{\rho}_{xpq}$, depending on the nature / complexity /...of x. The scheme to be developed in the analogue examples will reflect the links among these operators – often involving the **model(s)**, S_{x4}, to furnish some (partial) common framework(s). As usual, for Σ , this massive superstructure will seem over-elaborate, unless x is itself 'fairly complicated'; but this is mitigated by OIB facilities, and the overall **unification** fully justifies such representation.

With these preliminaries it is possible to formulate **prescriptions** for the formation of (potential) **direct** analogues – say, $\alpha_1(x)$; after which the further 'manoeuvres' to produce (potential) **inverse** analogues – say, $-\alpha_2[\alpha_1(x)]$ – may be investigated, and (partially) **systematized**, in terms of 'inverse associations' from scientific (or, mathematical) theories, as already outlined in **Problem 3** (where 'intertheoretic transformations' are used). The processes involved may be depicted as follows (in the notation already introduced):

$$x \to \mathbb{K}_x \to \mathcal{F}_x(\mathcal{P}_x) \to \mathbb{M}_x(\zeta, \xi_\zeta) \equiv \mathbb{CA}\mathcal{F}_x(\mathcal{P}_x)$$

where $\mathbb{K}_x$ denotes the (ordered) set of '**constituents**' of $\mathcal{F}_x$ – whose forms / interrelations may be inferred from the underlying **model(s)**, μ_x (which also furnish initial / boundary data, etc.). Similar considerations apply to the construction of the **operators** $\bar{\rho}_{xk}, \rho_{xk}^*$, and $\hat{\rho}_{xpq}$ – since all valid (partial) embodiments of x 'inhabit associated conceptual environments', such as ξ_ζ, whose characteristics depend on the nature of μ_x.

The role of **models** over Σ is discussed in several of the Essays – most notable (for the present context) in $\mathrm{E}_{11/I}$, in relation to the formation of (inverse) analogues over (compound) theories. Here, it is assumed that some specifier ('typically', $S_{x1} \equiv \Lambda_x$) of x has been ascertained (or, hypothesized), and the basic task is to obtain the other specifiers (typically, S_{x2}, S_{x3}) over the environment(s) governed by the model(s) $S_{x4} \equiv \mu_x$ – by developing the appropriate ρ-operators; and thence, quasi-optimal x-specifiers. The quest for **direct** analogues involves trial-perturbations of (some of) the specifier-ingredients – followed by heuristic interpretation of the resulting structure(s). At any point in this process, verifiable / plausible / conjectural **intertheoretic transformations** may be introduced – after which possible **novel structures** may be explored, with a view to ultimately rigorous development. In principle, such investigatory procedures may be pursued for arbitrary x, and it is an indication of the subtlety of (inverse) analogy, that 'interesting forms of $\pm\alpha(x)$' by no means **necessarily** stem from 'interesting forms of x' – a fascinating situation!

It may be argued, of course, that, by its nature, the discovery of analogues cannot be systematized; and, that only 'routine' / 'unstimulating' results may be 'derived'. This view, however, is vitiated by the multifarious discussions in these Essays. Indeed, the fundamental claim (already broadly justified) is that: **every ME/SE x is (synthetically) approachable over Σ_t**, 'through neighbourhoods in conceptual space(s)'. Far from inhibiting 'analogical awareness', the overall Σ -design exhibits analogy as **paradigmatic** of General Mathematical Activity! Up to now, most analogues have been identified incidentally, as offshoots of axiomatic-development / problem-solving – within formalized mathematical / scientific frameworks. The representations / procedures of E_{11} promote the **generation** of analogues, by introducing powerful techniques for analysis / synthesis – in many cases, constituting quasi-analogues of the original scientific synthesis tools: see, especially, $E_{11/II}$, $E_{11/III}$. This (standard!) Σ -circularity does not raise serious logical problems.

Notice that the formation of theory-environments (as considered in $E_{11/I}$) has basic similarities with the process of inverse-analogue formation – in the sense that there are sets (say, $\Lambda_{\tau_1(\tau_2)}$) of τ_1-elements at which **facets of τ_2 may** (consistently, etc.) **be 'injected'**. The parallel is only partial, however, since 'the injection of inter-theoretic transformations at the endpoints of direct-analogical segments' may be validated (if at all) only **retrospectively**, over some (partially) 'new structure(s)' – representable (in 'favourable cases') as examples of environment-extension, if the original sequence is reversed! The customary modes of presentation of proofs often mask the fortuitous nature of 'new ideas'. One of the basic aims of Σ is to reintroduce **serendipity** into written accounts of research projects – rather than the clinical published forms, totally lacking in insight. This may be achieved by printing records of interactive Σ -investigations, once the OIB is in regular use.

The possibilities for producing analogues correspond to the statements of Problems 1,2, and 3 – broadly understood – formulated over all admissible environments within Σ_t. For **syntheses**, see $E_{11/II}$, $E_{11/III}$, where selections of scientific, and mathematical, routines are considered. All other procedures contain **heuristic** elements, based on various types of **perturbations** of the constituents of the relevant specifiers – together with (possibly, conjectural) intertheoretic elements, for (quasi-)inverse analogues. Accordingly, each analogue example should be considered as **an exercise in solving Problems 1/2/3 over designated environments**. The detailed **properties** of the associated analogues may be studied after they have been incorporated into the OIB. It follows that the principal stress in $E_{11/IV}$ should be on the systematic design of quasi-templates for Problems 1/2/3, over (essentially) arbitrary environments – for all w.d. ME/SE. It turns out that such **templates** may be viewed as **specifiers** for the (quasi-)inverse analogues involved – which underlines the unifying power of this formalism.

The processes for producing near-optimal specifiers, through the 'stages' $S_x \twoheadrightarrow \bar{S}_x \twoheadrightarrow S_x^*$, have already been mentioned. In typical investigations, (near-)maximiza-
tion of the constituent-levels (in 'content', as well as in constructivity / operationality) is a primary aim. For certain types of analogues, or for very basic ME/SE x, the specifiers must be decomposed into more rudimentary 'factors'. This variation in the degrees of complexity / sophistication in pairs $(x, \pm\alpha(x))$ is characteristic of analogue exploration.

The formal problem templates (FPT) may be simple depicted as follows.

Here, v_x denotes any descriptive **name** for the (arbitrary) ME/SE x; $(S_x)_4 \equiv (S_{x1}, S_{x2}, S_{x3}, S_{x4})$, $S_{x3} \equiv \mathbb{M}_x$, represented by a diagonal matrix of operators, $m_{xi}, 1 \leqslant i \leqslant N; \gamma_\theta(\theta) \equiv$ conjectured form(s) of θ; $\gamma_B\mathbb{M}_x \equiv \{\gamma_i(m_{xi}) : i \in B \subset \bar{N}\}$ with $\bar{N} := (1, \ldots, N); \hat{p}_B\mathbb{M}_x \equiv \{\hat{P}_j m_{xi}(\text{VIA } \sigma_j(C_j) : j \in B)$, where $\sigma_l(C_l)$ is some 'segment' of the calculational / algorithmic procedure C_l.

This succinct representation of the FPT is based on the concepts / notation developed in $E_{11/I}$. When it is used in examples, the table-entries will, of course, vary greatly in intricacy / depth – but the clarity of the underlying structure(s) will make systematic investigation over Σ_t

	v	δ [DATA]	ρ [REQUIREMENTS]
P_1	x	$(S_x)_4; \gamma_\alpha(\alpha(x)); \gamma_B(\mathbb{M}_x)$	$(S_{\alpha(x)})_4; \models \alpha(x)$
P_2	x	$\mathbb{M}_x; \{\hat{P}_{B_ij}\mathbb{M}_x\}$	TO IDENTIFY / INTERPRET $(C_{\alpha(x)}, \alpha(x))$ in $\hat{P}_B\mathbb{M}_x \rightarrow$
P_3	x	$(S_x)_4; \sigma_j(C_j) : x \mapsto \alpha_j(x)$ in ξ_j	TO CONSTRUCT $\tau_{jk} : \xi_j \rightarrow \xi_k$, $\sigma_{jk}(C_k) : \alpha_j(x) \mapsto -\alpha_k(x)$ FROM $\gamma(-\alpha_k(x))$ in ξ_k

(relatively!) tractable, in the (quasi-iterative) accumulation of Σ -analogues. Indeed, it may be claimed (with utter impartiality!) that the overall Σ -framework – conceptual and notational – is outstandingly well adapted to the daunting task of organizing, and effectively extending, the corpus of Mathematical Knowledge!

The synthesis paradigms S-$P_k \equiv s_k$ – introduced in $E_{11/II}$, and discussed (quite extensively, in some cases) in $E_{11/III}$ – are best represented / formulated as flowcharts / block-diagrams – with the ultimate aim of developing quasi-universal representations over GST-frameworks[656], as considered (briefly) in $E_{11/II}$ and $E_{11/III}$, where possible structures for such representations are (tentatively) conjectured. In this quest, it is essential to produce preliminary flowcharts for the s_k within the underlying formalisms (of Network Theory, Control Theory, Dynamical Systems Theory, Automata Theory, etc.). These tasks are attempted here, on the basis of the accounts given in $E_{11/III}$, for fundamental procedures. Elaborations / variants will be covered, as Σ evolves, to produce collections, $\{s_k^i : i \in \bar{L}\}$, or, $\{s_k^\lambda : \lambda \in \Lambda\}$. One possible model for the s_k^i is furnished by the **T**opical **O**utlines / **F**low-**C**harts for the primary techniques used in Solid Mechanics (especially, in (LE)FM), linked to a list of (about 250) key references. This material is contained in a Farnborough Research Report (June, 1991) which (on account of its high relevance to the construction os s_k^i) is reproduced as an Appendix to this Essay ($E_{11/IV}$). Extracts (some, substantial) of other Farnborough Research Reports (all by JSN Elvey) are also appended (to earlier Essays). The inclusion of such voluminous accounts is justified (as partially explained in these Essays) by the evolution of 'the Σ -Project' as an approach / operational extension of the Farnborough IB – 'the ABC(FM)', which has been implemented over an IT/DBMS package, within a WINDOWS environment[657]. The 'ABC model' may be viewed as a paradigm for research-level (nonoperational) mathematical / scientific IB. The elaborateness / complexity of this ABC apparatus typifies the challenge of producing IB structures for arbitrary Scientific / Mathematical domains. The further imposition of **operational / approach criteria** – the principal concern of these Essays – will test Mathematics (and mathematicians) to the utmost limits – and **beyond** (an assertion requiring careful formulation / interpretation!). The ABC (MFM (Tax)) Manual is also reproduced, to illustrate the interplay of taxonomy / tagging / search procedures, at research level.

The combined reference lists (from $E_{11/II}$) for the topics #1–#16 – probably with some additions – correspond to the (preliminary) list of **B**asic **R**eferences of the ABC(FM). The **T**opical **O**utlines for the s_k may be produced through (extreme) condensation of the schemes outlined / cited in $E_{11/III}$, with all steps (broadly) operational. Eventually, each type of procedure (and all w.d. variants / extensions) must be incorporated. In this Essay, however, the main aim is to demonstrate the feasibility-in-principle of constructing SP in TO/FC form(s) – for arbitrary (validated) synthesis schemes with prescribed (partial) output governed by the 'environments' in #1–#16 – and, ultimately (through GST, or other unification) for arbitrary ME/SE, within 'the Σ -framework'.

The structure of a typical s_k^i is therefore as follows, for possible syntheses / analogues of β – w.d. over framework(s) / system(s) $\mathcal{F}_\beta/\mathcal{S}_\beta$

[656]**G**eneral **S**ystems **T**heory: see $E_{11/II}$, for some initial remarks.

[657]See: MFM (Tax) System: User Manual ©1993, J Mehers (Active Information Systems). The underlying software procedures of this (and later versions) are based on the taxonomic / tagging schemes covered by the Farnborough Research Reports.

(a) Specification of underlying framework(s) / system(s)

(b) Definitions ('over (a)') of β

(c) Synthesis scheme(s) for β

(d) Associated Σ -machines for β – say, $\mathbb{M}_\beta$

(e) Admissible perturbations of $\mathbb{M}_\beta$

(f) Exploratory ((quasi-)inverse-) analogues – say, $\pm\alpha(\beta)$

S_1 **: Electrical Networks (NW)**

A **network**, N, comprises (in)finitely many interconnected **elements** with a set of **ports** (i.e. accessible terminal pairs). A **circuit** is a network containing at least one **loop**. The elements represent physical devices - e.g. **resistors, capacitors, inductors; transformers and generators**. An **n-port** is any network with (exactly) n ports. This description holds for **all** networks, regardless of the detailed forms of interconnections/elements/ports. The diverse classes of networks correspond to **prescribed behaviour** of the various types of elements - for instance: **time-(in)dependence/(non)linearity/(non)-passivity** - as well as, to the **combinations of element types**. SeeA **network**, *N*, comprises (in)finitely many interconnected **elements** with a set of **ports** (i.e. accessible terminal pairs). A **circuit** is a network containing at least one **loop**. The elements represent physical devices - e.g. bf resistors/capacitors/inductors/transformers/generators. An **n-port** is any network with (exactly) n ports. This description holds for bf all networks, regardless of the detailed forms of interconnections/elements/ports. The diverse classes of networks correspond to **prescribed behaviour** of the variuous types of elements - for instance: bf time-(in)dependence/(non)linearity/(non)-passivity - as well as, to the **combinations of element types**. See $E_{11/III}$ for references to basic definitions, system-matrices, etc.

Of fundamental importance for most of the s_k are the concepts of (non)-**linearity** and time-**(in)variance** - defined (in Network Theory) at each terminal pair, where the I/O relation must be (un)changed by arbirary time-translation; and the whole network (for **linearity**) must admit **superposition** of admissible I/O pairs - which (collectively) therefore constitute (complex) **linear space**. A network is **nonlinear** IFF it is not linear; and it is **time-varying** IFF it is not time-invariant.

Although the effective synthesis procedures (as usually formulated) do have some limitations (e.g., to linear / time-invariant NW of class C_1; or, to nonlinear / time-varying NW of class C_3 – where the classes C_k correspond to the types / combinations / behaviour of the NW elements), significant extensions of scope continue to be made. Consequently, it is a reasonable assumption (for Σ) that broadly universal NW-synthesis schemes, of potentially arbitrary accuracy (over appropriate 'spaces'), will be developed – and may be incorporated (eventually) into the OIB. The sample schemes treated here are 'general within stipulated frameworks'; and hence, worthy of inclusion. Such levels of generality typify all of the s_k-presentations in this Essay (for TO/FC). As one would expect, the most efficient synthesis procedures hold for the most restricted NW – but, in Σ , the main requirement is that **some** constructive synthesis scheme should be valid for (essentially) arbitrary NW, subject to various consistency conditions, etc.. Such schemes correspond to I/O machines, whose component-perturbations allow analogue-exploration.

In most forms of synthesis, species of **system functions** are to be realized. It may be shown that such functions are (for **lumped** NW) **rational** – and so, amenable to canonical **representation**, and to effective **approximation**. The corresponding networks may also be **realized** via w.d. procedures: the TO/FC for the s_k. This broad description covers all forms of synthesis involving rational transfer (or comparable) functions, regardless of the field(s) of application – including all of the areas identified in $E_{11/II}$, and outlined in $E_{11/III}$. All of these procedures involve the successive imposition of (various) '**elementary operations**' (of which '**unit operations**' in Chemical Engineering furnish other examples) for the piecewise construction of some network(s) having the prescribed system function(s). The schemes are based on certain **realizability / passivity /...** criteria (centred on **Hurwitz polynomials** and **positive-real functions**). For **distributed** networks (e.g., transmission lines) the system function need not be rational: then, irrational / transcendental functions must be introduced. Examples of fairly general schemes will be indicated.

NOTE that the underlying procedures for the syntheses of diverse types of system functions (SF), Z, may be characterized very broadly, as follows, for NW $\mathcal{N}$.

(i) Z (as specified / conjectured), a **matrix** of order $n(\geqslant 1)$ of **multivariable / (ir)rational functions**, must be **positive-real**. Several sets of (necessary and) sufficient **conditions** for p.r. behaviour are known.

(ii) If Z **is p.r.**, then Z must be **decomposed** (reversibly) into **components** $C_1, \ldots,$ C_l, where each C_j **is realizable** as the SF, Z_j, of some **known** NW, $N_j, 1 \leqslant j \leqslant l$.

(iii) Then $N_1, \ldots, N_l$ must be suitable combined to produce the NW N with SF Z (as initially prescribed / guessed). This process is **effective**, since each w.d. assembly of NW selected from $\{N_1, \ldots, N_l\}$ has SF (ir)rationally expressible in terms of the corresponding Z_j.

Realizability criteria in domains **in**equivalent to NW-Theory (e.g., for other s_k) will, of course, differ from the p.r.-type conditions – but overall schemes comparable with {(i), (ii), (iii)}, will remain valid. This situation is reflected in the TO/FC presented in this Essay.

Further extensions of the domain of NW synthesis – especially, for classes of (partially) distributed systems – involve **p.r. functions of several** (complex) **variables**. See, e.g., Koga, T, IEE Trans. CT CT–15 (1968) 2–23; ibid. CT–18 (1971) 444–455. These schemes are also represented here. The longterm aim is to formulate all validated synthesis procedures within a maximally uniform framework – so that comparisons may be made efficiently, and redundancies may be eliminated. This (achievable) aim is in line with the overall design of the Σ -OIB: to maximize calculability (via **approach**) and to minimize the occurrence of trivial / repeated results – though **equivalence** of results is (mostly) important, and will always be identified / recorded. Thus, despite the immense diversity of SF / schemes, the small selection of TO/FC is adequate to illustrate the issues typically encountered.

The various types of synthesis correspond to properties such as: active / passive / lumped / distributed / time-(in)variant / stable / (non)linear; and, to the sort(s) / combination(s) of NW elements: resistors / inductors / capacitors / varactors /.... The resulting variety of (in)equivalent NW is considerable. Each class of broadly equivalent NW admits a set of characteristic synthesis procedures – whose generality / scope in Σ_t will increase with t. Indeed, **synthesis** (in all areas of Mathematics) remains an active field of research – especially in relation to the study of **Inverse Problems** (see SP-16 $\equiv s_{16}$, in $\mathrm{E}_{11/III}$, for introductory remarks). The schemes outlined here involve (often, sophisticated) techniques of (Linear) Algebra, (Complex) Analysis, and Functional Analysis (especially, for **non**linear NW). Other s_k make extensive use of Stochastic Processes (including Stochastic Calculus). Moreover, since syntheses are species of inverse problems, it follows that the s_k potentially span 'all of Mathematics'! This is, of course, merely an alternative statement of the Σ -claim that **every ME/SE is synthetically approachable / realizable** (within appropriate frameworks). It is also conjectured that: **the most general syntheses may** (eventually) **be representable over GST structures** for generalized I/O machines; and hence, that: possible ((quasi-)inverse) analogues of arbitrary ME/SE may be investigated – through perturbation of the components of the associated machine(s), and 'injection of suggestive non-algorithmic elements' (for inverse analogues).

The variety of established synthesis routines, s_k, is now considerable; and research in this area is still vigorous. Each routine is specified in terms of basic conditions – typically, as follows.

C$_1$. The set O of objects to be synthesized by some (in)finite network(s) $\mathcal{N}$ [e..g, TRANSFER FUNCTIONS, IMPEDANCES]

C$_2$. The set E of elements for which O is representable via characteristics of $\mathcal{N}$ [e.g., RESISTORS, INDUCTORS, CAPACITORS, TRANSFORMERS]

C$_3$. External / internal constraints on $\mathcal{N}$ [e.g., ACTIVE, PASSIVE, (IN)DEPENDENT SOURCES]

C$_4$. Modes of interconnection of members of E [e.g., SERIES, PARALLEL, SERIES-PARALLEL, LADDER]

C$_5$. Local / global topology of $\mathcal{N}$ [e.g., TREE, OPEN/CLOSED LOOP]

C$_6$. Algebraic / analytical (non)deterministic behaviour of elements of E [e.g., TIME-(IN)DEPENDENCE; PARAMETRIC (NON)LINEARITY; PRESCRIBED STOCHASTIC PROCESS(ES)]

C$_7$. Verification that $\mathcal{N}$ realizes O over E [e.g., EXACT; CONSTRUCTIVE; EFFECTIVE / APPROXIMATIVE – all, within w.d. (NON)DETERMINISTIC FRAMEWORKS]

The corresponding TO/FC versions may be produced by amalgamating / interrelating the conditions C_j (in maximally operational forms), with some additional properties.

The conditions C_j are broadly complemented by the 'typical structural characteristics, (a)–(f)', as already exhibited – covering **all** species of synthesis listed in $\mathsf{E}_{11/II}$, with key references, and discussed more fully in $\mathsf{E}_{11/III}$. Taken together, the appropriate forms of conditions (a)–(f) and $C_1, \ldots, C_7$ furnish the detailed information for the construction of TO/FC. Moreover, since {**I**nverse **P**roblems} formally subsumes {Syntheses}, as **sets**, all w.d. (future) synthesis procedures may be incorporated as IP – though (quasi-)**optimal** routines will (as usual) be developed individually. The aim here is to formulate at least one TP/FC for each species of synthesis identified in $\mathsf{E}_{11/II}$ – on the understanding that **all** validated syntheses will, eventually, be implemented in Σ – in t-max.-GST versions, potentially applicable to the 'generation' (as (partial) output) of classes of ME/SE, β, from the associated I/O machines, $\mathbb{M}_\beta$. Admissible perturbations of such $\mathbb{M}_\beta$ may then be used to explore possible ((quasi-)inverse) analogues of β.

At this stage it is important to recall that every **instance** of a w.d. SP (say, $\bar{s}_k$, for a specification of x) constitutes a **mode** of (quasi-)machine representation / realization of x, as (partial) $\mathbb{M}_x$-output. That is: x is (quasi-)**synthesized**, as $\bar{x}_k$, over $\bar{s}_k$.

Inversely: let $\delta(\beta)$ denote the defining conditions / characteristics of any ME/SE β. It is claimed in $\mathsf{E}_{11/I}$ that $\delta(\beta)$ – and hence, β – is (quasi-)**constructible** via some **machine(s)**, $\mathbb{M}_\beta$. Although there is no universal prescription 'to produce $\mathbb{M}_\beta$ from β', the operational extensions of the generalized **templates** for $\delta(\beta)$ – as introduced in $\mathsf{E}_{11/I}$, and subsequently discussed in detail – make this claim (at least) highly plausible – and 'essentially provable'.

Evidently, the modes of representation of β imply corresponding **stability / robustness** properties, under **perturbations** of $\mathbb{M}_\beta$. Moreover, the s_k themselves may have 'unobvious ramifications', significantly enlarging their original purviews. Consequently, it is enough (for the global schemes developed in $(\mathsf{E})_{11}$) to identify 'broadly typical s_k', for each k, and to exhibit corresponding TO/FC – **whose extended scope will** (potentially) **cover** M_t/S_t.

The **key insight** for the construction of **arbitrary syntheses** is that **G**eneric **C**ontrol **P**roblems are defined over a set, $\{A\}_N$, of **attributes**, A_j, including: **1-variable, n-variable, continuous, discrete, open-loop, closed-loop, deterministic, stochastic, feedback, sampled-data, optimal, adaptive, lumped, distributed,...** subject to **stability criteria**. The **techniques** deployed are broadly classified as **classical** (over **frequency domains**) and **modern / state-space** (over the **time domain**), though these classes are not disjoint. Other basic attributes are: **linear, nonlinear, stationary, time-invariant, time-varying, finite,** and **infinite**. Typically, $s_k \equiv s[C\langle W_k\rangle]$ is specified for $W_k \subset \{A\}_N$, and **G**eneric **F**undamental **M**athematical **P**rocedures, φ_m^W, are identified. Thus, the set of **generic** CP[W] is comparatively small; whereas, the profusion of synthesis **routines** stems from the wide range of **models** (in all areas of Pure / Applied / Applicable Mathematics) considered. The task of developing appropriate sets of φ_m^W is daunting (and will constitute a continuing process, as Σ evolves). Here, lists of references (with minimal outlines of the key procedures and their interrelations) must suffice.

The overriding priorities are for **stability** accurate **prediction** of I/O behaviour, and (extensions of) **controllability / observability** – for all types of control. The associated TO/FC must embody these mathematical characteristics – to be implemented (eventually) as I/O machines for arbitrary ME/SE, β, within GST-based frameworks – albeit, in relatively sophisticated / weakened formulations over appropriate approach domains.

Accordingly, the principal task is to develop TO/FC for the **Generic** FMP of Classical / Modern Control Theory – regardless of the field(s) of application; or, of the possible **modes** of control. On this basis the GFMP may now be listed. **NOTE.** it should be emphasized that the synthesis topics identified in $E_{11/II}$ are **all** representable via quasi-control processes – from Electrical NW to Inverse Problems. Consequently, the control schemes (in maximally general realizations) furnish a firm foundation for the Σ -OIB – with high potential for extension to the realms of 'arbitrary ME/SE', as discussed implicitly in much of $(E)_{10}$; and explicitly throughout E_{11}.

The treatment of **discrete** and **continuous** control systems within a **single framework** may be achieved through various **embedding** procedures. See, e.g., Porter, WA, '*Modern Foundations of Systems Engineering*' (Macmillan, 1966), where it is shown that S-D systems furnish a two-way link between discrete and continuous systems; and, Jacobs, OLR, '*Introduction to Control Theory*' (OUP, 1974), for an informal combined treatment of all types of discrete / continuous control problems (though each problem **type** has its own 'natural' framework(s)). More complete unifications may be attempted in terms of suitable defined **(non-)random measures** over various **function-spaces / algebraic-structures**. In this way, very general SP will emerge – eventually yielding **quasi-universal GST-schemes**, covering essentially arbitrary ME/SE, generated (as partial output) by associated I/O Σ -machines: see $E_{11/I}$.

Thus, if $\varphi_m, 1 \leqslant m \leqslant L$, denote the GFMP, then the TO/FC $T(s_k) \equiv T_k$ have the (schematic) form $T_k = T_k(\varphi_\sigma)_{m_k}$, where $(\sigma)_{m_k} \subset \bar{L}$. The φ_m are comparable with **unit operations** in **Chemical Engineering** – and the $T_k(\varphi_\sigma)_{m_k}$ (in $t_{\max}$-operational / approach versions) constitute **templates** for the study of **analogues**.

As a starting point for the synthesis of general β, the processes listed in $E_{11/II}$ (with some references), and discussed in $E_{11/III}$, may be formulated as **machine paradigms** – operational / approach **extensions** of the **problem templates** for Solid Mechanics introduced in Farnborough research reports (see $E_{11/I}$ and its Appendices for details). Hence: if the quintessential properties, $p(\beta)$, of β are prescribed, within the environment ξ_β, then the **core problem** is: **to produce** $C(\beta) \equiv \{p(\beta)\}$**, over** ξ_β – in terms of (typically, 'simpler') objects w.d. over ξ_β. Such processes are exemplified by the topics in $E_{11/II}$. On the basis of these schemes, more general formulations may be considered within **GST frameworks**. The (G)FMP may be constructed by examining some fundamental technique(s) for the SP S_k $(1 \leqslant k \leqslant K; K = 16$, so far) and then decomposing the underlying calculational scheme(s) into generic procedures – say, $\varphi_i^k, 1 \leqslant i \leqslant I_k$, for S_k. The set: $\Phi := \bigcup_{1 \leqslant k \leqslant K} \{\varphi_i^k : 1 \leqslant i \leqslant I_k\}$ may be regarded as a quasi-**basis** for synthesis algorithms.

Notice that the sets $\Phi^k := \{\varphi_i^k : 1 \leqslant i \leqslant I_k\}$ are typically **not** disjoint – since in many cases, the schemes for $S_{k'}, S_{k''}$ will have common subprocedures (possibly differently combined, etc.). Hence, the aim is to produce a list of all the **distinct** procedures φ_i^k, for the types of synthesis considered in $E_{11/II}$, and elaborated in $E_{11/III}$. This may be achieved by analysing some realization algorithm(s), ρ_k, for each of the S_k, to identify (and then formulate) the associated **sub**procedures and their **combinations** within ρ_k – that is, to construct the **template(s)** for S_k. Even for the topics covered in $E_{11/II}$, this is a highly nontrivial task, whose rigorous completion – and extension to GST frameworks – will form an important part of Σ -development. Here, attention is confined to indications, for each S_k, of possible φ_i^k – and, of templates, τ_k, assembled from appropriate combinations of these φ_i^k. From the τ_k, quasi-machines, $\mathbb{M}[S_k]$, may be introduced, embodying the algorithm(s) ρ_k.

The discovery of (quasi-)**inverse** analogues depends (often) on the capacity to **postulate** 'interesting behaviour' – say, $b(\theta)$ – for the so-far **un**known object(s), θ, to be **synthesized**, through various procedures, over w.d. domains, say, D_θ, of 'comparatively more basic' elements. For the SP in $E_{11/II}$, most of these elements are (non)linear, time-(in)variant, (non)deterministic,... components of classes of (in)finite **networks / systems** – for which such 'reconstruction problems' have long been studied intensively. The crucial issue, for Σ , is the possibility of **extending this formalism** to cover progressively more general species of behaviour – eventually, within the framework of **GST**. Although it is fundamental for most of this Essay, the **reconstruction paradigm** has been fully implemented only in limited mathematical / scientific areas (particularly, those of $E_{11/II}$). **The goal**, in the rest of $E_{11/IV}$, it **to systematize these implementations** through the introduction of **maximally uniform notation**. This, in turn, entails the prior **decomposition** of each synthesis algorithm, ρ_k, into (in)dependent **subprocedures**, φ_i^k, from which the **templates** τ_k may be **constructed**.

More generally, distinctions should be drawn between **direct**, and **indirect, syntheses** of the ME/SE θ.

(1) If θ admits LC/LD $l(\theta)$, then some I/O machine(s), $\mathbb{M}_\theta$, may be associated with θ – in principle, **directly**, since the components of $\mathbb{M}_\theta$ are **inherent** in the structure of algorithms, based on $l(\theta)$, for (approaches to) θ. Thereafter, explicit properties of θ may be investigated by **analysing** the procedures embodied in $\mathbb{M}_\theta$.

(2) By contrast, suppose that (as above) derived / observed / conjectured **characteristics**, $b(\theta)$, of θ, were **proposed**. Would such θ be **w.d. / unique / calculable**? If so, **how** could (approaches to) θ, exhibiting (approaches to) $b(\theta)$, be **(re)constructed** over specified domains, D_θ? Schemes for the resolution of this class of problems are predominantly of **indirect (inverse / iterative /...)** types.

Of course, the core motivation for the entire Σ -project is the fact that **'most ME' are non-constructively defined** – so the overriding aim is **to maximize constructivity through the universal development of approach formulations**.

To illustrate (no more!) the enormous range of **NW-synthesis** procedures, the following **S**ynthesis **R**eference **L**ist is now presented. Although these papers date from 1959–1982, they are still broadly representative – and they underline the vast potential scope of SP in all areas of Mathematics. From this list for NW Theory (and the shorter (!) lists for the other SP discussed in $E_{11/II}$, $E_{11/III}$), a small selection of archetypal TO/FC will be formulated – with the overall aim of developing GST-versions. All of the NW references are to the **IEE Trans. Circuit Theory**, and its successor, the **IEEE Trans. Circuits and Systems**. The (standard) forms **CT-m, CAS-n** are used for volumes m, n of these Transactions. (Some books and other journals may also be cited, as necessary.) The other SRL are comparatively short – comprising minimal selections of key papers / books, on which the associated TO/FC may be based. The combined SRL (of all species), say, $\mathcal{L}_S$, will be enlarged systematically in size / scope, as Σ evolves – to furnish asymptotic approaches to $\mathcal{L}_{M_t}$ – a typically modest Σ -aim! **NOTE.** The lists SRL: k should be augmented by the Key References for SP_k from $E_{11/II}$ – from whose Bibliographies most SRL:k-items are cited.

SRL:1 Electrical Networks.

CT–6 (288–291) NW t-realizability (AH Zemanian) 1959

CT–6 (340–344) Synthesis from Scattering Matrix (DC Youla)

CT–7 (251–260) n-Var. p.r. Functions / Applications (H Ozaki, et al.)

CT–7 (281–296) n-Port Brune Process (V Belevitch)

CT–7 (88–101) Takahasi Ladder-NW Schemes (L Weinberg, et al.)

CT–8 (153–164) Unified 2-Terminal NW Procedures (PM Kelley)

CT–9 (136–143) Lumped NW Equiv. of Transm^n Lines (V Ramachandran)

CT–9 (267–277) Pole-Zero Cond^{ns} for Rational p.r. Fn^{s} (K Steiglitz, et al.)

CT–10 (265–274) N-Port Realiz^{n} via Distribut^{n} Th. (AH Zemanian)

CT–10 (393–404) Funct^{nl} Anal. / Nonlin. Feedback Systems (G Zames)

CT–11 (214–224) 2-Var General^{zns} of CT / Applic^{ns} (HG Ansell)

CT–12 (150–157) Linear t-Var. NW Stability / State Space (ES Kuh)

CT–12 (164–170) Distribt^{n} Th. / General^{zd} Passive NW Anal. (AH Zemanian)

CT–12 (223–230) Scattering Matrix / Complex n-Port Load NW (RA Rohrer)

CT–13 (19–31) Filer Synthesis via p.r. Functions (DC Youla)

CT–13 (31–52) Passive n-Ports / Prescr. Reactances (T Koga)

CT–13 (244–258) NW Synthesis with Lossy Reactances (J Anderson, et al.)

CT–13 (331–332) State-Var. (In)dependence / Bashkow A. Matrix (SL Hakimi, et al.)

CT–13 (349–354) Generalized NW: Conference Abstracts [37 papers. . .]

CT–14 (129–139) TF-Realiz^{tn} for Grounded URC NW (JD Rhodes)

CT–14 (192–209) q-Valued Nonlin. NW without Memory (LO Chua)

CT–14 (394–403) Non-uniform Finite Transmission Lines (DS Heim, et al.)

CT–15 (2–23) Passive n-Ports / k-Var. p.r. Matrices (T Koga)

CT–15 (24–30) Lumped-NW Passivity Criteria (RA Rohrer)

CT–15 (139–143) q-Var. Cascaded Transmission-Line NW (I Shirakawa, et al.)

CT–15 (354–380) Simul. Realiz^{tn} of TF / Reflect^{n} Factors of n-Ports (NT Ming)

CT–16 (57–67) Tapered / Distrib^{td} RCG-NW (SC Lee)

CT–16 (313–318) Synthesis of Active t-Variable NW (R Saeks)

CT–17 (333-338) State-space Proc. / t-Invariant TF (P DeWilde, et al.)

CT–17 (575–584) q Nonlin. NW linked by Lossless TL (RJP de Figueiredo, et al.)

CT–17 (584–594) Lin. Transform^{n} Converters / Nonlin. Synthesis (LO Chua)

CT–18 (326–331) Infinite Resistive NW (H Flanders)

CT–18 (444–455) Terminated Cascade of TL / 2-parts (T Koga)

CT–19 (36–48) Generalized NW Hysteresis (LO Chua, et al.)

CT–19 (63–66) State. Eqs. of Lin. t-Invar. Systems (IC Göknar)

CT–19 (199–201) Global IFT (FF Wu, et al.)

CT–19 (227–232) State-Space NW Synthesis (R Yarlagadda)

CT–20 (239–246) State Rep^{rn} of Lin. Largescale DS (SP Singh, et al.)

CT–20 (370–382) Theory of Algebraic n-Ports (LO Chua, et al.)

CAS–21 (678–687) Linear t-Var. Passive NW (BDO Anderson, et al.)

CAS–21 (661–666) Decomp^{stn} / Synthesis of Nonlin. n-ports (LO Chua, et al.)

CAS–21 (709–721) Chemical Reaction NW (G Oster, et al.)

CAS–21 (598–605) $\mathcal{R}\{$p.r. q-Var F$^{\text{n}}\}$ / Applic^{ns} (V Ramachandran, et al.)

CAS–25 (88–98) Formulation of Nonline. Dynamic NW (Y Kawamura)

CAS–25 (31–40) Lin. t-Var. Passive Scattering Matrices (TG Liang)

CAS–26 (235–254) Hopf Bifurc^{tn} / Nonlin. Osc. in NW (AI Mees, et al.)

CAS–26 (105–111) Notes on n-Dim^{nl} System Th. (DC Youla)

CAS–29 (169–184) Stationary Lin. t-Var. Systems (TACM Claasen, et al.)

CAS–29 (220–233) Gen^{rlzd} Lagrange Form^{ltn} for Nonlin NW (HG Kwatny, et al.)

BOOKS

The items listed in $\mathrm{E}_{11/II}$ are also to be recorded here. In addition, the following references are relevant for the construction of (mathematical) TO/FC, in later sections, as well as for (electrical) networks.

Schwarz, RJ / Friedland, B, '*Linear Systems*' (McGr-H, 1965)

Porter, WA, '*Modern Foundations of Systems Engineering*' (MacMillan, 1966)

Takahashi, T, '*Mathematics of Automatic Control*' (H,R&W, Inc., 1966)

Anderson, BDO / Vongpanitlerd, S, '*Network Analysis and Synthesis...*' (P-H, 1973)

Kuo, BC, '*Automatic Control Systems*' (P-H; 3rd Ed., 1975)

Further Sigma Development

Although the Σ -Essays and various Appendices give a detailed account of the structure / function of the OIB, questions of practical **implementation** (over greatly enhanced SAM packages) are (mainly) deferred to the long-term (collective) phases of the project. This is inevitable, since the operational formulation of theorems over the Basic Spaces, ${}_SX$, is a cumulative activity, depending on 'new developments'. The apparent absence of detailed technical arguments / constructions, in most of the Essays – in spite of their scope: MK – is a consequence of the primary aim: **to demonstrate that some OIB of universal coverage may be developed systematically. Every result formulated in approach mode(s) presents a challenge to mathematical technique / ingenuity**. The resulting OT – of arbitrary type(s) – with be accumulated as the OIB evolves. **Most of the underlying procedures** (in all areas) **are indicated in references** to papers / books. **The resulting scheme** (over suitable computer-systems / SAM-packages) **will be vast and highly technical**; but such versions would be out of place in the 'Σ **-blueprint**' embodied in the Essays. Indeed, the minutiae of the proofs / constructions / computations (over the Basic Spaces, ${}_SX$, or their stochastic counterparts) constitute core material of the OIB, asymptotically encompassing MK. **The general formalism** for this cumulative process **is substantially created / exemplified within the Essays** – through specification of the component processes involved.

NOTE. The various Σ **-Lists** play an important part in the overall scheme. Apart from the S/M-List, there are lists for **C**ontiuum **M**echanics (± 3K items) and for miscellaneous papers (± 2.5K items) – reflecting previous research interests / projects from 1967 on. Moreover, the CM-List is closely linked to the '**ABC Project**' (for Fracture Mechanics), with its extensive **annotations** and associated **taggings** (identifying various strategies / techniques in considerable detail). The other lists have been partially tagged – and short annotations of some items on the S/M-List form a substantial part of $\text{E}_{10/II}$.

Eventually, **every item** on the S/M-List(s) will be annotated (with associated references). For the existing list(s) this is obviously a considerable task – to be completed as part of the initial implementations. More generally: every 'new' S/M items, contributed by users, will include the annotation(s) among the set of descriptors. As the system evolves, a 'standard formal for CD of S/M items' will emerge – to be imposed, retrospectively, on the earlier items in the lists, with associated (partial) taggings.

The role of reduced character-sets in the basic Σ -implementation.

Although LaTeXproduces elegant typography, etc., **it is fundamentally unsuitable for the effective implementation of Σ .** Rather, a (near-)minimal character-set of **WYSIWYG** gorm is essential – allowing arbitrary mathematical expressions to be **types directly** (as on the original dual-keyboard machines). This will certainly suffice for the coverage of all research activity.... The Σ -Essays constitute a **detailed plan for the development of an OIB covering (asymptotically) all of MK**. Ultimately, this resticted character-set should be accessible through **handwritten input** – with the possibility of gradual extension, as Σ evolves. All of the schemes discussed in the Σ -Essays may be implemented within this environment (through local (re-)definition of some symbols, if necessary). This includes **the formulation of arbitrary t-max.-operational theorems**, and the consequent **representation(s) of operational proofs.**

The ABC package, developed by Richard Bowden (Edinburgh, 1980–), has a WYSIWYG mathematical keyboard. However, less sophisticated apparatus will suffice to produce simpler – but adequte – input. One way of achieving this is to connect ASCII & basic-mathematical keyboards (u.c./l.c. Greek letters and a selection of symbols) to each user's screen / tablet, etc.. This would allow arbitrarily sophisticated theorems to be formulated / implemented through appropriately defined notation. The links with underlying SAM facilities would be forged by direct association – for instance[658]. $'\int f(x)dx \twoheadrightarrow$ INT$(x; f(x))$, with both f and the INT procedure ef-

[658]The notation $'\theta$ (discussed later) stands for the t-max. evaluation of θ over P_{Σ_t}.

fectively specified over P_Σ – the Σ -Mathematical processor. Although the resulting typography will mostly be somewhat basic / inelegant, the mathematical content will not be compromised, and it should be possible to allow bold / italic, as well as standard, type – and at least two sized of symbols / letters, to cover sub / superscripts, etc.. The (far more elaborate) LaTeXtypography is (no doubt) popular with publishers – who can reproduce it almost unchanged! – but: the crucial point is that proofs / constructions may be pursued, in conventional notations, within P_Σ, **interactively** – a **revolutionary advance**.

The distinction between mathematical expressions (wfe) and their (t-max.) Σ -evaluations must be maintained in all Σ -transactions. In this way, proofs / constructions / deductions may be developed naturally, as aspects of GMA. Every wfe is a combination of components – each of which may be displayed or (partially) evaluated, as appropriate. The distinction between display and evaluation (of a component θ, say) may be indicated as (e.g.) θ (display) and $'\theta$ (evaluation). The use of parentheses (e.g., $'(\theta_1\theta_2\theta_3)$), and similar devices, allows parts of wfe to be evaluated, with other parts merely displayed – as is often the case in GMA. For instance, $'(\int f(x)dx)$ produces the t-max. evaluated integral. An alternative scheme allows $'$ to act 'rightwards' until it is 'stopped'. Thus (for instance) $'\int f(x)dx + \sum_{1\leqslant n\leqslant L} a_n + \Phi(\zeta)$ leaves the w.f.e. $\Phi(\zeta)$ **un**evaluated. This minimal distortion of standard mathematical notation – with cognate forms of output from P_Σ – **produces genuine collaboration** between researchers and the OIB, to establish new / sharper / deeper results in all domains – to be incorporated within P_Σ, and used to produce yet more results, t-max.-ready for application.

For this to be feasible, all of the representations / calculations typical of arbitrarily sophisticated mathematics must be implemented over P_Σ. This process, in its multifarious aspects, is discussed in the Σ **-Essays**, for the most general **approach domains**. Every theorem (however ostensibly arcane) has some Σ -implementation(s), and so may be deployed in any Σ -investigation. The resulting quasi-dialogue facilitates the investigation of problems apparently beyond unaided human capacity – as has always been the case for (predominantly) **numerical** computations since computers were invented. The representation of **proofs** (of $T : p \Rightarrow q$) through Boolean combinations of compositions of suitably transformed OT (see, especially, E_8) with natural I/O, will cumulatively generate a universal research environment – at the highest level.

The Essays demonstrate conclusively that such global syntheses of Pure / Applied / Applicable Mathematics are not only theoretically, but also practically, attainable – as well as developing a comprehensive formalism to encompass essentially all w.d. GMA – quasi-asymptotically, if necessary. Most of these aims are (substantially) achievable through combinations of several types of **reduction processes:** to produce visible (but **not** pointlessly elaborate) I/O; to implement a growing range of analytical / algebraic / topological /... calculations over the (regularly extended) underlying SAM processor, P_Σ – coupled with extensive graphics facilities – and so to permit the application of (arbitrarily sophisticated, or technically complex) OT to wfe obtained within Σ .

All of these topics are discussed in considerable detail in the Σ -Essays, and the principal issues are effectively resolved. Accordinly, **the crucial, longterm tasks devolve on the construction of t-max.-effective** approach versions of fundamental theorems in all areas of Mathematics – to build up the 'initial system', Σ_{t_0}. In fact, the choice of Σ_{t_0} is somewhat open, since any 'large collection of operational facilities' covering 'most basic calculational domains' will suffice for the longterm systematic extension. This apparent vagueness is not problematical. It just emphasizes the cumulative nature of all largescale IKBS – which can aim only at **asymptotic** completeness.

The possibilities for **handwritten input** should be explored **seriously**. This is far from being an extreme luddite reaction to technological milieux; rather, it represents **the crucial importance of attaining symbiosis between people and computer systems for mathematical research**. Ultimately, a wide range of calligraphic styles should be (quasi-)reliably interpretable – and this aim **will** be realized, as AI and cognate techniques evolve. Initially, however, **printed** (u.c.) input may be used (for the 'English /... content') with the Greek alphabet and a substantial but limited range of mathematical symbols to be 'produced' within 'acceptable tolerances'. The existing computer interfaces (keyboards, screens,...) will be replaces by 'tablets', on which the input may be written directly. **NOTE.** A **checking facility** may be incorporated (at least, temporarily) for users to monitor how the computer interprets the input. Eventually, this should become superfluous! The corresponding **output** will be in conventional (printed) form, for the range of symbols currently incorporated. There will be no 'clash' between the written input and the printed output (whether on a screen, or else, on some (other) tablet).

NOTE. The **l.c. Roman alphabet** would still be used as input for symbols. This is not problematical.

A	B	C	D	E	F	G	H	I	J	K	L	M	N	O	P	Q	R
										S	T	U	V	W	X	Y	Z
a	b	c	d	e	f	g	h	i	j	k	l	m	n	o	p	q	r
										s	t	u	v	w	x	y	z
$\underline{\mathrm{A}}$	$\underline{\mathrm{B}}$	Γ	Δ	$\underline{\mathrm{E}}$	$\underline{\mathrm{Z}}$	$\underline{\mathrm{H}}$	Θ	$\underline{\mathrm{I}}$	$\underline{\mathrm{K}}$	Λ	$\underline{\mathrm{M}}$	$\underline{\mathrm{N}}$	Ξ	$\underline{\mathrm{O}}$	Π	$\underline{\mathrm{P}}$	Σ
												$\underline{\mathrm{T}}$	Υ	Φ	$\underline{\mathrm{X}}$	Ψ	Ω
$\therefore$	$\surd$	$\equiv$	$\int$	[	]	{	}	$\propto$	∞	$\Vert$	Σ	∇	$\bigvee$	$\bigcap$	$\cap$	$\cup$	$\sim$
$\approx$	$\div$	$<$	Γ	Δ	$>$	$'$	$\vert$	Θ	ϑ	$\hbar$	Λ	α	β	γ	δ	ϵ	ζ
η	θ	ι	κ	λ	χ	$\rightarrow$	Ξ	$\rightleftharpoons$	Π	$\vert$	Σ	l	$\bumpeq$	π	ρ	σ	τ
υ	ϕ	χ	Ψ	Ω	ψ	ω	$\leqslant$	$\geqslant$	(	)	$\exists$	$\forall$	$\doteqdot$	∂	/	\	..
$\wedge$																	

PLUS: all other characters on a standard 'QWERTY' keyboard.

NOTE: All of these 'symbols' could be accommodated on an enlarged / compound keyboard.

This (comparatively modest) set of alphabetical / mathematical symbols nevertheless suffices to express virtually all of the MK! Of course, a more extensive collection of fontes of characters would allow greated variety of notation; but 'locally', in any 'segment' of an exposition or calculation, the mathematical content may be specified, unambiguously and adequately, over this basic character-set – especially, if bold / italic forms and some size-variation are incorporated (as would be routine for modern equipment).

Since reliable recognition of handwritten input has so far not been prioritized in CS research, the extended / compound keyboard just indicated should be specified as the **standard** (input) **interface** for P_Σ – with associated notation for (un)evaluated (sub)expressions (i.e., wfe). In this way, typical Σ -dialogues will comprise the user's input, interspersed with (partially) evaluated Σ -output (where input / output could be, e.g., black / blue). In this way, arbitrary (O)T may be 'placed' within Σ , ready for application in the course of investigations – the resulting (suitable processed) output being returned to the user's tablet, to continue the investigation(s). With obvious 'matching' of input and output notations / styles, each investigation may be presented as a sequence (collection) of specifications / hypotheses / formal deductions / calculations, in a uniform style. The deails of (intermediate) 'computations' may be stored, along with all (interrelated) 'results' generated – to be designated as 'new theorems', if appropriate. **Thus:** the above set of symbols is found to be adequate both for **handwritten**, and for **keyboard, input**. These forms should be developed in parallel, until their relative merits have been properly explored.

In this context, P_Σ may be regarded as a (technically accomplished, but unimaginative) colleague, in (arbitrary) mathematical investigations, where the OT in P_{Σ_t} may be deployed in $P_{\Sigma_{t+\tau}}(\tau > 0)$. These OT will have varying degrees of operationality – from 'nil' to t-max. The ongoing implementation of OT, in **all** domains of Mathematics, is, of course, the primary research activity in the longterm development of Σ as an OIB – whose intricate, multi-faceted structure is considered, in great detail, in the Σ **-Essays**. The overall aim is to create a vast, interactive framework for the prosecution of mathematical researh, as embodied in the acronym SIGMA. The combination of I(/O) **tablets** (as above), with processed output, interspersed with input, inaugurates a radically new environment for developing maximally penetrating (new) results in Pure / Applied / Applicable Mathematics. Once Σ is properly established, the increase in investigative power (in all domains) will be **staggering**.

When all of the components discussed in the Essays are considered, the overall structure of Σ as a universal OIB becomes **pellucid. The titles of the Essays** (properly interpreted) collectively span the interrelated challenges to be met by the system – but **the crucial importance of easily-produced natural input, and natural output from P_Σ – in evocative mathematical notation** – cannot be exaggerated!

The development of t-max. constructive versions of fundamnetal OT constitutes the principal Σ -research challenge – a cumulative process, in which the current stock of OT may be used to produce further OT – an iterative / recursive scheme for the progressively constructive extension of P_Σ, at the highest level. The Σ -Essays furnish a **blueprint for a quasi-optimal mathematical-research environment** – regardless of possible innovations in computer design, affecting I/O procedures; or, even, human / computer interactions of more general types.

An International Research Institute for the initial construction, and longte- rm maintainance, of the Σ -OIB should, therefore, be established, as a matter of urgency – for the development / extension / application of all facets of MK. **The current, disconnected efforts** (even with internet facilities) **are ludicrously ineffectual in scope / power** (quite apart from the widespread duplication). Moreover, the primary aim – to attain t-max. operationality in all theorems – emphasizes constructive formulations, even in cases where such versions are highly elusive and hard to implement. The resulting (cumulative) OIB, with consistent / uniform notation, offers researchers **milieux** of unparalleled scope / penetration, for all aspects of mathematical investigation.

The elaborate framework gradually introduced in E_1–E_9, for exact / (standard-) approxiative MA, is diversely exemplified in $E_{10/I}$ – and illustrated by copious items in $E_{10/II}$. Also in $E_{10/II}$ is a vast collection of mathematical **analogues**, from all areas of Mathematics – as illustrated in the listed papers (c. 3,000) – some 10% of which have **C**apsule **D**escriptions.

In $E_{11/I--IV}$, a **general formalism for Mathematical Analogy** is developed – to a point where it may be extended, systematically, as the analogical content of the OIB grows. Although such a formalism cannot be definitive, **it is soundly defined**, and **effectively applicable** to arbitrary ME 'generated' in longterm Σ -practice. The material in E_{11} is, inevitably, rather elusive – since analogues (direct, or inverse) are, designedly, somewhat imprecise. The overall effect of E_1–$E_{11/IV}$ is to furnish **a near-blueprint for a universal mathematical-research environment**, into the indefinite future. **Nothing remotely similar has been propsed (still less, attempted) up to now.** The inevitable shortcomings of any initial version, Σ_{t_0}, will merely motivate the production of improved forms, $\Sigma_{t_1}, \Sigma_{t_2}\ldots$ The obvious imprecisions in this scheme of regular enhancement will (on informed consideration) prove to be insignificant, as impediments to productive research. Indeed, such imperfections are endemic in modern hardware / software packages in all areas of human activity. The over-riding aim – **to promote freewheeling MA, in all domains, at the highest level** – is incontrovertibly achieved within Σ .

Conventions and Notation

Stylistic Conventions & Terminology

In an innovative work of this kind, the adoption of some nonstandard usages is almost unavoidable. The 'local' introduction of special terms, etc., causes no problems, so attention is confined to 'global irregularities'.

(1) The construction '$A_1/\ldots/A_n$' ($n \geqslant 2$) is used to mean 'A_1 &/or $\ldots$ &/or A_n', for arbitrary 'single expressions', A_k (to include 'words', 'symbols', and 'formulae' – of all w.d. types). An expression is regarded as 'single' IFF it is treated as an **entity**, through hyphenation or specification.

This construction is ubiquitous here. It produces great economy of expression, and it should not be taken as indicating uncertainty! Rather, it reflects the richness and diversity of Mathematics[659], and the consequent necessity of using several descriptors (linguistic or symbolic) to convey the intended meanings with adequate precision.

Without such conventions these Essays would be longer (through tedious near-repetitions) and considerably harder to understand. Moreover, the formulations of the arguments would present many additional difficulties – and some of the most interesting ideas could scarcely have been developed at all.

(2) The (French) usage, of 'in equation' for 'inequality', is (often) followed, since, when both types of conditions may occur, the abbreviated designation (in)eq(s) covers all possibilities economically.

(3) The acronym SIGMA is (mostly) replace by the symbol, Σ ; as such it is used in many forms – especially as an adjective (as in 'Σ -operations').

(4) Several other acronyms are used throughout the Essays – after their introductions, in various contexts. A partial list is as follows.

(O)IB (Operational) Information Bank / Base

SAM Symbolic / Algebraic Manipulation

OT Operational Theorem(s) [T / T$_k$ / T′ /. . .]

I/O Input / Output

IT Information Technology

AI Artificial Intelligence

M_t Mathematics in Σ , at time t

L Linear (vector) space(s)

Γ Closure space(s)

T Topological space(s)

U Uniform space(s)

P Proximity space(s)

[659]and, of English!

C Contiguity space(s)

N Nearness space(s)

S Statistical-Metric space(s)

F Fuzzy space(s)

μ Measure space(s)

SA Scientific Analogue(s)

OP Operational Proof

BPA Basic Problem(s) for Analogues

TMA Theory of Mathematical Analogy

FPT Formal Problem Template(s)

ME Mathematical Entity / Entities

MA Mathematical Analogy / Analogies / Analogues

MP Mathematical Phenomenon / Phenomena

SP Scientific Phenomenon / Phenomena

$\mathbf{P}_\Sigma$ The Σ mathematical processor

LC Locational Conditions

Ab {Abstraction(s)}

Ap {Approximation(s)}

Apr {Approach(es)}

An {Analogues}

$-$**An** {Inverse-Analogues}

Σ^D Descriptive content of Σ

$\Sigma^{\mathbf{Op}}$ Operational content of Σ

$\mathbb{V}$ Variant(s)

Gen {Generalization(s)}

Spec {Specialization(s)}

$_SX$ Space of type X

DBMS Data-Base Management System

Σ^π Proofs in Σ (with subclassifications)

$\mathcal{J}$ Family of (O)T in Σ

wfe Well-formed expression(s)

wd Well-defined

κT (strict) converse of T

CB Calculational Base(s)

G Group(s)

R Ring(s)

Mod Module(s)

Mon Monoid(s)

V Vector Space(s)

A Algebras

D($\mathcal{S}$) Domain(s) with structure(s) $\mathcal{S}$

O Order

C Commutativity

Can Cancellations

Lin Linearity

D Differential

Char Characteristic

E Extension

F Factorization

(G)SF (General(ized)) Structured Framework

C^F Class of theorems underlying Formal algorithms

C^I Class of theorems underlying (efficient) Implementation

CF Calculational Framework(s)

$\delta(\zeta)$ Structural Decomposition for the Calculation ζ

AE Approach Environment(s)

GST General Systems Theory

$\mathcal{N}$ Nomenclature

$\mathcal{D}$ Definitions

$\mathcal{A}_x$ Axioms [Also: $Ax(S)$: axioms for structure S]

$\mathcal{R}$ Rudiments

$\mathcal{F}$ Foundation(s) (for Σ)

$(O)_q$ (Partially) Operational information

$(D)_n$ Descriptive items

$(P)_L$ t-max-Implemented Procedures

$(r)_m$ Rules (for the CB $B \equiv \{b\}_J$)

U_t Underlying objects in Σ_t; $U_t \equiv (u)_{L_t}$

OTC Operational-Theorem Calculus

UEF Universal Editorial Facility

CM Calculational Milieu(x)

$\mathbf{I}_t$ Information-set of Σ_t

CAD Computer-Aided Design

$\varphi_k(e)$ Facets of the ME e $(1 \leqslant k \leqslant 6)$

$\mathcal{D}(e)$ Diagram for the ME e

M Modification Operator(s) for diagrams $\mathcal{D}(e)$

$\mathbb{M}$ Modification operators for **classes** of diagrams

(C)V (Conceptual) Vicinities of ME

CPO Complete Partial Order

$\Sigma^{\,G}$ (Quasi-) Graphical Content of Σ

$\mathbb{A}(x)$ The set of admissible approaches to x

DC Diagram-Calculus (for Σ)

$\Sigma^{\,\mathbb{T}}$ Set of generic OT in Σ

$\rightarrow \Sigma^E$ Set of ME constituting Σ

π_T Proof-diagram for the theorem T

AE Approach Environment(s)

(G)CSF (Generalized) Combined Structural Framework(s)

EPCS Existence Problem for Combined Structures

SE/P Scientific Entities / Phenomena

Q Quintessence

LD Locational Description

${}^gT/{}^eT/{}^tT/{}^eT$ Generic / Explicit / Task / Cast forms of (O)T

$\pi(T)/'\pi(T)/''\pi(T)$ Full / Skeleton / Indicative proofs of T

MR Mathematical Reviews

AS Axiomatic System(s)

OC Operational Calculus

DC Diagram Calculus

OAC Operational Analogue Calculus

P/M Perturbation / Modification

MF-C Master Flow-Chart

[T**]**, $\langle T \rangle$ Premise(s), conclusion(s) of T

cat(e) Taxonomic Category of e

FH Formalized Heuristics

MAP Mathematical Analogies in Perspective

$\mathbb{P}$ {Positive Integers}

$\mathbb{R}$ {Real Numbers}

$\mathbb{Q}$ {Rational Numbers}

$\mathbb{C}$ {Complex Numbers}

$\mathbb{Z}$ {All Integers}$\cup\{0\}$

$e :: l(e) =: \bigwedge_{\alpha \in S} c_\alpha(\xi^{(\alpha)}) \equiv LC(e)$

$c^X \equiv \{c_\alpha : \alpha \in X\}$

$e \approx (\rho)_m$ e is characterized by $(\rho)_m$

BR Basic Reference(s)

CD Capsule Descriptions

$\mathcal{S}$ Specialization(s)

$\mathcal{G}$ Generalization(s)

$\mathcal{A}$ Standard Approximation(s) [all forms]

$\mathcal{Q}$ Abstraction(s) [Quintessence(s)]

$^{(+)}\mathcal{I}$ Analogy / Analogies [Imitations]

$^{-}\mathcal{I}$ Inverse-Analogy / Analogies

Ef**I** Environment for Investigation

$\delta\langle e\rangle$ Determination(s) [exact] / representations $\langle$of $e\rangle$

$\mathbb{A}\langle e\rangle$ {approaches to e} $=: (\delta \vee @ \vee \mathcal{A} \vee A)\langle e\rangle$

Lists / Arrays / Families / Functions

The (countable) **lists** / **arrays**, and the (**un**countable) families may be ordered or unordered. The term **function** is interpreted with great generality. Of particular importance for Σ are the applications of (all valid) Boolean combinations of compositions of transformed (generic) OT – to generate approaches to (arbitrary) w.d. ME. Since OT are regarded in Σ as functions of their 'constituent / cast arguments' (see, especially, E_4), and these collections of arguments may be represented as lists / arrays / families of elements – on which operators of all appropriate types may act – the need for a versatile / uniform notation is paramount.

Notation

$(\ldots)$ $\equiv$ ordered linear array

$\{\ldots\}$ $\equiv$ unordered linear array

$_{(n)}p \equiv (p, \ldots, p+n-1)$

$_{\{n\}}p \equiv \{p, \ldots, p+n-1\}$

$(x)_n \equiv (x_1, \ldots, x_n); (x)_{j,k} := (x_j, \ldots, x_k)$

$\{x\}_n \equiv \{x_1, \ldots, x_n\}; \{x\}_{j,k} := \{x_j, \ldots, x_k\}$

$(x_\sigma)_n \equiv (x_{\sigma_1}, \ldots, x_{\sigma_n})$

$\{x_\sigma\}_n \equiv \{x_{\sigma_1}, \ldots, x_{\sigma_n}\}$

$$({}_k x_\sigma)_n \equiv (\underbrace{x_{\sigma_1}, \ldots, x_{\sigma_1}}_{k_1}, \ldots, \underbrace{x_{\sigma_n}, \ldots, x_{\sigma_n}}_{k_n})$$

$$\{{}_k x_\sigma\}_n \equiv \{\underbrace{x_{\sigma_1}, \ldots, x_{\sigma_1}}_{k_1}, \ldots, \underbrace{x_{\sigma_n}, \ldots, x_{\sigma_n}}_{k_n}\}$$

$(x)_\Lambda \equiv (x_\lambda : \lambda \nearrow \Lambda)$, for '$\lambda$ increasing through the ordered family Λ'

$\{x\}_\Lambda \equiv \{x_\lambda : \lambda \nearrow \Lambda\}$, for an arbitrary family Λ

The notation '$\lambda \searrow \Lambda$' covers **de**crease in Λ. Here, $n, p, k_i \in \mathbb{P}$, and, $(\sigma)_n, \{\sigma\}_n \subset \mathbb{P}$. The $x_i, x_{\sigma_k}, x_\lambda$ are arbitrary wd elements – possibly, OT. For functions, one has:

$f(x)_n \equiv f((x)_n) = f(x_1, \ldots, x_n);$

$f\{x\}_n \equiv f(\{x\}_n) = f(\{x_1, \ldots, x_n\});$

$f[x]_{(n)} \equiv (f(x_1), \ldots, f(x_n));$

$f[x]_{\{n\}} \equiv \{f(x_1), \ldots, f(x_n)\}.$

More generally, if $x_{\langle m \rangle_k} \equiv x_{\langle m_1, \ldots, m_k \rangle}$, then one may define $f(x_{\langle m \rangle_k})_l$ by

$$f(x_{\langle m \rangle_k})_l := f(x_{\langle \langle m \rangle_{k-1}, m_{k_1} \rangle}, \ldots, x_{\langle \langle m \rangle_{k-1}, m_{k_l} \rangle}),$$

for $l \in \mathbb{P}$, and the case where $x_{\langle m \rangle_r} := x_{m_{1_{\ddots_{m_r}}}}$ is of special relevance in Σ .

NOTE The forms $f(x)^n, f\{x\}^n, f[x]^{(n)}, f[x]^{\{n\}}$ are defined analogously (for $(x)^n \equiv (x^1, \ldots, x^n), \{x\}^n \equiv x^1, \ldots, x^n\})$, with obvious extensions to $f(x^{p \cdots q} \equiv f(x^{p^1 \cdots q^1}, \ldots, x^{p^n \cdots q^n})$, etc.. The 'mixed forms', e.g., $T({}_k x^\sigma)_n, T({}^l x_\sigma)_m$ are also used.

Conspectus

The aim of this **Conspectus** is to demonstrate (in outline) that **the Σ -Scheme furnishes a comprehensive framework for the cumulative development/ synthesis of Mathematical Knowledge**

The collection of **Essays** and **Appendices** exhibits a variety of highly complex, but coherent, (non)-deterministic structures – diversely interrelated – unified by the pervasive concepts of **operational theorem** and approach. The power and scope of this 'apparatus' may be appreciated fully only after sustained effort.

The challenge is gargantuan, so the arguments are inescapably intricate and subtle. Above all, it becomes apparent that, **within Σ , calculability is maximized in all parts of Pure, Applied and Applicable Mathematics**; and, that the **boundaries between ostensibly disparate mathematical domains are virtually erased.**

Although the detailed superstructure is extremely complicated, **the essential features may** be understood by close inspection of the **Introduction** (E_0) and the **Contents Lists** of the Essays (backed by the **Conventions/Notation**, and **Glossary**).

Contents of Conspectus

• Prologue • Contents-Lists of $E_0, E_1, \ldots, E_{11}$
• C_0 : Introduction • Cellular-Automata/Science/Mathematics
• Conventions/Notation • Glossary

SIGMA : A Prototype Synthesis of Mathematics

Prologue

These Essays develop a comprehensive scheme for **the synthesis and continued extension of Mathematical Knowledge.** This gargantuan program is of crucial importance – for several reasons.

(1) The explosive growth of Pure/Applied/Applicable Mathematics cannot be halted, but the avalanche of 'new-results' **can** be organized/controlled/examined – to eliminate unsound/trivial/repeated items.

(2) The fragmentation of research domains (in all fields) drastically reduces the purview of Mathematics as a universal calculational system. Broad, 'pan-mathematical' views are now virtually unknown.

(3) Constructive paradigms in Pure Mathematics remain comparatively rare, and this limits the effectiveness of Applied/Applicable Mathematics.

(4) The extreme importance of **interrelations among ostensibly disparate structures** has still not been widely recognized. And so on!

The possibility of rectifying this situation devolves on the evolution of (computer-based) facilities for: **I**nformation-**T**echnology/**D**ata-**B**ase-**M**anagement/**A**rtificial **I**nt- elligence –and, above all, **S**ymbolic/**A**lgebra **M**anipulation – combined with **a growing appreciation of the roles of algorithms/constructivity in typical mathematical investigations.**

The **synthesis** proposed here involves a vast **O**perational **I**nformative **B**ank, whose primary constituents are **O**perational **T**heorems, capable of unlimited transformation and combination, for application to arbitrary (compatible) **M**athematical-**E**ntiti- es/**w**ell-**f**ormed-**e**xpressions.

The encompassing framework for these activities extends the most general notions of (non) deterministic approximation to those of APPROACH – subsuming **approximation, abstraction, analogy,** and **inverse-analogy**(from scientific theories). **The primary objective** is to outline a broad collection of **concepts, structures, procedures,** and **paradi- gms**, from which a (**U**niversal) **E**nvironment **f**or **I**nvestigation may be constructed – potentially adequate for **S**imulation/**I**mplementa- tion of **G**eneral **M**athematical **A**ctivity: SIGMA since the entire scheme seems to be unprecedented, with many novel/unfamiliar facets, all of the **components**, and their (mutual) **interactions**, must be discussed in considerable detail. Although the corpus, M_t^*, of 'Mathematical Knowledge at time t, cannot be precisely specified, it is argued that M_t (the **content of** Σ_t) may **approach** M_t^* **asymptotically**, as $t \to \infty$; so that **this prescription for** M_t^* **is susceptible of systematic development.**

The 'route' from any 'initial OIB', Σ_0, to Σ_t is, of course, far from unique; but it is **prototypical** for the establishment of a **superstructure for Mathematics.**

In these circumstances, the range and depth of topics discussed in the Essays may be regarded as close -to-minimal for indicating – however vaguely – the broad compass of M_t^*. Indeed, **the copious/diverse references** to topics in all recognized domains of Pure, Applied and Applicable Mathematics (as designated, for instance, in **Mathematical Reviews**, and in comparable **Science, Engineering, Social-Science . . . review journals** make no pretence to completeness, but are still adequate **to indicate the scale** of the massive tasks involved while, at the same time, demonstrating **the potential attainability** of the principal objectives.

The **universality** of SIGMA entails that the most powerful, intricate, sophisticated, arcane/. . . mathematical phenomena must be accommodated within a **unified descriptive, calculational milieu** – exhibiting structural versatility of a very high order. The formulations gradually elaborated in the Essays are, no doubt, far from optimal – deploying diverse procedures in various quasi-recursive (but fundamentally sound!) modes. This is inevitable, since even the concepts/techniques/. . .
used to define Σ_t will eventually be embedded in the OIB! In spite of formidable obstacles and recondite problems, it will become clear, from the material presented here, that **the essential synthesis of Mathematics is effectively attainable**; that the most profound issues have been **identified**, and **substantially investigated**; and, that systematic (cumulative) software implementations may now be explored.

The undoubted imperfections in this exposition are attributable (at least, in part!) to protracted research conducted in isolation. Indeed, the most desirable outcome of my efforts (during more than 30 years) would be the emergence of Σ /II: **concise, elegant, pellucid**, and **optimally implemented**! Meanwhile, the material is offered in this 'primitive version' – to start the ball rolling.

C_0 • Overall **aims** of $(E)_1 1$: **To maximize calculability, in all domains of Pure, Applied and Applicable Mathematics, on the basis of approach criteria**. • Taxonomic development of results algorithmic over Σ_t • Relation to SAM routines • Nature of t-dependence for Σ_t • Comments on cellular-automata representations of scientific/mathematical processes (as treated in: '**A New Kind of Science**', by **Stephen Wolfram** (Wolfram Media,2004)) • Evolution of Σ_t• Approach: neighbourhoods in 'Conceptual space' • Basic $_SX$• Mathematical information • Brief Comments on taxonomy design and DBMS structures • $(k-)$meta-mathematical aspects of Σ_t• Outlines of the E_k [• Abbreviated **Contents lists for the** E_k] → Conspectus

Comments on 'Cellular-Automata-Representations' of Scientific and Mathematical processes

The development of increasingly fast/powerful/small/cheap computers, with sophisticated graphics/IT capabilities – coupled with SAM packages – has facilitated **the simulation of diverse scientific/mathematical processes:** for instance, in **Biology**(evolution, genetics); **Chemistry** (reaction-paths, reaction-dynamics); **Continuum Mechanics** (viscous flow, turbulence, fracture); **Nonlinear Mechanics** (oscillations, chaotic behaviour). The impact on **Pure Mathematics** has been comparatively modest – except in primarily enumerative problems, where largescale 'trials' may suggest useful heuristics, even in well-established domains such as **Combinatorics** and **Number Theory**.

NOTE. Automatic theorem-proving pack-ages are still far from fulfilling their apparent potential in nontrivial procedures, though they have become efficient for routine – but totally unmotivated – deductions within various formal logical systems.

A highly speculative, hefty compendium of simulations (with cognate vignettes of the underlying 'standard science/computation) has been compiled by **Stephen Wolfram** in his book (/Software): '**A New Kind of Science**' (pp1157, Wolfram Media, Inc., 2002). Here, the extended (sometimes tenuous) argument devolves on the 'unexpectedly complex' evolution of (in particular) 256 **E**lementary **C**ellular **A**utomata, over tens, hundreds, thousands, millions /dots of iterations –using the SAM package **Mathematica**.

The key aim is to demonstrate that **'Simple CA rules' can produce 'arbitrary complicated system behaviour'**; and, moreover, that: **all CA (or comparable systems) whose evolutions 'are not obviously simple'** (i.e., not nested/periodic/ ...) **are essentially of the same degree of complexity** – the so called **principle of Computational Equivalence.** Connections with **N**eural **N**ets/**T**uning **M**achines/ (**F**inite) **A**utomata/**R**egister **M**achines/**T**ag **S**ystems ... are discussed at some leng- th and illustrated with copious examples of Mathematica graphics.

Another fundamental concept is that of (**Computational**) **Universality** of class- es of systems: a C U system can simulate arbitrary complex behaviour of all other (types of) systems. For instance, it is found that the 'rule-110 CA' is universal. The proof — entailing the checking of numerous conditions – is (so far) computer-based. The m-dimensional CA of order n are realized over lattices in $\mathbb{R}^m$, with n colours (or, in some 'totalistic' cases, a continuum of grey shades', from white to black).

Although the book is replete with 'guesses' and '(strong) suspicions' as to the (asymptotic) behaviour of multifarious systems (in their claimed CA/TM/... manifes- tations) there are virtually no **proofs** of such properties, at acceptable levels of rigour. The comparative analyses of computability, complexity, consistency, decidability ..., of classes of procedures over these systems, are probably well-formed, only the barest indications are given. **NOTE.** Systems are defined to be **computationally irreducible** if **simulation** is the only viable way to study their evolution.

The **underlying postulate –** that **every natural/scientific/mathematical process is in essence a computation** – mirrors the Σ -assumption that: **all mathematical entities (ME) are, in essence, (operational) theorems.** The crucial distinction is that the Σ -view requires no verification, since it may be formulated with full generality; where as, the PCE is a quasi-empirical axiom.

The graphics output is accompanied by various (slanted) interpretations. Processes from Biology, Chemistry, Quantum-Mechanics, Gravitation, High-Energy-Physics/, Patter-Analysis, Synthesis/... are discussed 'from the CA angle'. It is hypothesized that many basic processes are effectively 'screened out', by their essential computational irreducibility. (once again, there is an echo of Σ , where it is observed that countless significant (pure-) mathematical problems/investigations are overlooked/ ruled out on account of their intractability).

The exclusive use of **Mathematic** I/O (rather than, say, **Axiom**) is understandable; but the SAM-format **in the text** is distracting, and it significantly reduces the intuitive mathematical content of expressions.

NOTE. It is a basic premise in Σ that **the conceptual/operational sophistication of existing SAM packages is totally inadequate** – since the theorems typical of high-level mathematical discourse are not represented, even descriptively – let alone operationally. The difficulty of developing operational formulations of theorems poses a formidable challenge – which dominates these Essays. As things stand, the CA-scheme could offer insights at primitive theoretical levels; but it is inappropriate for the productive organization/extension of Mathematical Knowledge. The overall Σ -structure, however, allows the CA (and, also, the AI) routines to be consistently embedded in the OIB, as it evolves.

C_1: • P_{Sigma} as a **C**alculational **M**ilieu • Types of **Basic Spaces**, ${}_SX$• **Realizations if ME** over the ${}_SX$• Cumulative **axiom list** $(Ax)_M$ • List, $\mathcal{U}_t$, of **underlying ME** • ${}_SX$- approximative representations of ME • (inverse-) analogical modifications of the ${}_SX$• Initial lists (for Σ_0)L iterative development (from Handbooks, etc) • Pre-/sub-/skeleton/... taxonomies • Σ as a 'General System' • Inherent soundness of the overall Σ -scheme • Proofs as quasi-compositions of other (established) proofs • operator-function representations in terms of (other) ME/OT • Stability/Centrality/... • Brief comments on existing SAM packages • Main goals for Σ -operation • Σ as an extension of SAM via OIB • General representations of OT • Formal linguistic structure of Mathematics (for automatic theorem-proving) • Extension of structural features to Σ ... • Notation for lists, sequences, functions, ...— all involving OT with constituents in $\mathcal{U}_t$ • Descriptive/ nil-operational / t-max-operational/... forms of ME • General functions of OT • Action of OT on wfe • Calculational Bases • Basic ideas on 'constituents' and LC • Simple forms of Locational Neighbourhoods • Notions of Conceptual Space.

C_2 • Algebraic structures in **axiom** (and in other packages) • Reduction of analytical calculations to procedures over $\mathbb{R}$ or $\mathbb{C}$; and of algebraic calculations to realizable SAM routines • Indications of the Scope of **axiom** (and other systems) • Essence of 'algebraic activity' • **Claim** that all analytical-topological structures are realizable over 'suitable algebraic SF' • Reduction to SAM routines for polynomials/rational-functions/FPS (over general SF) • Fundamental procedures for n-variable functions • **Aim:** to embed arbitrary exact calculations 'naturally' within 'approach domains' (often **via** procedures in Commutative Algebra) • More sophisticated procedures (e.g., Hensel lifting, p-adic expansions) • Key references (books/papers) • Outlines of fundamental SAM algorithms • Outlines of routines for the integration of Elementary Functions, and for the solution of classes of ODE • List of underlying algebraic results in (e.g.) **axiom** • Key theorems for SAM algorithms • Brief remarks on SAM 'library routines' • **Claim** that arbitrary OT are (potentially) approachable over P_Σ • Synopses/Contents - Lists of seminal treatises on Algebra/Analysis (to support the 'calculational-reduction' claim) • Outline of a 'calculational-decomposition scheme' • The role of SAM in Σ • Selected 'advanced algorithms' (LLL-polynomial-factorization/Rothstein-Trager-Bronstein integration scheme/Risch-Davenport integration scheme).

C_3 • As structured, searchable collections of 'interacting items' • schemes for pre, skeletal, partial, full taxonomies: Heuristic aspects • Nomenclature/definitions/ axioms • Rudiments • Foundations • Detailed (hierarchical) proof -representations • Forms of definitions (explanatory/ inductive/recursive/implicit/...) • Types of Σ - information/Types of taxonomies • Taxonomic structures and techniques • t-max-operational information • Characterization of explicit/recursive/implicit definitions • Properties of Calculational Bases • Classes of wfe (e.g., k-atomic, n-variable predicate functions • Representations of wfe over $\mathcal{U}_t$ • Specification of CB • Notions of the OTC • The universal Editorial Facility • Roles of Network Theory /Information-Theory/Optimization-Theory/.../General-Systems-Theory in Σ -design.

C_4 • Notation for simple/multiple OT • Ingredients of OT • (Stylistic) formulations of OT • Task and cast representations of OT • Action of OT on wfe • Definition(s) of $f(T)$ • Class(es) of general transformations :$\{OT\} \to \{OT\}$ • Definitions of realizable, effective, constructive, implementable, operational, applicable • Interpretations and models/Overall aim: $M_t \sim M_t^*$ as $t \to \infty$ – within the most general approach environments • CB as elements of $\mathcal{P}B_t$ • $\mathcal{P}B_t$ as a Boolean algebra • Topological Boolean algebras • Cylindrical/Polyadic algebras • Classification of CB via fields covered • Fundamental approach – operations in CB • Sequential/Metric topologies /Formal languages/associated theories • Basic problems of the OTC • Diagrams as operational information • Some subject based classes of diagrams (with key references) • Facets of mathematical information • Modification of diagrams • quasi-addition theorems for diagrams, and for facets • Modification operators

C_5 • Specification of the Basic Spaces, ${}_SX$• Background references for the Basic Spaces • Comments on the references • Interrelations among analytical structures (including the ${}_SX$) • Remarks on structural interrelations • Types of **A**pproach **E**nvironments (i.e., of $L(V)/\Gamma/T/U/P$

$/C/N/S/F/mu)$ listed/briefly-discussed • Tables of references for all types of Basic Spaces • Some extra structures over the Basic Spaces • Representations of the analytical structures • Realizability problems for sets of axioms over specified base-sets • **E**xistence **P**roblems for **C**ombined **S**tructures • q realized over p – for classes $\{q\}, \{p\}$ of ME: general framework • Discussion of various cases (for the $_{S}X$) • Definitions of the fundamental AE/SF • Technically diverse specializations of generic AE/SF.

C_6 • LC/LD in conceptual space for arbitrary ME/SE • The classes $Ab(e)/An(e)/-An(s)$ • Quasi reflections and quasi -refractions and refractions • Quasi-scatteri- ng of ME/SE • LC-representations/criteria • LC-modification procedures • Further remarks on LC-modification • Internal /external ME behaviour • Classes of inner/outer conditions • Interactions of ME,e, with environment(s), ξ • Good analogues bu Closeness of reproduction (of e - conditions) • Inner conditions exemplified by 'SF-laws' • Outer conditions exemplified by optimization criteria • Interrelations among constituents of e • Simple and compound analogues • Analogues as mappings • The sets D, S, G, A, Q, I, and $\mathcal{M} := \{\mathcal{D}, \mathcal{S}, \mathcal{G}, \mathcal{A}, \mathcal{L}, \pm\mathcal{J}\}$ • LC- representations for abstractions • Relational -representations for analogues • Key/peripheral relations • Some comparison criteria for analogues • Analogues 'defects' • Quasi-morphism formulations • General binary/m-ary laws of combination • Tentative schemes for inverse analogues • LD-formalism • Comments.

C_7 • Approach as incomplete specification, or modification, of defining conditions over admissible SF • Forms of LC for e • Distinct modes of 'perturbation' of LC • Representation of the $_{S}X$via LC • Connections with Universal Algebra • 'Elementary perturbations' • Types of (standard) approximation • **C**onceptual **V**icinities • Quasi-neighbourhoods • More precise definitions for 'structured spaces' • LC-deformation-based neighbourhoods • quasi-metrics for predicate-function spaces • Results for classes of programming languages • Importance of **in**completely formalized specification languages • Other quasi-constructive models ... • Characterization of the (basic) modes of approach • Approaches to OT • Approaches to proof-diagrams (as listed) with specified components • Topical outlines of Σ -species of diagrams (with background notes/references) • Approaches to (general) Σ -diagrams: schematic framework(s) • Σ -**D**iagram **C**alculus • Classification of approaches to diagrams • Approaches to proof-diagrams

C_8 • Outline of 'linguistic structure of $\Sigma_t\mathcal{U}_t$' • 'Canonical versions' of OTL explicit/task/cast • Connections with 'environments for investigation' • General representations of OT **via** species-based decompositions of $\mathcal{U}_t$ • Combinations of transformed OT • OTC quasi-inverse(s) ($\equiv$ operational quasi-converse(s)) • Notions of quasi-convergence (for wfe) in proofs • Full/skeleton indicative representations of proofs • Role(s) of approach in the OTC • Interpretation(s) of functions $H(T)_m$ OTC-analogues of Riemann surfaces • OTC -analogues of eigen-elements • General discussion of composition(s) of OT • Compositional operational proofs • OT as computer programs • Martin-Löf's Type-Theory • Carnap's L-provability • Polyadic algebras • proof stability • Variations of proofs • Proof-preserving deformations • Operational quasi-converses • Estimates for sizes of proofs • Environments for investigation: taxonomic considerations.
[**NOTE.** All of the topics listed here are initially 'introduced' (with minimal definitions, etc.); after which each topic is briefly developed – to indicated possible modes of treatment.]

C_9 • Universality of approach within Σ • Notions of possibility, attainability, speed, efficiency, optimality, scope, stability ... of approach processes • Generic incompleteness of calculations • Exact calculations as 'exceptional elements' • Definitions, limitation-conditi- ons, quasi-varieties, abstract-boundaries as sources of approach criteria • The Σ -Lists (of examples0 • Possibility of calculational reduction to enhanced SAM/IT/AI/... packages • 'States', Σ_S of Σ 'generated by base sets S' • LC-representations as sources of all approach phenomena • Variants of ME • Stochastic LC • Interactions among ME • M_t as a quasi-space • (In)stability of LC-representations • ME-dependence on subsets of LC • 'Typical' C- changes • Irretrievable LC-properties • General 'LC-behaviour' • Conclusive proofs • Provisional proofs • Overall nature of Σ as a system • Brief discussions of possibility, attainability, speed, efficiency, optimality, scope, stability ... of approach processes • Outline of the MAP Scheme • List of MAP concepts: c_1(CONTINUITY), ..., c_63 (CALCULABILITY) • 'Specimen definitions' of $c_1, \ldots, c_63$

$C_{10/I}$ Limitations of current SAM packages • Some obvious enhancements • Notions of **S**ymbolic **I**nvestigation • Aims of Outlines from SCCM • Interpretation(s) of (partial) solutions in SI • Implicit Functions of n (Complex) variables • Implicit functions of n (complex) variables • Effective procedures in Function-Theory • Summability (and analogues) procedures • Differential Calculus in general spaces • Branching analysis for operator equations • Calculations in Celestial mechanics • Function- theoretic methods in Elasticity • Singular integral Equations • Ill-posed problems • Galois-field procedures for switching circuits • Problems in Robotics • Problems in Control Theory • Problems in Circuit Theory • Analysis of Nonlinear (O)DE • Basic calculations in Algebraic Topology • Calculations in Differential Geometry • Calculations with Generalized Functions • Procedures in Asymptotics • Operational Schema for (non)linear PDE • Topological approximation of fp of mappings • Calculations in Singularity Theory • BS-calculations: (quasi-)bases • Statistical Mechanics/ QFT • Group Theory • FEM/BEM routines • Solution of Stochastic functional operator equations • Group Representations • Perturbation Theory • Integral Operators (and some extensions) • Abstract Harmonic Analysis • Computational Geometry • Probability/Statistics • p-adic analysis • Complexity Theory • General Commentary

$C_{10/II}$ • Preliminary remarks on 'frameworks for approach' • Example: H/B for Nonparametric Statistics, Vols 1–3 by J.E. Walsh • Remarks on choice if topics for Σ -Lists • Minimal sets of item - characteristics • Primary/secondary/... sub-theories • Example (transplanted structures): Differential-Geometric techniques in Multivariate Analysis • Further remarks on the tagging of Σ -List items, $\langle i \rangle$ • Formal representation(s), $\langle i \rangle \approx \alpha(T_\sigma)_q$ • Item – profiles • Σ -superstructure • Explicatory versions of (O)T • Template - format for item- taggings • Item - descriptors, $(\Delta^m)_{11}$ • Nil/partially/fully -operational ME • Criteria for Σ -Lists • Some immediate consequences • Discussion of aims/consequences • Paths to mathematical progress • Discussion of key issues raised • Relation to the MAP scheme • Vignettes of **diverse examples of approach** (mainly, of (non) deterministic approximation). The **topics** correspond to selections from the S/M - List.

$C_{11/I}$ • Separation, $\sigma(Ae, e)$, of (inverse-)analogical approaches Ae, and e • Deterministic and stochastic versions of (inverse-)analogues • Example: Stochastic FEM • Inner (φ_I) and outer (φ_I) features of ME • Stages of analogue-formation • Recap: item-descriptors, $(\Delta^m)_1 1$ • Mechanisms, $\mu(Ae)$, of (inverse-) analogues of e • Mathematical/Scientific environments, E_m/E_S • Formulations, $F(i_m)$ of $\langle i_m \rangle$ • Comparison with Engineering Design Compendia • Behavioural properties, $b_{e\lambda}$, of e • (Partial) transfer of structures (from $\langle \zeta_1, \xi_1 \rangle$ to $\langle \zeta_2, \xi_2 \rangle$ • Recap: the list (c)$_6 3$ • Further comments on the MAP scheme • IllustrationL DIFFERENTIABILITY • Comments on the 'generation' of inverse analogues • (quasi-) graph representations of (inverse) analogues • Mechanisms of proofs/programs • Main Questions for Σ -analogues • Analogues as 'active ME': the OAC • Comments on 'Approximation in Physical Theories' (edited proceedings) • The ABC(MFM) templates • Examples of CD/annotations • Examples of CD-flowcharts • General schemes for taxonomy/tagging • Extension of the ABC scheme to general **OIB** • **M**aster **F**low **C**harts for ME • Variants of MF-C • Interim Summary • The roles if taxonomy/tagging in Σ • ME as (partial) output of (perturbed) MF-C • Interactions among (sub)theories • Links with **F**ormalized **M**athematical **T**heories • Matrix representations of ME/Machine transformation • **APPENDICES**: Calculational Flow Charts/ ABC - Problem Interface(s)/ **T**opical **O**utlines (for diverse mathematical techniques).

$C_{11/II}$ • Notation for (inverse) analogues of ME/SE • Basic Problems for (inverse) analogues • Contrast between the mechanisms for direct and for inverse, analogues • Preliminaries to (inverse-) analogue-formation • Perturbation operators for machines • Comparison of 'LC' and 'IM' representations • Interim Summary • Basic properties of assignment-perturbation operators • Machine analysis • Notions of machine synthesis • Some interrelations of analysis and synthesis (for IM) • Some fundamental (scientific/mathematical) synthesis schemes • Synthesis and quasi-constructive procedures • Aspects of chemical synthesis • **B**asic **R**eferences for **S**ynthesis **P**aradigms [16 lists, each prefaced by brief comments] • Interim Summary • CD of synthesis schemes: basic ideas • Connections with **G**eneral **S**ynthesis **T**heory • Brief remarks on **S**tochastic **C**ontrol **T**heory and **Control Theory • Global Aim: To specify arbitrary ME**

within GST; and, to develop GST-universal synthesis schemes for Σ

$C_{11/III}$ • Comparison of the 'tagging/descriptors' and 'LC/Machine' specifications of ME • Partial tagging format : $\{\delta, \mathcal{A}, A, -A, @\}\langle e\rangle$ • Interpretations of δ: examples • Basis for truncated tagging • Specification of e as $\varphi(u_\sigma)_L$ over U_t • Representations of arbitrary approaches to e • Reduced tagging format • Extension to **S**tandard **T**agging (by adjunction of brief outlines of model(s)/formulation(s)) • Basic Tagging: $\langle i_m\rangle \approx [m; (c_\psi m)_{P_m} ; f_m]$ • Operational specification(s) of e, e_λ via machines $\mathbb{M}, \mathbb{M}_\lambda$ • Connections with (general) engineering/biological/chemical/...
systems • Formalism for machine -perturbation • Interrelations of Σ with Theoretical CS • Σ as a 'multi- medium' of diversely interacting (sets of) ME • Analogue-exploration: a wide ranging collection of illustrations. • **APPENDICES:** the Σ -Lists

• **Mutual Influence** among (sets of) ME/SE, $\{\beta\}$ • **GST- Procedures** for (inverse) analogues • Typical **perturbations** of ME/SE • **Relation** of β-Machines to GST • **Quasi-equivalence** of realization and synthesis • **GST - representations** of β- machines • β-Machines as **I/O- Systems** • **Template-representations** of β- Machines • **Admissible perturbations** of β-machines • **P**erturbation/**M**odification Operations • β-Machines as constructive realizations of **models** • **Quasi-**inverse analogues of **ME** • Abstract/analogical **interpretations** of SE/SP • **Intuitive** responses to SE/SP: **Consistency criteria** • **Conceptual separation** of (inverse-) analogues from '**originals**' • **Canonical** P/M processes for direct analogues • **Non-**algorithmic nature of (quasi-) inverse analogues • **Incorporation** of 'new' (quasi) inverse analogues in the OIB • **RECAP:** β-machine structure • **RECAP: M**athematical **I**nformation Structures • Further **specification** of MI-structures **RECAP: internal structure** of β-machine components • **Basic Tasks:** Possibility/ adequacy of β=representation • **NOTES** on the **Basic Tasks**
• **NOTES on the individual $\mathbb{M}_\beta$-Components**

σ**:** formalized **scientific theories**

s**:** formalized **mathematical theories**

ζ**: frameworks** for ζ ('motivated by β')

d**:** LC/D for ζ over ξ_J

e**: auxiliary elements** for realization of $\zeta in \xi_J$

f**:** maximally **constructive forms** of ζ in ξ_J

$\mathcal{R}$**: representations** of ζ VIA auxiliary elements

g**:** formal **(in)equational conditions** on ζ

h**: formal auxiliary elements** for ζ in ξ_J

i**:** formal **realization** of ζ in ξ_J

M**:** mathematical **transformations**

j**: effective (in)equational conditions** on ζ in ξ_ζ

k**: effective auxiliary elements** for ζ in ξ_ζ

l**: effective realizations** of ζ (and so, of β)

• Comments on possible **CD/Flowcharts** for syntheses • Connections with **theoretical CS** • Connections with **program design/analysis** • **Algorithmic/Software** issues
• **Interim Summary**
Types of realization, $P_k(\beta)$, of β
Key features of the $P_k(\beta)$
$P_1(\beta) \equiv$ LC/D(β) [Locational Conditions/Descriptions]

$P_2(\beta) \equiv$ MF-C (β) [Master Flow-Chart form(s)]
$P_3(\beta) \equiv \mathbb{M}_\beta(\zeta, \xi_\zeta)$ [Machine-synthesis form(s)]
$P_4(\beta) \equiv \Pi_(\beta)$ [Computer program with specification(s)]
$_HP(\beta) \equiv$ [Hybrid form(s) of representation]

• Initial remarks on **formulations of syntheses** • General Comments on **STABILITY in syntheses**

Outlines for **S**ynthesis-**P**aradigms

• **SP**$_1$**:Electrical circuits** (and equivalent simulations) • **SP**$_2$**: Sampled-Data** control systems • **SP**$_3$**: Optimal/Feedback/Adaptive** Control Systems • **SP**$_4$**: Dynamical** Systems • **SP**$_5$**: Automata** • **SP**$_6$**: Threshold-Logic Gates** • **SP**$_7$**: Stoch- astic** control systems • **SP**$_8$**: Antenna** Design • **SP**$_9$**: Spectral-Function Theory** • **SP**$_10$: Integral Geometry • **SP**$_11$: Chemical Engineering Plant Design • **SP**$_12$: Multivariable Control Systems • **SP**$_13$: Chemical Structure Synthesis/Analysis • **SP**$_14$: Generalized Harmonic Analysis• **SP**$_15$: Machine Design • **SP**$_16$: Miscellaneous Inverse Problems • **Further Comments**

[$E_{11/IV}$ **Analogue Examples**]

Appendices

• Several (extracts from) Farnborough Research Reports • The Σ -Lists (reduced 'appro- ach-tagging')

• GST-justification for **general realization via** quasi-machine representations. • Models (m_e/m_s) of ME/SE • **P**erturbation/**M**odification of 'initial models' • Remarks on **inverse**-analogue formation: comparison of models and machines • Machines 'as' templates with operational components • (q-)inv analogues 'as' (partial) output of perturbed machines • RecapL Structure of **M**athematical **I**nformation • M_t 'as' a collection of theories: $\{\Theta_q : q \in \bar{Q}\}$ • Modes of combination of (sub) theories • The Overall Σ -Structure • The role of synthesis • The principal tasks of analogues formation • The representation $\mathbb{M}_\beta(\zeta) = \mathbb{CA}M(mathcalP_\beta)$, where $M(\mathcal{P}_\beta \equiv$ MF.C of 'the problem $\mathcal{P}_\beta$ for β'. • Specification of 'admissible components' for M($\mathcal{P}_\beta$) • The essence of inverse analogy • Discussion of the individual components of $M(\mathcal{P}_\beta)$ • Abstract synthesis frameworks: possible GST routines • The role of SAM in Σ • Links between Σ and **T**heoretical CS • Various 'calculational **levels**' in Σ • Some Key References for TCS in Σ • Interim Summary • Stability criteria in **S**ynthesis **P**aradigms • SP$_1$ Electrical Circuits /NW and simulations • SP$_2$ **S**ampled-**D**ata control systems • SP$_3$ Optimal/Feedback/Adaptive Control systems • SP$_4$ Dynamical systems • SP$_5$ Automata • SP$_6$ Threshold Logic Gates • SP$_7$ Stochastic Control Systems • SP$_8$ Antenna Design • SP$_9$ Spectral Function Theory • SP$_10$ Integral Geometry • SP$_{11}$ Chemical Engineering Plant Design • SP$_{12}$ Multivariable Control Systems • SP$_{13}$ Chemical-Structure Synthesis/Analysis • SP$_{14}$ Generalized Harmonic Analysis • SP$_{15}$ Machine Design • SP$_{16}$ Miscellaneous **I**nverse **P**roblems • Remarks on General Inverse Problems • Multifaceted Representations of ME/SE β• Bird's Eye View of the Σ -Project

$C_{11/IV}$ • Locators $\Lambda_x(:=$ LC(x) $\equiv$C$(x), x \in M;\ := LD(x) \equiv D(x), x \in S)$ • Implementations of Λ_x as machines, $\mathbb{M}_x$ • Extension of basic MFC (with constituents $\mathbb{K}_x$) to cover (descriptive) ME/SE over Σ • Development of (potentially) **operational** MFC • Conversion into quasi-I/O **Machines** • 'Logical'/Reproductive Conditions within $C(x), D(x)$ • Overall processes: $x \to \mathbb{K}_x \to \mathcal{F}_x(\mathcal{P}_x) \to \mathbb{M}_x(\zeta, \xi_\zeta) \to \mathbb{M}_{\lambda x}(\zeta_\lambda, \xi_{\zeta_\lambda})$ where $\mathbb{M}_{\lambda x}$ (for variants of x) involves perturbations of $\mathbb{K}_x$. • **Specifiers** (e.g. $\Lambda_x, \mathcal{F}_x, \mathbb{M}_x, \mu_x$) of x. **Notations:** S_x (i) $\forall$ w.d. $x \exists \mathbb{M}_x$; (ii) $\mathbb{M}_{\alpha(x)}$ yields analogues of $\alpha(x)$ • Forms of **decomposability** of $\mathbb{M}$ • 'Formal-heuristic' adequacy of Σ (for M_t) • Generation (via specifier-perturbation) of all analogues • Importance of **E**nvironments - **f**or - **I**nvestigation (in Σ) • Possible forms of Σ_0 (the 'initial (O)IB') • DBMS/GST aspects of Σ ; scope for AI facilities • Qualitative aspects of analogues: Comparisons. • **Basic Problems** (P_1, P_2, P_3) for Analogues • **Levels of detail** in specifiers S_x • Block-diagram forms of S_x • Reduction(s) of S_x to **quasi-finite type(s)**, $\bar{S}_x$ • Special character of **models**, μ_x, of x • Systematic notation for specifiers $S_x, \bar{S}_x, S_x^*$ – and, for maps, $\bar{P} : S \to \bar{S}, P^* : \bar{S} \to S^*$ • Insertion nodes **in** theory $\mathcal{J}_1$ **for** theory $\mathcal{J}_2$ • Block-diagram formulations of **S**ynthesis **P**aradigms • **T**opical **O**utlines as models for SP • Structure of typical SP • Recap on NW synthesis ((ir)rational system functions) • Some extensions for (partially) distributed systems • Specification of general synthesis problems: conditions $C_1, \ldots, C_7 \to \{C_j\} \equiv \{C\}_7$ • Derived TO/FC (via operational forms of the C_j, etc.) • Comparison of $\{C\}_7$ with the 'typical structural characteristics' $[(a)-(f)$, as above] • 'Arbitrary syntheses': **attributes**, $\{A\}_N$, of **G**eneric **C**ontrol **P**roblems; $s_k = Ls[C\langle W_k\rangle], W_k \subset \{A\}_N$ – **via G**eneric **F**undamental **M**athematical **P**rocedures, $\varphi_m^{W_k}$. • **Claim:** the set of GCP is 'effectively small'. • TO $[s_k] \equiv T_k \equiv T_k(\varphi_\sigma)_{m_k}, (\sigma)_{m_k} \subset \bar{L}$, where $(\varphi)_L$ is any (fixed) ordering of $\{\varphi\}_L \equiv$ {GFMP for SP in Σ_t} • Basic task: to fabricate $p(\beta/\xi_\beta) \equiv$ {quintessential properties of βin ξ_β} • Possible formulations for GST frameworks • Identification of $(\varphi_\sigma)_{m_k}$ **via** 'decomposition of s_k' • $\{\varphi\}_L$ 'as' a quasi-basis for syntheses over $\Sigma\ _t$ • (In)formal links between $(\varphi_\sigma)_{m_k}$ and templates, T_k, for $\mathbb{M}(s_k)$, with realizations, p_k • Postulation of 'behaviour', $b(\theta)$ – from which θ is to be (re)constructed over 'more basic domains', D_θ • Possible extension of this procedure to (some GST framework(s) • Distinction between direct and indirect syntheses – **direct**, via analysis of $\mathbb{M}_\theta$ (as derived from S_θ); **indirect**, by (re)construction from $b(\theta)$ bu Initial Illustration: NW synthesis procedures (A diverse list of sources for (non)linear/time -(in)-independent/(non)distributed/(non)deterministic models).

Glossary

CM – **C**alculational **M**ilieu – the overall mathematical environment, covering mathematical structures, IT representations, enhanced SAM facilities, etc...

SIGMA – Simulation/**I**mplementation of **G**eneral **M**athematical **A**ctivity

OT – Operational Theorems (t-max - implemented for application to wfe

SAM – Symbolic/Algebraic Manipulation

APPROACH – Closeness, as assessed **via** all wd combinations of (non) deterministic forms of approximation, abstraction, and (inverse-) analogy (over specified realms)

ME – **M**athematical **E**ntity(ies): arbitrary complex/sophisticated wd: 'objects'

OIB – **O**perational **I**nformation **B**ank: the DBMS containing all ME in Σ_t, with t-max operational/interactive facilities

$\mathbf{P}_\Sigma$ – The 'processor' for realizing all Σ -OT

BASIC SPACES – ${}_sX$: The structures over which all approximations in Σ may be formed

$\{A_x\}$ – Sets of **quasi-axioms** (subject only to mutual consistency) for the range of Σ -structures

${}_j\Sigma_0^D$ – quasi-recursively defined 'initial IB': ${}_j\Sigma_0^D =: \varphi_k({}_{j-1}\Sigma_0^D), 1 \leqslant j \leqslant 6$ (so far), where ${}_{\_0}\Sigma_0^D \equiv$ pure mathematical information, and the taxonomic/analytical/ operational/analogical characteristics are 'generated' successively, to produce Σ_t

$V_t(e)$ – Variants of the ME e over Σ_t

$L(v)$ – **L** (**V**ector) Space(s)

Γ – Closure Space(s)

T – Topological Spaecs

U – Uniform Space(s)

C – Contiguity Space(s)

P – Proximity Space(s)

N – Nearness Space(s)

S – Statistical (-Metric) Space(s)

μ – Measure Space(s)

L_Σ – Programming languages for P_Σ

$\mathcal{J}_t^D$ – Descriptive theorems in Σ_t

$\mathcal{J}_t^{Op}$ – OT in Σ_t

Σ_t^π – { proofs in Σ_t} descriptive (π^D)/formal(π^F)

Σ_t^E – { generic (formal) theorems in Σ_t}

CB$_t$ – Calculational Base $\equiv$ { generic OT in Σ_t}

LC(e) – { Locational Conditions for $e\} \equiv l(e)$ where $l(e) = \underset{\alpha\in A_e}{\Lambda} c_\alpha((u_{\sigma^e})_{h^e}; e)$

$^\mu$**LC**(e) – Stochastic fir $^\mu e$, involving measure μ

Functions of OT – $f(T(\theta_T)) := T(f(\theta_t))$ IFF RHS is w.d.

Operation of Tone – $Te := [\bigcup_{1\leqslant j\leqslant 3} b_j\alpha_j(T(\theta_T))]$, combining informative, substitutive and constructive effects

$\mathcal{F}$ – Fundamental algebraic framework(s)

$\mathcal{S}$ – Structural properties

SF – Structural Framework(s)

GSF – Generalized SF (for approach procedures)

ζ^F – {Formal Calculations }

ζ^I – {Implemented Calculations }

M_t^C – { 'ratified t-constructive ME' in Σ_t}

B-constructive – Bishop (-type) constructive

M_t^{*C} – { B-Constructive ME over Σ_t}

CF – Calculational Framework

N – Nomenclature

D – Definitions

A – Axioms

R – Rudiments

$\bar{\mathcal{T}}$ – (sub)theory

$\mathcal{T}$ – nonrudimentary part of $\bar{\mathcal{T}}$

Definitions-types – explicit/inductive/recursive/implicit

Taxonomies – pre/skeletal/partial/full

M_t^* – { mathematical results 'known at time t }

$\mathcal{J}$ – { nonrudimentary (sub)theories }

Formalized theories – $\{\mathcal{L}, \zeta, A_x\}$

t**-max-OI** – Information formulated t-max-operationally

Information flow – $\pm\dot{I}_t(S)$, in or out of 'domain S

OTC – Operational-Theorem Calculus

UEF – Universal Editorial Facility

Ef**I** – Environment for Investigation

$TLp \Rightarrow q$ – typical (generic) (OT); also, $T : p \Rightarrow q \vee r$

$T(\theta_T) \equiv \Gamma^T(u_\sigma)_h$ –formulations/specifications of OT

$T(\mathcal{T}_k$ – task - representation(s) of T

$i_d(T)T/i_{op}(T)$ – descriptive/operational indices of T

$T(\theta_T)e$ – action of T on e (as above)

$\mathcal{P}_T(e^*, e)$ – interrelation(s) for construction(s) in Te

$\hat{f}(T(u_\sigma)_h)$ – action of $\hat{f} \equiv (\hat{f})_h : \hat{f}(T) := T(\hat{f}(u_\sigma))_h$

$\{\varphi : \{OT\} \rightarrow \{OT\}\}$ – most general class of nongeneric mappings

π**(process)-type(s)** – realizable/effective/constructive/implementable/ operational/applicable

imp$_S(\pi)$ – implementation of π over SC Σ_t

ord$_S(\pi)$ – order of imp$_S(\pi)$

$\mathcal{U}_{t_k}, \mathcal{I}_{t_k}, \mathcal{B}_{t_k}$ – underlying /Information/CB sets fo Σ_{t_k}

Diagrams of ME – schematic/representational/calculational

Facets, $\varphi_k(e)$**, of** e – arithmetic/combinatorial/algebraic/analytical/ geometrical/topological

Modes, $m_k(e)$ **of** e – realizations of facets of e

k_S – { ME defined primarily w.r.t. φ_k}

$e \approx \langle \underset{1\leqslant k\leqslant b}{+} \rangle^k e$ –facet-resolutions/representations of e

Rules for e **- diagrams** – location/orientation/interconnection/ r-element-interaction

Components of e**-diagrams** – outline of environment/rules

$\{\mathcal{J}_k\}$ – basic theories for the facet(s) of φ_k

$\langle + \rangle$ – amalgamation of diagrams

$\{+\}$ –amalgamation/combination of facets

[$+$ **]** – combination(s) of ME

Ω_1 – $\{\Gamma, T, U, P, C, N\}$: primary set of Basic Spaces

Ω_2 – $\{L, \mu, S, F\}$: secondary set of Basic Spaces

$\mathcal{P}S$ – Power sets of S

AE – Approach Environments

$EPCS$ – Existence Problem for Combined Structures

CMS – Combined Mixed Structure(s)

QA – Qualifying Adjective(s)

CSF – Combined Structural Frameworks

q//p – q realized over p (e.g., $\Omega_2//\Omega_1$)

CV – Conceptual Vicinity

GCSF – Generalized CSF

SE/P – Scientific Entity/Phenomenon (or, plural)

M/S-reflection/refraction – Mechanisms of analogue-formation

Scattering of ME/SE –by collections of (other) ME/SE

MA – Mathematical Analogue(s)

SA – Scientific Analogue(s)

$l_\xi(e)$ – LC for e over environment ξ

Gen(e) – Generalization(s) of e

Spec(e) – Specialization(s) of e

Inner/Outer Conditions – for analogue-formation

$(P)_{m_e}$ – constituent interrelation(s) for e

$D/S/G/A/Q/I$ – Determination/Specialization/Generalization/Approximation/ Quintessence/Imitation

$\mathcal{D}, \mathcal{S}, \mathcal{G}, \mathcal{A}, \mathcal{L}, \mathcal{J}$ – classes of D/S/G/A/Q/I

$\mathbb{A}e$ – $[\mathcal{D} \cup \mathcal{S} \cup \mathcal{G} \cup \mathcal{A} \cup \mathcal{L} \cup \pm\mathcal{J}]e \equiv$ { approaches to e}

K_e/P_e – Key/Peripheral LC for e

${}_mL$ – m-**ary** laws of combination for (quasi-)morphisms

$LD(\alpha)$ – Locational Description(s) for the SE α

ξ_t – $\{wdME\ \mathcal{M}_t\}$; also denoted by Σ^E

$\mathbb{M}^{\lambda',\lambda''}_{D;E*}$ – Modification operator(s) for diagrams of ME

$\Gamma^e_{\lambda;j}$ – $(\gamma)_q$ – –{ Components of $D^e_{\lambda,j}$ (of type j, species λ)}

Proof-type/data-type – classes of diagrams

Component(s)-settings – context(s)/environment(s) for component-specification(s)

$\Sigma - DC$ – Σ- Diagram Calculus

$\Sigma^{\mathbb{T}}_t$ –{ generic OT in Σ_t}

${}^{e}T/{}^{\mathcal{T}}T/{}^{c}T/{}^{g}T$ – explicit/tast/cast versions of (O)T

${}^{k}\mathcal{U}_t$ – { underlying ME 'of species of s'_k}; $\mathcal{U}_t = \bigcup_{1 \leqslant k \leqslant N_t} {}^{k}\mathcal{U}_t$

${}^{k}\varphi$ – mapping, ${}^{k}\mathcal{U} \supset P \to Q \subset^{k} \mathcal{U}$

${}^{k,l}\psi$ – mapping, ${}^{k}\mathcal{U} \supset P \to Q \subset^{l} \mathcal{U}$

$\hat{\pi}$ – proof operator: $T : p \Rightarrow q IFF \hat{\pi} p - q \vee r$

${}^{-1}_{*}T$ – *- inverse of OT ($\approx$ converse, for composition)

$\pi/'\pi/''\pi$ – representations of full/skeleton/indicative proofs

$\pi(T)$ – $\Phi(\bigwedge_m C^{J_m})p = q \vee r$ (full); see E_4/E_8

$'\pi(T)$ – $\{T_j(f^j) : j \in M_k, 1 \leqslant k \leqslant K\}$ (skeleton)

$''\pi(T)$ –$\{T_j : j \in M_k, 1 \leqslant k \leqslant K\}$ (indicative)

Centrality – of results, over sets of proofs

[$AT\ \ (u_\sigma)_L$] – approximate formulation (of T)

$T((Au_\sigma)_L)$ – approximate realization (of T)

A**[**$Te\ \ /T[Ae]$] – approximate application (of T)

[$\mathbb{A}T\ \ (u)/T((\mathcal{A}u))/\mathcal{A}[Te]/T[\mathcal{A}e]$] – corresponding approach forms

$H(T)_m$ – representation of function(s) of $T_1, \ldots, T_m$

[T] – premise(s) of T

$\langle T \rangle$ – conclusion(s) of T

$V_\Sigma T$ – Variants of T, including $\mathbb{A}, A, W$(weak), S(strong), P(partial), G(generalized)

${}_K T/{}_k T$ – strict/full converse(s) of generic OT

${}_K ST/{}_k ST$ – strict/full converse(s) of Variants of T

mp, Mq – admissible modifications 'preserving' $T : p \Rightarrow q$

Σ_S – state of Σ 'generated by S'

s_0 – 'initial data'; conjecture: $\Sigma_{S_k} \sim \mathcal{M}(k \to \infty)$

CME – 'complete ME', via conceptual continuation

$V(e)$ – variant(s) of e

I**[**$e\ \ _k$] – Interactions among $e_1, \ldots, e_k$

SiMA/MaAT – Σ -lists of examples

$(\varphi)_6$ – Σ -facets of ME

CD – Capsule Description(s)

E_m/E_s – Mathematical/Scientific Environment(s)

T_A –approach types

F – formulations ($\equiv$ Models . . .)

$p[W]$ –admissible realizations of W

$\dot{I}(D)$ – flux of information (through D)

p_m –full profile of i_m on Σ -list(s)

$\langle i_m \rangle$ – formulation(s) of i_m

π_T – proof-diagram for T

$(\Delta^m)_{11}$ – tagging-descriptors for i_m on Σ -list(s)

$\mathcal{D}_t$ – $\{(\Delta^m)_1 1$ for all i_m on specified list(s) $\}$

$\pm A\mathcal{M}_t$ – (inverse-) analogical part of $\mathcal{M}_t$

$\sigma(\mathcal{A}e, e)/\sigma(\pm Ae, e)$ –separation of $\mathcal{A}\dot{e}/\pm Ae$ from e

ξ –environment for representation of e by LC(e)

φ_I/φ_O –inner/outer features of e

$L_{op}(e)$ – operational level (of e)

$F(m)$ – formulation(s) (of i_m)

$\mu(m)$ –analogue-mechanism(s) (of i_m)

$B(e)$ –$\{b_{e_\lambda} : \lambda \in \Lambda_e$ (of e)$\}$

$\langle S, \xi \rangle$ – structure(s) on environment(s) ξ

$\langle S_1, \xi_1 \rangle \hookrightarrow \langle S_2, \xi_2 \rangle$ – transfer of structure(s)

$\langle S_1', \xi_1 \rangle \hookrightarrow' \langle S_2, \xi_2 \rangle$ – 'suggestion' of 'new' mathematical structure(s) transferred from scientific structure(s)

$v_j(\mathcal{M}_t)$ –c_j - view (MAP) of $\mathcal{M}_t$

$V_S(\mathcal{M}_t$ – $U\{v_j(\mathcal{M}_t) : j \in SC\bar{N}\}$

$\{j_m\}_{k_j} \equiv \beta_j(t)$ – $\{c_j$ -basic ME $\}$

$(D)_l \equiv D_{\mathcal{M} \cup D^S}$ – Mathematical/scientific domains for (inverse-) analogue-formation

$G(\pm\alpha)$ – graphs of (inverse-) analogues α)

$p_\mathcal{T}$ – program-flow diagram(s) for task(s) $\mathcal{T}$

$\mathcal{J}_{\pm\alpha} \equiv \mathcal{J}_{\pm\alpha\mathcal{M}} \cup \mathcal{J}_{\alpha\mathcal{S}}$ –constituent subtheories for $\pm\alpha$

OAC – Operational-Analogue Calculus

$(D^*(t))_{L(t)}$ – permanently labelled mathematical (sub)domain(s)

$\mathcal{J}$ – quasi-formalized theory/theories, $\{langleS\}; \{W_j\}; \Omega; \alpha; \Lambda \rangle$, where $\{S\} \equiv$ $\{$ underlying space(s) $\}; \{W_j\} \subset \mathcal{P}\{S\}; W_j$ admits m-ary laws of combination, $2 \leqslant m \leqslant n_j$; $\Omega \equiv \{w_\mu\} \equiv$ prescribed 'operators' on certain 'objects' built over $\{S\} \cup \{W_j\}; \alpha \equiv$ axioms; $\Lambda \equiv \{$scientific laws, as specified $\}$

$M(\mathcal{J})$ – {meta-theories (for $\mathcal{J}$)}

$S_{\Lambda(\mathcal{J})}$ – { scientific theories constructible from $\mathcal{J}$ over Λ}

${}_{\eta}\mathcal{J}^*$ – $(\mathcal{J}_\eta, \lambda_\eta)$, for mathematical theory $\mathcal{J}_\eta$, 'laws' λ_η

${}_{\eta}s^*$ – $s({}_{\eta}\mathcal{J}^*) \equiv s(\mathcal{J}_\eta, \lambda_\eta)$

$\mathbb{M}_e(\xi, \xi)$ – **M**aster **F**low **C**hart/Machine for e over ξ

${}^{\mu}\mathbb{M}_e$ – μ based stochastic variants of $\mathbb{M}e$, for ${}^{\mu}e$ over ${}^{\mu}\xi$

$D(e)$ – Domain of $e \equiv$ { ME in definition(s) of e}

$p_\lambda(e_\lambda, D_\lambda, \xi_\lambda$ – representation of analogues, $\pm\alpha_\lambda(e)$, of e, via $\mathbb{M}_\lambda(\eta, \xi_\lambda)$

$\tilde{\mathbb{T}}\{\tilde{\mathcal{T}}\}$ – { scientific theories }

$\mathbb{T}\{\mathcal{T}\}$ – { mathematical }

${}_k\tilde{\mathcal{T}}/{}_k\mathcal{T}$ –arbitrary subtheories of ${}_{k-1}\hat{\mathcal{T}}/{}_{k-1}\mathcal{T}(k \geqslant 1)$

$\mathcal{T}_1 \circ_W \mathcal{T}_2$ – 'local' composition of subtheories

$G_{\mathcal{M}}$ – { generic ME in $\mathcal{M}$}

$G_{\mathcal{S}}$ – { generic SE in $\mathcal{S}$}

$Y_{G_{\mathcal{M}}}$ – $\{(LC(e), \mathbb{M}_e(\xi, \xi)) : e \in G_{\mathcal{M}}\} \equiv \{\eta\}$

$b\eta(d)$ – $(b_L L(d), b_{\mathbb{M}}\mathbb{M}(d))$: perturbations of η

$f({}_m L(\eta)^q_m)$ – perturbed quasi-morphisms (for analogues)

Comprehension Exercises for the Sigma Essays

The following Exercises are based on the Contents list in the Conspectus. The aim is to promote fluency in the interpretation / use of the many novel concepts introduced in the Σ -superstructure. There are no 'problems' (as such); but, rather, discussions / outlines / exemplifications / explications / justifications / demonstrations / explorations /. . . .

The set of model exercises for E_k is denoted by $[\varepsilon_k]$.

$\underline{[\varepsilon_1]}$

$*1$ Exemplify (distinctly) the Basic Spaces, ${}_SX$.

$*2$ Exemplify ${}_SX$-approximations of some ME.

$*3$ Exemplify some (inverse-)analogical modifications of the ${}_SX$.

$*4$ Exemplify some realizations of ME over the ${}_SX$.

$*5$ Illustrate the quasi-recursive development of Σ_0.

$*6$ Illustrate the terms: pre- / sub- / skeletal- / full TAXONOMIES.

$*7$ Justify the representation of Σ_t as a General System.

$*8$ Justify the claim of 'inherent soundness' of the Σ -Scheme.

$*9$ Illustrate: new proofs 'as' B-combinations of established subproofs.

$*10$ Explicate (briefly!) Σ 'as' an extension of SAM.

$*11$ Exemplify: Σ -representations of (O)T.

$*12$ Encapsulate the formal / logical / linguistic structure of M_t^*.

$*13$ Illustrate the Σ -notation for OT with 'arbitrary constituents'.

$*14$ Extend **A**utomatic-**T**heorem-**P**roving structures to Σ .

$*15$ Discuss: levels of operationality of ME in Σ .

$*16$ Discuss interpretations of $f(T)$, for functions f and OT, T.

$*17$ Explain / exemplify the action of $f(T)$ on the w.f.e. e.

$*18$ Discuss / illustrate the notion of **C**alculational **B**ase.

$*19$ Discuss the forms of LC for ME, and of LD for SE.

$*20$ Discuss / exemplify forms of **Conceptual V**icinities.

$[\varepsilon_2]$

*1 Summarise the algebraic structures underlying '**axiom**' (via the Manual and suitable references).

*2 Exemplify the reduction of 'arbitrary' analytical calculations to cases over $\mathbb{R}/\mathbb{C}$.

*3 Exemplify the reduction of 'arbitrary' algebraic calculations to cases over the '**axiom**' structures.

*4 Discuss the realizability of 'arbitrary' analytical-topological structures over 'suitable' algebraic SF.

*5 Exemplify calculational reductions to (potential) SAM forms.

*6 Discuss the fundamental analytical / algebraic processes for n-variable functions in SAM.

*7 Specify references for the basic SAM algorithms.

*8 Discuss claims that 'arbitrary OT are (potentially) approachable over P_Σ.

*9 Explicate the 'calculational reduction scheme, for (quasi-)arbitrary ME/SE.

$[\varepsilon_3]$

*1 Justify the characterization of CB as 'structured, searchable collections of interacting items'.

*2 Discuss the heuristic aspects of taxonomy schemes.

*3 Characterize (generally): nomenclature / definitions / axioms.

*4 Characterize (generally): foundations.

*5 Characterize (generally): distinct types of definitions.

*6 Characterize (generally): rudiments.

*7 Characterize (generally): types of taxonomies.

*8 Characterize (generally): types of Σ -information.

*9 Discuss general properties of CB.

*10 Outlive the roles of Network-Theory / Information-Theory / Optimization-Theory /General-Systems-Theory /... in Σ -design.

*11 Discuss the principal subdivisions of DBMS research. Exemplify!

$[\varepsilon_4]$

*1 Exemplify the notation for simple / multiple OT.

*2 Characterize: ingredients of OT.

*3 Characterize: the stylistic formulations of OT.

*4 Exemplify the 'task' and 'cast' versions of OT.

*5 Exemplify: the action of OT on w.f.e..

*6 Exemplify the definitions of $f(T)$.

$*7$ Discuss classes of transformations $\varphi : \{\text{OT}\} \to \{\text{OT}\}$.

$*8$ Exemplify the definitions of: realizable / effective / constructive / implementable / algorithmic / efficient / operational / applicable – identifying 'domains of (non-)coincidence'!

$*9$ Explicate the types of Topological Boolean Algebras; and, of cylindrical / polyadic algebras.

$*10$ Discuss: approach operations in CB.

$*11$ Formulate some 'basic problems' of the OTC.

$*12$ Explicate the OI-aspects of (classes of) Σ -diagrams.

$*13$ Exemplify the facets of ME, and of 'specified' MI.

$*14$ Illustrate the 'addition theorems for diagrams'.

$*15$ Discuss the 'admissible modifications' for diagrams.

$[\varepsilon_5]$

$*1$ Characterize / distinguish the Basic Spaces, ${}_SX$.

$*2$ Develop formalisms for interrelations among the ${}_SX$.

$*3$ Exemplify 'extra structures' over the ${}_SX$.

$*4$ Exemplify 'axiom-realizability problems' over base-sets.

$*5$ Formulate (in outline) a general existence theory for **C**ombined **S**tructural **F**rameworks.

$*6$ Exemplify the 'CSF-theory' for some ${}_SX$.

$*7$ List the 'main generic AE/SF'.

$*8$ Exemplify variants / specializations of generic AE/SF.

$*9$ Specify adequate references for the principal AE/SF.

$*10$ Discuss the general nature of 'stochastic versions of AE/SF'. How are such structures defined?

$[\varepsilon_6]$

$*1$ Exemplify LC/LD for some (collection of) ME/SE.

$*2$ Explicate the associated notions of 'conceptual spaces'.

$*3$ Discuss / characterize the classes Ab(e)m An(e), -An(s).

$*4$ Exemplify some quasi-reflections / refractions of ME/SE 'in various (sub)theories'.

$*5$ Exemplify the designated species of reflections / refractions.

$*6$ Discuss the concept of 'quasi-scattering' of some (collections of) ME/SE by others.

$*7$ Discuss 'admissible LC/LD-modifications.

$*8$ Exemplify: internal / external ME/SE behaviour.

$*9$ Exemplify: inner / outer conditions for ME/SE governed by 'Laws' / optimization criteria.

*10 Exemplify: interactions of ME/SE with environments ξ.

*11 Discuss: criteria for 'good analogues'.

*12 Examine 'reproduction criteria' for analogues of ME/SE.

*13 Explicate: simple / compound analogues.

*14 Explicate: 'analogues as mappings'.

*15 Explicate the set $\mathcal{M} := \{\mathcal{D}, \mathcal{S}, \mathcal{G}, \mathcal{A}, \mathcal{Q}, \pm\mathcal{J}\}$.

*16 Exemplify LC-forms for abstractions.

*17 Explicate 'the relational representation of analogues'.

*18 Exemplify: 'analogue defects'.

*19 Explicate (and exemplify!) the quasi-morphisms, and their approximate application for analogues.

*20 Explicate the general m-ary laws of combination (for quasi-morphisms involving the ME e).

$[\varepsilon_7]$

*1 Explicate the 'incomplete-specification / modification-of-LC' form of 'approach'.

*2 Explore possible forms of Lc for some ME.

*3 Discuss 'modes of perturbation' of LC.

*4 Demonstrate the LC-representation of the spaces $_X S$.

*5 Discuss links of Σ with aspects of Model Theory / Universal Algebra.

*6 Exemplify 'elementary perturbations'.

*7 Exemplify the modes of approximation listed as ##(i)–(xviii) in 'Varieties of Approximation' ($\equiv$ [ELVEY, 1994]).

*8 Exemplify 'conceptual vicinities', for predicate functions.

*9 Exemplify 'quasi-neighbourhoods' (for 'adequately structured spaces of ME).

*10 Discuss Lc-Deformations and quasi-metrics / norms for predicate-function spaces.

*11 Discuss possible representations of M_t over **in**completely formalized languages (say, $\mathcal{L}^-$).

*12 Compare properties of $\mathcal{L}^-$ with full-abstraction / semantic-equivalence for programming languages.

*13 Exemplify 'approaches to OT' (by sequences of OT).

*14 Discuss: approaches to diagrams (via approaches to their components).

*15 Outline possible forms of the Σ -Diagram Calculus.

*16 Discuss: approaches to proof-diagrams.

$[\varepsilon_8]$

$*1$ Outline the structure of Σ over $U_t \equiv \{$underlying ME$\}$.

$*2$ Discuss the various 'canonical forms' of OT.

$*3$ Explore the connections of canonical OT with Environments for Investigation.

$*4$ Examine representations of OT over U_t.

$*5$ Discuss w.d. forms of transformed OT.

$*6$ Discuss w.d. combinations of transformed OT.

$*7$ Exemplify some OTC-quasi-inverses (via converses).

$*8$ Discuss notions of quasi-convergence (for w.f.e. in proofs).

$*9$ Exemplify full / skeletal / indicative forms of proofs.

$*10$ Discuss the role(s) of approach in the OTC.

$*11$ Exemplify the interpretations of functions $H(T)_m$.

$*12$ Explore OTC-analogues of Riemann surfaces.

$*13$ Explore OTC-analogues of eigen-elements.

$*14$ Exemplify: composition (of collections of OT).

$*15$ Explore: OP as B-combinations of transformed OT.

$*16$ Explore: OT as computer programs.

$*17$ Explore OT-representations within Martin-Löf's Type Theory.

$*18$ Discuss Carnap's 'L-probability' as a possible framework for Σ .

$*19$ Discuss possible roles of polyadic algebras in Σ .

$*20$ Define: proof-stability / variation / preserving-deformations.

$*21$ Discuss possible forms of operational (quasi-)converses.

$*22$ Explore / exemplify estimates of proof-size / complexity.

$*23$ Discuss taxonomic aspects of environments-for-investigation.

$[\varepsilon_9]$

$*1$ Justify the claimed 'universality of approach' within Σ .

$*2$ Define and exemplify: possibility / attainability / speed / efficiency / optimality / scope / stability /. . . of approach.

$*3$ Discuss the 'generic incompleteness of calculations of Σ '.

$*4$ Exhibit $\{$exact calculations$\}$ as an 'exceptional subset' in Σ .

$*5$ Exemplify and discuss approach-criteria based on (i) definitions (ii) limitations (iii) quasi-varieties (iv) abstract boundaries.

$*6$ Characterize (if possible!) 'typical items' in the Σ -Lists.

$*7$ Outline (and exemplify) 'the calculational reduction problem'.

$*8$ Discuss the 'states, Σ_S' of Σ 'generated by base-sets S'.

$*9$ Explicate: 'LC as universal approach representations'.

$*10$ Discuss the formalism of 'variants of ME in Σ'.

$*11$ Discuss 'stochastic forms of ME' (via probability measures in LC).

$*12$ Exemplify: stochastic forms of ME.

$*13$ Discuss: forms of 'interactions among ME'.

$*14$ Outline: M_t as (i) a quasi-space (ii) a general system.

$*15$ Discuss forms of instability in LC-representations.

$*16$ Explore: ME-dependence on subsets of LC.

$*17$ Characterize 'typical LC-modifications'.

$*18$ Discuss: 'irretrievable LC-properties'.

$*19$ Discuss / exemplify: mutually / almost-isolated, and (in)separable, LC.

$*20$ Distinguish provisional proofs from conclusive proofs. Exemplify!

$*21$ Develop (heuristic) definitions for: possibility / attainability / speed / efficiency / optimality / scope / stability /... of approach processes.

$*22$ Outline the 'MAP scheme' for treating mathematical analogues.

$[\varepsilon_{10/I}]$

$*1$ Exemplify limitations (for Σ) in current SAM packages.

$*2$ List some key enhancements crucial for Σ-activity.

$*3$ Outline the essentials of 'Symbolic Investigation' in Σ.

$*4$ Discuss the (dis)connections between the proof(s) of any OT, T_1, and proofs of theorems used to prove T_1.

$*5$ For each of the (sets of) topics (##1–34) in SCCM ($\equiv$ [ELVEY, 1980]) summarised in $E_{10/I}$, develop Capsule Descriptions of the key concepts.

$[\varepsilon_{10/II}]$

$*1$ Outline the AE based on classes $\mathcal{D}, \mathcal{S}, \mathcal{G}, \mathcal{A}, \mathcal{Q}, \pm\mathcal{J}$.

$*2$ Discuss possible tagging schemes for Σ-examples.

$*3$ Explain how taggings yield 'approach outlines'.

$*4$ Sketch the structure of J. Walsh's 'H/B of Non-parametric Statistics' (as a quasi-IB).

$*5$ Explicate the 'template representation of ME/SE.

$*6$ Outline a classification scheme for 'primary mathematical / scientific subtheories.

$*7$ Suggest classifications for formulations of theorems.

$*8$ Give a general definition of 'transferred structures'.

$*9$ Discuss the set of item-descriptors for the Σ -Lists.

$*10$ Illustrate the approach characteristics of n items from the Σ -Lists.

$*11$ Discuss (briefly!) the selection criteria for the Σ -Lists.

$*12$ Criticize the cited 'strategies for mathematical progress'.

$*13$ Exemplify the (successful!) use of these strategies.

$[\varepsilon_{11/I}]$

$*1$ Discuss the 'separations' $\sigma(\mathcal{A}e, e)$ and $\sigma(Ae, e)$.

$*2$ Explore the 'stochastic FEM' over standard domains.

$*3$ Formulate the (stochastic) FEM over triangulable TS.

$*4$ Define / exemplify inner / outer features for LC/LD of e/s.

$*5$ Outline / exemplify the main stages in analogue formation.

$*6$ Explicate the mechanisms of ((quasi-)inverse) analogues.

$*7$ Exemplify mathematical / scientific environments for S/M-items.

$*8$ Explicate the role of 'formulations' in template representations.

$*9$ Compare $\{$analogues$\}$ with Compendia of engineering mechanisms.

$*10$ Discuss 'behavioural properties' $\{b_{e\lambda}\}$ of the ME e.

$*11$ Explicate 'structural transfer' (from $\langle S_1, \xi_1\rangle$ to $\langle S_2, \xi_2\rangle$).

$*12$ Explicate: reflections / refractions from mathematical / scientific theories.

$*13$ Discuss the MAP scheme (for treating ((quasi-)inverse) analogues.

$*14$ Discuss: 'interactions (of ME/SE constituents) in analogue formation.

$*15$ Sketch processes: $e \to \pm\alpha(e)$ via $(D)_e \subset \mathcal{M} \cup \mathcal{S}$ (with 'diagrams').

$*16$ Discuss OTC-operations for 'graphs', $G(\pm\alpha(e))$ and 'mechanisms', $\mu(\pm\alpha(e))$.

$*17$ Discuss: centrality of concepts / results under 'analogical mappings'.

$*18$ Discuss: forms of mechanisms for proofs / programs.

$*19$ Discuss: the 'main problems' for Σ -analogues.

$*20$ Outline: 'active ME' for analogues, and related OAC structures.

$*21$ Discuss: 'formalized heuristics' as a general framework for Σ .

$*22$ Exemplify: heuristic representations of mathematical / scientific theories.

$*23$ Discuss: models over informal scientific theories.

$*24$ Discuss: 'typical forms of approximation' in physical theories.

$*25$ Discuss: extension of 'ABC'-templates to operational forms for ME.

$*26$ Outline: the flowcharts for FM – and extensions to Σ .

$*27$ Outline: the approach / operational extensions of (i) ABC-flowcharts (ii) flowcharts for 'arbitrary ME/SE'.

$*28$ Outline links of MF-C(β) with 'machines' $\mathbb{M}_\beta$.

$*29$ Exemplify $\mathbb{M}_\beta$ for some ME/SE β.

$*30$ Outline procedures for 'developing $\mathbb{M}_{\{\beta\}}$'.

$*31$ Explicate the connections of perturbed $\mathbb{M}_\beta$ with $\pm\alpha(\beta)$.

$[\varepsilon_{11/II}]$

$*1$ Discuss: structure of $\mathbb{M}_\beta$ over {mathematical / scientific theories}.

$*2$ Is it possible to develop general synthesis schemes that cover (formally, at least) all of the paradigms ##1–16 sketched in $\mathrm{E}_{11/II}$?

$*3$ Is it possible to extend the GST-framework to include nonlinear / nondeterministic systems?

$*4$ Explicate links of synthesis schemes with GST.

$*5$ Explore the (formal) links of Circuit Theory with Control Theory.

$*6$ Discuss the formalism of assignment-perturbation operators.

$[\varepsilon_{11/III}]$

$*1$ Explicate: partial **tagging format** (for S/M-List items).

$*2$ Exemplify each of: $\delta, \mathcal{A}, A, -A, @$.

$*3$ Outline: **locators**, Λ_x, for the ME/SE x.

$*4$ Discuss the 'conversion' of Λ_x into the 'machine' $\mathbb{M}_x$.

$*5$ Outline the extension of the ABC **M**aster **F**low-**C**harts to Σ .

$*6$ Discuss the development of operational / approach MF-C.

$*7$ Explicate the conversion of MF-C into quasi-I/O machines.

$*8$ Explicate the **tagging**: $t_n \approx [n; (e_{\psi^n})_{p_n}; ({}_S X_{\chi^n})_{q_n}; V_n; W_n; Y_n; V_n'; Y_n']$.

$*9$ Exemplify the 'reduced tagging' specified in $*8$!

$*10$ Discuss interpretations of δ in the tagging formalism.

$*11$ Exemplify the representation: $e = \varphi(u_\sigma)_L$ over U_t.

$*12$ Exemplify $\mathbb{M}_x$-representations of ME/SE – identifying 'components', '(sub)mechanisms' and I/O processes.

$*13$ Indicate correspondences of Σ -machines with general engineering machines. Consider 'physical', 'chemical', 'biological', 'economic', ... machines similarly.

$*14$ Outline criteria / processes for machine-perturbation.

$*15$ Outline (formal) connections of Σ -machines with CS.

*16 Discuss the model of 'Σ as a multimedium of interacting ME'.

*17 Sketch the 'GST-realization / synthesis' argument – and indicate its potential applicability to general ME.

*18 Explicate the 'model formalism' for general ME/SE.

*19 Outline the perturbation-modification scheme.

*20 Outline the interpretation of Σ -machines as templates with operational components.

*21 Discuss: 'analogues 'as' (partial) output of perturbed Σ -machines'.

*22 Discuss: M_t 'as' a quasi-exhaustive collection of subtheories.

*23 Discuss (and exemplify) the Σ -combination of (disparate) theories.

*24 Explicate the roles of synthesis, and of analysis / predication in Σ .

*25 Show that synthesis and analysis are quasi-inverse processes.

*26 Outline the main stages in direct-analogue formation.

*27 Outline the key stages in ((quasi-)inverse) analogues formation.

*28 Outline the admissible specifications of M($\mathcal{P}_\beta$).

*29 Explicate: $\mathbb{M}_\beta(\zeta) = \mathbb{CA}M(\mathcal{P}_\beta)$.

*30 Discuss how inverse analogues may be 'fashioned' through 'temporary relaxation of rules'. Exemplify!

*31 Explicate: M($\mathcal{P}_\beta$) $\approx \sigma s\zeta d; e : f\mathcal{R}g; h : iMj; k : l$.

*32 Discuss: the (conjectured) 'universal (GST-based) synthesis block diagram'.

*33 Discuss: embedding / extension / modification of SAM procedures in Σ .

*34 Discuss: the longterm role(s) of Theoretical CS within Σ .

*35 Discuss: the longterm role(s) of 'constructive AI' within Σ .

*36 Outline the structures of (i) AUTOMATH (ii) nuperl (III) ONTIC.

*37 Discuss: the forms of **stability** crucial for sound **synthesis**.

*38 For each of the synthesis paradigms, s_k, $1 \leqslant k \leqslant 16$ listed / discussed in E$_{11/II}$/E$_{11/III}$, develop capsule descriptions of the key procedures.

$[\mathcal{E}_{11/IV}]$

*1 Outline the 'locator representation' of ME/SE, β.

*2 Exemplify: (i) finite (ii) countable (iii) uncountable locators.

*3 Specify the correspondence(s) of locators with Σ -machines.

*4 Outline the 'conversion' of Λ_β into $\mathbb{M}_\beta$ (via introduction of OT).

*5 Examine the role of (logical / reproductive) models in Σ .

*6 Explicate: $x \to \mathbb{K}_x \to \mathcal{F}_x(\mathcal{P}_x) \to \mathbb{M}_x(\zeta, \xi_\zeta) \to \mathbb{M}_{\lambda x}(\zeta_\lambda, \xi_{\zeta\lambda})$.

*7 Characterize: specifiers (of ME x).

$*8$ Outline: forms of decomposability for Σ -machines, $\mathbb{M}_x$.

$*9$ Indicate how general ME/SE specifiers 'stem from' ABC-forms.

$*10$ Discuss (and exemplify): admissible perturbations of $\mathbb{M}_x$.

$*11$ Discuss: 'new theories' via 'temporary violation of known rules'.

$*12$ Explicate: the Basic Problems, $\{\mathsf{P}_1, \mathsf{P}_2, \mathsf{P}_3\}$ for analogues.

$*13$ Discuss: key features of Environments-for-Investigation.

$*14$ Discuss: viable forms of Σ_0 (the 'initial (O)IB).

$*15$ Discuss: the DBMS / GST / AI aspects of Σ .

$*16$ Discuss: fundamental aspects (especially, qualitative) of ((quasi-)inverse) analogues.

$*17$ Exemplify (typical) levels of detail in ME/SE specifiers.

$*18$ Explicate the 'block-diagram' form(s) of ME specifiers, S_x.

$*19$ Discuss: reduction of S_x to $\bar{S}_x$ (of quasi-finite type(s)).

$*20$ Characterize: Models of x 'as' settings for x.

$*21$ Exemplify the notation $(S_x)_4$ and $\bar{\rho}_{xk}, \rho^*_{xk}$ for specifiers and associated mappings.

$*22$ Discuss: insertion nodes, in theory $\mathcal{I}_1$, for theory $\mathcal{I}_2$.

$*23$ Explicate: Problem templates (involving the Basic Problems $\{\mathsf{P}_1, \mathsf{P}_2, \mathsf{P}_3\}$.

$*24$ Discuss: synthesis-paradigms in block-diagrams / MFC forms.

$*25$ Explicate the s_k via processes (a)–(f) and conditions $(\mathrm{C})_7$.

$*26$ Discuss the conditions $(\mathrm{C})_7$ in general models.

$*27$ Show how TO/F-C are derivable from the (operational) C_j.

$*28$ Explicate: attributional representations of control problems.

$*29$ Explicate: $s_k =: s[C\langle W_k\rangle]$, $W_k \subset \{A\}_N$ (attributes) via: **G**eneric **F**undamental Procedures, $\varphi_m^{W_k}$.

$*30$ Justify: the set $\{$**G**eneric **C**ontrol **P**rocesses$\}$ is 'effective / small'.

$*31$ Explicate: $\mathrm{TO}(s_k) = T_k(\varphi_\sigma)_{m_k}, (\sigma)_{m_k} \subset \bar{L}, (\varphi)_L \equiv \{GFP\}$.

$*32$ Discuss: fabrication of $p(\beta|\xi_\beta) \equiv \{$key properties of β in $\xi_\beta\}$.

$*33$ Discuss: possible GST-formulations of SP.

$*34$ Discuss: the set $\Phi \equiv \{\varphi_i^k : i \in \bar{I}_k\}$ as a quasi-basis for the SP.

$*35$ Compare: $(\varphi_\sigma)_{m_k}$ and templates τ_k for $\mathbb{M}(s_k)$ with realizations ρ_k.

$*36$ Discuss: (re)construction of θ from 'behaviour $b(\theta)$' over D_θ.

$*37$ Explore: extension of scheme $*36$ to GST frameworks.

$*38$ Distinguish: direct synthesis (via analysis of $\mathbb{M}_\theta$) from **in**direct synthesis (based on behaviour $b(\theta)$).

References Web Sites

There are nearly 3,000 references occupying more than 300 pages. They are seen as an intergral part of the book but their length make it prohibitive to include in this book. So they are available instead on the following website.

http://www.personal.soton.ac.uk/rdi/References.pdf

www.ingramcontent.com/pod-product-compliance
Lightning Source LLC
Chambersburg PA
CBHW081715170526
45167CB00009B/3582
* 9 7 8 1 5 0 2 8 6 9 5 3 1 *